To the memory of my father and mother,
Albin John Carlson and Mildred Elizabeth Carlson
A. Bruce Carlson

To my parents,
Lois Crilly and I. Benjamin Crilly
Paul B. Crilly

To my son, Carter:
May your world be filled with the excitement of discovery
Janet C. Rutledge

Contents

The numbers in parentheses after section titles identify previous sections that contain the minimum prerequisite material. The symbol ★ identifies optional material.

Preface

This text, like its previous three editions, is an introduction to communication systems written at a level appropriate for advanced undergraduates and first-year graduate students in electrical or computer engineering. New features in this edition include the introduction of two other authors, Professors Rutledge and Crilly, to provide additional expertise for topics such as optical links and spread spectrum.

An initial study of signal transmission and the inherent limitations of physical systems establishes unifying concepts of communication. Attention is then given to analog communication systems, random signals and noise, digital systems, and information theory. However, as indicated in the table of contents, instructors may choose to skip over topics that have already been or will be covered elsewhere.

Mathematical techniques and models necessarily play an important role throughout the book, but always in the engineering context as means to an end. Numerous applications have been incorporated for their practical significance and as illustrations of concepts and design strategies. Some hardware considerations are also included to justify various communication methods, to stimulate interest, and to bring out connections with other branches of the field.

PREREQUISITE BACKGROUND

The assumed background is equivalent to the first two or three years of an electrical or computer engineering curriculum. Essential prerequisites are differential equations, steady-state and transient circuit analysis, and a first course in electronics. Students should also have some familiarity with operational amplifiers, digital logic, and matrix notation. Helpful but not required are prior exposure to linear systems analysis, Fourier transforms, and probability theory.

CONTENTS AND ORGANIZATION

A distinctive feature of this edition is the position and treatment of probability, random signals, and noise. These topics are located after the discussion of analog systems without noise. Other distinctive features are the new chapter on spread spectrum systems and the revised chapter on information and detection theory near the end of the book. The specific topics are listed in the table of contents and discussed further in Sect. 1.4.

Following an updated introductory chapter, this text has two chapters dealing with basic tools. These tools are then applied in the next four chapters to analog communication systems, including sampling and pulse modulation. Probability, random signals, and noise are introduced in the following three chapters and applied to analog systems. An appendix separately covers circuit and system noise. The remaining six chapters are devoted to digital communication and information theory, which require some knowledge of random signals and include coded pulse modulation.

All sixteen chapters can be presented in a year-long undergraduate course with minimum prerequisites. Or a one-term undergraduate course on analog communication might consist of material in the first seven chapters. If linear systems and probability theory are covered in prerequisite courses, then most of the last eight chapters can be included in a one-term senior/graduate course devoted primarily to digital communication.

The modular chapter structure allows considerable latitude for other formats. As a guide to topic selection, the table of contents indicates the minimum prerequisites for each chapter section. Optional topics within chapters are marked by the symbol ★.

INSTRUCTIONAL AIDS

Each chapter after the first one includes a list of instructional objectives to guide student study. Subsequent chapters also contain several examples and exercises. The exercises are designed to help students master their grasp of new material presented in the text, and exercise solutions are given at the back. The examples have been chosen to illuminate concepts and techniques that students often find troublesome.

Problems at the ends of chapters are numbered by text section. They range from basic manipulations and computations to more advanced analysis and design tasks. A manual of problem solutions is available to instructors from the publisher.

Several typographical devices have been incorporated to serve as aids for students. Specifically,

- Technical terms are printed in boldface type when they first appear.
- Important concepts and theorems that do not involve equations are printed inside boxes.
- Asterisks (*) after problem numbers indicate that answers are provided at the back of the book.
- The symbol ‡ identifies the more challenging problems.

Tables at the back of the book include transform pairs, mathematical relations, and probability functions for convenient reference. An annotated bibliography is also provided at the back in the form of a supplementary reading list.

Communication system engineers use many abbreviations, so the index lists common abbreviations and their meanings. Thus, the index additionally serves as a guide to many abbreviations in communications.

ACKNOWLEDGMENTS

We are indebted to the many people who contributed to previous editions. We also want to thank Profs. John Chaisson, Mostofa Howlader, Chaouki Abdallah, and

Mssrs. Joao Pinto and Steven Daniel for their assistance and the use of their libraries; the University of Tennessee Electrical and Computer Engineering department for support; Mssrs. Keith McKenzie, James Snively, Neil Troy, and Justin Acuff for their assistance in the manuscript preparation; the staff at McGraw-Hill, especially Michelle Flomenhoft and Mary Lee Harms, for assistance in the preparation of this edition; and the reviewers who helped shape the final manuscript. In particular, we want to thank:

Krishna Arora, Florida A&M University/The Florida State University
Tangul Basar, University of Illinois
Rajarathnam Chandramouli, Stevens Institute of Technology
John F. Doherty, Penn State University
Don R. Halverson, Texas A&M University
Ivan Howitt, University of Wisconsin-Milwaukee
Jacob Klapper, New Jersey Institute of Technology
Haniph Latchman, University of Florida
Harry Leib, McGill University
Mort Naraghi-Pour, Louisiana State University
Raghunathan Rajagopalan, University of Arizona
Rodney Roberts, Florida A&M University/The Florida State University
John Rulnick, Rulnick Engineering
Melvin Sandler, Cooper Union
Marvin Siegel, Michigan State University
Michael H. Thursby, Florida Institute of Technolgy

Special thanks for support, encouragement, and sense of humor go to our spouses and families.

A. Bruce Carlson
Paul B. Crilly
Janet C. Rutledge

chapter

1

Introduction

"**A**ttention, the Universe! By kingdoms, right wheel!" This prophetic phrase represents the first telegraph message on record. Samuel F. B. Morse sent it over a 16 km line in 1838. Thus a new era was born: the era of electrical communication.

Now, over a century and a half later, communication engineering has advanced to the point that earthbound TV viewers watch astronauts working in space. Telephone, radio, and television are integral parts of modern life. Long-distance circuits span the globe carrying text, data, voice, and images. Computers talk to computers via intercontinental networks, and control virtually every electrical appliance in our homes. Wireless personal communication devices keep us connected wherever we go. Certainly great strides have been made since the days of Morse. Equally certain, coming decades will usher in many new achievements of communication engineering.

This textbook introduces electrical communication systems, including analysis methods, design principles, and hardware considerations. We begin with a descriptive overview that establishes a perspective for the chapters that follow.

1.1   ELEMENTS AND LIMITATIONS OF COMMUNICATION SYSTEMS

A communication system conveys information from its source to a destination some distance away. There are so many different applications of communication systems that we cannot attempt to cover every type. Nor can we discuss in detail all the individual parts that make up a specific system. A typical system involves numerous components that run the gamut of electrical engineering—circuits, electronics, electromagnetics, signal processing, microprocessors, and communication networks, to name a few of the relevant fields. Moreover, a piece-by-piece treatment would obscure the essential point that a communication system is an integrated whole that really does exceed the sum of its parts.

We therefore approach the subject from a more general viewpoint. Recognizing that all communication systems have the same basic function of **information transfer,** we'll seek out and isolate the principles and problems of conveying information in electrical form. These will be examined in sufficient depth to develop analysis and design methods suited to a wide range of applications. In short, this text is concerned with communication systems as **systems.**

Information, Messages, and Signals

Clearly, the concept of **information** is central to communication. But information is a loaded word, implying semantic and philosophical notions that defy precise definition. We avoid these difficulties by dealing instead with the **message,** defined as the physical manifestation of information as produced by the source. Whatever form the message takes, the goal of a communication system is to reproduce at the destination an acceptable replica of the source message.

There are many kinds of information sources, including machines as well as people, and messages appear in various forms. Nonetheless, we can identify two distinct message categories, **analog** and **digital.** This distinction, in turn, determines the criterion for successful communication.

Figure 1.1–1   Communication system with input and output transducers.

An **analog** message is a physical quantity that varies with time, usually in a smooth and continuous fashion. Examples of analog messages are the acoustic pressure produced when you speak, the angular position of an aircraft gyro, or the light intensity at some point in a television image. Since the information resides in a time-varying waveform, an analog communication system should deliver this waveform with a specified degree of **fidelity.**

A **digital** message is an ordered sequence of symbols selected from a finite set of discrete elements. Examples of digital messages are the letters printed on this page, a listing of hourly temperature readings, or the keys you press on a computer keyboard. Since the information resides in discrete symbols, a digital communication system should deliver these symbols with a specified degree of **accuracy** in a specified amount of time.

Whether analog or digital, few message sources are inherently electrical. Consequently, most communication systems have input and output **transducers** as shown in Fig. 1.1–1. The input transducer converts the message to an electrical **signal,** say a voltage or current, and another transducer at the destination converts the output signal to the desired message form. For instance, the transducers in a voice communication system could be a microphone at the input and a loudspeaker at the output. We'll assume hereafter that suitable transducers exist, and we'll concentrate primarily on the task of **signal transmission.** In this context the terms **signal** and **message** will be used interchangeably since the signal, like the message, is a physical embodiment of information.

Elements of a Communication System

Figure 1.1–2 depicts the elements of a communication system, omitting transducers but including unwanted contaminations. There are three essential parts of any communication system, the transmitter, transmission channel, and receiver. Each part plays a particular role in signal transmission, as follows.

The **transmitter** processes the input signal to produce a transmitted signal suited to the characteristics of the transmission channel. Signal processing for transmission almost always involves **modulation** and may also include **coding.**

The **transmission channel** is the electrical medium that bridges the distance from source to destination. It may be a pair of wires, a coaxial cable, or a radio wave or laser beam. Every channel introduces some amount of transmission **loss** or **attenuation,** so the signal power progressively decreases with increasing distance.

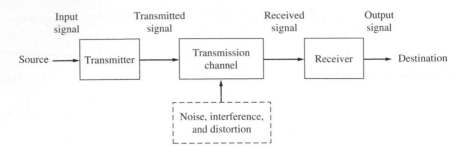

Figure 1.1–2   Elements of a communication system.

The **receiver** operates on the output signal from the channel in preparation for delivery to the transducer at the destination. Receiver operations include **amplification** to compensate for transmission loss, and **demodulation** and **decoding** to reverse the signal-processing performed at the transmitter. **Filtering** is another important function at the receiver, for reasons discussed next.

Various unwanted undesirable effects crop up in the course of signal transmission. Attenuation is undesirable since it reduces signal **strength** at the receiver. More serious, however, are distortion, interference, and noise, which appear as alterations of the signal **shape.** Although such contaminations may occur at any point, the standard convention is to blame them entirely on the channel, treating the transmitter and receiver as being ideal. Figure 1.1–2 reflects this convention.

Distortion is waveform perturbation caused by imperfect response of the system to the desired signal itself. Unlike noise and interference, distortion disappears when the signal is turned off. If the channel has a linear but distorting response, then distortion may be corrected, or at least reduced, with the help of special filters called **equalizers.**

Interference is contamination by extraneous signals from human sources— other transmitters, power lines and machinery, switching circuits, and so on. Interference occurs most often in radio systems whose receiving antennas usually intercept several signals at the same time. Radio-frequency interference (RFI) also appears in cable systems if the transmission wires or receiver circuitry pick up signals radiated from nearby sources. Appropriate filtering removes interference to the extent that the interfering signals occupy different frequency bands than the desired signal.

Noise refers to random and unpredictable electrical signals produced by natural processes both internal and external to the system. When such random variations are superimposed on an information-bearing signal, the message may be partially corrupted or totally obliterated. Filtering reduces noise contamination, but there inevitably remains some amount of noise that cannot be eliminated. This noise constitutes one of the fundamental system limitations.

Finally, it should be noted that Fig. 1.1–2 represents one-way or **simplex** (SX) transmission. Two-way communication, of course, requires a transmitter and receiver at each end. A **full-duplex** (FDX) system has a channel that allows simultaneous transmission in both directions. A **half-duplex** (HDX) system allows transmission in either direction but not at the same time.

Fundamental Limitations

An engineer faces two general kinds of constraints when designing a communication system. On the one hand are the **technological problems,** including such diverse considerations as hardware availability, economic factors, federal regulations, and so on. These are problems of feasibility that can be solved in theory, even though perfect solutions may not be practical. On the other hand are the **fundamental physical limitations,** the laws of nature as they pertain to the task in question. These limitations ultimately dictate what can or cannot be accomplished, irrespective of the technological problems. The fundamental limitations of information transmission by electrical means are **bandwidth** and **noise.**

The concept of bandwidth applies to both signals and systems as a measure of **speed.** When a signal changes rapidly with time, its frequency content, or **spectrum,** extends over a wide range and we say that the signal has a large bandwidth. Similarly, the ability of a system to follow signal variations is reflected in its usable frequency response or **transmission bandwidth.** Now all electrical systems contain energy-storage elements, and stored energy cannot be changed instantaneously. Consequently, every communication system has a finite bandwidth B that limits the rate of signal variations.

Communication under real-time conditions requires sufficient transmission bandwidth to accommodate the signal spectrum; otherwise, severe distortion will result. Thus, for example, a bandwidth of several megahertz is needed for a TV video signal, while the much slower variations of a voice signal fit into $B \approx 3$ kHz. For a digital signal with r symbols per second, the bandwidth must be $B \geq r/2$. In the case of information transmission without a real-time constraint, the available bandwidth determines the maximum signal speed. The time required to transmit a given amount of information is therefore inversely proportional to B.

Noise imposes a second limitation on information transmission. Why is noise unavoidable? Rather curiously, the answer comes from kinetic theory. At any temperature above absolute zero, thermal energy causes microscopic particles to exhibit random motion. The random motion of charged particles such as electrons generates random currents or voltages called **thermal noise.** There are also other types of noise, but thermal noise appears in every communication system.

We measure noise relative to an information signal in terms of the **signal-to-noise power ratio** S/N. Thermal noise power is ordinarily quite small, and S/N can be so large that the noise goes unnoticed. At lower values of S/N, however, noise degrades fidelity in analog communication and produces errors in digital communication. These problems become most severe on long-distance links when the transmission loss reduces the received signal power down to the noise level. Amplification at the receiver is then to no avail, because the noise will be amplified along with the signal.

Taking both limitations into account, Shannon (1948)[†] stated that the rate of information transmission cannot exceed the **channel capacity.**

$$C = B \log (1 + S/N)$$

This relationship, known as the **Hartley-Shannon law,** sets an upper limit on the performance of a communication system with a given bandwidth and signal-to-noise ratio.

1.2   MODULATION AND CODING

Modulation and coding are operations performed at the transmitter to achieve efficient and reliable information transmission. So important are these operations that they deserve further consideration here. Subsequently, we'll devote several chapters to modulating and coding techniques.

Modulation Methods

Modulation involves two waveforms: a **modulating signal** that represents the message, and a **carrier wave** that suits the particular application. A modulator systematically alters the carrier wave in correspondence with the variations of the modulating signal. The resulting modulated wave thereby "carries" the message information. We generally require that modulation be a *reversible* operation, so the message can be retrieved by the complementary process of **demodulation.**

Figure 1.2–1 depicts a portion of an analog modulating signal (part *a*) and the corresponding modulated waveform obtained by varying the amplitude of a **sinusoidal** carrier wave (part *b*). This is the familiar amplitude modulation (AM) used for radio broadcasting and other applications. A message may also be impressed on a sinusoidal carrier by frequency modulation (FM) or phase modulation (PM). All methods for sinusoidal carrier modulation are grouped under the heading of **continuous-wave** (CW) modulation.

Incidentally, you act as a CW modulator whenever you speak. The transmission of voice through air is accomplished by generating carrier tones in the vocal cords and modulating these tones with muscular actions of the oral cavity. Thus, what the ear hears as speech is a modulated acoustic wave similar to an AM signal.

Most long-distance transmission systems employ CW modulation with a carrier frequency much higher than the highest frequency component of the modulating signal. The spectrum of the modulated signal then consists of a band of frequency components clustered around the carrier frequency. Under these conditions, we say that CW modulation produces **frequency translation.** In AM broadcasting, for example, the message spectrum typically runs from 100 Hz to 5 kHz; if the carrier frequency is 600 kHz, then the spectrum of the modulated carrier covers 595–605 kHz.

[†]References are indicated in this fashion throughout the text. Complete citations are listed alphabetically by author in the References at the end of the book.

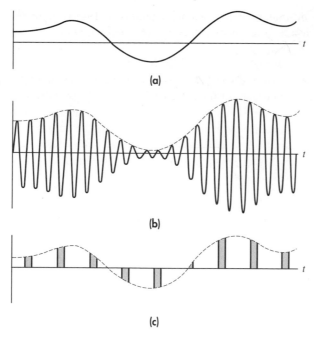

Figure 1.2–1   (*a*) Modulating signal; (*b*) sinusoidal carrier with amplitude modulation;
(*c*) pulse-train carrier with amplitude modulation.

Another modulation method, called **pulse modulation,** has a periodic train of
short pulses as the carrier wave. Figure 1.2–1*c* shows a waveform with pulse amplitude modulation (PAM). Notice that this PAM wave consists of short samples
extracted from the analog signal at the top of the figure. **Sampling** is an important
signal-processing technique and, subject to certain conditions, it's possible to
reconstruct an entire waveform from periodic samples.

But pulse modulation by itself does not produce the frequency translation
needed for efficient signal transmission. Some transmitters therefore combine pulse
and CW modulation. Other modulation techniques, described shortly, combine
pulse modulation with **coding.**

Modulation Benefits and Applications

The primary purpose of modulation in a communication system is to generate a modulated signal suited to the characteristics of the transmission channel. Actually, there
are several practical benefits and applications of modulation briefly discussed below.

Modulation for Efficient Transmission   Signal transmission over appreciable distance
always involves a traveling electromagnetic wave, with or without a guiding medium.

The efficiency of any particular transmission method depends upon the frequency of the signal being transmitted. By exploiting the frequency-translation property of CW modulation, message information can be impressed on a carrier whose frequency has been selected for the desired transmission method.

As a case in point, efficient line-of-sight ratio propagation requires antennas whose physical dimensions are at least 1/10 of the signal's wavelength. Unmodulated transmission of an audio signal containing frequency components down to 100 Hz would thus call for antennas some 300 km long. Modulated transmission at 100 MHz, as in FM broadcasting, allows a practical antenna size of about one meter. At frequencies below 100 MHz, other propagation modes have better efficiency with reasonable antenna sizes. Tomasi (1994, Chap. 10) gives a compact treatment of radio propagation and antennas.

For reference purposes, Fig. 1.2–2 shows those portions of the electromagnetic spectrum suited to signal transmission. The figure includes the free-space wavelength, frequency-band designations, and typical transmission media and propagation modes. Also indicated are representative applications authorized by the U.S. Federal Communications Commission.

Modulation to Overcome Hardware Limitations   The design of a communication system may be constrained by the cost and availability of hardware, hardware whose performance often depends upon the frequencies involved. Modulation permits the designer to place a signal in some frequency range that avoids hardware limitations. A particular concern along this line is the question of **fractional bandwidth,** defined as absolute bandwidth divided by the center frequency. Hardware costs and complications are minimized if the fractional bandwidth is kept within 1–10 percent. Fractional-bandwidth considerations account for the fact that modulation units are found in receivers as well as in transmitters.

It likewise follows that signals with large bandwidth should be modulated on high-frequency carriers. Since information rate is proportional to bandwidth, according to the Hartley-Shannon law, we conclude that a high information rate requires a high carrier frequency. For instance, a 5 GHz microwave system can accommodate 10,000 times as much information in a given time interval as a 500 kHz radio channel. Going even higher in the electromagnetic spectrum, one optical laser beam has a bandwidth potential equivalent to 10 million TV channels.

Modulation to Reduce Noise and Interference   A brute-force method for combating noise and interference is to increase the signal power until it overwhelms the contaminations. But increasing power is costly and may damage equipment. (One of the early transatlantic cables was apparently destroyed by high-voltage rupture in an effort to obtain a usable received signal.) Fortunately, FM and certain other types of modulation have the valuable property of suppressing both noise and interference.

This property is called **wideband noise reduction** because it requires the transmission bandwidth to be much greater than the bandwidth of the modulating signal. Wideband modulation thus allows the designer to exchange increased bandwidth for

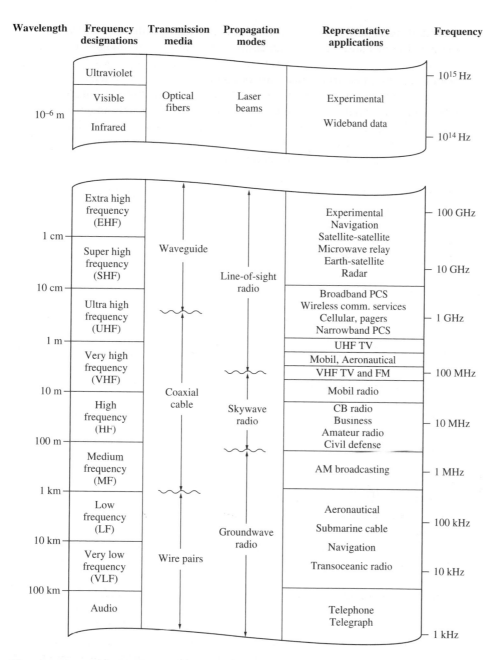

Figure 1.2–2 The electromagnetic spectrum.

decreased signal power, a trade-off implied by the Hartley-Shannon law. Note that a higher carrier frequency may be needed to accommodate wideband modulation.

Modulation for Frequency Assignment   When you tune a radio or television set to a particular station, you are selecting one of the many signals being received at that time. Since each station has a different assigned carrier frequency, the desired signal can be separated from the others by filtering. Were it not for modulation, only one station could broadcast in a given area; otherwise, two or more broadcasting stations would create a hopeless jumble of interference.

Modulation for Multiplexing   Multiplexing is the process of combining several signals for simultaneous transmission on one channel. **Frequency-division multiplexing** (FDM) uses CW modulation to put each signal on a different carrier frequency, and a bank of filters separates the signals at the destination. **Time-division multiplexing** (TDM) uses pulse modulation to put samples of different signals in nonoverlapping time slots. Back in Fig. 1.2–1c, for instance, the gaps between pulses could be filled with samples from other signals. A switching circuit at the destination then separates the samples for signal reconstruction. Applications of multiplexing include FM stereophonic broadcasting, cable TV, and long-distance telephone.

A variation of multiplexing is **multiple access** (MA). Whereas multiplexing involves a fixed assignment of the common communications resource (such as frequency spectrum) at the local level, MA involves the remote sharing of the resource. For example, code-division multiple access (CDMA) assigns a unique code to each digital cellular user, and the individual transmissions are separated by correlation between the codes of the desired transmitting and receiving parties. Since CDMA allows different users to share the same frequency band simultaneously, it provides another way of increasing communication efficiency.

Coding Methods and Benefits

We've described modulation as a *signal*-processing operation for effective transmission. **Coding** is a *symbol*-processing operation for improved communication when the information is digital or can be approximated in the form of discrete symbols. Both coding and modulation may be necessary for reliable long-distance digital transmission.

The operation of **encoding** transforms a digital message into a new sequence of symbols. **Decoding** converts an encoded sequence back to the original message with, perhaps, a few errors caused by transmission contaminations. Consider a computer or other digital source having $M \gg 2$ symbols. Uncoded transmission of a message from this source would require M different waveforms, one for each symbol. Alternatively, each symbol could be represented by a **binary codeword** consisting of K binary digits. Since there are 2^K possible codewords made up of K binary digits, we need $K \geq \log_2 M$ digits per codeword to encode M source symbols. If the source produces r symbols per second, the binary code will have Kr digits per second and the transmission bandwidth requirement is K times the bandwidth of an uncoded signal.

In exchange for increased bandwidth, binary encoding of M-ary source symbols offers two advantages. First, less complicated hardware is needed to handle a binary signal composed of just two different waveforms. Second, contaminating noise has less effect on a binary signal than it does on a signal composed of M different waveforms, so there will be fewer errors caused by the noise. Hence, this coding method is essentially a digital technique for wideband noise reduction.

Channel coding is a technique used to introduce controlled redundancy to further improve the performance reliability in a noisy channel. **Error-control coding** goes further in the direction of wideband noise reduction. By appending extra **check digits** to each binary codeword, we can detect, or even correct, most of the errors that do occur. Error-control coding increases both bandwidth and hardware complexity, but it pays off in terms of nearly error-free digital communication despite a low signal-to-noise ratio.

Now, let's examine the other fundamental system limitation: bandwidth. Many communication systems rely on the telephone network for transmission. Since the bandwidth of the transmission system is limited by decades-old design specifications, in order to increase the data rate, the signal bandwidth must be reduced. High-speed modems (data *mo*dulator/*dem*odulators) are one application requiring such data reduction. **Source-coding** techniques take advantage of the statistical knowledge of the source signal to enable efficient encoding. Thus, source coding can be viewed as the dual of channel coding in that it *reduces* redundancy to achieve the desired efficiency.

Finally, the benefits of digital coding can be incorporated in *analog* communication with the help of an analog-to-digital conversion method such as pulse-code-modulation (PCM). A PCM signal is generated by sampling the analog message, digitizing (quantizing) the sample values, and encoding the sequence of digitized samples. In view of the reliability, versatility, and efficiency of digital transmission, PCM has become an important method for analog communication. Furthermore, when coupled with high-speed microprocessors, PCM makes it possible to substitute **digital signal processing** for analog operations.

1.3   HISTORICAL PERSPECTIVE AND SOCIETAL IMPACT

In our daily lives we often take for granted the powerful technologies that allow us to communicate, nearly instantaneously, with people around the world. Many of us now have multiple phone numbers to handle our home and business telephones, facsimile machines, modems, and wireless personal communication devices. We send text, video, and music through electronic mail, and we "surf the Net" for information and entertainment. We have more television stations than we know what to do with, and "smart electronics" allow our household appliances to keep us posted on their health. It is hard to believe that most of these technologies were developed in the past 50 years.

Historical Perspective

The organization of this text is dictated by pedagogical considerations and does not necessarily reflect the evolution of communication systems. To provide at least some historical perspective, a chronological outline of electrical communication is presented in Table 1.3–1. The table lists key inventions, scientific discoveries, important papers, and the names associated with these events.

Table 1.3–1   A chronology of electrical communication

Year	Event
1800–1837	*Preliminary developments* Volta discovers the primary battery; the mathematical treatises by Fourier, Cauchy, and Laplace; experiments on electricity and magnetism by Oersted, Ampere, Faraday, and Henry; Ohm's law (1826); early telegraph systems by Gauss, Weber, and Wheatstone.
1838–1866	*Telegraphy* Morse perfects his system; Steinheil finds that the earth can be used for a current path; commercial service initiated (1844); multiplexing techniques devised; William Thomson (Lord Kelvin) calculates the pulse response of a telegraph line (1855); transatlantic cables installed by Cyrus Field and associates.
1845	Kirchhoff's circuit laws enunciated.
1864	Maxwell's equations predict electromagnetic radiation.
1876–1899	*Telephony* Acoustic transducer perfected by Alexander Graham Bell, after earlier attempts by Reis; first telephone exchange, in New Haven, with eight lines (1878); Edison's carbon-button transducer; cable circuits introduced; Strowger devises automatic step-by-step switching (1887); the theory of cable loading by Heaviside, Pupin, and Campbell.
1887–1907	*Wireless telegraphy* Heinrich Hertz verifies Maxell's theory; demonstrations by Marconi and Popov; Marconi patents a complete wireless telegraph system (1897); the theory of tuning circuits developed by Sir Oliver Lodge; commercial service begins, including ship-to-shore and transatlantic systems.
1892–1899	Oliver Heaviside's publications on operational calculus, circuits, and electromagnetics.
1904–1920	*Communication electronics* Lee De Forest invents the Audion (triode) based on Fleming's diode; basic filter types devised by G. A. Campbell and others; experiments with AM radio broadcasting; transcontinental telephone line with electronic repeaters completed by the Bell System (1915); multiplexed carrier telephony introduced; E. H. Armstrong perfects the superheterodyne radio receiver (1918); first commercial broadcasting station, KDKA, Pittsburgh.
1920–1928	*Transmission theory* Landmark papers on the theory of signal transmission and noise by J. R. Carson, H. Nyquist, J. B. Johnson, and R. V. L. Hartley.
1923–1938	*Television* Mechanical image-formation system demonstrated by Baird and Jenkins; theoretical analysis of bandwidth requirements; Farnsworth and Zworykin propose electronic systems; vacuum cathode-ray tubes perfected by DuMont and others; field tests and experimental broadcasting begin.
1927	Federal Communications Commission established.

Table 1.3–1   A chronology of electrical communication *(continued)*

Year	Event
1931	Teletypewriter service initiated.
1934	H. S. Black develops the negative-feedback amplifier.
1936	Armstrong's paper states the case for FM radio.
1937	Alec Reeves conceives pulse-code modulation.
1938–1945	*World War II* Radar and microwave systems developed; FM used extensively for military communications; improved electronics, hardware, and theory in all areas.
1944–1947	*Statistical communication theory* Rice develops a mathematical representation of noise; Weiner, Kolmogoroff, and Kotel'nikov apply statistical methods to signal detection.
1948–1950	*Information theory and coding* C. E. Shannon publishes the founding papers of information theory; Hamming and Golay devise error-correcting codes.
1948–1951	Transistor devices invented by Bardeen, Brattain, and Shockley.
1950	Time-division multiplexing applied to telephony.
1953	Color TV standards established in the United States.
1955	J. R. Pierce proposes satellite communication systems.
1956	First transoceanic telephone cable (36 voice channels).
1958	Long-distance data transmission system developed for military purposes.
1960	Maiman demonstrates the first laser.
1961	Integrated circuits go into commercial production; stereo FM broadcasts begin in the U.S.
1962	Satellite communication begins with Telstar I.
1962–1966	*High-speed digital communication* Data transmission service offered commercially; Touch-Tone telephone service introduced; wideband channels designed for digital signaling; pulse-code modulation proves feasible for voice and TV transmission; major breakthroughs in theory and implementation of digital transmission, including error-control coding methods by Viterbi and others, and the development of adaptive equalization by Lucky and coworkers.
1963	Solid-state microwave oscillators perfected by Gunn.
1964	Fully electronic telephone switching system (No. 1 ESS) goes into service.
1965	Mariner IV transmits pictures from Mars to Earth.
1966–1975	*Wideband communication systems* Cable TV systems; commercial satellite relay service becomes available; optical links using lasers and fiber optics.
1969	ARPANET created (precursor to Internet)
1971	Intel develops first single-chip microprocessor
1972	Motorola develops cellular telephone; first live TV broadcast across Atlantic ocean via satellite
1980	Compact disc developed by Philips and Sony
1981	FCC adopts rules creating commercial cellular telephone service; IBM PC is introduced (hard drives introduced two years later).
1982	AT&T agrees to divest 22 local service telephone companies; seven regional Bell system operating companies formed.

(continued)

Table 1.3–1   A chronology of electrical communication *(continued)*

Year	Event
1985	Fax machines widely available in offices.
1988–1989	Installation of trans-Pacific and trans-Atlantic optical cables for light-wave communications.
1990–2000	*Digital communication systems* Digital signal processing and communication systems in household appliances; digitally tuned receivers; direct-sequence spread spectrum systems; integrated services digital networks (ISDNs); high-definition digital television (HDTV) standards developed; digital pagers; handheld computers; digital cellular.
1994–1995	FCC raises $7.7 billion in auction of frequency spectrum for broadband personal communication devices
1998	Digital television service launched in U.S.

Several of the terms in the chronology have been mentioned already, while others will be described in later chapters when we discuss the impact and interrelationships of particular events. You may therefore find it helpful to refer back to this table from time to time.

Societal Impact

Our planet feels a little smaller in large part due to advances in communication. Multiple sources constantly provide us with the latest news of world events, and savvy leaders make great use of this to shape opinions in their own countries and abroad. Communication technologies change how we do business, and once-powerful companies, unable to adapt, are disappearing. Cable and telecommunications industries split and merge at a dizzying pace, and the boundaries between their technologies and those of computer hardware and software companies are becoming blurred. We are able (and expected) to be connected 24 hours a day, seven days a week, which means that we may continue to receive work-related E-mail, phone calls, and faxes, even while on vacation at the beach or in an area once considered remote.

These technology changes spur new public policy debates, chiefly over issues of personal privacy, information security, and copyright protection. New businesses taking advantage of the latest technologies appear at a faster rate than the laws and policies required to govern these issues. With so many computer systems connected to the Internet, malicious individuals can quickly spread computer viruses around the globe. Cellular phones are so pervasive that theaters and restaurants have created policies governing their use. For example, it was not so long ago that before a show an announcement would be made that smoking was not allowed in the auditorium. Now some theaters request that members of the audience turn off cell phones and beepers. State laws, municipal franchises, and public utility commissions must change to accommodate the telecommunications revolution. And the workforce must stay current with advances in technology via continuing education.

With new technologies developing at an exponential rate, we cannot say for certain what the world will be like in another 50 years. Nevertheless, a firm grounding in the basics of communication systems, creativity, commitment to ethical application of technology, and strong problem solving skills will equip the communications engineer with the capability to shape that future.

1.4   PROSPECTUS

This text provides a comprehensive introduction to analog and digital communications. A review of relevant background material precedes each major topic that is presented. Each chapter begins with an overview of the subjects covered and a listing of learning objectives. Throughout the text we rely heavily on mathematical models to cut to the heart of complex problems. Keep in mind, however, that such models must be combined with physical reasoning and engineering judgment.

Chapters 2 and 3 deal with deterministic signals, emphasizing time-domain and frequency-domain analysis of signal transmission, distortion, and filtering. Chapters 4 and 5 discuss the *how* and the *why* of various types of CW modulation. Particular topics include modulated waveforms, transmitters, and transmission bandwidth. Sampling and pulse modulation are introduced in Chapter 6, followed by analog modulation systems, including receivers, multiplexing systems, and television systems in Chapter 7. Before a discussion of the impact of noise on CW modulation systems in Chapter 10, Chapters 8 and 9 apply probability theory and statistics to the representation of random signals and noise.

Digital communication starts in Chapter 11 with baseband (unmodulated) transmission, so we can focus on the important concepts of digital signals and spectra, noise and errors, and synchronization. Chapter 12 then draws upon previous chapters for the study of coded pulse modulation, including PCM and digital multiplexing systems. A short survey of error-control coding is presented in Chapter 13. Chapter 14 analyzes digital transmission systems with CW modulation, culminating in a performance comparison of various methods. An expanded presentation of spread spectrum systems is presented in this edition in Chapter 15. Finally, an introduction to information theory in Chapter 16 provides a retrospective view of digital communication and returns us to the Hartley-Shannon law.

Each chapter contains several exercises designed to clarify and reinforce the concepts and analytic techniques. You should work these exercises as you come to them, checking your results with the answers provided at the back of the book. Also at the back you'll find tables containing handy summaries of important text material and mathematical relations pertinent to the exercises and to the problems at the end of each chapter.

Although we mostly describe communication systems in terms of "black boxes" with specified properties, we'll occasionally lift the lid to look at electronic circuits that carry out particular operations. Such digressions are intended to be illustrative rather than a comprehensive treatment of communication electronics.

Besides discussions of electronics, certain optional or more advanced topics are interspersed in various chapters and identified by the symbol ★. These topics may be omitted without loss of continuity. Other optional material of a supplementary nature is contained in the appendix.

Two types of references have been included. Books and papers cited within chapters provide further information about specific items. Additional references are collected in a supplementary reading list that serves as an annotated bibliography for those who wish to pursue subjects in greater depth.

Finally, as you have probably observed, communications engineers use many abbreviations and acronyms. Most abbreviations defined in this book are also listed in the index, to which you can refer if you happen to forget a definition.

chapter
2

Signals and Spectra

CHAPTER OUTLINE

Electrical communication signals are time-varying quantities such as voltage or current. Although a signal physically exists in the time domain, we can also represent it in the **frequency domain** where we view the signal as consisting of sinusoidal components at various frequencies. This frequency-domain description is called the **spectrum.**

Spectral analysis, using the Fourier series and transform, is one of the fundamental methods of communication engineering. It allows us to treat entire *classes* of signals that have similar properties in the frequency domain, rather than getting bogged down in detailed time-domain analysis of individual signals. Furthermore, when coupled with the frequency-response characteristics of filters and other system components, the spectral approach provides valuable insight for design work.

This chapter therefore is devoted to signals and spectral analysis, giving special attention to the frequency-domain interpretation of signal properties. We'll examine line spectra based on the Fourier series expansion of periodic signals, and continuous spectra based on the Fourier transform of nonperiodic signals. These two types of spectra will ultimately be merged with the help of the impulse concept.

As the first step in spectral analysis we must write equations representing signals as functions of time. But such equations are only mathematical *models* of the real world, and imperfect models at that. In fact, a completely faithful description of the simplest physical signal would be quite complicated and impractical for engineering purposes. Hence, we try to devise models that represent with minimum complexity the significant properties of physical signals. The study of many different signal models provides us with the background needed to choose appropriate models for specific applications. In many cases, the models will apply only to particular classes of signals. Throughout the chapter the major classifications of signals will be highlighted for their special properties.

OBJECTIVES

After studying this chapter and working the exercises, you should be able to do each of the following:

1. Sketch and label the one-sided or two-sided line spectrum of a signal consisting of a sum of sinusoids (Sect. 2.1).
2. Calculate the average value, average power, and total energy of a simple signal (Sects. 2.1 and 2.2).
3. Write the expressions for the exponential Fourier series and coefficients, the trigonometric Fourier series, and the direct and inverse Fourier transform (Sects. 2.1 and 2.2).
4. Identify the time-domain properties of a signal from its frequency-domain representation and vice versa (Sect. 2.2).
5. Sketch and label the spectrum of a rectangular pulse train, a single rectangular pulse, or a sinc pulse (Sects. 2.1 and 2.2).
6. State and apply Parseval's power theorem and Rayleigh's energy theorem (Sects. 2.1 and 2.2).
7. State the following transform theorems: superposition, time delay, scale change, frequency translation and modulation, differentiation and integration (Sect. 2.3).
8. Use transform theorems to find and sketch the spectrum of a signal defined by time-domain operations (Sect. 2.3).
9. Set up the convolution integral and simplify it as much as possible when one of the functions is a rectangular pulse (Sect. 2.4).
10. State and apply the convolution theorems (Sect. 2.4).
11. Evaluate or otherwise simplify expressions containing impulses (Sect. 2.5).
12. Find the spectrum of a signal consisting of constants, steps, impulses, sinusoids, and/or rectangular and triangular functions (Sect. 2.5).

2.1   LINE SPECTRA AND FOURIER SERIES

This section introduces and interprets the frequency domain in terms of rotating phasors. We'll begin with the line spectrum of a sinusoidal signal. Then we'll invoke the Fourier series expansion to obtain the line spectrum of any periodic signal that has finite average power.

Phasors and Line Spectra

Consider the familiar sinusoidal or ac (alternating-current) waveform $v(t)$ plotted in Fig. 2.1–1. By convention, we express sinusoids in terms of the cosine function and write

$$v(t) = A \cos (\omega_0 t + \phi) \qquad \text{[1]}$$

where A is the peak value or **amplitude** and ω_0 is the **radian frequency.** The **phase angle** ϕ represents the fact that the peak has been shifted away from the time origin and occurs at $t = -\phi/\omega_0$. Equation (1) implies that $v(t)$ repeats itself for all time, with repetition **period** $T_0 = 2\pi/\omega_0$. The reciprocal of the period equals the **cyclical frequency**

$$f_0 \triangleq \frac{1}{T_0} = \frac{\omega_0}{2\pi} \qquad \text{[2]}$$

measured in cycles per second or hertz.

Obviously, no real signal goes on forever, but Eq. (1) could be a reasonable model for a sinusoidal waveform that lasts a long time compared to the period. In particular, ac steady-state circuit analysis depends upon the assumption of an eternal sinusoid—usually represented by a complex exponential or **phasor.** Phasors also play a major role in the spectral analysis.

The phasor representation of a sinusoidal signal comes from **Euler's theorem**

$$e^{\pm j\theta} = \cos \theta \pm j \sin \theta \qquad \text{[3]}$$

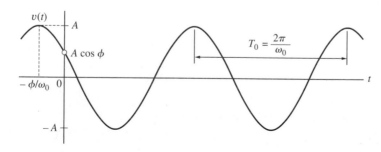

Figure 2.1–1   A sinusoidal waveform $v(t) = A \cos (\omega_0 t + \phi)$.

where $j \triangleq \sqrt{-1}$ and θ is an arbitrary angle. If we let $\theta = \omega_0 t + \phi$, we can write any sinusoid as the real part of a complex exponential, namely

$$A \cos (\omega_0 t + \phi) = A \operatorname{Re} \left[e^{j(\omega_0 t + \phi)} \right] \qquad \text{[4]}$$
$$= \operatorname{Re} \left[A e^{j\phi} e^{j\omega_0 t} \right]$$

This is called a phasor representation because the term inside the brackets may be viewed as a rotating vector in a complex plane whose axes are the real and imaginary parts, as Fig. 2.1–2a illustrates. The phasor has length A, rotates counterclockwise at a rate of f_0 revolutions per second, and at time $t = 0$ makes an angle ϕ with respect to the positive real axis. The projection of the phasor on the real axis equals the sinusoid in Eq. (4).

Now observe that only three parameters completely specify a phasor: amplitude, phase angle, and rotational frequency. To describe the same phasor in the **frequency domain,** we must associate the corresponding amplitude and phase with the particular frequency f_0. Hence, a suitable frequency-domain description would be the **line spectrum** in Fig. 2.1–2b, which consists of two plots: amplitude versus frequency and phase versus frequency. While this figure appears simple to the point of being trivial, it does have great conceptual value when extended to more complicated signals. But before taking that step, four conventions regarding line spectra should be stated.

1. In all our spectral drawings the independent variable will be **cyclical frequency** f hertz, rather than radian frequency ω, and any specific frequency such as f_0 will be identified by a subscript. (We'll still use ω with or without subscripts as a shorthand notation for $2\pi f$ since that combination occurs so often.)

2. Phase angles will be measured with respect to **cosine** waves or, equivalently, with respect to the positive real axis of the phasor diagram. Hence, sine waves need to be converted to cosines via the identity

$$\sin \omega t = \cos (\omega t - 90°) \qquad \text{[5]}$$

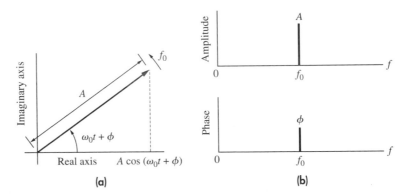

(a)   (b)

Figure 2.1–2   Representations of $A \cos (\omega_0 t + \phi)$. (a) Phasor diagram; (b) line spectrum.

3. We regard amplitude as always being a **positive** quantity. When negative signs appear, they must be absorbed in the phase using

$$-A \cos \omega t = A \cos (\omega t \pm 180°)$$ [6]

It does not matter whether you take $+180°$ or $-180°$ since the phasor ends up in the same place either way.

4. Phase angles usually are expressed in degrees even though other angles such as ωt are inherently in radians. No confusion should result from this mixed notation since angles expressed in degrees will always carry the appropriate symbol.

To illustrate these conventions and to carry further the idea of line spectrum, consider the signal

$$w(t) = 7 - 10 \cos (40\pi t - 60°) + 4 \sin 120\pi t$$

which is sketched in Fig. 2.1–3a. Converting the constant term to a zero frequency or dc (direct-current) component and applying Eqs. (5) and (6) gives the sum of cosines

$$w(t) = 7 \cos 2\pi 0t + 10 \cos (2\pi 20t + 120°) + 4 \cos (2\pi 60t - 90°)$$

whose spectrum is shown in Fig. 2.1–3b.

Drawings like Fig. 2.1–3b, called **one-sided** or *positive*-frequency line spectra, can be constructed for any linear combination of sinusoids. But another spectral representation turns out to be more valuable, even though it involves *negative* frequencies. We obtain this representation from Eq. (4) by recalling that $\mathrm{Re}[z] = \frac{1}{2}(z + z^*)$, where z is any complex quantity with complex conjugate z^*. Hence, if $z = Ae^{j\phi}e^{j\omega_0 t}$ then $z^* = Ae^{j\phi}e^{-j\omega_0 t}$ and Eq. (4) becomes

$$A \cos (\omega_0 t + \phi) = \frac{A}{2} e^{j\phi}e^{j\omega_0 t} + \frac{A}{2} e^{-j\phi}e^{-j\omega_0 t}$$ [7]

so we now have a *pair of conjugate phasors.*

The corresponding phasor diagram and line spectrum are shown in Fig. 2.1–4. The phasor diagram consists of two phasors with equal lengths but opposite angles and directions of rotation. The phasor sum always falls along the real axis to yield $A \cos (\omega_0 t + \phi)$. The line of spectrum is **two-sided** since it must include negative frequencies to allow for the opposite rotational directions, and one-half of the original amplitude is associated with each of the two frequencies $\pm f_0$. The amplitude spectrum has **even symmetry** while the phase spectrum has **odd symmetry** because we are dealing with conjugate phasors. This symmetry appears more vividly in Fig. 2.1–5, which is the two-sided version of Fig. 2.1–3b.

It should be emphasized that these line spectra, one-sided or two-sided, are just pictorial ways of representing sinusoidal or phasor time functions. A single line in the one-sided spectrum represents a *real cosine wave,* whereas a single line in the two-sided spectrum represents a *complex exponential* and the conjugate term must be added to get a real cosine wave. Thus, whenever we speak of some frequency interval such as f_1 to f_2 in a two-sided spectrum, we should also include the corresponding

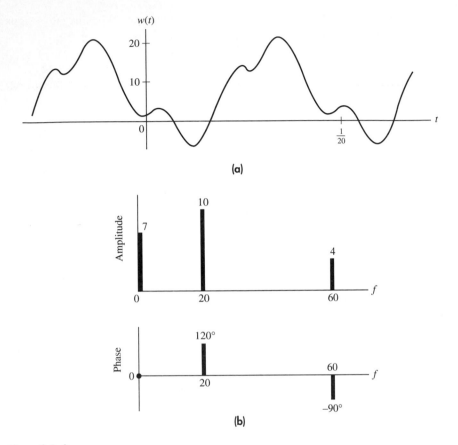

(a)

(b)

Figure 2.1–3

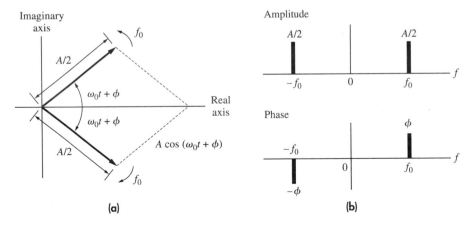

(a)

(b)

Figure 2.1–4   (a) Conjugate phasors; (b) two-sided spectrum.

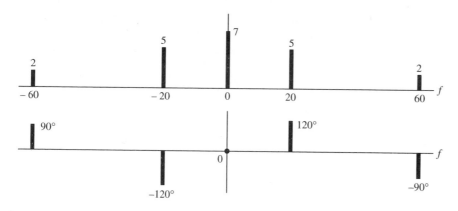

Figure 2.1–5

negative-frequency interval $-f_1$ to $-f_2$. A simple notation for specifying both inter-
vals is $f_1 \leq |f| \leq f_2$.

Finally, note that

> The amplitude spectrum in either version conveys more information than the phase spectrum. Both
> parts are required to define the time-domain function, but the amplitude spectrum by itself tells us
> what frequencies are present and in what proportion.

Putting this another way, the amplitude spectrum displays the signal's *frequency
content.*

Construct the one-sided and two-sided spectrum of $v(t) = -3 - 4 \sin 30\pi t$.   **EXERCISE 2.1–1**

Periodic Signals and Average Power

Sinusoids and phasors are members of the general class of *periodic signals.* These
signals obey the relationship

$$v(t \pm mT_0) = v(t) \qquad -\infty < t < \infty \qquad \text{[8]}$$

where m is any integer. This equation simply says that shifting the signal by an integer
number of periods to the left or right leaves the waveform unchanged. Consequently, a
periodic signal is fully described by specifying its behavior over any one period.

The frequency-domain representation of a periodic signal is a line spectrum
obtained by Fourier series expansion. The expansion requires that the signal have

finite average power. Because average power and other time averages are important signal properties, we'll formalize these concepts here.

Given any time function $v(t)$, its **average value** over all time is defined as

$$\langle v(t) \rangle \triangleq \lim_{T \to \infty} \frac{1}{T} \int_{-T/2}^{T/2} v(t)\, dt \qquad [9]$$

The notation $\langle v(t) \rangle$ represents the averaging operation on the right-hand side, which comprises three steps: integrate $v(t)$ to get the net area under the curve from $-T/2 \leq t \leq T/2$; divide that area by the duration T of the time interval; then let $T \to \infty$ to encompass all time. In the case of a **periodic** signal, Eq. (9) reduces to the average over any interval of duration T_0. Thus

$$\langle v(t) \rangle = \frac{1}{T_0} \int_{t_1}^{t_1 + T_0} v(t)\, dt = \frac{1}{T_0} \int_{T_0} v(t)\, dt \qquad [10]$$

where the shorthand symbol $\int_{T_0}$ stands for an integration from any time t_1 to $t_1 + T_0$.

If $v(t)$ happens to be the voltage across a resistance R, it produces the current $i(t) = v(t)/R$ and we could compute the resulting average power by averaging the instantaneous power $v(t)i(t) = v^2(t)/R = Ri^2(t)$. But we don't necessarily know whether a given signal is a voltage or current, so let's **normalize** power by assuming henceforth that $R = 1\ \Omega$. Our definition of the **average power** associated with an arbitrary periodic signal then becomes

$$P \triangleq \langle |v(t)|^2 \rangle = \frac{1}{T_0} \int_{T_0} |v(t)|^2\, dt \qquad [11]$$

where we have written $|v(t)|^2$ instead of $v^2(t)$ to allow for the possibility of *complex* signal models. In any case, the value of P will be *real* and *nonnegative*.

When the integral in Eq. (11) exists and yields $0 < P < \infty$, the signal $v(t)$ is said to have well-defined average power, and will be called a *periodic power signal*. Almost all periodic signals of practical interest fall in this category. The average value of a power signal may be positive, negative, or zero.

Some signal averages can be found by inspection, using the physical interpretation of averaging. As a specific example take the sinusoid

$$v(t) = A \cos (\omega_0 t + \phi)$$

which has

$$\langle v(t) \rangle = 0 \qquad P = \frac{A^2}{2} \qquad [12]$$

You should have no trouble confirming these results if you sketch one period of $v(t)$ and $|v(t)|^2$.

Fourier Series

The signal $w(t)$ back in Fig. 2.1–3 was generated by summing a dc term and two sinusoids. Now we'll go the other way and decompose periodic signals into sums of sinusoids or, equivalently, rotating phasors. We invoke the **exponential Fourier series** for this purpose.

Let $v(t)$ be a power signal with period $T_0 = 1/f_0$. Its exponential Fourier series expansion is

$$v(t) = \sum_{n=-\infty}^{\infty} c_n e^{j2\pi n f_0 t} \qquad n = 0, 1, 2, \ldots \tag{13}$$

The series coefficients are related to $v(t)$ by

$$c_n = \frac{1}{T_0} \int_{T_0} v(t) e^{-j2\pi n f_0 t}\, dt \tag{14}$$

so c_n equals the average of the product $v(t)e^{-j2\pi n f_0 t}$. Since the coefficients are complex quantities in general, they can be expressed in the polar form

$$c_n = |c_n|\, e^{j \arg c_n}$$

where arg c_n stands for the **angle** of c_n. Equation (13) thus expands a periodic power signal as an infinite sum of phasors, the nth term being

$$c_n e^{j2\pi n f_0 t} = |c_n|\, e^{j \arg c_n} e^{j2\pi n f_0 t}$$

The series convergence properties will be discussed after considering its spectral implications.

Observe that $v(t)$ in Eq. (13) consists of phasors with amplitude $|c_n|$ and angle arg c_n at the frequencies $nf_0 = 0, \pm f_0, \pm 2f_0, \ldots$. Hence, the corresponding frequency-domain picture is a two-sided line spectrum defined by the series coefficients. We emphasize the spectral interpretation by writing

$$c(nf_0) \triangleq c_n$$

so that $|c(nf_0)|$ represents the **amplitude spectrum** as a function of f, and arg $c(nf_0)$ represents the **phase spectrum.** Three important spectral properties of periodic power signals are listed below.

1. All frequencies are integer multiples or **harmonics** of the **fundamental frequency** $f_0 = 1/T_0$. Thus the spectral lines have uniform spacing f_0.

2. The dc component equals the **average value** of the signal, since setting $n = 0$ in Eq. (14) yields

$$c(0) = \frac{1}{T_0} \int_{T_0} v(t)\, dt = \langle v(t) \rangle \tag{15}$$

Calculated values of $c(0)$ may be checked by inspecting $v(t)$—a wise practice when the integration gives an ambiguous result.

3. If $v(t)$ is a **real** (noncomplex) function of time, then

$$c_{-n} = c_n^* = |c_n| \, e^{-j \arg c_n} \qquad [16a]$$

which follows from Eq. (14) with n replaced by $-n$. Hence

$$|c(-nf_0)| = |c(nf_0)| \qquad \arg c(-nf_0) = -\arg c(nf_0) \qquad [16b]$$

which means that the amplitude spectrum has even symmetry and the phase spectrum has odd symmetry.

When dealing with real signals, the property in Eq. (16) allows us to regroup the exponential series into **complex-conjugate pairs,** except for c_0. Equation (13) then becomes

$$v(t) = c_0 + \sum_{n=1}^{\infty} |2c_n| \, \cos\left(2\pi nf_0\, t + \arg c_n\right) \qquad [17]$$

which is the **trigonometric Fourier series** and suggests a one-sided spectrum. Most of the time, however, we'll use the exponential series and two-sided spectra.

One final comment should be made before taking up an example. The integration for c_n often involves a phasor average in the form

$$\frac{1}{T} \int_{-T/2}^{T/2} e^{j2\pi ft}\, dt = \frac{1}{j2\pi f\, T}\left(e^{j\pi fT} - e^{-j\pi fT}\right) \qquad [18]$$

$$= \frac{1}{\pi f\, T}\, \sin \pi f\, T$$

Since this expression occurs time and again in spectral analysis, we'll now introduce the **sinc function** defined by

$$\text{sinc}\ \lambda \triangleq \frac{\sin \pi \lambda}{\pi \lambda} \qquad [19]$$

where λ represents the independent variable. Some authors use the related **sampling function** defined as $\text{Sa}\,(x) \triangleq (\sin x)/x$ so that $\text{sinc}\ \lambda = \text{Sa}\,(\pi \lambda)$. Figure 2.1–6 shows that $\text{sinc}\ \lambda$ is an even function of λ having its peak at $\lambda = 0$ and zero crossings at all other integer values of λ, so

$$\text{sinc}\ \lambda = \begin{cases} 1 & \lambda = 0 \\ 0 & \lambda = \pm 1, \pm 2, \dots \end{cases}$$

Numerical values of $\text{sinc}\ \lambda$ and $\text{sinc}^2\ \lambda$ are given in Table T.4 at the back of the book, while Table T.3 includes several mathematical relations that you'll find helpful for Fourier analysis.

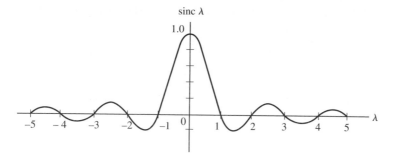

Figure 2.1–6   The function sinc $\lambda = (\sin \pi\lambda)/\pi\lambda$.

Rectangular Pulse Train   **EXAMPLE 2.1–1**

Consider the periodic train of rectangular pulses in Fig. 2.1–7. Each pulse has height
or amplitude A and width or duration τ. There are stepwise **discontinuities** at each
pulse-edge location $t = \pm\tau/2$, and so on, so the values of $v(t)$ are *undefined* at these
points of discontinuity. This brings out another possible difference between a physical signal and its mathematical model, for a physical signal never makes a perfect
stepwise transition. However, the model may still be reasonable if the actual transition times are quite small compared to the pulse duration.

To calculate the Fourier coefficients, we'll take the range of integration in
Eq. (14) over the central period $-T_0/2 \le t \le T_0/2$, where

$$v(t) = \begin{cases} A & |t| < \tau/2 \\ 0 & |t| > \tau/2 \end{cases}$$

Thus

$$c_n = \frac{1}{T_0} \int_{-T_0/2}^{T_0/2} v(t) e^{-j2\pi nf_0 t}\, dt = \frac{1}{T_0} \int_{-\tau/2}^{\tau/2} A e^{-j2\pi nf_0 t}\, dt$$

$$= \frac{A}{-j2\pi nf_0 T_0} \left(e^{-j\pi nf_0 \tau} - e^{+j\pi nf_0 \tau} \right)$$

$$= \frac{A}{T_0} \frac{\sin \pi nf_0 \tau}{\pi nf_0}$$

Multiplying and dividing by τ finally gives

$$c_n = \frac{A\tau}{T_0} \operatorname{sinc} nf_0 \tau \qquad\qquad\qquad \textbf{[20]}$$

which follows from Eq. (19) with $\lambda = nf_0 \tau$.

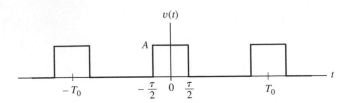

Figure 2.1-7   Rectangular pulse train.

The amplitude spectrum obtained from $|c(nf_0)| = |c_n| = Af_0\,\tau|\text{sinc } nf_0\,\tau|$ is shown in Fig. 2.1–8a for the case of $\tau/T_0 = f_0\,\tau = 1/4$. We construct this plot by drawing the continuous function $Af_0\,\tau|\text{sinc } f\tau|$ as a dashed curve, which becomes the **envelope** of the lines. The spectral lines at $\pm4f_0$, $\pm8f_0$, and so on, are "missing" since they fall precisely at multiples of $1/\tau$ where the envelope equals zero. The dc component has amplitude $c(0) = A\tau/T_0$ which should be recognized as the average value of $v(t)$ by inspection of Fig. 2.1–7. Incidentally, τ/T_0 equals the ratio of "on" time to period, frequently designated as the **duty cycle** in pulse electronics work.

The phase spectrum in Fig. 2.1–8b is obtained by observing that c_n is always real but sometimes negative. Hence, arg $c(nf_0)$ takes on the values $0°$ and $\pm180°$, depending on the sign of sinc $nf_0\,\tau$. Both $+180°$ and $-180°$ were used here to bring out the odd symmetry of the phase.

Having decomposed the pulse train into its frequency components, let's build it back up again. For that purpose, we'll write out the trigonometric series in Eq. (17),

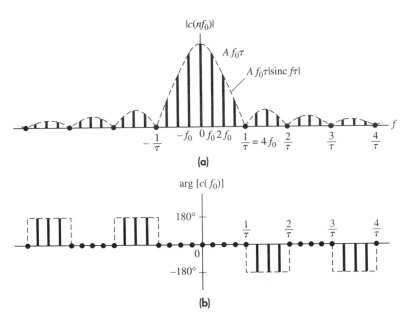

Figure 2.1-8   Spectrum of rectangular pulse train with $f_c\tau = 1/4$. (a) Amplitude; (b) phase.

still taking $\tau/T_0 = f_0\,\tau = 1/4$ so $c_0 = A/4$ and $|2c_n| = (2A/4)\,|\text{sinc } n/4| = (2A/\pi n)|\sin \pi n/4|$. Thus

$$v(t) = \frac{A}{4} + \frac{\sqrt{2}\,A}{\pi}\cos \omega_0\,t + \frac{A}{\pi}\cos 2\omega_0\,t + \frac{\sqrt{2}\,A}{3\pi}\cos 3\omega_0\,t + \cdots$$

Summing terms through the third harmonic gives the approximation of $v(t)$ sketched in Fig. 2.1–9a. This approximation contains the gross features of the pulse train but lacks sharp corners. A more accurate approximation shown in Fig. 2.1–9b comprises all components through the seventh harmonic. Note that the small-amplitude higher harmonics serve primarily to square up the corners. Also note that the series is converging toward the midpoint value $A/2$ at $t = \pm\tau/2$ where $v(t)$ has discontinuities.

Sketch the amplitude spectrum of a rectangular pulse train for each of the following cases: $\tau = T_0/5$, $\tau = T_0/2$, $\tau = T_0$. In the last case the pulse train degenerates into a constant for all time; how does this show up in the spectrum?   **EXERCISE 2.1–2**

Convergence Conditions and Gibbs Phenomenon

We've seen that a periodic signal can be approximated with a finite number of terms of its Fourier series. But does the infinite series **converge** to $v(t)$? The study of convergence involves subtle mathematical considerations that we'll not go into here. Instead, we'll state without proof some of the important results. Further details are given by Ziemer, Tranter and Fannin (1998) or Stark, Tuteur and Anderson (1988).

The **Dirichlet conditions** for Fourier series expansion are as follows: If a periodic function $v(t)$ has a finite number of maxima, minima, and discontinuities per period, and if $v(t)$ is **absolutely integrable,** so that $v(t)$ has a finite area per period, then the Fourier series exists and converges uniformly wherever $v(t)$ is continuous. These conditions are sufficient but not strictly necessary.

An alternative condition is that $v(t)$ be **square integrable,** so that $|v(t)|^2$ has finite area per period—equivalent to a power signal. Under this condition, the series **converges in the mean** such that if

$$v_N(t) = \sum_{n=-N}^{N} c_n\, e^{j2\pi n f_0 t}$$

then

$$\lim_{N\to\infty} \int_{T_0} |v(t) - v_N(t)|^2\, dt = 0$$

In other words, the mean square difference between $v(t)$ and the partial sum $v_N(t)$ vanishes as more terms are included.

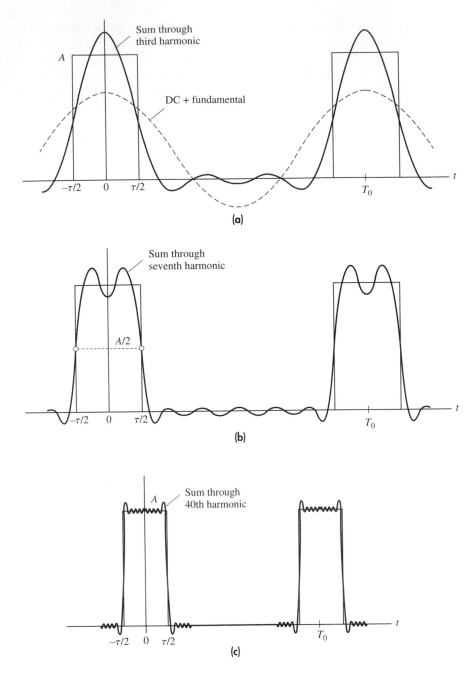

Figure 2.1–9 Fourier-series reconstruction of a rectangular pulse train.

Regardless of whether $v(t)$ is absolutely integrable or square integrable, the series exhibits a behavior known as **Gibbs phenomenon** at points of discontinuity. Figure 2.1–10 illustrates this behavior for a stepwise discontinuity at $t = t_0$. The partial sum $v_N(t)$ converges to the **midpoint** at the discontinuity, which seems quite reasonable. However, on each side of the discontinuity, $v_N(t)$ has oscillatory overshoot with period $T_0/2N$ and peak value of about 9 percent of the step height, independent of N. Thus, as $N \rightarrow \infty$, the oscillations collapse into nonvanishing spikes called "Gibbs ears" above and below the discontinuity as shown in Fig. 2.1–9c. Kamen and Heck (1997, Chap. 4) provide Matlab examples to further illustrate Gibbs phenomenon.

Since a real signal must be continuous, Gibbs phenomenon does not occur and we're justified in treating the Fourier series as being *identical* to $v(t)$. But idealized signal models like the rectangular pulse train often do have discontinuities. You therefore need to pay attention to convergence when working with such models.

Gibbs phenomenon also has implications for the shapes of the filters used with real signals. An ideal filter that is shaped like a rectangular pulse will result in discontinuities in the spectrum that will lead to distortions in the time signal. Another way to view this is that multiplying a signal in the frequency domain by a rectangular filter results in the time signal being convolved with a sinc function. Therefore, real applications use other window shapes with better time-frequency characteristics, such as Hamming or Hanning windows. See Oppenheim, Schafer and Buck (1999) for a more complete discussion on the effects of window shape.

Parseval's Power Theorem

Parseval's theorem relates the average power P of a periodic signal to its Fourier coefficients. To derive the theorem, we start with

$$P = \frac{1}{T_0} \int_{T_0} |v(t)|^2 \, dt = \frac{1}{T_0} \int_{T_0} v(t)v^*(t) \, dt$$

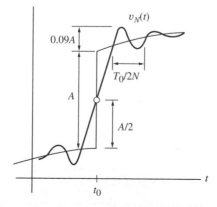

Figure 2.1–10   Gibbs phenomenon at a step discontinuity.

Now replace $v*(t)$ by its exponential series

$$v*(t) = \left[\sum_{n=-\infty}^{\infty} c_n e^{j2\pi n f_0 t} \right]^* = \sum_{n=-\infty}^{\infty} c_n^* e^{-j2\pi n f_0 t}$$

so that

$$P = \frac{1}{T_0} \int_{T_0} v(t) \left[\sum_{n=-\infty}^{\infty} c_n^* e^{-j2\pi n f_0 t} \right] dt$$

$$= \sum_{n=-\infty}^{\infty} \left[\frac{1}{T_0} \int_{T_0} v(t) e^{-j2\pi n f_0 t} \, dt \right] c_n^*$$

and the integral inside the sum equals c_n. Thus

$$P = \sum_{n=-\infty}^{\infty} c_n c_n^* = \sum_{n=-\infty}^{\infty} |c_n|^2 \qquad \text{[21]}$$

which is Parseval's theorem.

The spectral interpretation of this result is extraordinarily simple:

The average power can be found by squaring and adding the heights $|c_n| = |c(nf_0)|$ of the amplitude lines.

Observe that Eq. (21) does not involve the phase spectrum, underscoring our prior comment about the dominant role of the amplitude spectrum relative to a signal's frequency content. For further interpretation of Eq. (21) recall that the exponential Fourier series expands $v(t)$ as a sum of phasors of the form $c_n e^{j2\pi n f_0 t}$. You can easily show that the average power of each phasor is

$$\langle |c_n e^{j2\pi n f_0 t}|^2 \rangle = |c_n|^2 \qquad \text{[22]}$$

Therefore, Parseval's theorem implies **superposition of average power,** since the total average power of $v(t)$ is the sum of the average powers of its phasor components.

Several other theorems pertaining to Fourier series could be stated here. However, they are more conveniently treated as special cases of Fourier transform theorems covered in Sect. 2.3. Table T.2 lists some of the results, along with the Fourier coefficients for various periodic waveforms encountered in communication systems.

EXERCISE 2.1–3   Use Eq. (21) to calculate P from Fig. 2.1–5.

2.2   FOURIER TRANSFORMS AND CONTINUOUS SPECTRA

Now let's turn from periodic signals that last forever (in theory) to nonperiodic signals concentrated over relatively short-time durations. If a nonperiodic signal has finite total energy, its frequency-domain representation will be a continuous spectrum obtained from the Fourier transform.

Fourier Transforms

Figure 2.2–1 shows two typical nonperiodic signals. The single rectangular pulse (Fig. 2.2–1a) is **strictly timelimited** since $v(t)$ is identically zero outside the pulse duration. The other signal is **asymptotically timelimited** in the sense that $v(t) \to 0$ as $t \to \pm \infty$. Such signals may also be described loosely as "pulses." In either case, if you attempt to average $v(t)$ or $|v(t)|^2$ over all time you'll find that these averages equal zero. Consequently, instead of talking about average power, a more meaningful property of a nonperiodic signal is its energy.

If $v(t)$ is the voltage across a resistance, the total delivered energy would be found by integrating the instantaneous power $v^2(t)/R$. We therefore define normalized **signal energy** as

$$E \triangleq \int_{-\infty}^{\infty} |v(t)|^2 \, dt \qquad\qquad [1]$$

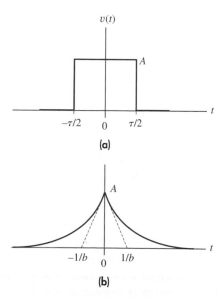

Figure 2.2–1

Some energy calculations can be done by inspection, since E is just the total area under the curve of $|v(t)|^2$. For instance, the energy of a rectangular pulse with amplitude A is simply $E = A^2\tau$.

When the integral in Eq. (1) exists and yields $0 < E < \infty$, the signal $v(t)$ is said to have well-defined energy and is called a **nonperiodic energy signal.** Almost all timelimited signals of practical interest fall in this category, which is the essential condition of spectral analysis using the Fourier transform.

To introduce the Fourier transform, we'll start with the Fourier series representation of a periodic power signal

$$v(t) = \sum_{n=-\infty}^{\infty} c(nf_0)e^{j2\pi nf_0 t} \qquad [2]$$

$$= \sum_{n=-\infty}^{\infty} \left[\frac{1}{T_0} \int_{T_0} v(t)e^{-j2\pi nf_0 t} \, dt \right] e^{j2\pi nf_0 t}$$

where the integral expression for $c(nf_0)$ has been written out in full. According to the **Fourier integral theorem** there's a similar representation for a nonperiodic energy signal that may be viewed as a limiting form of the Fourier series of a signal as the period goes to infinity. Example 2.1–1 showed that the spectral components of a pulse train are spaced at intervals of $nf_0 = n/T_0$, so they become closer together as the period of the pulse train increased. However, the shape of the spectrum remains unchanged if the pulse width τ stays constant. Let the frequency spacing $f_0 = T_0^{-1}$ approach zero (represented in Eq. 3 as df) and the index n approach infinity such that the product nf_0 approaches a continuous frequency variable f. Then

$$v(t) = \int_{-\infty}^{\infty} \left[\int_{-\infty}^{\infty} v(t)e^{-j2\pi ft} \, dt \right] e^{j2\pi ft} \, df \qquad [3]$$

The bracketed term is the **Fourier transform** of $v(t)$ symbolized by $V(f)$ or $\mathcal{F}[v(t)]$ and defined as

$$V(f) = \mathcal{F}[v(t)] \triangleq \int_{-\infty}^{\infty} v(t)e^{-j2\pi ft} \, dt \qquad [4]$$

an integration over all time that yields a function of the continuous variable f.

The time function $v(t)$ is recovered from $V(f)$ by the **inverse Fourier transform**

$$v(t) = \mathcal{F}^{-1}[V(f)] \triangleq \int_{-\infty}^{\infty} V(f)e^{j2\pi ft} \, df \qquad [5]$$

an integration over all frequency f. To be more precise, it should be stated that $\mathcal{F}^{-1}[V(f)]$ **converges in the mean** to $v(t)$, similar to Fourier series convergence, with Gibbs phenomenon occurring at discontinuities. But we'll regard Eq. (5) as being an equality for most purposes. A proof that $\mathcal{F}^{-1}[V(f)] = v(t)$ will be outlined in Sect. 2.5.

Equations (4) and (5) constitute the pair of **Fourier integrals**†. At first glance, these integrals seem to be a closed circle of operations. In a given problem, however, you usually know either $V(f)$ or $v(t)$. If you know $V(f)$, you can find $v(t)$ from Eq. (5); if you know $v(t)$, you can find $V(f)$ from Eq. (4).

Turning to the frequency-domain picture, a comparison of Eqs. (2) and (5) indicates that $V(f)$ plays the same role for nonperiodic signals that $c(nf_0)$ plays for periodic signals. Thus, $V(f)$ is the **spectrum** of the nonperiodic signal $v(t)$. But $V(f)$ is a continuous function defined for all values of f whereas $c(nf_0)$ is defined only for discrete frequencies. Therefore, a nonperiodic signal will have a **continuous spectrum** rather than a line spectrum. Again, comparing Eqs. (2) and (5) helps explain this difference: in the periodic case we return to the time domain by *summing* discrete-frequency phasors, while in the nonperiodic case we *integrate* a continuous frequency function. Three major properties of $V(f)$ are listed below.

1. The Fourier transform is a complex function, so $|V(f)|$ is the amplitude spectrum of $v(t)$ and arg $V(f)$ is the phase spectrum.

2. The value of $V(f)$ at $f = 0$ equals the **net area** of $v(t)$, since

$$V(0) = \int_{-\infty}^{\infty} v(t)\, dt \qquad [6]$$

which compares with the periodic case where $c(0)$ equals the average value of $v(t)$.

3. If $v(t)$ is **real**, then

$$V(-f) = V^*(f) \qquad [7a]$$

and

$$|V(-f)| = |V(f)| \qquad \text{arg } V(-f) = -\text{arg } V(f) \qquad [7b]$$

so again we have even amplitude symmetry and odd phase symmetry. The term **hermitian symmetry** describes complex functions that obey Eq. (7).

Rectangular Pulse   EXAMPLE 2.2–1

In the last section we found the line spectrum of a rectangular pulse train. Now consider the single rectangular pulse in Fig. 2.2–1a. This is so common a signal model that it deserves a symbol of its own. Let's adopt the pictorial notation

$$\Pi(t/\tau) \triangleq \begin{cases} 1 & |t| < \tau/2 \\ 0 & |t| > \tau/2 \end{cases} \qquad [8]$$

† Other definitions take ω for the frequency variable and therefore include $1/2\pi$ or $1/\sqrt{2\pi}$ as multiplying terms.

which stands for a rectangular function with unit amplitude and duration τ centered at $t = 0$. The pulse in the figure is then written

$$v(t) = A\Pi(t/\tau) \tag{9a}$$

Inserting $v(t)$ in Eq. (4) yields

$$V(f) = \int_{-\tau/2}^{\tau/2} Ae^{-j2\pi ft}\,dt = \frac{A}{\pi f}\sin \pi f\tau \tag{9b}$$

$$= A\tau \operatorname{sinc} f\tau$$

so $V(0) = A\tau$, which clearly equals the pulse's area. The corresponding spectrum, plotted in Fig. 2.2–2, should be compared with Fig. 2.1–8 to illustrate the similarities and differences between line spectra and continuous spectra.

Further inspection of Fig. 2.2–2 reveals that the significant portion of the spectrum is in the range $|f| < 1/\tau$ since $|V(f)| \ll |V(0)|$ for $|f| > 1/\tau$. We therefore may take $1/\tau$ as a measure of the spectral "width." Now if the pulse duration is reduced (small τ), the frequency width is increased, whereas increasing the duration reduces the spectral width. Thus, short pulses have broad spectra, and long pulses have narrow spectra. This phenomenon, called **reciprocal spreading,** is a general property of all signals, pulses or not, because high-frequency components are demanded by rapid time variations while smoother and slower time variations require relatively little high-frequency content.

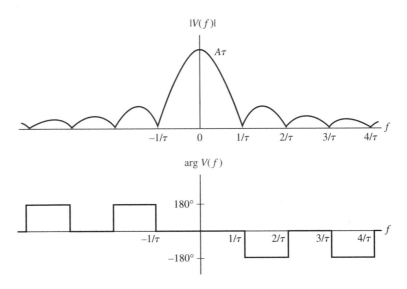

Figure 2.2–2   Rectangular pulse spectrum $V(f) = A\tau \operatorname{sinc} f\tau$.

Symmetric and Causal Signals

When a signal possesses symmetry with respect to the time axis, its transform integral can be simplified. Of course any signal symmetry depends upon both the wave-shape and the location of the time origin. But we're usually free to choose the time origin since it's not physically unique—as contrasted with the frequency-domain origin which has a definite physical meaning.

To develop the time-symmetry properties, we'll write ω in place of $2\pi f$ for notational convenience and expand Eq. (4) using $e^{-j2\pi ft} = \cos \omega t - j \sin \omega t$. Thus, in general

$$V(f) = V_e(f) + jV_o(f) \qquad\qquad \text{[10a]}$$

where

$$V_e(f) \triangleq \int_{-\infty}^{\infty} v(t) \cos \omega t \, dt \qquad\qquad \text{[10b]}$$

$$V_o(f) \triangleq - \int_{-\infty}^{\infty} v(t) \sin \omega t \, dt$$

which are the **even** and **odd** parts of $V(f)$, regardless of $v(t)$. Incidentally, note that if $v(t)$ is **real,** then

$$\text{Re}\,[V(f)] = V_e(f) \qquad \text{Im}\,[V(f)] = V_o(f)$$

so $V^*(f) = V_e(f) - jV_o(f) = V(-f)$, as previously asserted in Eq. (7).

When $v(t)$ has time symmetry, we simplify the integrals in Eq. (10b) by applying the general relationship

$$\int_{-\infty}^{\infty} w(t) \, dt = \int_{0}^{\infty} w(t) \, dt + \int_{-\infty}^{0} w(t) \, dt \qquad\qquad \text{[11]}$$

$$= \begin{cases} 2 \int_{0}^{\infty} w(t) \, dt & w(t) \text{ even} \\ 0 & w(t) \text{ odd} \end{cases}$$

where $w(t)$ stands for either $v(t) \cos \omega t$ or $v(t) \sin \omega t$. If $v(t)$ has **even symmetry** so that

$$v(-t) = v(t) \qquad\qquad \text{[12a]}$$

then $v(t) \cos \omega t$ is even whereas $v(t) \sin \omega t$ is odd. Hence, $V_o(f) = 0$ and

$$V(f) = V_e(f) = 2 \int_{0}^{\infty} v(t) \cos \omega t \, dt \qquad\qquad \text{[12b]}$$

Conversely, if $v(t)$ has **odd symmetry** so that

$$v(-t) = -v(t) \qquad\qquad \text{[13a]}$$

then

$$V(f) = jV_o(f) = -j2 \int_0^\infty v(t) \sin \omega t \, dt \qquad \text{[13b]}$$

and $V_e(f) = 0$.

Equations (12) and (13) further show that the spectrum of a **real symmetrical** signal will be either purely real and even or purely imaginary and odd. For instance, the rectangular pulse in Example 2.2–1 is a real and even time function and its spectrum was found to be a real and even frequency function.

Now consider the case of a **causal** signal, defined by the property that

$$v(t) = 0 \qquad t < 0 \qquad \text{[14a]}$$

This simply means that the signal "starts" at or after $t = 0$. Since causality precludes any time symmetry, the spectrum consists of both real and imaginary parts computed from

$$V(f) = \int_0^\infty v(t)e^{-j2\pi ft} \, dt \qquad \text{[14b]}$$

This integral bears a resemblance to the **Laplace transform** commonly used for the study of transients in linear circuits and systems. Therefore, we should briefly consider the similarities and differences between these two types of transforms.

The unilateral or one-sided Laplace transform is a function of the complex variable $s = \sigma + j\omega$ defined by

$$\mathscr{L}[v(t)] \triangleq \int_0^\infty v(t)e^{-st} \, dt$$

which implies that $v(t) = 0$ for $t < 0$. Comparing $\mathscr{L}[v(t)]$ with Eq. (14b) shows that if $v(t)$ is a **causal energy** signal, you can get $V(f)$ from the Laplace transform by letting $s = j2\pi f$. But a typical table of Laplace transforms includes many **nonenergy** signals whose Laplace transforms exist only with $\sigma > 0$ so that $|v(t)e^{-st}| = |v(t)e^{-\sigma t}| \to 0$ as $t \to \infty$. Such signals do not have a Fourier transform because $s = \sigma + j\omega$ falls outside the frequency domain when $\sigma \neq 0$. On the other hand, the Fourier transform exists for noncausal energy signals that do not have a Laplace transform. See Kamen and Heck (1997, Chap. 7) for further discussion.

EXAMPLE 2.2–2   **Causal Exponential Pulse**

Figure 2.2–3a shows a causal waveform that decays exponentially with time constant $1/b$, so

$$v(t) = \begin{cases} Ae^{-bt} & t > 0 \\ 0 & t < 0 \end{cases} \qquad \text{[15a]}$$

The spectrum can be obtained from Eq. (14b) or from the Laplace transform $\mathcal{L}[v(t)] = A/(s + b)$, with the result that

$$V(f) = \frac{A}{b + j2\pi f} \qquad \text{[15b]}$$

which is a complex function in unrationalized form. Multiplying numerator and denominator of Eq. (15b) by $b - j2\pi f$ yields the rationalized expression

$$V(f) = \frac{b - j2\pi f}{b^2 + (2\pi f)^2} A$$

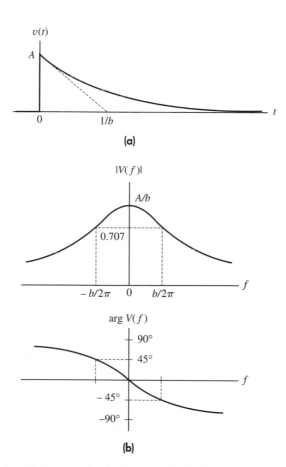

Figure 2.2–3   Causal exponential pulse. (a) Waveform; (b) spectrum.

and we see that

$$V_e(f) = \text{Re}\,[V(f)] = \frac{bA}{b^2 + (2\pi f)^2}$$

$$V_o(f) = \text{Im}\,[V(f)] = -\frac{2\pi f A}{b^2 + (2\pi f)^2}$$

Conversion to polar form then gives the amplitude and phase spectrum

$$|V(f)| = \sqrt{V_e^2(f) + V_o^2(f)} = \frac{A}{\sqrt{b^2 + (2\pi f)^2}}$$

$$\arg V(f) = \arctan \frac{V_o(f)}{V_e(f)} = -\arctan \frac{2\pi f}{b}$$

which are plotted in Fig. 2.2–3b.

The phase spectrum in this case is a smooth curve that includes all angles from $-90°$ to $+90°$. This is due to the signal's lack of time symmetry. But $V(f)$ still has hermitian symmetry since $v(t)$ is a real function. Also note that the spectral width is proportional to b, whereas the time "width" is proportional to the time constant $1/b$—another illustration of reciprocal spreading.

EXERCISE 2.2–1   Find and sketch $V(f)$ for the symmetrical decaying exponential $v(t) = Ae^{-b|t|}$ in Fig. 2.2–1b. (You must use a definite integral from Table T.3.) Compare your result with $V_e(f)$ in Example 2.2–2. Confirm the reciprocal-spreading effect by calculating the frequency range such that $|V(f)| \geq (1/2)|V(0)|$.

Rayleigh's Energy Theorem

Rayleigh's energy theorem is analogous to Parseval's power theorem. It states that the energy E of a signal $v(t)$ is related to the spectrum $V(f)$ by

$$E = \int_{-\infty}^{\infty} V(f)V^*(f)\,df = \int_{-\infty}^{\infty} |V(f)|^2\,df \qquad [16]$$

Therefore,

Integrating the square of the amplitude spectrum over all frequency yields the total energy.

The value of Eq. (16) lies not so much in computing E, since the time-domain integration of $|v(t)|^2$ often is easier. Rather, it implies that $|V(f)|^2$ gives the distribution of energy in the frequency domain, and therefore may be termed the **energy spectral density.** By this we mean that the energy in any differential frequency band df equals $|V(f)|^2\,df$, an interpretation we'll further justify in Sect. 3.6. That interpretation, in turn, lends quantitative support to the notion of spectral width in the sense that most of the energy of a given signal should be contained in the range of frequencies taken to be the spectral width.

By way of illustration, Fig. 2.2–4 is the energy spectral density of a rectangular pulse, whose spectral width was claimed to be $|f| < 1/\tau$. The energy in that band is the shaded area in the figure, namely

$$\int_{-1/\tau}^{1/\tau} |V(f)|^2\,df = \int_{-1/\tau}^{1/\tau} (A\tau)^2\,\text{sinc}^2\,f\tau\,df = 0.92A^2\tau$$

a calculation that requires numerical methods. But the total pulse energy is $E \approx A^2\tau$, so the asserted spectral width encompasses more than 90 percent of the total energy.

Rayleigh's theorem is actually a special case of the more general integral relationship

$$\int_{-\infty}^{\infty} v(t)w^*(t)\,dt = \int_{-\infty}^{\infty} V(f)W^*(f)\,df \qquad \textbf{[17]}$$

where $v(t)$ and $w(t)$ are arbitrary energy signals with transforms $V(f)$ and $W(f)$. Equation (17) yields Eq. (16) if you let $w(t) = v(t)$ and note that $\int_{-\infty}^{\infty} v(t)v^*(t)\,dt = E$. Other applications of Eq. (17) will emerge subsequently.

The proof of Eq. (17) follows the same lines as our derivation of Parseval's theorem. We substitute for $w^*(t)$ the inverse transform

$$w^*(t) = \left[\int_{-\infty}^{\infty} W(f)e^{j\omega t}\,df \right]^* = \int_{-\infty}^{\infty} W^*(f)e^{-j\omega t}\,df$$

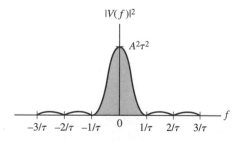

Figure 2.2–4   Energy spectral density of a rectangular pulse.

Interchanging the order of time and frequency integrations then gives

$$\int_{-\infty}^{\infty} v(t)w^*(t)\,dt = \int_{-\infty}^{\infty} v(t)\left[\int_{-\infty}^{\infty} W^*(f)e^{-j\omega t}\,df\right]dt$$

$$= \int_{-\infty}^{\infty}\left[\int_{-\infty}^{\infty} v(t)e^{-j\omega t}\,dt\right]W^*(f)\,df$$

which completes the proof since the bracketed term equals $V(f)$.

The *interchange of integral operations* illustrated here is a valuable technique in signal analysis, leading to many useful results. However, you should not apply the technique willy-nilly without giving some thought to the validity of the interchange. As a pragmatic guideline, you can assume that the interchange is valid if the results make sense. If in doubt, test the results with some simple cases having known answers.

EXERCISE 2.2–2   Calculate the energy of a causal exponential pulse by applying Rayleigh's theorem to $V(f)$ in Eq. (15*b*). Then check the result by integrating $|v(t)|^2$.

Duality Theorem

If you reexamine the pair of Fourier integrals, you'll see that they differ only by the variable of integration and the sign in the exponent. A fascinating consequence of this similarity is the **duality theorem.** The theorem states that if $v(t)$ and $V(f)$ constitute a known transform pair, and if there exists a **time function** $z(t)$ related to the function $V(f)$ by

$$z(t) = V(t) \tag{18a}$$

then

$$\mathcal{F}[z(t)] = v(-f) \tag{18b}$$

where $v(-f)$ equals $v(t)$ with $t = -f$.

Proving the duality theorem hinges upon recognizing that Fourier transforms are definite integrals whose variables of integration are *dummy* variables. Therefore, we may replace f in Eq. (5) with the dummy variable λ and write

$$v(t) = \int_{-\infty}^{\infty} V(\lambda)e^{j2\pi\lambda t}\,d\lambda$$

Furthermore, since t is a dummy variable in Eq. (4) and since $z(t) = V(t)$ in the theorem,

$$\mathcal{F}[z(t)] = \int_{-\infty}^{\infty} z(\lambda)e^{-j2\pi f\lambda}\,d\lambda = \int_{-\infty}^{\infty} V(\lambda)e^{j2\pi\lambda(-f)}\,d\lambda$$

Comparing these integrals then confirms that $\mathcal{F}[z(t)] = v(-f)$.

Although the statement of duality in Eq. (18) seems somewhat abstract, it turns out to be a handy way of generating new transform pairs without the labor of integration. The theorem works best when $v(t)$ is real and even so $z(t)$ will also be real and even, and $Z(f) = \mathcal{F}[z(t)] = v(-f) = v(f)$. The following example should clarify the procedure.

Sinc Pulse   **EXAMPLE 2.2–3**

A rather strange but important time function in communication theory is the **sinc pulse** plotted in Fig. 2.2–5a and defined by

$$z(t) = A \operatorname{sinc} 2Wt \qquad [19a]$$

We'll obtain $Z(f)$ by applying duality to the transform pair

$$v(t) = B\Pi(t/\tau) \qquad V(f) = B\tau \operatorname{sinc} f\tau$$

Rewriting Eq. (19a) as

$$z(t) = \left(\frac{A}{2W}\right)(2W) \operatorname{sinc} t(2W)$$

brings out the fact that $z(t) - V(t)$ with $\tau = 2W$ and $B = A/2W$. Duality then says that $\mathcal{F}[z(t)] = v(-f) = B\Pi(-f/\tau) = (A/2W)\Pi(-f/2W)$ or

$$Z(f) = \frac{A}{2W} \Pi\left(\frac{f}{2W}\right) \qquad [19b]$$

since the rectangle function has even symmetry.

The plot of $Z(f)$, given in Fig. 2.2–5b, shows that the spectrum of a sinc pulse equals zero for $|f| > W$. Thus, the spectrum has clearly defined width W, measured in terms of positive frequency, and we say that $Z(f)$ is **bandlimited**. Note, however, that the signal $z(t)$ goes on forever and is only asymptotically timelimited.

Find the transform of $z(t) = B/[1 + (2\pi t)^2]$ by applying duality to the result of Exercise 2.2–1.   **EXERCISE 2.2–3**

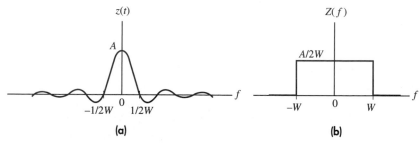

(a)   (b)

Figure 2.2–5   A sinc pulse and its bandlimited spectrum.

Transform Calculations

Except in the case of a very simple waveform, brute-force integration should be viewed as the method of last resort for transform calculations. Other, more practical methods are discussed here.

When the signal in question is defined mathematically, you should first consult a table of Fourier transforms to see if the calculation has been done before. Both columns of the table may be useful, in view of the duality theorem. A table of Laplace transforms also has some value, as mentioned in conjunction with Eq. (14).

Besides duality, there are several additional transform theorems covered in Sect. 2.3. These theorems often help you decompose a complicated waveform into simpler parts whose transforms are known. Along this same line, you may find it expedient to *approximate* a waveform in terms of idealized signal models. Suppose $\tilde{z}(t)$ approximates $z(t)$ and magnitude-squared error $|z(t) - \tilde{z}(t)|^2$ is a small quantity. If $Z(f) = \mathcal{F}[z(t)]$ and $\tilde{Z}(f) = \mathcal{F}[\tilde{z}(t)]$ then

$$\int_{-\infty}^{\infty} |Z(f) - \tilde{Z}(f)|^2 \, df = \int_{-\infty}^{\infty} |z(t) - \tilde{z}(t)|^2 \, dt \qquad [20]$$

which follows from Rayleigh's theorem with $v(t) = z(t) - \tilde{z}(t)$. Thus, the integrated approximation error has the same value in the time and frequency domains.

The above methods are easily modified for the calculation of Fourier series coefficients. Specifically, let $v(t)$ be a periodic signal and let $z(t) = v(t)\Pi(t/T_0)$, a nonperiodic signal consisting of one period of $v(t)$. If you can obtain

$$Z(f) = \mathcal{F}[v(t)\Pi(t/T_0)] \qquad [21a]$$

then, from Eq. (14), Sect. 2.1, the coefficients of $v(t)$ are given by

$$c_n = \frac{1}{T_0} Z(nf_0) \qquad [21b]$$

This relationship facilitates the application of transform theorems to Fourier series calculations.

Finally, if the signal is defined in numerical form, its transform can be found via numerical calculations. For this purpose, the FFT computer algorithm is especially well suited. For details on the algorithm and the supporting theory of discrete Fourier transforms, see Oppenheim, Schafer and Buck (1999).

2.3   TIME AND FREQUENCY RELATIONS

Rayleigh's theorem and the duality theorem in the previous section helped us draw useful conclusions about the frequency-domain representation of energy signals. Now we'll look at some of the many other theorems associated with Fourier transforms. They are included not just as manipulation exercises but for two very practical reasons. First, the theorems are invaluable when interpreting spectra, for they express relationships between time-domain and frequency-domain operations. Second, we can build up an extensive catalog of transform pairs by applying the theorems to known pairs—and such a catalog will be useful as we seek new signal models.

In stating the theorems, we indicate a signal and its transform (or spectrum) by lowercase and uppercase letters, as in $V(f) = \mathcal{F}[v(t)]$ and $v(t) = \mathcal{F}^{-1}[V(f)]$. This is also denoted more compactly by $v(t) \leftrightarrow V(f)$. Table T.1 at the back lists the theorems and transform pairs covered here, plus a few others.

Superposition

Superposition applies to the Fourier transform in the following sense. If a_1 and a_2 are constants and

$$v(t) = a_1 v_1(t) + a_2 v_2(t)$$

then

$$\mathcal{F}[v(t)] = a_1 \mathcal{F}[v_1(t)] + a_2 \mathcal{F}[v_2(t)]$$

Generalizing to sums with an arbitrary number of terms, we write the superposition (or linearity) theorem as

$$\sum_k a_k v_k(t) \leftrightarrow \sum_k a_k V_k(f) \qquad [1]$$

This theorem simply states that linear combinations in the time domain become linear combinations in the frequency domain.

Although proof of the theorem is trivial, its importance cannot be overemphasized. From a practical viewpoint Eq. (1) greatly facilitates spectral analysis when the signal in question is a linear combination of functions whose individual spectra are known. From a theoretical viewpoint it underscores the applicability of the Fourier transform for the study of linear systems.

Time Delay and Scale Change

Given a time function $v(t)$, various other waveforms can be generated from it by modifying the argument of the function. Specifically, replacing t by $t - t_d$ produces the **time-delayed** signal $v(t - t_d)$. The delayed signal has the same shape as $v(t)$ but shifted t_d units to the right along the time axis. In the frequency domain, time delay causes an added linear phase with slope $-2\pi t_d$, so that

$$v(t - t_d) \leftrightarrow V(f)e^{-j2\pi f t_d} \qquad [2]$$

If t_d is a negative quantity, the signal is *advanced* in time and the added phase has positive slope. The amplitude spectrum remains unchanged in either case, since $|V(f)e^{-j2\pi f t_d}| = |V(f)||e^{-j2\pi f t_d}| = |V(f)|$.

Proof of the time-delay theorem is accomplished by making the change of variable $\lambda = t - t_d$ in the transform integral. Thus, using $\omega = 2\pi f$ for compactness, we have

$$\mathcal{F}[v(t - t_d)] = \int_{-\infty}^{\infty} v(t - t_d)e^{-j\omega t}\, dt$$

$$= \int_{-\infty}^{\infty} v(\lambda)e^{-j\omega(\lambda + t_d)}\, d\lambda$$

$$= \left[\int_{-\infty}^{\infty} v(\lambda)e^{-j\omega\lambda}\, d\lambda\right]e^{-j\omega t_d}$$

The integral in brackets is just $V(f)$, so $\mathcal{F}[v(t - t_d)] = V(f)e^{-j\omega t_d}$.

Another time-axis operation is *scale change*, which produces a horizontally scaled image of $v(t)$ by replacing t with αt. The scale signal $v(\alpha t)$ will be *expanded* if $|\alpha| < 1$ or *compressed* if $|\alpha| > 1$; a negative value of α yields time *reversal* as well as expansion or compression. These effects may occur during playback of recorded signals, for instance.

Scale change in the time domain becomes *reciprocal* scale change in the frequency domain, since

$$v(\alpha t) \leftrightarrow \frac{1}{|\alpha|}\, V\!\left(\frac{f}{\alpha}\right) \qquad \alpha \neq 0 \qquad\qquad [3]$$

Hence, compressing a signal expands its spectrum, and vice versa. If $\alpha = -1$, then $v(-t) \leftrightarrow V(-f)$ so both the signal and spectrum are reversed.

We'll prove Eq. (3) for the case $\alpha < 0$ by writing $\alpha = -|\alpha|$ and making the change of variable $\lambda = -|\alpha|t$. Therefore, $t = \lambda/\alpha$, $dt = -d\lambda/|\alpha|$, and

$$\mathcal{F}[v(-|\alpha|t)] = \int_{-\infty}^{+\infty} v(-|\alpha|t)e^{-j\omega t}\, dt$$

$$= \frac{-1}{|\alpha|} \int_{+\infty}^{-\infty} v(\lambda)e^{-j\omega\lambda/\alpha}\, d\lambda$$

$$= \frac{1}{|\alpha|} \int_{-\infty}^{+\infty} v(\lambda)e^{-j2\pi(f/\alpha)\lambda}\, d\lambda$$

$$= \frac{1}{|\alpha|}\, V\!\left(\frac{f}{\alpha}\right)$$

Observe how this proof uses the general relationship

$$\int_{a}^{b} x(\lambda)\, d(-\lambda) = -\int_{-a}^{-b} x(\lambda)\, d\lambda = \int_{-b}^{-a} x(\lambda)\, d\lambda$$

Hereafter, the intermediate step will be omitted when this type of manipulation occurs.

EXAMPLE 2.3–1

The signal in Fig. 2.3–1a has been constructed using two rectangular pulses $v(t) = A\Pi(t/\tau)$ such that

$$z_a(t) = v(t - t_d) + (-1)v[t - (t_d + T)]$$

Application of the superposition and time-delay theorems yields

$$Z_a(f) = V(f)e^{-j2\pi f t_d} + (-1)V(f)e^{-j2\pi f(t_d+T)}$$
$$= V(f)[e^{-j2\pi f t_d} - e^{-j2\pi f(t_d+T)}]$$

where $V(f) = A\tau \, \text{sinc} \, f\tau$.

The bracketed term in $Z_a(f)$ is a particular case of the expression $e^{j2\theta_1} \pm e^{j2\theta_2}$ which often turns up in Fourier analysis. A more informative version of this expression is obtained by factoring and using Euler's theorem, as follows:

$$e^{j2\theta_1} \pm e^{j2\theta_2} = [e^{j(\theta_1-\theta_2)} \pm e^{-j(\theta_1-\theta_2)}]e^{j(\theta_1+\theta_2)} \quad [4]$$
$$= \begin{cases} 2\cos(\theta_1 - \theta_2)e^{j(\theta_1+\theta_2)} \\ j2\sin(\theta_1 - \theta_2)e^{j(\theta_1+\theta_2)} \end{cases}$$

The upper result in Eq. (4) corresponds to the upper $(+)$ sign and the lower result to the lower $(-)$ sign.

In the problem at hand we have $\theta_1 = -\pi f t_d$ and $\theta_2 = -\pi f(t_d + T)$, so $\theta_1 - \theta_2 = \pi f T$ and $\theta_1 + \theta_2 = -2\pi f t_0$ where $t_0 = t_d + T/2$ as marked in Fig. 2.3–1a. Therefore, after substituting for $V(f)$, we obtain

$$Z_a(f) = (A\tau \, \text{sinc} \, f\tau)(j2 \sin \pi f T \, e^{-j2\pi f t_0})$$

Note that $Z_a(0) = 0$, agreeing with the fact that $z_a(t)$ has zero net area.

If $t_0 = 0$ and $T = \tau$, $z_a(t)$ degenerates to the waveform in Fig. 2.3–1b where

$$z_b(t) = A\Pi\left(\frac{t + \tau/2}{\tau}\right) - A\Pi\left(\frac{t - \tau/2}{\tau}\right)$$

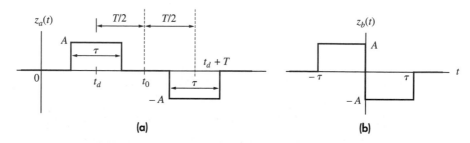

Figure 2.3–1 Signals in Example 2.3–1.

The spectrum then becomes

$$Z_b(f) = (A\tau \operatorname{sinc} f\tau)(j2 \sin \pi f\tau)$$

$$= (j2\pi f\tau)A\tau \operatorname{sinc}^2 f\tau$$

This spectrum is purely imaginary because $z_b(t)$ has odd symmetry.

EXERCISE 2.3-1   Let $v(t)$ be a *real* but otherwise arbitrary energy signal. Show that if

$$z(t) = a_1 v(t) + a_2 v(-t) \qquad [5a]$$

then

$$Z(f) = (a_1 + a_2)V_e(f) + j(a_1 - a_2)V_o(f) \qquad [5b]$$

where $V_e(f)$ and $V_o(f)$ are the real and imaginary parts of $V(f)$.

Frequency Translation and Modulation

Besides generating new transform pairs, duality can be used to generate transform theorems. In particular, a dual of the time-delay theorem is

$$v(t)e^{j\omega_c t} \leftrightarrow V(f - f_c) \qquad \omega_c = 2\pi f_c \qquad [6]$$

We designate this as **frequency translation** or **complex modulation,** since multiplying a time function by $e^{j\omega_c t}$ causes its spectrum to be translated in frequency by $+f_c$.

To see the effects of frequency translation, let $v(t)$ have the bandlimited spectrum of Fig. 2.3–2a, where the amplitude and phase are plotted on the same axes using solid and broken lines, respectively. Also let $f_c > W$. Inspection of the translated spectrum $V(f - f_c)$ in Fig. 2.3–2b reveals the following:

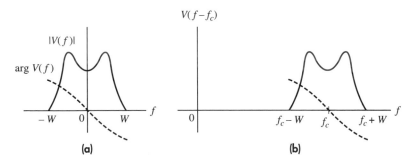

(a)   (b)

Figure 2.3–2   Frequency translation of a bandlimited spectrum.

1. The significant components are concentrated around the frequency f_c.

2. Though $V(f)$ was bandlimited in W, $V(f - f_c)$ has a spectral width of $2W$. Translation has therefore doubled spectral width. Stated another way, the negative-frequency portion of $V(f)$ now appears at positive frequencies.

3. $V(f - f_c)$ is not hermitian but does have symmetry with respect to translated origin at $f = f_c$.

These considerations may appear somewhat academic in view of the fact that $v(t)e^{j\omega_c t}$ is not a real time function and cannot occur as a communication signal. However, signals of the form $v(t) \cos(\omega_c t + \phi)$ are common—in fact, they are the basis of carrier modulation—and by direct extension of Eq. (6) we have the following **modulation theorem**:

$$v(t) \cos(\omega_c t + \phi) \leftrightarrow \frac{e^{j\phi}}{2} V(f - f_c) + \frac{e^{-j\phi}}{2} V(f + f_c) \qquad [7]$$

In words, multiplying a signal by a sinusoid translates its spectrum *up and down* in frequency by f_c. All the comments about complex modulation also apply here. In addition, the resulting spectrum is hermitian, which it must be if $v(t) \cos(\omega_c t + \phi)$ is a real function of time. The theorem is easily proved with the aid of Euler's theorem and Eq. (6).

RF Pulse   EXAMPLE 2.3–2

Consider the finite-duration sinusoid of Fig. 2.3–3a, sometimes referred to as an RF pulse when f_c falls in the radio-frequency band. (See Fig.1.1–2 for the range of frequencies that supports radio waves.) Since

$$z(t) = A\Pi\left(\frac{t}{\tau}\right) \cos \omega_c t$$

we have immediately

$$Z(f) = \frac{A\tau}{2} \text{sinc} \, (f - f_c)\tau + \frac{A\tau}{2} \text{sinc} \, (f + f_c)\tau.$$

obtained by setting $v(t) = A\Pi(t/\tau)$ and $V(f) = A\tau \, \text{sinc} \, f\tau$ in Eq. (7). The resulting amplitude spectrum is sketched in Fig. 2.3–3b for the case of $f_c \gg 1/\tau$ so the two translated sinc functions have negligible overlap.

Because this is a sinusoid of finite duration, its spectrum is continuous and contains more than just the frequencies $f = \pm f_c$. Those other frequencies stem from the fact that $z(t) = 0$ for $|t| > \tau/2$, and the smaller τ is, the larger the spectral spread around $\pm f_c$—reciprocal spreading, again. On the other hand, had we been dealing with a sinusoid of *infinite* duration, the frequency-domain representation would be a two-sided *line* spectrum containing only the discrete frequencies $\pm f_c$.

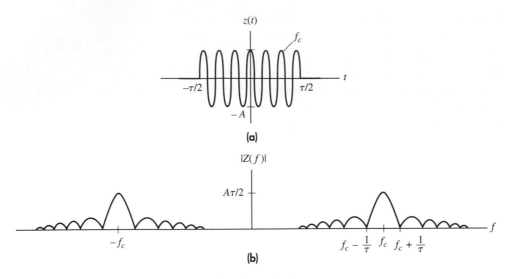

Figure 2.3–3   (a) RF pulse; (b) amplitude spectrum.

Differentiation and Integration

Certain processing techniques involve differentiating or integrating a signal. The frequency-domain effects of these operations are indicated in the theorems below. A word of caution, however: The theorems should not be applied before checking to make sure that the differentiated or integrated signal is Fourier-transformable; the fact that $v(t)$ has finite energy is not a guarantee that the same holds true for its derivative or integral.

To derive the differentiation theorem, we replace $v(t)$ by the inverse transform integral and interchange the order of operations, as follows:

$$
\frac{d}{dt} v(t) = \frac{d}{dt} \left[\int_{-\infty}^{\infty} V(f) e^{j2\pi ft} \, df \right]
$$

$$
= \int_{-\infty}^{\infty} V(f) \left(\frac{d}{dt} e^{j2\pi ft} \right) df
$$

$$
= \int_{-\infty}^{\infty} [j2\pi f V(f)] e^{j2\pi ft} \, df
$$

Referring back to the definition of the inverse transform reveals that the bracketed term must be $\mathcal{F}[dv(t)/dt]$, so

$$
\frac{d}{dt} v(t) \leftrightarrow j2\pi f V(f)
$$

and by iteration we get

$$\frac{d^n}{dt^n} v(t) \leftrightarrow (j2\pi f)^n V(f) \qquad [8]$$

which is the **differentiation theorem.**

Now suppose we generate another function from $v(t)$ by integrating it over all past time. We write this operation as $\int_{-\infty}^{t} v(\lambda)\, d\lambda$, where the dummy variable λ is needed to avoid confusion with the independent variable t in the upper limit. The **integration theorem** says that if

$$V(0) = \int_{-\infty}^{\infty} v(\lambda)\, d\lambda = 0 \qquad [9a]$$

then

$$\int_{-\infty}^{t} v(\lambda)\, d\lambda \leftrightarrow \frac{1}{j2\pi f} V(f) \qquad [9b]$$

The zero net area condition in Eq. (9a) ensures that the integrated signal goes to zero as $t \to \infty$. (We'll relax this condition in Sect. 2.5.)

To interpret these theorems, we see that

> Differentiation enhances the high-frequency components of a signal, since $|j2\pi f V(f)| > |V(f)|$ for $|f| > 1/2\pi$. Conversely, integration suppresses the high-frequency components.

Spectral interpretation thus agrees with the time-domain observation that differentiation accentuates time variations while integration smoothes them out.

Triangular Pulse   EXAMPLE 2.3–3

The waveform $z_b(t)$ in Fig. 2.3–1b has zero net area, and integration produces a **triangular** pulse shape. Specifically, let

$$w(t) = \frac{1}{\tau} \int_{-\infty}^{t} z_b(\lambda)\, d\lambda = \begin{cases} A\left(1 - \dfrac{|t|}{\tau}\right) & |t| < \tau \\ 0 & |t| > \tau \end{cases}$$

which is sketched in Fig. 2.3–4a. Applying the integration theorem to $Z_b(f)$ from Example 2.3–1, we obtain

$$W(f) = \frac{1}{\tau} \frac{1}{j2\pi f} Z_b(f) = A\tau \operatorname{sinc}^2 f\tau$$

as shown in Fig. 2.3–4b. A comparison of this spectrum with Fig. 2.2–2 reveals that the triangular pulse has less high-frequency content than a rectangular pulse with

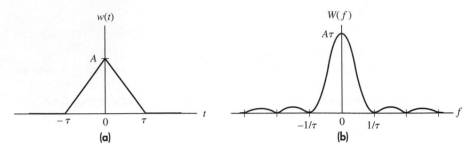

Figure 2.3–4   A triangular pulse and its spectrum.

amplitude A and duration τ, although they both have area $A\tau$. The difference is traced to the fact that the triangular pulse is spread over 2τ seconds and does not have the sharp, stepwise time variations of the rectangular shape.

This transform pair can be written more compactly by defining the **triangular function**

$$\Lambda\left(\frac{t}{\tau}\right) \triangleq \begin{cases} 1 - \dfrac{|t|}{\tau} & |t| < \tau \\ 0 & |t| > \tau \end{cases} \qquad [10]$$

Then $w(t) = A\Lambda(t/\tau)$ and

$$A\Lambda\left(\frac{t}{\tau}\right) \leftrightarrow A\tau \, \text{sinc}^2 f\tau \qquad [11]$$

It so happens that triangular functions can be generated from rectangular functions by another mathematical operation, namely, convolution. And convolution happens to be the next item on our agenda.

EXERCISE 2.3–2

A dual of the differentiation theorem is

$$t^n v(t) \leftrightarrow \frac{1}{(-j2\pi)^n} \frac{d^n}{df^n} V(f) \qquad [12]$$

Derive this relationship for $n = 1$ by differentiating the transform integral $\mathcal{F}[v(t)]$ with respect to f.

2.4   CONVOLUTION

The mathematical operation known as convolution ranks high among the tools used by communication engineers. Its applications include system analysis and probability theory as well as transform calculations. Here we are concerned with convolution in the time and frequency domains.

Convolution Integral

The convolution of two functions of the same variable, say $v(t)$ and $w(t)$, is defined by

$$v * w(t) \triangleq \int_{-\infty}^{\infty} v(\lambda)w(t - \lambda)\, d\lambda \qquad \text{[1]}$$

The notation $v * w(t)$ merely stands for the operation on the right-hand side of Eq. (1) and the asterisk ($*$) has nothing to do with complex conjugation. Equation (1) is the **convolution integral,** often denoted $v * w$ when the independent variable is unambiguous. At other times the notation $[v(t)] * [w(t)]$ is necessary for clarity. Note carefully that the independent variable here is t, the same as the independent variable of the functions being convolved; the integration is always performed with respect to a dummy variable (such as λ) and t is a constant insofar as the integration is concerned.

Calculating $v * w(t)$ is no more difficult than ordinary integration when the two functions are continuous for all t. Often, however, one or both of the functions is defined in a piecewise fashion, and the graphical interpretation of convolution becomes especially helpful.

By way of illustration, take the functions in Fig. 2.4–1a where

$$v(t) = Ae^{-t} \qquad 0 < t < \infty$$
$$w(t) = t/T \qquad 0 < t < T$$

For the integrand in Eq. (1), $v(\lambda)$ has the same shape as $v(t)$ and

$$w(t - \lambda) = \frac{t - \lambda}{T} \qquad 0 < t - \lambda < T$$

But obtaining the picture of $w(t - \lambda)$ as a function of λ requires two steps: first, we reverse $w(t)$ in time and replace t with λ to get $w(-\lambda)$; second, we shift $w(-\lambda)$ to the right by t units to get $w[-(\lambda - t)] = w(t - \lambda)$ for a given value of t. Figure 2.4–1b shows $v(\lambda)$ and $w(t - \lambda)$ with $t < 0$. The value of t always equals the distance from the origin of $v(\lambda)$ to the shifted origin of $w(-\lambda)$ indicated by the dashed line.

As $v * w(t)$ is evaluated for $-\infty < t < \infty$, $w(t - \lambda)$ slides from left to right with respect to $v(\lambda)$, so the convolution integrand changes with t. Specifically, we see in Fig. 2.4–1b that the functions don't overlap when $t < 0$; hence, the integrand equals zero and

$$v * w(t) = 0 \qquad t < 0$$

When $0 < t < T$ as in Fig. 2.4–1c, the functions overlap for $0 < \lambda < t$, so t becomes the upper limit of integration and

$$v * w(t) = \int_0^t Ae^{-\lambda}\left(\frac{t - \lambda}{T}\right) d\lambda$$

$$= \frac{A}{T}(t - 1 + e^{-t}) \qquad 0 < t < T$$

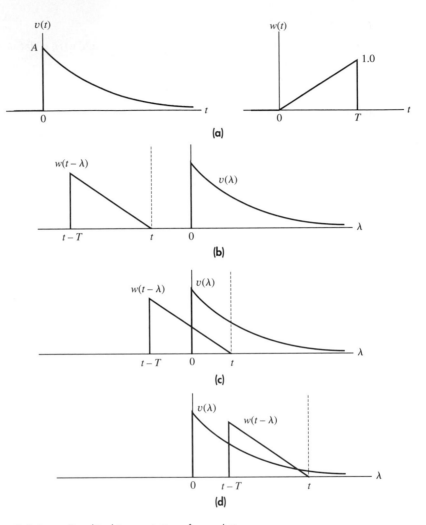

Figure 2.4–1   Graphical interpretation of convolution.

Finally, when $t > T$ as in Fig. 2.4–1d, the functions overlap for $t - T < \lambda < t$ and

$$v * w(t) = \int_{t-T}^{t} Ae^{-\lambda}\left(\frac{t-\lambda}{T}\right) d\lambda$$

$$= \frac{A}{T}(T - 1 + e^{-T})e^{-(t-T)} \qquad t > T$$

The complete result plotted in Fig. 2.4–2 shows that convolution is a *smoothing* operation in the sense that $v * w(t)$ is "smoother" than either of the original functions.

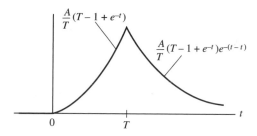

Figure 2.4–2   Result of the convolution in Fig. 2.4–1.

Convolution Theorems

The convolution operation satisfies a number of important and useful properties. They can all be derived from the convolution integral in Eq. (1). In some cases they are also apparent from graphical analysis. For example, further study of Fig. 2.4–1 should reveal that you get the same result by reversing v and sliding it past w, so convolution is **commutative.** This property is listed below along with the **associative** and **distributive** properties.

$$v * w = w * v \qquad\qquad\text{[2a]}$$

$$v * (w * z) = (v * w) * z \qquad\qquad\text{[2b]}$$

$$v * (w + z) = (v * w) + (v * z) \qquad\qquad\text{[2c]}$$

All of these can be derived from Eq. (1).

Having defined and examined the convolution operation, we now list the two **convolution theorems:**

$$v * w(t) \leftrightarrow V(f)W(f) \qquad\qquad\text{[3]}$$

$$v(t)w(t) \leftrightarrow V * W(f) \qquad\qquad\text{[4]}$$

These theorems state that convolution in the time domain becomes multiplication in the frequency domain, while multiplication in the time domain becomes convolution in the frequency domain. Both of these relationships are important for future work.

The proof of Eq. (3) uses the time-delay theorem, as follows:

$$\mathcal{F}[v * w(t)] = \int_{-\infty}^{\infty}\left[\int_{-\infty}^{\infty} v(\lambda)w(t-\lambda)\,d\lambda\right]e^{-j\omega t}\,dt$$

$$= \int_{-\infty}^{\infty} v(\lambda)\left[\int_{-\infty}^{\infty} w(t-\lambda)e^{-j\omega t}\,dt\right]d\lambda$$

$$= \int_{-\infty}^{\infty} v(\lambda)[W(f)e^{-j\omega\lambda}]\,d\lambda$$

$$= \left[\int_{-\infty}^{\infty} v(\lambda)e^{-j\omega\lambda} \, d\lambda \right] W(f) = V(f)W(f)$$

Equation (4) can be proved by writing out the transform of $v(t)w(t)$ and replacing $w(t)$ by the inversion integral $\mathcal{F}^{-1}[W(f)]$.

EXAMPLE 2.4.1   **Trapezoidal Pulse**

To illustrate the convolution theorem—and to obtain yet another transform pair—let's convolve the rectangular pulses in Fig. 2.4–3a. This is a relatively simple task using the graphical interpretation and symmetry considerations. If $\tau_1 > \tau_2$, the problem breaks up into three cases: no overlap, partial overlap, and full overlap. Fig. 2.4–3b shows $v(\lambda)$ and $w(t - \lambda)$ in one case where there is no overlap and $v * w(t) = 0$. For this region

$$t + \frac{\tau_2}{2} < -\frac{\tau_1}{2}$$

or

$$t < -\frac{(\tau_1 + \tau)}{2}$$

There is a corresponding region with no overlap where $t - \tau_2/2 > \tau_1/2$, or $t > (\tau_1 + \tau_2)/2$. Combining these together yields the region of no overlap as $|t| > (\tau_1 + \tau_2)/2$. In the region where there is partial overlap $t + \tau_2/2 > -\tau_1/2$ and $t - \tau_2/2 < -\tau_1/2$, which yields

$$v * w(t) = \int_{-\frac{\tau_1}{2}}^{t+\frac{\tau_2}{2}} A_1 A_2 \, d\lambda = A_1 A_2 \left(t + \frac{\tau_1 + \tau_2}{2} \right) \qquad -\frac{\tau_1 + \tau_2}{2} < t < -\frac{\tau_1 - \tau_2}{2}$$

By properties of symmetry the other region of partial overlap can be found to be

$$v * w(t) = \int_{t-\frac{\tau_s}{2}}^{\frac{\tau_1}{2}} A_1 A_2 \, d\lambda = A_1 A_2 \left(-t + \frac{\tau_1 + \tau_2}{2} \right) \qquad \frac{\tau_1 - \tau_2}{2} < t < \frac{\tau_1 + \tau_2}{2}$$

Finally, the convolution in the region of total overlap is

$$v * w(t) = \int_{t-\frac{\tau_2}{2}}^{t+\frac{\tau_2}{2}} A_1 A_2 \, d\lambda = A_1 A_2 \tau_2 \qquad |t| < \frac{\tau_1 - \tau_2}{2}$$

The result is the *trapezoidal* pulse shown in Fig. 2.4–3c, whose transform will be the product $V(f)W(f) = (A_1\tau_1 \operatorname{sinc} f\tau_1)(A_2\tau_2 \operatorname{sinc} f\tau_2)$.

Now let $\tau_1 = \tau_2 = \tau$ so the trapezoidal shape reduces to the *triangular* pulse back in Fig. 2.3–4a with $A = A_1 A_2 \tau$. Correspondingly, the spectrum becomes $(A_1\tau \operatorname{sinc} f\tau)(A_2\tau \operatorname{sinc} f\tau) = A\tau \operatorname{sinc}^2 f\tau$, which agrees with our prior result.

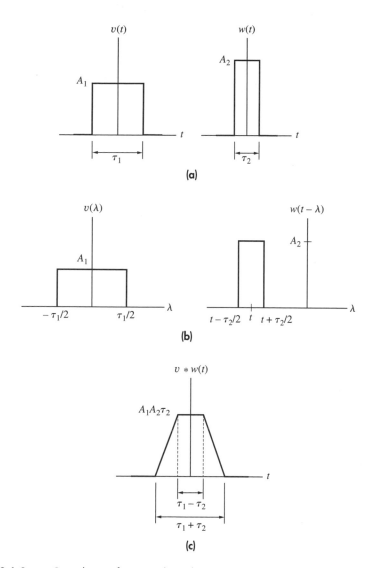

Figure 2.4–3   Convolution of rectangular pulses.

Ideal Lowpass Filter

EXAMPLE 2.4–2

In Section 2.1 we mentioned the impact of the discontinuities introduced in a signal as a result of filtering with an ideal filter. We will examine this further by taking the rectangular function from Example 2.2–1 $v(t) = A\Pi(t/\tau)$ whose transform, $V(f) = A\tau$ sinc $f\tau$, exists for all values of f. We can lowpass filter this signal at $f = 1/\tau$ by multiplying $V(f)$ by the rectangular function

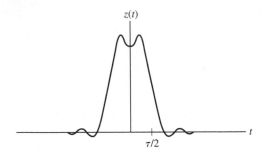

Figure 2.4–4

$$W(f) = \Pi\left(\frac{f}{2/\tau}\right) \leftrightarrow \frac{2}{\tau}\operatorname{sinc}\left(\frac{2t}{\tau}\right)$$

The output function is

$$z(t) = v(t) * w(t) = w(t) * v(t) = \int_{t-\frac{\tau}{2}}^{t+\frac{\tau}{2}} \frac{2A}{\tau}\operatorname{sinc}\frac{2\lambda}{\tau}\, d\lambda$$

This integral cannot be evaluated in closed form; however, it can be evaluated numerically using Table T.4 to obtain the result shown in Fig. 2.4–4. Note the similarity to the result in Fig. 2.1–9b.

EXERCISE 2.4–1   Let $v(t) = A\operatorname{sinc} 2Wt$, whose spectrum is bandlimited in W. Use Eq. (4) with $w(t) = v(t)$ to show that the spectrum of $v^2(t)$ will be bandlimited in $2W$.

2.5   IMPULSES AND TRANSFORMS IN THE LIMIT

So far we've maintained a distinction between two spectral classifications: line spectra that represent periodic power signals and continuous spectra that represent nonperiodic energy signals. But the distinction poses something of a quandary when you encounter a signal consisting of periodic *and* nonperiodic terms. We'll resolve this quandary here by allowing **impulses** in the frequency domain for the representation of discrete frequency components. The underlying notion of **transforms in the limit** also permits the spectral representation of time-domain impulses and other signals whose transforms don't exist in the usual sense.

Properties of the Unit Impulse

The **unit impulse** or **Dirac delta function** $\delta(t)$ is not a function in the strict mathematical sense. Rather, it belongs to a special class known as **generalized functions** or **distributions** whose definitions are stated by assignment rules. In particular, the properties of $\delta(t)$ will be derived from the defining relationship

$$\int_{t_1}^{t_2} v(t)\,\delta(t)\,dt = \begin{cases} v(0) & t_1 < 0 < t_2 \\ 0 & \text{otherwise} \end{cases} \qquad [1]$$

where $v(t)$ is any ordinary function that's continuous at $t = 0$. This rule assigns a number—either $v(0)$ or 0—to the expression on the left-hand side. Equation (1) and all subsequent expressions will also apply to the frequency-domain impulse $\delta(f)$ by replacing t with f.

If $v(t) = 1$ in Eq. (1), it then follows that

$$\int_{-\infty}^{\infty} \delta(t)\,dt = \int_{-\epsilon}^{\epsilon} \delta(t)\,dt = 1 \qquad [2]$$

with ϵ being arbitrarily small. We interpret Eq. (2) by saying that $\delta(t)$ has **unit area** concentrated at the discrete point $t = 0$ and no net area elsewhere. Carrying this argument further suggests that

$$\delta(t) = 0 \qquad t \neq 0 \qquad [3]$$

Equations (2) and (3) are the more familiar definitions of the impulse, and lead to the common graphical representation. For instance, the picture of $A\,\delta(t - t_d)$ is shown in Fig. 2.5–1, where the letter A next to the arrowhead means that $A\,\delta(t - t_d)$ has area or weight A located at $t = t_d$.

Although an impulse does not exist physically, there are numerous conventional functions that have all the properties of $\delta(t)$ in the limit as some parameter ϵ goes to zero. In particular, if the function $\delta_\epsilon(t)$ is such that

$$\lim_{\epsilon \to 0} \int_{-\infty}^{\infty} v(t)\,\delta_\epsilon(t)\,dt = v(0) \qquad [4a]$$

then we say that

$$\lim_{\epsilon \to 0} \delta_\epsilon(t) = \delta(t) \qquad [4b]$$

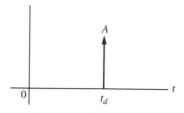

Figure 2.5–1   Graphical representation of $A\delta(t - t_d)$.

Two functions satisfying Eq. (4a) are

$$\delta_\epsilon(t) = \frac{1}{\epsilon} \Pi\left(\frac{t}{\epsilon}\right) \qquad [5]$$

$$\delta_\epsilon(t) = \frac{1}{\epsilon} \text{sinc}\, \frac{t}{\epsilon} \qquad [6]$$

which are plotted in Fig. 2.5–2. You can easily show that Eq. (5) satisfies Eq. (4a) by expanding $v(t)$ in a Maclaurin series prior to integrating. An argument for Eq. (6) will be given shortly when we consider impulses and transforms.

By definition, the impulse has no mathematical or physical meaning unless it appears under the operation of integration. Two of the most significant integration properties are

$$v(t) * \delta(t - t_d) = v(t - t_d) \qquad [7]$$

$$\int_{-\infty}^{\infty} v(t)\, \delta(t - t_d)\, dt = v(t_d) \qquad [8]$$

both of which can derived from Eq. (1). Equation (7) is a **replication** operation, since convolving $v(t)$ with $\delta(t - t_d)$ reproduces the entire function $v(t)$ delayed by t_d. In contrast, Eq. (8) is a **sampling** operation that picks out or samples the value of $v(t)$ at $t = t_d$ — the point where $\delta(t - t_d)$ is "located."

Given the stipulation that any impulse expression must eventually be integrated, you can use certain nonintegral relations to simplify expressions *before* integrating. Two such relations are

$$v(t)\, \delta(t - t_d) = v(t_d)\, \delta(t - t_d) \qquad [9a]$$

$$\delta(\alpha t) = \frac{1}{|\alpha|} \delta(t) \qquad \alpha \neq 0 \qquad [9b]$$

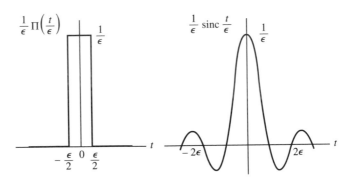

Figure 2.5–2   Two functions that become impulses as $\epsilon \to 0$.

which are justified by integrating both sides over $-\infty < t < \infty$. The product relation in Eq. (9a) simply restates the sampling property. The scale-change relation in Eq. (9b) says that, relative to the independent variable t, $\delta(\alpha t)$ acts like $\delta(t)/|\alpha|$. Setting $\alpha = -1$ then brings out the even-symmetry property $\delta(-t) = \delta(t)$.

Evaluate or simplify each of the following expressions with $v(t) = (t - 3)^2$:

EXERCISE 2.5-1

(a) $\displaystyle\int_{-\infty}^{\infty} v(t)\, \delta(t + 4)\, dt$; (b) $v(t) * \delta(t + 4)$; (c) $v(t)\, \delta(t + 4)$; (d) $v(t) * \delta(-t/4)$.

Impulses in Frequency

Impulses in frequency represent phasors or constants. In particular, let $v(t) = A$ be a constant for all time. Although this signal has infinite energy, we can obtain its transform in a limiting sense by considering that

$$v(t) = \lim_{W \to 0} A \text{ sinc } 2Wt = A \qquad [10a]$$

Now we already have the transform pair A sinc $2Wt \leftrightarrow (A/2W)\Pi(f/2W)$, so

$$\mathcal{F}[v(t)] = \lim_{W \to 0} \frac{A}{2W} \Pi\left(\frac{f}{2W}\right) = A\, \delta(f) \qquad [10b]$$

which follows from Eq. (5) with $\epsilon = 2W$ and $t = f$. Therefore,

$$A \leftrightarrow A\, \delta(f) \qquad [11]$$

and the spectrum of a constant in the time domain is an impulse in the frequency domain at $f = 0$.

This result agrees with intuition in that a constant signal has no time variation and its spectral content ought to be confined to $f = 0$. The impulsive form results simply because we use integration to return to the time domain, via the inverse transform, and an impulse is required to concentrate the nonzero area at a discrete point in frequency. Checking this argument mathematically using Eq. (1) gives

$$\mathcal{F}^{-1}[A\, \delta(f)] = \int_{-\infty}^{\infty} A\, \delta(f)e^{j2\pi ft}\, dt = Ae^{j2\pi ft}\Big|_{f=0} = A$$

which justifies Eq. (11) for our purposes. Note that the impulse has been integrated to obtain a physical quantity, namely the signal $v(t) = A$.

As an alternative to the above procedure, we could have begun with a rectangular pulse, $A\Pi(t/\tau)$, and let $\tau \to \infty$ to get a constant for all time. Then, since $\mathcal{F}[A\Pi(t/\tau)] = A\tau$ sinc $f\tau$, agreement with Eq. (11) requires that

$$\lim_{\tau \to \infty} A\tau \text{ sinc } f\tau = A\, \delta(f)$$

And this supports the earlier assertion in Eq. (6) that a sinc function becomes an impulse under appropriate limiting conditions.

To generalize Eq. (11), direct application of the frequency-translation and modulation theorems yields

$$Ae^{j\omega_c t} \leftrightarrow A\,\delta(f - f_c) \tag{12}$$

$$A\cos(\omega_c t + \phi) \leftrightarrow \frac{Ae^{j\phi}}{2}\delta(f - f_c) + \frac{Ae^{-j\phi}}{2}\delta(f + f_c) \tag{13}$$

Thus, the spectrum of a single phasor is an impulse at $f = f_c$ while the spectrum of a sinusoid has two impulses, shown in Fig. 2.5–3. Going even further in this direction, if $v(t)$ is an arbitrary periodic signal whose exponential Fourier series is

$$v(t) = \sum_{n=-\infty}^{\infty} c(nf_0)e^{j2\pi nf_0 t} \tag{14a}$$

then its Fourier transform is

$$V(f) = \sum_{n=-\infty}^{\infty} c(nf_0)\,\delta(f - nf_0) \tag{14b}$$

where superposition allows us to transform the sum term by term.

By now it should be obvious from Eqs. (11)–(14) that any two-sided line spectrum can be converted to a "continuous" spectrum using this rule: convert the spectral lines to impulses whose weights equal the line heights. The phase portion of the line spectrum is absorbed by letting the impulse weights be complex numbers. Hence, with the aid of transforms in the limit, we can represent both periodic and nonperiodic signals by continuous spectra. That strange beast the impulse function thereby emerges as a key to unifying spectral analysis.

But you may well ask: What's the difference between the line spectrum and the "continuous" spectrum of a period signal? Obviously there can be no physical difference; the difference lies in the mathematical conventions. To return to the time domain from the line spectrum, we sum the phasors which the lines represent. To return to the time domain from the continuous spectrum, we integrate the impulses to get phasors.

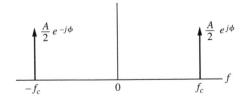

Figure 2.5–3   Spectrum of $A\cos(\omega_c t + \phi)$.

The sinusoidal waveform in Fig. 2.5–4a has constant frequency f_c except for the
interval $-1/f_c < t < 1/f_c$ where the frequency jumps to $2f_c$. Such a signal might be
produced by the process of **frequency modulation,** to be discussed in Chap. 5. Our
interest here is the spectrum, which consists of both impulsive and nonimpulsive
components.

EXAMPLE 2.5–1

For analysis purposes, we'll let $\tau = 2/f_c$ and decompose $v(t)$ into a sum of three
terms as follows:

$$v(t) = A \cos \omega_c t - A\Pi(t/\tau) \cos \omega_c t + A\Pi(t/\tau) \cos 2\omega_c t$$

The first two terms represent a cosine wave with a "hole" to make room for an RF
pulse at frequency $2f_c$ represented by the third term. Transforming $v(t)$ term by term
then yields

$$V(f) = \frac{A}{2} [\delta(f - f_c) + \delta(f + f_c)]$$

$$-\frac{A\tau}{2} [\text{sinc } (f - f_c)\tau + \text{sinc } (f + f_c)\tau]$$

$$+\frac{A\tau}{2} [\text{sinc } (f - 2f_c)\tau + \text{sinc } (f + 2f_c)\tau]$$

where we have drawn upon Eq.(13) and the results of Example 2.3–2. The amplitude spectrum is sketched in Fig. 2.5–4b, omitting the negative-frequency portion.
Note that $|V(f)|$ is not symmetric about $f = f_c$ because the nonimpulsive component
must include the term at $2f_c$.

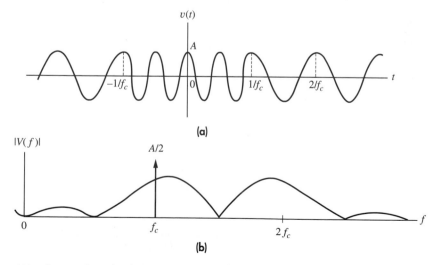

(a)

(b)

Figure 2.5–4   Waveform and amplitude spectrum in Example 2.5–1.

Step and Signum Functions

We've seen that a constant for all time becomes a dc impulse in the frequency domain. Now consider the **unit step function** in Fig. 2.5–5a which steps from "off" to "on" at $t = 0$ and is defined as

$$u(t) \triangleq \begin{cases} 1 & t > 0 \\ 0 & t < 0 \end{cases} \qquad [15]$$

This function has several uses in Fourier theory, especially with regard to **causal** signals since any time function multiplied by $u(t)$ will equal zero for $t < 0$. However, the lack of symmetry creates a problem when we seek the transform in the limit, because limiting operations are equivalent to **contour integrations** and must be performed in a symmetrical fashion—as we did in Eq. (10). To get around this problem, we'll start with the **signum function** (also called the **sign function**) plotted in Fig. 2.5–5b and defined as

$$\text{sgn } t \triangleq \begin{cases} +1 & t > 0 \\ -1 & t < 0 \end{cases} \qquad [16]$$

which clearly has odd symmetry.

The signum function is a limited case of the energy signal $z(t)$ in Fig. 2.5–6 where $v(t) = e^{-bt}u(t)$ and

$$z(t) = v(t) - v(-t) = \begin{cases} +e^{-bt} & t > 0 \\ -e^{+bt} & t < 0 \end{cases}$$

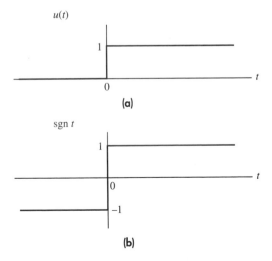

Figure 2.5–5   (a) Unit step function; (b) signum function.

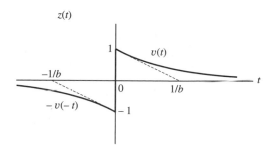

Figure 2.5–6

so that $z(t) \rightarrow \operatorname{sgn} t$ if $b \rightarrow 0$. Combining the results of Example 2.2–2 and Exercise 2.3–1 yields

$$\mathcal{F}[z(t)] = Z(f) = j2V_o(f) = \frac{-j4\pi f}{b^2 + (2\pi f)^2}$$

Therefore,

$$\mathcal{F}[\operatorname{sgn} t] = \lim_{b \to 0} Z(f) = \frac{-j}{\pi f}$$

and we have the transform pair

$$\operatorname{sgn} t \leftrightarrow \frac{1}{j\pi f} \qquad\qquad [17]$$

We then observe from Fig. 2.5–5 that the step and signum functions are related by

$$u(t) = \tfrac{1}{2}(\operatorname{sgn}\ t + 1) = \tfrac{1}{2}\operatorname{sgn} t + \tfrac{1}{2}$$

Hence,

$$u(t) \leftrightarrow \frac{1}{j2\pi f} + \frac{1}{2}\delta(f) \qquad\qquad [18]$$

since $\mathcal{F}[1/2] = \tfrac{1}{2}\delta(f)$.

Note that the spectrum of the signum function does not include a dc impulse. This agrees with the fact that sgn t is an odd function with *zero average value* when averaged over all time, as in Eq. (9), Sect. 2.1. In contrast, the average value of the unit step is $<u(t)> = 1/2$ so its spectrum includes $\tfrac{1}{2}\delta(f)$—just as the transform of a periodic signal with average value $c(0)$ would include the dc term $c(0)\ \delta(f)$.

An impulsive dc term also appears in the **integration theorem** when the signal being integrated has nonzero net area. We derive this property by convolving $u(t)$ with an arbitrary energy signal $v(t)$ to get

$$v * u(t) = \int_{-\infty}^{\infty} v(\lambda)u(t - \lambda)\ d\lambda \qquad\qquad [19]$$

$$= \int_{-\infty}^{t} v(\lambda)\, d\lambda$$

since $u(t - \lambda) = 0$ for $\lambda > t$. But from the convolution theorem and Eq. (18)

$$\mathcal{F}[v * u(t)] = V(f)\left[\frac{1}{j2\pi f} + \frac{1}{2}\delta(f)\right]$$

so

$$\int_{-\infty}^{t} v(\lambda)\, d\lambda \leftrightarrow \frac{1}{j2\pi f} V(f) + \frac{1}{2} V(0)\, \delta(f) \qquad [20]$$

where we have used $V(f)\, \delta(f) = V(0)\, \delta(f)$. Equation (20) reduces to our previous statement of the integration theorem when $V(0) = 0$.

EXERCISE 2.5–2   Apply the modulation theorem to obtain the spectrum of the *causal sinusoid* $v(t) = Au(t) \cos \omega_c t$.

Impulses in Time

Although the time-domain impulse $\delta(t)$ seems a trifle farfetched as a signal model, we'll run into meaningful practical applications in subsequent chapters. Equally important is the value of $\delta(t)$ as an analytic device. To derive its transform, we let $\tau \to 0$ in the known pair

$$\frac{A}{\tau} \Pi\left(\frac{t}{\tau}\right) \leftrightarrow A \operatorname{sinc} f\tau$$

which becomes

$$A\, \delta(t) \leftrightarrow A \qquad [21]$$

Hence, the transform of a time impulse has **constant amplitude,** meaning that its spectrum contains **all frequencies in equal proportion.**

You may have noticed that $A\, \delta(t) \leftrightarrow A$ is the dual of $A \leftrightarrow A\, \delta(f)$. This dual relationship embraces the two extremes of reciprocal spreading in that

An impulsive signal with "zero" duration has infinite spectral width, whereas a constant signal with infinite duration has "zero" spectral width.

Applying the time-delay theorem to Eq. (21) yields the more general pair

$$A\, \delta(t - t_d) \leftrightarrow Ae^{-j2\pi ft_d} \qquad [22]$$

It's a simple matter to confirm the direct transform relationship $\mathcal{F}[A\delta(t - t_d)] = Ae^{-j2\pi ft_d}$; consistency therefore requires that $\mathcal{F}^{-1}[Ae^{-j2\pi ft_d}] = A\,\delta(t - t_d)$, which leads to a significant integral expression for the unit impulse. Specifically, since

$$\mathcal{F}^{-1}[e^{-j2\pi ft_d}] = \int_{-\infty}^{\infty} e^{-j2\pi ft_d}e^{j2\pi ft}\,df$$

we conclude that

$$\int_{-\infty}^{\infty} e^{j2\pi f(t-t_d)}\,df = \delta(t - t_d) \qquad \text{[23]}$$

Thus, the integral on the left side may be evaluated in the limiting form of the unit impulse—a result we'll put immediately to work in a proof of the Fourier integral theorem.

Let $v(t)$ be a continuous time function with a well-defined transform $V(f) = \mathcal{F}[v(t)]$. Our task is to show that the inverse transform does, indeed, equal $v(t)$. From the definitions of the direct and inverse transforms we can write

$$\mathcal{F}^{-1}[V(f)] = \int_{-\infty}^{\infty}\left[\int_{-\infty}^{\infty} v(\lambda)e^{-j2\pi f\lambda}\,d\lambda\right]e^{j2\pi ft}\,df$$

$$= \int_{\infty}^{\infty} v(\lambda)\left[\int_{-\infty}^{\infty} e^{j2\pi(t-\lambda)f}\,df\right]d\lambda$$

But the bracketed integral equals $\delta(t - \lambda)$, from Eq. (23), so

$$\mathcal{F}^{-1}[V(f)] = \int_{-\infty}^{\infty} v(\lambda)\,\delta(t - \lambda)\,d\lambda = v(t) * \delta(t) \qquad \text{[24]}$$

Therefore $\mathcal{F}^{-1}[V(f)]$ equals $v(t)$ in the same sense that $v(t) * \delta(t) = v(t)$. A more rigorous proof, including Gibbs' phenomenon at points of discontinuity, is given by Papoulis (1962, Chap. 2).

Lastly, we relate the unit impulse to the **unit step** by means of the integral

$$\int_{-\infty}^{t} \delta(\lambda - t_d)\,d\lambda = \begin{cases} 1 & t > t_d \\ 0 & t < t_d \end{cases} \qquad \text{[25]}$$

$$= u(t - t_d)$$

Differentiating both sides then yields

$$\delta(t - t_d) = \frac{d}{dt}u(t - t_d) \qquad \text{[26]}$$

which provides another interpretation of the impulse in terms of the derivative of a step discontinuity.

Equations (26) and (22), coupled with the differentiation theorem, expedite certain transform calculations and help us predict a signal's high-frequency spectral rolloff. The method is as follows. Repeatedly differentiate the signal in question until one or more stepwise discontinuities first appear. The next derivative, say the nth, then includes an impulse $A_k \, \delta(t - t_k)$ for each discontinuity of height A_k at $t = t_k$, so

$$\frac{d^n}{dt^n} v(t) = w(t) + \sum_k A_k \, \delta(t - t_k) \qquad \text{[27a]}$$

where $w(t)$ is a nonimpulsive function. Transforming Eq. (27a) gives

$$(j2\pi f)^n V(f) = W(f) + \sum_k A_k \, e^{-j2\pi f t_k} \qquad \text{[27b]}$$

which can be solved for $V(f)$ if we know $W(f) = \mathcal{F}\,[w(t)]$.

Furthermore, if $|W(f)| \to 0$ as $f \to \infty$, the high-frequency behavior of $|V(f)|$ will be proportional to $|f|^{-n}$ and we say that the spectrum has an **nth-order rolloff**. A large value of n thus implies that the signal has very little high-frequency content—an important consideration in the design of many communication systems.

EXAMPLE 2.5–2   **Raised Cosine Pulse**

Figure 2.5–7a shows a waveform called the *raised cosine pulse* because

$$v(t) = \frac{A}{2}\left(1 + \cos\frac{\pi t}{\tau}\right)\Pi\left(\frac{t}{2\tau}\right)$$

We'll use the differentiation method to find the spectrum $V(f)$ and the high-frequency rolloff. The first three derivatives of $v(t)$ are sketched in Fig. 2.5–7b, and we see that

$$\frac{dv(t)}{dt} = -\left(\frac{\pi}{\tau}\right)\frac{A}{2}\sin\frac{\pi t}{\tau}\Pi\left(\frac{t}{2\tau}\right)$$

which has no discontinuities. However, $d^2v(t)/dt^2$ is discontinuous at $t = \pm\tau$ so

$$\frac{d^3}{dt^3}v(t) = \left(\frac{\pi}{\tau}\right)^3\frac{A}{2}\sin\frac{\pi t}{\tau}\,\Pi\left(\frac{t}{2\tau}\right) + \left(\frac{\pi}{\tau}\right)^2\frac{A}{2}\delta(t + \tau) - \left(\frac{\pi}{\tau}\right)^2\frac{A}{2}\delta(t - \tau)$$

This expression has the same form as Eq. (27a), but we do not immediately know the transform of the first term.

Fortunately, a comparison of the first and third derivatives reveals that the first term of $d^3v(t)/dt^3$ can be written as $w(t) = -(\pi/\tau)^2 \, dv(t)/dt$. Therefore, $W(f) = -(\pi/\tau)^2(j2\pi f)V(f)$ and Eq. (27b) gives

$$(j2\pi f)^3 V(f) = -\left(\frac{\pi}{\tau}\right)^2 (j2\pi f)V(f) + \left(\frac{\pi}{\tau}\right)^2\frac{A}{2}\left(e^{j2\pi f\tau} - e^{-j2\pi f\tau}\right)$$

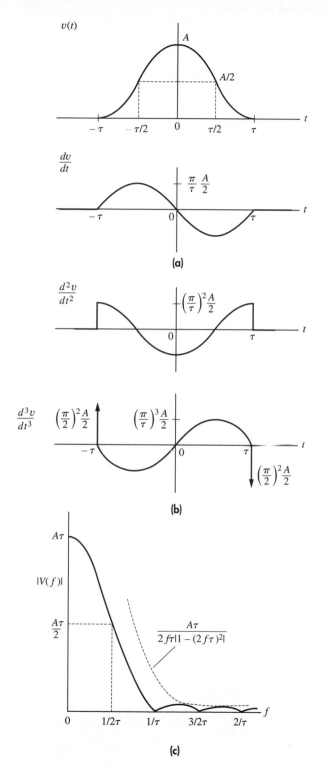

Figure 2.5–7   Raised cosine pulse. (a) Waveform; (b) derivatives; (c) amplitude spectrum.

Routine manipulations finally produce the result

$$V(f) = \frac{jA \sin 2\pi f\tau}{j2\pi f + (\tau/\pi)^2(j2\pi f)^3} = \frac{A\tau \, \text{sinc} \, 2f\tau}{1 - (2f\tau)^2}$$

whose amplitude spectrum is sketched in Fig. 2.5–7c for $f \geq 0$. Note that $|V(f)|$ has a third-order rolloff ($n = 3$), whereas a rectangular pulse with $|V(f)| = |\text{sinc} \, f\tau| = |(\sin \pi f\tau)/(\pi f\tau)|$ would have only a first-order rolloff.

EXERCISE 2.5–3   Let $v(t) = (2At/\tau)\Pi(t/\tau)$. Sketch $dv(t)/dt$ and use Eq. (27) to find $V(f)$.

2.6   PROBLEMS

2.1–1   Consider the phasor signal $v(t) = Ae^{j\phi}e^{j2\pi f_0 t}$. Confirm that Eq. (14) yields just one nonzero coefficient c_m having the appropriate amplitude and phase.

2.1–2   If a periodic signal has the **even-symmetry** property $v(-t) = v(t)$, then Eq. (14) may be written as

$$c_n = \frac{2}{T_0} \int_0^{T_0/2} v(t) \cos (2\pi nt/T_0) \, dt$$

Use this expression to find c_n when $v(t) = A$ for $|t| < T_0/4$ and $v(t) = -A$ for $T_0/4 < |t| < T_0/2$. As a preliminary step you should sketch the waveform and determine c_0 directly from $\langle v(t) \rangle$. Then sketch and label the spectrum after finding c_n.

2.1–3   Do Prob. 2.1–2 with $v(t) = A - 2A|t|/T_0$ for $|t| < T_0/2$.

2.1–4   Do Prob. 2-1–2 with $v(t) = A \cos (2\pi t/T_0)$ for $|t| < T_0/2$.

2.1–5   If a periodic signal has the **odd-symmetry** property $v(-t) = -v(t)$, then Eq. (14) may be written as

$$c_n = -j\frac{2}{T_0} \int_0^{T_0/2} v(t) \sin (2\pi nt/T_0) \, dt$$

Use this expression to find c_n when $v(t) = A$ for $0 < t < T_0/2$ and $v(t) = -A$ for $-T_0/2 < t < 0$. As a preliminary step you should sketch the waveform and determine c_0 directly from $\langle v(t) \rangle$. Then sketch and label the spectrum after finding c_n.

2.1–6   Do Prob. 2.1–5 with $v(t) = A \sin(2\pi t/T_0)$ for $|t| < T_0/2$.

*Indicates answer given in the back of the book.

2.1–7‡   Consider a periodic signal with the **half-wave symmetry** property $\langle v(t \pm T_0/2) = -v(t)\rangle$, so the second half of any period looks like the first half inverted. Show that $c_n = 0$ for all *even* harmonics.

2.1–8*   Use Parseval's power theorem to calculate the average power in the rectangular pulse train with $\tau/T_0 = 1/4$ if all frequencies above $|f| > 1/\tau$ are removed. Repeat for the cases where all frequencies above $|f| > 2/\tau$ and $|f| > 1/2\tau$ are removed.

2.1–9   Let $v(t)$ be the triangular wave with even symmetry listed in Table T.2, and let $v'(t)$ be the approximating obtained with the first three nonzero terms of the trigonometric Fourier series. (*a*) What percentage of the total signal power is contained in $v'(t)$? (*b*) Sketch $v'(t)$ for $|t| < T_0/2$.

2.1–10   Do Prob. 2.1–9 for the square wave in Table T.2.

2.1–11‡   Calculate P for the sawtooth wave listed in Table T.2. Then apply Parseval's power theorem to show that the infinite sum $1/1^2 + 1/2^2 + 1/3^2 + \cdots$ equals $\pi^2/6$.

2.1–12‡   Calculate P for the triangular wave listed in Table T.2. Then apply Parseval's power theorem to show that the infinite sum $1/1^4 + 1/3^4 + 1/5^4 + \cdots$ equals $\pi^4/96$.

2.2–1   Consider the **cosine pulse** $v(t) = A\cos(\pi t/\tau)\Pi(t/\tau)$. Show that $V(f) = (A\tau/2)$ $[\text{sinc}(f\tau - 1/2) + \text{sinc}(f\tau + 1/2]$. Then sketch and label $|V(f)|$ for $f \geq 0$ to verify reciprocal spreading.

2.2–2   Consider the **sine pulse** $v(t) = A\sin(2\pi t/\tau)\Pi(t/\tau)$. Show that $V(f) = -j(A\tau/2)$ $[\text{sinc}(f\tau - 1) - \text{sinc}(f\tau + 1)]$. Then sketch and label $|V(f)|$ for $f \geq 0$ to verify reciprocal spreading.

2.2–3   Find $V(f)$ when $v(t) = (A - A|t|/\tau)\Pi(t/2\tau)$. Express your result in terms of the sinc function.

2.2–4*   Find $V(f)$ when $v(t) = (At/\tau)\Pi(t/2\tau)$. Express your result in terms of the sinc function.

2.2–5   Use Rayleigh's theorem to calculate the energy in the signal $v(t) = \text{sinc}2Wt$.

2.2–6*   Let $v(t)$ be the causal exponential pulse in Example 2.2–2. Use Rayleigh's theorem to calculate the percentage of the total energy contained in $|f| < W$ when $W = b/2\pi$ and $W = 2b/\pi$.

2.2–7   Suppose the left-hand side of Eq. (17) had been written as

$$\int_{-\infty}^{\infty} v(t)w(t)\,dt$$

Find the resulting right-hand side and simplify for the case when $v(t)$ is real and $w(t) = v(t)$.

2.2–8   Show that $\mathcal{F}[w^*(t)] = W^*(-f)$. Then use Eq. (17) to obtain a frequency-domain expression for $\int_{-\infty}^{\infty} v(t)z(t)dt$.

2.2–9   Use the duality theorem to find the Fourier transform of $v(t) = \text{sinc}2t/\tau$.

2.2–10*   Apply duality to the result of Prob. 2.2–1 to find $z(t)$ when $Z(f) = A\cos(\pi f/2W)$ $\Pi(f/2W)$.

2.2–11   Apply duality to the result of Prob. 2.2–2 to find $z(t)$ when $Z(f) = -jA\sin(\pi f/W)$ $\Pi(f/2W)$.

2.2–12‡   Use Eq. (16) and a known transform pair to show that

$$\int_0^\infty (a^2 + x^2)^{-2} \, dx = \pi/4a^3$$

2.3–1*   Let $v(t)$ be the rectangular pulse in Fig. 2.2–1a. Find and sketch $Z(f)$ for $z(t) = v(t - T) + v(t + T)$ taking $\tau \ll T$.

2.3–2   Repeat Prob. 2.3–1 for $z(t) = v(t - 2T) + 2v(t) + v(t + 2T)$.

2.3–3   Repeat Prob. 2.3–1 for $z(t) = v(t - 2T) - 2v(t) + v(t + 2T)$.

2.3–4   Sketch $v(t)$ and find $V(f)$ for

$$v(t) = A\Pi\left(\frac{t - T/2}{T}\right) + B\Pi\left(\frac{t - 3T/2}{T}\right)$$

2.3–5   Sketch $v(t)$ and find $V(f)$ for

$$v(t) = A\Pi\left(\frac{t - 2T}{4T}\right) + B\Pi\left(\frac{t - 2T}{2T}\right)$$

2.3–6*   Find $Z(f)$ in terms of $V(f)$ when $z(t) = v(at - t_d)$.

2.3–7   Prove Eq. (6) (p. 48).

2.3–8   Use Eq. (7) to obtain the transform pair in Prob. 2.2–1.

2.3–9   Use Eq. (7) to obtain the transform pair in Prob. 2.2–2.

2.3–10   Use Eq. (7) to find $Z(f)$ when $z(t) = Ae^{-|t|}\cos\omega_c t$.

2.3–11   Use Eq. (7) to find $Z(f)$ when $z(t) = Ae^{-t}\sin\omega_c t$ for $t \geq 0$ and $z(t) = 0$ for $t < 0$.

2.3–12   Use Eq. (12) to do Prob. 2.2–4.

2.3–13*   Use Eq. (12) to find $Z(f)$ when $z(t) = Ate^{-b|t|}$.

2.3–14   Use Eq. (12) to find $Z(f)$ when $z(t) = At^2 e^{-t}$ for $t \geq 0$ and $z(t) = 0$ for $t < 0$.

2.3–15   Consider the Gaussian pulse listed in Table T.1. Generate a new transform pair by: (a) applying Eq. (8) with $n = 1$; (b) applying Eq. (12) with $n = 1$.

2.4–1   Find and sketch $y(t) = v^*w(t)$ when $v(t) = t$ for $0 < t < 2$ and $w(t) = A$ for $t > 0$. Both signals equal zero outside the specified ranges.

2.4–2*   Do Prob. 2.4–1 with $w(t) = A$ for $0 < t < 3$.

2.4–3   Do Prob. 2.4–1 with $w(t) = A$ for $0 < t < 1$.

2.4–4   Find and sketch $y(t) = v^*w(t)$ when $v(t) = 2\Pi(\frac{t-1}{2})$, $w(t) = A$ for $t \geq 4$, and $w(t) = 0$ otherwise.

2.4–5 Do Prob. 2.4–4 with $w(t) = e^{-2t}$ for $t > 0$ and $w(t) = 0$ otherwise.

2.4–6 Do Prob. 2.4–4 with $w(t) = \Lambda(\frac{t}{\tau})$.

2.4–7* Find $y(t) = v^*w(t)$ for $v(t) = Ae^{-at}$ for $t > 0$ and $w(t) = Be^{-bt}$ for $t > 0$. Both signals equal zero outside the specified ranges.

2.4–8 Do Prob. 2.4–7 with $w(t) = \sin \pi t$ for $0 \le t \le 2$, $w(t) = 0$ otherwise. (Hint: express sinusoid as a sum of exponentials)

2.4–9 Prove Eq. (2a) from Eq. (1) (p. 53).

2.4–10 Let $v(t)$ and $w(t)$ have even symmetry. Show from Eq. (1) that $v^*w(t)$ will have even symmetry.

2.4–11 Let $v(t)$ and $w(t)$ have odd symmetry. Show from Eq. (1) that $v^*w(t)$ will have odd symmetry.

2.4–12 Find and sketch v^*v^*v when $v(t) = \Pi(\frac{t}{\tau})$. You may use the symmetry property stated in Prob. 2.4–10.

2.4–13 Use Eq. (3) to prove Eq. (2b) (p. 55).

2.4–14* Find and sketch $y(t) = v^*w(t)$ when $v(t) = \text{sinc } 4t$ and $w(t) = 2 \text{ sinc } \frac{t}{2}$.

2.5–1 Consider the signal $z(t)$ and its transform $Z(f)$ from Example 2.3–2. Find $z(t)$ and $Z(f)$ as $\tau \to 0$.

2.5–2 Let $v(t)$ be a periodic signal whose Fourier series coefficients are denoted by $c_v(nf_0)$. Use Eq. (14) and an appropriate transform theorem to express $c_w(nf_0)$ in terms of $c_v(nf_0)$ when $w(t) = v(t - t_d)$.

2.5–3 Do Prob. 2.5–2 with $w(t) = dv(t)/dt$.

2.5–4‡ Do Prob. 2.5–2 with $w(t) = v(t) \cos m\omega_0 t$.

2.5–5* Let $v(t) = A$ for $0 < t < 2\tau$ and $v(t) = 0$ otherwise. Use Eq. (18) to find $V(f)$. Check your result by writing $v(t)$ in terms of the rectangle function.

2.5–6 Let $v(t) = A$ for $|t| > \tau$ and $v(t) = 0$ otherwise. Use Eq. (18) to find $V(f)$. Check your result by writing $v(t)$ in terms of the rectangle function.

2.5–7 Let $v(t) = A$ for $t < -T$, and $v(t) = -A$ for $t > T$, and $v(t) = 0$ otherwise. Use Eq. (18) to find $V(f)$. Check your result by letting $T \to 0$.

2.5–8 Let

$$w(t) = \int_{-\infty}^{t} v(\lambda) \, d\lambda$$

with $v(t) = (1/\epsilon)\Pi(t/\epsilon)$. Sketch $w(t)$ and use Eq. (20) to find $W(f)$. Then let $\epsilon \to 0$ and compare your results with Eq. (18).

2.5–9 Do Prob. 2.5–8 with $v(t) = (1/\epsilon)e^{-t/\epsilon} u(t)$.

2.5–10 Obtain the transform of the signal in Prob. 2.3–1 by expressing $z(t)$ as the convolution of $v(t)$ with impulses.

2.5–11*   Do Prob. 2.5–10 for the signal in Prob. 2.3–2.

2.5–12   Do Prob. 2.5–10 for the signal in Prob. 2.3–3.

2.5–13*   Find and sketch the signal $v(t) = \sum_{n=0}^{8} \sin{(2\pi t)}\delta(t - 0.5n)$ using Eq. (9a).

2.5–14   Find and sketch the signal $v(t) = \sum_{n=-10}^{10} \cos{(2\pi t)}\delta(t - 0.1n)$ using Eq. (9a).

chapter

3

Signal Transmission and Filtering

CHAPTER OUTLINE

Signal transmission is the process whereby an electrical waveform gets from one location to another, ideally arriving without distortion. In contrast, signal filtering is an operation that purposefully distorts a waveform by altering its spectral content. Nonetheless, most transmission systems and filters have in common the properties of linearity and time invariance. These properties allow us to model both transmission and filtering in the time domain in terms of the impulse response, or in the frequency domain in terms of the frequency response.

This chapter begins with a general consideration of system response in both domains. Then we'll apply our results to the analysis of signal transmission and distortion for a variety of media and systems such as fiber optics and satellites. We'll examine the use of various types of filters and filtering in communication systems. Some related topics—notably transmission loss, Hilbert transforms, and correlation—are included as starting points for subsequent development.

OBJECTIVES

After studying this chapter and working the exercises, you should be able to do each of the following:

1. State and apply the input–output relations for an LTI system in terms of its impulse response $h(t)$, step response $g(t)$, or transfer function $H(f)$ (Sect. 3.1).
2. Use frequency-domain analysis to obtain an exact or approximate expression for the output of a system (Sect. 3.1).
3. Find $H(f)$ from the block diagram of a simple system (Sect. 3.1).
4. Distinguish between amplitude distortion, delay distortion, linear distortion, and nonlinear distortion (Sect. 3.2).
5. Identify the frequency ranges that yield distortionless transmission for a given channel, and find the equalization needed for distortionless transmission over a specified range (Sect. 3.2).
6. Use dB calculations to find the signal power in a cable transmission system with amplifiers (Sect. 3.3).
7. Discuss the characteristics of and requirements for transmission over fiber optic and satellite systems (Sect. 3.3).
8. Identify the characteristics and sketch $H(f)$ and $h(t)$ for an ideal LPF, BPF, or HPF (Sect. 3.4).
9. Find the 3-dB bandwidth of a real LPF, given $H(f)$ (Sect. 3.4).
10. State and apply the bandwidth requirements for pulse transmission (Sect. 3.4).
11. State and apply the properties of the Hilbert transform (Sect. 3.5).
12. Define the crosscorrelation and autocorrelation functions for power or energy signals, and state their properties (Sect. 3.6).
13. State the Wiener-Kinchine theorem and the properties of spectral density functions (Sect. 3.6).
14. Given $H(f)$ and the input correlation or spectral density function, find the output correlation or spectral density (Sect. 3.6).

3.1   RESPONSE OF LTI SYSTEMS

Figure 3.1–1 depicts a system inside a "black box" with an external **input signal** $x(t)$ and an **output signal** $y(t)$. In the context of electrical communication, the system usually would be a two-port network driven by an applied voltage or current at the input port, producing another voltage or current at the output port. Energy stor-

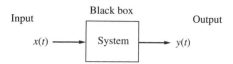

Figure 3.1–1

age elements and other internal effects may cause the output waveform to look quite different from the input. But regardless of what's in the box, the system is character-ized by an **excitation-and-response** relationship between input and output.

Here we're concerned with the special but important class of **linear time-invariant** systems—or LTI systems for short. We'll develop the input–output rela-tionship in the time domain using the superposition integral and the system's impulse response. Then we'll turn to frequency-domain analysis expressed in terms of the system's transfer function.

Impulse Response and the Superposition Integral

Let Fig. 3.1–1 be an LTI system having no internal stored energy at the time the input $x(t)$ is applied. The output $y(t)$ is then the **forced response** due entirely to $x(t)$, as represented by

$$y(t) = F[x(t)] \tag{1}$$

where $F[x(t)]$ stands for the functional relationship between input and output. The **linear** property means that Eq. (1) obeys the **principle of superposition.** Thus, if

$$x(t) = \sum_k a_k x_k(t) \tag{2a}$$

where a_k are constants, then

$$y(t) = \sum_k a_k F[x_k(t)] \tag{2b}$$

The **time-invariance** property means that the system's characteristics remain **fixed with time.** Thus, a time-shifted input $x(t - t_d)$ produces

$$F[x(t - t_d)] = y(t - t_d) \tag{3}$$

so the output is time-shifted but otherwise unchanged.

Most LTI systems consist entirely of **lumped-parameter** elements (such as resistors capacitors, and inductors), as distinguished from elements with spatially distributed phenomena (such as transmission lines). Direct analysis of a lumped-parameter system starting with the element equations leads to the input–output rela-tion as a linear differential equation in the form

$$a_n \frac{d^n y(t)}{dt^n} + \cdots + a_1 \frac{dy(t)}{dt} + a_0 y(t) = b_m \frac{d^m x(t)}{dt^m} + \cdots + b_1 \frac{dx(t)}{dt} + b_0 x(t) \tag{4}$$

where the a's and b's are constant coefficients involving the element values. The number of independent energy-storage elements determines the value of n, known as the *order* of the system. Unfortunately, Eq. (4) doesn't provide us with a direct expression for $y(t)$.

To obtain an explicit input–output equation, we must first define the system's **impulse response**

$$h(t) \triangleq F[\delta(t)] \tag{5}$$

which equals the forced response when $x(t) = \delta(t)$. But any continuous input signal can be written as the convolution $x(t) = x(t) * \delta(t)$, so

$$y(t) = F\left[\int_{-\infty}^{\infty} x(\lambda)\delta(t - \lambda) \, d\lambda \right]$$

$$= \int_{-\infty}^{\infty} x(\lambda)F[\delta(t - \lambda)] \, d\lambda$$

in which the interchange of operations is allowed by virtue of the system's linearity. Now, from the time-invariance property, $F[\delta(t - \lambda)] = h(t - \lambda)$ and hence

$$y(t) = \int_{-\infty}^{\infty} x(\lambda)h(t - \lambda) \, d\lambda \tag{6a}$$

$$= \int_{-\infty}^{\infty} h(\lambda)x(t - \lambda) \, d\lambda \tag{6b}$$

where we have drawn upon the commutivity of convolution.

Either form of Eq. (6) is called the **superposition integral.** It expresses the forced response as a convolution of the input $x(t)$ with the impulse response $h(t)$. System analysis in the time domain therefore requires knowledge of the impulse response along with the ability to carry out the convolution.

Various techniques exist for determining $h(t)$ from a differential equation or some other system model. But you may be more comfortable taking $x(t) = u(t)$ and calculating the system's **step response**

$$g(t) \triangleq F[u(t)] \tag{7a}$$

from which

$$h(t) = \frac{dg(t)}{dt} \tag{7b}$$

This derivative relation between the impulse and step response follows from the general convolution property

$$\frac{d}{dt}[v * w(t)] = v(t) * \left[\frac{dw(t)}{dt} \right]$$

Thus, since $g(t) = h * u(t)$ by definition, $dg(t)/dt = h(t) * [du(t)/dt] = h(t) * \delta(t) = h(t)$.

Time Response of a First-Order System

EXAMPLE 3.1–1

The simple RC circuit in Fig. 3.1–2 has been arranged as a two-port network with input voltage $x(t)$ and output voltage $y(t)$. The reference voltage polarities are indicated by the $+/-$ notation where the assumed higher potential is indicated by the $+$ sign. This circuit is a first-order system governed by the differential equation

$$RC\frac{dy(t)}{dt} + y(t) = x(t)$$

Similar expressions describe certain transmission lines and cables, so we're particularly interested in the system response.

From either the differential equation or the circuit diagram, the step response is readily found to be

$$g(t) = (1 - e^{-t/RC})u(t) \qquad \text{[8a]}$$

Interpreted physically, the capacitor starts at zero initial voltage and charges toward $y(\infty) = 1$ with time constant RC when $x(t) = u(t)$. Figure 3.1–3a plots this behavior, while Fig. 3.1–3b shows the corresponding impulse response

$$h(t) = \frac{1}{RC}e^{-t/RC}u(t) \qquad \text{[8b]}$$

obtained by differentiating $g(t)$. Note that $g(t)$ and $h(t)$ are *causal* waveforms since the input equals zero for $t < 0$.

The response to an arbitrary input $x(t)$ can now be found by putting Eq. (8b) in the superposition integral. For instance, take the case of a rectangular pulse applied at $t = 0$, so $x(t) = A$ for $0 < t < \tau$. The convolution $y(t) = h * x(t)$ divides into three parts, like the example back in Fig. 2.4–1 with the result that

$$y(t) = \begin{cases} 0 & t < 0 \\ A(1 - e^{-t/RC}) & 0 < t < \tau \\ A(1 - e^{-\tau/RC})e^{-(t-\tau)/RC} & t > \tau \end{cases} \qquad \text{[9]}$$

as sketched in Fig. 3.1–4 for three values of τ/RC.

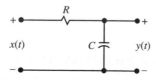

Figure 3.1–2   RC lowpass filter.

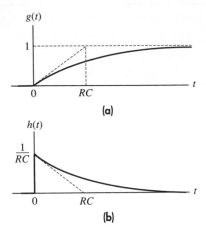

Figure 3.1–3   Output of an RC lowpass filter. (*a*) Step response; (*b*) impulse response.

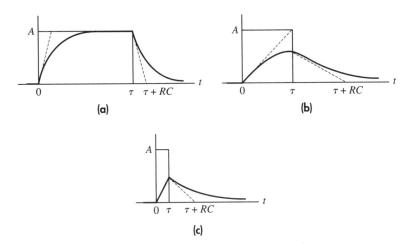

Figure 3.1–4   Rectangular pulse response of an RC lowpass filter. (*a*) $\tau \gg$ RC; (*b*) $\tau \approx$ RC; (*c*) $\tau \ll$ RC.

EXERCISE 3.1–1   Let the resistor and the capacitor be interchanged in Fig. 3.1–2. Find the step and impulse response.

Transfer Functions and Frequency Response

Time-domain analysis becomes increasingly difficult for higher-order systems, and the mathematical complications tend to obscure significant points. We'll gain a different and often clearer view of system response by going to the frequency domain. As a first step in this direction, we define the system **transfer function** to be the Fourier transform of the impulse response, namely,

$$H(f) \triangleq \mathcal{F}[h(t)] = \int_{-\infty}^{\infty} h(t)e^{-j2\pi ft}\, dt \qquad [10]$$

This definition requires that $H(f)$ exists, at least in a limiting sense. In the case of an **unstable** system, $h(t)$ **grows with time** and $H(f)$ does not exist.

When $h(t)$ is a *real* time function, $H(f)$ has the hermitian symmetry

$$H(-f) = H^*(f) \qquad [11a]$$

so that

$$|H(-f)| = |H(f)| \qquad \arg H(-f) = -\arg H(f) \qquad [11b]$$

We'll assume this property holds unless otherwise stated.

The frequency-domain interpretation of the transfer function comes from $y(t) = h * x(t)$ with a *phasor* input, say

$$x(t) = A_x e^{j\phi_x} e^{j2\pi f_0 t} \qquad -\infty < t < \infty \qquad [12a]$$

The stipulation that $x(t)$ persists for all time means that we're dealing with *steady-state* conditions, like the familiar case of ac steady-state circuit analysis. The steady-state forced response is

$$
\begin{aligned}
y(t) &= \int_{-\infty}^{\infty} h(\lambda) A_x e^{j\phi_x} e^{j2\pi f_0(t-\lambda)}\, d\lambda \\
&= \left[\int_{-\infty}^{\infty} h(\lambda) e^{-j2\pi f_0 \lambda}\, d\lambda \right] A_x e^{j\phi_x} e^{j2\pi f_0 t} \\
&= H(f_0) A_x e^{j\phi_x} e^{j2\pi f_0 t}
\end{aligned}
$$

where, from Eq. (10), $H(f_0)$ equals $H(f)$ with $f = f_0$. Converting $H(f_0)$ to polar form then yields

$$y(t) = A_y e^{j\phi_y} e^{j2\pi f_0 t} \qquad -\infty < t < \infty \qquad [12b]$$

in which we have identified the output phasor's amplitude and angle

$$A_y = |H(f_0)|A_x \qquad \phi_y = \arg H(f_0) + \phi_x \qquad [13]$$

Using conjugate phasors and superposition, you can similarly show that if

$$x(t) = A_x \cos\,(2\pi f_0 t + \phi_x)$$

then

$$y(t) = A_y \cos\,(2\pi f_0 t + \phi_y)$$

with A_y and ϕ_y as in Eq. (13).

Since $A_y/A_x = |H(f_0)|$ at any frequency f_0, we conclude that $|H(f)|$ represents the system's **amplitude ratio** as a function of frequency (sometimes called the **amplitude response** or **gain**). By the same token, arg $H(f)$ represents the **phase shift,** since $\phi_y - \phi_x = $ arg $H(f_0)$. Plots of $|H(f)|$ and arg $H(f)$ versus frequency give us the frequency-domain representation of the system or, equivalently, the system's **frequency response.** Henceforth, we'll refer to $H(f)$ as either the transfer function or frequency-response function.

Now let $x(t)$ be any signal with spectrum $X(f)$. Calling upon the convolution theorem, we take the transform of $y(t) = h * x(t)$ to obtain

$$Y(f) = H(f)X(f) \qquad\qquad [14]$$

This elegantly simple result constitutes the basis of frequency-domain system analysis. It says that

The output spectrum Y(f) equals the input spectrum X(f) multiplied by the transfer function H(f).

The corresponding amplitude and phase spectra are

$$|Y(f)| = |H(f)||X(f)|$$
$$\text{arg } Y(f) = \text{arg } H(f) + \text{arg } X(f)$$

which compare with the single-frequency expressions in Eq. (13). If $x(t)$ is an energy signal, then $y(t)$ will be an energy signal whose spectral density and total energy are given by

$$|Y(f)|^2 = |H(f)|^2|X(f)|^2 \qquad\qquad [15a]$$

$$E_y = \int_{-\infty}^{\infty} |H(f)|^2|X(f)|^2 \, df \qquad\qquad [15b]$$

as follows from Rayleigh's energy theorem.

Equation (14) sheds new light on the meaning of the system transfer function and the transform pair $h(t) \leftrightarrow H(f)$. For if we let $x(t)$ be a unit impulse, then $X(f) = 1$ and $Y(f) = H(f)$—in agreement with the definition $y(t) = h(t)$ when $x(t) = \delta(t)$. From the frequency-domain viewpoint, the "flat" input spectrum $X(f) = 1$ contains all frequencies in equal proportion and, consequently, the output spectrum takes on the shape of the transfer function $H(f)$.

Figure 3.1–5 summarizes our input–output relations in both domains. Clearly, when $H(f)$ and $X(f)$ are given, the output spectrum $Y(f)$ is much easier to find than the output signal $y(t)$. In principle, we could compute $y(t)$ from the inverse transform

$$y(t) = \mathscr{F}^{-1}[H(f)X(f)] = \int_{-\infty}^{\infty} H(f)X(f)e^{j2\pi ft} \, df$$

But this integration does not necessarily offer any advantages over time-domain convolution. Indeed, the power of frequency-domain system analysis largely

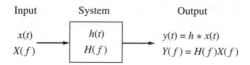

Figure 3.1–5   Input–output relations for an LTI system.

depends on staying in that domain and using our knowledge of spectral properties to draw inferences about the output signal.

Finally, we point out two ways of determining $H(f)$ that don't involve $h(t)$. If you know the *differential equation* for a lumped-parameter system, you can immediately write down its transfer function as the ratio of polynomials

$$H(f) = \frac{b_m(j2\pi f)^m + \cdots + b_1(j2\pi f) + b_0}{a_n(j2\pi f)^n + \cdots + a_1(j2\pi f) + a_0} \qquad [16]$$

whose coefficients are the same as those in Eq. (4). Equation (16) follows from Fourier transformation of Eq. (4).

Alternatively, if you can calculate a system's **steady-state phasor response,** Eqs. (12) and (13) show that

$$H(f) = \frac{y(t)}{x(t)} \qquad \text{when} \qquad x(t) = e^{j2\pi ft} \qquad [17]$$

This method corresponds to impedance analysis of electrical circuits, but is equally valid for any LTI system. Furthermore, Eq. (17) may be viewed as a special case of the s domain transfer function $H(s)$ used in conjunction with Laplace transforms. Since $s = \sigma + j\omega$ in general, $H(f)$ is obtained from $H(s)$ simply by letting $s = j2\pi f$. These methods assume, of course, that the system is stable.

Frequency Response of a First-Order System   EXAMPLE 3.1–2

The RC circuit from Example 3.1–1 has been redrawn in Fig. 3.1–6a with the impedances $Z_R = R$ and $Z_C = 1/j\omega C$ replacing the elements. Since $y(t)/x(t) = Z_C/(Z_C + Z_R)$ when $x(t) = e^{j\omega t}$, Eq. (17) gives

$$H(f) = \frac{(1/j2\pi fC)}{(1/j2\pi fC) + R} = \frac{1}{1 + j2\pi fRC}$$

$$= \frac{1}{1 + j(f/B)} \qquad [18a]$$

where we have introduced the system parameter

$$B \triangleq \frac{1}{2\pi RC} \qquad [18b]$$

Identical results would have been obtained from Eq. (16), or from $H(f) = \mathcal{F}[h(t)]$. (In fact, the system's impulse response has the same form as the causal exponential pulse discussed in Example 2.2–2.) The amplitude ratio and phase shift are

$$|H(f)| = \frac{1}{\sqrt{1 + (f/B)^2}} \qquad \arg H(f) = -\arctan\frac{f}{B} \qquad \text{[18c]}$$

as plotted in Fig. 3.1–6b for $f \geq 0$. The hermitian symmetry allows us to omit $f < 0$ without loss of information.

The amplitude ratio $|H(f)|$ has special significance relative to any *frequency-selective* properties of the system. We call this particular system a **lowpass filter** because it has almost no effect on the amplitude of low-frequency components, say $|f| \ll B$, while it drastically reduces the amplitude of high-frequency components, say $|f| \gg B$. The parameter B serves as a measure of the filter's **passband** or **bandwidth.**

To illustrate how far you can go with frequency-domain analysis, let the input $x(t)$ be an arbitrary signal whose spectrum has negligible content for $|f| > W$. There are three possible cases to consider, depending on the relative values of B and W:

1. If $W \ll B$, as shown in Fig. 3.1–7a, then $|H(f)| \approx 1$ and $\arg H(f) \approx 0$ over the signal's frequency range $|f| < W$. Thus, $Y(f) = H(f)X(f) \approx X(f)$ and $y(t) \approx x(t)$ so we have **undistorted transmission** through the filter.

2. If $W \approx B$, as shown in Fig. 3.1–7b, then $Y(f)$ depends on both $H(f)$ and $X(f)$. We can say that the output is **distorted,** since $y(t)$ will differ significantly from $x(t)$, but time-domain calculations would be required to find the actual waveform.

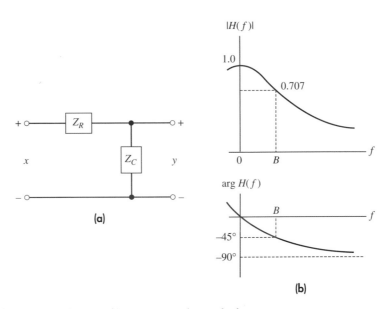

Figure 3.1–6   RC lowpass filter.(a) circuit; (b) transfer function.

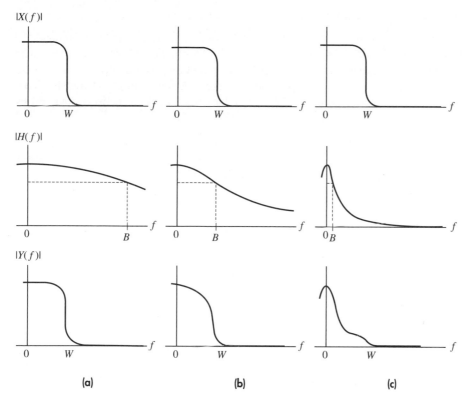

Figure 3.1–7   Frequency-domain analysis of a first-order lowpass filter. (a) $B \gg W$; (b) $B \approx$ W, (c) $B \ll W$.

3. If $W \gg B$, as shown in Fig. 3.1–7c, the input spectrum has a nearly constant value $X(0)$ for $|f| < B$ so $Y(f) \approx X(0)H(f)$. Thus, $y(t) \approx X(0)h(t)$, and the output signal now looks like the filter's *impulse response*. Under this condition, we can reasonably model the input signal as an impulse.

Our previous time-domain analysis with a rectangular input pulse confirms these conclusions since the nominal spectral width of the pulse is $W = 1/\tau$. The case $W \ll B$ thus corresponds to $1/\tau \ll 1/2\pi RC$ or $\tau/RC \gg 1$, and we see in Fig. 3.1–4a that $y(t) \approx x(t)$. Conversely, $W \gg B$ corresponds to $\tau/RC \ll 1$ as in Fig. 3.1–4c where $y(t)$ looks more like $h(t)$.

Find $H(f)$ when $Z_L = j\omega L$ replaces Z_C in Fig. 3.1–6a. Express your result in terms of the system parameter $f_\ell = R/2\pi L$, and justify the name "highpass filter" by sketching $|H(f)|$ versus f.   **EXERCISE 3.1–2**

Block-Diagram Analysis

More often than not, a communication system comprises many interconnected building blocks or subsystems. Some blocks might be two-port networks with known transfer functions, while other blocks might be given in terms of their time-domain operations. Any LTI operation, of course, has an equivalent transfer function. For reference purposes, Table 3.1–1 lists the transfer functions obtained by applying transform theorems to four primitive time-domain operations.

Table 3.1–1

Time-Domain Operation		Transfer Function
Scalar multiplication	$y(t) = \pm Kx(t)$	$H(f) = \pm K$
Differentiation	$y(t) = \dfrac{dx(t)}{dt}$	$H(f) = j2\pi f$
Integration	$y(t) = \displaystyle\int_{-\infty}^{t} x(\lambda)\, d\lambda$	$H(f) = \dfrac{1}{j2\pi f}$
Time delay	$y(t) = x(t - t_d)$	$H(f) = e^{-j2\pi f t_d}$

When the subsystems in question are described by individual transfer functions, it is possible and desirable to lump them together and speak of the overall system transfer function. The corresponding relations are given below for two blocks connected in parallel, cascade, and feedback. More complicated configurations can be analyzed by successive application of these basic rules. One essential assumption must be made, however, namely, that any interaction or *loading* effects have been accounted for in the individual transfer functions so that they represent the actual response of the subsystems in the context of the overall system. (A simple op-amp voltage follower might be used to provide isolation between blocks and prevent loading.)

Figure 3.1–8a diagrams two blocks in **parallel:** both units have the same input and their outputs are summed to get the system's output. From superposition it follows that $Y(f) = [H_1(f) + H_2(f)]X(f)$ so the overall transfer function is

$$H(f) = H_1(f) + H_2(f) \qquad \text{Parallel connection} \qquad [19a]$$

In the **cascade** connection, Fig. 3.1–8b, the output of the first unit is the input to the second, so $Y(f) = H_2(f)[H_1(f)X(f)]$ and

$$H(f) = H_1(f)H_2(f) \qquad \text{Cascade connection} \qquad [19b]$$

The **feedback** connection, Fig. 3.1–8c, differs from the other two in that the output is sent back through $H_2(f)$ and subtracted from the input. Thus,

$$Y(f) = H_1(f)[X(f) - H_2(f)Y(f)]$$

and rearranging yields $Y(f) = \{H_1(f)/[1 + H_1(f)H_2(f)]\}X(f)$ so

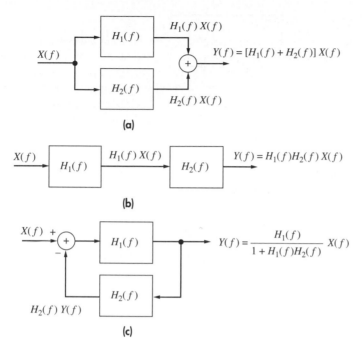

$$Y(f) = [H_1(f) + H_2(f)]\, X(f)$$

$$Y(f) = H_1(f)H_2(f)\, X(f)$$

$$Y(f) = \frac{H_1(f)}{1 + H_1(f)H_2(f)}\, X(f)$$

Figure 3.1–8   (a) Parallel connection; (b) cascade connection; (c) feedback connection.

$$H(f) = \frac{H_1(f)}{1 + H_1(f)H_2(f)} \qquad \text{Feedback connection} \qquad \text{[19c]}$$

This case is more properly termed the *negative* feedback connection as distinguished from positive feedback, where the returned signal is added to the input instead of subtracted.

Zero-Order Hold   EXAMPLE 3.1–3

The zero-order hold system in Fig. 3.1–9a has several applications in electrical communication. Here we take it as an instructive exercise of the parallel and cascade relations. But first we need the individual transfer functions, determined as follows: the upper branch of the parallel section is a straight-through path so, trivially, $H_1(f) = 1$; the lower branch produces pure time delay of T seconds followed by sign inversion, and lumping them together gives $H_2(f) = -e^{-j2\pi fT}$; the integrator in the final block has $H_3(f) = 1/j2\pi f$. Figure 3.1–9b is the equivalent block diagram in terms of these transfer functions.

Having gotten this far, the rest of the work is easy. We combine the parallel branches in $H_{12}(f) = H_1(f) + H_2(f)$ and use the cascade rule to obtain

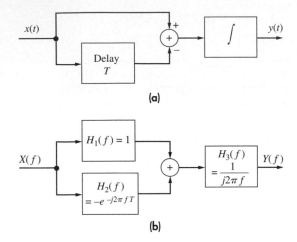

(a)

(b)

Figure 3.1–9   Block diagrams of a zero-order hold. (a) Time domain; (b) frequency domain.

$$H(f) = H_{12}(f)H_3(f) = \left[H_1(f) + H_2(f)\right]H_3(f)$$

$$= \left[1 - e^{-j2\pi fT}\right]\frac{1}{j2\pi f}$$

$$= \frac{e^{j\pi fT} - e^{-j\pi fT}}{j2\pi f} e^{-j\pi fT} = \frac{\sin \pi fT}{\pi f} e^{-j\pi fT}$$

$$= T \operatorname{sinc} fT e^{-j\pi fT}$$

Hence we have the unusual result that the amplitude ratio of this system is a *sinc function* in frequency!

To confirm this result by another route, let's calculate the impulse response $h(t)$ drawing upon the definition that $y(t) = h(t)$ when $x(t) = \delta(t)$. Inspection of Fig. 3.1–9a shows that the input to the integrator then is $x(t) - x(t - T) = \delta(t) - \delta(t - T)$, so

$$h(t) = \int_{-\infty}^{t} \left[\delta(\lambda) - \delta(\lambda - T)\right] d\lambda = u(t) - u(t - T)$$

which represents a rectangular pulse starting at $t = 0$. Rewriting the impulse response as $h(t) = \Pi[(t - T/2)/T]$ helps verify the transform relation $h(t) \leftrightarrow H(f)$.

EXERCISE 3.1–3   Let $x(t) = A\Pi(t/\tau)$ be applied to the zero-order hold. Use frequency-domain analysis to find $y(t)$ when $\tau \ll T$, $\tau = T$, and $\tau \gg T$.

3.2   SIGNAL DISTORTION IN TRANSMISSION

A signal transmission system is the electrical channel between an information source and destination. These systems range in complexity from a simple pair of wires to a sophisticated laser-optics link. But all transmission systems have two physical attributes of particular concern in communication: internal power dissipation that reduces the size of the output signal, and energy storage that alters the shape of the output.

Our purpose here is to formulate the conditions for distortionless signal transmission, assuming an LTI system so we can work with its transfer function. Then we'll define various types of distortion and address possible techniques for minimizing their effects.

Distortionless Transmission

Distortionless transmission means that the output signal has the same "shape" as the input. More precisely, given an input signal $x(t)$, we say that

The output is undistorted if it differs form the input only by a multiplying constant and a finite time delay.

Analytically, we have distortionless transmission if

$$y(t) = Kx(t - t_d) \qquad\qquad [1]$$

where K and t_d are constants.

The properties of a distortionless system are easily found by examining the output spectrum

$$Y(f) = \mathscr{F}[y(t)] = Ke^{-j\omega t_d}X(f)$$

Now by definition of transfer function, $Y(f) = H(f)X(f)$, so

$$H(f) = Ke^{-j\omega t_d} \qquad\qquad [2a]$$

In words, a system giving distortionless transmission must have *constant amplitude response* and negative *linear phase shift*, so

$$|H(f)| = |K| \qquad \arg H(f) = -2\pi t_d f \pm m180° \qquad [2b]$$

Note that arg $H(f)$ must pass through the origin or intersect at an integer multiple of $\pm180°$. We have added the term $\pm m180°$ to the phase to account for K being positive or negative. In the case of zero time delay, the phase is constant at 0 or $\pm180°$.

An important and rather obvious qualification to Eq. (2) should be stated immediately. The conditions on $H(f)$ are required only over those frequencies where the input signal has significant spectral content. To underscore this point, Fig. 3.2–1 shows the energy spectral density of an average voice signal obtained from laboratory measurements. Since the spectral density is quite small for $f < 200$ Hz and

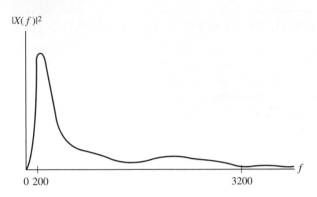

Figure 3.2–1   Energy spectral density of an average voice signal.

$f > 3200$ Hz, we conclude that a system satisfying Eq. (2) over $200 \leq |f| \leq 3200$ Hz would yield nearly distortion-free voice transmission. Similarly, since the human ear only processes sounds between about 20 Hz and 20,000 Hz, an audio system that is distortion free in this range is sufficient.

However, the stringent demands of distortionless transmission can only be satisfied approximately in practice, so transmission systems always produce some amount of signal distortion. For the purpose of studying distortion effects on various signals, we'll define three major types of distortion:

1. *Amplitude distortion*, which occurs when

$$|H(f)| \neq |K|$$

2. *Delay distortion*, which occurs when

$$\arg H(f) \neq -2\pi t_d f \pm m180°$$

3. *Nonlinear distortion*, which occurs when the system includes nonlinear elements

The first two types can be grouped under the general designation of *linear* distortion, described in terms of the transfer function of a linear system. For the third type, the nonlinearity precludes the existence of a transfer function.

EXAMPLE 3.2–1   Suppose a transmission system has the frequency response plotted in Fig. 3.2–2. This system satisfies Eq. (2) for $20 \leq |f| \leq 30$ kHz. Otherwise, there's amplitude distortion for $|f| < 20$ kHz and $|f| > 50$ kHz, and delay distortion for $|f| > 30$ kHz.

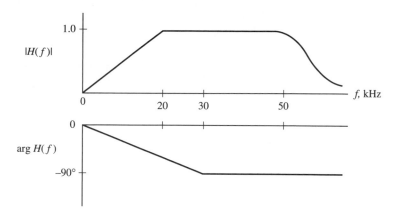

Figure 3.2–2

Linear Distortion

Linear distortion includes any amplitude or delay distortion associated with a linear transmission system. Amplitude distortion is easily described in the frequency domain; it means simply that the output frequency components are not in correct proportion. Since this is caused by $|H(f)|$ not being constant with frequency, amplitude distortion is sometimes called **frequency distortion.**

The most common forms of amplitude distortion are excess attenuation or enhancement of extreme high or low frequencies in the signal spectrum. Less common, but equally bothersome, is disproportionate response to a band of frequencies within the spectrum. While the frequency-domain description is easy, the effects in the time domain are far less obvious, save for very simple signals. For illustration, a suitably simple test signal is $x(t) = \cos \omega_0 t - 1/3 \cos 3\omega_0 t + 1/5 \cos 5\omega_0 t$, a rough approximation to a square wave sketched in Fig. 3.2–3. If the low-frequency or high-frequency component is attenuated by one-half, the resulting outputs are as shown in Fig. 3.2–4. As expected, loss of the high-frequency term reduces the "sharpness" of the waveform.

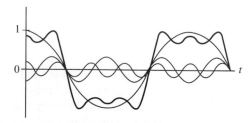

Figure 3.2–3   Test signal $x(t) = \cos \omega_0 t - 1/3 \cos 3\omega_0 t + 1/5 \cos 5\omega_0 t$.

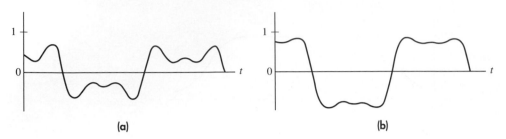

Figure 3.2–4   Test signal with amplitude distortion. *(a)* Low frequency attenuated; *(b)* high frequency attenuated.

Beyond qualitative observations, there's little more we can say about amplitude distortion without experimental study of specific signal types. Results of such studies are usually couched in terms of required "flat" frequency response—meaning the frequency range over which $|H(f)|$ must be constant to within a certain tolerance so that the amplitude distortion is sufficiently small.

We now turn our attention to phase shift and time delay. If the phase shift is not linear, the various frequency components suffer different amounts of time delay, and the resulting distortion is termed **phase** or **delay distortion.** For an arbitrary phase shift, the time delay is a function of frequency and can be found by writing arg $H(f) = -2\pi f t_d(f)$ with all angles expressed in *radians*. Thus

$$t_d(f) = -\frac{\arg H(f)}{2\pi f} \qquad\qquad \text{[3]}$$

which is independent of frequency only if arg $H(f)$ is linear with frequency.

A common area of confusion is **constant time delay** versus **constant phase shift.** The former is desirable and is required for distortionless transmission. The latter, in general, causes distortion. Suppose a system has the constant phase shift θ not equal to $0°$ or $\pm m180°$. Then each signal frequency component will be delayed by $\theta/2\pi$ *cycles* of its own frequency; this is the meaning of constant phase shift. But the time delays will be different, the frequency components will be scrambled in time, and distortion will result.

That constant phase shift does give distortion is simply illustrated by returning to the test signal of Fig. 3.2–3 and shifting each component by one-fourth cycle, $\theta = -90°$. Whereas the input was roughly a square wave, the output will look like the triangular wave in Fig. 3.2–5. With an arbitrary nonlinear phase shift, the deterioration of waveshape can be even more severe.

You should also note from Fig. 3.2–5 that the *peak* excursions of the phase-shifted signal are substantially greater (by about 50 percent) than those of the input test signal. This is not due to amplitude response, since the output amplitudes of the three frequency components are, in fact, unchanged; rather, it is because the components of the distorted signal all attain maximum or minimum values at the same time, which was not true of the input. Conversely, had we started with Fig. 3.2–5 as

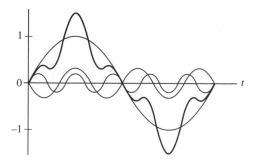

Figure 3.2–5   Test signal with constant phase shift $\theta = -90°$.

the test signal, a constant phase shift of $+90°$ would yield Fig. 3.2–3 for the output waveform. Thus we see that *delay distortion alone* can result in an increase or decrease of peak values as well as other waveshape alterations.

Clearly, delay distortion can be critical in pulse transmission, and much labor is spent *equalizing* transmission delay for digital data systems and the like. On the other hand, the human ear is curiously insensitive to delay distortion; the waveforms of Figs. 3.2–3 and 3.2–5 would sound just about the same when driving a loudspeaker. Thus, delay distortion is seldom of concern in voice and music transmission.

Let's take a closer look at the impact of phase delay on a modulated signal. The transfer function of an arbitrary channel can be expressed as

$$H(f) = Ae^{j(-2\pi f t_g + \phi_0)} = (Ae^{j\phi_0})e^{-j2\pi f t_g} \qquad [4]$$

where $\arg H(f) = -2\pi f t_g + \phi_0$ leads to $t_d(f) = t_g - \phi_0/2\pi f$ from Eq. (3). If the input to this bandpass channel is

$$x(t) = x_1(t) \cos \omega_c t - x_2(t) \sin \omega_c t \qquad [5]$$

then by the time delay property of Fourier transforms, the output will be delayed by t_g. Since $e^{j\phi_0}$ can be incorporated into the sine and cosine terms, the output of the channel is

$$y(t) = Ax_1(t - t_g) \cos \left[\omega_c(t - t_g) + \phi_0 \right] - Ax_2(t - t_g) \sin \left[\omega_c(t - t_g) + \phi_0 \right]$$

We observe that $\arg H(f_c) = -\omega_c t_g + \phi_0 = -\omega_c t_d$ so that

$$y(t) = Ax_1(t - t_g) \cos \left[\omega_c(t - t_d) \right] - Ax_2(t - t_g) \sin \left[\omega_c(t - t_d) \right] \qquad [6]$$

From Eq. (6) we see that the carrier has been delayed by t_d and the signals that modulate the carrier, x_1 and x_2, are delayed by t_g. The time delay t_d corresponding to the phase shift in the carrier is called the **phase delay** of the channel. This delay is also sometimes referred to as the **carrier** delay. The delay between the envelope of the input signal and that of the received signal, t_g, is called the **envelope** or **group delay** of the channel. In general, $t_d \neq t_g$.

This leads to a set of conditions under which a linear bandpass channel is distortionless. As in the general case of distortionless transmission described earlier, the amplitude response must be constant. For the channel in Eq. (4) this implies $|H(f)| = |A|$. In order to recover the original signals x_1 and x_2, the group delay must be constant. Therefore, from Eq. (4) this implies that t_g can be found directly from the derivative of arg $H(f) = \theta(f)$ as

$$t_g = -\frac{1}{2\pi} \frac{d\theta(f)}{df} \qquad\qquad [7]$$

Note that this condition on arg $H(f)$ is less restrictive than in the general case presented earlier. If $\phi_0 = 0$ then the general conditions of distortionless transmission are met and $t_d = t_g$.

EXERCISE 3.2–1   Use Eq. (3) to plot $t_d(f)$ from arg $H(f)$ given in Fig. 3.2–2.

Equalization

Linear distortion—both amplitude and delay—is theoretically curable through the use of *equalization* networks. Figure 3.2–6 shows an equalizer $H_{eq}(f)$ in cascade with a distorting transmission channel $H_C(f)$. Since the overall transfer function is $H(f) = H_C(f)H_{eq}(f)$ the final output will be distortionless if $H_C(f)H_{eq}(f) = Ke^{-j\omega t_d}$, where K and t_d are more or less arbitrary constants. Therefore, we require that

$$H_{eq}(f) = \frac{Ke^{-j\omega t_d}}{H_C(f)} \qquad\qquad [8]$$

wherever $X(f) \neq 0$.

Rare is the case when an equalizer can be designed to satisfy Eq. (8) exactly—which is why we say that equalization is a *theoretical* cure. But excellent approximations often are possible so that linear distortion can be reduced to a tolerable level. Probably the oldest equalization technique involves the use of *loading coils* on twisted-pair telephone lines. These coils are lumped inductors placed in shunt across the line every kilometer or so, giving the improved amplitude ratio typically illustrated in Fig. 3.2–7. Other lumped-element circuits have been designed for specific equalization tasks.

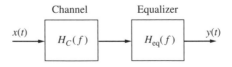

Figure 3.2–6   Channel with equalizer for linear distortion.

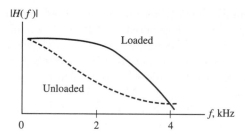

Figure 3.2–7   Amplitude ratio of a typical telephone line with and without loading coils for equalization.

More recently, the **tapped-delay-line equalizer,** or **transversal filter,** has emerged as a convenient and flexible device. To illustrate the principle, Fig. 3.2–8 shows a delay line with total time delay 2Δ having taps at each end and the middle. The tap outputs are passed through adjustable gains, c_{-1}, c_0, and c_1, and summed to form the final output. Thus

$$y(t) = c_{-1}x(t) + c_0\,x(t - \Delta) + c_1 x(t - 2\Delta) \qquad [9a]$$

and

$$H_{eq}(f) = c_{-1} + c_0\,e^{-j\omega\Delta} + c_1 e^{-j\omega2\Delta}$$
$$= (c_{-1}e^{+j\omega\Delta} + c_0 + c_1 e^{-j\omega\Delta})e^{-j\omega\Delta} \qquad [9b]$$

Generalizing Eq. (9b) to the case of a $2M\Delta$ delay line with $2M + 1$ taps yields

$$H_{eq}(f) = \left(\sum_{m=-M}^{M} c_m e^{-j\omega m\Delta} \right)e^{-j\omega M\Delta} \qquad [10]$$

which has the form of an *exponential Fourier series* with frequency periodicity $1/\Delta$. Therefore, given a channel $H_C(f)$ to be equalized over $|f| < W$, you can approximate the right-hand side of Eq. (8) by a Fourier series with frequency periodicity $1/\Delta \geq W$ (thereby determining Δ), estimate the number of significant terms (which determines M), and match the tap gains to the series coefficients.

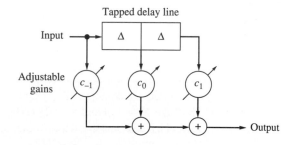

Figure 3.2–8   Transversal filter with three taps.

In many applications, the tap gains must be readjusted from time to time to compensate for changing channel characteristics. Adjustable equalization is especially important in switched communication networks, such as a telephone system, since the route between source and destination cannot be determined in advance. Sophisticated **adaptive equalizers** have therefore been designed with provision for automatic readjustment.

Adaptive equalization is usually implemented with digital circuitry and microprocessor control, in which case the delay line may be replaced by a shift register or charge-coupled device (CCD). For fixed (nonadjustable) equalizers, the transversal filter can be fabricated in an integrated circuit using a surface-acoustic-wave (SAW) device.

EXAMPLE 3.2–2   **Multipath Distortion**

Radio systems sometimes suffer from multipath distortion caused by two (or more) propagation paths between transmitter and receiver. Reflections due to mismatched impedance on a cable system produce the same effect. As a simple example, suppose the channel output is

$$y(t) = K_1 x(t - t_1) + K_2 x(t - t_2)$$

whose second term corresponds to an *echo* of the first if $t_2 > t_1$. Then

$$H_C(f) = K_1 e^{-j\omega t_1} + K_2 e^{-j\omega t_2} \qquad [11]$$
$$= K_1 e^{-j\omega t_1}(1 + k e^{-j\omega t_0})$$

where $k = K_2/K_1$ and $t_0 = t_2 - t_1$.

If we take $K = K_1$ and $t_d = t_1$ for simplicity in Eq. (8), the required equalizer characteristic becomes

$$H_{eq}(f) = \frac{1}{1 + k e^{-j\omega t_0}} = 1 - k e^{-j\omega t_0} + k^2 e^{-j2\omega t_0} + \cdots$$

The binomial expansion has been used here because, in this case, it leads to the form of Eq. (10) without any Fourier-series calculations. Assuming a small echo, so that $k^2 \ll 1$, we drop the higher-power terms and rewrite $H_{eq}(f)$ as

$$H_{eq}(f) \approx (e^{+j\omega t_0} - k + k^2 e^{-j\omega t_0}) e^{-j\omega t_0}$$

Comparison with Eqs. (9b) or (10) now reveals that a three-tap transversal filter will do the job if $c_{-1} = 1$, $c_0 = -k$, $c_1 = k^2$, and $\Delta = t_0$.

EXERCISE 3.2–2   Sketch $|H_{eq}(f)|$ and arg $H_{eq}(f)$ needed to equalize the frequency response in Fig. 3.2–2 over $5 \le |f| \le 50$ kHz. Take $K = 1/4$ and $t_d = 1/120$ ms in Eq. (8).

Nonlinear Distortion and Companding

A system having nonlinear elements cannot be described by a transfer function. Instead, the instantaneous values of input and output are related by a curve or function $y(t) = T[x(t)]$, commonly called the **transfer characteristic.** Figure 3.2–9 is a representative transfer characteristic; the flattening out of the output for large input excursions is the familiar saturation-and-cutoff effect of transistor amplifiers. We'll consider only *memoryless* devices, for which the transfer characteristic is a complete description.

Under small-signal input conditions, it may be possible to linearize the transfer characteristic in a piecewise fashion, as shown by the thin lines in the figure. The more general approach is a polynomial approximation to the curve, of the form

$$y(t) = a_1 x(t) + a_2 x^2(t) + a_3 x^3(t) + \cdots \qquad \text{[12a]}$$

and the higher powers of $x(t)$ in this equation give rise to the nonlinear distortion.

Even though we have no transfer function, the output spectrum can be found, at least in a formal way, by transforming Eq. (12a). Specifically, invoking the convolution theorem,

$$Y(f) = a_1 X(f) + a_2 X * X(f) + a_3 X * X * X(f) + \cdots \qquad \text{[12b]}$$

Now if $x(t)$ is bandlimited in W, the output of a linear network will contain no frequencies beyond $|f| < W$. But in the nonlinear case, we see that the output includes $X * X(f)$, which is bandlimited in $2W$, $X * X * X(f)$, which is bandlimited in $3W$, and so on. The nonlinearities have therefore created output frequency components that were not present in the input. Furthermore, since $X * X(f)$ may contain components for $|f| < W$, this portion of the spectrum overlaps that of $X(f)$. Using filtering techniques, the added components at $|f| > W$ can be removed, but there is no convenient way to get rid of the added components at $|f| < W$. These, in fact, constitute the nonlinear distortion.

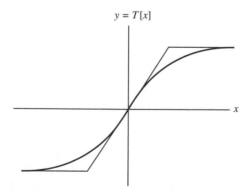

Figure 3.2–9   Transfer characteristic of a nonlinear device.

A quantitative measure of nonlinear distortion is provided by taking a simple cosine wave, $x(t) = \cos \omega_0 t$, as the input. Inserting in Eq. (12a) and expanding yields

$$y(t) = \left(\frac{a_2}{2} + \frac{3a_4}{8} + \cdots \right) + \left(a_1 + \frac{3a_3}{4} + \cdots \right) \cos \omega_0 t$$

$$+ \left(\frac{a_2}{2} + \frac{a_4}{4} + \cdots \right) \cos 2\omega_0 t + \cdots$$

Therefore, the nonlinear distortion appears as **harmonics** of the input wave. The amount of second-harmonic distortion is the ratio of the amplitude of this term to that of the fundamental, or in percent:

$$\text{Second-harmonic distortion} = \left| \frac{a_2/2 + a_4/4 + \cdots}{a_1 + 3a_3/4 + \cdots} \right| \times 100\%$$

Higher-order harmonics are treated similarly. However, their effect is usually much less, and many can be removed entirely by filtering.

If the input is a sum of two cosine waves, say $\cos \omega_1 t + \cos \omega_2 t$, the output will include all the harmonics of f_1 and f_2, plus crossproduct terms which yield $f_2 - f_1$, $f_2 + f_1, f_2 - 2f_1$, etc. These sum and difference frequencies are designated as **intermodulation distortion.** Generalizing the intermodulation effect, if $x(t) = x_1(t) + x_2(t)$, then $y(t)$ contains the **crossproduct** $x_1(t)x_2(t)$ (and higher-order products, which we ignore here). In the frequency domain $x_1(t)x_2(t)$ becomes $X_1 * X_2(f)$; and even though $X_1(f)$ and $X_2(f)$ may be separated in frequency, $X_1 * X_2(f)$ can overlap both of them, producing one form of **cross talk.** This aspect of nonlinear distortion is of particular concern in telephone transmission systems. On the other hand the crossproduct term is the desired result when nonlinear devices are used for modulation purposes.

It is important to note the difference between cross talk and other types of interference. Cross talk occurs when one signal crosses over to the frequency band of another signal due to nonlinear distortion in the channel. Picking up a conversation on a cordless phone or baby monitor occurs because the frequency spectrum allocated to such devices is too crowded to accommodate all of the users on separate frequency carriers. Therefore some "sharing" may occur from time to time. While cross talk resulting from nonlinear distortion is now rare in telephone transmission due to advances in technology, it was a major problem at one time.

The crossproduct term is the desired result when nonlinear devices are used for modulation purposes. In Sect. 4.3 we will examine how nonlinear devices can be used to achieve amplitude modulation. In Chap. 5, carefully controlled nonlinear distortion again appears in both modulation and detection of FM signals.

Although nonlinear distortion has no perfect cure, it can be minimized by careful design. The basic idea is to make sure that the signal does not exceed the linear operating range of the channel's transfer characteristic. Ironically, one strategy along this line utilizes two nonlinear signal processors, a *compressor* at the input and an *expander* at the output, as shown in Fig. 3.2–10.

Figure 3.2–10   Companding system.

A compressor has greater amplification at low signal levels than at high signal levels, similar to Fig. 3.2–9, and thereby compresses the range of the input signal. If the compressed signal falls within the linear range of the channel, the signal at the channel output is proportional to $T_{\text{comp}}[x(t)]$ which is distorted by the compressor but not the channel. Ideally, then, the expander has a characteristic that perfectly complements the compressor so the expanded output is proportional to $T_{\text{exp}}\{T_{\text{comp}}[x(t)]\} = x(t)$, as desired.

The joint use of compressing and expanding is called **companding** (surprise?) and is of particular value in telephone systems. Besides reducing nonlinear distortion, companding tends to compensate for the signal-level difference between loud and soft talkers. Indeed, the latter is the key advantage of companding compared to the simpler technique of linearly attenuating the signal at the input (to keep it in the linear range of the channel) and linearly amplifying it at the output.

3.3   TRANSMISSION LOSS AND DECIBELS

In addition to any signal distortion, a transmission system also reduces the power level or "strength" of the output signal. This signal-strength reduction is expressed in terms of transmission power loss. Although transmission loss can be compensated by power amplification, the ever-present electrical noise may prevent successful signal recovery in the face of large transmission loss.

This section describes transmission loss encountered on cable and radio communication systems. We'll start with a brief review of the more familiar concept of power gain, and we'll introduce decibels as a handy measure of power ratios used by communication engineers.

Power Gain

Let Fig. 3.3–1 represent an LTI system whose input signal has average power P_{in}. If the system is distortionless, the average signal power at the output will be proportional to P_{in}. Thus, the system's **power gain** is

$$g \triangleq P_{\text{out}}/P_{\text{in}} \qquad [1]$$

a constant parameter not to be confused with our step-response notation $g(t)$. Systems that include amplification may have very large values of g, so we'll find it convenient to express power gain in **decibels** (dB) defined as

$$g_{\text{dB}} \triangleq 10 \log_{10} g \qquad [2]$$

Figure 3.3–1   LTI system with power gain g.

The "B" in dB is capitalized in honor of Alexander Graham Bell who first used logarithmic power measurements.

Since the decibel is a logarithmic unit, it converts powers of 10 to products of 10. For instance, $g = 10^m$ becomes $g_{dB} = m \times 10$ dB. Power gain is always positive, of course, but negative dB values occur when $g \leq 1.0 = 10^0$ and hence $g_{dB} \leq 0$ dB. Note carefully that 0 dB corresponds to *unity* gain $(g = 1)$. Given a value in dB, the ratio value is

$$g = 10^{(g_{dB}/10)} \qquad [3]$$

obtained by inversion of Eq. (2).

While decibels always represent *power ratios*, signal power itself may be expressed in dB if you divide P by one watt or one milliwatt, as follows:

$$P_{dBW} = 10 \log_{10} \frac{P}{1\ W} \qquad P_{dBm} = 10 \log_{10} \frac{P}{1\ mW} \qquad [4]$$

Rewriting Eq. (1) as $(P_{out}/1\ mW) = g(P_{in}/1\ mW)$ and taking the logarithm of both sides then yields the dB equation

$$P_{out_{dBm}} = g_{dB} + P_{in_{dBm}}$$

Such manipulations have particular advantages for the more complicated relations encountered subsequently, where multiplication and division become addition and subtraction of known dB quantities. Communication engineers usually work with dBm because the signal powers are quite small at the output of a transmission system.

Now consider a system described by its transfer function $H(f)$. A sinusoidal input with amplitude A_x produces the output amplitude $A_y = |H(f)|A_x$, and the **normalized** signal powers are $P_x = A_x^2/2$ and $P_y = A_y^2/2 = |H(f)|^2 P_x$. These normalized powers do not necessarily equal the actual powers in Eq. (1). However, when the system has the same impedance level at input and output, the ratio P_y/P_x does equal P_{out}/P_{in}. Therefore, if $H(f) = Ke^{-j\omega t_d}$, then

$$g = |H(f)|^2 = K^2 \qquad [5]$$

In this case, the power gain also applies to *energy* signals in the sense that $E_y = gE_x$. When the system has unequal input and output impedances, the power (and energy) gain is proportional to K^2.

If the system is frequency-selective, Eq. (5) does not hold but $|H(f)|^2$ still tells us how the gain varies as a function of frequency. For a useful measure of frequency dependence in terms of signal power we take

$$|H(f)|_{dB} \triangleq 10 \log_{10} |H(f)|^2 \qquad\qquad [6]$$

which represents the **relative gain** in dB.

(a) Verify that $P_{dBm} = P_{dBW} + 30$ dB. (b) Show that if $|H(f)|_{dB} = -3$ dB then $|H(f)| \approx 1/\sqrt{2}$ and $|H(f)|^2 \approx \frac{1}{2}$. The significance of this result is discussed in the section on real filters.

EXERCISE 3.3–1

Transmission Loss and Repeaters

Any passive transmission medium has power loss rather than gain, since $P_{out} < P_{in}$. We therefore prefer to work with the transmission **loss,** or **attenuation**

$$L \triangleq 1/g = P_{in}/P_{out}$$

$$L_{dB} = -g_{dB} = 10 \log_{10} P_{in}/P_{out} \qquad\qquad [7]$$

Hence, $P_{out} = P_{in}/L$ and $P_{out_{dBm}} = P_{in_{dBm}} - L_{dB}$.

In the case of transmission lines, coaxial and fiber-optic cables, and wave-guides, the output power decreases *exponentially* with distance. We'll write this relation in the form

$$P_{out} = 10^{-(\alpha\ell/10)}P_{in}$$

where ℓ is the **path length** between source and destination and α is the **attenuation coefficient** in dB per unit length. Equation (7) then becomes

$$L = 10^{(\alpha\ell/10)} \qquad L_{dB} = \alpha\ell \qquad\qquad [8]$$

Table 3.3–1   Typical values of transmission loss

Transmission Medium	Frequency	Loss dB/km
Open-wire pair (0.3 cm diameter)	1 kHz	0.05
Twisted-wire pair (16 gauge)	10 kHz	2
	100 kHz	3
	300 kHz	6
Coaxial cable (1 cm diameter)	100 kHz	1
	1 MHz	2
	3 MHz	4
Coaxial cable (15 cm diameter)	100 MHz	1.5
Rectangular waveguide (5 × 2.5 cm)	10 GHz	5
Helical waveguide (5 cm diameter)	100 GHz	1.5
Fiber-optic cable	3.6×10^{14} Hz	2.5
	2.4×10^{14} Hz	0.5
	1.8×10^{14} Hz	0.2

showing that the dB loss is proportional to the length. Table 3.3–1 lists some typical values of α for various transmission media and signal frequencies.

Attenuation values in dB somewhat obscure the dramatic decrease of signal power with distance. To bring out the implications of Eq. (8) more clearly, suppose you transmit a signal on a 30 km length of cable having $\alpha = 3$ dB/km. Then $L_{dB} = 3 \times 30 = 90$ dB, $L = 10^9$, and $P_{out} = 10^{-9} P_{in}$. Doubling the path length doubles the attenuation to 180 dB, so that $L = 10^{18}$ and $P_{out} = 10^{-18} P_{in}$. This loss is so great that you'd need an input power of one megawatt (10^6 W) to get an output power of one picowatt (10^{-12} W)!

Large attenuation certainly calls for *amplification* to boost the output signal. As an example, Fig. 3.3–2 represents a cable transmission system with an output amplifier and a **repeater amplifier** inserted near the middle of the path. (Any *preamplification* at the input would be absorbed in the value of P_{in}.) Since power gains multiply in a cascade connection like this,

$$P_{out} = (g_1 \, g_2 \, g_3 \, g_4)P_{in} = \frac{g_2 \, g_4}{L_1 \, L_3} P_{in} \qquad [9a]$$

which becomes the dB equation

$$P_{out} = (g_2 + g_4) - (L_1 + L_3) + P_{in} \qquad [9b]$$

We've dropped the dB subscripts here for simplicity, but the addition and subtraction in Eq. (9b) unambiguously identifies it as a dB equation. Of course, the units of P_{out} (dBW or dBm) will be the same as those of P_{in}.

The repeater in Fig. 3.3–2 has been placed near the middle of the path to prevent the signal power from dropping down into the noise level of the amplifier. Long-haul cable systems have repeaters spaced every few kilometers for this reason, and a transcontinental telephone link might include more than 2000 repeaters. The signal-power analysis of such systems follows the same lines as Eq. (9). The noise analysis is presented in the Appendix.

Fiber Optics

Optical communication systems have become increasingly popular over the last two decades with advances in laser and fiber-optic technologies. Because optical systems use carrier frequencies in the range of 2×10^{14} Hz, the transmitted signals can have much larger bandwidth than is possible with metal cables such as twisted-wire

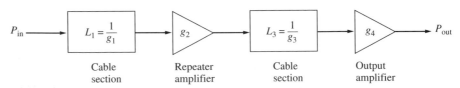

Figure 3.3–2   Cable transmission system with a repeater amplifier.

pair and coaxial cable. We will see in the next chapter that the theoretical maximum bandwidth for that carrier frequency is on the order of 2×10^{13} Hz! While we may never need that much bandwidth, it is nice to have extra if we need it.

In the 1960s fiber-optic cables were extremely lossy, with losses around 1000 dB/km, and were impractical for commercial use. Today these losses are on the order of 0.2 to 2 dB/km depending on the type of fiber used and the wavelength of the signal. This is lower than most twisted-wire pair and coaxial cable systems. There are many advantages to using fiber-optic channels in addition to large bandwidth and low loss. The dielectric waveguide property of the optical fiber makes it less susceptible to interference from external sources. Since the transmitted signal is light rather than current, there is no electromagnetic field to generate cross talk and no radiated RF energy to interfere with other communication systems. In addition, since moving photons do not interact, there is no noise generated inside the optical fiber. Fiber-optic channels are safer to install and maintain since there is no large current or voltage to worry about. Furthermore, since it is virtually impossible to tap into a fiber-optic channel without the user detecting it, they are secure enough for military applications. They are rugged and flexible, and operate over a larger temperature variation than metal cable. The small size (about the diameter of a human hair) and weight mean they take up less storage space and are cheaper to transport. Finally, they are fabricated from sand, which is a plentiful resource. While the up-front installation costs are higher, it is predicted that the long-term costs will ultimately be lower than with metal-based cables.

Most optical communication systems are digital since system limitations on the amplitude of analog modulation make it impractical. The system is a hybrid of electrical and optical components, since the signal sources and final receivers are still made up of electronics. Optical transmitters use either LEDs or solid-state lasers to generate light pulses. The choice between these two is driven by design constraints. LEDs, which produce noncoherent (multiple wavelengths) light, are rugged, inexpensive, and have low power output ($\sim$0.5 mW). Lasers are much higher in cost and have a shorter lifetime; however they produce coherent (single wavelength) light and have a power output of around 5 mW. The receivers are usually PIN diodes or avalanche photodiodes (APD), depending on the wavelength of the transmitted signal. In the remainder of this discussion we will concentrate our attention on the fiber-optic channel itself.

Fiber-optic cables have a **core** made of pure silica glass surrounded by a **cladding** layer also usually made of silica glass, but sometimes made of plastic. There is an outer, thin protective jacket made of plastic in most cases. In the core the signal traverses the fiber. The cladding reduces losses by keeping the signal power within the core. There are three main types of fiber-optic cable: single-mode fibers, multimode step-index fibers, and multimode graded-index fibers. Figure 3.3–3*a* shows three light rays traversing a single-mode fiber. Because the diameter of the core is sufficiently small ($\sim$8 μm), there is only a single path for each of the rays to follow as they propagate down the length of the fiber. The difference in the index of refraction between the core and cladding layers causes the light to be reflected back

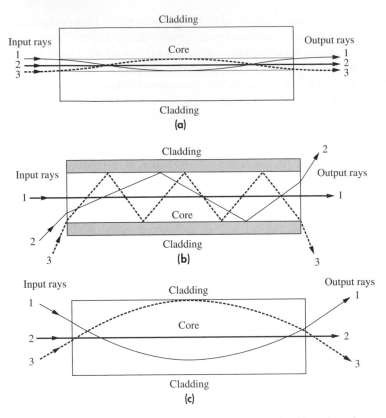

Figure 3.3–3   (a) Light propagation down a single-mode step-index fiber. (b) Light propagation down a multimode step-index fiber. (c) Light propagation down a multimode graded-index fiber.

into the channel, and thus the rays follow a straight path through the fiber. Consequently, each ray of light travels the same distance in a given period of time, and a pulse input would have essentially the same shape at the output. Therefore single-mode fibers have the capacity for large transmission bandwidths, which makes them very popular for commercial applications. However, the small core diameter makes it difficult to align cable section boundaries and to couple the source to the fiber, and thus losses can occur.

Multimode fibers allow multiple paths through the cable. Because they have a larger core diameter ($\sim 50~\mu$m) it is easier to splice and couple the fiber segments, resulting in less loss. In addition, more light rays at differing angles can enter the channel. In a multimode step-index fiber there is a step change between the index of refraction of the core and cladding, as there is with single-mode fibers. Fig. 3.3–3*b* shows three rays entering a multimode step-index fiber at various angles. It is clear that the paths of the rays will be quite different. Ray 1 travels straight through as in the case of the single-mode fiber. Ray 2 is reflected off of the core-cladding boundary a few times and thus takes a longer path through the cable. Ray 3, with multiple reflections, has a much longer path. As Fig. 3.3–3b shows, the angle of incidence impacts the time to reach the receiver. We can define two terms to describe this channel delay. The average time difference between the arrivals of the various rays is termed **mean-time delay,** and the standard deviation is called the **delay spread.** The impact on a narrow pulse would be to broaden the pulse width as the signal propagates down the channel. If the broadening exceeds the gap between the pulses, overlap may result and the pulses will not be distinguishable at the output. Therefore the maximum bandwidth of the transmitted signal in a multimode step-index channel is much lower than in the single mode case.

Multimode graded-index fibers give us the best of both worlds in performance. The large central core has an index of refraction that is not uniform. The refractive index is greatest at the center and tapers gradually toward the outer edge. As shown in Fig. 3.3–3c, the rays again propagate along multiple paths; however since they are constantly refracted there is a continuous bending of the light rays. The velocity of the wave is inversely proportional to the refractive index so that those waves farthest from the center propagate fastest. The refractive index profile can be designed so that all of the waves have approximately the same delay when they reach the output. Therefore the lower dispersion permits higher transmission bandwidth. While the bandwidth of a multimode graded-index fiber is lower than that of a single-mode fiber, the benefits of the larger core diameter are sufficient to make it suitable for long-distance communication applications.

With all of the fiber types there are several places where losses occur, including where the fiber meets the transmitter or receiver, where the fiber sections connect to each other, and within the fiber itself. Attenuation within the fiber results primarily from **absorption** losses due to impurities in the silica glass, and **scattering** losses due to imperfections in the waveguide. Losses increase exponentially with distance traversed and also vary with wavelength. There are three wavelength regions where there are relative minima in the attenuation curve, and they are given in Table 3.3–1. The smallest amount of loss occurs around 1300 and 1500 nm, so those frequencies are used most often for long-distance communication systems.

Current commercial applications require repeaters approximately every 40 km. However, each year brings technology advances, so this spacing continues to increase. Conventional repeater amplifiers convert the light wave to an electrical signal, amplify it, and convert it back to an optical signal for retransmission. However, light wave amplifiers are being developed and may be available soon.

Fiber-optic communication systems are quickly becoming the standard for long-distance telecommunications. Homes and businesses are increasingly wired internally and externally with optical fibers. Long-distance telephone companies advertise the clear, quiet channels with claims that listeners can hear a pin drop. Underwater fiber cables now cover more than two-thirds of the world's circumference and can handle over 100,000 telephone conversations at one time. Compare that to the first transoceanic cable that was a technological breakthrough in 1956 and carried just 36 voice channels. While current systems can handle 90 Mbits/sec to 2.5 Gbits/sec, there have been experimental results as high as 1000 Gbits/sec. At current transmission rates of 64 kbits/sec, this represents 15 million telephone conversations over a single optical fiber. As capacity continues to expand, we will no doubt find new ways to fill it.

Radio Transmission★

Signal transmission by radiowave propagation can reduce the required number of repeaters, and has the additional advantage of eliminating long cables. Although radio involves modulation processes described in later chapters, it seems appropriate here to examine the transmission loss for **line-of-sight propagation** illustrated in Fig. 3.3–4 where the radio wave travels a direct path from transmitting to receiving antenna. This propagation mode is commonly employed for long-distance communication at frequencies above about 100 MHz.

The **free-space loss** on a line-of-sight path is due to spherical dispersion of the radio wave. This loss is given by

$$L = \left(\frac{4\pi\ell}{\lambda}\right)^2 = \left(\frac{4\pi f\ell}{c}\right)^2 \qquad [10a]$$

in which λ is the wavelength, f the signal frequency, and c the speed of light. If we express ℓ in kilometers and f in gigahertz (10^9 Hz), Eq. (10a) becomes

$$L_{dB} = 92.4 + 20 \log_{10} f_{GHz} + 20 \log_{10} \ell_{km} \qquad [10b]$$

We see that L_{dB} increases as the logarithm of ℓ, rather than in direct proportion to path length. Thus, for instance, doubling the path length increases the loss by only 6 dB.

Furthermore, directional antennas have a focusing effect that acts like amplification in the sense that

$$P_{out} = \frac{g_T g_R}{L} P_{in} \qquad [11]$$

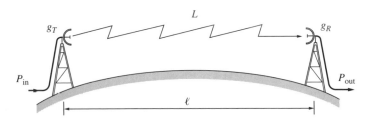

Figure 3.3–4   Line-of-sight radio transmission.

where g_T and g_R represent the antenna gains at the transmitter and receiver. The maximum transmitting or receiving gain of an antenna with effective **aperture area** A_e is

$$g = \frac{4\pi A_e}{\lambda^2} = \frac{4\pi A_e f^2}{c^2} \qquad [12]$$

where $c \approx 3 \times 10^5$ km/s. The value of A_e for a horn or dish antenna approximately equals its physical area, and large parabolic dishes may provide gains in excess of 60 dB.

Commercial radio stations often use compression to produce a transmitted signal that has higher power but doesn't exceed the system's linear operating region. As mentioned in Sect. 3.2, compression provides greater amplification of low-level signals, and can raise them above the background noise level. However since your home radio does not have a built-in expander to complete the companding process, some audible distortion may be present. To cope with this, music production companies often preprocess the materials sent to radio stations to ensure the integrity of the desired sound.

Satellites employ line-of-sight radio transmission over very long distances. They have a broad coverage area and can reach areas that are not covered by cable or fiber, including mobile platforms such as ships and planes. Even though fiber-optic systems are carrying an increasing amount of transoceanic telephone traffic (and may make satellites obsolete for many applications), satellite relays still handle the bulk of very long distance telecommunications. Satellite relays also make it possible to transmit TV signals across the ocean. They have a wide bandwidth of about 500 MHz that can be subdivided for use by individual transponders. Most satellites are in geostationary orbit. This means that they are synchronous with Earth's rotation and are located directly above the equator, and thus they appear stationary in the sky. The main advantage is that antennas on Earth pointing at the satellite can be fixed.

A typical C band satellite has an uplink frequency of 6 GHz, a downlink frequency of 4 GHz, and 12 transponders each having a bandwidth of 36 MHz. The advantages in using this frequency range are that it allows use of relatively inexpensive microwave equipment, has low attenuation due to rainfall (the primary atmospheric cause of signal loss), and has a low sky background noise. However, there can be severe interference from terrestrial microwave systems, so many satellites now use the Ku band. The Ku band frequencies are 14 GHz for uplink and 12 GHz for downlink. This allows smaller and less expensive antennas. C band satellites are most commonly used for commercial cable TV systems, whereas Ku band is used for videoconferencing. A newer service that allows direct broadcast satellites (DBS) for home television service uses 17 GHz for uplink and 12 GHz for downlink.

By their nature, satellites require multiple users to access them from different locations at the same time. A variety of **multiple access** techniques have been developed, and will be discussed further in a later chapter. Personal communication devices such as cellular phones rely on multiple access techniques such as time division multiple access (TDMA) and code division multiple access (CDMA). Propagation delay can be a problem over long distances for voice communication, and may require echo cancellation in the channel.

Current technology allows portable satellite uplink systems to travel to where news or an event is happening. In fact, all equipment can fit in a van or in several large trunks that can be shipped on an airplane. For a more complete but practical overview of satellites, see Tomasi (1994, Chap. 18).

EXAMPLE 3.3–1   **Satellite Relay System**

Figure 3.3–5 shows a simplified transoceanic television system with a satellite relay serving as a repeater. The satellite is in geostationary orbit and is about 22,300 miles (36,000 km) above the equator. The uplink frequency is 6 GHz, and the downlink frequency is 4 GHz. Equation (10b) gives an uplink path loss

$$L_u = 92.4 + 20 \log_{10} 6 + 20 \log_{10} 3.6 \times 10^4 = 199.1 \text{ dB}$$

and a downlink loss

$$L_d = 92.4 + 20 \log_{10} 4 + 20 \log_{10} 3.6 \times 10^4 = 195.6 \text{ dB}$$

since the distance from the transmitter and receiver towers to the satellite is approximately the same as the distance from Earth to the satellite. The antenna gains in dB are given on the drawing with subscripts identifying the various functions—for example, g_{RU} stands for the receiving antenna gain on the uplink from ground to satellite. The satellite has a repeater amplifier that produces a typical output of 18 dBW. If the transmitter input power is 35 dBW, the power received at the satellite

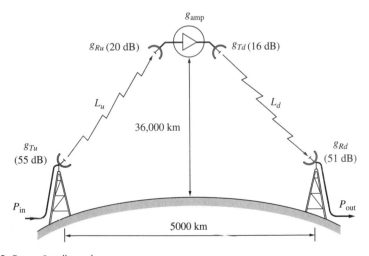

Figure 3.3–5   Satellite relay system.

is 35 dBW + 55 dB − 199.1 dB + 20 dB = −144.1 dBW. The power output at the receiver is 18 dBW + 16 dB − 195.6 dB + 51 dB = −110.6 dBW. Inverting Eq. (4) gives

$$P_{\text{out}} = 10^{(-100.6/10)} \times 1 \text{ W} = 8.7 \times 10^{-12} \text{ W}$$

Such minute power levels are typical for satellite systems.

A 40 km cable system has $P_{\text{in}} = 2$ W and a repeater with 64 dB gain is inserted 24 km from the input. The cable sections have $\alpha = 2.5$ dB/km. Use dB equations to find the signal power at: (*a*) the repeater's input; (*b*) the final output.

EXERCISE 3.3–2

3.4   FILTERS AND FILTERING

Virtually every communication system includes one or more filters for the purpose of separating an information-bearing signal from unwanted contaminations such as interference, noise, and distortion products. In this section we'll define ideal filters, describe the differences between real and ideal filters, and examine the effect of filtering on pulsed signals.

Ideal Filters

By definition, an ideal filter has the characteristics of distortionless transmission over one or more specified frequency bands and zero response at all other frequencies. In particular, the transfer function of an ideal **bandpass** filter (BPF) is

$$H(f) = \begin{cases} Ke^{-j\omega t_d} & f_\ell \leq |f| \leq f_u \\ 0 & \text{otherwise} \end{cases} \qquad [1]$$

as plotted in Fig. 3.4–1. The parameters f_ℓ and f_u are the lower and upper **cutoff frequencies,** respectively, since they mark the end points of the **passband.** The filter's **bandwidth** is

$$B = f_u - f_\ell$$

which we measure in terms of the positive-frequency portion of the passband.

In similar fashion, an ideal **lowpass** filter (LPF) is defined by Eq. (1) with $f_\ell = 0$, so $B = f_u$, while an ideal **highpass** filter (HPF) has $f_\ell > 0$ and $f_u = \infty$. Ideal **band-rejection** or **notch** filters provide distortionless transmission over all frequencies except some **stopband,** say $f_\ell \leq |f| \leq f_u$, where $H(f) = 0$.

But all such filters are physically **unrealizable** in the sense that their characteristics cannot be achieved with a finite number of elements. We'll skip the general proof of this assertion. Instead, we'll give an instructive plausibility argument based on the impulse response.

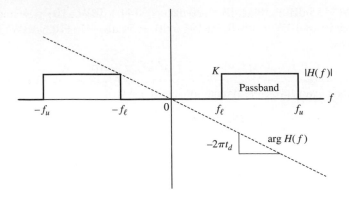

Figure 3.4–1   Transfer function of an ideal bandpass filter.

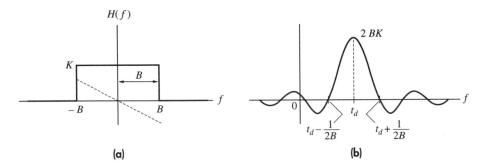

(a)   **(b)**

Figure 3.4–2   Ideal lowpass filter. (a) Transfer function; (b) impulse response.

Consider an ideal LPF whose transfer function, shown in Fig. 3.4–2a, can be written as

$$H(f) = Ke^{-j\omega t_d}\Pi\left(\frac{f}{2B}\right) \qquad \text{[2a]}$$

Its impulse response will be

$$h(t) = \mathcal{F}^{-1}[H(f)] = 2BK \text{ sinc } 2B(t - t_d) \qquad \text{[2b]}$$

which is sketched in Fig. 3.4–2b. Since $h(t)$ is the response to $\delta(t)$ and $h(t)$ has nonzero values for $t < 0$, the *output appears before the input is applied.* Such a filter is said to be **anticipatory** or **noncausal,** and the portion of the output appearing before the input is called a **precursor.** Without doubt, such behavior is physically impossible, and hence the filter must be unrealizable. Like results hold for the ideal BPF and HPF.

Fictitious though they may be, ideal filters have great conceptual value in the study of communication systems. Furthermore, many real filters come quite close to ideal behavior.

Show that the impulse response of an ideal BPF is

$$h(t) = 2BK \, \text{sinc} \, B(t - t_d) \, \cos \, \omega_c(t - t_d)$$

where $\omega_c = \pi(f_\ell + f_u)$.

EXERCISE 3.4–1

Bandlimiting and Timelimiting

Earlier we said that a signal $v(t)$ is *bandlimited* if there exists some constant W such that

$$V(f) = 0 \qquad |f| > W$$

Hence, the spectrum has no content outside $|f| > W$. Similarly, a **timelimited** signal is defined by the property that, for the constants $t_1 < t_2$,

$$v(t) = 0 \qquad t < t_1 \text{ and } t > t_2$$

Hence, the signal "starts" at $t \geq t_1$ and "ends" at $t \leq t_2$. Let's further examine these two definitions in the light of real versus ideal filters.

The concepts of ideal filtering and bandlimited signals go hand in hand, since applying a signal to an ideal LPF produces a bandlimited signal at the output. We've also seen that the impulse response of an ideal LPF is a sinc pulse lasting for all time. We now assert that any signal emerging from an ideal LPF will exist for all time. Consequently, a strictly bandlimited signal cannot be timelimited. Conversely, by duality, a strictly timelimited signal cannot be bandlimited. Every transform pair we've encountered supports these assertions, and a general proof is given in Wozencraft and Jacobs (1965, App. 5B). Thus,

Perfect bandlimiting and timelimiting are mutually incompatible.

This observation raises concerns about the signal and filter models used in the study of communication systems. Since a signal cannot be both bandlimited and timelimited, we should either abandon bandlimited signals (and ideal filters) or else we must accept signal models that exist for all time. On the one hand, we recognize that any real signal is timelimited, having starting and ending times. On the other hand, the concepts of bandlimited spectra and ideal filters are too useful and appealing to be dismissed entirely.

The resolution of our dilemma is really not so difficult, requiring but a small compromise. Although a strictly timelimited signal is not strictly bandlimited, its spectrum may be negligibly small above some upper frequency limit W. Likewise, a strictly bandlimited signal may be negligibly small outside a certain time interval $t_1 \leq t \leq t_2$. Therefore, we will often assume that signals are essentially both bandlimited and timelimited for most practical purposes.

Real Filters

The design of realizable filters that approach ideal behavior is an advanced topic outside the scope of this book. But we should at least look at the major differences between real and ideal filters to gain some understanding of the approximations implied by the assumption of an ideal filter. Further information on filter design and implementation can be found in texts such as Van Valkenburg (1982).

To begin our discussion, Fig. 3.4–3 shows the amplitude ratio of a typical real bandpass filter. Compared with the ideal BPF in Fig. 3.4–1, we see a passband where $|H(f)|$ is relatively large (but not constant) and stopbands where $|H(f)|$ is quite small (but not zero). The end points of the passband are usually defined by

$$|H(f)| = \frac{1}{\sqrt{2}}\, |H(f)|_{\max} = \frac{K}{\sqrt{2}} \qquad f = f_\ell, f_u \qquad \text{[3]}$$

so that $|H(f)|^2$ falls no lower than $K^2/2$ for $f_\ell \leq |f| \leq f_u$. The bandwidth $B = f_u - f_\ell$ is then called the **half-power** or 3 dB bandwidth. Similarly, the end points of the stopbands can be taken where $|H(f)|$ drops to a suitably small value such as $K/10$ or $K/100$.

Between the passband and stopbands are **transition** regions, shown shaded, where the filter neither "passes" nor "rejects" frequency components. Therefore, effective signal filtering often depends on having a filter with very narrow transition regions. We'll pursue this aspect by examining one particular class of filters in some detail. Then we'll describe other popular designs.

The simplest of the standard filter types is the nth-order **Butterworth** LPF, whose circuit contains n reactive elements (capacitors and inductors). The transfer function with $K = 1$ has the form

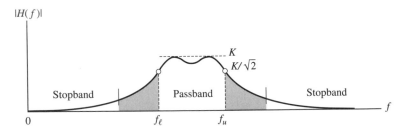

Figure 3.4–3   Typical amplitude ratio of a real bandpass filter.

$$H(f) = \frac{1}{P_n(jf/B)} \tag{4a}$$

where B equals the 3 dB bandwidth and $P_n(jf/B)$ is a complex polynomial. The family of Butterworth polynomials is defined by the property

$$|P_n(jf/B)|^2 = 1 + (f/B)^{2n}$$

so that

$$|H(f)| = \frac{1}{\sqrt{1 + (f/B)^{2n}}} \tag{4b}$$

Consequently, the first n derivatives of $|H(f)|$ equal zero at $f = 0$ and we say that $|H(f)|$ is **maximally flat.** Table 3.4–1 lists the Butterworth polynomials for $n = 1$ through 4, using the normalized variable $p = jf/B$.

   A first-order Butterworth filter has the same characteristics as an RC lowpass filter and would be a poor approximation of an ideal LPF. But the approximation improves as you increase n by adding more elements to the circuit. For instance, the impulse response of a third-order filter sketched in Fig. 3.4–4a bears obvious resemblance to that of an ideal LPF—without the precursors, of course. The frequency-response curves of this filter are plotted in Fig. 3.4–4b. Note that the phase shift has a reasonably linear slope over the passband, implying time delay plus some delay distortion.

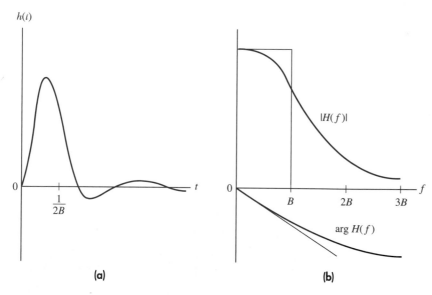

Figure 3.4–4   Third-order Butterworth LPF. (a) Impulse response; (b) transfer function.

Table 3.4–1   Butterworth polynomials

n	$P_n(p)$
1	$1 + p$
2	$1 + \sqrt{2}\,p + p^2$
3	$(1 + p)(1 + p + p^2)$
4	$(1 + 0.765p + p^2)(1 + 1.848p + p^2)$

A clearer picture of the amplitude ratio in the transition region is obtained from a **Bode diagram,** constructed by plotting $|H(f)|$ in dB versus f on a logarithmic scale. Figure 3.4–5 shows the Bode diagram for Butterworth lowpass filters with various values of n. If we define the edge of the stopband at $|H(f)| = -20$ dB, the width of the transition region when $n = 1$ is $10B - B = 9B$ but only $1.25B - B = 0.25B$ when $n = 10$. Clearly, $|H(f)|$ approaches the ideal square characteristic in the limit as $n \rightarrow \infty$. At the same time, however, the slope of the phase shift (not shown) increases with n and the delay distortion may become intolerably large.

In situations where potential delay distortion is a major concern, a **Bessel-Thomson** filter would be the preferred choice. This class of filters is characterized by **maximally linear phase shift** for a given value of n, but has a wider transition

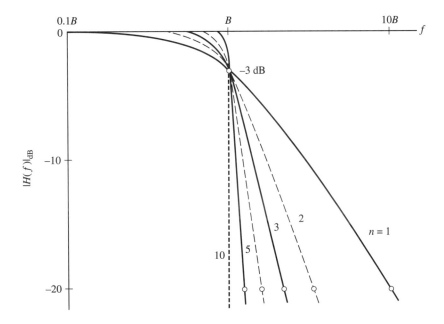

Figure 3.4–5   Bode diagram for Butterworth LPFs.

region. At the other extreme, the class of **equiripple** filters (including Chebyshev and elliptic filters) provides the sharpest transition for a given value of n; but these filters have small amplitude ripples in the passband and significantly nonlinear phase shift. Equiripple filters would be satisfactory in audio applications, for instance, whereas pulse applications might call for the superior transient performance of Bessel-Thomson filters.

All three filter classes can be implemented with active devices (such as operational amplifiers) that eliminate the need for bulky inductors. **Switched-capacitor filter** designs go even further and eliminate resistors that would take up too much space in a large-scale integrated circuit. All three classes can also be modified to obtain highpass or bandpass filters. However, some practical implementation problems do arise when you want a bandpass filter with a narrow but reasonably square passband. Special designs that employ electromechanical phenomena have been developed for such applications. For example, Fig. 3.4–6 shows the amplitude ratio of a seventh-order monolithic crystal BPF intended for use in an AM radio.

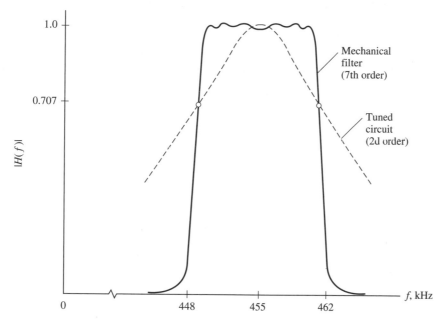

Figure 3.4–6   Amplitude ratio of a mechanical filter.

The circuit in Fig. 3.4–7 is one implementation of a second-order Butterworth LPF with   **EXAMPLE 3.4–1**

$$B = \frac{1}{2\pi\sqrt{LC}}$$

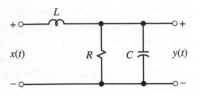

Figure 3.4–7   Second-order Butterworth LPF.

We can obtain an expression for the transfer function as

$$H(f) = \frac{Z_{RC}}{Z_{RC} + j\omega L}$$

where

$$Z_{RC} = \frac{R/j\omega C}{R + 1/j\omega C} = \frac{R}{1 + j\omega RC}$$

Thus

$$H(f) = \frac{1}{1 + j\omega L/R - \omega^2 LC}$$

$$= \left[1 + j\frac{2\pi L}{R}f - (2\pi\sqrt{LC}f)^2 \right]^{-1}$$

From Table 3.4–1 with $p = jf/B$, we want

$$H(f) = \left[1 + j\sqrt{2}\frac{f}{B} - \left(\frac{f}{B}\right)^2 \right]^{-1}$$

The required relationship between R, L, and C that satisfies the equation can be found by setting

$$\frac{2\pi L}{R} = \frac{\sqrt{2}}{B} = \sqrt{2}\,2\pi\sqrt{LC}$$

which yields $R = \sqrt{\dfrac{L}{2C}}$.

EXERCISE 3.4–2   Show that a Butterworth LPF has $|H(f)|_{dB} \approx -20n \log_{10}(f/B)$ when $f > B$. Then find the minimum value of n needed so that $|H(f)| \leq 1/10$ for $f \geq 2B$.

Pulse Response and Risetime

A rectangular pulse, or any other signal with an abrupt transition, contains significant high-frequency components that will be attenuated or eliminated by a lowpass filter. Pulse filtering therefore produces a smoothing or smearing effect that must be studied in the time domain. The study of pulse response undertaken here leads to useful information about pulse transmission systems.

Let's begin with the unit step input signal $x(t) = u(t)$, which could represent the leading edge of a rectangular pulse. In terms of the filter's impulse response $h(t)$, the step response will be

$$g(t) \triangleq \int_{-\infty}^{\infty} h(\lambda) u(t - \lambda) \, d\lambda = \int_{-\infty}^{t} h(\lambda) \, d\lambda \qquad [5]$$

since $u(t - \lambda) = 0$ for $\lambda > t$. We saw in Examples 3.1–1 and 3.1–2 for instance, that a first-order lowpass filter has

$$g(t) = (1 - e^{-2\pi Bt}) u(t)$$

where B is the 3 dB bandwidth.

Of course a first-order LPF doesn't severely restrict high-frequency transmission. So let's go to the extreme case of an ideal LPF, taking unit gain and zero time delay for simplicity. From Eq. (2b) we have $h(t) = 2B \operatorname{sinc} 2Bt$ and Eq. (5) becomes

$$g(t) = \int_{-\infty}^{t} 2B \operatorname{sinc} 2B\lambda \, d\lambda$$

$$= \int_{-\infty}^{0} \operatorname{sinc} \mu \, d\mu + \int_{0}^{2Bt} \operatorname{sinc} \mu \, d\mu$$

where $\mu = 2B\lambda$. The first integral is known to equal 1/2, but the second requires numerical evaluation. Fortunately, the result can be expressed in terms of the tabulated *sine integral function*

$$\operatorname{Si}(\theta) \triangleq \int_{0}^{\theta} \frac{\sin \alpha}{\alpha} \, d\alpha = \pi \int_{0}^{\theta/\pi} \operatorname{sinc} \mu \, d\mu \qquad [6]$$

which is plotted in Fig. 3.4–8 for $\theta > 0$ and approaches the value $\pi/2$ as $\theta \to \infty$. The function is also defined for $\theta < 0$ by virtue of the odd-symmetry property $\operatorname{Si}(-\theta) = -\operatorname{Si}(\theta)$. Using Eq. (6) in the problem at hand we get

$$g(t) = \frac{1}{2} + \frac{1}{\pi} \operatorname{Si}(2\pi Bt) \qquad [7]$$

obtained by setting $\theta/\pi = 2Bt$.

For comparison purposes, Fig. 3.4–9 shows the step response of an ideal LPF along with that of a first-order LPF. The ideal LPF completely removes all high frequencies $|f| > B$, producing *precursors, overshoot,* and *oscillations* in the step response. (This behavior is the same as Gibbs's phenomenon illustrated in Fig. 2.1–10 and in Example 2.4–2.) None of these effects appears in the response of the first-order LPF, which gradually attenuates but does not eliminate high frequencies.

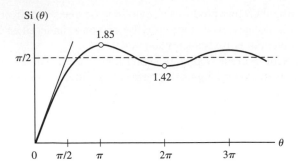

Figure 3.4–8   The sine integral function.

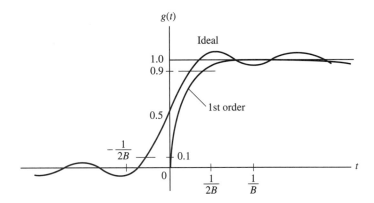

Figure 3.4–9   Step response of ideal and first-order LPFs.

The step response of a more selective filter—a third-order Butterworth LPF, for example—would more nearly resemble a time-delayed version of the ideal LPF response.

Before moving on to pulse response per se, there's an important conclusion to be drawn from Fig. 3.4–9 regarding **risetime**. Risetime is a measure of the "speed" of a step response, usually defined as the time interval t_r between $g(t) = 0.1$ and $g(t) = 0.9$ and known as the 10–90% risetime. The risetime of a first-order lowpass filter can be computed from $g(t)$ as $t_r \approx 0.35/B$, while the ideal filter has $t_r \approx 0.44/B$. Both values are reasonably close to $0.5/B$ so we'll use the approximation

$$t_r \approx \frac{1}{2B} \qquad [8]$$

for the risetime of an arbitrary lowpass filter with bandwidth B.

Our work with step response pays off immediately in the calculation of pulse response if we take the input signal to be a unit-height rectangular pulse with duration τ starting at $t = 0$. Then we can write

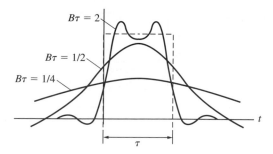

Figure 3.4–10   Pulse response of an ideal LPF.

$$x(t) = u(t) - u(t - \tau)$$

and hence

$$y(t) = g(t) - g(t - \tau)$$

which follows from superposition.

Using $g(t)$ from Eq. (7), we obtain the pulse response of an ideal LPF as

$$y(t) = \frac{1}{\pi} \{ \mathrm{Si}\, (2\pi Bt) - \mathrm{Si}\, [2\pi B(t - \tau)] \}$$ [9]

which is plotted in Fig. 3.4–10 for three values of the product $B\tau$. The response has a more-or-less rectangular shape when $B\tau \geq 2$, whereas it becomes badly smeared and spread out if $B\tau \leq \frac{1}{4}$. The intermediate case $B\tau = \frac{1}{2}$ gives a recognizable but not rectangular output pulse. The same conclusions can be drawn from the pulse response of a first-order lowpass filter previously sketched in Fig. 3.1–3, and similar results would hold for other input pulse shapes and other lowpass filter characteristics.

Now we're in a position to make some general statements about *bandwidth requirements for pulse transmission*. Reproducing the actual pulse shape requires a large bandwidth, say

$$B \gg \frac{1}{\tau_{\min}}$$

where $\tau_{\min}$ represents the smallest output pulse duration. But if we only need to *detect* that a pulse has been sent, or perhaps measure the pulse amplitude, we can get by with the smaller bandwidth

$$B \geq \frac{1}{2\tau_{\min}}$$ [10]

an important and handy rule of thumb.

Equation (10) also gives the condition for distinguishing between, or *resolving*, output pulses spaced by $\tau_{\min}$ or more. Figure 3.4–11 shows the resolution condition

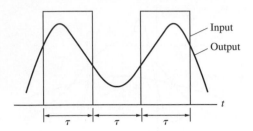

Figure 3.4–11   Pulse resolution of an ideal LPF. $B = 1/2\tau$.

for an ideal lowpass channel with $B = \frac{1}{2}\tau$. A smaller bandwidth or smaller spacing would result in considerable overlap, making it difficult to identify separate pulses.

Besides pulse detection and resolution, we'll occasionally be concerned with *pulse position* measured relative to some reference time. Such measurements have inherent ambiguity due to the rounded output pulse shape and nonzero risetime of leading and trailing edges. For a specified minimum risetime, Eq. (8) yields the bandwidth requirement

$$B \geq \frac{1}{2t_{r_{min}}} \qquad [11]$$

another handy rule of thumb.

Throughout the foregoing discussion we've tacitly assumed that the transmission channel has satisfactory phase-shift characteristics. If not, the resulting delay distortion could render the channel useless for pulse transmission, regardless of the bandwidth. Therefore, our bandwidth requirements in Eqs. (10) and (11) imply the additional stipulation of nearly linear phase shift over $|f| \leq B$. A phase equalization network may be needed to achieve this condition.

EXERCISE 3.4–3   A certain signal consists of pulses whose durations range from 10 to 25 μs; the pulses occur at random times, but a given pulse always starts at least 30 μs after the starting time of the previous pulse. Find the minimum transmission bandwidth required for pulse detection and resolution, and estimate the resulting risetime at the output.

3.5   QUADRATURE FILTERS AND HILBERT TRANSFORMS

The Fourier transform serves most of our needs in the study of filtered signals since, in most cases, we are interested in the separation of signals based on their frequency content. However, there are times when separating signals on the basis of phase is more convenient. For these applications we'll use the Hilbert transform, which we'll introduce in conjunction with quadrature filtering. In Chap. 4 we will make use of

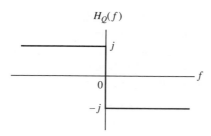

Figure 3.5–1   Transfer function of a quadrature phase shifter.

the Hilbert transform in the study of two important applications: the generation of single-sideband amplitude modulation and the mathematical representation of band-pass signals.

A **quadrature filter** is an allpass network that merely shifts the phase of positive frequency components by $-90°$ and negative frequency components by $+90°$. Since a $\pm90°$ phase shift is equivalent to multiplying by $e^{\pm j90°} = \pm j$, the transfer function can be written in terms of the signum function as

$$H_Q(f) = -j\,\mathrm{sgn}\,f = \begin{cases} -j & f > 0 \\ +j & f < 0 \end{cases} \qquad\qquad \text{[1a]}$$

which is plotted in Fig. 3.5–1. The corresponding impulse response is

$$h_Q(t) = \frac{1}{\pi t} \qquad\qquad \text{[1b]}$$

We obtain this result by applying duality to $\mathscr{F}[\mathrm{sgn}\,t] = 1/j\pi f$ which yields $\mathscr{F}[1/j\pi t] = \mathrm{sgn}\,(-f) = -\mathrm{sgn}\,f$, so $\mathscr{F}^{-1}[-j\,\mathrm{sgn}\,f] = j/j\pi t = 1/\pi t$.

Now let an arbitrary signal $x(t)$ be the input to a quadrature filter. The output signal $y(t) = x(t) * h_Q(t)$ will be defined as the **Hilbert transform** of $x(t)$, denoted by $\hat{x}(t)$. Thus

$$\hat{x}(t) \triangleq x(t) * \frac{1}{\pi t} = \frac{1}{\pi} \int_{-\infty}^{\infty} \frac{x(\lambda)}{t - \lambda}\, d\lambda \qquad\qquad \text{[2]}$$

Note that Hilbert transformation is a *convolution* and does not change the domain, so both $x(t)$ and $\hat{x}(t)$ are functions of time. Even so, we can easily write the spectrum of $\hat{x}(t)$, namely

$$\mathscr{F}[\hat{x}(t)] = (-j\,\mathrm{sgn}\,f)X(f) \qquad\qquad \text{[3]}$$

since phase shifting produces the output spectrum $H_Q(f)X(f)$.

The catalog of Hilbert transform pairs is quite short compared to our Fourier transform catalog, and the Hilbert transform does not even exist for many common signal models. Mathematically, the trouble comes from potential singularities in Eq. (2) when $\lambda = t$ and the integrand becomes undefined. Physically, we see from Eq. (1b) that $h_Q(t)$

is *noncausal*, which means that the quadrature filter is *unrealizable*—although its behavior can be approximated over a finite frequency band using a real network.

Although the Hilbert transform operates exclusively in the time domain, it has a number of useful properties. Those applicable to our interests are discussed here. In all cases we will assume that the signal $x(t)$ is real.

1. A signal $x(t)$ and its Hilbert transform $\hat{x}(t)$ have the same amplitude spectrum. In addition, the energy or power in a signal and its Hilbert transform are also equal. These follow directly from Eq. (3) since $|-j \operatorname{sgn} f| = 1$ for all f.

2. If $\hat{x}(t)$ is the Hilbert transform of $x(t)$, then $-x(t)$ is the Hilbert transform of $\hat{x}(t)$. The details of proving this property are left as an exercise; however, it follows that two successive shifts of 90° result in a total shift of 180°.

3. A signal $x(t)$ and its Hilbert transform $\hat{x}(t)$ are orthogonal. In Sect. 3.6 we will show that this means

$$\int_{-\infty}^{\infty} x(t)\hat{x}(t)\, dt = 0 \text{ for energy signals}$$

and

$$\lim_{T \to \infty} \frac{1}{2T} \int_{-T}^{T} x(t)\hat{x}(t)\, dt = 0 \text{ for power signals}$$

EXAMPLE 3.5–1

Hilbert Transform of a Cosine Signal

The simplest and most obvious Hilbert transform pair follows directly from the phase-shift property of the quadrature filter. Specifically, if the input is

$$x(t) = A \cos(\omega_0 t + \phi)$$

then

$$\hat{X}(f) = -j \operatorname{sgn} f X(f) = \frac{-jA}{2} [\delta(f - f_0) + \delta(f + f_0)] \operatorname{sgn} f$$

$$= \frac{A}{2j} [\delta(f - f_0) + \delta(f + f_0)]$$

and thus $\hat{x}(t) = A \sin(\omega_0 t + \phi)$.

This transform pair can be used to find the Hilbert transform of any signal that consists of a sum of sinusoids. However, most other Hilbert transforms involve performing the convolution operation in Eq. (2), as illustrated by the following example.

EXAMPLE 3.5–2

Hilbert Transform of a Rectangular Pulse

Consider the delayed rectangular pulse $x(t) = A[u(t) - u(t - \tau)]$. The Hilbert transform is

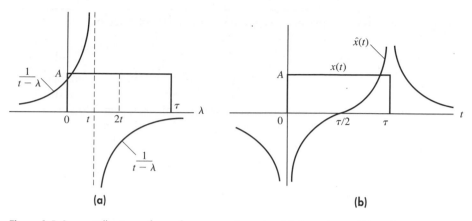

Figure 3.5–2   Hilbert transform of a rectangular pulse. (a) Convolution; (b) result.

$$\hat{x}(t) = \frac{A}{\pi} \int_0^\tau \frac{1}{t - \lambda} \, d\lambda$$

whose evaluation requires graphical interpretation. Figure 3.5–2a shows the case $0 < t < \tau/2$ and we see that the areas cancel out between $\lambda = 0$ and $\lambda = 2t$, leaving

$$\hat{x}(t) = \frac{A}{\pi} \int_{2t}^\tau \frac{d\lambda}{t - \lambda} = \frac{A}{\pi} \left[\ln(-t) - \ln(t - \tau) \right]$$

$$= \frac{A}{\pi} \ln\left(\frac{-t}{t - \tau} \right) = \frac{A}{\pi} \ln\left(\frac{t}{\tau - t} \right)$$

This result also holds for $\tau/2 < t < \tau$, when the areas cancel out between $\lambda = 2t - \tau$ and $\lambda = \tau$. There is no area cancellation for $t < 0$ or $t > \tau$, and

$$\hat{x}(t) = \frac{A}{\pi} \int_0^\tau \frac{d\lambda}{t - \lambda} = \frac{A}{\pi} \ln\left(\frac{t}{t - \tau} \right)$$

These separate cases can be combined in one expression

$$\hat{x}(t) = \frac{A}{\pi} \ln \left| \frac{t}{t - \tau} \right| \qquad\qquad \text{[4]}$$

which is plotted in Fig. 3.5–2b along with $x(t)$.

The infinite spikes in $\hat{x}(t)$ at $t = 0$ and $t = \tau$ can be viewed as an extreme manifestation of *delay distortion*. See Fig. 3.2–5 for comparison.

The inverse Hilbert transform recovers $x(t)$ from $\hat{x}(t)$. Use spectral analysis to show that $\hat{x}(t) * (-1/\pi t) = x(t)$.

EXERCISE 3.5–1

3.6   CORRELATION AND SPECTRAL DENSITY

This section introduces **correlation functions** as another approach to signal and system analysis. Correlation focuses on time averages and signal power or energy. Taking the Fourier transform of a correlation function leads to frequency-domain representation in terms of **spectral density functions,** equivalent to energy spectral density in the case of an energy signal. In the case of a power signal, the spectral density function tells us the power distribution over frequency.

But the signals themselves need not be Fourier transformable. Hence, spectral density allows us to deal with a broader range of signal models, including the important class of *random* signals. We develop correlation and spectral density here as analytic tools for nonrandom signals. You should then feel more comfortable with them when we get to random signals in Chap. 9.

Correlation of Power Signals

Let $v(t)$ be a power signal, but not necessarily real nor periodic. Our only stipulation is that it must have well-defined **average power**

$$P_v \triangleq \langle |v(t)|^2 \rangle = \langle v(t)v^*(t) \rangle \geq 0 \qquad [1]$$

The time-averaging operation here is interpreted in the general form

$$\langle z(t) \rangle = \lim_{T \to \infty} \frac{1}{T} \int_{-T/2}^{T/2} z(t)\, dt$$

where $z(t)$ is an arbitrary time function. For reference purposes, we note that this operation has the following properties:

$$\langle z^*(t) \rangle = \langle z(t) \rangle^* \qquad [2a]$$

$$\langle z(t - t_d) \rangle = \langle z(t) \rangle \qquad \text{any } t_d \qquad [2b]$$

$$\langle a_1 z_1(t) + a_2 z_2(t) \rangle = a_1 \langle z_1(t) \rangle + a_2 \langle z_2(t) \rangle \qquad [2c]$$

We'll have frequent use for these properties in conjunction with correlation.

If $v(t)$ and $w(t)$ are power signals, the average $\langle v(t)w^*(t) \rangle$ is called the **scalar product** of $v(t)$ and $w(t)$. The scalar product is a number, possibly complex, that serves as a measure of *similarity* between the two signals. **Schwarz's inequality** relates the scalar product to the signal powers P_v and P_w, in that

$$|\langle v(t)w^*(t) \rangle|^2 \leq P_v P_w \qquad [3]$$

You can easily confirm that the equality holds when $v(t) = aw(t)$, with a being an arbitrary constant. Hence, $|\langle v(t)w^*(t) \rangle|$ is maximum when the signals are *proportional*. We'll soon define correlation in terms of the scalar product.

First, however, let's further interpret $\langle v(t)w^*(t) \rangle$ and prove Schwarz's inequality by considering

$$z(t) = v(t) - aw(t) \qquad [4a]$$

The average power of $z(t)$ is

$$P_z = \langle z(t)z^*(t)\rangle = \langle [v(t) - aw(t)][v^*(t) - a^*w^*(t)]\rangle \qquad [4b]$$

$$= \langle v(t)v^*(t)\rangle + aa^*\langle w(t)w^*(t)\rangle - a^*\langle v(t)w^*(t)\rangle - a\langle v^*(t)w(t)\rangle$$

$$= P_v + aa^*P_w - 2\,\text{Re}\,[a^*\langle v(t)w^*(t)\rangle]$$

where Eqs. (2a) and (2c) have been used to expand and combine terms. If $a = 1$, then $z(t) = v(t) - w(t)$ and

$$P_z = P_v + P_w - 2\,\text{Re}\,\langle v(t)w^*(t)\rangle$$

A large value of the scalar product thus implies similar signals, in the sense that the difference signal $v(t) - w(t)$ has small average power. Conversely, a small scalar product implies dissimilar signals and $P_z \approx P_v + P_w$.

To prove Schwarz's inequality from Eq. (4b), let $a = \langle v(t)w^*(t)\rangle / P_w$ so

$$aa^*P_w = a^*\langle v(t)w^*(t)\rangle = |\langle v(t)w^*(t)\rangle|^2/P_w$$

Then $P_z = P_v - |\langle v(t)w^*(t)\rangle|^2/P_w \geq 0$, which reduces to Eq. (3) and completes the preliminary work.

Now we define the **crosscorrelation** of two power signals as[†]

$$R_{vw}(\tau) \overset{\triangle}{=} \langle v(t)w^*(t - \tau)\rangle = \langle v(t + \tau)w^*(t)\rangle \qquad [5]$$

This is a scalar product with the second signal delayed by τ relative to the first or, equivalently, the first signal advanced by τ relative to the second. The relative displacement τ is the *independent variable* in Eq. (5), the variable t having been washed out in the time average. General properties of $R_{vw}(\tau)$ are

$$|R_{vw}(\tau)|^2 \leq P_v P_w \qquad [6a]$$

$$R_{wv}(\tau) = R_{vw}^*(-\tau) \qquad [6b]$$

Equation (6a) simply restates Schwarz's inequality, while Eq. (6b) points out that $R_{wv}(\tau) \neq R_{vw}(\tau)$.

We conclude from our previous observations that $R_{vw}(\tau)$ measures the similarity between $v(t)$ and $w(t - \tau)$ as a function of τ. Crosscorrelation is thus a more sophisticated measure than the ordinary scalar product since it detects time-shifted similarities or differences that would be ignored in $\langle v(t)w^*(t)\rangle$.

But suppose we correlate a signal with itself, generating the **autocorrelation function**

$$R_v(\tau) \overset{\triangle}{=} R_{vv}(\tau) = \langle v(t)v^*(t - \tau)\rangle = \langle v(t + \tau)v^*(t)\rangle \qquad [7]$$

This autocorrelation tells us something about the *time variation* of $v(t)$, at least in an averaged sense. If $|R_v(\tau)|$ is large, we infer that $v(t - \tau)$ is very similar to $v(t)$ for

---

[†]Another definition used by some authors is $\langle v^*(t)w(t + \tau)\rangle$, equivalent to interchanging the subscripts on $R_{vw}(\tau)$ in Eq. (5).

that particular value of τ; whereas if $|R_v(\tau)|$ is small, then $v(t)$ and $v(t-\tau)$ must look quite different.

Properties of the autocorrelation function include

$$R_v(0) = P_v \tag{8a}$$

$$|R_v(\tau)| \leq R_v(0) \tag{8b}$$

$$R_v(-\tau) = R_v^*(\tau) \tag{8c}$$

Hence, $R_v(\tau)$ has hermitian symmetry and a maximum value at the origin equal to the signal power. If $v(t)$ is real, then $R_v(\tau)$ will be real and even. If $v(t)$ happens to be periodic, $R_v(\tau)$ will have the same periodicity.

Lastly, consider the sum or difference signal

$$z(t) = v(t) \pm w(t) \tag{9a}$$

Upon forming its autocorrelation, we find that

$$R_z(\tau) = R_v(\tau) + R_w(\tau) \pm [R_{vw}(\tau) + R_{wv}(\tau)] \tag{9b}$$

If $v(t)$ and $w(t)$ are **uncorrelated** for all τ, so

$$R_{vw}(\tau) = R_{wv}(\tau) = 0$$

then $R_z(\tau) = R_v(\tau) + R_w(\tau)$ and setting $\tau = 0$ yields

$$P_z = P_v + P_w$$

Superposition of average power therefore holds for uncorrelated signals.

EXAMPLE 3.6–1   **Correlation of Phasors and Sinusoids**

The calculation of correlation functions for phasors and sinusoidal signals is expedited by calling upon Eq. (18), Sect. 2.1, written as

$$\langle e^{j\omega_1 t} e^{-j\omega_2 t} \rangle = \lim_{T\to\infty} \frac{1}{T} \int_{-T/2}^{T/2} e^{j(\omega_1-\omega_2)t}\, dt \tag{10}$$

$$= \lim_{T\to\infty} \text{sinc}\,\frac{(\omega_1-\omega_2)T}{2\pi} = \begin{cases} 0 & \omega_2 \neq \omega_1 \\ 1 & \omega_2 = \omega_1 \end{cases}$$

We'll apply this result to the phasor signals

$$v(t) = C_v e^{j\omega_v t} \qquad w(t) = C_w e^{j\omega_w t} \tag{11a}$$

where C_v and C_w are complex constants incorporating the amplitude and phase angle. The crosscorrelation is

$$R_{vw}(\tau) = \langle [C_v e^{j\omega_v t}][C_w e^{j\omega_w(t-\tau)}]^* \rangle$$

$$= C_v C_w^* e^{j\omega_w \tau} \langle e^{j\omega_v t} e^{-j\omega_w t} \rangle$$

$$= \begin{cases} 0 & \omega_w \neq \omega_v \\ C_v C_w^* e^{j\omega_v \tau} & \omega_w = \omega_v \end{cases}$$ [11b]

Hence, the phasors are uncorrelated unless they have identical frequencies. The autocorrelation function is

$$R_v(\tau) = |C_v|^2 e^{j\omega_v \tau}$$ [11c]

which drops out of Eq. (11b) when $w(t) = v(t)$.

Now it becomes a simple task to show that the sinusoidal signal

$$z(t) = A \cos (\omega_0 t + \phi)$$ [12a]

has

$$R_z(\tau) = \frac{A^2}{2} \cos \omega_0 \tau$$ [12b]

Clearly, $R_z(\tau)$ is real, even, and periodic, and has the maximum value $R_z(0) = A^2/2 = P_z$. This maximum also occurs whenever $\omega_0 \tau$ equals a multiple of 2π radians, so $z(t \pm \tau) = z(t)$. On the other hand, $R_z(\tau) = 0$ when $z(t \pm \tau)$ and $z(t)$ are in phase quadrature.

But notice that the phase angle ϕ does not appear in $R_z(\tau)$, owing to the averaging effect of correlation. This emphasizes the fact that the autocorrelation function does not uniquely define a signal.

Derive Eq. (12b) by writing $z(t)$ as a sum of conjugate phasors and applying Eqs. (9) and (11).

EXERCISE 3.6–1

Correlation of Energy Signals

Averaging products of energy signals over all time yields zero. But we can meaningfully speak of the **total energy**

$$E_v \triangleq \int_{-\infty}^{\infty} v(t) v^*(t) \, dt \geq 0$$ [13]

Similarly, the correlation functions for energy signals can be defined as

$$R_{vw}(\tau) \triangleq \int_{-\infty}^{\infty} v(t) w^*(t - \tau) \, dt$$ [14a]

$$R_v(\tau) \triangleq R_{vv}(\tau)$$ [14b]

Since the integration operation $\int_{-\infty}^{\infty} z(t) \, dt$ has the same mathematical properties as the time-average operation $\langle z(t) \rangle$, all of our previous correlation relations hold for the case of energy signals if we replace average power P_v with total energy E_v. Thus, for instance, we have the property

$$|R_{vw}(\tau)|^2 \leq E_v E_w \qquad [15]$$

as the energy-signal version of Eq. (6a).

Closer examination of Eq. (14) reveals that energy-signal correlation is a type of *convolution*. For with $z(t) = w^*(-t)$ and $t = \lambda$, the right-hand side of Eq. (14a) becomes

$$\int_{-\infty}^{\infty} v(\lambda)z(\tau - \lambda) \, d\lambda = v(\tau) * z(\tau)$$

and therefore

$$R_{vw}(\tau) = v(\tau) * w^*(-\tau) \qquad [16]$$

Likewise, $R_v(\tau) = v(\tau) * v^*(-\tau)$.

Some additional relations are obtained in terms of the Fourier transforms $V(f) = \mathcal{F}[v(t)]$, etc. Specifically, from Eqs. (16) and (17), Sect. 2.2,

$$R_v(0) = E_v = \int_{-\infty}^{\infty} |V(f)|^2 \, df$$

$$R_{vw}(0) = \int_{-\infty}^{\infty} v(t)w^*(t) \, dt = \int_{-\infty}^{\infty} V(f)W^*(f) \, df$$

Combining these integrals with $|R_{vw}(0)|^2 \leq E_v E_w = R_v(0)R_w(0)$ yields

$$\left| \int_{-\infty}^{\infty} V(f)W^*(f) \, df \right|^2 \leq \int_{-\infty}^{\infty} |V(f)|^2 \, df \int_{-\infty}^{\infty} |W(f)|^2 \, df \qquad [17]$$

Equation (17) is a frequency-domain statement of Schwarz's inequality. The equality holds when $V(f)$ and $W(f)$ are proportional.

EXAMPLE 3.6–2   **Pattern Recognition**

Crosscorrelation can be used in pattern recognition tasks. If the crosscorrelation of objects A and B is similar to the autocorrelation of A, then B is assumed to match A. Otherwise B does not match A. For example, the autocorrelation of $x(t) = \Pi(t)$ can be found from performing the graphical correlation in Eq. (14b) as $R_x(\tau) = \Lambda(\tau)$. If we examine the similarity of $y(t) = 2\Pi(t)$ to $x(t)$ by finding the crosscorrelation $R_{xy}(\tau) = 2\Lambda(\tau)$, we see that $R_{xy}(\tau)$ is just a scaled version of $R_x(\tau)$. Therefore $y(t)$ matches $x(t)$. However, if we take the crosscorrelation of $z(t) = u(t)$ with $x(t)$, we obtain

$$R_{xz}(\tau) = \begin{cases} 1 & \text{for } \tau < -1/2 \\ 1/2 - \tau & \text{for } -1/2 \leq \tau \leq 1/2 \\ 0 & \text{for } \tau > 1/2 \end{cases}$$

and conclude that $z(t)$ doesn't match $x(t)$.

This type of graphical correlation is particularly effective for signals that do not have a closed-form solution. For example, autocorrelation can find the pitch (fundamental frequency) of speech signals. The crosscorrelation can determine if two speech samples have the same pitch, and thus may have come from the same individual.

Let $v(t) = A[u(t) - u(t - D)]$ and $w(t) = v(t - t_d)$. Use Eq. (16) with $z(\tau) = w^*(-\tau)$ to sketch $R_{vw}(\tau)$. Confirm from your sketch that $|R_{vw}(\tau)|^2 \leq E_v E_w$ and that $|R_{vw}(\tau)|^2_{\max} = E_v E_w$ at $\tau = -t_d$.

EXERCISE 3.6–2

We next investigate system analysis in the "τ domain," as represented by Fig. 3.6–1. A signal $x(t)$ having known autocorrelation $R_x(\tau)$ is applied to an LTI system with impulse response $h(t)$, producing the output signal

$$y(t) = h(t) * x(t) = \int_{-\infty}^{\infty} h(\lambda)\, x(t - \lambda)\, d\lambda$$

We'll show that the input-output crosscorrelation function is

$$R_{yx}(\tau) = h(\tau) * R_x(\tau) = \int_{-\infty}^{\infty} h(\lambda)\, R_x(\tau - \lambda)\, d\lambda \qquad \text{[18]}$$

and that the output autocorrelation function is

$$R_y(\tau) = h^*(-\tau) * R_{yx}(\tau) = \int_{-\infty}^{\infty} h^*(-\mu) R_{yx}(\tau - \mu)\, d\mu \qquad \text{[19a]}$$

Substituting Eq. (18) into (19a) then gives

$$R_y(\tau) = h^*(-\tau) * h(\tau) * R_x(\tau) \qquad \text{[19b]}$$

Note that these τ-domain relations are convolutions, similar to the time-domain relation.

For derivation purposes, let's assume that $x(t)$ and $y(t)$ are power signals so we can use the compact time-averaged notation. Obviously, the same results will hold when $x(t)$ and $y(t)$ are both energy signals. The assumption of a stable system ensures that $y(t)$ will be the same type of signal as $x(t)$.

Starting with the crosscorrelation $R_{yx}(\tau) = \langle y(t)x^*(t - \tau)\rangle$, we insert the superposition integral $h(t)x^*(t)$ for $y(t)$ and interchange the order of operations to get

$$R_{yx}(\tau) = \int_{-\infty}^{\infty} h(\lambda)\langle x(t - \lambda)x^*(t - \tau)\rangle\, d\lambda$$

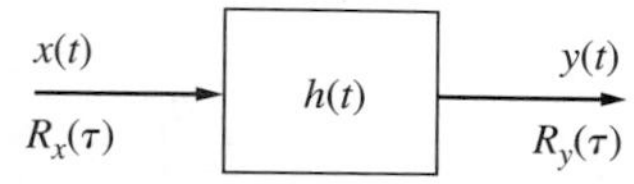

Figure 3.6–1

But since $\langle z(t) \rangle = \langle z(t + \lambda) \rangle$ for any λ,

$$\langle x(t - \lambda)x^*(t - \tau) \rangle = \langle x(t + \lambda - \lambda)x^*(t + \lambda - \tau) \rangle$$
$$= \langle x(t)x^*[t - (\tau - \lambda)] \rangle$$
$$= R_x(\tau - \lambda)$$

Hence,

$$R_{yx}(\tau) = \int_{-\infty}^{\infty} h(\lambda)R_x(\tau - \lambda)\, d\lambda$$

Proceeding in the same fashion for $R_y(\tau) = \langle y(t)y^*(t - \tau) \rangle$ we arrive at

$$R_y(\tau) = \int_{-\infty}^{\infty} h^*(\lambda)\langle y(t)x^*(t - \tau - \lambda) \rangle d\lambda$$

in which $\langle y(t)x^*(t - \tau - \lambda) \rangle = R_{yx}(\tau + \lambda)$. Equation (19a) follows from the change of variable $\mu = -\lambda$.

Spectral Density Functions

At last we're prepared to discuss spectral density functions. Given a power or energy signal $v(t)$, its spectral density function $G_v(f)$ represents the distribution of power or energy in the frequency domain and has two essential properties. First, the area under $G_v(f)$ equals the average power or total energy, so

$$\int_{-\infty}^{\infty} G_v(f)\, df = R_v(0) \tag{20}$$

Second, if $x(t)$ is the input to an LTI system with $H(f) = \mathcal{F}[h(t)]$, then the input and output spectral density functions are related by

$$G_y(f) = |H(f)|^2 G_x(f) \tag{21}$$

since $|H(f)|^2$ is the power or energy gain at any f. These two properties are combined in

$$R_y(0) = \int_{-\infty}^{\infty} |H(f)|^2 G_x(f)\, df \tag{22}$$

which expresses the output power or energy $R_y(0)$ in terms of the input spectral density.

Equation (22) leads to a physical interpretation of spectral density with the help of Fig. 3.6–2. Here, $G_x(f)$ is arbitrary and $|H(f)|^2$ acts like a narrowband filter with unit gain, so

$$G_y(f) = \begin{cases} G_x(f) & f_c - \dfrac{\Delta f}{2} < f < f_c + \dfrac{\Delta f}{2} \\ 0 & \text{otherwise} \end{cases}$$

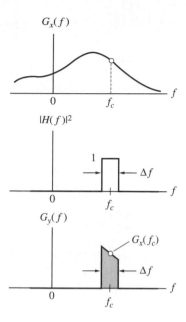

Figure 3.6–2   Interpretation of spectral density functions.

If Δf is sufficiently small, the area under $G_y(f)$ will be $R_y(0) \approx G_x(f_c)\Delta f$ and

$$G_x(f_c) \approx R_y(0)/\Delta f$$

We conclude that at any frequency $f = f_c$, $G_x(f_c)$ equals the signal *power or energy per unit frequency*. We further conclude that any spectral density function must be *real* and *nonnegative* for all values of f.

But how do you determine $G_v(f)$ from $v(t)$? The *Wiener-Kinchine theorem* states that you first calculate the autocorrelation function and then take its Fourier transform. Thus,

$$G_v(f) = \mathscr{F}_\tau[R_v(\tau)] \triangleq \int_{-\infty}^{\infty} R_v(\tau)e^{-j2\pi f\tau}\, d\tau \qquad \text{[23a]}$$

where $\mathscr{F}_\tau$ stands for the Fourier transform operation with τ in place of t. The inverse transform is

$$R_v(\tau) = \mathscr{F}_\tau^{-1}[G_v(f)] \triangleq \int_{-\infty}^{\infty} G_v(f)e^{j2\pi f\tau}\, df \qquad \text{[23b]}$$

so we have the Fourier transform pair

$$R_v(\tau) \leftrightarrow G_v(f)$$

All of our prior transform theorems therefore may be invoked to develop relationships between autocorrelation and spectral density.

If $v(t)$ is an *energy* signal with $V(f) = \mathcal{F}[v(t)]$, application of Eqs. (16) and (23a) shows that

$$G_v(f) = |V(f)|^2 \qquad \text{[24]}$$

and we have the **energy spectral density.** If $v(t)$ is a *periodic power* signal with the Fourier series expansion

$$v(t) = \sum_{n=-\infty}^{\infty} c(nf_0) e^{j2\pi n f_0 t} \qquad \text{[25a]}$$

the Wiener-Kinchine theorem gives the **power spectral density,** or **power spectrum,** as

$$G_v(f) = \sum_{n=-\infty}^{\infty} |c(nf_0)|^2 \delta(f - nf_0) \qquad \text{[25b]}$$

This power spectrum consists of *impulses* representing the average phasor power $|c(nf_0)|^2$ concentrated at each harmonic frequency $f = nf_0$. Substituting Eq. (25b) into Eq. (20) then yields a restatement of Parseval's power theorem. In the special case of a sinusoidal signal

$$z(t) = A \cos(\omega_0 t + \phi)$$

we use $R_z(\tau)$ from Eq. (12b) to get

$$G_v(f) = \mathcal{F}_\tau[(A^2/2) \cos 2\pi f_0 \tau]$$
$$= \frac{A^2}{4} \delta(f - f_0) + \frac{A^2}{4} \delta(f + f_0)$$

which is plotted in Fig. 3.6–3.

All of the foregoing cases lend support to the Wiener-Kinchine theorem but do not constitute a general proof. To prove the theorem, we must confirm that taking $G_v(f) = \mathcal{F}_\tau[R_v(\tau)]$ satisfies the properties in Eqs. (20) and (21). The former immediately follows from the inverse transform in Eq. (23b) with $\tau = 0$. Now recall the output autocorrelation expression

$$R_y(\tau) = h^*(-\tau) * h(\tau) * R_x(\tau)$$

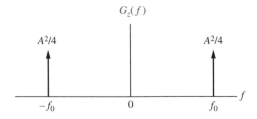

Figure 3.6–3   Power spectrum of $z(t) = A \cos(\omega_0 t + \phi)$.

Since

$$\mathscr{F}_\tau[h(\tau)] = H(f) \qquad \mathscr{F}_\tau[h^*(-\tau)] = H^*(f)$$

the convolution theorem yields

$$\mathscr{F}_\tau[R_y(\tau)] = H^*(f)H(f)\mathscr{F}_\tau[R_x(\tau)]$$

and thus $G_y(f) = |H(f)|^2 G_x(f)$ if we take $\mathscr{F}_\tau[R_y(\tau)] = G_y(f)$, etc.

The signal $x(t) = \operatorname{sinc} 10t$ is input to the system in Fig. 3.6–1 having the transfer function

EXAMPLE 3.6–3

$$H(f) = 3\Pi\left(\frac{f}{4}\right)e^{-j4\pi f}$$

We can find the energy spectral density of $x(t)$ from Eq. (24)

$$G_x(f) = |X(f)|^2 = \frac{1}{100}\,\Pi\left(\frac{f}{10}\right)$$

and the corresponding spectral density of the output $y(t)$

$$G_y(f) = |H(f)|^2 G_x(f)$$

$$= \left[9\Pi\left(\frac{f}{4}\right)\right]\left[\frac{1}{100}\,\Pi\left(\frac{f}{10}\right)\right]$$

$$= \frac{9}{100}\,\Pi\left(\frac{f}{4}\right)$$

since the amplitudes multiply only in the region where the functions overlap. There are several ways to find the total energies E_x and E_y. We know that

$$E_x = \int_{-\infty}^{\infty} |x(t)|^2\, dt = \int_{-\infty}^{\infty} |X(f)|^2\, df = \int_{-\infty}^{\infty} G_x(f)\, df$$

$$= \int_{-5}^{5} \frac{1}{100}\, df = \frac{1}{10}$$

Or we can find $R_x(\tau) = \mathscr{F}_\tau^{-1}\{G_x(f)\} = \frac{1}{10}\operatorname{sinc} 10t$ from which $E_x = R_x(0) = \frac{1}{10}$. Similarly,

$$E_y = \int_{-\infty}^{\infty} |y(t)|^2\, dt = \int_{-\infty}^{\infty} |Y(f)|^2\, df = \int_{-\infty}^{\infty} G_y(f)\, df$$

$$= \int_{-2}^{2} \frac{9}{100}\, df = \frac{9}{25}$$

And correspondingly $R_y(\tau) = \mathcal{F}_\tau^{-1}\{G_y(f)\} = \frac{9}{25}$ sinc $4t$ which leads to the same result that $E_y = R_y(0) = \frac{9}{25}$. We can find the output signal $y(t)$ directly from the relationship

$$Y(f) = X(f)H(f) = \frac{3}{10}\Pi\left(\frac{f}{4}\right)e^{-j4\pi f}$$

by doing the same type of multiplication between rectangular functions as we did earlier for the spectral density. Using the Fourier transform theorems, $y(t) = \frac{6}{5}$ sinc $4(t - 2)$.

EXAMPLE 3.6–4   **Comb Filter**

Consider the **comb filter** in Fig. 3.6–4a. The impulse response is

$$h(t) = \delta(t) - \delta(t - T)$$

so

$$H(f) = 1 - e^{-j2\pi fT}$$

and

$$|H(f)|^2 = 2 - e^{-j2\pi fT} - e^{j2\pi fT}$$

$$= 4 \sin^2 2\pi(f/f_c) \qquad f_c = 2/T$$

The sketch of $|H(f)|^2$ in Fig. 3.6–4b explains the name of this filter.

If we know the input spectral density, the output density and autocorrelation can be found from

$$G_y(f) = 4 \sin^2 2\pi(f/f_c)G_x(f)$$

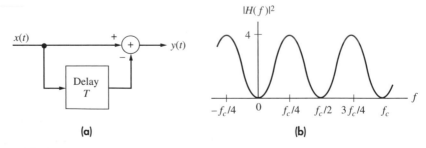

(a)   (b)

Figure 3.6–4   Comb filter.

$$R_y(\tau) = \mathcal{F}_\tau^{-1}[G_y(f)]$$

If we also know the input autocorrelation, we can write

$$R_y(\tau) = \mathcal{F}_\tau^{-1}[|H(f)|^2] * R_x(f)$$

where, using the exponential expression for $|H(f)|^2$,

$$\mathcal{F}_\tau^{-1}[|H(f)|^2] = 2\delta(\tau) - \delta(\tau - T) - \delta(\tau + T)$$

Therefore,

$$R_y(\tau) = 2R_x(\tau) - R_x(\tau - T) - R_x(\tau + T)$$

and the output power or energy is $R_y(0) = 2R_x(0) - R_x(-T) - R_x(T)$.

Let $v(t)$ be an energy signal. Show that $\mathcal{F}_\tau[v^*(-\tau)] = V^*(f)$. Then derive $G_v(f) = |V(f)|^2$ by applying Eq. (23a) to Eq. (16).   **EXERCISE 3.6–3**

3.7   PROBLEMS

3.1–1 A given system has impulse response $h(t)$ and transfer function $H(f)$. Obtain expressions for $y(t)$ and $Y(f)$ when $x(t) = A[\delta(t + t_d) - \delta(t - t_d)]$.

3.1–2 Do Prob. 3.1–1 with $x(t) = A[\delta(t + t_d) + \delta(t)]$

3.1–3* Do Prob. 3.1–1 with $x(t) = Ah(t - t_d)$.

3.1–4 Do Prob. 3.1–1 with $x(t) = Au(t - t_d)$.

3.1–5 Justify Eq. (7b) from Eq. (14) with $x(t) = u(t)$.

3.1–6 Find and sketch $|H(f)|$ and arg $H(f)$ for a system described by the differential equation $dy(t)/dt + 4\pi y(t) = dx(t)/dt + 16\pi x(t)$.

3.1–7 Do Prob. 3.1–6 with $dy(t)/dt + 16\pi y(t) = dx(t)/dt + 4\pi x(t)$.

3.1–8 Do Prob. 3.1–6 with $dy(t)/dt - 4\pi y(t) = -dx(t)/dt + 4\pi x(t)$.

3.1–9* Use frequency-domain analysis to obtain an approximate expression for $y(t)$ when $H(f) = B/(B + jf)$ and $x(t)$ is such that $X(f) \approx 0$ for $|f| < W$ with $W \gg B$.

3.1–10 Use frequency-domain analysis to obtain an approximate expression for $y(t)$ when $H(f) = jf/(B + jf)$ and $x(t)$ is such that $X(f) \approx 0$ for $|f| > W$ with $W \ll B$.

3.1–11 The input to an RC lowpass filter is $x(t) = 2 \operatorname{sinc} 4Wt$. Plot the energy ratio E_y/E_x versus B/W.

3.1–12 Sketch and label the impulse response of the cascade system in Fig. 3.1–8b when the blocks represent zero-order holds with time delays $T_1 > T_2$.

3.1–13 Sketch and label the impulse response of the cascade system in Fig. 3.1–8b when $H_1(f) = [1 + j(f/B)]^{-1}$ and the second block represents a zero-order hold with time delay $T \gg 1/B$.

3.1–14* Find the step and impulse response of the feedback system in Fig. 3.1–8c when $H_1(f)$ is a differentiator and $H_2(f)$ is a gain K.

3.1–15 Find the step and impulse response of the feedback system in Fig. 3.1–8c when $H_1(f)$ is a gain K and $H_2(f)$ is a differentiator.

3.1–16‡ If $H(f)$ is the transfer function of a physically realizable system, then $h(t)$ must be real and causal. As a consequence, for $t \geq 0$ show that

$$h(t) = 4 \int_0^\infty H_r(f) \cos \omega t \, df = 4 \int_0^\infty H_i(f) \cos \omega t \, df$$

where $H_r(f) = \text{Re}[H(f)]$ and $H_i(f) = \text{Im}[H(f)]$.

3.2–1 Show that a first-order lowpass system yields essentially distortionless transmission if $x(t)$ is bandlimited to $W \ll B$.

3.2–2 Find and sketch $y(t)$ when the test signal $x(t) = 4 \cos \omega_0 t + \frac{4}{9} \cos 3\omega_0 t + \frac{4}{25} \cos 5\omega_0 t$, which approximates a triangular wave, is applied to a first-order lowpass system with $B = 3f_0$.

3.2–3* Find and sketch $y(t)$ when the test signal from Prob. 3.2–2 is applied to a first-order highpass system with $H(f) = jf/(B + jf)$ and $B = 3f_0$.

3.2–4 The signal 2 sinc $40t$ is to be transmitted over a channel with transfer function $H(f)$. The output is $y(t) = 20$ sinc $(40t - 200)$. Find $H(f)$ and sketch its magnitude and phase over $|f| \leq 30$.

3.2–5 Evaluate $t_d(f)$ at $f = 0, 0.5, 1$, and 2 kHz for a first-order lowpass system with $B = 2$ kHz.

3.2–6 A channel has the transfer function

$$H(f) = \begin{cases} 4\Pi\left(\dfrac{f}{40}\right) e^{-j\pi f/30} & \text{for}\, |f| \leq 15 \text{ Hz} \\[3mm] 4\Pi\left(\dfrac{f}{40}\right) e^{-j\pi/2} & \text{for}\, |f| > 15 \text{ Hz} \end{cases}$$

Sketch the phase delay $t_d(f)$ and group delay $t_g(f)$. For what values of f does $t_d(f) = t_g(f)$?

3.2–7 Consider a transmission channel with $H_C(f) = (1 + 2\alpha \cos \omega T)e^{-j\omega T}$, which has *amplitude ripples*. (a) Show that $y(t) = \alpha x(t) + x(t - T) + \alpha x(t - 2T)$, so the output includes a leading and trailing echo. (b) Let $x(t) = \Pi(t/\tau)$ and $\alpha = 1/2$. Sketch $y(t)$ for $\tau = 2T/3$ and $4T/3$.

3.2–8* Consider a transmission channel with $H_C(f) = \exp[-j(\omega T - \alpha \sin \omega T)]$, which has *phase ripples*. Assume $|\alpha| \ll \pi/2$ and use a series expansion to show that the output includes a leading and trailing echo.

3.2–9   Design a tapped-delay line equalizer for $H_c(f)$ in Prob. 3.2–8 with $\alpha = 0.4$.

3.2–10   Design a tapped-delay line equalizer for $H_c(f)$ in Prob. 3.2–7 with $\alpha = 0.4$.

3.2–11   Suppose $x(t) = A \cos \omega_0 t$ is applied to a nonlinear system with $y(t) = 2x(t) - 3x^3(t)$. Write $y(t)$ as a sum of cosines. Then evaluate the second-harmonic and third-harmonic distortion when $A = 1$ and $A = 2$.

3.2–12   Do Prob. 3.2–11 with $y(t) = 5x(t) - 2x^2(t) + 4x^3(t)$.

3.3–1*   Let the repeater system in Fig. 3.3–2 have $P_{in} = 0.5$ W, $\alpha = 2$ dB/km, and a total path length of 50 km. Find the amplifier gains and the location of the repeater so that $P_{out} = 50$ mW and the signal power at the input to each amplifier equals 20μW.

3.3–2   Do Prob. 3.3–1 with $P_{in} = 100$ mW and $P_{out} = 0.1$ W.

3.3–3   A 400 km repeater system consists of m identical cable sections with $\alpha = 0.4$ dB/km and m identical amplifiers with 30 dB maximum gain. Find the required number of sections and the gain per amplifier so that $P_{out} = 50$ mW when $P_{in} = 2$W.

3.3–4   A 3000 km repeater system consists of m identical fiber-optic cable sections with $\alpha = 0.5$ dB/km and m identical amplifiers. Find the required number of sections and the gain per amplifier so that $P_{out} = P_{in} = 5$ mW and the input power to each amplifier is at least 67 μW.

3.3–5   Do Prob. 3.3–4 with $\alpha = 2.5$ dB/km.

3.3–6*   Suppose the radio link in Fig. 3.3–4 has $f = 3$ GHz, $\ell = 40$ km, and $P_{in} = 5$W. If both antennas are circular dishes with the same radius r, find the value of r that yields $P_{out} = 2$ μW.

3.3–7   Do Prob. 3.3–6 with $f = 200$ MHz and $\ell = 10$ km.

3.3–8   The radio link in Fig. 3.3–4 is used to transmit a metropolitan TV signal to a rural cable company 50 km away. Suppose a radio repeater with a total gain of g_{rpt} (including antennas and amplifier) is inserted in the middle of the path. Obtain the condition on the value of g_{rpt} so that P_{out} is increased by 20 percent.

3.3–9   A direct broadcast satellite (DBS) system uses 17 GHz for the uplink and 12 GHz for the downlink. Using the values of the amplifiers from Example 3.3–1, find P_{out} assuming $P_{in} = 30$ dBW.

3.4–1   Find and sketch the impulse response of the ideal HPF defined by Eq. (1) with $f_u = \infty$.

3.4–2*   Find and sketch the impulse response of an ideal band-rejection filter having $H(f) = 0$ for $f_c - B/2 < |f| < f_c + B/2$ and distortionless transmission for all other frequencies.

3.4–3   Find the minimum value of n such that a Butterworth filter has $|H(f)| \geq -1$ dB for $|f| < 0.7B$. Then calculate $|H(3B)|$ in dB.

3.4–4   Find the minimum value of n such that a Butterworth filter has $|H(f)| \geq -1$ dB for $|f| < 0.9B$. Then calculate $|H(3B)|$ in dB.

3.4–5   The impulse response of a second-order Butterworth LPF is $h(t) = 2be^{-bt} \sin bt \, u(t)$ with $b = 2\pi B/\sqrt{2}$. Derive this result using a table of Laplace transforms by taking $p = s/2\pi B$ in Table 3.4–1.

3.4–6   Let $R = \sqrt{L/C}$ in Fig. 3.4–7. (a) Show that $|H(f)|^2 = [1 - (f/f_0)^2 + (f/f_0)^4]^{-1}$ with $f_0 = 1/2\pi\sqrt{LC}$. (b) Find the 3 dB bandwidth in terms of f_0. Then sketch $|H(f)|$ and compare with a second-order Butterworth response.

3.4–7   Show that the 10–90% risetime of a first-order LPF equals $1/2.87B$.

3.4–8*   Use $h(t)$ given in Prob. 3.4–5 to find the step response of a second-order Butterworth LPF. Then plot $g(t)$ and estimate the risetime in terms of B.

3.4–9   Let $x(t) = A \operatorname{sinc} 4Wt$ be applied to an ideal LPF with bandwidth B. Taking the duration of sinc at to be $\tau = 2/a$, plot the ratio of output to input pulse duration as a function of B/W.

3.4–10‡   The **effective bandwidth** of an LPF and the **effective duration** of its impulse response are defined by

$$B_{\text{eff}} \triangleq \frac{\int_{-\infty}^{\infty} |H(f)| \, df}{2H(0)} \qquad \tau_{\text{eff}} \triangleq \frac{\int_{-\infty}^{\infty} h(t) \, dt}{|h(t)|_{\max}}$$

Obtain expressions for $H(0)$ and $|h(t)|$ from $\mathcal{F}[h(t)]$ and $\mathcal{F}^{-1}[H(f)]$, respectively. Then show that $\tau_{\text{eff}} \geq 1/2 \, B_{\text{eff}}$.

3.4–11‡   Let the impulse response of an ideal LPF be truncated to obtain the causal function

$$h(t) = 2KB \operatorname{sinc} 2B(t - t_d) \qquad 0 < t < 2t_d$$

and $h(t) = 0$ elsewhere. (a) Show by Fourier transformation that

$$H(f) = \frac{K}{\pi} e^{-j\omega t_d} \{ \operatorname{Si}[2\pi(f + B)t_d] - \operatorname{Si}[2\pi(f - B)t_d] \}$$

(b) Sketch $h(t)$ and $|H(f)|$ for $t_d \gg 1/B$ and $t_d = 1/2B$.

3.5–1   Let $x(t) = \delta(t)$. (a) Find $\hat{x}(t)$ from Eq. (2) and use your result to confirm that $\mathcal{F}^{-1}[-j \operatorname{sgn} f] = 1/\pi t$. (b) Then derive another Hilbert transform pair from the property $\hat{x}(t)*(-1/\pi t) = x(t)$.

3.5–2*   Use Eq. (3), Sect. 3.1, and the results in Example 3.5–2 to obtain the Hilbert transform of $A\Pi(t/\tau)$. Now show that if $v(t) = A$ for all time, then $\hat{v}(t) = 0$.

3.5–3   Use Eq. (3) to show that if $x(t) = \operatorname{sinc} 2Wt$ then $\hat{x}(t) = \pi Wt \operatorname{sinc}^2 Wt$.

3.5–4   Find the Hilbert transform of the signal in Fig. 3.2–3 using the results of Example 3.5–1.

3.5–5*   Find the Hilbert transform of the signal

$$x(t) = 4 \cos \omega_0 t + \tfrac{4}{9} \cos 3\omega_0 t + \tfrac{4}{25} \cos 5\omega_0 t.$$

3.5–6   Show that the functions that form the Hilbert transform pair in Prob. 3.5–3 have the same amplitude spectrum by finding the magnitude of the Fourier transform of each. (*Hint:* Express the sinc^2 term as the product of a sine function and sinc function.)

3.5–7   Show that $\int_{-\infty}^{\infty} x(t)\hat{x}(t)dt = 0$ for $x(t) = A \cos \omega_0 t$.

3.5–8‡   Let the transfer function of a filter be written in the form $H(f) = H_e(f) + jH_o(f)$, as in Eq. (10), Sect. 2.2. If the filter is physically realizable, then its impulse response must have the causal property $h(t) = 0$ for $t < 0$. Hence, we can write $h(t) = (1 + \mathrm{sgn}\, t)h_e(t)$ where $h_e(t) = \tfrac{1}{2}h(|t|)$ for $-\infty < t < \infty$. Show that $\mathcal{F}[h_e(t)] = H_e(f)$ and thus causality requires that $H_o(f) = -\hat{H}_e(f)$.

3.6–1   Prove Eq. (6b).

3.6–2   Let $v(t)$ be periodic with period T_0. Show from Eq. (7) that $R_v(\tau)$ has the same periodicity.

3.6–3   Derive Eq. (8b) by taking $w(t) = v(t - \tau)$ in Eq. (3).

3.6–4   Use the method of pattern recognition demonstrated in Example 3.6–2 to determine whether $y(t) = \sin 2\omega_0 t$ is similar to $x(t) = \cos 2\omega_0 t$.

3.6–5*   Use Eq. (24) to obtain the spectral density, autocorrelation, and signal energy when $v(t) = A\Pi[(t - t_d)/D]$.

3.6–6   Do Prob. 3.6–5 with $v(t) = A \,\mathrm{sinc}\, 4W(t + t_d)$.

3.6–7   Do Prob. 3.6–5 with $v(t) = Ae^{-bt}u(t)$.

3.6–8   Use Eq. (25) to obtain the spectral density, autocorrelation, and signal power when $v(t) = A_0 + A_1 \sin(\omega_0 t + \phi)$.

3.6–9   Do Prob. 3.6–8 with $v(t) = A_1 \cos(\omega_0 t + \phi_1) + A_2 \sin(2\omega_0 t + \phi_1)$.

3.6–10*   Obtain the autocorrelation of $v(t) = Au(t)$ from Eq. (7). Use your result to find the signal power and spectral density.

3.6–11   The energy signal $x(t) = \Pi(10t)$ is input to an ideal lowpass filter system with $K = 3$, $B = 20$, and $t_d = 0.05$, producing the output signal $y(t)$. Write and simplify an expression for $R_y(\tau)$.

Linear CW Modulation

CHAPTER OUTLINE

The several purposes of modulation were itemized in Chap. 1 along with a qualitative description of the process. To briefly recapitulate: *modulation* is the systematic alteration of one waveform, called the *carrier*, according to the characteristics of another waveform, the modulating signal or message. The fundamental goal is to produce an information-bearing modulated wave whose properties are best suited to the given communication task.

We now embark on a tour of **continuous-wave** (CW) **modulation systems.** The carrier in these systems is a sinusoidal wave modulated by an analog signal—AM and FM radio being familiar examples. The abbreviation CW also refers to on-off keying of a sinusoid, as in radio telegraphy, but that process is more accurately termed *interrupted* continuous wave (ICW).

This chapter deals specifically with **linear** CW modulation, which involves direct frequency translation of the message spectrum. Double-sideband modulation (DSB) is precisely that. Minor modifications of the translated spectrum yield conventional amplitude modulation (AM), single-sideband modulation (SSB), or vestigial-sideband modulation (VSB). Each of these variations has its own distinct advantages and significant practical applications. Each will be given due consideration, including such matters as waveforms and spectra, modulation methods, transmitters, and demodulation. The chapter begins with a general discussion of bandpass signals and systems, pertinent to all forms of CW modulation.

OBJECTIVES

After studying this chapter and working the exercises, you should be able to do each of the following:

1. Given a bandpass signal, find its envelope and phase, in-phase and quadrature components, and lowpass equivalent signal and spectrum (Sect. 4.1).
2. State and apply the fractional-bandwidth rule of thumb for bandpass systems (Sect. 4.1).
3. Sketch the waveform and envelope of an AM or DSB signal, and identify the spectral properties of AM, DSB, SSB, and VSB (Sects. 4.2 and 4.4).
4. Construct the line spectrum and phasor diagram, and find the sideband power and total power of an AM, DSB, SSB or VSB signal with tone modulation (Sects. 4.2 and 4.4).
5. Distinguish between product, power-law, and balanced modulators, and analyze a modulation system (Sect. 4.3).
6. Identify the characteristics of synchronous, homodyne, and envelope detection (Sect. 4.5).

4.1   BANDPASS SIGNALS AND SYSTEMS

Effective communication over appreciable distance usually requires a high-frequency sinusoidal carrier. Consequently, by applying the frequency translation (or modulation) property of the Fourier transform from Sect. 2.3 to a bandlimited message signal, we can see that most long-haul transmission systems have a **bandpass** frequency response. The properties are similar to those of a bandpass filter, and any signal transmitted on such a system must have a bandpass spectrum. Our purpose here is to present the characteristics and methods of analysis unique to bandpass systems and signals. Before plunging into the details, let's establish some conventions regarding the message and modulated signals.

Analog Message Conventions

Whenever possible, our study of analog communication will be couched in terms of an arbitrary message waveform $x(t)$—which might stand for a sample function from the ensemble of possible messages produced by an information source. The one essential condition imposed on $x(t)$ is that it must have a reasonably well-defined **message bandwidth W,** so there's negligible spectral content for $|f| > W$. Accordingly, Fig. 4.1–1 represents a typical message spectrum $X(f) = \mathcal{F}[x(t)]$ assuming the message is an energy signal.

For mathematical convenience, we'll also scale or normalize all messages to have a magnitude not exceeding unity, so

$$|x(t)| \leq 1 \qquad [1]$$

This normalization puts an upper limit on the average message power, namely

$$S_x = \langle x^2(t) \rangle \leq 1 \qquad [2]$$

when we assume $x(t)$ is a deterministic power signal. Both energy-signal and power-signal models will be used for $x(t)$, depending on which one best suits the circumstances at hand.

Occasionally, analysis with arbitrary $x(t)$ turns out to be difficult if not impossible. As a fall-back position we may then resort to the specific case of sinusoidal or **tone** modulation, taking

$$x(t) = A_m \cos \omega_m t \qquad A_m \leq 1 \qquad f_m < W \qquad [3]$$

Tone modulation allows us to work with one-sided line spectra and simplifies power calculations. Moreover, if you can find the response of the modulation system at a particular frequency f_m, you can infer the response for all frequencies in the message band—barring any nonlinearities. To reveal potential nonlinear effects, you must use **multitone** modulation such as

$$x(t) = A_1 \cos \omega_1 t + A_2 \cos \omega_2 t + \cdots$$

with $A_1 + A_2 + \cdots \leq 1$ to satisfy Eq. (1).

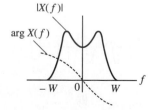

Figure 4.1–1   Message spectrum with bandwidth W.

Bandpass Signals

We next explore the characteristics unique to bandpass signals and establish some useful analysis tools that will aid our discussions of bandpass transmission. Consider a real energy signal $v_{bp}(t)$ whose spectrum $V_{bp}(f)$ has the bandpass characteristic sketched in Fig. 4.1–2a. This spectrum exhibits hermitian symmetry, because $v_{bp}(t)$ is real, but $V_{bp}(f)$ is not necessarily symmetrical with respect to $\pm f_c$. We define a **bandpass signal** by the frequency domain property

$$V_{bp}(f) = 0 \quad |f| < f_c - W \tag{4}$$

$$|f| > f_c + W$$

which simply states that the signal has no spectral content outside a band of width $2W$ centered at f_c. The values of f_c and W may be somewhat arbitrary, as long as they satisfy Eq. (4) with $W < f_c$.

The corresponding bandpass waveform in Fig. 4.1–2b looks like a sinusoid at frequency f_c with slowly changing amplitude and phase angle. Formally we write

$$v_{bp}(t) = A(t) \cos \left[\omega_c t + \phi(t) \right] \tag{5}$$

where $A(t)$ is the **envelope** and $\phi(t)$ is the **phase**, both functions of time. The envelope, shown as a dashed line, is defined as nonnegative, so that $A(t) \geq 0$. Negative "amplitudes," when they occur, are absorbed in the phase by adding $\pm 180°$.

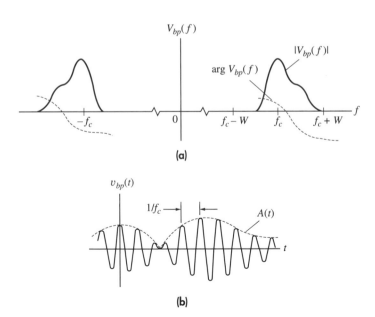

(a)

(b)

Figure 4.1–2   Bandpass signal. (a) Spectrum; (b) waveform.

Figure 4.1–3a depicts $v_{bp}(t)$ as a complex-plane vector whose length equals $A(t)$ and whose angle equals $\omega_c t + \phi(t)$. But the angular term $\omega_c t$ represents a steady counterclockwise rotation at f_c revolutions per second and can just as well be suppressed, leading to Fig. 4.1–3b. This phasor representation, used regularly hereafter, relates to Fig. 4.1–3a in the following manner: If you pin the origin of Fig. 4.1–3b and rotate the entire figure counterclockwise at the rate f_c, it becomes Fig. 4.1–3a.

Further inspection of Fig. 4.1–3a suggests another way of writing $v_{bp}(t)$. If we let

$$v_i(t) \triangleq A(t) \cos \phi(t) \qquad v_q(t) \triangleq A(t) \sin \phi(t) \qquad \text{[6]}$$

then

$$v_{bp}(t) = v_i(t) \cos \omega_c t - v_q(t) \sin \omega_c t \qquad \text{[7]}$$

$$= v_i(t) \cos \omega_c t + v_q(t) \cos (\omega_c t + 90°)$$

Equation (7) is called the **quadrature-carrier** description of a bandpass signal, as distinguished from the **envelope-and-phase** description in Eq. (5). The functions $v_i(t)$ and $v_q(t)$ are named the **in-phase** and **quadrature components,** respectively. The quadrature-carrier designation comes about from the fact that the two terms in Eq. (7) may be represented by phasors with the second at an angle of $+90°$ compared to the first.

While both descriptions of a bandpass signal are useful, the quadrature-carrier version has advantages for the frequency-domain interpretation. Specifically, Fourier transformation of Eq. (7) yields

$$V_{bp}(f) = \frac{1}{2}[V_i(f - f_c) + V_i(f + f_c)] + \frac{j}{2}[V_q(f - f_c) - V_q(f + f_c)] \qquad \text{[8]}$$

where

$$V_i(f) = \mathcal{F}[v_i(t)] \qquad V_q(f) = \mathcal{F}[v_q(t)]$$

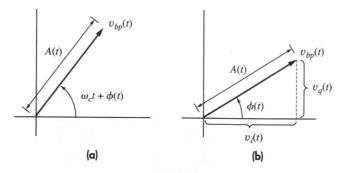

(a)   (b)

Figure 4.1–3   (a) Rotating phasor; (b) phasor diagram with rotation suppressed.

To obtain Eq. (8) we have used the modulation theorem from Eq. (7), Sect. 2.3, along with $e^{\pm j90°} = \pm j$. The envelope-and-phase description does not readily convert to the frequency domain since, from Eq. (6) or Fig. 4.1–3b,

$$A(t) = \sqrt{v_i^2(t) + v_q^2(t)} \qquad \phi(t) = \arctan \frac{v_q(t)}{v_i(t)} \tag{9}$$

which are not Fourier-transformable expressions.

An immediate implication of Eq. (8) is that, in order to satisfy the bandpass condition in Eq. (4), the in-phase and quadrature functions must be **lowpass** signals with

$$V_i(f) = V_q(f) = 0 \qquad |f| > W$$

In other words,

> $V_{bp}(f)$ consists of two lowpass spectra that have been translated and, in the case of $V_q(f)$, quadrature phase shifted.

We'll capitalize upon this property in the definition of the **lowpass equivalent spectrum**

$$V_{\ell p}(f) \triangleq \tfrac{1}{2}[V_i(f) + jV_q(f)] \tag{10a}$$

$$= V_{bp}(f - f_c)u(f + f_c) \tag{10b}$$

As shown in Fig. 4.1–4, $V_{\ell p}(f)$ simply equals the positive-frequency portion of $V_{bp}(f)$ translated down to the origin.

Going from Eq. (10) to the time domain, we obtain the **lowpass equivalent signal**

$$v_{\ell p}(t) = \mathcal{F}^{-1}[V_{\ell p}(f)] = \tfrac{1}{2}[v_i(t) + jv_q(t)] \tag{11a}$$

Thus, $v_{\ell p}(t)$ is a fictitious *complex* signal whose real part equals $\tfrac{1}{2}v_i(t)$ and whose imaginary part equals $\tfrac{1}{2}v_q(t)$. Alternatively, rectangular-to-polar conversion yields

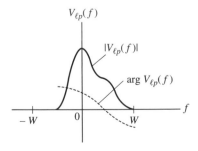

Figure 4.1–4   Lowpass equivalent spectrum.

$$v_{\ell p}(t) = \tfrac{1}{2}A(t)e^{j\,\phi(t)} \qquad [11b]$$

where we've drawn on Eq. (9) to write $v_{\ell p}(t)$ in terms of the envelope and phase functions. The complex nature of the lowpass equivalent signal can be traced back to its spectrum $V_{\ell p}(f)$, which lacks the hermitian symmetry required for the transform of a real time function. Nonetheless, $v_{\ell p}(t)$ does represent a *real* bandpass signal.

The connection between $v_{\ell p}(t)$ and $v_{bp}(t)$ is derived from Eqs. (5) and (11b) as follows:

$$\begin{aligned}
v_{bp}(t) &= \mathrm{Re}\left\{A(t)e^{j[\omega_c t + \phi(t)]}\right\} & [12]\\
&= 2\,\mathrm{Re}\left[\tfrac{1}{2}A(t)e^{j\omega_c t}e^{j\phi(t)}\right]\\
&= 2\,\mathrm{Re}\left[v_{\ell p}(t)e^{j\omega_c t}\right]
\end{aligned}$$

This result expresses the **lowpass-to-bandpass transformation** in the time domain. The corresponding frequency-domain transformation is

$$V_{bp}(f) = V_{\ell p}(f - f_c) + V_{\ell p}^*(-f - f_c) \qquad [13a]$$

whose first term constitutes the positive-frequency portion of $V_{bp}(f)$ while the second term constitutes the negative-frequency portion. Since we'll deal only with real bandpass signals, we can keep the hermitian symmetry of $V_{bp}(f)$ in mind and use the simpler expression

$$V_{bp}(f) = V_{\ell p}(f - f_c) \qquad f > 0 \qquad [13b]$$

which follows from Figs. 4.1–2a and 4.1–4.

Let $z(t) = v_{\ell p}(t)e^{j\omega_c t}$ and use $2\,\mathrm{Re}\,[z(t)] = z(t) + z^*(t)$ to derive Eq. (13a) from Eq. (12). **EXERCISE 4.1–1**

Bandpass Transmission

Now we have the tools needed to analyze bandpass transmission represented by Fig. 4.1–5a where a bandpass signal $x_{bp}(t)$ applied to a bandpass system with transfer function $H_{bp}(f)$ produces the bandpass output $y_{bp}(t)$. Obviously, you could attempt direct bandpass analysis via $Y_{bp}(f) = H_{bp}(f)X_{bp}(f)$. But it's usually easier to work with the lowpass equivalent spectra related by

$$Y_{\ell p}(f) = H_{\ell p}(f)\,X_{\ell p}(f) \qquad [14a]$$

where

$$H_{\ell p}(f) = H_{bp}(f - f_c)u(f + f_c) \qquad [14b]$$

which is the **lowpass equivalent transfer function.**

Equation (14) permits us to replace a bandpass system with the lowpass equivalent model in Fig. 4.1–5b. Besides simplifying analysis, the lowpass model provides

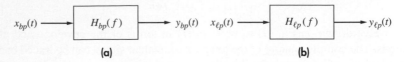

Figure 4.1–5   (a) Bandpass system; (b) lowpass model.

valuable insight to bandpass phenomena by analogy with known lowpass relationships. We move back and forth between the bandpass and lowpass models with the help of our previous results for bandpass signals.

In particular, after finding $Y_{\ell p}(f)$ from Eq. (14), you can take its inverse Fourier transform

$$y_{\ell p}(t) = \mathcal{F}^{-1}[Y_{\ell p}(f)] = \mathcal{F}^{-1}[H_{\ell p}(f)X_{\ell p}(f)]$$

The lowpass-to-bandpass transformation in Eq. (12) then yields the output signal $y_{bp}(t)$. Or you can get the output quadrature components or envelope and phase immediately from $y_{\ell p}(t)$ as

$$y_i(t) = 2\,\mathrm{Re}\,[y_{\ell p}(t)] \qquad y_q(t) = 2\,\mathrm{Im}\,[y_{\ell p}(t)] \qquad \text{[15]}$$
$$A_y(t) = 2\,|y_{\ell p}(t)| \qquad \phi_y(t) = \arg\,[y_{\ell p}(t)]$$

which follow from Eq. (10). The example below illustrates an important application of these techniques.

EXAMPLE 4.1–1   **Carrier and Envelope Delay**

Consider a bandpass system having constant amplitude ratio but *nonlinear* phase shift $\theta(f)$ over its passband. Thus,

$$H_{bp}(f) = Ke^{j\theta(f)} \qquad f_\ell < |f| < f_u$$

and

$$H_{\ell p}(f) = Ke^{j\theta(f+f_c)}u(f + f_c) \qquad f_\ell - f_c < f < f_u - f_c$$

as sketched in Fig. 4.1–6. Assuming the phase nonlinearities are relatively smooth, we can write the approximation

$$\theta(f + f_c) \approx -2\pi(t_0 f_c + t_1 f)$$

where

$$t_0 \triangleq -\frac{\theta(f_c)}{2\pi f_c} \qquad t_1 \triangleq -\frac{1}{2\pi}\frac{d\theta(f)}{df}\bigg|_{f=f_c} \qquad \text{[16]}$$

This approximation comes from the first two terms of the Taylor series expansion of $\theta(f + f_c)$.

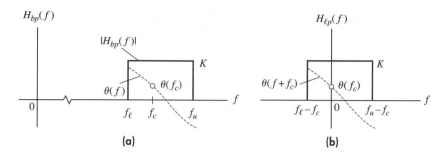

Figure 4.1–6   (a) Bandpass transfer function; (b) lowpass equivalent.

To interpret the parameters t_0 and t_1, let the input signal have zero phase so that $x_{bp}(t) = A_x(t) \cos \omega_c t$ and $x_{\ell p}(t) = \frac{1}{2}A_x(t)$. If the input spectrum $X_{bp}(f)$ falls entirely within the system's passband, then, from Eq. (14),

$$Y_{\ell p}(f) = Ke^{j\theta(f+f_c)}X_{\ell p}(f) \approx Ke^{-j2\pi(t_0 f_c + t_1 f)}X_{\ell p}(f)$$

$$\approx Ke^{-j\omega_c t_0}[X_{\ell p}(f)e^{-j2\pi f t_1}]$$

Recalling the time-delay theorem, we see that the second term corresponds to $x_{\ell p}(t)$ delayed by t_1. Hence,

$$y_{\ell p}(t) \approx Ke^{-j\omega_c t_0}x_{\ell p}(t - t_1) = Ke^{-j\omega_c t_0} \tfrac{1}{2}A_x(t - t_1)$$

and Eq. (12) yields the bandpass output

$$y_{bp}(t) \approx KA_x(t - t_1) \cos \omega_c(t - t_0)$$

Based on this result, we conclude that t_0 is the **carrier delay** while t_1 is the **envelope delay** of the system. And since t_1 is independent of frequency, at least to the extent of our approximation for $\theta(f + f_c)$, the envelope has not suffered delay distortion. Envelope delay is also called the **group** delay.

We'll later describe **multiplexing** systems in which several bandpass signals at different carrier frequencies are transmitted over a single channel. Plots of $d\theta/df$ versus f are used in this context to evaluate the channel's delay characteristics. If the curve is not reasonably flat over a proposed band, phase equalization may be required to prevent excessive envelope distortion.

Suppose a bandpass system has zero phase shift but $|H_{bp}(f)| = K_0 + (K_1/f_c)(f - f_c)$ for $f_\ell < f < f_u$, where $K_0 > (K_1/f_c)(f_\ell - f_c)$. Sketch $H_{\ell p}(f)$ taking $f_\ell < f_c$ and $f_u > f_c$. Now show that if $x_{bp}(t) = A_x(t) \cos \omega_c t$ then the quadrature components of $y_{bp}(t)$ are

EXERCISE 4.1–2

$$y_i(t) = K_0 A_x(t) \qquad y_q(t) = -\frac{K_1}{2\pi f_c}\frac{dA_x(t)}{dt}$$

provided that $X_{bp}(f)$ falls entirely within the bandpass of the system.

The simplest bandpass system is the parallel resonant or *tuned* circuit represented by Fig. 4.1–7a. The voltage transfer function plotted in Fig. 4.1–7b can be written as

$$H(f) = \frac{1}{1 + jQ\left(\dfrac{f}{f_0} - \dfrac{f_0}{f}\right)} \qquad [17a]$$

in which the resonant frequency f_0 and quality factor Q are related to the element values by

$$f_0 = \frac{1}{2\pi\sqrt{LC}} \qquad Q = R\sqrt{\frac{C}{L}}$$

The 3 dB bandwidth between the lower and upper cutoff frequencies is

$$B = f_u - f_\ell = \frac{f_0}{Q} \qquad [17b]$$

Since practical tuned circuits usually have $10 < Q < 100$, the 3 dB bandwidth falls between 1 and 10 percent of the center-frequency value.

A complete bandpass system consists of the transmission channel plus tuned amplifiers and coupling devices connected at each end. Hence, the overall frequency response has a more complicated shape than that of a simple tuned circuit. Nonetheless, various physical effects result in a loose but significant connection between the system's bandwidth and the carrier frequency f_c—similar to Eq. (17b).

For instance, the antennas in a radio system produce considerable distortion unless the frequency range is small compared to f_c. Moreover, designing a reasonably distortionless bandpass amplifier turns out to be quite difficult if B is either

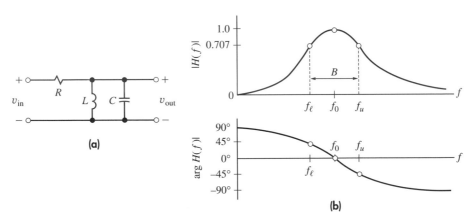

Figure 4.1–7   (a) Tuned circuit; (b) transfer function.

very large or very small compared to f_c. As a rough rule of thumb, the **fractional bandwidth** B/f_c should be kept within the range

$$0.01 < \frac{B}{f_c} < 0.1 \qquad \text{[18]}$$

Otherwise, the signal distortion may be beyond the scope of practical equalizers.

From Eq. (18) we see that

> Large bandwidths require high carrier frequencies.

This observation is underscored by Table. 4.1–1, which lists selected carrier frequencies and the corresponding nominal bandwidth $B \approx 0.02 f_c$ for different frequency bands. Larger bandwidths can be achieved, of course, but at substantially greater cost. As a further consequence of Eq. (18), the terms **bandpass** and **narrowband** are virtually synonymous in signal transmission.

Table 4.1–1   Selected carrier frequencies and nominal bandwidth

Frequency Band	Carrier Frequency	Bandwidth
Longwave radio	100 kHz	2 kHz
Shortwave radio	5 MHz	100 kHz
VHF	100 MHz	2 MHz
Microwave	5 GHz	100 MHz
Millimeterwave	100 GHz	2 GHz
Optical	5×10^{14} Hz	10^{13} Hz

Bandpass Pulse Transmission   **EXAMPLE 4.1–2**

We found in Sect. 3.4 that transmitting a pulse of duration τ requires a lowpass bandwidth $B \geq 1/2\tau$. We also found in Example 2.3–2 that frequency translation converts a pulse to a bandpass waveform and *doubles* its spectral width. Putting these two observations together, we conclude that bandpass pulse transmission requires

$$B \geq 1/\tau$$

Since Eq. (18) imposes the additional constraint $0.1f_c > B$, the carrier frequency must satisfy

$$f_c > 10/\tau$$

These relations have long served as useful guidelines in radar work and related fields. To illustrate, if $\tau = 1\ \mu s$ then bandpass transmission requires $B \geq 1$ MHz and $f_c > 10$ MHz.

4.2   DOUBLE-SIDEBAND AMPLITUDE MODULATION

There are two types of double-sideband amplitude modulation: standard amplitude modulation (AM), and suppressed-carrier double-sideband modulation (DSB). We'll examine both types and show that the minor theoretical difference between them has major repercussions in practical applications.

AM Signals and Spectra

The unique property of AM is that the *envelope* of the modulated carrier has the same *shape* as the message. If A_c denotes the unmodulated carrier amplitude, modulation by $x(t)$ produces the modulated envelope

$$A(t) = A_c[1 + \mu x(t)] \qquad [1]$$

where μ is a positive constant called the **modulation index.** The complete AM signal $x_c(t)$ is then

$$x_c(t) = A_c[1 + \mu x(t)] \cos \omega_c t \qquad [2]$$
$$= A_c \cos \omega_c t + A_c \mu x(t) \cos \omega_c t$$

Since $x_c(t)$ has no time-varying phase, its in-phase and quadrature components are

$$x_{ci}(t) = A(t) \qquad\qquad x_{cq}(t) = 0$$

as obtained from Eqs. (5) and (6), Sect. 4.1, with $\phi(t) = 0$. Actually, we should include a constant carrier phase shift to emphasize that the carrier and message come from independent and unsynchronized sources. However, putting a constant phase in Eq. (2) increases the notational complexity without adding to the physical understanding.

Figure 4.2–1 shows part of a typical message and the resulting AM signal with two values of μ. The envelope clearly reproduces the shape of $x(t)$ if

$$f_c \gg W \qquad \mu \leq 1 \qquad [3]$$

When these conditions are satisfied, the message $x(t)$ is easily extracted from $x_c(t)$ by use of a simple **envelope detector** whose circuitry will be described in Sect. 4.5.

The condition $f_c \gg W$ ensures that the carrier oscillates rapidly compared to the time variation of $x(t)$; otherwise, an envelope could not be visualized. The condition $\mu \leq 1$ ensures that $A_c[1 + \mu x(t)]$ does not go negative. With 100-percent modulation ($\mu = 1$), the envelope varies between $A_{min} = 0$ and $A_{max} = 2A_c$. **Overmodulation** ($\mu > 1$), causes **phase reversals** and **envelope distortion** illustrated by Fig. 4.2–1c.

Going to the frequency domain, Fourier transformation of Eq. (2) yields

$$X_c(f) = \frac{1}{2} A_c \, \delta(f - f_c) + \frac{\mu}{2} A_c X(f - f_c) \qquad f > 0 \qquad [4]$$

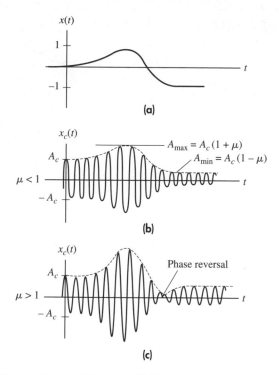

Figure 4.2–1   AM waveforms. (a) Message; (b) AM wave with $\mu < 1$; (c) AM wave with $\mu > 1$.

where we've written out only the positive-frequency half of $X_c(f)$. The negative-frequency half will be the hermitian image of Eq. (4) since $x_c(t)$ is a real bandpass signal. Both halves of $X_c(f)$ are sketched in Fig. 4.2–2 with $X(f)$ from Fig. 4.1–1. The AM spectrum consists of carrier-frequency impulses and **symmetrical sidebands** centered at $\pm f_c$. The presence of **upper** and **lower sidebands** accounts for the name **double-sideband** amplitude modulation. It also accounts for the AM **transmission bandwidth**

$$B_T = 2W \qquad\qquad\qquad [5]$$

Note that AM requires twice the bandwidth needed to transmit $x(t)$ at baseband without modulation.

   Transmission bandwidth is an important consideration for the comparison of modulation systems. Another important consideration is the **average transmitted power**

$$S_T \triangleq \langle x_c^2(t) \rangle$$

Upon expanding $x_c^2(t)$ from Eq. (2), we have

$$S_T = \tfrac{1}{2}A_c^2 \langle 1 + 2\mu x(t) + \mu^2 x^2(t) \rangle + \tfrac{1}{2}A_c^2 \langle [1 + \mu x(t)]^2 \cos 2\omega_c t \rangle$$

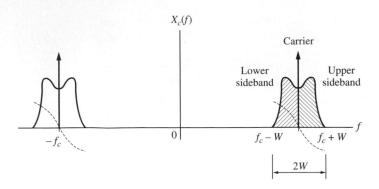

Figure 4.2–2   AM spectrum.

whose second term averages to zero under the condition $f_c \gg W$. Thus, if $\langle x(t) \rangle = 0$ and $\langle x^2(t) \rangle = S_x$ then

$$S_T = \tfrac{1}{2}A_c^2(1 + \mu^2 S_x) \tag{6}$$

The assumption that the message has zero average value (or no dc component) anticipates the conclusion from Sect. 4.5 that ordinary AM is not practical for transmitting signals with significant low-frequency content.

   We bring out the interpretation of Eq. (6) by putting it in the form

$$S_T = P_c + 2P_{sb}$$

where

$$P_c = \tfrac{1}{2}A_c^2 \qquad P_{sb} = \tfrac{1}{4}A_c^2\mu^2 S_x = \tfrac{1}{2}\mu^2 S_x P_c \tag{7}$$

The term P_c represents the **unmodulated carrier power,** since $S_T = P_c$ when $\mu = 0$; the term P_{sb} represents the **power per sideband** since, when $\mu \neq 0$, S_T consists of the power in the carrier plus two symmetric sidebands. The modulation constraint $|\mu x(t)| \leq 1$ requires that $\mu^2 S_x \leq 1$, so $P_{sb} \leq \tfrac{1}{2}P_c$ and

$$P_c = S_T - 2P_{sb} \geq \tfrac{1}{2}S_T \qquad P_{sb} \leq \tfrac{1}{4}S_T \tag{8}$$

Consequently, at least 50 percent of the total transmitted power resides in a carrier term that's independent of $x(t)$ and thus conveys no message information.

DSB Signals and Spectra

The "wasted" carrier power in amplitude modulation can be eliminated by setting $\mu = 1$ and suppressing the unmodulated carrier-frequency component. The resulting modulated wave becomes

$$x_c(t) = A_c x(t) \cos \omega_c t \tag{9}$$

which is called **double-sideband–suppressed-carrier modulation**—or DSB for short. (The abbreviations DSB–SC and DSSC are also used.) The transform of Eq. (9) is simply

$$X_c(f) = \tfrac{1}{2}A_c X(f - f_c) \qquad f > 0$$

and the DSB spectrum looks like an AM spectrum without the unmodulated carrier impulses. The transmission bandwidth thus remains unchanged at $B_T = 2W$.

Although DSB and AM are quite similar in the frequency domain, the time-domain picture is another story. As illustrated by Fig. 4.2–3 the DSB envelope and phase are

$$A(t) = A_c|x(t)| \qquad \phi(t) = \begin{cases} 0 & x(t) > 0 \\ \pm 180° & x(t) < 0 \end{cases} \tag{10}$$

The envelope here takes the shape of $|x(t)|$, rather than $x(t)$, and the modulated wave undergoes a **phase reversal** whenever $x(t)$ crosses zero. Full recovery of the message requires knowledge of these phase reversals, and could not be accomplished by an envelope detector. Suppressed-carrier DSB thus involves more than just "amplitude" modulation and, as we'll see in Sect. 4.5, calls for a more sophisticated demodulation process.

However, carrier suppression does put all of the average transmitted power into the information-bearing sidebands. Thus

$$S_T = 2P_{sb} = \tfrac{1}{2}A_c^2 S_x \tag{11}$$

which holds even when $x(t)$ includes a dc component. From Eqs. (11) and (8) we see that DSB makes better use of the total average power available from a given transmitter. Practical transmitters also impose a limit on the **peak envelope power** A_{max}^2.

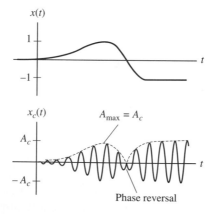

Figure 4.2–3   DSB waveforms.

We'll take account of this peak-power limitation by examining the ratio P_{sb}/A_{max}^2 under maximum modulation conditions. Using Eq. (11) with $A_{max} = A_c$ for DSB and using Eq. (7) with $A_{max} = 2A_c$ for AM, we find that

$$P_{sb}/A_{max}^2 = \begin{cases} S_x/4 & \text{DSB} \\ S_x/16 & \text{AM with } \mu = 1 \end{cases} \qquad [12]$$

Hence, if A_{max}^2 is fixed and other factors are equal, a DSB transmitter produces four times the sideband power of an AM transmitter.

The foregoing considerations suggest a trade-off between power efficiency and demodulation methods.

> DSB conserves power but requires complicated demodulation circuitry, whereas AM requires increased power to permit simple envelope detection.

EXAMPLE 4.2–1

Consider a radio transmitter rated for $S_T \leq 3$ kW and $A_{max}^2 \leq 8$ kW. Let the modulating signal be a tone with $A_m = 1$ so $S_x = A_m^2/2 = \frac{1}{2}$. If the modulation is DSB, the maximum possible power per sideband equals the lesser of the two values determined from Eqs. (11) and (12). Thus

$$P_{sb} = \tfrac{1}{2}S_T \leq 1.5 \text{ kW} \qquad P_{sb} = \tfrac{1}{8}A_{max}^2 \leq 1.0 \text{ kW}$$

which gives the upper limit $P_{sb} = 1.0$ kW.

If the modulation is AM with $\mu = 1$, then Eq. (12) requires that $P_{sb} = A_{max}^2/32 \leq 0.25$ kW. To check on the average-power limitation, we note from Eq. (7) that $P_{sb} = P_c/4$ so $S_T = P_c + 2P_{sb} = 6P_{sb}$ and $P_{sb} = S_T/6 \leq 0.5$ kW. Hence, the peak power limit again dominates and the maximum sideband power is $P_{sb} = 0.25$ kW. Since transmission range is proportional to P_{sb}, the AM path length would be only 25 percent of the DSB path length with the same transmitter.

EXERCISE 4.2–1

Let the modulating signal be a *square wave* that switches periodically between $x(t) = +1$ and $x(t) = -1$. Sketch $x_c(t)$ when the modulation is AM with $\mu = 0.5$, AM with $\mu = 1$, and DSB. Indicate the envelopes by dashed lines.

EXERCISE 4.2–2

Suppose a voice signal has $|x(t)|_{max} = 1$ and $S_x = 1/5$. Calculate the values of S_T and A_{max}^2 needed to get $P_{sb} = 10$ W for DSB and for AM with $\mu = 1$.

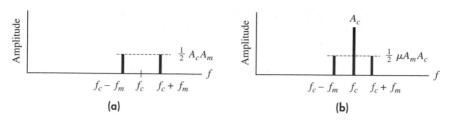

Figure 4.2–4   Line spectra for tone modulation. (a) DSB; (b) AM.

Tone Modulation and Phasor Analysis

Setting $x(t) = A_m \cos \omega_m t$ in Eq. (9) gives the tone-modulated DSB waveform

$$x_c(t) = A_c A_m \cos \omega_m t \cos \omega_c t \qquad \text{[13a]}$$

$$= \frac{A_c A_m}{2} \cos (\omega_c - \omega_m)t + \frac{A_c A_m}{2} \cos (\omega_c + \omega_m)t$$

where we have used the trigonometric expansion for the product of cosines. Similar expansion of Eq. (2) yields the tone-modulated AM wave

$$x_c(t) = A_c \cos \omega_c t + \frac{A_c \mu A_m}{2} \cos (\omega_c - \omega_m)t + \frac{A_c \mu A_m}{2} \cos (\omega_c + \omega_m)t \quad \text{[13b]}$$

Figure 4.2–4 shows the positive-frequency line spectra obtained from Eqs. (13a) and (13b).

It follows from Fig. 4.2–4 that tone modulated DSB or AM can be viewed as a sum of ordinary *phasors*, one for each spectral line. This viewpoint prompts the use of phasor analysis to find the envelope-and-phase or quadrature-carrier terms. Phasor analysis is especially helpful for studying the effects of transmission distortion, interference, and so on, as demonstrated in the example below.

Let's take the case of tone-modulated AM with $\mu A_m = \frac{2}{3}$ for convenience. The phasor diagram is constructed in Fig. 4.2–5a by adding the sideband phasors to the tip of the horizontal carrier phasor. Since the carrier frequency is f_c, the sideband phasors at $f_c \pm f_m$ rotate with speeds of $\pm f_m$ relative to the carrier phasor. The resultant of the sideband phasors is seen to be colinear with the carrier, and the phasor sum equals the envelope $A_c(1 + \frac{2}{3} \cos \omega_m t)$.

But suppose a transmission channel completely removes the lower sideband, so we get the diagram in Fig. 4.2–5b. Now the envelope becomes

$$A(t) = [(A_c + \tfrac{1}{3}A_c \cos \omega_m t)^2 + (\tfrac{1}{3}A_c \sin \omega_m t)^2]^{1/2}$$

$$= A_c \sqrt{\tfrac{10}{9} + \tfrac{2}{3} \cos \omega_m t}$$

EXAMPLE 4.2–2

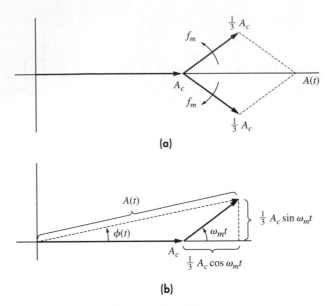

(a)

(b)

Figure 4.2–5   Phasor diagrams for Example 4.2–2.

from which the *envelope distortion* can be determined. Also note that the transmission amplitude distortion has produced a *time-varying phase* $\phi(t)$.

EXERCISE 4.2–3   Draw the phasor diagram for tone-modulated DSB with $A_m = 1$. Then find $A(t)$ and $\phi(t)$ when the amplitude of the lower sideband is cut in half.

4.3   MODULATORS AND TRANSMITTERS

The sidebands of an AM or DSB signal contain new frequencies that were not present in the carrier or message. The modulator must therefore be a **time-varying** or **nonlinear** system, because LTI systems never produce new frequency components. This section describes the operating principles of modulators and transmitters that employ product, square-law, or switching devices. Detailed circuit designs are given in the references cited in the Supplementary Reading.

Product Modulators

Figure 4.3–1*a* is the block diagram of a **product modulator** for AM based on the equation $x_c(t) = A_c \cos \omega_c t + \mu x(t) A_c \cos \omega_c t$. The schematic diagram in Fig. 4.3–1*b* implements this modulator with an analog multiplier and an op-amp summer. Of course, a DSB product modulator needs only the multiplier to produce

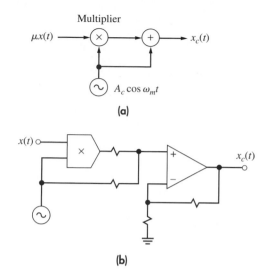

(a)

(b)

Figure 4.3–1   (a) Product modulator for AM; (b) schematic diagram with analog multiplier.

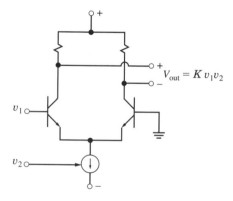

Figure 4.3–2   Circuit for variable transconductance multiplier.

$x_c(t) = x(t)A_c \cos \omega_c t$. In either case, the crucial operation is *multiplying* two analog signals.

Analog multiplication can be carried out electronically in a number of different
ways. One popular integrated-circuit design is the **variable transconductance multiplier** illustrated by Fig. 4.3–2. Here, input voltage v_1 is applied to a differential
amplifier whose gain depends on the transconductance of the transistors which, in
turn, varies with the total emitter current. Input v_2 controls the emitter current by
means of a voltage-to-current converter, so the differential output equals Kv_1v_2.
Other circuits achieve multiplication directly with **Hall-effect** devices, or indirectly

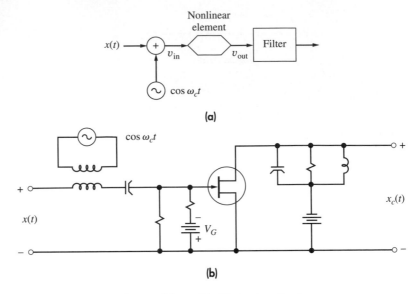

Figure 4.3–3   (a) Square-law modulator; (b) FET circuit realization.

with **log** and **antilog amplifiers** arranged to produce antilog (log v_1 + log v_2) = $v_1 v_2$. However, most analog multipliers are limited to low power levels and relatively low frequencies.

Square-Law and Balanced Modulators

Signal multiplication at higher frequencies can be accomplished by the **square-law modulator** diagrammed in Fig. 4.3–3a. The circuit realization in Fig. 4.3–3b uses a field-effect transistor as the nonlinear element and a parallel RLC circuit as the filter. We assume the nonlinear element approximates the square-law transfer curve

$$v_{out} = a_1 v_{in} + a_2 v_{in}^2$$

Thus, with $v_{in}(t) = x(t) + \cos \omega_c t$,

$$v_{out}(t) = a_1 x(t) + a_2 x^2(t) + a_2 \cos^2 \omega_c t + a_1 \left[1 + \frac{2a_2}{a_1} x(t) \right] \cos \omega_c t \qquad [1]$$

The last term is the desired AM wave, with $A_c = a_1$ and $\mu = 2a_2/a_1$, provided it can be separated from the rest.

As to the feasibility of separation, Fig. 4.3–4 shows the spectrum $V_{out}(f) = \mathscr{F}[v_{out}(t)]$ taking $X(f)$ as in Fig. 4.1–1. Note that the $x^2(t)$ term in Eq. (1) becomes $X * X(f)$, which is bandlimited in $2W$. Therefore, if $f_c > 3W$, there is no spectral overlapping and the required separation can be accomplished by a bandpass

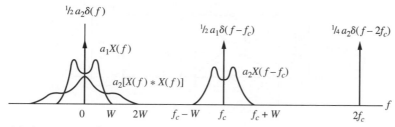

Figure 4.3–4   Spectral components in Eq. (1).

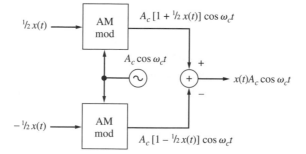

Figure 4.3–5
Balanced
modulator.

filter of bandwidth $B_T = 2W$ centered at f_c. Also note that the carrier-frequency impulse disappears and we have a DSB wave if $a_1 = 0$—corresponding to the *perfect* square-law curve $v_{out} = a_2 v_{in}^2$.

Unfortunately, perfect square-law devices are rare, so high-frequency DSB is obtained in practice using *two* AM modulators arranged in a balanced configuration to cancel out the carrier. Figure 4.3–5 shows such a **balanced modulator** in block-diagram form. Assuming the AM modulators are identical, save for the reversed sign of one input, the outputs are $A_c[1 + \frac{1}{2}x(t)]$ cos $\omega_c t$ and $A_c[1 - \frac{1}{2}x(t)]$ cos $\omega_c t$. Subtracting one from the other yields $x_c(t) = x(t)A_c$ cos $\omega_c t$, as required. Hence, a balanced modulator is a multiplier. You should observe that if the message has a *dc term*, that component is *not* canceled out in the modulator, even though it appears at the carrier frequency in the modulated wave.

Another modulator that is commonly used for generating DSB signals is the **ring modulator** shown in Fig. 4.3–6. A square-wave carrier $c(t)$ with frequency f_c causes the diodes to switch on and off. When $c(t) > 0$, the top and bottom diodes are switched on, while the two inner diodes in the cross-arm section are off. In this case, $v_{out} = x(t)$. Conversely, when $c(t) < 0$, the inner diodes are switched on and the top and bottom diodes are off, resulting in $v_{out} = -x(t)$. Functionally, the ring modulator can be thought of as multiplying $x(t)$ and $c(t)$. However because $c(t)$ is a periodic function, it can be represented by a Fourier series expansion. Thus

$$v_{out}(t) = \frac{4}{\pi} x(t) \cos \omega_c t - \frac{4}{3\pi} x(t) \cos 3\omega_c t + \frac{4}{5\pi} x(t) \cos 5\omega_c t - \cdots$$

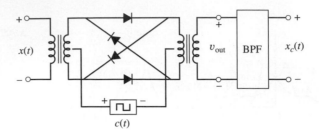

Figure 4.3–6   Ring modulator.

Observe that the DSB signal can be obtained by passing $v_{out}(t)$ through a bandpass filter having bandwidth $2W$ centered at f_c. This modulator is often referred to as a **double-balanced** modulator since it is balanced with respect to both $x(t)$ and $c(t)$.

A balanced modulator using switching circuits is discussed in Chap. 6 under the heading of **bipolar choppers.** Other circuit realizations can be found in the literature.

EXERCISE 4.3–1   Suppose the AM modulators in Fig. 4.3–5 are constructed with identical nonlinear elements having $v_{out} = a_1 v_{in} + a_2 v_{in}^2 + a_3 v_{in}^3$. Take $v_{in} = \pm x(t) + A_c \cos \omega_c t$ and show that the AM signals have second-harmonic distortion but, nonetheless, the final output is undistorted DSB.

Switching Modulators

In view of the heavy filtering required, square-law modulators are used primarily for *low-level* modulation, i.e., at power levels lower than the transmitted value. Substantial linear amplification is then necessary to bring the power up to S_T. But RF power amplifiers of the required linearity are not without problems of their own, and it often is better to employ *high-level* modulation if S_T is to be large.

Efficient high-level modulators are arranged so that undesired modulation products never fully develop and need not be filtered out. This is usually accomplished with the aid of a *switching* device, whose detailed analysis is postponed to Chap. 6. However, the basic operation of the supply-voltage modulated class C amplifier is readily understood from its idealized equivalent circuit and waveforms in Fig. 4.3–7.

The active device, typically a transistor, serves as a switch driven at the carrier frequency, closing briefly every $1/f_c$ seconds. The RLC load, called a **tank** circuit, is tuned to resonate at f_c, so the switching action causes the tank circuit to "ring" sinusoidally. The steady-state load voltage in absence of modulation is then $v(t) = V \cos \omega_c t$. Adding the message to the supply voltage, say via transformer, gives $v(t) = [V + Nx(t)] \cos \omega_c t$, where N is the transformer turns ratio. If V and N are correctly proportioned, the desired modulation has been accomplished without appreciable generation of undesired components.

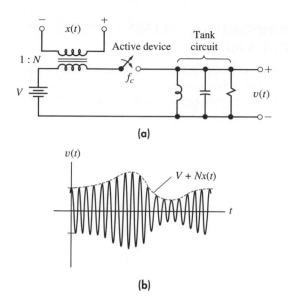

(a)

(b)

Figure 4.3–7 Class C amplifier with supply-voltage modulation. (a) Equivalent circuit; (b) output waveform.

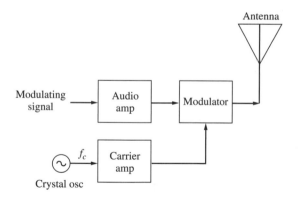

Figure 4.3–8 AM transmitter with high-level modulation.

A complete AM transmitter is diagrammed in Fig. 4.3–8 for the case of high-level modulation. The carrier wave is generated by a crystal-controlled oscillator to ensure stability of the carrier frequency. Because high-level modulation demands husky input signals, both the carrier and message are amplified before modulation. The modulated signal is then delivered directly to the antenna.

4.4   SUPPRESSED-SIDEBAND AMPLITUDE MODULATION

Conventional amplitude modulation is wasteful of both transmission power and bandwidth. Suppressing the carrier reduces the transmission power. Suppressing one sideband, in whole or part, reduces transmission bandwidth and leads to single-sideband modulation (SSB) or vestigial-sideband modulation (VSB) discussed in this section.

SSB Signals and Spectra

The upper and lower sidebands of DSB are uniquely related by symmetry about the carrier frequency, so either one contains *all* the message information. Hence, transmission bandwidth can be cut in half if one sideband is suppressed along with the carrier.

Figure 4.4–1*a* presents a conceptual approach to single-sideband modulation. Here, the DSB signal from a balanced modulator is applied to a sideband filter that suppresses one sideband. If the filter removes the lower sideband, the output spectrum

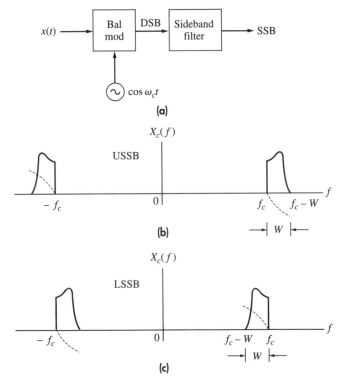

Figure 4.4–1   Single-sideband modulation. (*a*) Modulator; (*b*) USSB spectrum; (*c*) LSSB spectrum.

$X_c(f)$ consists of the upper sideband alone, as illustrated by Fig. 4.4–1*b*. We'll label this a USSB spectrum to distinguish it from the LSSB spectrum containing just the lower sideband, as illustrated by Fig. 4.4–1*c*. The resulting signal in either case has

$$B_T = W \qquad S_T = P_{sb} = \tfrac{1}{4}A_c^2 S_x \qquad\qquad [1]$$

which follow directly from our DSB results.

Although SSB is readily visualized in the frequency domain, the time-domain description is not immediately obvious—save for the special case of tone modulation. By referring back to the DSB line spectrum in Fig. 4.4–4*a*, we see that removing one sideband line leaves only the other line. Hence,

$$x_c(t) = \tfrac{1}{2}A_c A_m \cos(\omega_c \pm \omega_m)t \qquad\qquad [2]$$

in which the upper sign stands for USSB and the lower for LSSB, a convention employed hereafter. Note that the frequency of a tone-modulated SSB wave is offset from f_c by $\pm f_m$ and the envelope is a *constant* proportional to A_m. Obviously, envelope detection won't work for SSB.

To analyze SSB with an arbitrary message $x(t)$, we'll draw upon the fact that the sideband filter in Fig. 4.4–1*a* is a *bandpass* system with a bandpass DSB input $x_{bp}(t) = A_c x(t) \cos \omega_c t$ and a bandpass SSB output $y_{bp}(t) = x_c(t)$. Hence, we'll find $x_c(t)$ by applying the *equivalent lowpass method* from Sect. 4.1. Since $x_{bp}(t)$ has no quadrature component, the lowpass equivalent input is simply

$$x_{\ell p}(t) = \tfrac{1}{2}A_c x(t) \qquad X_{\ell p}(f) = \tfrac{1}{2}A_c X(f)$$

The bandpass filter transfer function for USSB is plotted in Fig. 4.4–2*a* along with the equivalent lowpass function

$$H_{\ell p}(f) = H_{bp}(f + f_c)u(f + f_c) = u(f) - u(f - W)$$

The corresponding transfer functions for LSSB are plotted in Fig. 4.4–2*b*, where

$$H_{\ell p}(f) = u(f + W) - u(f)$$

Both lowpass transfer functions can be represented by

$$H_{\ell p}(f) = \tfrac{1}{2}(1 \pm \operatorname{sgn} f) \qquad |f| \le W \qquad\qquad [3]$$

You should confirm for yourself that this rather strange expression does include both parts of Fig. 4.4–2.

Multiplying $H_{\ell p}(f)$ and $X_{\ell p}(f)$ yields the lowpass equivalent spectrum for either USSB or LSSB, namely

$$Y_{\ell p}(f) = \tfrac{1}{4}A_c(1 \pm \operatorname{sgn} f)X(f) = \tfrac{1}{4}A_c[X(f) \pm (\operatorname{sgn} f)X(f)]$$

Now recall that $(-j \operatorname{sgn} f)X(f) = \mathscr{F}[\hat{x}(t)]$, where $\hat{x}(t)$ is the *Hilbert transform* of $x(t)$ defined in Sect. 3.5. Therefore, $\mathscr{F}^{-1}[(\operatorname{sgn} f)X(f)] = j\hat{x}(t)$ and

$$y_{\ell p}(t) = \tfrac{1}{4}A_c[x(t) \pm j\hat{x}(t)]$$

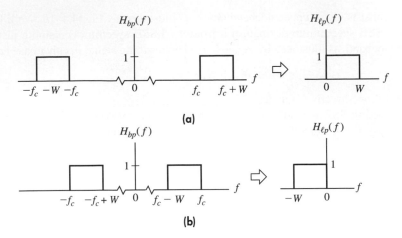

Figure 4.4–2   Ideal sideband filters and lowpass equivalents. (a) USSB; (b) LSSB.

Finally, we perform the lowpass-to-bandpass transformation $x_c(t) = y_{bp}(t) = 2 \, \text{Re}[y_{\ell p}(t) e^{j\omega_c t}]$ to obtain

$$x_c(t) = \tfrac{1}{2} A_c [x(t) \cos \omega_c t \mp \hat{x}(t) \sin \omega_c t] \qquad [4]$$

This is our desired result for the SSB waveform in terms of an arbitrary message $x(t)$.

Closer examination reveals that Eq. (4) has the form of a quadrature-carrier expression. Hence, the in-phase and quadrature components are

$$x_{ci}(t) = \tfrac{1}{2} A_c x(t) \qquad x_{cq}(t) = \pm \tfrac{1}{2} A_c \hat{x}(t)$$

while the SSB envelope is

$$A(t) = \tfrac{1}{2} A_c \sqrt{x^2(t) + \hat{x}^2(t)} \qquad [5]$$

The complexity of Eqs. (4) and (5) makes it a difficult task to sketch SSB waveforms or to determine the peak envelope power. Instead, we must infer time-domain properties from simplified cases such as tone modulation or pulse modulation.

EXAMPLE 4.4–1   **SSB with Pulse Modulation**

Whenever the SSB modulating signal has abrupt transitions, the Hilbert transform $\hat{x}(t)$ contains sharp peaks. These peaks then appear in the envelope $A(t)$, giving rise to the effect known as envelope *horns*. To demonstrate this effect, let's take the rectangular pulse $x(t) = u(t) - u(t - \tau)$ so we can use $\hat{x}(t)$ found in Example 3.5–2. The resulting SSB envelope plotted in Fig. 4.4–3 exhibits infinite peaks at $t = 0$ and $t = \tau$, the instants when $x(t)$ has stepwise discontinuities. Clearly, a transmitter

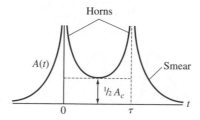

Figure 4.4–3   Envelope of SSB with pulse modulation.

couldn't handle the peak envelope power needed for these infinite horns. Also note the *smears* in $A(t)$ before and after each peak.

    We thus conclude that

> SSB is not appropriate for pulse transmission, digital data, or similar applications, and more suitable modulating signals (such as audio waveforms) should still be lowpass filtered before modulation in order to smooth out any abrupt transitions that might cause excessive horns or smearing.

Show that Eqs. (4) and (5) agree with Eq. (2) when $x(t) = A_m \cos \omega_m t$ so $\hat{x}(t) = A_m \sin \omega_m t$.

EXERCISE 4.4–1

SSB Generation

Our conceptual SSB generation system (Fig. 4.4–1a) calls for the ideal filter functions in Fig. 4.4–2. But a perfect cutoff at $f = f_c$ cannot be synthesized, so a real sideband filter will either pass a portion of the undesired sideband or attenuate a portion of the desired sideband. (Doing both is tantamount to vestigial-sideband modulation.) Fortunately, many modulating signals of practical interest have little or no low-frequency content, their spectra having "holes" at zero frequency as shown in Fig. 4.4–4a. Such spectra are typical of audio signals (voice and music), for example. After translation by the balanced modulator, the zero-frequency hole appears as a vacant space centered about the carrier frequency into which the transition region of a practical sideband filter can be fitted. Figure 4.4–4b illustrates this point.

    As a rule of thumb, the width 2β of the transition region cannot be much smaller than 1 percent of the nominal cutoff frequency, which imposes the limit $f_{co} < 200\beta$. Since 2β is constrained by the width of the spectral hole and f_{co} should equal f_c, it may not be possible to obtain a sufficiently high carrier frequency with a given message spectrum. For these cases the modulation process can be carried out in two (or more) steps using the system in Fig. 4.4–5 (see Prob. 4.4–5).

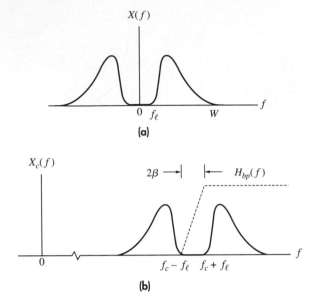

Figure 4.4–4   (*a*) Message spectrum with zero-frequency hole; (*b*) practical sideband filter.

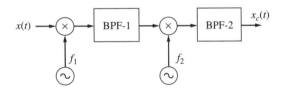

Figure 4.4–5   Two-step SSB generation.

Another method for SSB generation is based on writing Eq. (4) in the form

$$x_c(t) = \frac{A_c}{2} x(t) \cos \omega_c t \pm \frac{A_c}{2} \hat{x}(t) \cos (\omega_c t - 90°) \qquad [6]$$

This expression suggests that an SSB signal consists of *two* DSB waveforms with quadrature carriers and modulating signals $x(t)$ and $\hat{x}(t)$. Figure 4.4–6 diagrams a system that implements Eq. (6) and produces either USSB or LSSB, depending upon the sign at the summer. This system, known as the **phase-shift method,** bypasses the need for sideband filters. Instead, the DSB sidebands are phased such that they cancel out on one side of f_c and add on the other side to create a single-sideband output.

However, the quadrature phase shifter $H_Q(f)$ is itself an unrealizable network that can only be approximated — usually with the help of additional but identical phase networks in both branches of Fig. 4.4–6. Approximation imperfections gener-

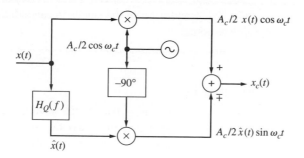

Figure 4.4–6   Phase-shift method for SSB generation.

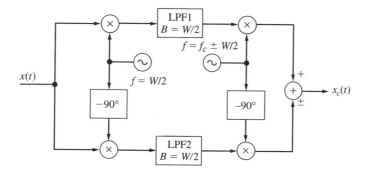

Figure 4.4–7 Weaver's SSB modulator.

ally cause low-frequency signal distortion, and the phase-shift system works best with message spectra of the type in Fig. 4.4–4a. A third method for SSB generation that avoids both sideband filters and quadrature phase shifters is considered in Example 4.4–2.

Weaver's SSB Modulator   EXAMPLE 4.4–2

Consider the modulator in Fig. 4.4–7 taking $x(t) = \cos 2\pi f_m t$ with $f_m < W$. Then $x_c(t) = v_1 \pm v_2$ where v_1 is the signal from the upper part of the loop and v_2 is from the lower part. Taking these separately, the input to the upper LPF is $\cos 2\pi f_m t \cos 2\pi \frac{W}{2} t$. The output of LPF1 is multiplied by $\cos 2\pi (f_c \pm \frac{W}{2})t$, resulting in $v_1 = \frac{1}{4}[\cos 2\pi(f_c \pm \frac{W}{2} - \frac{W}{2} + f_m)t + \cos 2\pi(f_c \pm \frac{W}{2} + \frac{W}{2} - f_m)t]$. The input to the lower LPF is $\cos 2\pi f_m t \sin 2\pi \frac{W}{2} t$. The output of LPF2 is multiplied by $\sin 2\pi(f_c \pm \frac{W}{2})t$, resulting in $v_2 = \frac{1}{4}[\cos 2\pi(f_c \pm \frac{W}{2} - \frac{W}{2} + f_m)t - \cos 2\pi(f_c \pm \frac{W}{2} + \frac{W}{2} - f_m)t]$. Taking the upper signs, $x_c(t) = 2 \times \frac{1}{4} \cos 2\pi(f_c + \frac{W}{2} - \frac{W}{2} + f_m)t = \frac{1}{2} \cos (\omega_c + \omega_m)t$, which corresponds to USSB. Similarly, we achieve LSSB by taking the lower signs, resulting in $x_c(t) = \frac{1}{2} \cos (\omega_c - \omega_m)t$.

EXERCISE 4.4–2   Take $x(t) = \cos \omega_m t$ in Fig. 4.4–6 and confirm the sideband cancellation by sketching line spectra at appropriate points.

VSB Signals and Spectra★

Consider a modulating signal of very large bandwidth having significant low-frequency content. Principal examples are television video, facsimile, and high-speed data signals. Bandwidth conservation argues for the use of SSB, but practical SSB systems have poor low-frequency response. On the other hand, DSB works quite well for low message frequencies but the transmission bandwidth is twice that of SSB. Clearly, a compromise modulation scheme is desired; that compromise is VSB.

VSB is derived by filtering DSB (or AM) in such a fashion that one sideband is passed almost completely while just a trace, or **vestige,** of the other sideband is included. The key to VSB is the sideband filter, a typical transfer function being that of Fig. 4.4–8a. While the exact shape of the response is not crucial, it must have odd symmetry about the carrier frequency and a relative response of 1/2 at f_c. Therefore, taking the upper sideband case, we have

$$H(f) = u(f - f_c) - H_\beta(f - f_c) \qquad f > 0 \qquad\qquad [7a]$$

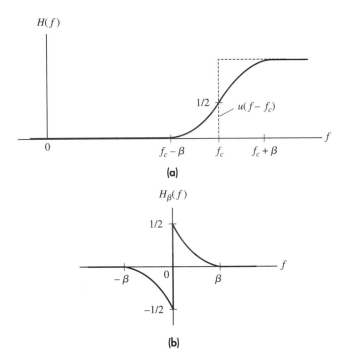

(a)

(b)

Figure 4.4–8   VSB filter characteristics.

where

$$H_\beta(-f) = -H_\beta(f) \qquad \text{and} \qquad H_\beta(f) = 0 \qquad |f| > \beta \qquad \text{[7b]}$$

as shown in Fig. 4.4–8b.

The VSB filter is thus a *practical* sideband filter with transition width 2β. Because the width of the partial sideband is one-half the filter transition width, the transmission bandwidth is

$$B_T = W + \beta \approx W \qquad \text{[8]}$$

However, in some applications the vestigial filter symmetry is achieved primarily at the receiver, so the transmission bandwidth must be slightly larger than $W + \beta$.

When $\beta \ll W$, which is usually true, the VSB spectrum looks essentially like an SSB spectrum. The similarity also holds in the time domain, and a VSB waveform can be expressed as a modification of Eq. (4). Specifically,

$$x_c(t) = \tfrac{1}{2}A_c[x(t) \cos \omega_c t - x_q(t) \sin \omega_c t] \qquad \text{[9a]}$$

where $x_q(t)$ is the quadrature message component defined by

$$x_q(t) = \hat{x}(t) + x_\beta(t) \qquad \text{[9b]}$$

with

$$x_\beta(t) = j2 \int_{-\beta}^{\beta} H_\beta(f)X(f)e^{j\omega t}\, df \qquad \text{[9c]}$$

If $\beta \ll W$, VSB approximates SSB and $x_\beta(t) \approx 0$; conversely, for large β, VSB approximates DSB and $\hat{x}(t) + x_\beta(t) \approx 0$. The transmitted power S_T is not easy to determine exactly, but is bounded by

$$\tfrac{1}{4}A_c^2 S_x \leq S_T \leq \tfrac{1}{2}A_c^2 S_x \qquad \text{[10]}$$

depending on the vestige width β.

Finally, suppose an AM wave is applied to a vestigial sideband filter. This modulation scheme, termed VSB plus carrier (VSB + C), is used for television video transmission. The unsuppressed carrier allows for *envelope detection*, as in AM, while retaining the bandwidth conservation of suppressed sideband. Distortionless envelope modulation actually requires symmetric sidebands, but VSB + C can deliver a fair approximation.

To analyze the envelope of VSB + C, we incorporate a carrier term and modulation index μ in Eq. (9) which becomes

$$x_c(t) = A_c\{[1 + \mu x(t)] \cos \omega_c t - \mu x_q(t) \sin \omega_c t\} \qquad \text{[11]}$$

The in-phase and quadrature components are then

$$x_{ci}(t) = A_c[1 + \mu x(t)] \qquad x_{cq}(t) = A_c \mu x_q(t)$$

so the envelope is $A(t) = [x_{ci}^2(t) + x_{cq}^2(t)]^{1/2}$ or

$$A(t) = A_c[1 + \mu x(t)]\left\{1 + \left[\frac{\mu x_q(t)}{1 + \mu x(t)}\right]^2\right\}^{1/2} \qquad [12]$$

Hence, if μ is not too large and β not too small, then $|\mu x_q(t)| \ll 1$ and

$$A(t) \approx A_c[1 + \mu x(t)]$$

as desired. Empirical studies with typical signals are needed to find values for μ and β that provide a suitable compromise between the conflicting requirements of distortionless envelope modulation, power efficiency, and bandwidth conservation.

4.5   FREQUENCY CONVERSION AND DEMODULATION

Linear CW modulation—be it AM, DSB, SSB, or VSB—produces upward translation of the message spectrum. **Demodulation** therefore implies *downward frequency translation* in order to recover the message from the modulated wave. Demodulators that perform this operation fall into the two broad categories of **synchronous detectors** and **envelope detectors.**

Frequency translation, or **conversion,** is also used to shift a modulated signal to a new carrier frequency (up or down) for amplification or other processing. Thus, translation is a fundamental concept in linear modulation systems and includes modulation and detection as special cases. Before examining detectors, we'll look briefly at the general process of frequency conversion.

Frequency Conversion

Frequency conversion starts with multiplication by a sinusoid. Consider, for example, the DSB wave $x(t) \cos \omega_1 t$. Multiplying by $\cos \omega_2 t$, we get

$$x(t) \cos \omega_1 t \cos \omega_2 t = \tfrac{1}{2}x(t) \cos (\omega_1 + \omega_2)t + \tfrac{1}{2}x(t) \cos (\omega_1 - \omega_2)t \qquad [1]$$

The product consists of the *sum* and *difference frequencies*, $f_1 + f_2$ and $|f_1 - f_2|$, each modulated by *x(t)*. We write $|f_1 - f_2|$ for clarity, since $\cos (\omega_2 - \omega_1)t = \cos (\omega_1 - \omega_2)t$. Assuming $f_2 \neq f_1$, multiplication has translated the signal spectra to *two* new carrier frequencies. With appropriate filtering, the signal is up-converted or down-converted. Devices that carry out this operation are called **frequency converters** or **mixers.** The operation itself is termed **heterodyning** or **mixing.**

Figure 4.5–1 diagrams the essential components of a frequency converter. Implementation of the multiplier follows the same line as the modulator circuits discussed in Sect. 4.3. Converter applications include beat-frequency oscillators, regenerative frequency dividers, speech scramblers, and spectrum analyzers, in addition to their roles in transmitters and receivers.

EXAMPLE 4.5–1   Figure 4.5–2 represents a simplified **transponder** in a satellite relay that provides two-way communication between two ground stations. Different carrier frequencies,

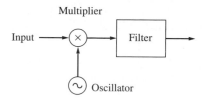

Figure 4.5–1   Frequency converter.

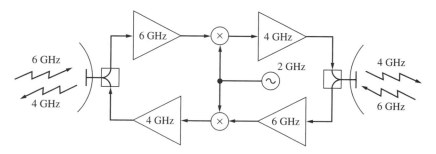

Figure 4.5–2   Satellite transponder with frequency conversion.

6 GHz and 4 GHz, are used on the uplink and downlink to prevent self-oscillation due to positive feedback from the transmitting side to the receiving side. A frequency converter translates the spectrum of the amplified uplink signal to the passband of the downlink amplifier.

Sketch the spectrum of Eq. (1) for $f_2 < f_1$, $f_2 = f_1$, and $f_2 > f_1$, taking $X(f)$ as in Fig. 4.1–1.   **EXERCISE 4.5–1**

Synchronous Detection

All types of linear modulation can be detected by the *product* demodulator of Fig. 4.5–3. The incoming signal is first multiplied with a locally generated sinusoid and then lowpass-filtered, the filter bandwidth being the same as the message bandwidth W or somewhat larger. It is assumed that the local oscillator (LO) is *exactly synchronized* with the carrier, in both phase and frequency, accounting for the name **synchronous** or **coherent detection.**

For purposes of analysis, we'll write the input signal in the generalized form

$$x_c(t) = [K_c + K_\mu x(t)] \cos \omega_c t - K_\mu x_q(t) \sin \omega_c t \qquad [2]$$

which can represent any type of linear modulation with proper identification of K_c, K_μ, and $x_q(t)$—i.e., take $K_c = 0$ for suppressed carrier, $x_q(t) = 0$ for double sideband, and so on. The filter input is thus the product

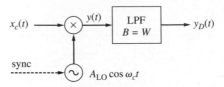

Figure 4.5–3   Synchronous product detection.

$$x_c(t)A_{LO} \cos \omega_c t$$

$$= \frac{A_{LO}}{2} \{[K_c + K_\mu x(t)] + [K_c + K_\mu x(t)] \cos 2\omega_c t - K_\mu x_q(t) \sin 2\omega_c t\}$$

Since $f_c > W$, the double-frequency terms are rejected by the lowpass filter, leaving only the leading term

$$y_D(t) = K_D[K_c + K_\mu x(t)] \tag{3}$$

where K_D is the detection constant. The DC component $K_D K_c$ corresponds to the translated carrier if present in the modulated wave. This can be removed from the output by a blocking capacitor or transformer—which also removes any DC term in $x(t)$ as well. With this minor qualification we can say that the message has been fully recovered from $x_c(t)$.

Although perfectly correct, the above manipulations fail to bring out what goes on in the demodulation of VSB. This is best seen in the frequency domain with the message spectrum taken to be constant over W (Fig. 4.5–4a) so the modulated spectrum takes the form of Fig. 4.5–4b. The downward-translated spectrum at the filter input will then be as shown in Fig. 4.5–4c. Again, high-frequency terms are elimi-

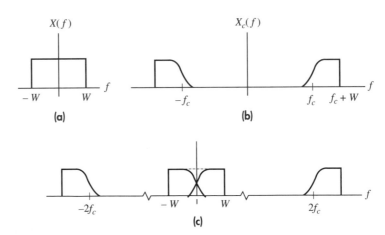

Figure 4.5–4   VSB spectra. (a) Message; (b) modulated signal; (c) frequency-translated signal before lowpass filtering.

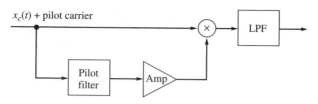

Figure 4.5–5   Homodyne detection.

nated by filtering, while the down-converted sidebands overlap around zero frequency. Recalling the symmetry property of the vestigial filter, we find that the portion removed from the upper sideband is exactly restored by the corresponding vestige of the lower sideband, so $X(f)$ has been reconstructed at the output and the detected signal is proportional to $x(t)$.

Theoretically, product demodulation borders on the trivial; in practice, it can be rather tricky. The crux of the problem is **synchronization**—synchronizing an oscillator to a sinusoid that is not even present in the incoming signal if carrier is suppressed. To facilitate the matter, suppressed-carrier systems may have a small amount of carrier reinserted in $x_c(t)$ at the transmitter. This **pilot carrier** is picked off at the receiver by a narrow bandpass filter, amplified, and used in place of an LO. The system, shown in Fig. 4.5–5, is called **homodyne detection.** (Actually, the amplified pilot more often serves to synchronize a separate oscillator rather than being used directly.)

A variety of other techniques are possible for synchronization, including **phase-lock loops** (to be covered in Sect. 7.3) or the use of highly stable, crystal-controlled oscillators at transmitter and receiver. Nonetheless, some degree of asynchronism must be expected in synchronous detectors. It is therefore important to investigate the effects of phase and frequency drift in various applications. This we'll do for DSB and SSB in terms of tone modulation.

Let the local oscillator wave be $\cos(\omega_c t + \omega' t + \phi')$, where ω' and ϕ' represent slowly drifting frequency and phase errors compared to the carrier. For double sideband with tone modulation, the detected signal becomes

$$y_D(t) = K_D \cos \omega_m t \cos (\omega' t + \phi') \qquad \text{[4]}$$

$$= \begin{cases} \dfrac{K_D}{2} [\cos (\omega_m + \omega')t + \cos (\omega_m - \omega')t] & \phi' = 0 \\ K_D \cos \omega_m t \cos \phi' & \omega' = 0 \end{cases}$$

Similarly, for single sideband with $x_c(t) = \cos(\omega_c \pm \omega_m)t$, we get

$$y_D(t) = K_D \cos [\omega_m t \mp (\omega' t + \phi')] \qquad \text{[5]}$$

$$= \begin{cases} K_D \cos (\omega_m \mp \omega')t & \phi' = 0 \\ K_D \cos (\omega_m t \mp \phi') & \omega' = 0 \end{cases}$$

All of the foregoing expressions come from simple trigonometric expansions.

Clearly, in both DSB and SSB, a frequency drift that's not small compared to W will substantially alter the detected tone. The effect is more severe in DSB since a *pair* of tones, $f_m + f'$ and $f_m - f'$, is produced. If $f' \ll f_m$, this sounds like warbling or the beat note heard when two musical instruments play in unison but slightly out of tune. While only one tone is produced with SSB, this too can be disturbing, particularly for music transmission. To illustrate, the major triad chord consists of three notes whose frequencies are related as the integers 4, 5, and 6. Frequency error in detection shifts each note by the same absolute amount, destroying the harmonic relationship and giving the music an East Asian flavor. (Note that the effect is *not* like playing recorded music at the wrong speed, which preserves the frequency ratios.) For voice transmission, subjective listener tests have shown that frequency drifts of less than ± 10 Hz are tolerable, otherwise, everyone sounds rather like Donald Duck.

As to phase drift, again DSB is more sensitive, for if $\phi' = \pm 90°$ (LO and carrier in quadrature), the detected signal vanishes entirely. With slowly varying ϕ', we get an apparent **fading** effect. Phase drift in SSB appears as **delay distortion,** the extreme case being when $\phi' = \pm 90°$ and the demodulated signal becomes $\hat{x}(t)$. However, as was remarked before, the human ear can tolerate sizeable delay distortion, so phase drift is not so serious in voice-signal SSB systems.

To summarize,

> Phase and frequency synchronization requirements are rather modest for voice transmission via SSB. But in data, facsimile, and video systems with suppressed carrier, careful synchronization is a necessity. Consequently, television broadcasting employs VSB + C rather than suppressed-carrier VSB.

Envelope Detection

Very little was said earlier in Sect. 4.5 about synchronous demodulation of AM for the simple reason that it's almost never used. True, synchronous detectors work for AM, but so does an **envelope detector,** which is much simpler. Because the envelope of an AM wave has the same shape as the message, independent of carrier frequency and phase, demodulation can be accomplished by extracting the envelope with no worries about synchronization.

A simplified envelope detector and its waveforms are shown in Fig. 4.5–6, where the diode is assumed to be piecewise-linear. In absence of further circuitry, the voltage v would be just the half-rectified version of the input v_{in}. But $R_1 C_1$ acts as a lowpass filter, responding only to variations in the peaks of v_{in} provided that

$$W \ll \frac{1}{R_1 C_1} \ll f_c \qquad\qquad [6]$$

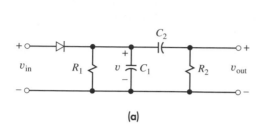

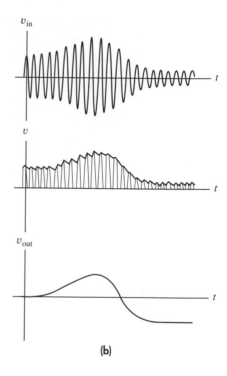

Figure 4.5–6 Envelope detection (*a*) Circuit; (*b*) waveforms.

Thus, as noted earlier, we need $f_c \gg W$ so the envelope is clearly defined. Under these conditions, C_1 discharges only slightly between carrier peaks, and v approximates the envelope of v_{in}. More sophisticated filtering produces further improvement if needed. Finally, $R_2 C_2$ acts as a dc block to remove the bias of the unmodulated carrier component. Since the dc block distorts low-frequency message components, conventional envelope detectors are inadequate for signals with important low-frequency content.

The voltage v may also be filtered to remove the *envelope* variations and produce a dc voltage proportional to the carrier amplitude. This voltage in turn is fed back to earlier stages of the receiver for **automatic volume control** (AVC) to compensate for fading. Despite the nonlinear element, Fig. 4.5–6 is termed a **linear envelope detector;** the output is linearly proportional to the input envelope. Power-law diodes can also be used, but then v will include terms of the form v_{in}^2, v_{in}^3, and so on, and there may be appreciable second-harmonic distortion unless $\mu \ll 1$.

Some DSB and SSB demodulators employ the method of **envelope reconstruction** diagrammed in Fig. 4.5–7. The addition of a large, locally generated carrier to the incoming signal reconstructs the envelope for recovery by an envelope detector. This method eliminates signal multiplication but does not get around the synchronization problem, for the local carrier must be as well synchronized as the LO in a product demodulator.

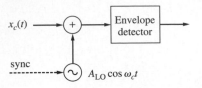

Figure 4.5–7   Envelope reconstruction for suppressed-carrier modulation.

EXERCISE 4.5–2   Let the input in Fig. 4.5–7 be SSB with tone modulation, and let the LO have a phase error ϕ' but no frequency error. Use a phasor diagram to obtain an expression for the resulting envelope. Then show that $A(t) \approx A_{\mathrm{LO}} + \frac{1}{2} A_c A_m \cos (\omega_m t \mp \phi')$ if $A_{\mathrm{LO}} \gg A_c A_m$.

4.6   PROBLEMS

4.1–1   Use a phasor diagram to obtain expressions for $v_i(t)$, $v_q(t)$, $A(t)$, and $\phi(t)$ when $v_{bp}(t) = v_1(t) \cos \omega_c t + v_2(t) \cos (\omega_c t + \alpha)$. Then simplify $A(t)$ and $\phi(t)$ assuming $|v_2(t)| \ll |v_1(t)|$.

4.1–2   Do Prob. 4.1–1 with $v_{bp}(t) = v_1(t) \cos (\omega_c - \omega_0)t + v_2(t) \cos (\omega_c + \omega_0)t$.

4.1–3   Let $v_i(t)$ and $v_q(t)$ in Eq. (7) be lowpass signals with energy E_i and E_q, respectively, and bandwidth $W < f_c$. (a) Use Eq. (17), Sect. 2.2, to prove that

$$\int_{-\infty}^{\infty} v_{bp}(t)dt = 0$$

(b) Now show that the bandpass signal energy equals $(E_i + E_q)/2$.

4.1–4*   Find $v_{\ell p}(t)$, $v_i(t)$ and $v_q(t)$ when $f_c = 1200$ Hz and

$$V_{bp}(f) = \begin{cases} 1 & 900 \leq |f| < 1300 \\ 0 & \text{otherwise} \end{cases}$$

4.1–5   Do Prob. 4.1–4 with

$$V_{bp}(f) = \begin{cases} 1 & 1100 \leq |f| < 1200 \\ 1/2 & 1200 \leq |f| < 1350 \\ 0 & \text{otherwise} \end{cases}$$

4.1–6   Let $v_{bp}(t) = 2z(t) \cos [(\omega_c \pm \omega_0)t + \alpha]$. Find $v_i(t)$ and $v_q(t)$ to obtain

$$v_{\ell p}(t) = z(t) \exp j(\pm \omega_0 t + \alpha)$$

4.1–7   Derive Eq. (17b) by obtaining expressions for f_ℓ and f_u from Eq. (17a).

4.1–8 Let $f = (1 + \delta)f_0$ in Eq. (17a) and assume that $|\delta| \ll 1$. Derive the handy approximation

$$H(f) \approx 1/[1 + j2Q(f - f_0)/f_0]$$

which holds for $f > 0$ and $|f - f_0| \ll f_0$.

4.1–9 A *stagger-tuned* bandpass system centered at $f = f_c$ has $H(f) = 2H_1(f)H_2(f)$, where $H_1(f)$ is given by Eq. (17a) with $f_0 = f_c - b$ and $Q = f_0/2b$ while $H_2(f)$ is given by Eq. (17a) with $f_0 = f_c + b$ and $Q = f_0/2b$. Use the approximation in Prob. 4.1–8 to plot $|H(f)|$ for $f_c - 2b < f < f_c + 2b$ and compare it with a simple tuned circuit having $f_0 = f_c$ and $B = 2\sqrt{2b}$.

4.1–10* Use lowpass *time-domain* analysis to find and sketch $y_{bp}(t)$ when $x_{bp}(t) = A \cos \omega_c t\, u(t)$ and $H_{bp}(f) = 1/[1 + j2(f - f_c)/B]$ for $f > 0$, which corresponds to the tuned-circuit approximation in Prob. 4.1–8.

4.1–11 Do Prob. 4.1–10 with $H_{bp}(f) = \Pi[(f - f_c)/B]e^{-j\omega t_d}$ for $f > 0$, which corresponds to an ideal BPF. *Hint:* See Eq. (9), Sect. 3.4.

4.1–12‡ The bandpass signal in Prob. 4.1–6 has $z(t) = 2u(t)$ and is applied to an ideal BPF with unit gain, zero time delay, and bandwidth B centered at f_c. Use lowpass frequency-domain analysis to obtain an approximation for the bandpass output signal when $B \ll f_0$.

4.1–13‡ Consider a BPF with bandwidth B centered at f_c, unit gain, and parabolic phase shift $\theta(f) = (f - f_c)^2/b$ for $f > 0$. Obtain a quadrature-carrier approximation for the output signal when $|b| \gg (B/2)^2$ and $x_{bp}(t) = z(t) \cos \omega_c t$, where $z(t)$ has a band-limited lowpass spectrum with $W \le \frac{B}{2}$.

4.2–1 Let $x(t) = \cos 2\pi f_m t\, u(t)$ with $f_m \ll f_c$. Sketch $x_c(t)$ and indicate the envelope when the modulation is AM with $\mu < 1$, AM with $\mu > 1$, and DSB. Identify locations where any phase reversals occur.

4.2–2 Do Prob. 4.2–1 with $x(t) = 0.5u(t) - 1.5u(t - T)$ with $T \gg 1/f_c$.

4.2–3* If $x(t) = \cos 200\pi t$, find B_T and S_T for the AM modulated signal assuming $A_c = 10$ and $\mu = 0.6$. Repeat for DSB transmission.

4.2–4 The signal $x(t) = \text{sinc}^2 40t$ is to be transmitted using AM with $\mu < 1$. Sketch the double-sided spectrum of $x_c(t)$ and find B_T.

4.2–5 Calculate the transmitted power of an AM wave with 100 percent tone modulation and peak envelope power 32 kW.

4.2–6 Consider a radio transmitter rated for 4 kW peak envelope power. Find the maximum allowable value of μ for AM with tone modulation and $S_T = 1$ kW.

4.2–7 The multitone modulating signal $x(t) = 3K(\cos 8\pi t + 2 \cos 20\pi t)$ is input to an AM transmitter with $\mu = 1$ and $f_c = 1000$. Find K so that $x(t)$ is properly normalized, draw the positive-frequency line spectrum of the modulated wave, and calculate the upper bound on $2P_{sb}/S_T$.

4.2–8   Do Prob. 4.2–7 with $x(t) = 2K(\cos 8\pi t + 1) \cos 20\pi t$.

4.2–9*   The signal $x(t) = 4 \sin \frac{\pi}{2} t$ is transmitted by DSB. What range of carrier frequencies can be used?

4.2–10   The signal in Prob. 4.2–9 is transmitted by AM with $\mu = 1$. Draw the phasor diagram. What is the minimum amplitude of the carrier such that phase reversals don't occur?

4.2–11   The signal $x(t) = \cos 2\pi 40t + \frac{1}{2} \cos 2\pi 90t$ is transmitted using DSB. Sketch the positive-frequency line spectrum and the phasor diagram.

4.3–1   The signal $x(t) = \frac{1}{2} \cos 2\pi 70t + \frac{1}{3} \cos 2\pi 120t$ is input to the square-law modulator system given in Fig. 4.3–3a (p. 160) with a carrier frequency of 10 kHz. Assume $v_{out} = a_1 v_{in} + a_2 v_{in}^2$. (a) Give the center frequency and bandwidth of the filter such that this system will produce a standard AM signal. (b) Determine values of a_1 and a_2 such that $A_c = 10$ and $\mu = \frac{1}{2}$.

4.3–2*   A modulation system with nonlinear elements produces the signal $x_c(t) = aK^2(v(t) + A \cos \omega_c t)^2 - b(v(t) - A \cos \omega_c t)^2$. If the carrier has frequency f_c and $v(t) = x(t)$, show that an appropriate choice of K produces DSB modulation without filtering. Draw a block diagram of the modulation system.

4.3–3   Find K and $v(t)$ so that the modulation system from Prob. 4.3–2 produces AM without filtering. Draw a block diagram of the modulation system.

4.3–4   A modulator similar to the one in Fig. 4.3–3a (p. 160) has a nonlinear element of the form $v_{out} = a_1 v_{in} + a_3 v_{in}^3$. Sketch $V_{out}(f)$ for the input signal in Fig. 4.1–1 (p. 143). Find the parameters of the oscillator and BPF to produce a DSB signal with carrier frequency f_c.

4.3–5   Design in block-diagram form an AM modulator using the nonlinear element from Prob. 4.3–4 and a frequency doubler. Carefully label all components and find a required condition on f_c in terms of W to realize this system.

4.3–6   Find the output signal in Fig. 4.3–5 (p. 161) when the AM modulators are **unbalanced,** so that one nonlinear element has $v_{out} = a_1 v_{in} + a_2 v_{in}^2 + a_3 v_{in}^3$ while the other has $v_{out} = b_1 v_{in} + b_2 v_{in}^2 + b_3 v_{in}^3$.

4.3–7*   The signal $x(t) = 20\text{sinc}^2 400t$ is input to the ring modulator in Fig. 4.3–6 (p. 162). Sketch the spectrum of v_{out} and find the range of values of f_c that can be used to transmit this signal.

4.4–1   Derive Eq. (4) from $y_{\ell p}(t)$.

4.4–2   Take the transform of Eq. (4) to obtain the SSB spectrum

$$X_c(f) = \frac{1}{4} A_c \{ [1 \pm \text{sgn}(f - f_c)] X(f - f_c) + [1 \mp \text{sgn}(f + f_c)] X(f + f_c) \}.$$

4.4–3   Confirm that the expression for $X_c(f)$ in Prob. 4.4–2 agrees with Figs. 4.4–1b and 4.4–1c (p. 164).

4.4–4   Find the SSB envelope when $x(t) = \cos \omega_m t + \frac{1}{9} \cos 3\omega_m t$ which approximates a triangular wave. Sketch $A(t)$ taking $A_c = 81$ and compare with $x(t)$.

4.4–5   The system in Fig. 4.4–5 produces USSB with $f_c = f_1 + f_2$ when the lower cutoff frequency of the first BPF equals f_1 and the lower cutoff frequency of the second BPF equals f_2. Demonstrate the system's operation by taking $X(f)$ as in Fig. 4.4–4a and sketching spectra at appropriate points. How should the system be modified to produce LSSB?

4.4–6   Suppose the system in Fig. 4.4–5 is designed for USSB as described in Prob. 4.4–5. Let $x(t)$ be a typical voice signal, so $X(f)$ has negligible content outside $200 < |f| < 3200$ Hz. Sketch the spectra at appropriate points to find the maximum permitted value of f_c when the transition regions of the BPFs must satisfy $2\beta \geq 0.01 f_{co}$.

4.4–7*   The signal $x(t) = \cos 2\pi 100t + 3 \cos 2\pi 200t + 2 \cos 2\pi 400t$ is input to an LSSB amplitude modulation system with a carrier frequency of 10 kHz. Sketch the double-sided spectrum of the transmitted signal. Find the transmitted power S_T and bandwidth B_T.

4.4–8   Draw the block diagram of a system that would generate the LSSB signal in Prob. 4.4–7, giving exact values for filter cutoff frequencies and oscillators. Make sure your filters meet the fractional bandwidth rule.

4.4–9   Suppose the carrier phase shift in Fig. 4.4–6 is actually $-90° + \delta$, where δ is a small angular error. Obtain approximate expressions for $x_c(t)$ and $A(t)$ at the output.

4.4–10   Obtain an approximate expression for $x_c(t)$ at the output in Fig. 4.4–6 when $x(t) = \cos \omega_m t$ and the quadrature phase shifter has $|H_Q(f_m)| = 1 - \epsilon$ and $\arg H_Q(f_m) = -90° + \delta$, where ϵ and δ are small errors. Write your answer as a sum of two sinusoids.

4.4–11   The tone signal $x(t) = A_m \cos 2\pi f_m t$ is input to a VSB + C modulator. The resulting transmitted signal is

$$x_c(t) = A_c \cos 2\pi f_c t + \frac{1}{2} a A_m A_c \cos [2\pi(f_c + f_m)t]$$
$$+ \frac{1}{2}(1 - a)A_m A_c \cos [2\pi(f_c - f_m)t].$$

Sketch the phasor diagram assuming $a > \frac{1}{2}$. Find the quadrature component $x_{cq}(t)$.

4.4–12*   Obtain an expression for VSB with tone modulation taking $f_m < \beta$ so the VSB filter has $H(f_c \pm f_m) = 0.5 \pm a$. Then show that $x_c(t)$ reduces to DSB when $a = 0$ or SSB when $a = \pm 0.5$.

4.4–13   Obtain an expression for VSB + C with tone modulation taking $f_m > \beta$. Construct the phasor diagram and find $A(t)$.

4.5–1   Given a bandpass amplifier centered at 66 MHz, design a television transponder that receives a signal on Channel 11 (199.25 MHz) and transmits it on Channel 4 (67.25 MHz). Use only one oscillator.

4.5–2   Do Prob. 4.5–1 with the received signal on Channel 44 (651.25 MHz) and the transmitted signal on Channel 22 (519.25 MHz).

4.5–3   The system in Fig. 4.4–5 becomes a *scrambler* when the first BPF passes only the upper sideband, the second oscillator frequency is $f_2 = f_1 + W$, and the second BPF is replaced by an LPF with $B = W$. Sketch the output spectrum taking $X(f)$ as in Fig. 4.4–4a, and explain why this output would be unintelligible when $x(t)$ is a voice signal. How can the output signal be unscrambled?

4.5–4   Take $x_c(t)$ as in Eq. (2) and find the output of a synchronous detector whose local oscillator produces $2 \cos (\omega_c t + \phi)$, where ϕ is a constant phase error. Then write separate answers for AM, DSB, SSB, and VSB by appropriate substitution of the modulation parameters.

4.5–5*   The transmitted signal in Prob. 4.4–11 is demodulated using envelope detection. Assuming $0 \le a \le 1$, what values of a minimize and maximize the distortion at the output of the envelope detector?

4.5–6   The signal $x(t) = 2 \cos 4\pi t$ is transmitted by DSB. Sketch the output signal if envelope detection is used for demodulation.

4.5–7   Suppose the DSB waveform from Prob. 4.5–6 is demodulated using a synchronous detector that has a square wave with a fundamental frequency of f_c as the local oscillator. Will the detector properly demodulate the signal? Will the same be true if periodic signals other than the square wave are substituted for the oscillator?

4.5–8   Sketch a half-rectified AM wave having tone modulation with $\mu A_m = 1$ and $f_m = W$. Use your sketch to determine upper and lower limits on the time constant $R_1 C_1$ of the envelope detector in Fig. 4.5–6. From these limits find the minimum practical value of f_c / W.

chapter

5

Exponential CW Modulation

CHAPTER OUTLINE

Two properties of *linear* CW modulation bear repetition at the outset of this chapter: the modulated spectrum is basically the translated message spectrum and the transmission bandwidth never exceeds twice the message bandwidth. A third property, derived in Chap. 10, is that the destination signal-to-noise ratio $(S/N)_D$ is no better than baseband transmission and can be improved only by increasing the transmitted power. **Exponential** modulation differs on all three counts.

In contrast to linear modulation, exponential modulation is a *nonlinear* process; therefore, it should come as no surprise that the modulated spectrum is not related in a simple fashion to the message spectrum. Moreover, it turns out that the transmission bandwidth is usually much greater than twice the message bandwidth. Compensating for the bandwidth liability is the fact that exponential modulation can provide increased signal-to-noise ratios without increased transmitted power. Exponential modulation thus allows you to *trade bandwidth for power* in the design of a communication system.

We begin our study of exponential modulation by defining the two basic types, **phase modulation** (PM) and **frequency modulation** (FM). We'll examine signals and spectra, investigate the transmission bandwidth and distortion problem, and describe typical hardware for generation and detection. The analysis of interference at the end of the chapter brings out the value of FM for radio broadcasting and sets the stage for our consideration of noise in Chap. 10.

OBJECTIVES

After studying this chapter and working the exercises, you should be able to do each of the following:

1.   Find the instantaneous phase and frequency of a signal with exponential modulation (Sect. 5.1).
2.   Construct the line spectrum and phasor diagram for FM or PM with tone modulation (Sect. 5.1).
3.   Estimate the bandwidth required for FM or PM transmission (Sect. 5.2).
4.   Identify the effects of distortion, limiting, and frequency multiplication on an FM or PM signal (Sect. 5.2).
5.   Design an FM generator and detector appropriate for an application (Sect. 5.3).
6.   Use a phasor diagram to analyze interference in AM, FM, and PM (Sect. 5.4).

5.1   PHASE AND FREQUENCY MODULATION

This section introduces the concepts of instantaneous phase and frequency for the definition of PM and FM signals. Then, since the nonlinear nature of exponential modulation precludes spectral analysis in general terms, we must work instead with the spectra resulting from particular cases such as narrowband modulation and tone modulation.

PM and FM Signals

Consider a CW signal with constant envelope but time-varying phase, so

$$x_c(t) = A_c \cos\left[\omega_c t + \phi(t)\right] \tag{1}$$

Upon defining the **total instantaneous angle**

$$\theta_c(t) \triangleq \omega_c t + \phi(t)$$

we can express $x_c(t)$ as

$$x_c(t) = A_c \cos \theta_c(t) = A_c \operatorname{Re}\left[e^{j\theta_c(t)}\right]$$

Hence, if $\theta_c(t)$ contains the message information $x(t)$, we have a process that may be termed either **angle** modulation or **exponential** modulation. We'll use the latter name because it emphasizes the nonlinear relationship between $x_c(t)$ and $x(t)$.

As to the specific dependence of $\theta_c(t)$ on $x(t)$, **phase modulation** (PM) is defined by

$$\phi(t) \triangleq \phi_\Delta x(t) \qquad \phi_\Delta \le 180° \tag{2}$$

so that

$$x_c(t) = A_c \cos\left[\omega_c t + \phi_\Delta x(t)\right] \tag{3}$$

These equations state that the instantaneous phase varies directly with the modulating signal. The constant ϕ_Δ represents the **maximum phase shift** produced by $x(t)$, since we're still keeping our normalization convention $|x(t)| \le 1$. The upper bound $\phi_\Delta \le 180°$ (or π radians) limits $\phi(t)$ to the range $\pm 180°$ and prevents phase ambiguities—after all, there's no physical distinction between angles of $+270$ and $-90°$, for instance. The bound on ϕ_Δ is analogous to the restriction $\mu \le 1$ in AM, and ϕ_Δ can justly be called the **phase modulation index,** or the **phase deviation.**

The rotating-phasor diagram in Fig. 5.1–1 helps interpret phase modulation and leads to the definition of frequency modulation. The total angle $\theta_c(t)$ consists of the constant rotational term $\omega_c t$ plus $\phi(t)$, which corresponds to angular shifts relative to the dashed line. Consequently, the phasor's instantaneous rate of rotation in cycles per second will be

$$f(t) \triangleq \frac{1}{2\pi}\dot{\theta}_c(t) = f_c + \frac{1}{2\pi}\dot{\phi}(t) \tag{4}$$

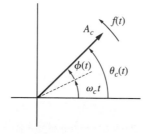

Figure 5.1–1   Rotating-phasor representation of exponential modulation.

in which the dot notation stands for the time derivative, that is, $\dot{\phi}(t) = d\phi(t)/dt$, and so on. We call $f(t)$ the **instantaneous frequency** of $x_c(t)$. Although $f(t)$ is measured in *hertz*, it should *not* be equated with spectral frequency. Spectral frequency f is the independent variable of the frequency domain, whereas instantaneous frequency $f(t)$ is a time-dependent property of waveforms with exponential modulation.

In the case of **frequency modulation** (FM), the instantaneous frequency of the modulated wave is defined to be

$$f(t) \triangleq f_c + f_\Delta x(t) \qquad f_\Delta < f_c \qquad [5]$$

so $f(t)$ varies in proportion with the modulating signal. The proportionality constant f_Δ, called the **frequency deviation,** represents the maximum shift of $f(t)$ relative to the carrier frequency f_c. The upper bound $f_\Delta < f_c$ simply ensures that $f(t) > 0$. However, we usually want $f_\Delta \ll f_c$ in order to preserve the bandpass nature of $x_c(t)$.

Equations (4) and (5) show that an FM wave has $\dot{\phi}(t) = 2\pi f_\Delta \, x(t)$, and integration yields the phase modulation

$$\phi(t) = 2\pi f_\Delta \int_{t_0}^{t} x(\lambda) \, d\lambda + \phi(t_0) \qquad t \geq t_0 \qquad [6a]$$

If t_0 is taken such that $\phi(t_0) = 0$, we can drop the lower limit of integration and use the informal expression

$$\phi(t) = 2\pi f_\Delta \int^{t} x(\lambda) \, d\lambda \qquad [6b]$$

The FM waveform is then written as

$$x_c(t) = A_c \, \cos \left[\omega_c t + 2\pi f_\Delta \int^{t} x(\lambda) \, d\lambda \right] \qquad [7]$$

But it must be assumed that the message has no *dc component* so the above integrals do not diverge when $t \to \infty$. Physically, a dc term in $x(t)$ would produce a constant carrier-frequency shift equal to $f_\Delta \langle x(t) \rangle$.

A comparison of Eqs. (3) and (7) implies little difference between PM and FM, the essential distinction being the integration of the message in FM. Moreover, nomenclature notwithstanding, both FM and PM have both time-varying phase and frequency, as underscored by Table 5.1–1. These relations clearly indicate that, with the help of integrating and differentiating networks, a phase modulator can produce frequency modulation and vice versa. In fact, in the case of tone modulation it's nearly impossible visually to distinguish FM and PM waves.

On the other hand, a comparison of exponential modulation with linear modulation reveals some pronounced differences. For one thing,

The amplitude of an exponentially modulated wave is *constant*.

Table 5.1–1   Comparison of PM and FM

	Instantaneous phase $\phi(t)$	Instantaneous frequency $f(t)$
PM	$\phi_\Delta x(t)$	$f_c + \dfrac{1}{2\pi}\,\phi_\Delta \dot{x}(t)$
FM	$2\pi f_\Delta \displaystyle\int^t x(\lambda)\,d\lambda$	$f_c + f_\Delta x(t)$

Therefore, regardless of the message $x(t)$, the average transmitted power is

$$S_T = \tfrac{1}{2}A_c^2 \tag{8}$$

For another, the zero crossings of an exponentially modulated wave are *not periodic,* whereas they are always periodic in linear modulation. Indeed, because of the constant-amplitude property of FM and PM, it can be said that

> The message resides in the *zero crossings alone,* providing the carrier frequency is large.

Finally, since exponential modulation is a nonlinear process,

> The modulated wave does not look at all like the message waveform.

Figure 5.1–2 illustrates some of these points by showing typical AM, FM, and PM waves. As a mental exercise you may wish to check these waveforms against the corresponding modulating signals. For FM and PM this is most easily done by considering the instantaneous frequency rather than by substituting $x(t)$ in Eqs. (3) and (7).

Despite the many similarities of PM and FM, frequency modulation turns out to have superior noise-reduction properties and thus will receive most of our attention. To gain a qualitative appreciation of FM noise reduction, suppose a demodulator simply extracts the instantaneous frequency $f(t) = f_c + f_\Delta x(t)$ from $x_c(t)$. The demodulated output is then proportional to the frequency deviation f_Δ, which can be increased without increasing the transmitted power S_T. If the noise level remains constant, increased signal output is equivalent to reduced noise. However, noise reduction does require increased transmission bandwidth to accommodate large frequency deviations.

Ironically, frequency modulation was first conceived as a means of **bandwidth reduction,** the argument going somewhat as follows: If, instead of modulating the

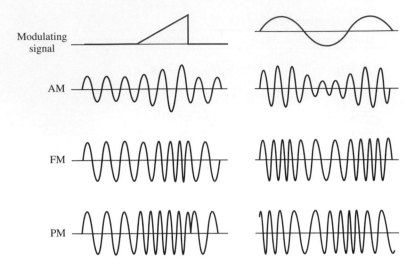

Figure 5.1–2   Illustrative AM, FM, and PM waveforms.

carrier amplitude, we modulate the frequency by swinging it over a range of, say, ±50 Hz, then the transmission bandwidth will be 100 Hz regardless of the message bandwidth. As we'll soon see, this argument has a serious flaw, for it ignores the distinction between **instantaneous** and **spectral frequency.** Carson (1922) recognized the fallacy of the bandwidth-reduction notion and cleared the air on that score. Unfortunately, he and many others also felt that exponential modulation had no advantages over linear modulation with respect to noise. It took some time to overcome this belief but, thanks to Armstrong (1936), the merits of exponential modulation were finally appreciated. Before we can understand them quantitatively, we must address the problem of spectral analysis.

EXERCISE 5.1–1   Suppose FM had been defined in direct analogy to AM by writing $x_c(t) = A_c \cos \omega_c(t)\, t$ with $\omega_c(t) = \omega_c[1 + \mu x(t)]$. Demonstrate the physical impossibility of this definition by finding $f(t)$ when $x(t) = \cos \omega_m t$.

Narrowband PM and FM

Our spectral analysis of exponential modulation starts with the quadrature-carrier version of Eq. (1), namely

$$x_c(t) = x_{ci}(t) \cos \omega_c t - x_{cq}(t) \sin \omega_c t \qquad [9]$$

where

$$x_{ci}(t) = A_c \cos \phi(t) = A_c\left[1 - \frac{1}{2!}\phi^2(t) + \cdots\right] \qquad [10]$$

$$x_{cq}(t) = A_c \sin \phi(t) = A_c \left[\phi(t) - \frac{1}{3!} \phi^3(t) + \cdots \right]$$

Now we impose the simplifying condition

$$|\phi(t)| \ll 1 \text{ rad} \qquad\qquad\qquad [11a]$$

so that

$$x_{ci}(t) \approx A_c \qquad x_{cq}(t) \approx A_c \phi(t) \qquad\qquad [11b]$$

Then it becomes an easy task to find the spectrum $X_c(f)$ of the modulated wave in terms of an arbitrary message spectrum $X(f)$.

Specifically, the transforms of Eqs. (9) and (11b) yield

$$X_c(f) = \frac{1}{2} A_c \delta(f - f_c) + \frac{j}{2} A_c \Phi(f - f_c) \qquad f > 0 \qquad [12a]$$

in which

$$\Phi(f) = \mathcal{F}[\phi(t)] = \begin{cases} \phi_\Delta X(f) & \text{PM} \\ -jf_\Delta X(f)/f & \text{FM} \end{cases} \qquad [12b]$$

The FM expression comes from the integration theorem applied to $\phi(t)$ in Eq. (6).

Based on Eq. (12), we conclude that if $x(t)$ has message bandwidth $W \ll f_c$, then $x_c(t)$ will be a bandpass signal with bandwidth $2W$. But this conclusion holds only under the conditions of Eq. (11). For larger values of $|\phi(t)|$, the terms $\phi^2(t)$, $\phi^3(t)$, . . . cannot be ignored in Eq. (10) and will *increase* the bandwidth of $x_c(t)$. Hence, Eqs. (11) and (12) describe the special case of **narrowband** phase or frequency modulation (NBPM or NBFM).

An informative illustration of Eq. (12) is provided by taking $x(t) = \text{sinc } 2Wt$, so $X(f) = (1/2W)\Pi(f/2W)$. The resulting NBPM and NBFM spectra are depicted in Fig. 5.1–3. Both spectra have carrier-frequency impulses and bandwidth $2W$. However, the lower sideband in NBFM is 180° out of phase (represented by the negative sign), whereas both NBPM sidebands have a 90° phase shift (represented by j). Except for the phase shift, the NBPM spectrum looks just like an AM spectrum with the same modulating signal.

EXAMPLE 5.1–1

Use the second-order approximations $x_{ci}(t) \approx A_c[1 - \frac{1}{2}\phi^2(t)]$ and $x_{cq}(t) \approx A_c \phi(t)$ to find and sketch the components of the PM spectrum when $x(t) = \text{sinc } 2Wt$.

EXERCISE 5.1–2

Tone Modulation

The study of FM and PM with tone modulation can be carried out jointly by the simple expedient of allowing a 90° difference in the modulating tones. For if we take

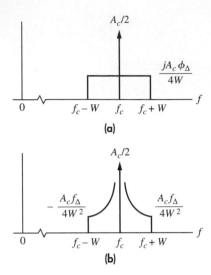

Figure 5.1-3   Narrowband modulated spectra with $x(t) = \text{sinc } 2Wt$. (a) PM: (b) FM.

$$x(t) = \begin{cases} A_m \sin \omega_m t & \text{PM} \\ A_m \cos \omega_m t & \text{FM} \end{cases}$$

then Eqs. (2) and (6) both give

$$\phi(t) = \beta \sin \omega_m t \qquad\qquad [13a]$$

where

$$\beta \triangleq \begin{cases} \phi_\Delta A_m & \text{PM} \\ (A_m/f_m)f_\Delta & \text{FM} \end{cases} \qquad\qquad [13b]$$

The parameter β serves as the **modulation index** for PM or FM with tone modulation. This parameter equals the **maximum phase deviation** and is proportional to the tone amplitude A_m in both cases. Note, however, that β for FM is inversely proportional to the tone frequency f_m since the integration of $\cos \omega_m t$ yields $(\sin \omega_m t)/\omega_m$.

Narrowband tone modulation requires $\beta \ll 1$, and Eq. (9) simplifies to

$$x_c(t) \approx A_c \cos \omega_c t - A_c \beta \sin \omega_m t \sin \omega_c t \qquad\qquad [14]$$

$$\approx A_c \cos \omega_c t - \frac{A_c \beta}{2} \cos (\omega_c - \omega_m)t + \frac{A_c \beta}{2} \cos (\omega_c + \omega_m)t$$

The corresponding line spectrum and phasor diagram are shown in Fig. 5.1–4. Observe how the phase reversal of the lower sideband line produces a component perpendicular or *quadrature* to the carrier phasor. This quadrature relationship is precisely what's needed to create phase or frequency modulation instead of amplitude modulation.

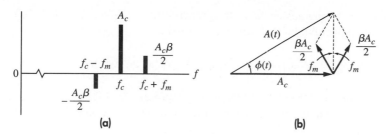

Figure 5.1–4   NBFM with tone modulation. (a) Line spectrum; (b) phasor diagram.

Now, to determine the line spectrum with an arbitrary value of the modulation index, we drop the narrowband approximation and write

$$x_c(t) = A_c[\cos \phi(t) \cos \omega_c t - \sin \phi(t) \sin \omega_c t] \qquad [15]$$
$$= A_c[\cos (\beta \sin \omega_m t) \cos \omega_c t - \sin (\beta \sin \omega_m t) \sin \omega_c t]$$

Then we use the fact that, even though $x_c(t)$ is not necessarily periodic, the terms $\cos (\beta \sin \omega_m t)$ and $\sin (\beta \sin \omega_m t)$ are periodic and each can be expanded as a trigonometric Fourier series with $f_0 = f_m$. Indeed, a well-known result from applied mathematics states that

$$\cos (\beta \sin \omega_m t) = J_0(\beta) + \sum_{\substack{n=... \\ n \text{ even}}}^{\infty} 2\, J_n(\beta) \cos n\omega_m t \qquad [16]$$

$$\sin (\beta \sin \omega_m t) = \sum_{\substack{n=... \\ n \text{ odd}}}^{\infty} 2\, J_n(\beta) \sin n\omega_m t$$

where n is positive and

$$J_n(\beta) \triangleq \frac{1}{2\pi} \int_{-\pi}^{\pi} e^{j(\beta \sin \lambda - n\lambda)}\, d\lambda \qquad [17]$$

The coefficients $J_n(\beta)$ are **Bessel functions** of the first kind, of order n and argument β. With the aid of Eq. (17), you should encounter little difficulty in deriving the trigonometric expansions given in Eq. (16).

Substituting Eq. (16) into Eq. (15) and expanding products of sines and cosines finally yields

$$x_c(t) = A_c J_0(\beta) \cos \omega_c t \qquad [18a]$$

$$+ \sum_{\substack{n=... \\ n \text{ odd}}}^{\infty} A_c J_n(\beta)[\cos (\omega_c + n\omega_m)t - \cos (\omega_c - n\omega_m)t]$$

$$+ \sum_{\substack{n=... \\ n \text{ even}}}^{\infty} A_c J_n(\beta)[\cos (\omega_c + n\omega_m)t + \cos (\omega_c - n\omega_m)t]$$

Alternatively, taking advantage of the property that $J_{-n}(\beta) = (-1)^n J_n(\beta)$, we get the more compact but less informative expression

$$x_c(t) = A_c \sum_{n=-\infty}^{\infty} J_n(\beta) \cos (\omega_c + n\omega_m)t \qquad \text{[18b]}$$

In either form, Eq. (18) is the mathematical representation for a constant-amplitude wave whose instantaneous frequency varies sinusoidally. A phasor interpretation, to be given shortly, will shed more light on the matter.

Examining Eq. (18), we see that

> The FM spectrum consists of a carrier-frequency line plus an *infinite* number of sideband lines at frequencies $f_c \pm nf_m$. All lines are equally spaced by the modulating frequency and the odd-order lower sideband lines are reversed in phase or inverted relative to the unmodulated carrier. In a positive-frequency line spectrum, any apparent negative frequencies $(f_c + nf_m < 0)$ must be folded back to the positive values $|f_c + nf_m|$.

A typical spectrum is illustrated in Fig. 5.1–5. Note that negative frequency components will be negligible as long as $\beta f_m \ll f_c$. In general, the relative amplitude of a line at $f_c + nf_m$ is given by $J_n(\beta)$, so before we can say more about the spectrum, we must examine the behavior of Bessel functions.

Figure 5.1–6a shows a few Bessel functions of various order plotted versus the argument β. Several important properties emerge from this plot.

1. The relative amplitude of the carrier line $J_0(\beta)$ varies with the modulation index and hence depends on the modulating signal. Thus, in contrast to linear modulation, the carrier-frequency component of an FM wave "contains" part of the message information. Nonetheless, there will be spectra in which the carrier line has zero amplitude since $J_0(\beta) = 0$ when $\beta = 2.4, 5.5,$ and so on.

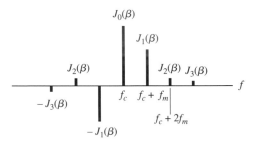

Figure 5.1–5   Line spectrum of FM with tone modulation.

2. The number of sideband lines having appreciable relative amplitude also depends on β. With $\beta \ll 1$ only J_0 and J_1 are significant, so the spectrum will consist of carrier and two sideband lines as in Fig. 5.1–4a. But if $\beta \gg 1$, there will be many sideband lines, giving a spectrum quite unlike linear modulation.

3. Large β implies a large bandwidth to accommodate the extensive sideband structure, agreeing with the physical interpretation of large frequency deviation.

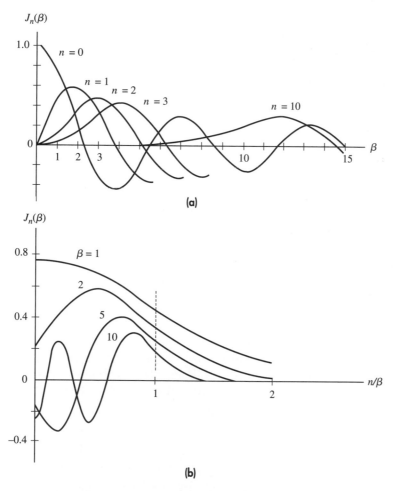

(a)

(b)

Figure 5.1–6 Plots of Bessel functions. (a) Fixed order n, variable argument β; (b) fixed argument β, variable order n.

Table 5.1–2   Selected values of $J_n(\beta)$

n	$J_n(0.1)$	$J_n(0.2)$	$J_n(0.5)$	$J_n(1.0)$	$J_n(2.0)$	$J_n(5.0)$	$J_n(10)$	n
0	1.00	0.99	0.94	0.77	0.22	−0.18	−0.25	0
1	0.05	0.10	0.24	0.44	0.58	−0.33	0.04	1
2			0.03	0.11	0.35	0.05	0.25	2
3				0.02	0.13	0.36	0.06	3
4					0.03	0.39	−0.22	4
5						0.26	−0.23	5
6						0.13	−0.01	6
7						0.05	0.22	7
8						0.02	0.32	8
9							0.29	9
10							0.21	10
11							0.12	11
12							0.06	12
13							0.03	13
14							0.01	14

Some of the above points are better illustrated by Fig. 5.1–6b, which gives $J_n(\beta)$ as a function of n/β for various *fixed* values of β. These curves represent the "envelope" of the sideband lines if we multiply the horizontal axis by βf_m to obtain the line position nf_m relative to f_c. Observe in particular that all $J_n(\beta)$ decay monotonically for $n/\beta > 1$ and that $|J_n(\beta)| \ll 1$ if $|n/\beta| \gg 1$. Table 5.1–2 lists selected values of $J_n(\beta)$, rounded off at the second decimal place. Blanks in the table correspond to $|J_n(\beta)| < 0.01$.

Line spectra drawn from the data in Table 5.1–2 are shown in Fig. 5.1–7, omitting the sign inversions. Part *a* of the figure has β increasing with f_m held fixed, and applies to FM and PM. Part *b* applies only to FM and illustrates the effect of increasing β by *decreasing* f_m with $A_m f_\Delta$ held fixed. The dashed lines help bring out the concentration of significant sideband lines within the range $f_c \pm \beta f_m$ as β becomes large.

For the phasor interpretation of $x_c(t)$ in Eq. (18), we first return to the narrowband approximation and Fig. 5.1–4. The envelope and phase constructed from the carrier and first pair of sideband lines are seen to be

$$A(t) \approx \sqrt{A_c^2 + \left(2\frac{\beta}{2}A_c\sin\omega_m t\right)^2} \approx A_c\left[1 + \frac{\beta^2}{4} - \frac{\beta^2}{4}\cos 2\omega_m t\right]$$

$$\phi(t) \approx \arctan\left[\frac{2(\beta/2)A_c\sin\omega_m t}{A_c}\right] \approx \beta\sin\omega_m t$$

Thus the phase variation is approximately as desired, but there is an additional *amplitude* variation at twice the tone frequency. To cancel out the latter we should

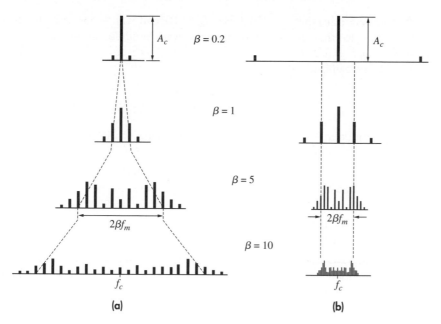

Figure 5.1–7   Tone-modulated line spectra. (a) FM or PM with f_m fixed; (b) FM with $A_m f_\Delta$ fixed.

include the second-order pair of sideband lines that rotate at $\pm 2f_m$ relative to the carrier and whose resultant is collinear with the carrier. While the second-order pair virtually wipes out the undesired amplitude modulation, it also distorts $\phi(t)$. The phase distortion is then corrected by adding the third-order pair, which again introduces amplitude modulation, and so on ad infinitum.

When all spectral lines are included, the odd-order pairs have a resultant in quadrature with the carrier that provides the desired frequency modulation plus unwanted amplitude modulation. The resultant of the even-order pairs, being collinear with the carrier, corrects for the amplitude variations. The net effect is then as illustrated in Fig. 5.1–8. The tip of the resultant sweeps through a circular arc reflecting the constant amplitude A_c.

The narrowband FM signal $x_c(t) = 100 \cos \left[2\pi\, 5000t + 0.05 \sin 2\pi\, 200t \right]$ is transmitted. To find the instantaneous frequency $f(t)$ we take the derivative of $\theta(t)$

EXAMPLE 5.1–2

$$f(t) = \frac{1}{2\pi} \dot{\theta}(t)$$

$$= \frac{1}{2\pi} \left[2\pi\, 5000 + 0.05(2\pi\, 200) \cos 2\pi\, 200\, t \right]$$

$$= 5000 + 10 \cos 2\pi\, 200\, t$$

From $f(t)$ we determine that $f_c = 5000$ Hz, $f_\Delta = 10$, and $x(t) = \cos 2\pi 200t$. There are two ways to find β. For NBFM with tone modulation we know that $\phi(t) = \beta \sin \omega_m t$. Since $x_c(t) = A_c \cos [\omega_c t + \phi(t)]$, we can see that $\beta = 0.05$. Alternatively we can calculate

$$\beta = \frac{A_m}{f_m} f_\Delta$$

From $f(t)$ we find that $A_m f_\Delta = 10$ and $f_m = 200$ so that $\beta = 10/200 = 0.05$ just as we found earlier. The line spectrum has the form of Fig. 5.1–4a with $A_c = 100$ and sidelobes $A_c \beta/2 = 2.5$. The minor distortion from the narrowband approximation shows up in the transmitted power. From the line spectrum we get $S_T = \frac{1}{2}(-2.5)^2 + \frac{1}{2}(100)^2 + \frac{1}{2}(2.5)^2 = 5006.25$ versus $S_T = \frac{1}{2}A_c^2 = \frac{1}{2}(100)^2 = 5000$ when there are enough sidelobes so that there is no amplitude distortion.

EXERCISE 5.1–3   Consider tone-modulated FM with $A_c = 100$, $A_m f_\Delta = 8$ kHz, and $f_m = 4$ kHz. Draw the line spectrum for $f_c = 30$ kHz and for $f_c = 11$kHz.

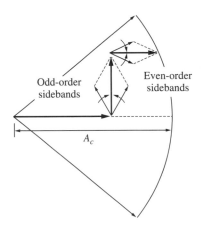

Figure 5.1–8   FM phasor diagram for arbitrary β.

Multitone and Periodic Modulation★

The Fourier series technique used to arrive at Eq. (18) also can be applied to the case of FM with multitone modulation. For instance, suppose that $x(t) = A_1 \cos \omega_1 t + A_2 \cos \omega_2 t$, where f_1 and f_2 are not harmonically related. The modulated wave is first written as

$$x_c(t) = A_c[(\cos \alpha_1 \cos \alpha_2 - \sin \alpha_1 \sin \alpha_2) \cos \omega_c t$$

$$-(\sin \alpha_1 \cos \alpha_2 + \cos \alpha_1 \sin \alpha_2) \sin \omega_c t]$$

where $\alpha_1 = \beta_1 \sin \omega_1 t$, $\beta_1 = A_1 f_\Delta / f_1$, and so on. Terms of the form $\cos \alpha_1$, $\sin \alpha_1$, and so on, are then expanded according to Eq. (16), and after some routine manipulations we arrive at the compact result

$$x_c(t) = A_c \sum_{n=-\infty}^{\infty} \sum_{m=-\infty}^{\infty} J_n(\beta_1) J_m(\beta_2) \cos (\omega_c + n\omega_1 + m\omega_2)t \qquad [19]$$

This technique can be extended to include three or more nonharmonic tones; the procedure is straightforward but tedious.

To interpret Eq. (19) in the frequency domain, the spectral lines can be divided into four categories: (1) the carrier line of amplitude $A_c J_0(\beta_1) J_0(\beta_2)$; (2) sideband lines at $f_c \pm nf_1$ due to one tone alone; (3) sideband lines at $f_c \pm mf_2$ due to the other tone alone; and (4) sideband lines at $f_c \pm nf_1 \pm mf_2$ which appear to be beat-frequency modulation at the sum and difference frequencies of the modulating tones and their harmonics. (This last category would not occur in linear modulation where simple superposition of sideband lines is the rule.) A double-tone FM spectrum showing the various types of spectral lines is given in Fig. 5.1–9 for $f_1 \ll f_2$ and $\beta_1 > \beta_2$. Under these conditions there exists the curious property that each sideband line at $f_c \pm mf_2$ looks like another FM carrier with tone modulation of frequency f_1.

When the tone frequencies are harmonically related—meaning that $x(t)$ is a *periodic* waveform—then $\phi(t)$ is periodic and so is $e^{j\phi(t)}$. The latter can be expanded in an exponential Fourier series with coefficients

$$c_n = \frac{1}{T_0} \int_{T_0} \exp j[\phi(t) - n\omega_0 t]\, dt \qquad [20a]$$

Therefore

$$x_c(t) = A_c \operatorname{Re}\left[\sum_{n=-\infty}^{\infty} c_n e^{j(\omega_c + n\omega_0)t} \right] \qquad [20b]$$

and $A_c |c_n|$ equals the magnitude of the spectral line at $f = f_c + nf_0$.

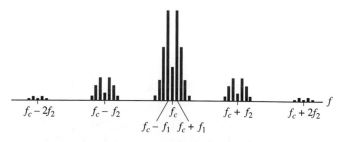

Figure 5.1–9   Double-tone FM line spectrum with $f_1 \ll f_2$ and $\beta_1 > \beta_2$.

EXAMPLE 5.1–3   **FM with Pulse-Train Modulation**

Let $x(t)$ be a unit-amplitude rectangular pulse train with period T_0, pulse duration τ, and duty cycle $d = \tau/T_0$. After removing the dc component $\langle x(t) \rangle = d$, the instantaneous frequency of the resulting FM wave is as shown in Fig. 5.1–10a. The time origin is chosen such that $\phi(t)$ plotted in Fig. 5.1–10b has a peak value $\phi_\Delta = 2\pi f_\Delta \tau$ at $t = 0$. We've also taken the constant of integration such that $\phi(t) \geq 0$. Thus

$$\phi(t) = \begin{cases} \phi_\Delta(1 + t/\tau) & -\tau < t < 0 \\ \phi_\Delta[1 - t/(T_0 - \tau)] & 0 < t < T_0 - \tau \end{cases}$$

which defines the range of integration for Eq. (20a).

The evaluation of c_n is a nontrivial exercise involving exponential integrals and trigonometric relations. The final result can be written as

$$c_n = \left[\frac{\sin \pi(\beta - n)d}{\pi(\beta - n)} + \frac{(1 - d) \sin \pi(\beta - n)d}{\pi(\beta - n)d + \pi n} \right] e^{j\pi(\beta+n)d}$$

$$= \frac{\beta d}{(\beta - n)d + n} \operatorname{sinc}(\beta - n)d \, e^{j\pi(\beta+n)d}$$

where we've let

$$\beta = f_\Delta T_0 = f_\Delta/f_0$$

which plays a role similar to the modulation index for single-tone modulation.

Figure 5.1–10c plots the magnitude line spectrum for the case of $d = 1/4$, $\beta = 4$, and $A_c = 1$. Note the absence of symmetry here and the peaking around

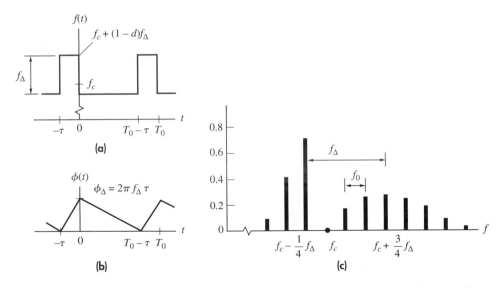

Figure 5.1–10   FM with pulse-train modulation. (a) Instantaneous frequency; (b) phase; (c) line spectrum for $d = 1/4$.

$f = f_c - \frac{1}{4}f_\Delta$ and $f = f_c + \frac{3}{4}f_\Delta$, the two values taken on by the instantaneous frequency. The fact that the spectrum contains other frequencies as well underscores the difference between spectral frequency and instantaneous frequency. The same remarks apply for the *continuous* spectrum of FM with a single modulating pulse— demonstrated by our results in Example 2.5–1.

5.2   TRANSMISSION BANDWIDTH AND DISTORTION

The spectrum of a signal with exponential modulation has infinite extent, in general. Hence, generation and transmission of *pure* FM requires *infinite bandwidth*, whether or not the message is bandlimited. But practical FM systems having finite bandwidth do exist and perform quite well. Their success depends upon the fact that, sufficiently far away from the carrier frequency, the spectral components are quite small and may be discarded. True, omitting any portion of the spectrum will cause *distortion* in the demodulated signal; but the distortion can be minimized by keeping all significant spectral components.

We'll formulate in this section estimates of transmission bandwidth requirements by drawing upon results from Sect. 5.1. Then we'll look at distortion produced by linear and nonlinear systems. Topics encountered in passing include the concept of **wideband** FM and that important piece of FM hardware known as a **limiter.** We'll concentrate primarily on FM, but minor modifications make the analyses applicable to PM.

Transmission Bandwidth Estimates

Determination of FM transmission bandwidth boils down to the question: How much of the modulated signal spectrum is *significant?* Of course, significance standards are not absolute, being contingent upon the amount of distortion that can be tolerated in a specific application. However, rule-of-thumb criteria based on studies of tone modulation have met with considerable success and lead to useful approximate relations. Our discussion of FM bandwidth requirements therefore begins with the significant sideband lines for tone modulation.

Figure 5.1–6 indicated that $J_n(\beta)$ falls off rapidly for $|n/\beta| > 1$, particularly if $\beta \gg 1$. Assuming that the modulation index β is large, we can say that $|J_n(\beta)|$ is significant only for $|n| \leq \beta = A_m f_\Delta/f_m$. Therefore, all significant lines are contained in the frequency range $f_c \pm \beta f_m = f_c \pm A_m f_\Delta$, a conclusion agreeing with intuitive reasoning. On the other hand, suppose the modulation index is small; then *all* sideband lines are small compared to the carrier, since $J_0(\beta) \gg J_{n\neq0}(\beta)$ when $\beta \ll 1$. But we must retain at least the first-order sideband pair, else there would be no frequency modulation at all. Hence, for small β, the significant sideband lines are contained in $f_c \pm f_m$.

To put the above observations on a quantitative footing, all sideband lines having relative amplitude $|J_n(\beta)| > \epsilon$ are *defined* as being significant, where ϵ ranges from 0.01 to 0.1 according to the application. Then, if $|J_M(\beta)| > \epsilon$ and $|J_{M+1}(\beta)| < \epsilon$, there are M significant sideband *pairs* and $2M + 1$ significant lines all told. The bandwidth is thus written as

$$B = 2M(\beta)f_m \qquad M(\beta) \geq 1 \tag{1}$$

since the lines are spaced by f_m and M depends on the modulation index β. The condition $M(\beta) \geq 1$ has been included in Eq. (1) to account for the fact that B cannot be less than $2f_m$.

Figure 5.2–1 shows M as a continuous function of β for $\epsilon = 0.01$ and 0.1. Experimental studies indicate that the former is often overly conservative, while the latter may result in small but noticeable distortion. Values of M between these two bounds are acceptable for most purposes and will be used hereafter.

But the bandwidth B is not the transmission bandwidth B_T; rather it's the minimum bandwidth necessary for modulation by a tone of specified amplitude and frequency. To estimate B_T, we should calculate the *maximum* bandwidth required when the tone parameters are constrained by $A_m \leq 1$ and $f_m \leq W$. For this purpose, the dashed line in Fig. 5.2–1 depicts the approximation

$$M(\beta) \approx \beta + 2 \tag{2}$$

which falls midway between the solid lines for $\beta \geq 2$. Inserting Eq. (2) into Eq. (1) gives

$$B \approx 2(\beta + 2)f_m = 2\left(\frac{A_m f_\Delta}{f_m} + 2\right)f_m = 2(A_m f_\Delta + 2f_m)$$

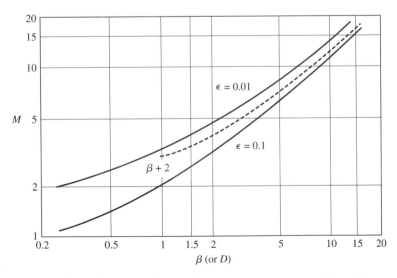

Figure 5.2–1   The number of significant sideband pairs as a function of β (or D).

Now, bearing in mind that f_Δ is a property of the modulator, what tone produces the maximum bandwidth? Clearly, it is the *maximum-amplitude–maximum-frequency* tone having $A_m = 1$ and $f_m = W$. The worst-case tone-modulation bandwidth is then

$$B_T \approx 2(f_\Delta + 2W) \qquad \text{if } \beta > 2$$

Note carefully that the corresponding modulation index $\beta = f_\Delta/W$ is not the maximum value of β but rather the value which, combined with the maximum modulating frequency, yields the maximum bandwidth. Any other tone having $A_m < 1$ or $f_m < W$ will require less bandwidth even though β may be larger.

Finally, consider a reasonably smooth but otherwise *arbitrary modulating signal* having the message bandwidth W and satisfying the normalization convention $|x(t)| \leq 1$. We'll estimate B_T directly from the worst-case tone-modulation analysis, assuming that any component in $x(t)$ of smaller amplitude or frequency will require a smaller bandwidth than B_T. Admittedly, this procedure ignores the fact that superposition is not applicable to exponential modulation. However, our investigation of multitone spectra has shown that the beat-frequency sideband pairs are contained primarily within the bandwidth of the dominating tone alone, as illustrated by Fig. 5.1–9.

Therefore, extrapolating tone modulation to an arbitrary modulating signal, we define the **deviation ratio**

$$D \triangleq \frac{f_\Delta}{W} \qquad\qquad \text{[3]}$$

which equals the maximum deviation divided by the maximum modulating frequency, analogous to the modulation index of worst-case tone modulation. The transmission bandwidth required for $x(t)$ is then

$$B_T = 2\,M(D)W \qquad\qquad \text{[4]}$$

where D is treated just like β to find $M(D)$, say from Fig. 5.2–1.

Lacking appropriate curves or tables for $M(D)$, there are several approximations to B_T that can be invoked. With extreme values of the deviation ratio we find that

$$B_T = \begin{cases} 2DW = 2f_\Delta & D \gg 1 \\ 2W & D \ll 1 \end{cases}$$

paralleling our results for tone modulation with β very large or very small. Both of these approximations are combined in the convenient relation

$$B_T \approx 2(f_\Delta + W) = 2(D + 1)W \qquad \begin{matrix} D \gg 1 \\ D \ll 1 \end{matrix} \qquad \text{[5]}$$

known as **Carson's rule.** Perversely, the majority of actual FM systems have $2 < D < 10$, for which Carson's rule somewhat underestimates the transmission bandwidth. A better approximation for equipment design is then

$$B_T \approx 2(f_\Delta + 2W) = 2(D + 2)W \qquad D > 2 \qquad \text{[6]}$$

which would be used, for example, to determine the 3 dB bandwidths of FM amplifiers. Note that Carson's rule overestimates B_T for some applications using the
narrowband approximation. The bandwidth of the transmitted signal in Example
5.1–2 is 400 Hz, whereas Eq. (5) estimates $B_T \approx 420$ Hz.

Physically, the deviation ratio represents the **maximum phase deviation** of an
FM wave under worst-case bandwidth conditions. Our FM bandwidth expressions
therefore apply to **phase** modulation if we replace D with the maximum phase deviation ϕ_Δ of the PM wave. Accordingly, the transmission bandwidth for PM with
arbitrary $x(t)$ is estimated to be

$$B_T = 2M(\phi_\Delta)W \qquad M(\phi_\Delta) \geq 1 \qquad\qquad \text{[7a]}$$

or

$$B_T \approx 2(\phi_\Delta + 1)W \qquad\qquad \text{[7b]}$$

which is the approximation equivalent to Carson's rule. These expressions differ
from the FM case in that ϕ_Δ is independent of W.

You should review our various approximations and their conditions of validity.
In deference to most of the literature, we'll usually take B_T as given by Carson's rule
in Eqs. (5) and (7b). But when the modulating signal has discontinuities—a rectangular pulse train, for instance—the bandwidth estimates become invalid and we
must resort to brute-force spectral analysis.

EXAMPLE 5.2–1   **Commercial FM Bandwidth**

Commercial FM broadcast stations in the United States are limited to a maximum
frequency deviation of 75 kHz, and modulating frequencies typically cover 30 Hz to
15 kHz. Letting $W = 15$ kHz, the deviation ratio is $D = 75$ kHz/15 kHz $= 5$ and Eq.
(6) yields $B_T \approx 2(5 + 2) \times 15$ kHz $= 210$ kHz. High-quality FM radios have bandwidths of at least 200 kHz. Carson's rule in Eq. (5) underestimates the bandwidth,
giving $B_T \approx 180$ kHz.

If a single modulating tone has $A_m = 1$ and $f_m = 15$ kHz, then $\beta = 5$, $M(\beta) \approx 7$,
and Eq. (1) shows that $B = 210$ kHz. A lower-frequency tone, say 3 kHz, would
result in a larger modulation index ($\beta = 25$), a greater number of significant sideband
pairs ($M = 27$), but a smaller bandwidth since $B = 2 \times 27 \times 3$ kHz $= 162$ kHz.

EXERCISE 5.2–1   Calculate B_T/W for $D = 0.3, 3$, and 30 using Eqs. (5) and (6) where applicable.

Linear Distortion

The analysis of distortion produced in an FM or PM wave by a linear network is an
exceedingly knotty problem—so much so that several different approaches to it

have been devised, none of them easy. Panter (1965) devotes three chapters to the subject and serves as a reference guide. Since we're limited here to a few pages, we can only view the "tip of the iceberg." Nonetheless, we'll gain some valuable insights regarding linear distortion of FM and PM.

Figure 5.2–2 represents an exponentially modulated bandpass signal $x_c(t)$ applied to a linear system with transfer function $H(f)$, producing the output $y_c(t)$. The constant-amplitude property of $x_c(t)$ allows us to write the lowpass equivalent input

$$x_{\ell p}(t) = \tfrac{1}{2}A_c e^{j\phi(t)} \qquad [8]$$

where $\phi(t)$ contains the message information. In terms of $X_{\ell p}(f)$, the lowpass equivalent output spectrum is

$$Y_{\ell p}(f) = H(f + f_c)u(f + f_c)X_{\ell p}(f) \qquad [9]$$

Lowpass-to-bandpass transformation finally gives the output as

$$y_c(t) = 2\,\mathrm{Re}\left[y_{\ell p}(t)e^{j\omega_c t}\right] \qquad [10]$$

While this method appears simple on paper, the calculations of $X_{\ell p}(f) = \mathscr{F}[x_{\ell p}(t)]$ and $y_{\ell p}(t) = \mathscr{F}^{-1}[Y_{\ell p}(f)]$ generally prove to be major stumbling blocks. Computer-aided numerical techniques are then necessary.

One of the few cases for which Eqs. (8)–(10) yield closed-form results is the transfer function plotted in Fig. 5.2–3. The gain $|H(f)|$ equals K_0 at f_c and increases (or decreases) linearly with slope K_1/f_c; the phase-shift curve corresponds to carrier

Figure 5.2–2

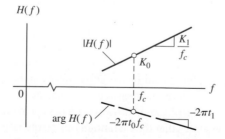

Figure 5.2–3

delay t_0 and group delay t_1, as discussed in Example 4.1–1. The lowpass equivalent of $H(f)$ is

$$H(f + f_c)\, u\, (f + f_c) = \left(K_0 + \frac{K_1}{f_c} f \right) e^{-j2\pi(t_0 f_c + t_1 f)}$$

and Eq. (9) becomes

$$Y_{\ell p}(f) = K_0 e^{-j\omega_c t_0}\left[X_{\ell p}(f)\, e^{-j2\pi t_1 f} \right] + \frac{K_1}{j\omega_c} e^{-j\omega_c t_0}\left[(j2\pi f) X_{\ell p}(f) e^{-j2\pi t_1 f} \right]$$

Invoking the time-delay and differentiation theorems for $\mathcal{F}^{-1}[Y_{\ell p}(f)]$ we see that

$$y_{\ell p}(t) = K_0 e^{-j\omega_c t_0} x_{\ell p}(t - t_1) + \frac{K_1}{j\omega_c} e^{-j\omega_c t_0}\, \dot{x}_{\ell p}(t - t_1)$$

where

$$\dot{x}_{\ell p}(t - t_1) = \frac{d}{dt}\left[\frac{1}{2} A_c\, e^{j\phi(t-t_1)} \right] = \frac{j}{2} A_c \dot{\phi}(t - t_1) e^{j\phi(t-t_1)}$$

obtained from Eq. (8).

Inserting these expressions into Eq. (10) gives the output signal

$$y_c(t) = A(t) \cos\left[\omega_c(t - t_0) + \phi(t - t_1) \right] \tag{11a}$$

which has a time-varying amplitude

$$A(t) = A_c\left[K_0 + \frac{K_1}{\omega_c} \dot{\phi}(t - t_1) \right] \tag{11b}$$

In the case of an FM input, $\dot{\phi}(t) = 2\pi f_\Delta\, x(t)$ so

$$A(t) = A_c\left[K_0 + \frac{K_1 f_\Delta}{f_c} x(t - t_1) \right] \tag{12}$$

Equation (12) has the same form as the envelope of an AM wave with $\mu = K_1 f_\Delta / K_0 f_c$. We thus conclude that $|H(f)|$ in Fig. 5.2–3 produces **FM-to-AM conversion,** along with the carrier delay t_0 and group delay t_1 produced by arg $H(f)$. (By the way, a second look at Example 4.2–2 reveals that amplitude distortion of an AM wave can produce AM-to-PM conversion.)

FM-to-AM conversion does not present an insurmountable problem for FM or PM transmission, as long as $\phi(t)$ suffers no ill effects other than time delay. We therefore ignore the amplitude distortion from any reasonably smooth gain curve. But delay distortion from a nonlinear phase-shift curve can be quite severe and must be equalized in order to preserve the message information.

A simplified approach to phase-distortion effects is provided by the *quasi-static approximation* which assumes that the instantaneous frequency of an FM wave with $f_\Delta \gg W$ varies so slowly compared to $1/W$ that $x_c(t)$ looks more or less like an ordinary sinusoid at frequency $f(t) = f_c + f_\Delta x(t)$. For if the system's response to a carrier-frequency sinusoid is

$$y_c(t) = A_c|H(f_c)| \cos [\omega_c t + \arg H(f_c)]$$

and if $x_c(t)$ has a slowly changing instantaneous frequency $f(t)$, then

$$y_c(t) \approx A_c|H[f(t)]| \cos \{\omega_c t + \phi(t) + \arg H[f(t)]\} \qquad \text{[13]}$$

It can be shown that this approximation requires the condition

$$|\ddot{\phi}(t)|_{max} \left| \frac{1}{H(f)} \frac{d^2 H(f)}{df^2} \right|_{max} \ll 8\pi^2 \qquad \text{[14]}$$

in which $|\ddot{\phi}(t)| \leq 4\pi^2 f_\Delta W$ for tone-modulated FM with $f_m \leq W$. If $H(f)$ represents a single-tuned bandpass filter with 3 dB bandwidth B, then the second term in Eq. (14) equals $8/B^2$ and the condition becomes $4f_\Delta W/B^2 \ll 1$ which is satisfied by the transmission bandwidth requirement $B \geq B_T$.

Now suppose that Eq. (14) holds and the system has a nonlinear phase shift such as $\arg H(f) = \alpha f^2$, where α is a constant. Upon substituting $f(t) = f_c + \dot{\phi}(t)/2\pi$ we get

$$\arg H[f(t)] = \alpha f_c^2 + \frac{\alpha f_c}{\pi} \dot{\phi}(t) + \frac{\alpha}{4\pi^2} \dot{\phi}^2(t)$$

Thus, the total phase in Eq. (13) will be distorted by the addition of $\dot{\phi}(t)$ and $\dot{\phi}^2(t)$.

Let $|H(f)| = 1$ and $\arg H(f) = -2\pi t_1 (f - f_c)$. Show that Eqs. (11) and (13) give the same result with $\phi(t) = \beta \sin \omega_m t$ provided that $\omega_m t_1 \ll \pi$.

EXERCISE 5.2–2

Nonlinear Distortion and Limiters

Amplitude distortion of an FM wave produces FM-to-AM conversion. Here we'll show that the resulting AM can be eliminated through the use of controlled *nonlinear* distortion and filtering.

For purposes of analysis, let the input signal in Fig. 5.2–4 be

$$v_{in}(t) = A(t) \cos \theta_c(t)$$

where $\theta_c(t) = \omega_c t + \phi(t)$ and $A(t)$ is the amplitude. The nonlinear element is assumed to be **memoryless**—meaning no energy storage—so the input and output are related by an instantaneous nonlinear transfer characteristic $v_{out} = T[v_{in}]$. We'll also assume for convenience that $T[0] = 0$.

Although $v_{in}(t)$ is not necessarily periodic in time, it may be viewed as a *periodic function of θ_c* with period 2π. (Try to visualize plotting v_{in} versus θ_c with time

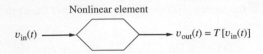

Nonlinear element

$$v_{in}(t) \longrightarrow \longrightarrow v_{out}(t) = T[v_{in}(t)]$$

Figure 5.2–4

held fixed.) Likewise, the output is a periodic function of θ_c and can be expanded in the trigonometric Fourier series

$$v_{out} = \sum_{n=1}^{\infty} |2a_n| \cos(n\,\theta_c + \arg a_n) \qquad [15a]$$

where

$$a_n = \frac{1}{2\pi} \int_{2\pi} T[v_{in}]e^{-jn\theta_c}\,d\theta_c \qquad [15b]$$

The time variable t does not appear explicitly here, but v_{out} depends on t via the time-variation of θ_c. Additionally, the coefficients a_n may be functions of time when the amplitude of v_{in} has time variations.

But we'll first consider the case of an undistorted FM input, so $A(t)$ equals the constant A_c and all the a_n are constants. Hence, writing out Eq. (15a) term by term with t explicitly included, we have

$$\begin{aligned} v_{out}(t) = {} & |2a_1| \cos[\omega_c t + \phi(t) + \arg a_1] \qquad [16] \\ & + |2a_2| \cos[2\omega_c t + 2\phi(t) + \arg a_2] \\ & + \cdots \end{aligned}$$

This expression reveals that the nonlinear distortion produces additional FM waves at *harmonics* of the carrier frequency, the nth harmonic having constant amplitude $|2a_n|$ and phase modulation $n\phi(t)$ plus a constant phase shift $\arg a_n$.

If these waves don't overlap in the frequency domain, the *undistorted* input can be recovered by applying the distorted output to a *bandpass filter*. Thus, we say that FM enjoys considerable immunity from the effects of memoryless nonlinear distortion.

Now let's return to FM with unwanted amplitude variations $A(t)$. Those variations can be flattened out by an **ideal hard limiter** or **clipper** whose transfer characteristic is plotted in Fig. 5.2–5a. Figure 5.2–5b shows a clipper circuit employing back-to-back Zener diodes with breakdown voltage V_0 at the output of a high-gain amplifier.

The clipper output looks essentially like a square wave, since $T[v_{in}] = V_0 \operatorname{sgn} v_{in}$ and

$$v_{out} = \begin{cases} +V_0 & v_{in} > 0 \\ -V_0 & v_{in} < 0 \end{cases}$$

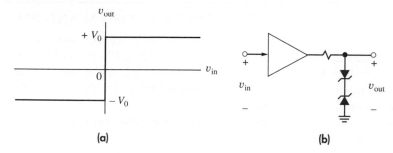

Figure 5.2–5   Hard limiter. (a) Transfer characteristic; (b) circuit realization with Zener diodes.

The coefficients are then found from Eq. (15b) to be

$$
a_n = \begin{cases} 4V_0/\pi n & n = 1, 5, 9, \ldots \\ -4V_0/\pi n & n = 3, 7, 11, \ldots \\ 0 & n = 2, 4, 6, \ldots \end{cases}
$$

which are independent of time because $A(t) \geq 0$ does not affect the sign of v_{in}. Therefore,

$$
v_{\text{out}}(t) = \frac{8V_0}{\pi} \cos[\omega_c t + \phi(t)] - \frac{8V_0}{3\pi} \cos[3\omega_c t + 3\phi(t)] + \cdots \qquad \text{[17]}
$$

and bandpass filtering yields a constant-amplitude FM wave if the components of $v_{\text{out}}(t)$ have no spectral overlap. Incidentally, this analysis lends support to the previous statement that message information resides entirely in the *zero-crossings* of an FM or PM wave.

Figure 5.2–6 summarizes our results. The limiter plus BPF in part *a* removes unwanted amplitude variations from an AM or PM wave, and would be used in a receiver. The nonlinear element in part *b* distorts a constant-amplitude wave, but the BPF passes only the undistorted term at the *n*th harmonic. This combination acts as a *frequency multiplier* if $n > 1$, and is used in certain types of transmitters.

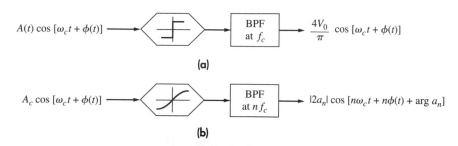

Figure 5.2–6   Nonlinear processing circuits. (a) Amplitude limiter; (b) frequency multiplier.

5.3   GENERATION AND DETECTION OF FM AND PM

The operating principles of several methods for the generation and detection of exponential modulation are presented in this section. Other FM and PM systems that involve phase-lock loops will be mentioned in Sect. 7.3. Additional methods and information regarding specific circuit designs can be found in the radio electronics texts cited at the back of the book.

When considering equipment for exponential modulation, you should keep in mind that the instantaneous phase or frequency varies linearly with the message waveform. Devices are thus required that produce or are sensitive to phase or frequency variation in a linear fashion. Such characteristics can be approximated in a variety of ways, but it is sometimes difficult to obtain a suitably linear relationship over a wide operating range.

On the other hand, the constant-amplitude property of exponential modulation is a definite advantage from the hardware viewpoint. For one thing, the designer need not worry about excessive power dissipation or high-voltage breakdown due to extreme envelope peaks. For another, the relative immunity to nonlinear distortion allows the use of nonlinear electronic devices that would hopelessly distort a signal with linear modulation. Consequently, considerable latitude is possible in the design and selection of equipment. As a case in point, the microwave repeater links of long-distance telephone communications employ FM primarily because the wideband linear amplifiers required for amplitude modulation are unavailable at microwave frequencies.

Direct FM and VCOs

Conceptually, direct FM is straightforward and requires nothing more than a **voltage-controlled oscillator** (VCO) whose oscillation frequency has a linear dependence on applied voltage. It's possible to modulate a conventional tuned-circuit oscillator by introducing a **variable-reactance** element as part of the LC parallel resonant circuit. If the equivalent capacitance has a time dependence of the form

$$C(t) = C_0 - Cx(t)$$

and if $Cx(t)$ is "small enough" and "slow enough," then the oscillator produces $x_c(t) = A_c \cos \theta_c(t)$ where

$$\dot{\theta}_c(t) = \frac{1}{\sqrt{LC(t)}} = \frac{1}{\sqrt{LC_0}}\left[1 - \frac{C}{C_0} x(t) \right]^{-1/2}$$

Letting $\omega_c = 1/\sqrt{LC_0}$ and assuming $|(C/C_0)x(t)| \ll 1$, the binomial series expansion gives $\dot{\theta}_c(t) \approx \omega_c[1 + (C/2C_0)x(t)]$, or

$$\theta_c(t) \approx 2\pi f_c t + 2\pi \frac{C}{2C_0} f_c \int^t x(\lambda) d\lambda \qquad [1]$$

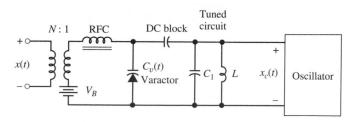

Figure 5.3–1   VCO circuit with varactor diode for variable reactance.

which constitutes frequency modulation with $f_\Delta = (C/2C_0)f_c$. Since $|x(t)| \le 1$, the approximation is good to within 1 percent when $C/C_0 < 0.013$ so the attainable frequency deviation is limited by

$$f_\Delta = \frac{C}{2C_0} f_c \le 0.006\, f_c \qquad\qquad [2]$$

This limitation quantifies our meaning of $Cx(t)$ being "small" and seldom imposes a design hardship. Similarly, the usual condition $W \ll f_c$ ensures that $Cx(t)$ is "slow enough."

Figure 5.3–1 shows a tuned-circuit oscillator with a varactor diode biased to get $Cx(t)$. The input transformer, RF choke (RFC), and dc block serve to isolate the low-frequency, high-frequency, and dc voltages. The major disadvantage with this type of circuit is that the carrier frequency tends to drift and must be stabilized by rather elaborate feedback frequency control. For this reason, many older FM transmitters are of the indirect type.

Linear integrated-circuit (IC) voltage-controlled oscillators can generate a direct FM output waveform that is relatively stable and accurate. However, in order to operate, IC VCOs require several additional external components to function. Because of their low output power, they are most suitable for applications such as cordless telephones. Figure 5.3–2 shows the schematic diagram for a direct FM transmitter using the Motorola MC1376, an 8-pin IC FM modulator. The MC1376 operates with carrier frequencies between 1.4 and 14 MHz. The VCO is fairly linear between 2 and 4 volts and can produce a peak frequency deviation of approximately 150 kHz. Higher power outputs can be achieved by utilizing an auxiliary transistor connected to a 12-V power supply.

Phase Modulators and Indirect FM

Although we seldom transmit a PM wave, we're still interested in phase modulators because: (1) the implementation is relatively easy; (2) the carrier can be supplied by a stable frequency source, such as a crystal-controlled oscillator; and (3) integrating the input signal to a phase modulator produces a *frequency*-modulated output.

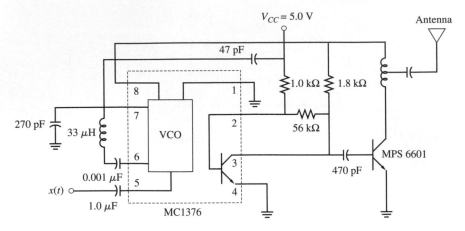

Figure 5.3–2   Schematic diagram of IC VCO direct FM generator utilizing the Motorola MC1376.

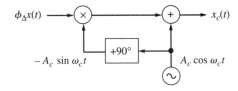

Figure 5.3–3   Narrowband phase modulator.

Figure 5.3–3 depicts a narrowband phase modulator derived from the approximation $x_c(t) \approx A_c \cos \omega_c t - A_c \phi_\Delta x(t) \sin \omega_c t$—see Eqs. (9) and (11), Sect. 5.1. The evident simplicity of this modulator depends upon the approximation condition $|\phi_\Delta x(t)| \ll 1$ radian, and phase deviations greater than $10°$ result in distorted modulation.

Larger phase shifts can be achieved by the **switching-circuit** modulator in Fig. 5.3–4. The typical waveforms shown in Fig. 5.3–4 help explain the operation. The modulating signal and a sawtooth wave at twice the carrier frequency are applied to a comparator. The comparator's output voltage goes high whenever $x(t)$ exceeds the sawtooth wave, and the flip-flop switches states at each rising edge of a comparator pulse. The flip-flop thus produces a phase-modulated square wave (like the output of a hard limiter), and bandpass filtering yields $x_c(t)$.

Now consider the indirect FM transmitter diagrammed in Fig. 5.3–5. The integrator and phase modulator constitute a *narrowband frequency modulator* that generates an initial NBFM signal with instantaneous frequency

$$f_1(t) = f_{c_1} + \frac{\phi_\Delta}{2\pi T} \, x(t)$$

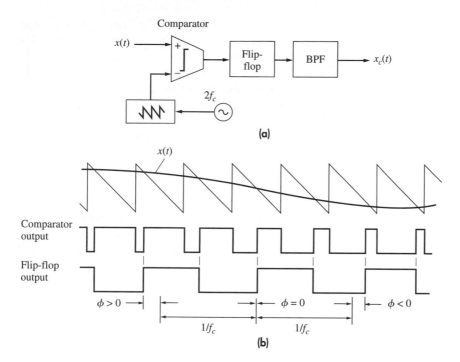

Figure 5.3–4   Switching-circuit phase modulator. (a) Schematic diagram (b) waveforms.

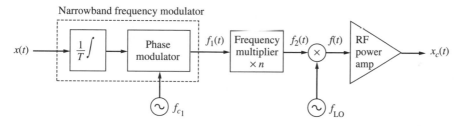

Figure 5.3–5   Indirect FM transmitter.

where T is the integrator's proportionality constant. The initial frequency deviation therefore equals $\phi_\Delta/2\pi T$ and must be increased to the desired value f_Δ by a **frequency multiplier.**

The frequency multiplier produces n-fold multiplication of *instantaneous frequency*, so

$$f_2(t) = nf_1(t) = nf_{c_1} + f_\Delta x(t) \qquad\qquad \textbf{[3]}$$

where

$$f_\Delta = n\left(\frac{\phi_\Delta}{2\pi T}\right)$$

Typical frequency multipliers consist of a chain of doublers and triplers, each unit constructed as shown in Fig. 5.2–6b. Note that this multiplication is a subtle process, affecting the *range* of frequency variation but not the *rate*. Frequency multiplication of a tone-modulated signal, for instance, increases the carrier frequency and modulation index but not the modulation frequency, so the amplitude of the sideband lines is altered while the line spacing remains the same. (Compare the spectra in Fig. 5.1–7a with $\beta = 5$ and $\beta = 10$.)

The amount of multiplication required to get f_Δ usually results in nf_{c_1} being much higher than the desired carrier frequency. Hence, Fig. 5.3–5 includes a *frequency converter* that translates the spectrum intact down to $f_c = |nf_{c_1} \pm f_{LO}|$ and the final instantaneous frequency becomes $f(t) = f_c + f_\Delta x(t)$. (The frequency conversion may actually be performed in the middle of the multiplier chain to keep the frequencies at reasonable values.) The last system component is a *power amplifier*, since all of the previous operations must be carried out at low power levels. Note the similarity to the ring modulator discussed in Sect. 4.3 that is used to generate DSB signals.

EXAMPLE 5.3–1

The indirect FM system originally designed by Armstrong employed a narrowband phase modulator in the form of Fig. 5.3–3 and produced a minute initial frequency deviation. As an illustration with representative numbers, suppose that $\phi_\Delta/2\pi T \approx 15$ Hz (which ensures negligible modulation distortion) and that $f_{c_1} = 200$ kHz (which falls near the lower limit of practical crystal-oscillator circuits). A broadcast FM output with $f_\Delta = 75$ kHz requires frequency multiplication by the factor $n \approx 75{,}000 \div 15 = 5000$. This could be achieved with a chain of four triplers and six doublers, so $n = 3^4 \times 2^6 = 5184$. But $nf_{c_1} \approx 5000 \times 200$ kHz $= 1000$ MHz, and a downconverter with $f_{LO} \approx 900$ MHz is needed to put f_c in the FM band of 88–108 MHz.

EXERCISE 5.3–1

Show that the phase at the output of Fig. 5.3–3 is given by

$$\phi(t) = \phi_\Delta x(t) - \tfrac{1}{3}\phi_\Delta^3 x^3(t) + \tfrac{1}{5}\phi_\Delta^5 x^5(t) + \cdots \qquad [4]$$

Hence, $\phi(t)$ contains *odd-harmonic distortion* unless ϕ_Δ is quite small.

Triangular-Wave FM★

Triangular-wave FM is a modern and rather novel method for frequency modulation that overcomes the inherent problems of conventional VCOs and indirect FM systems. The method generates virtually distortionless modulation at carrier frequencies up to 30 MHz, and is particularly well suited for instrumentation applications.

We'll define triangular FM by working backwards from $x_c(t) = A_c \cos \theta_c(t)$ with

$$\theta_c(t) = \omega_c t + \phi(t) - \phi(0)$$

where the initial phase shift $-\phi(0)$ has been included so that $\theta_c(0) = 0$. This phase shift does not affect the instantaneous frequency

$$f(t) = \frac{1}{2\pi} \dot{\theta}_c(t) = f_c + f_\Delta x(t)$$

Expressed in terms of $\theta_c(t)$, a unit-amplitude triangular FM signal is

$$x_\Lambda(t) = \frac{2}{\pi} \arcsin \left[\cos \theta_c(t) \right] \qquad\qquad \text{[5a]}$$

which defines a triangular waveform when $\phi(t) = 0$. Even with $\phi(t) \neq 0$, Eq. (5a) represents a *periodic triangular function of* θ_c, as plotted in Fig. 5.3–6a. Thus,

$$x_\Lambda = \begin{cases} 1 - \dfrac{2}{\pi} \theta_c & 0 < \theta_c < \pi \\[2mm] -3 + \dfrac{2}{\pi} \theta_c & \pi < \theta_c < 2\pi \end{cases} \qquad\qquad \text{[5b]}$$

and so forth for $\theta_c > 2\pi$.

Figure 5.3–6b shows the block diagram of a system that produces $x_\Lambda(t)$ from the voltage

$$v(t) = \frac{2}{\pi} \dot{\theta}_c(t) = 4[f_c + f_\Delta x(t)]$$

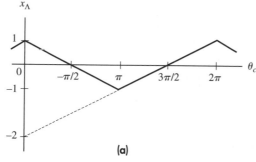

(a)

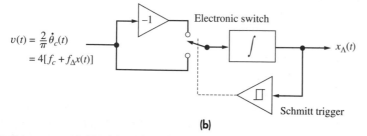

(b)

Figure 5.3–6   Triangular-wave FM. (a) Waveform; (b) modulation system.

which is readily derived from the message waveform $x(t)$. The system consists of an analog inverter, an integrator, and a Schmitt trigger controlling an electronic switch. The trigger puts the switch in the upper position whenever $x_\Lambda(t)$ increases to $+1$ and puts the switch in the lower position whenever $x_\Lambda(t)$ decreases to -1.

Suppose the system starts operating at $t = 0$ with $x_\Lambda(0) = +1$ and the switch in the upper position. Then, for $0 < t < t_1$,

$$x_\Lambda(t) = 1 - \int_0^t v(\lambda)\, d\lambda = 1 - \frac{2}{\pi}[\theta_c(t) - \theta_c(0)]$$

$$= 1 - \frac{2}{\pi}\theta_c(t) \qquad 0 < t < t_1$$

so $x_\Lambda(t)$ traces out the downward ramp in Fig. 5.3–6a until time t_1 when $x_\Lambda(t_1) = -1$, corresponding to $\theta_c(t_1) = \pi$. Now the trigger throws the switch to the lower position and

$$x_\Lambda(t) = -1 + \int_{t_1}^t v(\lambda)\, d\lambda = -1 + \frac{2}{\pi}[\theta_c(t) - \theta_c(t_1)]$$

$$= -3 + \frac{2}{\pi}\theta_c(t) \qquad t_1 < t < t_2$$

so $x_\Lambda(t)$ traces out the upward ramp in Fig. 5.3–6a. The upward ramp continues until time t_2 when $\theta_c(t_2) = 2\pi$ and $x_\Lambda(t_2) = +1$. The switch then triggers back to the upper position, and the operating cycle goes on periodically for $t > t_2$.

A sinusoidal FM wave is obtained from $x_\Lambda(t)$ using a nonlinear waveshaper with transfer characteristics $T[x_\Lambda(t)] = A_c \sin[(\pi/2)x_\Lambda(t)]$, which performs the inverse of Eq. (5a). Or $x_\Lambda(t)$ can be applied to a hard limiter to produce **square-wave** FM. A laboratory test generator might have all three outputs available.

Frequency Detection

A **frequency detector,** often called a **discriminator,** produces an output voltage that should vary linearly with the instantaneous frequency of the input. There are perhaps as many different circuit designs for frequency detection as there are designers who have considered the problem. However, almost every circuit falls into one of the following four operational categories:

1. FM-to-AM conversion
2. Phase-shift discrimination
3. Zero-crossing detection
4. Frequency feedback

We'll look at illustrative examples from the first three categories, postponing frequency feedback to Sect. 7.3. Analog *phase* detection is not discussed here because

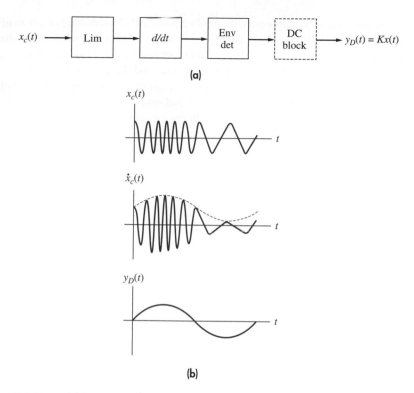

(a)

$x_c(t)$

$\dot{x}_c(t)$

$y_D(t)$

(b)

Figure 5.3–7   (a) Frequency detector with limiter and FM-to-AM conversion; (b) waveforms.

it's seldom needed in practice and, if needed, can be accomplished by integrating the output of a frequency detector.

Any device or circuit whose output equals the *time derivative* of the input produces **FM-to-AM conversion.** To be more specific, let $x_c(t) = A_c \cos \theta_c(t)$ with $\dot{\theta}_c(t) = 2\pi[f_c + f_\Delta x(t)]$; then

$$\dot{x}_c(t) = -A_c \dot{\theta}_c(t) \sin \theta_c(t) \qquad\qquad [6]$$
$$= 2\pi A_c[f_c + f_\Delta\, x(t)] \sin [\theta_c(t) \pm 180°]$$

Hence, an **envelope detector** with input $\dot{x}_c(t)$ yields an output proportional to $f(t) = f_c + f_\Delta x(t)$.

Figure 5.3–7*a* diagrams a conceptual frequency detector based on Eq. (6). The diagram includes a limiter at the input to remove any spurious amplitude variations from $x_c(t)$ before they reach the envelope detector. It also includes a dc block to remove the constant carrier-frequency offset from the output signal. Typical waveforms are sketched in Fig. 5.3–7*b* taking the case of tone modulation.

For actual hardware implementation of FM-to-AM conversion, we draw upon the fact that an ideal differentiator has $|H(f)| = 2\pi f$. Slightly above or below resonance, the transfer function of an ordinary tuned circuit shown in Fig. 5.3–8a approximates the desired linear amplitude response over a small frequency range. Thus, for instance, a detuned AM receiver will roughly demodulate FM via **slope detection.**

Extended linearity is achieved by the **balanced discriminator** circuit in Fig. 5.3–8b. A balanced discriminator includes two resonant circuits, one tuned above f_c and the other below, and the output equals the difference of the two envelopes. The resulting frequency-to-voltage characteristic takes the form of the well-known S curve in Fig. 5.3–8c. No dc block is needed, since the carrier-frequency

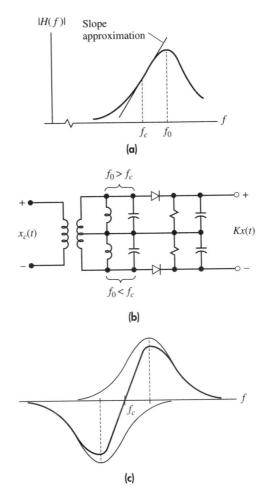

Figure 5.3–8   (a) Slope detection with a tuned circuit; (b) balanced discriminator circuit; (c) frequency-to-voltage characteristic.

offset cancels out, and the circuit has good performance at low modulating frequencies. The balanced configuration easily adapts to the microwave band, with resonant cavities serving as tuned circuits and crystal diodes for envelope detectors.

Phase-shift discriminators involve circuits with linear *phase* response, in contrast to the linear amplitude response of slope detection. The underlying principle comes from an approximation for time differentiation, namely

$$\dot{v}(t) \approx \frac{1}{t_1}\left[v(t) - v(t - t_1)\right] \qquad\qquad [7]$$

providing that t_1 is small compared to the variation of $v(t)$. Now an FM wave has $\dot{\phi}(t) = 2\pi f_\Delta x(t)$ so

$$\phi(t) - \phi(t - t_1) \approx t_1\dot{\phi}(t) = 2\pi f_\Delta t_1 x(t) \qquad\qquad [8]$$

The term $\phi(t - t_1)$ can be obtained with the help of a delay line or, equivalently, a linear phase-shift network.

Figure 5.3–9 represents a phase-shift discriminator built with a network having group delay t_1 and carrier delay t_0 such that $\omega_c t_0 = 90°$—which accounts for the name **quadrature detector.** From Eq. (11), Sect. 5.2, the phase-shifted signal is proportional to $\cos[\omega_c t - 90° + \phi(t - t_1)] = \sin[\omega_c t + \phi(t - t_1)]$. Multiplication by $\cos[\omega_c t + \phi(t)]$ followed by lowpass filtering yields an output proportional to

$$\sin[\phi(t) - \phi(t - t_1)] \approx \phi(t) - \phi(t - t_1)$$

assuming t_1 is small enough that $|\phi(t) - \phi(t - t_1)| \ll \pi$. Therefore,

$$y_D(t) \approx K_D f_\Delta x(t)$$

where the detection constant K_D includes t_1. Despite these approximations, a quadrature detector provides better linearity than a balanced discriminator and is often found in high-quality receivers.

Other phase-shift circuit realizations include the **Foster-Seely discriminator** and the popular **ratio detector.** The latter is particularly ingenious and economical, for it combines the operations of limiting and demodulation into one unit. See Tomasi (1998, Chap. 7) for further details.

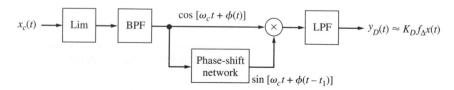

Figure 5.3–9   Phase-shift discriminator or quadrature detector.

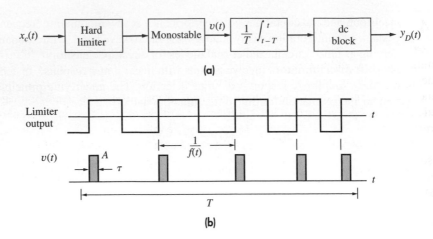

Figure 5.3–10   Zero-crossing detector. (a) Diagram; (b) waveforms.

Lastly, Fig. 5.3–10 gives the diagram and waveforms for a simplified **zero-crossing detector.** The square-wave FM signal from a hard limiter triggers a monostable pulse generator, which produces a short pulse of fixed amplitude A and duration τ at each upward (or downward) zero crossing of the FM wave. If we invoke the quasi-static viewpoint and consider a time interval T such that $W \ll 1/T \ll f_c$, the monostable output $v(t)$ looks like a rectangular pulse train with nearly constant period $1/f(t)$. Thus, there are $n_T \approx Tf(t)$ pulses in this interval, and continually integrating $v(t)$ over the past T seconds yields

$$\frac{1}{T}\int_{t-T}^{t} v(\lambda)\,d\lambda = \frac{1}{T}n_T A\tau \approx A\tau f(t)$$

which becomes $y_D(t) \approx K_D f_\Delta x(t)$ after the dc block.

Commercial zero-crossing detectors may have better than 0.1 percent linearity and operate at center frequencies from 1 Hz to 10 MHz. A divide-by-ten counter inserted after the hard limiter extends the range up to 100 MHz.

Today most FM communication devices utilize linear integrated circuits for FM detection. Their reliability, small size, and ease of design have fueled the growth of portable two-way FM and cellular radio communications systems. Phase-lock loops and FM detection will be discussed in Sect. 7.3.

EXERCISE 5.3–2   Given a delay line with time delay $t_0 \ll 1/f_c$, devise a frequency detector based on Eqs. (6) and (7).

5.4 INTERFERENCE

Interference refers to the contamination of an information-bearing signal by another similar signal, usually from a human source. This occurs in radio communication when the receiving antenna picks up two or more signals in the same frequency band. Interference may also result from multipath propagation, or from electromagnetic coupling between transmission cables. Regardless of the cause, severe interference prevents successful recovery of the message information.

Our study of interference begins with the simple but nonetheless informative case of interfering sinusoids, representing unmodulated carrier waves. This simplified case helps bring out the differences between interference effects in AM, FM, and PM. Then we'll see how the technique of **deemphasis filtering** improves FM performance in the face of interference. We conclude with a brief examination of the FM **capture effect.**

Interfering Sinusoids

Consider a receiver tuned to some carrier frequency f_c. Let the total received signal be

$$v(t) = A_c \cos \omega_c t + A_i \cos \left[(\omega_c + \omega_i)t + \phi_i \right]$$

The first term represents the desired signal as an unmodulated carrier, while the second term is an interfering carrier with amplitude A_i, frequency $f_c + f_i$, and relative phase angle ϕ_i.

To put $v(t)$ in the envelope-and-phase form $v(t) = A_v(t) \cos \left[\omega_c t + \phi_v(t) \right]$, we'll introduce

$$\rho \triangleq A_i/A_c \qquad \theta_i(t) \triangleq \omega_i t + \phi_i \qquad \text{[1]}$$

Hence, $A_i = \rho A_c$ and the phasor construction in Fig. 5.4–1 gives

$$A_v(t) = A_c \sqrt{1 + \rho^2 + 2\rho \cos \theta_i(t)} \qquad \text{[2]}$$

$$\phi_v(t) = \arctan \frac{\rho \sin \theta_i(t)}{1 + \rho \cos \theta_i(t)}$$

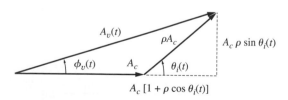

Figure 5.4–1   Phasor diagram of interfering carriers.

These expressions show that interfering sinusoids produce both *amplitude* and *phase modulation*. In fact, if $\rho \ll 1$ then

$$A_v(t) \approx A_c[1 + \rho \cos(\omega_i t + \phi_i)] \tag{3}$$

$$\phi_v(t) \approx \rho \sin(\omega_i t + \phi_i)$$

which looks like *tone modulation* at frequency f_i with AM modulation index $\mu = \rho$ and FM or PM modulation index $\beta = \rho$. At the other extreme, if $\rho \gg 1$ then

$$A_v(t) \approx A_i[1 + \rho^{-1} \cos(\omega_i t + \phi_i)]$$

$$\phi_v(t) \approx \omega_i t + \phi_i$$

so the envelope still has tone modulation but the phase corresponds to a shifted carrier frequency $f_c + f_i$ plus the constant ϕ_i.

Next we investigate what happens when $v(t)$ is applied to an ideal envelope, phase, or frequency demodulator with detection constant K_D. We'll take the weak interference case ($\rho \ll 1$) and use the approximation in Eq. (3) with $\phi_i = 0$. Thus, the demodulated output is

$$y_D(t) \approx \begin{cases} K_D(1 + \rho \cos \omega_i t) & \text{AM} \\ K_D \rho \sin \omega_i t & \text{PM} \\ K_D \rho f_i \cos \omega_i t & \text{FM} \end{cases} \tag{4}$$

provided that $|f_i| \leq W$—otherwise, the lowpass filter at the output of the demodulator would reject $|f_i| > W$. The constant term in the AM result would be removed if the demodulator includes a dc block. As written, this result also holds for *synchronous* detection in DSB and SSB systems since we've assumed $\phi_i = 0$. The multiplicative factor f_i in the FM result comes from the instantaneous frequency deviation $\dot{\phi}_v(t)/2\pi$.

Equation (4) reveals that weak interference in a linear modulation system or phase modulation system produces a spurious output tone with amplitude proportional to $\rho = A_i/A_c$, independent of f_i. But the tone amplitude is proportional to ρf_i in an FM system. Consequently, FM will be less vulnerable to interference from a *cochannel* signal having the same carrier frequency, so $f_i \approx 0$, and more vulnerable to *adjacent-channel* interference ($f_i \neq 0$). Figure 5.4–2 illustrates this difference in

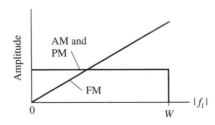

Figure 5.4–2   Amplitude of demodulated interference from a carrier at frequency $f_c + f_i$.

the form of a plot of demodulated interference amplitude versus $|f_i|$. (The crossover point would correspond to $|f_i| = 1$ Hz if all three detector constants had the same numerical value.)

The analysis of demodulated interference becomes a much more difficult task with arbitrary values of ρ and/or modulated carriers. We'll return to that problem after exploring the implications of Fig. 5.4–2.

Let $A_i = A_c$ so $\rho = 1$ in Eq. (2). Take $\phi_i = 0$ and use trigonometric identities to show that

EXERCISE 5.4–1

$$A_v(t) = 2A_c \, |\cos (\omega_i t/2)| \qquad \phi_v(t) = \omega_i t/2$$

Then sketch the demodulated output waveform for envelope, phase, and frequency detection assuming $f_i \ll W$.

Deemphasis and Preemphasis Filtering

The fact that detected FM interference is most severe at large values of $|f_i|$ suggests a method for improving system performance with selective postdetection filtering, called **deemphasis filtering.** Suppose the demodulator is followed by a lowpass filter having an amplitude ratio that begins to decrease gradually *below W;* this will *deemphasize* the high-frequency portion of the message band and thereby reduce the more serious interference. A sharp-cutoff (ideal) lowpass filter is still required to remove any residual components above W, so the complete demodulator consists of a frequency detector, deemphasis filter, and lowpass filter, as in Fig. 5.4–3.

Obviously deemphasis filtering also attenuates the high-frequency components of the message itself, causing distortion of the output signal unless corrective measures are taken. But it's a simple matter to compensate for deemphasis distortion by **predistorting** or **preemphasizing** the modulating signal at the transmitter before modulation. The preemphasis and deemphasis filter characteristics should be related by

$$H_{\text{pe}}(f) = \frac{1}{H_{\text{de}}(f)} \qquad |f| \leq W \tag{5}$$

to yield net undistorted transmission. In essence,

Figure 5.4–3   Complete FM demodulator.

We preemphasize the message before modulation (where the interference is absent) so we can deemphasize the interference relative to the message after demodulation.

Preemphasis/deemphasis filtering offers potential advantages whenever undesired contaminations tend to predominate at certain portions of the message band. For instance, the *Dolby* system for tape recording dynamically adjusts the amount of preemphasis/deemphasis in inverse proportion to the high-frequency signal content; see Stremler (1990, App. F) for details. However, little is gained from deemphasizing phase modulation or linear modulation because the demodulated interference amplitude does not depend on the frequency.

The FM deemphasis filter is usually a simple first-order network having

$$H_{de}(f) = \left[1 + j\left(\frac{f}{B_{de}}\right) \right]^{-1} \approx \begin{cases} 1 & |f| \ll B_{de} \\ \dfrac{B_{de}}{jf} & |f| \gg B_{de} \end{cases} \qquad [6]$$

where the 3 dB bandwidth B_{de} is considerably less than the message bandwidth W. Since the interference amplitude increases linearly with $|f_i|$ in the absence of filtering, the deemphasized interference response is $|H_{de}(f_i)| \times |f_i|$, as sketched in Fig. 5.4–4. Note that, like PM, this becomes constant for $|f_i| \gg B_{de}$. Therefore, FM can be superior to PM for both adjacent-channel and cochannel interference.

At the transmitting end, the corresponding preemphasis filter function should be

$$H_{pe}(f) = \left[1 + j\left(\frac{f}{B_{de}}\right) \right] \approx \begin{cases} 1 & |f| \ll B_{de} \\ \dfrac{jf}{B_{de}} & |f| \gg B_{de} \end{cases} \qquad [7]$$

which has little effect on the lower message frequencies. At higher frequencies, however, the filter acts as a *differentiator*, the output spectrum being proportional to $fX(f)$ for $|f| \gg B_{de}$. But differentiating a signal before frequency modulation is equivalent to *phase modulation*! Hence, preemphasized FM is actually a combination of FM and PM, combining the advantages of both with respect to interference.

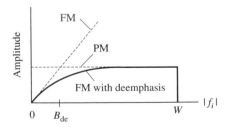

Figure 5.4–4   Demodulated interference amplitude with FM deemphasis filtering.

As might be expected, this turns out to be equally effective for reducing *noise*, as will be discussed in more detail in Chap. 10.

Referring to $H_{pe}(f)$ as given above, we see that the amplitude of the maximum modulating frequency is increased by a factor of W/B_{de}, which means that the frequency deviation is increased by this same factor. Generally speaking, the increased deviation requires a greater transmission bandwidth, so the preemphasis-deemphasis improvement is not without price. Fortunately, many modulating signals of interest, particularly audio signals, have relatively little energy in the high-frequency end of the message band, and therefore the higher frequency components do not develop maximum deviation, the transmission bandwidth being dictated by lower components of larger amplitude. Adding high-frequency preemphasis tends to equalize the message spectrum so that all components require the same bandwidth. Under this condition, the transmission bandwidth need not be increased.

Typical deemphasis and preemphasis networks for commercial FM are shown in Fig. 5.4–5 along with their Bode diagrams. The RC time constant in both circuits equals 75 μs, so $B_{de} = 1/2\pi RC \approx 2.1$ kHz. The preemphasis filter has an upper break frequency at $f_u = (R + r)/2\pi RrC$, usually chosen to be well above the audio range, say $f_u \geq 30$ kHz.

EXAMPLE 5.4–1

Suppose an audio signal is modeled as a sum of tones with low-frequency amplitudes $A_m \leq 1$ for $f_m \leq 1$ kHz and high-frequency amplitudes $A_m \leq 1$ kHz$/f_m$ for $f_m > 1$ kHz. Use Eqs. (1) and (2), Sect. 5.2, to estimate the bandwidth required for a single tone at $f_m = 15$ kHz whose amplitude has been preemphasized by $|H_{pe}(f)|$ given in Eq. (7) with $B_{de} = 2$ kHz. Assume $f_\Delta = 75$ kHz and compare your result with $B_T \approx 210$ kHz.

EXERCISE 5.4–2

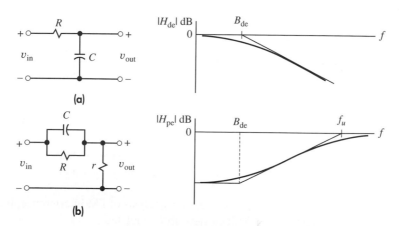

Figure 5.4–5   (a) Deemphasis filter; (b) preemphasis filter.

FM Capture Effect★

Capture effect is a phenomenon that takes place in FM systems when two signals have nearly equal amplitudes at the receiver. Small variations of relative amplitude then cause the stronger of the two to dominate the situation, suddenly displacing the other signal at the demodulated output. You may have heard the annoying results when listening to a distant FM station.

For a reasonably tractable analysis of capture effect, we'll consider an unmodulated carrier with modulated cochannel interference ($f_i = 0$). The resultant phase $\phi_v(t)$ is then given by Eq. (2) with $\theta_i(t) = \phi_i(t)$, where $\phi_i(t)$ denotes the phase modulation of the interfering signal. Thus, if $K_D = 1$ for simplicity, the demodulated signal becomes

$$y_D(t) = \dot{\phi}_v(t) = \frac{d}{dt}\left[\arctan \frac{\rho \sin \phi_i(t)}{1 + \rho \cos \phi_i(t)} \right]$$ [8a]

$$= \alpha(\rho, \phi_i)\,\dot{\phi}_i(t)$$

where

$$\alpha(\rho,\phi_i) \triangleq \frac{\rho^2 + \rho \cos \phi_i}{1 + \rho^2 + 2\rho \cos \phi_i}$$ [8b]

The presence of $\dot{\phi}_i(t)$ in Eq. (8a) indicates potentially **intelligible** interference (or **cross talk**) to the extent that $\alpha(\rho, \phi_i)$ remains constant with time. After all, if $\rho \gg 1$ then $\alpha(\rho, \phi_i) \approx 1$ and $y_D(t) \approx \dot{\phi}_i(t)$.

But capture effect occurs when $A_i \approx A_c$, so $\rho \approx 1$ and Eq. (8b) does not immediately simplify. Instead, we note that

$$\alpha(\rho, \phi_i) = \begin{cases} \rho/(1 + \rho) & \phi_i = 0,\ \pm\, 2\pi, \ldots \\ \rho^2/(1 + \rho^2) & \phi_i = \pm\, \pi/2,\ \pm\, 3\pi/2, \ldots \\ -\rho/(1 - \rho) & \phi_i = \pm\, \pi,\ \pm\, 3\pi, \ldots \end{cases}$$

and we resort to plots of $\alpha(\rho, \phi_i)$ versus ϕ_i as shown in Fig. 5.4–6a. Except for the negative spikes, these plots approach $\alpha(\rho, \phi_i) = 0.5$ as $\rho \to 1$, and thus $y_D(t) = 0.5$ $\dot{\phi}_i(t)$. For $\rho < 1$, the strength of the demodulated interference essentially depends on the peak-to-peak value

$$\alpha_{pp} = \alpha(\rho, 0) - \alpha(\rho, \pi) = 2\rho/(1 - \rho^2)$$

which is plotted versus ρ in Fig. 5.4–6b. This knee-shaped curve reveals that if transmission fading causes ρ to vary around a nominal value of about 0.7, the interference almost disappears when $\rho < 0.7$ whereas it takes over and "captures" the output when $\rho > 0.7$.

Panter (1965, Chap. 11) presents a detailed analysis of FM interference, including waveforms that result when both carriers are modulated.

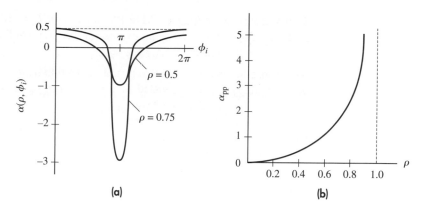

Figure 5.4–6

5.5   PROBLEMS

5.1–1   Sketch and label $\phi(t)$ and $f(t)$ for PM and FM when $x(t) = A\Lambda(t/\tau)$. Take $\phi(-\infty) = 0$ in the FM case.

5.1–2   Do Prob. 5.1–1 with $x(t) = A\cos(\pi t/\tau)\Pi(t/2\tau)$.

5.1–3   Do Prob. 5.1–1 with $x(t) = \dfrac{4At}{t^2 - 16}$ for $t > 4$.

5.1–4*   A *frequency-sweep generator* produces a sinusoidal output whose instantaneous frequency increases linearly from f_1 at $t = 0$ to f_2 at $t = T$. Write $\theta_c(t)$ for $0 \le t \le T$.

5.1–5   Besides PM and FM, two other possible forms of exponential modulation are *phase-integral modulation*, with $\phi(t) = K\, dx(t)/dt$, and *phase-acceleration modulation*, with

$$f(t) = f_c + K \int^{t} x(\lambda)d\lambda$$

Add these to Table 5.1–1 and find the maximum values of $\phi(t)$ and $f(t)$ for all four types when $x(t) = \cos 2\pi f_m t$.

5.1–6   Use Eq. (16) to obtain Eq. (18a) from Eq. (15).

5.1–7‡   Derive Eq. (16) by finding the exponential Fourier series of the complex periodic function $\exp(j\beta \sin \omega_m t)$.

5.1–8   Tone modulation is applied simultaneously to a frequency modulator and a phase modulator and the two output spectra are identical. Describe how these two spectra will change when: (*a*) the tone amplitude is increased or decreased; (*b*) the tone frequency is increased or decreased; (*c*) the tone amplitude and frequency are increased or decreased in the same proportion.

5.1–9   Consider a tone-modulated FM or PM wave with $f_m = 10$ kHz, $\beta = 2.0$, $A_c = 100$, and $f_c = 30$ kHz. (a) Write an expression for $f(t)$. (b) Draw the line spectrum and show therefrom that $S_T < A_c^2/2$.

5.1–10* Do Prob. 5.1–9 with $f_m = 20$ kHz and $f_c = 40$ kHz, in which case $S_T > A_c^2/2$.

5.1–11   Construct phasor diagrams for tone-modulated FM with $A_c = 10$ and $\beta = 0.5$ when $\omega_m t = 0$, $\pi/4$, and $\pi/2$. Calculate A and ϕ from each diagram and compare with the theoretical values.

5.1–12   Do Prob. 5.1–11 with $\beta = 1.0$.

5.1–13   A tone-modulated FM signal with $\beta = 1.0$ and $f_m = 100$ Hz is applied to an ideal BPF with $B = 250$ Hz centered at $f_c = 500$. Draw the line spectrum, phasor diagram, and envelope of the output signal.

5.1–14   Do Prob. 5.1–13 with $\beta = 5.0$.

5.1–15   One implementation of a music synthesizer exploits the harmonic structure of FM tone modulation. The violin note C_2 has a frequency of $f_0 = 405$ Hz with harmonics at integer multiples of f_0 when played with a bow. Construct a system using FM tone modulation and frequency converters to synthesize this note with f_0 and three harmonics.

5.1–16‡ Consider FM with periodic square-wave modulation defined by $x(t) = 1$ for $0 < t < T_0/2$ and $x(t) = -1$ for $-T_0/2 < t < 0$. (a) Take $\phi(0) = 0$ and plot $\phi(t)$ for $-T_0/2 < t < T_0/2$. Then use Eq. (20a) to obtain

$$c_n = \frac{1}{2} e^{j\pi\beta} \left[\operatorname{sinc}\left(\frac{n+\beta}{2}\right) e^{j\pi n/2} + \operatorname{sinc}\left(\frac{n-\beta}{2}\right) e^{-j\pi n/2} \right]$$

where $\beta = fT_0$. (b) Sketch the resulting magnitude line spectrum when β is a large integer.

5.2–1   A message has $W = 15$ kHz. Estimate the FM transmission bandwidth for $f_\Delta = 0.1$, 0.5, 1, 5, 10, 50, 100, and 500 kHz.

5.2–2   Do Prob. 5.2–1 with $W = 5$ kHz.

5.2–3   An FM system has $f_\Delta = 10$ kHz. Use Table 9.4–1 and Fig. 5.2–1 to estimate the bandwidth for: (a) barely intelligible voice transmission; (b) telephone-quality voice transmission: (c) high-fidelity audio transmission.

5.2–4   A video signal with $W = 5$ MHz is to be transmitted via FM with $f_\Delta = 25$ MHz. Find the minimum carrier frequency consistent with fractional bandwidth considerations. Compare your results with transmission via DSB amplitude modulation.

5.2–5*   Your new wireless headphones use infrared FM transmission and have a frequency response of 30–15,000 Hz. Find B_T and f_Δ consistent with fractional bandwidth considerations, assuming $f_c = 5 \times 10^{14}$ Hz.

5.2–6   A commercial FM radio station alternates between music and talk show/call-in formats. The broadcasted CD music is bandlimited to 15 kHz based on convention. Assuming $D = 5$ is used for both music and voice, what percentage of the available transmission bandwidth is used during the talk show if we take $W = 5$ kHz for voice signals?

5.2–7   An FM system with $f_\Delta = 30$ kHz has been designed for $W = 10$ kHz. Approximately what percentage of B_T is occupied when the modulating signal is a unit-amplitude tone at $f_m = 0.1$, 1.0, or 5.0 kHz? Repeat your calculations for a PM system with $\phi_\Delta = 3$ rad.

5.2–8   Consider phase-integral and phase-acceleration modulation defined in Prob. 5.1–5. Investigate the bandwidth requirements for tone modulation, and obtain transmission bandwidth estimates. Discuss your results.

5.2–9*   The transfer function of a single-tuned BPF is $H(f) \approx 1/[1 + j2Q\,(f - f_c)/f_c]$ over the positive-frequency passband. Use Eq. (10) to obtain an expression for the output signal and its instantaneous phase when the input is an NBPM signal.

5.2–10   Use Eq. (10) to obtain an expression for the output signal and its amplitude when an FM signal is distorted by a system having $H(f) = K_0 - K_3(f - f_c)^3$ over the positive-frequency passband.

5.2–11   Use Eq. (13) to obtain an expression for the output signal and its instantaneous frequency when an FM signal is distorted by a system having $|H(f)| = 1$ and $\arg H(f) = \alpha_1(f - f_c) + \alpha_3(f - f_c)^3$ over the positive-frequency passband.

5.2–12   An FM signal is applied to the BPF in Prob. 5.2–9. Let $\alpha = 2Qf_\Delta/f_c \ll 1$ and use Eq. (13) to obtain an approximate expression for the output signal and its instantaneous frequency.

5.2–13   Let the input to the system in Fig. 5.2–6a be an FM signal with $D = f_\Delta/W$ and spurious amplitude variations. Sketch the spectrum at the output of the limiter and show that successful operation requires $f_\Delta < (f_c - W)/2$.

5.2–14   The input to the system in Fig. 5.2–6b is an FM signal with $D = f_\Delta/W$ and the BPF is centered at $3f_c$, corresponding to a frequency tripler. Sketch the spectrum at the filter's input and obtain a condition on f_Δ in terms of f_c and W that ensures successful operation.

5.2–15* Do Prob. 5.2–14 with the BPF centered at $4f_c$, corresponding to a frequency quadrupler.

5.3–1   The equivalent tuning capacitance in Fig. 5.3–1 is $C(t) = C_1 + C_v(t)$ where $C_v(t) = C_2/\sqrt{V_B + x(t)/N}$. Show that $C(t) \approx C_0 - C_x(t)$ with 1 percent accuracy if $NV_B \geq 300/4$. Then show that the corresponding limitation on the frequency deviation is $f_\Delta < f_c/300$.

5.3–2   The direct FM generator in Fig. 5.3–2 is used for a remote-controlled toy car. Find the range of allowable values for W so that B_T satisfies the fractional bandwidth requirements, assuming the maximum frequency deviation of 150 kHz is used.

5.3–3   Confirm that $x_c(t) = A_c \cos \theta_c(t)$ is a solution of the integrodifferential equation $\dot{x}_c(t) = -\dot{\theta}_c(t) \int \dot{\theta}_c(t) x_c(t)\, dt$. Then draw the block diagram of a direct FM generator based on this relationship.

5.3–4   Suppose an FM detector receives the transmitted signal that was generated by the phase modulator in Fig. 5.3–3. Describe the distortion in the output message signal. (Hint: Consider the relationship between the message signal amplitude and frequency, and the modulation index.)

5.3–5*   An audio message signal is transmitted using frequency modulation. Describe the distortion on the output message signal if it is received by a PM detector. (Hint: Consider the relationship between the message signal amplitude and frequency, and the modulation index.)

5.3–6   Design a wireless stereo speaker system using indirect FM. Assuming $W = 15$ kHz, $D = 5, f_{c1} = 500$ kHz, $f_c = 915$ MHz, and $\phi_\Delta/2\pi T < 20$, determine the number of triplers needed in your multiplier stage, and find the value of f_{LO} needed to design your system.

5.3–7   The audio portion of a television transmitter is an indirect FM system having $W = 10$ kHz, $D = 2.5$, and $f_c = 4.5$ MHz. Devise a block diagram of this system with $\phi_\Delta/2\pi T < 20$ Hz and $f_c = 200$ kHz. Use the shortest possible multiplier chain consisting of frequency triplers and doublers, and locate the down-converter such that no frequency exceeds 100 MHz.

5.3–8   A signal with $W = 4$ kHz is transmitted using indirect FM with $f_c = 1$ MHz and $f_\Delta = 12$ kHz. If $\phi_\Delta/2\pi T < 100$ and $f_{c1} = 10$ kHz, how many doublers will be needed to achieve the desired output parameters? Draw the block diagram of the system indicating the value and location of the local oscillator such that no frequency exceeds 10 MHz.

5.3–9   Suppose the phase modulator in Fig. 5.3–5 is implemented as in Fig. 5.3–3. Take $x(t) = A_m \cos \omega_m t$ and let $\beta = (\phi_\Delta/2\pi T)(A_m/f_m)$. (a) Show that if $\beta \ll 1$, then

$$f_1(t) \approx f_c + \beta f_m \left[\cos \omega_m t + (\beta/2)^2 \cos 3\omega_m t\right]$$

(b) Obtain a condition on $\phi_\Delta/2\pi T$ so the third-harmonic distortion does not exceed 1 percent when $A_m \le 1$ and 30 Hz $\le f_m \le 15$ kHz, as in FM broadcasting.

5.3–10   Let the input to Fig. 5.3–7a be an FM signal with $f_\Delta \ll f_c$ and let the differentiator be implemented by a tuned circuit with $H(f) = 1/[1 + j(2Q/f_0)(f - f_0)]$ for $f \approx f_0$. Use the quasi-static method to show that $y_D(t) \approx K_D f_\Delta x(t)$ when $f_0 = f_c + b$ provided that $f_\Delta \ll b \ll f_0/2Q$.

5.3–11* Let the input to Fig. 5.3–7a be an FM signal with $f_\Delta \ll f_c$ and let the differentiator be implemented by a first-order lowpass filter with $B = f_c$. Use quasi-static analysis to show that $y_D(t) \approx -K_1 f_\Delta x(t) + K_2 f_\Delta^2 x^2(t)$. Then take $x(t) = \cos \omega_m t$ and obtain a condition on f_Δ/f_c so the second-harmonic distortion is less than 1%.

5.3–12   The tuned circuits in Fig. 5.3–8*b* have transfer functions of the form $H(f) = 1/[1 + j(2Q/f_0)(f - f_0)]$ for $f \approx f_0$. Let the two center frequencies be $f_0 = f_c \pm b$ with $f_\Delta \le b \ll f_c$. Use quasi-static analysis to show that if both circuits have $(2Q/f_0)b = \alpha \ll 1$, then $y_D(t) \approx K_1 x(t) - K_3 x^3(t)$ where $K_3/K_1 \ll 1$.

5.4–1   Obtain an approximate expression for the output of an amplitude demodulator when the input is an AM signal with 100 percent modulation plus an interfering signal $A_i[1 + x_i(t)] \cos[(\omega_c + \omega_i)t + \phi_i]$ with $\rho = A_i/A_c \ll 1$. Is the demodulated interference intelligible?

5.4–2   Obtain an approximate expression for the output of a phase demodulator when the input is an NBPM signal with 100 percent modulation plus an interfering signal $A_i \cos[(\omega_c + \omega_i)t + \phi_i(t)]$ with $\rho = A_i/A_c \ll 1$. Is the demodulated interference intelligible?

5.4–3   Investigate the performance of envelope detection versus synchronous detection of AM in the presence of multipath propagation, so that $v(t) = x_c(t) + \alpha x_c(t - t_d)$ with $\alpha^2 < 1$. Consider the special cases $\omega_c t_d \approx \pi/2$ and $\omega_c t_d \approx \pi$.

5.4–4   You are talking on your cordless phone, which uses amplitude modulation, when someone turns on a motorized appliance, causing static on the phone. You switch to your new FM cordless phone, and the call is clear. Explain.

5.4–5*   In World War II they first used preemphasis/deemphasis in amplitude modulation for mobile communications to make the high-frequency portion of speech signals more intelligible. Assuming that the amplitude of the speech spectrum is bandlimited to 3.5 kHz and rolls off at about 6 dB per decade (factor of 10 on a log-frequency scale) above 500 Hz, draw the Bode diagrams of the preemphasis and deemphasis filters so that the message signal has a flattened spectrum prior to transmission. Discuss the impact on the transmitted power for DSB versus standard AM with $\mu = 1$.

5.4–6   Preemphasis filters can also be used in hearing aid applications. Suppose a child has a hearing loss that gets worse at high frequencies. A preemphasis filter can be designed to be the approximate inverse of the high frequency deemphasis that takes place in the ear. In a noisy classroom it is often helpful to have the teacher speak into a microphone and have the signal transmitted by FM to a receiver that the child is wearing. Is it better to have the preemphasis filter at the microphone end prior to FM transmission or at the receiver worn by the child? Discuss your answer in terms of transmitted power, transmitted bandwidth, and susceptibility to interference.

5.4–7   A message signal $x(t)$ has an energy or power spectrum that satisfies the condition

$$G_x(f) \le (B_{de}/f)^2 G_{max} \qquad |f| > B_{de}$$

where G_{max} is the maximum of $G_x(f)$ in $|f| < B_{de}$. If the preemphasis filter in Eq. (7) is applied to $x(t)$ before FM transmission, will the transmitted bandwidth be increased?

5.4–8   Equation (8) also holds for the case of unmodulated adjacent-channel interference if we let $\phi_i(t) = \omega_i t$. Sketch the resulting demodulated waveform when $\rho = 0.4, 0.8,$ and 1.2.

5.4–9   If the amplitude of an interfering sinusoid and the amplitude of the sinusoid of interest are approximately equal, $\rho = A_i/A_c \approx 1$ and Eq. (8b) appears to reduce to $\alpha(\rho, \phi_i) = 1/2$ for all ϕ_i, resulting in cross talk. However, large spikes will appear at the demodulator output when $\phi_i = \pm\pi$. Show that if $\phi_i = \pi$ and $\rho = 1 \pm \epsilon$, then $\alpha(\rho, \pi) \to \pm\infty$ as $\epsilon \to 0$. Conversely, show that if ρ is slightly less than 1 and $\phi_i = \pi \pm \epsilon$, then $\alpha(\rho, \phi_i) \to -\infty$ as $\epsilon \to 0$.

5.4–10‡*   Develop an expression for the demodulated signal when an FM signal with instantaneous phase $\phi(t)$ has interference from an unmodulated adjacent-channel carrier. Write your result in terms of $\phi(t)$, $\rho = A/A_c$, and $\theta_i(t) = \omega_i t + \phi_i$.

Sampling and Pulse Modulation

Experimental data and mathematical functions are frequently displayed as *continuous* curves, even though a finite number of *discrete points* was used to construct the graphs. If these points, or *samples*, have sufficiently close spacing, a smooth curve drawn through them allows you to interpolate intermediate values to any reasonable degree of accuracy. It can therefore be said that the continuous curve is adequately described by the sample points alone.

In similar fashion, an electric *signal* satisfying certain requirements can be reproduced from an appropriate set of *instantaneous samples*. Sampling therefore makes it possible to transmit a message in the form of *pulse modulation*, rather than a continuous signal. Usually the pulses are quite short compared to the time between them, so a pulse-modulated wave has the property of being "off" most of the time.

This property of pulse modulation offers two potential advantages over CW modulation. First, the transmitted power can be concentrated into short bursts instead of being generated continuously. The system designer then has greater latitude for equipment selection, and may choose devices such as lasers and high-power microwave tubes that operate only on a pulsed basis. Second, the time interval between pulses can be filed with sample values from other signals, a process called *time-division multiplexing* (*TDM*).

But pulse modulation has the disadvantage of requiring very large transmission bandwidth compared to the message bandwidth. Consequently, the methods of *analog* pulse modulation discussed in this chapter are used primarily as *message processing* for TDM and/or prior to CW modulation. *Digital* or *coded* pulse modulation has additional advantages that compensate for the increased bandwidth, as we'll see in Chapter 12.

OBJECTIVES

After studying this chapter and working the exercises, you should be able to do each of the following:

1.   Draw the spectrum of a sampled signal (Sect. 6.1).
2.   Define the minimum sampling frequency to adequately represent a signal given the maximum value of aliasing error, message bandwidth, LPF characteristics, and so forth (Sect. 6.1).
3.   Know what is meant by the *Nyquist rate* and know where it applies (Sect. 6.1).
4.   Describe the implications of practical sampling versus ideal sampling (Sect. 6.1).
5.   Reconstruct a signal from its samples using an ideal LPF (Sect. 6.1).
6.   Explain the operation of pulse-amplitude modulation, pulse-duration modulation, and pulse-position modulation; sketch their time domain waveforms; and calculate their respective bandwidths (Sects. 6.2 and 6.3).

6.1   SAMPLING THEORY AND PRACTICE

The theory of sampling presented here sets forth the conditions for signal sampling and reconstruction from sample values. We'll also examine practical implementation of the theory and some related applications.

Chopper Sampling

A simple but highly informative approach to sampling theory comes from the switching operation of Fig. 6.1–1a. The switch periodically shifts between two contacts at a rate of $f_s = 1/T_s$ Hz, dwelling on the input signal contact for τ seconds and on the grounded contact for the remainder of each period. The output $x_s(t)$ then consists of short segments for the input $x(t)$, as shown in Fig. 6.1–1b. Figure 6.1–1c is an electronic version of Fig. 6.1–1a; the output voltage equals the input voltage except when the clock signal forward-biases the diodes and thereby clamps the output to zero. This operation, variously called *single-ended* or **unipolar chopping,** is not instantaneous sampling in the strict sense. Nonetheless, $x_s(t)$ will be designated the sampled wave and f_s the sampling frequency.

We now ask: Are the sampled segments sufficient to describe the original input signal and, if so, how can $x(t)$ be retrieved from $x_s(t)$? The answer to this question lies in the frequency domain, in the spectrum of the sampled wave.

As a first step toward finding the spectrum, we introduce a **switching function** $s(t)$ such that

$$x_s(t) = x(t)s(t) \qquad\qquad [1]$$

Thus the sampling operation becomes multiplication by $s(t)$, as indicated schematically in Fig. 6.1–2a, where $s(t)$ is nothing more than the periodic pulse train of Fig. 6.1–2b. Since $s(t)$ is periodic, it can be written as a Fourier series. Using the results of Example 2.1–1 we have

$$s(t) = \sum_{n=-\infty}^{\infty} f_s \tau \operatorname{sinc} n f_s \tau\, e^{j2\pi n f_s t} = c_0 + \sum_{n=1}^{\infty} 2c_n \cos n\omega_s t \qquad [2]$$

where

$$c_n = f_s \tau \operatorname{sinc} n f_s \tau \qquad \omega_s = 2\pi f_s$$

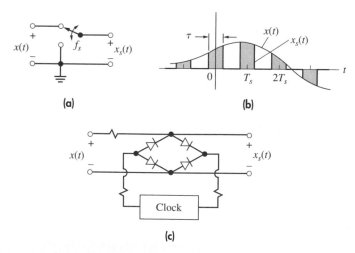

(a)

(b)

(c)

Figure 6.1–1   Switching sampler. (a) Functional diagram; (b) waveforms; (c) circuit realization with diode bridge.

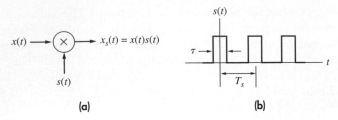

Figure 6.1–2   Sampling as multiplication. (a) Functional diagram; (b) switching function.

Combining Eq. (2) with Eq. (1) yields the term-by-term expansion

$$x_s(t) = c_0\, x(t) + 2c_1\, x(t)\cos \omega_s t + 2c_2\, x(t)\cos 2\omega_c t + \cdots \qquad [3]$$

Thus, if the input message spectrum is $X(f) = \mathscr{F}[x(t)]$, the output spectrum is

$$X_s(f) = c_0\, X(f) + c_1[X(f - f_s) + X(f + f_s)]$$
$$+ c_2[X(f - 2f_s) + X(f + 2f_s)] \qquad [4]$$
$$+ \cdots$$

which follows directly from the modulation theorem.

While Eq. (4) appears rather messy, the spectrum of a sampled wave is readily sketched if the input signal is assumed to be *bandlimited*. Figure 6.1–3 shows a convenient $X(f)$ and the corresponding $X_s(f)$ for two cases, $f_s > 2W$ and $f_s < 2W$. This figure reveals something quite surprising: the sampling operation has left the message spectrum *intact*, merely repeating it periodically in the frequency domain with a spacing of f_s. We also note that the first term of Eq. (4) is precisely the message spectrum, attenuated by the **duty cycle** $c_0 = f_s \tau = \tau/T_s$.

If sampling preserves the message spectrum, it should be possible to recover or reconstruct $x(t)$ from the sampled wave $x_s(t)$. The reconstruction technique is not at all obvious from the time-domain relations in Eqs. (1) and (3). But referring again to Fig. 6.1–3, we see that $X(f)$ can be separated from $X_s(f)$ by *lowpass filtering*, provided that the spectral sidebands don't overlap. And if $X(f)$ alone is filtered from $X_s(f)$, we have recovered $x(t)$. Two conditions obviously are necessary to prevent overlapping spectral bands: the message must be bandlimited, and the sampling frequency must be sufficiently great that $f_s - W \geq W$. Thus we require

$$X(f) = 0 \quad |f| > W$$

and in the case of nonsinusoidal signals

$$f_s \geq 2W \quad \text{or} \quad T_s \leq \frac{1}{2W} \qquad [5a]$$

If the sampled signal is sinusoidal, its frequency spectrum will consist of impulses and equality of Eq. (5a) does not hold, and we thus require

$$f_s > 2W \quad \text{or} \quad T_s < \frac{1}{2W} \qquad \text{(sinusoidal signals)} \qquad \textbf{[5b]}$$

The minimum sampling frequency $f_{s_{\min}} = 2W$, or in the case of sinusoidal signals $f_{s_{\min}} = 2W^+$, is called the **Nyquist rate.** When Eq. (5) is satisfied and $x_s(t)$ is filtered by an ideal LPF, the output signal will be proportional to $x(t)$; thus message reconstruction from the sampled signal has been achieved. The exact value of the filter bandwidth B is unimportant as long as

$$W < B < f_s - W \qquad \textbf{[6]}$$

so the filter passes $X(f)$ and rejects all higher components in Fig. 6.1–3b. Sampling at $f_s > 2W$ creates a *guard band* into which the transition region of a practical LPF can be fitted.

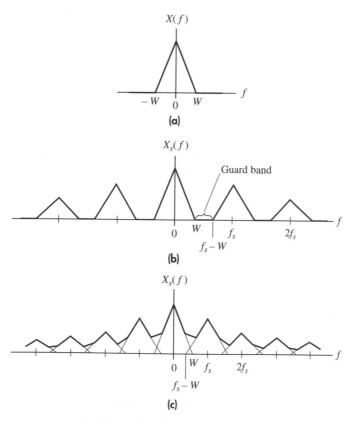

Figure 6.1–3   Spectra for switching sampling. (a) Message; (b) sampled message, $f_s > 2W$; (c) sampled message, $f_s < 2W$.

This analysis has shown that if a bandlimited signal is sampled at a frequency greater than the Nyquist rate, it can be *completely* reconstructed from the sampled wave. Reconstruction is accomplished by lowpass filtering. These conclusions may be difficult to believe at first exposure; they certainly test our faith in spectral analysis. Nonetheless, they are quite correct.

Finally, it should be pointed out that our results are independent of the sample-pulse duration, save as it appears in the duty cycle. If τ is made very small, $x_s(t)$ approaches a string of *instantaneous sample points*, which corresponds to *ideal sampling*. We'll pursue ideal sampling theory after a brief look at the **bipolar chopper**, which has $\tau = T_s/2$.

EXAMPLE 6.1–1   **Bipolar Choppers**

Figure 6.1–4a depicts the circuit and waveforms for a bipolar chopper. The equivalent switching function is a *square wave* alternating between $s(t) = +1$ and -1. From the series expansion of $s(t)$ we get

$$x_s(t) = \frac{4}{\pi} x(t) \cos \omega_s t - \frac{4}{3\pi} x(t) \cos 3\omega_s t + \frac{4}{5\pi} x(t) \cos 5\omega_s t - \cdots \qquad [7]$$

whose spectrum is sketched in Fig. 6.1–4b for $f \geq 0$. Note that $X_s(f)$ contains no dc component and only the odd harmonics of f_s. Clearly, we can't recover $x(t)$ by lowpass filtering. Instead, the practical applications of bipolar choppers involve *bandpass* filtering.

If we apply $x_s(t)$ to a BPF centered at some odd harmonic nf_s, the output will be proportional to $x(t) \cos n\omega_s t$—a double-sideband suppressed-carrier waveform.

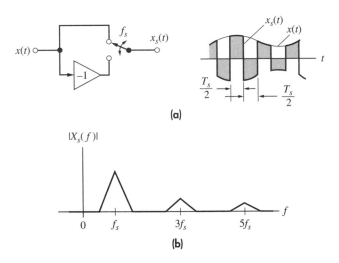

(a)

(b)

Figure 6.1–4   Bipolar chopper. (a) Circuit and waveforms; (b) spectrum.

Thus, a bipolar chopper serves as a *balanced modulator*. It also serves as a *synchronous detector* when the input is a DSB or SSB signal and the output is lowpass filtered. These properties are combined in the **chopper-stabilized amplifier,** which makes possible dc and low-frequency amplification using a high-gain ac amplifier. Additionally, Prob. 6.1–4 indicates how a bipolar chopper can be modified to produce the baseband multiplexed signal for FM stereo.

Ideal Sampling and Reconstruction

By definition, ideal sampling is *instantaneous* sampling. The switching device of Fig. 6.1–1a yields instantaneous values only if $\tau \to 0$; but then $f_s\tau \to 0$, and so does $x_s(t)$. Conceptually, we overcome this difficulty by multiplying $x_s(t)$ by $1/\tau$ so that, as $\tau \to 0$ and $1/\tau \to \infty$, the sampled wave becomes a train of *impulses* whose *areas* equal the instantaneous sample values of the input signal. Formally, we write the rectangular pulse train as

$$s(t) = \sum_{k=-\infty}^{\infty} \Pi\left(\frac{t - kT_s}{\tau}\right)$$

from which we define the **ideal sampling function**

$$s_\delta(t) \triangleq \lim_{\tau\to 0} \frac{1}{\tau} s(t) = \sum_{k=-\infty}^{\infty} \delta(t - kT_s) \qquad [8]$$

The **ideal sampled wave** is then

$$x_\delta(t) \triangleq x(t)s_\delta(t) \qquad [9a]$$

$$= x(t) \sum_{k=-\infty}^{\infty} \delta(t - kT_s)$$

$$= \sum_{k=-\infty}^{\infty} x(kT_s)\,\delta(t - kT_s) \qquad [9b]$$

since $x(t)\,\delta(t - kT_s) = x(kT_s)\,\delta(t - kT_s)$.

To obtain the corresponding spectrum $X_\delta(f) = \mathcal{F}[x_\delta(t)]$ we note that $(1/\tau)x_s(t) \to x_\delta(t)$ as $\tau \to 0$ and, likewise, $(1/\tau)X_s(f) \to X_\delta(f)$. But each coefficient in Eq. (4) has the property $c_n/\tau = f_s$ sinc $nf_s\tau = f_s$ when $\tau = 0$. Therefore,

$$X_\delta(f) = f_s X(f) + f_s[X(f - f_s) + X(f + f_s)] + \cdots$$

$$= f_s \sum_{n=-\infty}^{\infty} X(f - nf_s) \qquad [10]$$

which is illustrated in Fig. 6.1–5 for the message spectrum of Fig. 6.1–3a taking $f_s > 2W$. We see that $X_\delta(f)$ is periodic in frequency with period f_s, a crucial observation in the study of sampled-data systems.

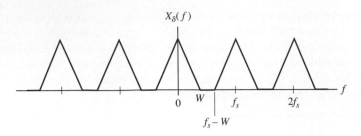

Figure 6.1–5   Spectrum of ideally sampled message.

Somewhat parenthetically, we can also develop an expression for $S_\delta(f) = \mathscr{F}[s_\delta(t)]$ as follows. From Eq. (9a) and the convolution theorem, $X_\delta(f) = X(f) * S_\delta(f)$ whereas Eq. (10) is equivalent to

$$X_\delta(f) = X(f) * \left[\sum_{n=-\infty}^{\infty} f_s\, \delta(f - nf_s) \right]$$

Therefore, we conclude that

$$S_\delta(f) = f_s \sum_{n=-\infty}^{\infty} \delta(f - nf_s) \qquad [11]$$

so the spectrum of a periodic string of unit-weight impulses in the time domain is a periodic string of impulses in the frequency domain with spacing $f_s = 1/T_s$; in both domains we have a function that looks like a picket fence.

Returning to the main subject and Fig. 6.1–5, it's immediately apparent that if we invoke the same conditions as before—$x(t)$ bandlimited in W and $f_s \geq 2W$—then a filter of suitable bandwidth will reconstruct $x(t)$ from the ideal sampled wave. Specifically, for an ideal LPF of gain K, time delay t_d, and bandwidth B, the transfer function is

$$H(f) = K\Pi\left(\frac{f}{2B}\right)e^{-j\omega t_d}$$

so filtering $x_\delta(t)$ produces the output spectrum

$$Y(f) = H(f)X_\delta(f) = Kf_s X(f)e^{-j\omega t_d}$$

assuming B satisfies Eq. (6). The output time function is then

$$y(t) = \mathscr{F}^{-1}[Y(f)] = Kf_s\, x(t - t_d) \qquad [12]$$

which is the original signal amplified by Kf_s and delayed by t_d.

Further confidence in the sampling process can be gained by examining reconstruction in the time domain. The impulse response of the LPF is

$$h(t) = 2BK \text{ sinc } 2B(t - t_d)$$

And since the input $x_\delta(t)$ is a train of weighted impulses, the output is a train of weighted *impulse responses*, namely,

$$y(t) = h(t) * x_\delta(t) = \sum_k x(kT_s)h(t - kT_s) \qquad [13]$$

$$= 2BK \sum_{k=-\infty}^{\infty} x(kT_s) \operatorname{sinc} 2B(t - t_d - kT_s)$$

Now suppose for simplicity that $B = f_s/2$, $K = 1/f_s$, and $t_d = 0$, so

$$y(t) = \sum_k x(kT_s) \operatorname{sinc} (f_s t - k)$$

We can then carry out the reconstruction process graphically, as shown in Fig. 6.1–6. Clearly the correct values are reconstructed at the sampling instants $t = kT_s$, for all sinc functions are zero at these times save one, and that one yields $x(kT_s)$. Between sampling instants $x(t)$ is *interpolated* by summing the precursors and postcursors from *all* the sinc functions. For this reason the LPF is often called an **interpolation filter,** and its impulse is called the **interpolation function.**

The above results are well summarized by stating the important theorem of **uniform (periodic) sampling.** While there are many variations of this theorem, the following form is best suited to our purposes.

> If a signal contains no frequency components for $|f| \geq W$, it is completely described by instantaneous sample values uniformly spaced in time with period $T_s \leq 1/2W$. If a signal has been sampled at the Nyquist rate or greater $(f_s \geq 2W)$ and the sample values are represented as weighted impulses, the signal can be exactly reconstructed from its samples by an ideal LPF of bandwidth B, where $W \leq B \leq f_s - W$.

Another way to express the theorem comes from Eqs. (12) and (13) with $K = T_s$ and $t_d = 0$. Then $y(t) = x(t)$ and

$$x(t) = 2BT_s \sum_{k=-\infty}^{\infty} x(kT_s) \operatorname{sinc} 2B(t - kT_s) \qquad [14]$$

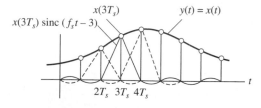

Figure 6.1–6   Ideal reconstruction.

provided $T_s \leq 1/2W$ and B satisfies Eq. (6). Therefore, just as a periodic signal is completely described by its Fourier series coefficients, a bandlimited signal is completely described by its instantaneous sample values *whether or not the signal actually is sampled.*

EXERCISE 6.1–1   Consider a sampling pulse train of the general form

$$s_p(t) = \sum_{k=-\infty}^{\infty} p(t - kT_s) \qquad \text{[15a]}$$

whose pulse type $p(t)$ equals zero for $|t| > T_s/2$ but is otherwise arbitrary. Use an exponential Fourier series and Eq. (21), Sect. 2.2, to show that

$$S_p(f) = f_s \sum_{n=-\infty}^{\infty} P(nf_s)\, \delta(f - nf_s) \qquad \text{[15b]}$$

where $P(f) = \mathcal{F}[p(t)]$. Then let $p(t) = \delta(t)$ to obtain Eq. (11).

Practical Sampling and Aliasing

Practical sampling differs from ideal sampling in three obvious aspects:

1. The sampled wave consists of pulses having finite amplitude and duration, rather than impulses.
2. Practical reconstruction filters are not ideal filters.
3. The messages to be sampled are *timelimited* signals whose spectra are not and cannot be strictly bandlimited.

The first two differences may present minor problems, while the third leads to the more troublesome effect known as **aliasing.**

Regarding pulse-shape effects, our investigation of the unipolar chopper and the results of Exercise 6.1–1 correctly imply that almost any pulse shape $p(t)$ will do when sampling takes the form of a *multiplication* operation $x(t)s_p(t)$. Another operation produces *flat-top sampling* described in the next section. This type of sampling may require equalization, but it does not alter our conclusion that pulse shapes are relatively inconsequential.

Regarding practical reconstruction filters, we consider the typical filter response superimposed in a sampled-wave spectrum in Fig. 6.1–7. As we said earlier, reconstruction can be done by interpolating between samples. The ideal LPF does a perfect interpolation. With practical systems, we can reconstruct the signal using a **zero-order hold (ZOH)** with

$$y(t) = \sum_k x(kT_s)\, \Pi\!\left(\frac{t - kT_s}{T_s}\right) \qquad \text{[16]}$$

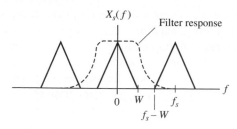

$X_s(f)$

Filter response

$0 \quad W \quad f_s$

$f_s - W$

f

Figure 6.1–7   Practical reconstruction filter.

or a **first-order hold (FOH)** which performs a linear interpolation using

$$y(t) = \sum_k x(kT_s)\Lambda\left(\frac{t - kT_s}{T_s}\right) \qquad [17]$$

The reconstruction process for each of these is shown in Fig. 6.1–8. Both the ZOH and FOH functions are lowpass filters with transfer function magnitudes of $|H_{ZOH}(f)| = |T_s \operatorname{sinc}(fT_s)|$ and $|H_{FOH}(f)| = |T_s\sqrt{1 + (2\pi fT_s)^2} \operatorname{sinc}^2(fT_s)|$, respectively. See Problems 6.1–11 and 6.1–12 for more insight.

If the filter is reasonably flat over the message band, its output will consist of $x(t)$ plus spurious frequency components at $|f| > f_s - W$ outside the message band. In audio systems, these components would sound like high-frequency hissing or "noise." However, they are considerably attenuated and their strength is proportional to $x(t)$, so they disappear when $x(t) = 0$. When $x(t) \neq 0$, the message tends to mask their presence and render them more tolerable. The combination of careful filter design and an adequate guard band created by taking $f_s > 2W$ makes practical reconstruction filtering nearly equivalent to ideal reconstruction. In the case of ZOH and FOH reconstruction, their frequency response shape $\operatorname{sinc}(fT_s)$ and $\operatorname{sinc}^2(fT_s)$ will distort the spectra of $x(t)$. We call this **aperature error,** which can be minimized by either increasing the sampling rate or compensating with the appropriate inverse filter.

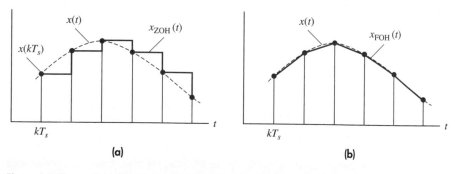

(a)

(b)

Figure 6.1–8   Signal reconstruction from samples using (a) ZOH, (b) FOH.

Regarding the timelimited nature of real signals, a message spectrum like Fig. 6.1–9a may be viewed as a bandlimited spectrum if the frequency content above W is small and presumably unimportant for conveying the information. When such a message is sampled, there will be unavoidable overlapping of spectral components as shown in Fig. 6.1–9b. In reconstruction, frequencies originally outside the normal message band will appear at the filter output in the form of much *lower* frequencies. Thus, for example, $f_1 > W$ becomes $f_s - f_1 < W$, as indicated in the figure.

This phenomenon of downward frequency translation is given the descriptive name of **aliasing.** The aliasing effect is far more serious than spurious frequencies passed by nonideal reconstruction filters, for the latter fall *outside* the message band, whereas aliased components fall *within* the message band. Aliasing is combated by filtering the message as much as possible *before* sampling and, if necessary, sampling at higher than the Nyquist rate. This is often done when the antialiasing filter does not have a sharp cutoff characteristic, as is the case of RC filters. Let's consider a broadband signal whose message content has a bandwidth of W but is corrupted by other frequency components such as noise. This signal is filtered using the simple first-order RC LPF antialiasing filter that has bandwidth $B = 1/2\pi RC$ with $W \ll B$ and is shown in Figure 6.1–9a. It is then sampled to produce the spectra shown in Fig. 6.1–9b. The shaded area represents the aliased components that have spilled into the filter's passband. Observe that the shaded area decreases if f_s increases or if we employ a more ideal LPF. Assuming reconstruction is done with the first-order Butterworth LPF, the maximum percent aliasing error in the passband is

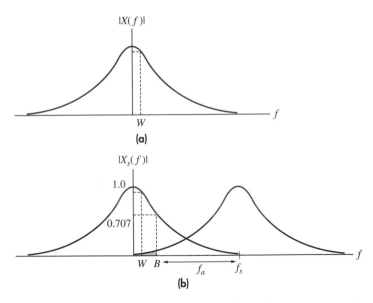

Figure 6.1–9   Message spectrum. (a) Output of RC filter; (b) after sampling. Shaded area represents aliasing spillover into passband.

$$Error\% = \left(\frac{1/0.707}{\sqrt{1 + (f_a/B)^2}} \right) \times 100\% \qquad [18]$$

with $f_a = f_s - B$ and the 0.707 factor is due to the filter's gain at its half-power frequency, B. See Ifeachor and Jervis (1993).

Oversampling

EXAMPLE 6.1–2

When using VLSI technology for *digital signal processing* (DSP) of analog signals, we must first sample the signal. Because it is relatively difficult to fabricate integrated circuit chips with large values of R and C we use the most feasible RC LPF and then *oversample* the signal at several times its Nyquist rate. We follow with a digital filter to reduce frequency components above the information bandwidth W. We then reduce the effective sampling frequency to its Nyquist rate using a process called *downsampling*. Both the digital filtering and downsampling processes are readily done with VLSI technology.

Let's say the maximum values of R and C we can put on a chip are 10 kΩ and 100 pF, respectively, and we want to sample a telephone quality voice such that the aliased components will be at least 30 dB below the desired signal. Using Eq. (18) with

$$B = \frac{1}{2\pi RC} = \frac{1}{2\pi \times 10^4 \times 100^{-12}} = 159 \text{ kHz}$$

we get

$$5\% - \left(\frac{1/0.707}{\sqrt{1 + (f_a/159 \text{ kHz})^2}} \right) \times 100\%.$$

Solving yields $f_a = 4.49$ MHz, and therefore the sampling frequency is $f_s = f_a + B = 4.65$ MHz. With our RC LPF, and $f_s = 4.49$ MHZ, any aliased components at 159 kHz will be at least 5 percent below the signal level at the half-power frequency. Of course the level of aliasing will be considerably less than 5 percent at frequencies below the telephone bandwidth of 3.2 kHz.

Sampling Oscilloscopes

EXAMPLE 6.1–3

A practical application of aliasing occurs in the sampling oscilloscope, which exploits undersampling to display high-speed *periodic* waveforms that would otherwise be beyond the capability of the electronics. To illustrate the principle, consider the periodic waveform $x(t)$ with period $T_x = 1/f_x$ in Fig. 6.1–10a. If we use a sampling interval T_s slightly greater than T_x and interpolate the sample points, we get the *expanded* waveform $y(t) = x(\alpha t)$ shown as a dashed curve. The corresponding sampling frequency is

$$f_s = (1 - \alpha)f_x \qquad 0 < \alpha < 1$$

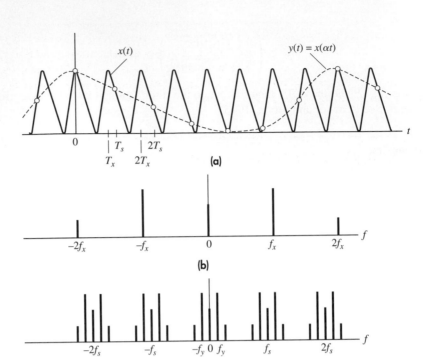

Figure 6.1–10   (a) Periodic waveform with undersampling; (b) spectrum of x(t); (c) spectrum of y(t) = x(αt), α<1.

so $f_s < f_x$ and even the fundamental frequency of $x(t)$ will be undersampled. Now let's find out if and how this system actually works by going to the frequency domain.

We assume that $x(t)$ has been prefiltered to remove any frequency components higher than the mth harmonic. Figure 6.1–10b shows a typical two-sided line spectrum of $x(t)$, taking $m = 2$ for simplicity. Since sampling translates all frequency components up and down by nf_s, the fundamental will appear in the spectrum of the sampled signal at

$$\pm f_y = \pm |f_x - f_s| = \pm \alpha f_x$$

as well as at $\pm f_x$ and at $f_x \pm nf_s = (1 + n)f_x \pm nf_y$. Similar translations applied to the dc component and second harmonic yield the spectrum in Figure 6.1–10c, which contains a *compressed* image of the original spectrum centered at each multiple of f_s. Therefore, a lowpass filter with $B = f_s/2$ will construct $y(t) = x(\alpha t)$ from $x_s(t)$ provided that

$$\alpha < \frac{1}{2m + 1}$$

which prevents spectral overlap.

Demonstrate the aliasing effect for yourself by making a careful sketch of $\cos 2\pi 10t$ and $\cos 2\pi 70t$ for $0 \leq t \leq \frac{1}{10}$. Put both sketches on the same set of axes and find the sample values at $t = 0, \frac{1}{80}, \frac{2}{80}, \ldots, \frac{8}{80}$, which corresponds to $f_s = 80$. Also, convince yourself that no other waveform bandlimited in $10 < W < 40$ can be interpolated from the sample values of $\cos 2\pi 10t$.

EXERCISE 6.1–2

6.2   PULSE-AMPLITUDE MODULATION

If a message waveform is adequately described by periodic sample values, it can be transmitted using analog pulse modulation wherein the sample values modulate the amplitude of a pulse train. This process is called **pulse-amplitude modulation (PAM).** An example of a message waveform and corresponding PAM signal are shown in Figure 6.2–1.

As Figure 6.2–1 indicates, the pulse amplitude varies in direct proportion to the sample values of $x(t)$. For clarity, the pulses are shown as rectangular and their durations have been grossly exaggerated. Actual modulated waves would also be delayed slightly compared to the message because the pulses can't be generated before the sampling instances.

It should be evident from the waveform that a PAM signal has significant dc content and that the bandwidth required to preserve the pulse shape far exceeds the message bandwidth. Consequently you seldom encounter a single-channel communication system with PAM or, for that matter, other analog pulse-modulated methods. But analog pulse modulation deserves attention for its major roles in time-division multiplexing, data telemetry, and instrumentation systems.

Flat-Top Sampling and PAM

Although a PAM wave could be obtained from a chopper circuit, a more popular method employs the **sample-and-hold** (S/H) technique. This operation produces flat-top pulses, as in Fig. 6.2–1, rather than curved-top chopper pulses. We therefore begin here with the properties of *flat-top sampling*.

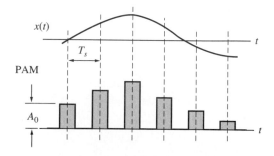

Figure 6.2–1

A rudimentary S/H circuit consists of two FET switches and a capacitor, connected as shown in Fig. 6.2–2a. A gate pulse at G1 briefly closes the sampling switch and the capacitor holds the sampled voltage until discharged by a pulse applied to G2. (Commercial integrated-circuit S/H units have further refinements, including isolating op-amps at input and output). Periodic gating of the sample-and-hold circuit generates the sampled wave

$$x_p(t) = \sum_k x(kT_s)p(t - kT_s) \tag{1}$$

illustrated by Fig. 6.2–2b. Note that each output pulse of duration τ represents a single instantaneous sample value.

To analyze flat-top sampling, we'll draw upon the relation $p(t - kT_s) = p(t) *$ $\delta(t - kT_s)$ and write

$$x_p(t) = p(t) * \left[\sum_k x(kT_s)\, \delta(t - kT_s) \right] = p(t) * x_\delta(t)$$

Fourier transformation of this *convolution* operation yields

$$X_p(f) = P(f)\left[f_s \sum_n X(f - nf_s) \right] = P(f)X_\delta(f) \tag{2}$$

Figure 6.2–3 provides a graphical interpretation of Eq. (2), taking $X(f) = \Pi(f/2W)$. We see that flat-top sampling is equivalent to passing an ideal sampled wave through a network having the transfer function $P(f) = \mathcal{F}[p(t)]$.

The high-frequency rolloff characteristic of a typical $P(f)$ acts like a *lowpass* filter and attenuates the upper portion of the message spectrum. This loss of high-frequency content is called **aperture effect.** The larger the pulse duration or aperture τ, the larger the effect. Aperture effect can be corrected in reconstruction by including an *equalizer* with

$$H_{eq}(f) = Ke^{-j\omega t_d}/P(f) \tag{3}$$

However, little if any equalization is needed when $\tau/T_s \ll 1$.

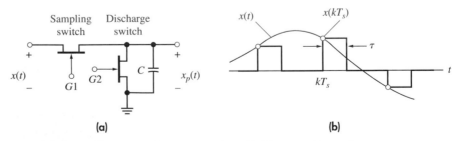

(a)	(b)

Figure 6.2–2   Flat-top sampling. (a) Sample-and-hold circuit; (b) waveforms.

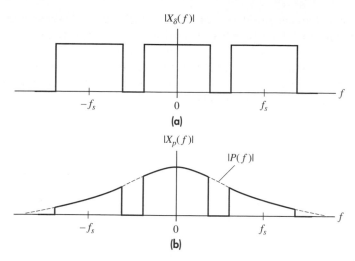

Figure 6.2–3   (a) Spectrum for ideal sampling when $X(f) = \Pi(f/2W)$; (b) aperture effect in flat-top sampling.

Now consider a unipolar flat-top PAM signal defined by

$$x_p(t) = \sum_k A_0[1 + \mu x(kT_s)]p(t - kT_s)$$ [4]

The constant A_0 equals the unmodulated pulse amplitude, and the *modulation index* μ controls the amount of amplitude variation. The condition

$$1 + \mu x(t) > 0$$ [5]

ensures a unipolar (single-polarity) waveform with no missing pulses. The resulting constant pulse rate f_s is particularly important for synchronization in time-division multiplexing.

Comparison of Eqs. (1) and (4) shows that a PAM signal can be obtained from a sample-and-hold circuit with input $A_0[1 + \mu x(t)]$. Correspondingly, the PAM spectrum will look like Fig. 6.2–3b with $X(f)$ replaced by

$$\mathscr{F}\{A_0[1 + \mu x(t)]\} = A_0[\delta(f) + \mu X(f)],$$

which results in spectral impulses at all harmonics of f_s and at $f = 0$. Reconstruction of $x(t)$ from $x_p(t)$ therefore requires a *dc block* as well as lowpass filtering and equalization.

Clearly, PAM has many similarities to AM CW modulation—modulation index, spectral impulses, and dc blocks. (In fact, an AM wave could be derived from PAM by bandpass filtering). But the PAM spectrum extends from dc up through several harmonics of f_s, and the estimate of required transmission bandwidth B_T must be based on time-domain considerations. For this purpose, we assume a small pulse duration τ compared to the time between pulses, so

$$\tau \ll T_s \leq \frac{1}{2W}$$

Adequate pulse resolution then requires

$$B_T \geq \frac{1}{2\tau} \gg W \qquad\qquad [6]$$

Hence, practical applications of PAM are limited to those situations in which the advantages of a pulsed waveform outweigh the disadvantages of large bandwidth.

EXERCISE 6.2–1   Consider PAM transmission of a voice signal with $W \approx 3$ kHz. Calculate B_T if $f_s = 8$ kHz and $\tau = 0.1\, T_s$.

6.3   PULSE-TIME MODULATION

The sample values of a message can also modulate the time parameters of a pulse train, namely the pulse width or its position. The corresponding processes are designated as **pulse-duration (PDM)** and **pulse-position modulation (PPM)** and are illustrated in Fig. 6.3–1. PDM is also called **pulse-width modulation (PWM).** Note the pulse width or pulse position varies in direct proportion to the sample values of $x(t)$.

Pulse-Duration and Pulse-Position Modulation

We lump PDM and PPM together under one heading for two reasons. First, in both cases a *time* parameter of the pulse is being modulated, and the pulses have *constant amplitude*. Second, a close relationship exists between the modulation methods for PDM and PPM.

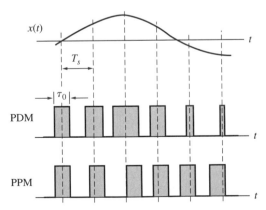

Figure 6.3–1   Types of pulse-time modulation.

To demonstrate these points, Fig. 6.3–2 shows the block diagram and wave-forms of a system that combines the sampling and modulation operations for either PDM or PPM. The system employs a comparator and a sawtooth-wave generator with period T_s. The output of the comparator is zero except when the message wave-form $x(t)$ exceeds the sawtooth wave, in which case the output is a positive constant A. Hence, as seen in the figure, the comparator produces a PDM signal with *trailing-edge modulation* of the pulse duration. (Reversing the sawtooth results in leading-edge modulation while replacing the sawtooth with a triangular wave results in modulation on both edges.) Position modulation is obtained by applying the PDM signal to a monostable pulse generator that triggers on trailing edges at its input and produces short output pulses of fixed duration.

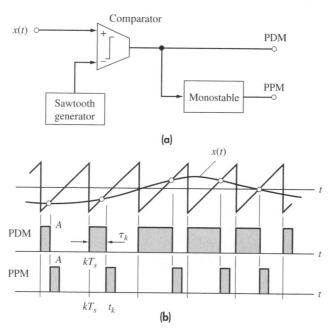

Figure 6.3–2   Generation of PDM or PPM. (*a*) Block diagram; (*b*) waveforms.

Careful examination of Fig. 6.3–2*b* reveals that the modulated duration or posi-tion depends on the message value at the time location t_k of the pulse edge, rather than the apparent sample time kT_s. Thus, the sample values are *nonuniformly* spaced. Inserting a sample-and-hold circuit at the input of the system gives uniform sampling if desired, but there's little difference between uniform and nonuniform sampling in the practical case of small amounts of time modulation such that $t_k - kT_s \ll T_s$.

If we assume nearly uniform sampling, the duration of the kth pulse in the PDM signal is

$$\tau_k = \tau_0[1 + \mu x(kT_s)] \qquad\qquad \textbf{[1]}$$

in which the unmodulated duration τ_0 represents $x(kT_s) = 0$ and the modulation index μ controls the amount of duration modulation. Our prior condition on μ in Eq. (5), Sect. 6.2, applies here to prevent missing pulses and "negative" durations when $x(kT_s) \leq 0$. The PPM pulses have fixed duration and amplitude so, unlike PAM and PDM, there's no potential problem of missing pulses. The kth pulse in a PPM signal begins at time

$$t_k = kT_s + t_d + t_0\, x(kT_s) \tag{2}$$

in which the unmodulated position $kT_s + t_d$ represents $x(kT_s) = 0$ and the constant t_0 controls the displacement of the modulated pulse.

The variable time parameters in Eqs. (1) and (2) make the expressions for $x_p(t)$ rather awkward. However, an informative approximation for the PDM waveform is derived by taking rectangular pulses with amplitude A centered at $t = kT_s$ and assuming that τ_k varies slowly from pulse to pulse. Series expansion then yields

$$x_p(t) \approx A f_s \tau_0 [1 + \mu x(t)] + \sum_{n=1}^{\infty} \frac{2A}{\pi n} \sin n\phi(t) \cos n\omega_s t \tag{3}$$

where $\phi(t) = \pi f_s \tau_0 [1 + \mu x(t)]$. Without attempting to sketch the corresponding spectrum, we see from Eq. (3) that the PDM signal contains the message $x(t)$ plus a dc component and *phase-modulated* waves at the harmonics of f_s. The phase modulation has negligible overlap in the message band when $\tau_0 \ll T_s$, so $x(t)$ can be recovered by lowpass filtering with a dc block.

Another message reconstruction technique converts pulse-*time* modulation into pulse-*amplitude* modulation, and works for PDM and PPM. To illustrate this technique the middle waveform in Fig. 6.3–3 is produced by a *ramp generator* that starts at time kT_s, stops at t_k, restarts at $(k + 1)T_s$, and so forth. Both the start and stop commands can be extracted from the edges of a PDM pulse, whereas PPM reconstruction must have an auxiliary *synchronization* signal for the start command.

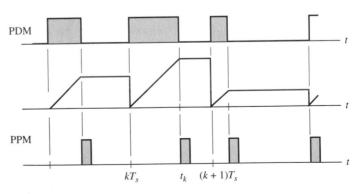

Figure 6.3–3   Conversion of PDM or PPM into PAM.

Regardless of the particular details, demodulation of PDM or PPM requires received pulses with short *risetime* in order to preserve accurate message information. For a specified risetime $t_r \ll T_s$, the transmission bandwidth must satisfy

$$B_T \geq \frac{1}{2t_r} \qquad \qquad \text{[4]}$$

which will be substantially greater than the PAM transmission bandwidth. In exchange for the extra bandwidth, we gain the benefit of constant-amplitude pulses that suffer no ill effects from *nonlinear distortion* in transmission since nonlinear distortion does not alter pulse duration or position.

Additionally, like PM and FM CW modulation, PDM and PPM have the potential for *wideband noise reduction*—a potential more fully realized by PPM than by PDM. To appreciate why this is so, recall that the information resides in the time location of the pulse *edges*, not in the pulses themselves. Thus, somewhat like the carrier-frequency power of AM, the pulse power of pulse-time modulation is "wasted" power, and it would be more efficient to suppress the pulses and just transmit the edges! Of course we cannot transmit edges without transmitting pulses to define them. But we can send very short pulses indicating the position of the edges, a process equivalent to PPM. The reduced power required for PPM is a fundamental advantage over PDM, an advantage that becomes more apparent when we examine the signal-to-noise ratios.

Derive Eq. (3) by the following procedure. First, assume constant pulse duration τ and write $x_p(t) = As(t)$ with $s(t)$ given by Eq. (2), Sect. 6.1. Then apply the *quasi-static approximation* $\tau \approx \tau_0[1 + \mu x(t)]$.

EXERCISE 6.3–1

PPM Spectral Analysis★

Because PPM with nonuniform sampling is the most efficient type of analog pulse modulation for message transmission, we should take the time to analyze its spectrum. The analysis method itself is worthy of examination.

Let the kth pulse be centered at time t_k. If we ignore the constant time delay t_d in Eq. (2), nonuniform sampling extracts the sample value at t_k, rather than kT_s, so

$$t_k = kT_s + t_0 x(t_k) \qquad \qquad \text{[5]}$$

By definition, the PPM wave is a summation of constant-amplitude position-modulated pulses, and can be written as

$$x_p(t) = \sum_k Ap(t - t_k) = Ap(t) * \left[\sum_k \delta(t - t_k) \right]$$

where A is the pulse amplitude and $p(t)$ the pulse shape. A simplification at this point is made possible by noting that $p(t)$ will (or should) have a very small duration compared to T_s. Hence, for our purposes, the pulse shape can be taken as impulsive, and

$$x_p(t) \approx A \sum_k \delta(t - t_k) \qquad [6]$$

If desired, Eq. (6) can later be convolved with $p(t)$ to account for the nonimpulsive shape.

In their present form, Eqs. (5) and (6) are unsuited to further manipulation; the trouble is the position term t_k, which cannot be solved for explicitly. Fortunately, t_k can be eliminated entirely. Consider any function $g(t)$ having a single first-order zero at $t = \lambda$, such that $g(\lambda) = 0$, $g(t) \neq 0$ for $t \neq \lambda$, and $\dot{g}(t) \neq 0$ at $t = \lambda$. The distribution theory of impulses then shows that

$$\delta(t - \lambda) = |\dot{g}(t)| \, \delta[g(t)] \qquad [7]$$

whose right-hand side is independent of λ. Equation (7) can therefore be used to remove t_k from $\delta(t - t_k)$ if we can find a function $g(t)$ that satisfies $g(t_k) = 0$ and the other conditions but does not contain t_k.

Suppose we take $g(t) = t - kT_s - t_0 x(t)$, which is zero at $t = kT_s + t_0 x(t)$. Now, for a given value of k, there is only one PPM pulse, and it occurs at $t_k = kT_s + t_0 x(t_k)$. Thus $g(t_k) = t_k - kT_s - t_0 x(t_k) = 0$, as desired. Inserting $\lambda = t_k$, $\dot{g}(t) = 1 - t_0 \dot{x}(t)$, etc., into Eq. (7) gives

$$\delta(t - t_k) = |1 - t_0 \dot{x}(t)| \, \delta[t - kT_s - t_0 x(t)]$$

and the PPM wave of Eq. (6) becomes

$$x_p(t) = A[1 - t_0 \dot{x}(t)] \sum_k \delta[t - t_0 x(t) - kT_s]$$

The absolute value is dropped since $|t_0 \dot{x}(t)| < 1$ for most signals of interest if $t_0 \ll T_s$. We then convert the sum of impulses to a sum of exponentials via

$$\sum_{k=-\infty}^{\infty} \delta(t - kT_s) = f_s \sum_{n=-\infty}^{\infty} e^{jn\omega_s t} \qquad [8]$$

which is **Poisson's sum formula.** Thus, we finally obtain

$$x_p(t) = Af_s[1 - t_0 \dot{x}(t)] \sum_{n=-\infty}^{\infty} e^{jn\omega_s[t - t_0 x(t)]}$$

$$= Af_s[1 - t_0 \dot{x}(t)] \left\{ 1 + \sum_{n=1}^{\infty} 2 \cos [n\omega_s t - n\omega_s t_0 x(t)] \right\} \qquad [9]$$

The derivation of Eq. (8) is considered in Prob. 6.3–6.

Interpreting Eq. (9), we see that PPM with nonuniform sampling is a combination of linear and exponential carrier modulation, for each harmonic of f_s is phase-modulated by the message $x(t)$ and amplitude-modulated by the derivative $\dot{x}(t)$. The spectrum therefore consists of AM and PM sidebands centered at all multiples of f_s, plus a dc impulse and the spectrum of $\dot{x}(t)$. Needless to say, sketching such a spectrum is a tedious exercise even for tone modulation. The leading term of Eq. (9) suggests that the message can be retrieved by lowpass filtering and *integrating*. However, the integration method does not take full advantage of the noise-reduction properties of PPM, so the usual procedure is conversion to PAM or PDM followed by lowpass filtering.

6.4   PROBLEMS

6.1-1   Consider the chopper-sampled waveform in Eq. (3) with $\tau = T_s/2, f_s = 100$ Hz, and $x(t) = 2 + 2 \cos 2\pi 30t + \cos 2\pi 80t$. Draw and label the one-sided line spectrum of $x_s(t)$ for $0 \leq f \leq 300$ Hz. Then find the output waveform when $x_s(t)$ is applied to an ideal LPF with $B = 75$ Hz.

6.1-2   Do Prob. 6.1-1 with $x(t) = 2 + 2 \cos 2\pi 30t + \cos 2\pi 140t$.

6.1-3   The usable frequency range of a certain amplifier is f_ℓ to $f_\ell + B$, with $B \gg f_\ell$. Devise a system that employs bipolar choppers and allows the amplifier to handle signals having significant dc content and bandwidth $W \ll B$.

6.1-4*   The baseband signal for FM stereo is

$$x_b(t) = [x_L(t) + x_R(t)] + [x_L(t) - x_R(t)] \cos \omega_s t + A \cos \omega_s t/2$$

with $f_s = 38$ kHz. The chopper system in Fig. 6.1-4 is intended to generate this signal. The LPF has gain K_1 for $|f| \leq 15$ kHZ, gain K_2 for $23 \leq |f| \leq 53$ kHz, and rejects $|f| \geq 99$ kHz. Use a sketch to show that $x_s(t) = x_L(t)s(t) + x_R(t)[1 - s(t)]$, where $s(t)$ is a unipolar switching function with $\tau = T_s/2$. Then find the necessary values of K_1 and K_2.

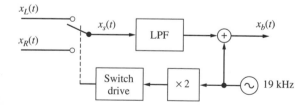

Figure P6.1-4

6.1-5   A popular *stereo decoder* circuit employs transistor switches to generate $v_L(t) = x_1(t) - Kx_2(t)$ and $v_R(t) = x_2(t) - Kx_1(t)$ where K is a constant, $x_1(t) = x_b(t)s(t)$, $x_2(t) = x_b(t)[1 - s(t)]$, $x_b(t)$ is the FM stereo baseband signal in Prob. 6.1-4, and $s(t)$ is a unipolar switching function with $\tau = T_s/2$. (a) Determine K such that lowpass filtering of $v_L(t)$ and $v_R(t)$ yields the desired left- and right-channel signals. (b) What's the disadvantage of a simpler switching circuit that has $K = 0$?

6.1–6   Derive Eq. (11) using Eq. (14), Sect. 2.5.

6.1–7   Suppose $x(t)$ has the spectrum in Fig. P6.1–7 with $f_u = 25$ kHz and $W = 10$ kHz. Sketch $x_\delta(f)$ for $f_s = 60$, 45, and 25 kHz. Comment in each case on the possible reconstruction of $x(t)$ from $x_\delta(t)$.

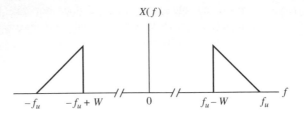

Figure P6.1–7

6.1–8   Consider the bandpass signal spectrum in Fig. P6.1–7 whose Nyquist rate is $f_s = 2f_u$. However, the *bandpass sampling theorem* states that $x(t)$ can be reconstructed from $x_\delta(t)$ by *bandpass* filtering if $f_s = 2f_u/m$ and the integer m satisfies $(f_u/W) - 1 < m \le f_u/W$. (a) Find m and plot f_s/W versus f_u/W for $0 < f_u/W \le 5$. (b) Check the theorem by plotting $X_\delta(f)$ when $f_u = 2.5W$ and $f_s = 2.5W$. Also show that the higher rate $f_s = 4W$ would not be acceptable.

6.1–9   The signal $x(t) = \operatorname{sinc}^2 5t$ is ideally sampled at $t = 0, \pm0.1, \pm0.2, \ldots$, and reconstructed by an ideal LPF with $B = 5$, unit gain, and zero time delay. Carry out the reconstruction process graphically, as in Fig. 6.1–6 for $|t| \le 0.2$.

6.1–10   A rectangular pulse with $\tau = 2$ is ideally sampled and reconstructed using an ideal LPF with $B = f_s/2$. Sketch the resulting output waveforms when $T_s = 0.8$ and 0.4, assuming one sample time is at the center of the pulse.

6.1–11   Suppose an ideally sampled wave is reconstructed using a *zero-order hold* with time delay $T = T_s$. (a) Find and sketch $y(t)$ to show that the reconstructed waveform is a *staircase* approximation of $x(t)$. (b) Sketch $|Y(f)|$ for $X(f) = \Pi(f/2W)$ with $W \ll f_s$. Comment on the significance of your result.

6.1–12‡   The reconstruction system in Fig. P6.1–12 is called a *first-order hold*. Each block labeled ZOH is a zero-order hold with time delay $T = T_s$. (a) Find $h(t)$ and sketch $y(t)$ to interpret the reconstruction operation. (b) Show that $H(f) = T_s(1 + j2\pi fT_s)$ $(\operatorname{sinc}^2 fT_s) \exp(-j2\pi fT_s)$. Then sketch $|Y(f)|$ for $X(f) = \Pi(f/2W)$ with $W < f_s/2$.

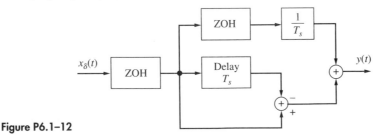

Figure P6.1–12

6.1–13‡ Use Parseval's theorem and Eq. (14) with $T_s = 1/2W$ and $B = W$ to show that the energy of a bandlimited signal is related to its sample values by

$$E = (1/2W) \sum_{k=-\infty}^{\infty} |x(k/2W)|^2$$

6.1–14 The **frequency-domain sampling theorem** says that if $x(t)$ is a *timelimited* signal, such that $x(t) = 0$ for $|t| \geq T$, then $X(f)$ is completely determined by its sample values $X(nf_0)$ with $f_0 \leq 1/2T$. Prove this theorem by writing the Fourier series for the periodic signal $v(t) = x(t) * [\sum_k \delta(t - kT_0)]$, where $T_0 \geq 2T$, and using the fact that $x(t) = v(t)\Pi(t/2T)$.

6.1–15* A signal with period $T_x = 0.08 \ \mu s$ is to be displayed using a sampling oscilloscope whose internal high-frequency response cuts off at $B = 6$ MHz. Determine maximum values for the sampling frequency and the bandwidth of the presampling LPF.

6.1–16 Explain why the sampling oscilloscope in Prob. 6.1–15 will not provide a useful display when $T_x < 1/3B$.

6.1–17* A $W = 15$ kHz signal has been sampled at 150 kHz. What will be the maximum percent aperature error if the signal is reconstructed using a (*a*) ZOH, (*b*) FOH?

6.1–18 A $W = 15$ kHz signal is sampled at 150 kHz with a first-order Butterworth antialiasing filter. What will be the maximum percent aliasing error in the passband?

6.1–19 Show that the equality in Eq. (5) of Sect. 6.1 does not hold for a sinusoidal signal.

6.1–20* What is the Nyquist rate to adequately sample the following signals: (*a*) sinc (100*t*), (*b*) sinc^2 (100*t*), (*c*) 10 cos$^3(2\pi 10^5 t)$?

6.1–21 Repeat Example 6.1–2 such that aliased components will be least 40 dB below the signal level at the half-power frequency of 159 kHz.

6.2–1 Sketch $|X_p(f)|$ and find $H_{eq}(f)$ for flat-top sampling with $\tau = T_s/2$, $f_s = 2.5W$, and $p(t) = \Pi(t/\tau)$. Is equalization essential in this case?

6.2–2 Do Prob. 6.2–1 for $p(t) = (\cos \pi t/\tau)\Pi(t/\tau)$.

6.2–3‡* Some sampling devices extract from $x(t)$ its *average value* over the sampling duration, so $x(kT_s)$ in Eq. (1) is replaced by

$$\bar{x}(kT_s) \triangleq \frac{1}{\tau} \int_{kT_s-\tau}^{kT_s} x(\lambda) \, d\lambda$$

(*a*) Devise a frequency-domain model of this process using an averaging filter, with input $x(t)$ and output $\bar{x}(t)$, followed by instantaneous flat-top sampling. Then obtain the inpulse response of the averaging filter and write the resulting expression for $X_p(f)$. (*b*) Find the equalizer needed when $p(t)$ is a rectangular pulse.

6.2–4 Consider the PAM signal in Eq. (4). (*a*) Show that its spectrum is

$$X_p(f) = A_0 f_s P(f) \left\{ \sum_n [\delta(f - nf_s) + \mu X(f - nf_s)] \right\}$$

(b) Sketch $|X_p(f)|$ when $p(t) = \Pi(t/\tau)$ with $\tau = T_s/2$ and $\mu x(t) = \cos 2\pi f_m/t$ with $f_m < f_s/2$.

6.2–5   Suppose the PAM signal in Eq. (4) is to be transmitted over a transformer-coupled channel, so the pulse shape is taken as $p(t) = \Pi[(t - \tau/2)/\tau] - \Pi[(t + \tau/2)/\tau]$ to eliminate the dc component of $x_p(t)$. (a) Use the expression in Prob. 6.2–4a to sketch $|X_p(f)|$ when $\tau = T_s/4$, $X(f) = \Pi(f/2W)$, and $f_s > 2W$. (b) Find an appropriate equalizer, assuming that $x(t)$ has negligible frequency content for $|f| < f_\ell < W$. Why is this assumption necessary?

6.2–6   Show how a PAM signal can be demodulated using a product detector. Be sure to describe frequency parameters for the LO and the LPF.

6.3–1*   Calculate the transmission bandwidth needed for voice PDM with $f_s = 8$ kHz, $|\mu x(t)| \leq 0.8$, and $\tau_0 = T_s/5$ when we want $t_r \leq 0.25\tau_{min}$.

6.3–2   A voice PDM signal with $f_s = 8$ kHz and $|\mu x(t)| \leq 0.8$ is to be transmitted over a channel having $B_T = 500$ kHz. Obtain bounds on τ_0 such that $\tau_{max} \leq T_s/3$ and $\tau_{min} \geq 3t_r$.

6.3–3   A pulse-modulated wave is generated by uniformly sampling the signal $x(t) = \cos 2\pi t/T_m$ at $t = kT_s$, where $T_s = T_m/3$. Sketch and label $x_p(t)$ when the modulation is: (a) PDM with $\mu = 0.8$, $\tau_0 = 0.4T_s$, and leading edges fixed at $t = kT_s$; (b) PPM with $t_d = 0.5T_s$ and $t_0 = \tau = 0.2T_s$.

6.3–4   Do Prob. 6.3–3 with $T_s = T_m/6$.

6.3–5   Use Eq. (9) to devise a system that employs a PPM generator and produces narrowband phase modulation with $f_c = mf_s$.

6.3–6   **Poisson's sum formula** states in general that

$$\sum_{n=-\infty}^{\infty} e^{\pm j2\pi n\lambda/L} = L \sum_{m=-\infty}^{\infty} \delta(\lambda - mL)$$

where λ is an independent variable and L is a constant. (a) Derive the time-domain version as given in Eq. (8) by taking $\mathscr{F}^{-1}[S_\delta(f)]$. (b) Derive the frequency-domain version by taking $\mathscr{F}[s_\delta(t)]$.

6.3-7‡   Let $g(t)$ be any continuous function that monotonically increases or decreases over $a \leq t \leq b$ and crosses zero at $t = \lambda$ within this range. Justify Eq. (7) by making the change-of-variable $v = g(t)$ in

$$\int_a^b \delta[g(t)]\, dt$$

chapter

7

Analog Communication Systems

CHAPTER OUTLINE

Communication systems that employ linear or exponential CW modulation may differ in many respects—type of modulation, carrier frequency, transmission medium, and so forth. But they have in common the property that a sinusoidal bandpass signal with time-varying envelope and/or phase conveys the message information. Consequently, generic hardware items such as oscillators, mixers, and bandpass filters are important building blocks for *all* CW modulation systems. Furthermore, many systems involve both linear and exponential CW modulation techniques.

This chapter therefore takes a broader look at CW modulation systems and hardware, using concepts and results from Chapters 4 through 6. Specific topics include CW receivers, frequency- and time-division multiplexing, phase-lock loops, and television systems.

OBJECTIVES

After studying this chapter and working the exercises, you should be able to do each of the following:

1. Design, in block-diagram form, a superheterodyne receiver that satisfies stated specifications (Sect. 7.1).
2. Predict at what frequencies a superheterodyne is susceptible to spurious inputs (Sect. 7.1).
3. Draw the block diagram of either an FDM or TDM system, given the specifications, and calculate the various bandwidths (Sect. 7.2).
4. Identify the phase-locked loop structures used for pilot filtering, frequency synthesis, and FM detection (Sect. 7.3).
5. Analyze a simple phase-lock-loop system and determine the condition for locked operation (Sect. 7.3).
6. Explain the following TV terms: *scanning raster, field, frame, retrace, luminance, chrominance,* and *color compatibility* (Sect. 7.4).
7. Estimate the bandwidth requirement for image transmission given the vertical resolution, active line time, and aspect ratio (Sect. 7.4).
8. Describe the significant performance differences of NTSC versus HDTV systems (Sect. 7.4).

7.1   RECEIVERS FOR CW MODULATION

All that is really essential in a CW receiver is some tuning mechanism, demodulation, and amplification. With a sufficiently strong received signal, you may even get by without amplification—witness the historic crystal radio set. However, most receivers operate on a more sophisticated **superheterodyne** principle, which we'll discuss first. Then we'll consider other types of receivers and the related scanning spectrum analyzer.

Superheterodyne Receivers

Beside demodulation, a typical broadcast receiver must perform three other operations: (1) carrier-frequency *tuning* to select the desired signal, (2) *filtering* to separate that signal from others received with it, and (3) *amplification* to compensate for transmission loss. And at least some of the amplification should be provided before demodulation to bring the signal up to a level useable by the demodulator circuitry.

For example, if the demodulator is based on a diode envelope detector, the input signal must overcome the diode's forward voltage drop. In theory, all of the foregoing requirements could be met with a high-gain tunable bandpass amplifier. In practice, fractional-bandpass and stability problems make such an amplifier expensive and difficult to build. Armstrong devised the superheterodyne or "superhet" receiver to circumvent these problems.

The superhet principle calls for two distinct amplification and filtering sections prior to demodulation, as diagrammed in Fig. 7.1–1. The incoming signal $x_c(t)$ is first selected and amplified by a **radio-frequency** (**RF**) section tuned to the desired carrier frequency f_c. The amplifier has a relatively broad bandwidth B_{RF} that partially passes adjacent-channel signals along with $x_c(t)$. Next a **frequency converter** comprised of a mixer and local oscillator translates the RF output down to an **intermediate-frequency** (**IF**) at $f_{IF} < f_c$. The adjustable LO frequency tracks with the RF tuning such that

$$f_{LO} = f_c + f_{IF} \qquad \text{or} \qquad f_{LO} = f_c - f_{IF} \qquad [1]$$

and hence

$$|f_c - f_{LO}| = f_{IF} \qquad [2]$$

An IF section with bandwidth $B_{IF} \approx B_T$ now removes the adjacent-channel signals. This section is a fixed bandpass amplifier, called the **IF strip,** which provides most of the gain. Finally, the IF output goes to the demodulator for message recovery and baseband amplification. The parameters for commercial broadcast AM and FM receivers are given in Table 7.1–1.

Table 7.1–1   Parameters of AM and FM radios

	AM	FM
Carrier frequency	540–1600 kHz	88.1–107.9 MHz
Carrier spacing	10 kHz	200 kHz
Intermediate frequency	455 kHz	10.7 MHz
IF bandwidth	6–10 kHz	200–250 kHz
Audio bandwidth	3–5 kHz	15 kHz

The spectral drawings of Fig. 7.1–2 help clarify the action of a superhet receiver. Here we assume a modulated signal with symmetric sidebands (as distinguished from SSB or VSB), and we take $f_{LO} = f_c + f_{IF}$ so

$$f_c = f_{LO} - f_{IF}$$

The RF input spectrum in Fig. 7.1–2a includes our desired signal plus adjacent-channel signals on either side and another signal at the **image frequency**

$$f_c' = f_c + 2f_{IF} = f_{LO} + f_{IF} \qquad [3]$$

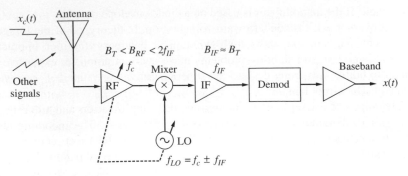

Figure 7.1–1   Superheterodyne receiver.

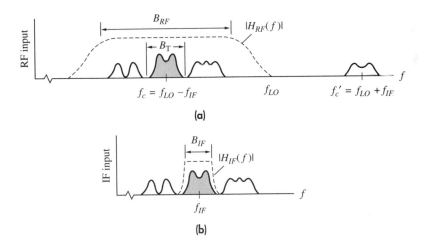

Figure 7.1–2

The main task of the RF section is to pass $f_c \pm B_T/2$ while rejecting the image frequency signal. For f_c' to reach the mixer, it would be down-converted to

$$f_c' - f_{LO} = (f_{LO} + f_{IF}) - f_{LO} = f_{IF}$$

and the image frequency signal would produce an effect similar to cochannel interference. Hence, we want an RF response $|H_{RF}(f)|$ like the dashed line, with

$$B_T < B_{RF} < 2f_{IF} \qquad\qquad \text{[4]}$$

The filtered and down-converted spectrum at the IF input is shown in Fig 7.1–2b. The indicated IF response $|H_{IF}(f)|$ with $B_{IF} \approx B_T$ completes the task of adjacent-channel rejection.

The superheterodyne structure results in several practical benefits. First, tuning takes place entirely in the "front" end so the rest of the circuitry, including the demodulator, requires no adjustment to change f_c. Second, the separation between f_c and f_{IF} eliminates potential instability due to stray feedback from the amplified output to the receiver's inputs. Third, most of the gain and selectivity is concentrated in the fixed-frequency IF strip. Since f_{IF} is an internal design parameter, it can be chosen to obtain a reasonable fractional bandwidth B_{IF}/f_{IF} for ease of implementation. Taken together, these benefits make it possible to build superhets with extremely high gain—75 dB or more in the IF strip alone. We can also employ high-Q mechanical, ceramic, crystal, and SAW bandpass filters and thus achieve tremendous reductions in adjacent channel interference.

Additionally, when the receiver must cover a wide carrier-frequency range, the choice of $f_{LO} = f_c + f_{IF}$ may result in a smaller and more readily achieved LO **tuning ratio.** For example, with AM broadcast radios, where $540 < f_c < 1600$ kHz and $f_{IF} = 455$ kHz using $f_{LO} = f_c + f_{IF}$ results in $995 < f_{LO} < 2055$ kHz and thus a LO tuning range of 2:1. On the other hand, if we chose $f_{LO} = f_c - f_{IF}$, then for the same IF and input frequency range, we get $85 < f_{LO} < 1145$ kHz or a LO tuning range of 13:1. We should point out, however, that taking $f_{LO} > f_c$ in an SSB superhet causes **sideband reversal** in the down-converted signal, so USSB at RF becomes LSSB at IF, and vice versa.

A major disadvantage of the superhet structure is its potential for spurious responses at frequencies beside f_c. Image-frequency response is the most obvious problem. The radio of Figure 7.1–1 employs a tunable BPF for image rejection. Given today's integrated electronics technology, high-Q tunable BPFs may not be economical and thus some other means of image rejection must be employed. Raising f_{IF} will increase the spacing between f_c and f_c' and thus reduce the requirements for the RF amplifier's BPF. In fact, if we set f_{IF} high enough, we could use a more economical LPF for image rejection.

But images are not the only problem superhets face with respect to spurious responses. Any distortion in the LO signal will generate harmonics that get mixed with a spurious input and be allowed to pass to the IF strip. That's why the LO must be a "clean" sine wave. The nonsinusoidal shape of digital signals is loaded with harmonics and thus if a receiver contains digital circuitry, special care must be taken to prevent these signals from "leaking" into the mixer stage. Further problems come from signal feedthrough and nonlinearities. For example, when a strong signal frequency near $\frac{1}{2}f_{IF}$ gets to the IF input, its second harmonic may be produced if the first stage of the IF amplifier is nonlinear. This second harmonic, approximately f_{IF}, will then be amplified by later stages and appear at the detector input as interference.

Superheterodyne receivers often contain an **automatic gain control (AGC)** such that the receiver's gain is automatically adjusted according to the input signal level. AGC is accomplished by rectifying the receiver's audio signal, thus calculating its average value. This dc value is then fed back to the IF or RF stage to increase or decrease the stage's gain. An AM radio usually includes an **automatic volume control (AVC)** signal from the demodulator back to the IF, while an FM receiver has **automatic frequency control (AFC)** fed back to the LO to correct small frequency drifts.

EXAMPLE 7.1–1

Superhets and spurious signal response

A superhet receiver with $f_{IF} = 500$ kHz and $3.5 < f_{LO} < 4.0$ MHz has a tuning dial calibrated to receive signals from 3 to 3.5 MHz. It is set to receive a 3.0-MHz signal. The receiver has a broadband RF amplifier, and it has been found that the LO has a significant third harmonic output. If a signal is heard, what are all its possible carrier frequencies? With $f_{LO} = 3.5$ MHz, $f_c = f_{LO} - f_{IF} = 3.5 - 0.5 = 3.0$ MHz, and the image frequency is $f_c' = f_c + 2f_{IF} = 4.0$ MHz. But the oscillator's third harmonic is 10.5 MHz and thus $f_c'' = 3f_{LO} - f_{IF} = 10.5 - 0.5 = 10.0$ MHz. The corresponding image frequency is then $f_c''' = f_c'' + 2f_{IF} = 10 + 1 = 11$ MHz. Therefore, with this receiver, even though the dial states the station is transmitting at 3.0 MHz, it actually may also be 4, 10, or 11 MHz.

EXERCISE 7.1–1

Determine the spurious frequencies for the receiver of Example 7.1–1 if $f_{IF} = 7.0$ MHz with $10 \leq f_{LO} \leq 10.5$ MHz and the local oscillator outputs a third harmonic. What would the minimum spurious input rejection be in dB, if the receiver's input was preceded by a first-order Butterworth LPF with $B = 4$ MHz.

Direct Conversion Receivers

Direct conversion receivers (**DC**) are a class of **tuned-RF** (**TRF**) receivers that consist of an RF amplifier followed by a product detector and suitable message amplification. They are called **homodyne** receivers. A DC receiver is diagrammed in Fig. 7.1–3. Adjacent channel interference rejection is accomplished by the LPF after the mixer. The DC receiver does not suffer from the *same* image problem that affects the superhet and because of improved circuit technology, particularly with higher gain and more stable RF amplifiers, it is capable of good performance. The DC's simplicity lends itself to subminiature wireless sensor applications.

The DC's chief drawback is that it does not reject the image signal that is present in the opposite sideband and is thus more susceptible to noise and interference. Fig. 7.1–4 illustrates the output from two single-tone SSB signals, one transmitting at the upper sideband, or $f_c + f_1$, and an interfering signal at the lower sideband, or $f_c - f_2$. Both f_1 and f_2 will appear at the receiver's output. However, the system shown in Figure 7.1–4, which was originally developed by Campbell (1993), eliminates the other sideband. If the nodes in Figure 7.1–4 are studied, the receiver's output only contains the upper sideband $f_c + f_1$ signal.

Special-Purpose Receivers

Other types of receivers used for special purposes include the heterodyne, the TRF, and the double-conversion structure. A **heterodyne receiver** is a superhet without the RF section, which raises potential image-frequency problems. Such receivers

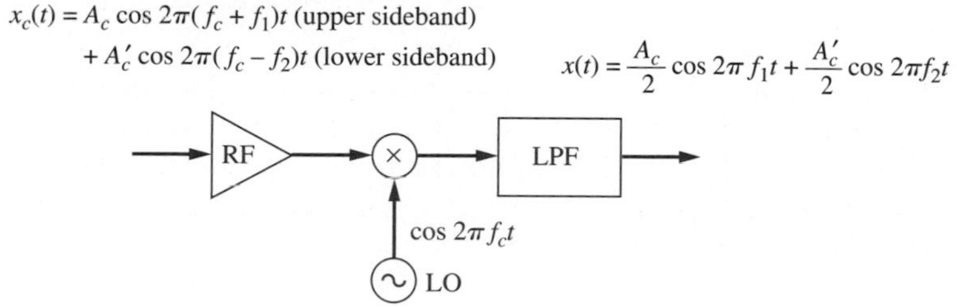

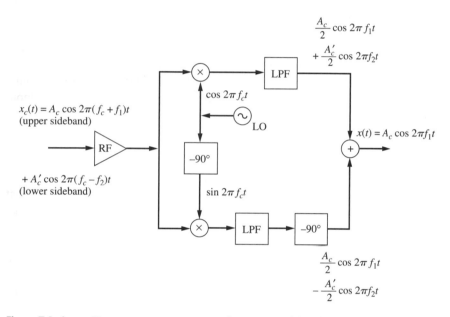

$x_c(t) = A_c \cos 2\pi(f_c + f_1)t$ (upper sideband)
$+ A_c' \cos 2\pi(f_c - f_2)t$ (lower sideband)
$x(t) = \frac{A_c}{2} \cos 2\pi f_1 t + \frac{A_c'}{2} \cos 2\pi f_2 t$
RF
LPF
$\cos 2\pi f_c t$
LO

Figure 7.1–3 Direct conversion receiver.

$\frac{A_c}{2} \cos 2\pi f_1 t$
$+ \frac{A_c'}{2} \cos 2\pi f_2 t$
LPF
$\cos 2\pi f_c t$
$x_c(t) = A_c \cos 2\pi(f_c + f_1)t$
(upper sideband)
LO
$x(t) = A_c \cos 2\pi f_1 t$
RF
–90°
$+ A_c' \cos 2\pi(f_c - f_2)t$
(lower sideband)
$\sin 2\pi f_c t$
LPF
–90°
$\frac{A_c}{2} \cos 2\pi f_1 t$
$- \frac{A_c'}{2} \cos 2\pi f_2 t$

Figure 7.1–4 Direct conversion receiver with opposite sideband rejection.

can be built at microwave frequencies with a diode mixer preceded by a fixed microwave filter to reject images. In addition to the DC TRF receiver, we can also have a TRF using a tunable RF amplifier and envelope detector. The classic crystal radio is the simplest TRF.

A **double-conversion receiver** in Fig. 7.1–5 takes the superhet principle one step further by including two frequency converters and two IF sections. The second IF is always fixed-tuned, while the first IF and second LO may be fixed or tunable. In either case, double conversion permits a larger value of $f_{IF\text{-}1}$ to improve image rejection in the RF section, and a smaller value of $f_{IF\text{-}2}$ to improve adjacent-channel rejection in the second IF. High-performance receivers for SSB and shortwave AM usually improve this design strategy.

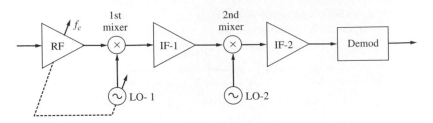

Figure 7.1–5   Double-conversion receiver.

Notice that a double-conversion SSB receiver with synchronous detection requires *three* stable oscillators plus automatic frequency control and synchronization circuitry. Fortunately IC technology has made the **frequency synthesizer** available for this application. We'll discuss frequency synthesis using phase-lock loops in Sect. 7.3.

Receiver Specifications

We now want to consider several parameters that determine the ability of a receiver to successfully demodulate a radio signal. Receiver **sensitivity** is the minimum input voltage that produces a specified signal-to-noise ratio at the output of the IF section. A good quality short-wave radio typically has a sensitivity of 1μV for a 40 dB signal-to-noise ratio. **Dynamic range** is the difference between the largest input signal that will not become distorted and the smallest signal that can be discerned, and is an important parameter. It is measured in dB. The maximum dynamic range of most analog amplifiers is 100 dB. Let's say we are listening to a weak AM broadcast signal and there is a strong station transmitting at a significantly different frequency, but within the RF amplifier's passband. The strong signal can overload the RF amplifier and thus wipe out the weak signal. **Selectivity** specifies a receiver's ability to discriminate against adjacent channel signals. It is a function of the IF strip's BPF. The **noise figure** indicates how much the receiver degrades the input signal's signal-to-noise ratio and is calculated by

$$Nf = \frac{(S/N)_{input}}{(S/N)_{output}} \qquad [5]$$

Typical noise figures are 5–10 dB. Finally **image rejection** is defined as

$$RR = 10 \log|H_{RF}(f_c)/H_{RF}(f_c')|^2 \qquad \text{dB} \qquad [6]$$

A typical value of image rejection is 50 dB. This equation may apply to other types of *spurious inputs* as well.

EXERCISE 7.1–2

Suppose a superhet's RF section is a typical tuned circuit described by Eq. (17), Sect. 4.1, with $f_o = f_c$ and $Q = 50$. Show that achieving $RR = 60$ dB requires $f_c/f_c' \approx 20$ when $f_c' = f_c + 2f_{IF}$. This requirement could easily be satisfied by a double conversion receiver with $f_{IF-1} \approx 9.5f_c$.

Scanning Spectrum Analyzers★

If the LO in a superhet is replaced by a VCO, then the predetection portion acts like a *voltage-tunable bandpass amplifier* with center frequency $f_0 = f_{LO} \pm f_{IF}$ and bandwidth $B = B_{IF}$. This property is at the heart of the *scanning spectrum analyzer* in Fig. 7.1–6a—a useful laboratory instrument that displays the spectral magnitude of an input signal over some selected frequency range.

The VCO is driven by a periodic ramp generator that sweeps the instantaneous frequency $f_{LO}(t)$ linearly from f_1 to f_2 in T seconds. The IF section has a narrow bandwidth B, usually adjustable, and the IF output goes to an envelope detector. Hence, the system's amplitude response at any instant t looks like Fig. 7.1–6b, where $f_0(t) = f_{LO}(t) - f_{IF}$. A fixed BPF (or LPF) at the input passes $f_1 \leq f \leq f_2$ while rejecting the image at $f_0(t) + 2f_{IF}$.

As $f_0(t)$ repeatedly scans past the frequency components of an input signal $v(t)$, its spectrum is displayed by connecting the envelope detector and ramp generator to the vertical and horizontal deflections of an oscilloscope. Obviously, a transient signal would not yield a fixed display, so $v(t)$ must be either a *periodic* or *quasi-periodic signal* or a *stationary random signal* over the time of observation. Correspondingly,

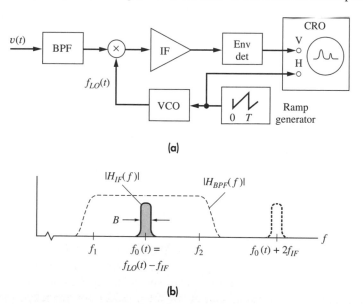

(a)

(b)

Figure 7.1–6   Scanning spectrum analyzer, (a) Block diagram; (b) amplitude response.

the display represents the amplitude line spectrum or the power spectral density. (A square-law envelope detector would be used for the latter.)

Some of the operational subtleties of this system are best understood by assuming that $v(t)$ consists of two or more sinusoids. To resolve one spectral line from the others, the IF bandwidth must be smaller than the line spacing. Hence, we call B the *frequency resolution*, and the maximum number of resolvable lines equals $(f_2 - f_1)/B$. The IF output produced by a single line takes the form of a *bandpass pulse* with time duration

$$\tau = BT/(f_2 - f_1) = B/\dot{f}_0$$

where $\dot{f}_0 = (f_2 - f_1)/T$ represents the *frequency sweep rate* in hertz per second.

But a rapid sweep rate may exceed the IF pulse response. Recall that our guideline for bandpass pulses requires $B \geq 1/\tau = \dot{f}_o/B$, or

$$\dot{f}_0 = \frac{f_2 - f_1}{T} \leq B^2 \qquad \qquad [7]$$

This important relation shows that accurate resolution (small B) calls for a slow rate and correspondingly long observation time. Also note that Eq. (7) involves four parameters adjustable by the user. Some scanning spectrum analyzers have built-in hardware that prevents you from violating Eq. (7); others simply have a warning light.

7.2   MULTIPLEXING SYSTEMS

When several communication channels are needed between the same two points, significant economies may be realized by sending all the messages on one transmission facility—a process called *multiplexing*. Applications of multiplexing range from the vital, if prosaic, telephone network, to the glamour of FM stereo and space-probe telemetry systems. There are three basic multiplexing techniques: frequency-division multiplexing (FDM), time-division multiplexing (TDM), and code-division multiplexing, treated in Chapter 15. The goal of these techniques is to enable multiple users to share a channel, and hence they are referred to as *frequency-division multiple access* (FDMA), *time-division multiple access* (TDMA), and *code-division multiple access* (CDMA).

Frequency-Division Multiplexing

The principle of FDM is illustrated by Fig. 7.2–1a, where several input messages (three are shown) individually modulate the *subcarriers* f_{c1}, f_{c2}, and so forth, after passing through LPFs to limit the message bandwidths. We show the subcarrier modulation as SSB as it often is, but any of the CW modulation techniques could be employed, or a mixture of them. The modulated signals are then summed to produce the *baseband* signal, with spectrum $X_b(f)$ as shown in Fig. 7.2–1b. (The designation "baseband" indicates that final carrier modulation has not yet taken place.) The baseband time function $x_b(t)$ is left to your imagination.

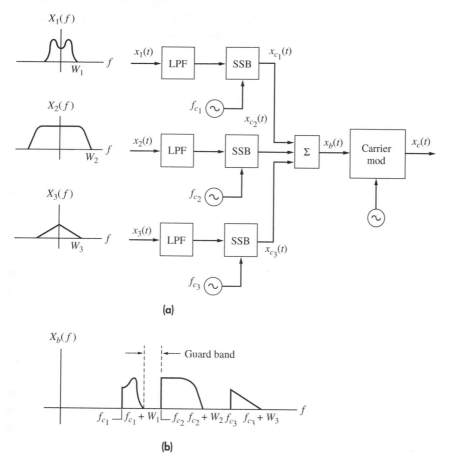

Figure 7.2–1   Typical FDM transmitter. (a) Input spectra and block diagram; (b) baseband FDM spectrum.

Assuming that the subcarrier frequencies are properly chosen, the multiplexing operation has assigned a slot in the frequency domain for each of the individual messages in modulated form, hence the name **frequency-division multiplexing.** The baseband signal may then be transmitted directly or used to modulate a transmitted carrier of frequency f_c. We are not particularly concerned here with the nature of the final carrier modulation, since the baseband spectrum tells the story.

Message recovery or demodulation of FDM is accomplished in three steps portrayed by Fig. 7.2–2. First, the carrier demodulator reproduces the baseband signal $x_b(t)$. Then the modulated subcarriers are separated by a bank of bandpass filters in parallel, following which the messages are individually detected.

The major practical problem of FDM is **cross talk,** the unwanted coupling of one message into another. Intelligible cross talk (cross-modulation) arises primarily because of nonlinearities in the system which cause one message signal to appear as

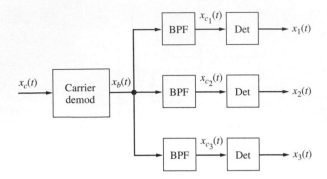

Figure 7.2–2   Typical FDM receiver.

modulation on another subcarrier. Consequently, standard practice calls for negative feedback to minimize amplifier nonlinearity in FDM systems. (As a matter of historical fact, the FDM cross talk problem was a primary motivator for the development of negative-feedback amplifiers.)

Unintelligible cross talk may come from nonlinear effects or from imperfect spectral separation by the filter bank. To reduce the latter, the modulated message spectra are spaced out in frequency by **guard bands** into which the filter transition regions can be fitted. For example, the guard band marked in Fig. 7.2–1b is of width $f_{c2} - (f_{c1} + W_1)$. The net **baseband bandwidth** is therefore the sum of the modulated message bandwidths plus the guard bands. But the scheme in Fig. 7.2–2 is not the only example of FDM. The commercial AM or FM broadcast bands are everyday examples of FDMA, where several broadcasters can transmit simultaneously in the same band, but at slightly different frequencies.

EXAMPLE 7.2–1   **FDMA Satellite Systems**

The Intelsat global network adds a third dimension to long-distance communication. Since a particular satellite links several ground stations in different countries, various *access* methods have been devised for international telephony. One scheme, known as *frequency-division multiple access* (FDMA), assigns a fixed number of voice channels between pairs of ground stations. These channels are grouped with standard FDM hardware, and relayed through the satellite using FM carrier modulation.

For the sake of example, suppose a satellite over the Atlantic Ocean serves ground stations in the United States, Brazil, and France. Further suppose that 36 channels (three groups) are assigned to the US–France route and 24 channels (two groups) to the US–Brazil route. Figure 7.2–3 shows the arrangement of the US transmitter and the receivers in Brazil and France. Not shown are the French and Brazilian transmitters and the US receiver needed for two-way conversations. Addi-

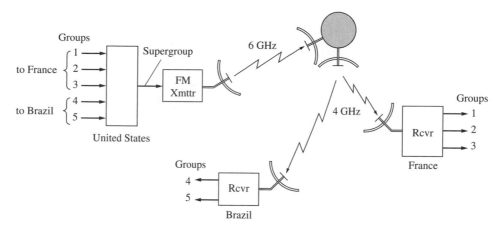

Figure 7.2–3   Simplified FDMA satellite system.

tional transmitters and receivers at slightly different carrier frequencies would provide a Brazil–France route.

The FDMA scheme creates at the satellite a *composite FDM signal* assembled with the FM signals from all ground stations. The satellite equipment consists of a bank of transponders. Each transponder has 36-MHz bandwidth accommodating 336 to 900 voice channels, depending on the ground-pair assignments. More details and other access schemes can be found in the literature.

Suppose an FDM baseband amplifier has cube-law nonlinearity which produces a baseband component proportional to $(v_2 \cos \omega_2 t)^2 v_1 \cos \omega_1 t$, where f_1 and f_2 are two subcarrier frequencies. Show that AM subcarrier modulation with $v_1 = 1 + x_1(t)$ and $v_2 = 1 + x_2(t)$ results in both intelligible and unintelligible cross talk on subcarrier f_1. Compare with the DSB case $v_1 = x_1(t)$ and $v_2 = x_2(t)$.

EXERCISE 7.2–1

FM Stereo Multiplexing

EXAMPLE 7.2–2

Figure 7.2–4a diagrams the FDM system that generates the baseband signal for FM stereophonic broadcasting. The left-speaker and right-speaker signals are first matrixed and preemphasized to produce $x_L(t) + x_R(t)$ and $x_L(t) - x_R(t)$. The sum signal is heard with a monophonic receiver; matrixing is required so the monaural listener will not be subjected to sound gaps in program material having stereophonic Ping-Pong effects. The $x_L(t) + x_R(t)$ signal is then inserted directly into the baseband, while $x_L(t) - x_R(t)$ DSB modulates a 38-kHz subcarrier. Double-sideband modulation is employed to preserve fidelity at low frequencies, and a 19-kHz pilot tone is added for receiver synchronization.

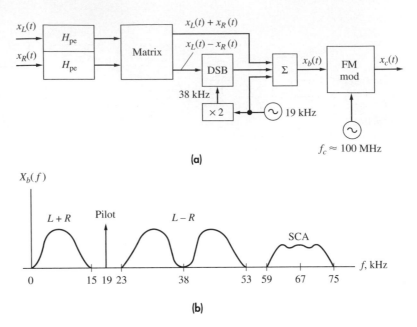

(a)

(b)

Figure 7.2–4   FM stereo multiplexing. (a) Transmitter; (b) baseband spectrum.

The resulting baseband spectrum is sketched in Fig. 7.2–4b. Also shown is another spectral component labeled SCA, which stands for Subsidiary Communication Authorization. The SCA signal has NBFM subcarrier modulation and is transmitted by some FM stations for the use of private subscribers who pay for commercial-free program material—the so-called "wallpaper music" heard in stores and offices.

For stereo broadcasting without SCA, the pilot carrier is allocated 10 percent of the peak frequency deviation and the seesaw relationship between $L + R$ and $L - R$ permits each to achieve nearly 90 percent deviation. The fact that the baseband spectrum extends to 53 kHz (or 75 kHz with SCA) does not appreciably increase the transmission bandwidth requirement because the higher frequencies produce smaller deviation ratios. High-fidelity stereo receivers typically have $B_{IF} \approx 250$ kHz.

The stereo *demultiplexing* or *decoding* system is diagrammed in Fig. 7.2–5. Notice how the pilot tone is used to actuate the stereo indicator as well as for synchronous detection. Integrated-circuit decoders employ switching circuits or phase-lock loops to carry out the functional operations.

Incidentally, discrete four-channel (quadraphonic) disk recording takes a logical extension of the FM stereo strategy to multiplex four independent signals on the two channels of a stereophonic record. Let's denote the four signals as L_F, L_R, R_F, and R_R (for left-front, left-rear, etc.). The matrixed signal $L_F + L_R$ is recorded directly on one channel along with $L_F - L_R$ multiplexed via frequency modulation of a 30-kHz subcarrier. The matrixed signals $R_F + R_R$ and $R_F - R_R$ are likewise multiplexed on the other channel. Because the resulting baseband spectrum goes up to 45 kHz, discrete quadraphonic signals cannot be transmitted in full on stereo FM.

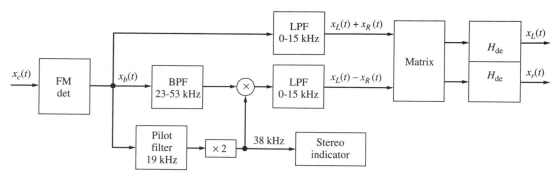

Figure 7.2–5   FM stereo multiplex receiver.

Other quadraphonic systems have only two independent channels and are compatible with FM stereo.

Quadrature-Carrier Multiplexing

Quadrature-carrier multiplexing, also known as *quadrature amplitude modulation* (QAM), utilizes carrier phase shifting and synchronous detection to permit two DSB signals to occupy the same frequency band. Figure 7.2–6 illustrates the multiplexing and demultiplexing arrangement. The transmitted signal has the form

$$x_c(t) = A_c[x_1(t) \cos \omega_c t \pm x_2(t) \sin \omega_c t] \qquad \text{[1]}$$

Since the modulated spectra overlap each other, this technique is more properly characterized as frequency-*domain* rather than frequency-*division* multiplexing.

From our prior study of synchronous detection for DSB and SSB, you should readily appreciate the fact that QAM involves more stringent synchronization than, say, an FDM system with SSB subcarrier modulation. Hence, QAM is limited to specialized applications, notably color television and digital data transmission.

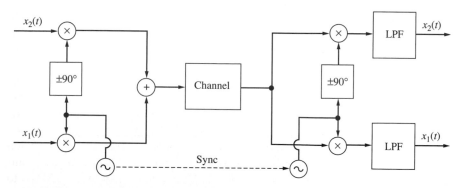

Figure 7.2–6   Quadrature-carrier multiplexing.

Time-Division Multiplexing

A sampled waveform is "off" most of the time, leaving the time between samples available for other purposes. In particular, sample values from several different signals can be interlaced into a single waveform. This is the principle of **time-division multiplexing** (TDM) discussed here.

The simplified system in Fig. 7.2–7 demonstrates the essential features of time-division multiplexing. Several input signals are prefiltered by the bank of input LPFs and sampled sequentially. The rotating sampling switch or *commutator* at the transmitter extracts one sample from each input per revolution. Hence, its output is a PAM waveform that contains the individual samples periodically interlaced in time. A similar rotary switch at the receiver, called a *decommutator* or *distributor*, separates the samples and distributes them to another bank of LPFs for reconstruction of the individual messages.

If all inputs have the same message bandwidth W, the commutator should rotate at the rate $f_s \geq 2W$ so that successive samples from any one input are spaced by $T_s = 1/f_s \leq 1/2W$. The time interval T_s containing one sample from each input is called a *frame*. If there are M input channels, the pulse-to-pulse spacing within a frame is $T_s/M = 1/Mf_s$. Thus, the total number of pulses per second will be

$$r = Mf_s \geq 2MW \tag{2}$$

which represents the pulse rate or *signaling rate* of the TDM signal.

Our primitive system shows mechanical switching to generate multiplexed PAM. But almost all practical TDM systems employ electronic switching. Further-

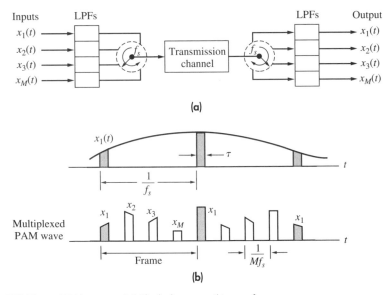

(a)

(b)

Figure 7.2–7   TDM system. (*a*) Block diagram; (*b*) waveforms.

more, other types of pulse modulation can be used instead of PAM. Therefore, a more generalized commutator might have the structure diagrammed in Fig. 7.2–8, where pulse-modulation *gates* process the individual inputs to form the TDM output. The gate control signals come from a flip-flop chain (a broken-ring counter) driven by a digital clock at frequency Mf_s. The decommutator would have a similar structure.

Regardless of the type of pulse modulation, TDM systems require careful *synchronization* between commutator and decommutator. Synchronization is a critical consideration in TDM, because each pulse must be distributed to the correct output line at the appropriate time. A popular brute-force synchronization technique devotes one time slot per frame to a distinctive *marker* pulse or nonpulse, as illustrated in Fig. 7.2–9. These markers establish the frame frequency f_s at the receiver, but the number of signal channels is reduced to $M - 1$. Other synchronization methods involve auxiliary pilot tones or the statistical properties of the TDM signal itself.

Radio-frequency transmission of TDM necessitates the additional step of CW modulation to obtain a *bandpass* waveform. For instance, a TDM signal composed of duration or position-modulated pulses could be applied to an AM transmitter with 100 percent modulation, thereby producing a train of constant-amplitude RF pulses.

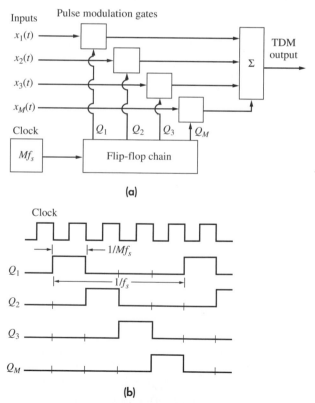

(a)

(b)

Figure 7.2–8   (a) Electronic commutator for TDM; (b) timing diagram.

The compound process would be designated PDM/AM or PPM/AM, and the required transmission bandwidth would be twice that of the baseband TDM signal. The relative simplicity of this technique suits low-speed multichannel applications such as radio control for model airplanes.

More sophisticated TDM systems may use PAM/SSB for bandwidth conservation or PAM/FM for wideband noise reduction. The complete transmitter diagram in Fig. 7.2–10a now includes a lowpass *baseband filter* with bandwidth

$$B_b = \tfrac{1}{2}r = \tfrac{1}{2}Mf_s \qquad \text{[3]}$$

Baseband filtering prior to CW modulation produces a smooth modulating waveform $x_b(t)$ having the property that it passes through the individual sample values at the corresponding sample times, as portrayed in Fig. 7.2–10b. Since the interlaced sample spacing equals $1/Mf_s$, the baseband filter constructs $x_b(t)$ in the same way

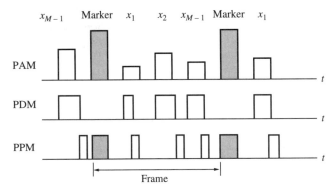

Figure 7.2–9   TDM synchronization markers.

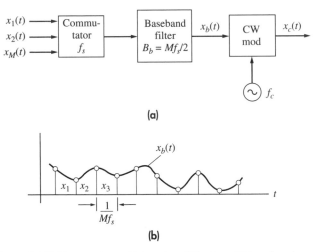

(a)

(b)

Figure 7.2–10   (a) TDM transmitter with baseband filtering; (b) baseband waveform.

that an LPF with $B = f_s/2$ would reconstruct a waveform $x(t)$ from its periodic samples $x(kT_s)$ with $T_s = 1/f_s$.

If baseband filtering is employed, and if the sampling frequency is close to the Nyquist rate $f_{s_{min}} = 2W$ for the individual inputs, then the transmission bandwidth for PAM/SSB becomes

$$B_T = \tfrac{1}{2}M \times 2W = MW$$

Under these conditions, TDM approaches the theoretical *minimum bandwidth* of frequency-division multiplexing with SSB subcarrier modulation.

Although we've assumed so far that all input signals have the same bandwidth, this restriction is not essential and, moreover, would be unrealistic for the important case of analog data telemetry. The purpose of a telemetry system is to combine and transmit physical measurement data from different sources at some remote location. The sampling frequency required for a particular measurement depends on the physical process involved and can range from a fraction of one hertz up to several kilohertz. A typical telemetry system has a *main multiplexer* plus *submultiplexers* arranged to handle 100 or more data channels with various sampling rates.

TDM Telemetry   EXAMPLE 7.2–3

For the sake of illustration, suppose we need 5 data channels with minimum sampling rates of 3000, 700, 600, 300, and 200 Hz. If we used a 5-channel multiplexer with $f_s = 3000$ Hz for all channels, the TDM signaling rate would be $r = 5 \times 3000 = 15$ kHz— not including synchronization markers. A more efficient scheme involves an 8-channel main multiplexer with $f_s = 750$ Hz and a 2-channel submultiplexer with $f_s = 375$ Hz connected as shown in Fig. 7.2–11.

The two lowest-rate signals $x_4(t)$ and $x_5(t)$ are combined by the submultiplexer to create a pulse rate of $2 \times 375 = 750$ Hz for insertion into one channel of the

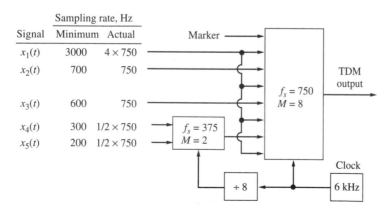

Figure 7.2–11   TDM telemetry system with main multiplexer and submultiplexer.

main multiplexer. Hence, the samples of $x_4(t)$ and $x_5(t)$ will appear in *alternate* frames. On the other hand, the highest-rate signal $x_1(t)$ is applied to four inputs on the main multiplexer. Consequently, its samples appear in four equispaced slots within each frame, for an equivalent sampling rate of $4 \times 750 = 3000$ Hz. The total output signaling rate, including a marker, is $r = 8 \times 750$ Hz $= 6$ kHz. Baseband filtering would yield a smoothed signal whose bandwidth $B_b = 3$ kHz fits nicely into a voice telephone channel!

EXERCISE 7.2–2

Suppose the output in Fig. 7.2–11 is an unfiltered PAM signal with 50 percent duty cycle. Sketch the waveform for two successive frames, labeling each pulse with its source signal. Then calculate the required transmission bandwidth B_T from Eq. (6), Sect. 6.2.

Cross Talk and Guard Times

When a TDM system includes baseband filtering, the filter design must be done with extreme care to avoid interchannel *cross talk* from one sample value to the next in the frame. Digital signals suffer a similar problem called *intersymbol interference*, and we defer the treatment of baseband waveform shaping to Sect. 11.3.

A TDM signal without baseband filtering also has cross talk if the transmission channel results in pulses whose tails or postcursors overlap into the next time slot of the frame. Pulse overlap is controlled by establishing *guard times* between pulses, analogous to the guard bands between channels in an FDM system. Practical TDM systems have both guard times and guard bands, the former to suppress cross talk, the latter to facilitate message reconstruction with nonideal filters.

For a quantitative estimate of cross talk, let's assume that the transmission channel acts like a first-order lowpass filter with 3-dB bandwidth B. The response to a rectangular pulse then decays exponentially, as sketched in Fig. 7.2–12. The guard time T_g represents the *minimum* pulse spacing, so the pulse tail decays to a value no larger than $A_{ct} = Ae^{-2\pi BT_g}$ by the time the next pulse arrives. Accordingly, we define the *cross-talk reduction factor*.

$$k_{ct} \triangleq 10 \log (A_{ct}/A)^2 \approx -54.5 \, BT_g \qquad \text{dB} \qquad \qquad [4]$$

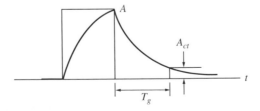

Figure 7.2–12   Cross talk in TDM.

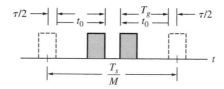

Figure 7.2–13   TDM/PPM with guard time.

Keeping the cross talk below -30 dB calls for $T_g > 1/2B$.

Guard times are especially important in TDM with pulse-duration or pulse-position modulation because the pulse edges move around within their frame slots. Consider the PPM case in Fig. 7.2–13: here, one pulse has been position-modulated forward by an amount t_0 and the next pulse backward by the same amount. The allowance for guard time T_g requires that $T_g + 2t_0 + 2(\tau/2) \leq T_s/M$ or

$$t_0 \leq \frac{1}{2}\left(\frac{T_s}{M} - \tau - T_g\right) \qquad \text{[5]}$$

A similar modulation limit applies in the case of PDM.

Nine voice signals plus a marker are to be transmitted via PPM on a channel having $B = 400$ kHz. Calculate T_g such that $k_{ct} \approx -60$ dB. Then find the maximum permitted value of t_0 if $f_s = 8$ kHz and $\tau = \frac{1}{5}(T_s/M)$.

EXERCISE 7.2–3

Comparison of TDM and FDM

Time-division and frequency-division multiplexing accomplish the same end by different means. Indeed, they may be classified as *dual* techniques. Individual TDM channels are assigned to distinct *time slots* but jumbled together in the frequency domain; conversely, individual FDM channels are assigned to distinct *frequency slots* but jumbled together in the time domain. What advantages then does each offer over the other?

Many of the TDM advantages are technology driven. TDM is readily implemented with high-density VLSI circuitry where digital switches are extremely economical. Recall that FDM requires an analog subcarrier modulator, bandpass filter, and demodulator for every message channel. These are relatively expensive to implement in VLSI. But all of these are replaced by a TDM commutator and decommutator switching circuits, easily put on a chip. However, TDM synchronization is only slightly more demanding than that of suppressed-carrier FDM.

Second, TDM is invulnerable to the usual causes of cross talk in FDM, namely, imperfect bandpass filtering and nonlinear cross-modulation. However, TDM cross-talk immunity does depend on the transmission bandwidth and the absence of delay distortion.

Third, the use of submultiplexers allows a TDM system to accommodate different signals whose bandwidths or pulse rates may differ by more than an order of magnitude. This flexibility has particular value for multiplexing digital signals, as we'll see in Sect. 12.5.

Finally, TDM may or may not be advantageous when the transmission medium is subject to **fading.** Rapid wideband fading might strike only occasional pulses in a given TDM channel, whereas all FDM channels would be affected. But slow narrowband fading wipes out all TDM channels, whereas it might hurt only one FDM channel.

Many systems such as satellite relay are a hybrid of FDMA and TDMA. For example, we have FDMA where specific frequency channels will be allocated to various services. In turn then, each channel may be shared by individual users using TDMA.

7.3   PHASE-LOCK LOOPS

The phase-lock loop (PLL) is undoubtedly the most versatile building block available for CW modulation systems. PLLs are found in modulators, demodulators, frequency synthesizers, multiplexers, and a variety of signal processors. We'll illustrate some of these applications after discussing PLL operation and lock-in conditions. Our introductory study provides a useful working knowledge of PLLs but does not go into detailed analysis of nonlinear behavior and transients. Treatments of these advanced topics are given in Blanchard (1976), Gardner (1979), Meyr and Ascheid (1990), and Lindsey (1972).

PLL Operation and Lock-In

The basic aim of a PLL is to *lock* or *synchronize* the instantaneous angle (i.e., phase and frequency) of a VCO output to the instantaneous angle of an external bandpass signal that may have some type of CW modulation. For this purpose, the PLL must perform *phase comparison*. We therefore begin with a brief look at phase comparators.

The system in Fig. 7.3–1a is an **analog phase comparator.** It produces an output $y(t)$ that depends on the instantaneous angular difference between two bandpass input signals, $x_c(t) = A_c \cos \theta_c(t)$ and $v(t) = A_v \cos \theta_v(t)$. Specifically, if

$$\theta_v(t) = \theta_c(t) - \epsilon(t) + 90° \qquad [1]$$

and if the LPF simply extracts the difference-frequency term from the product $x_c(t)v(t)$, then

$$y(t) = \tfrac{1}{2} A_c A_v \cos [\theta_c(t) - \theta_v(t)]$$
$$= \tfrac{1}{2} A_c A_v \cos [\epsilon(t) - 90°] = \tfrac{1}{2} A_c A_v \sin \epsilon(t)$$

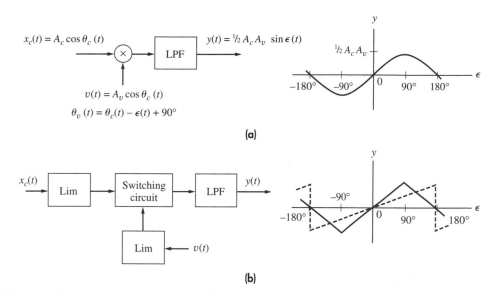

Figure 7.3–1   Phase comparators. (a) Analog; (b) digital.

We interpret $\epsilon(t)$ as the *angular error*, and the plot of y versus ϵ emphasizes that $y(t) = 0$ when $\epsilon(t) = 0$. Had we omitted the 90° shift in Eq. (1), we would get $y(t) = 0$ at $\epsilon(t) = \pm 90°$. Thus, zero output from the phase comparator corresponds to a *quadrature* phase relationship.

Also note that $y(t)$ depends on $A_c A_v$ when $\epsilon(t) \neq 0$, which could cause problems if the input signals have amplitude modulation. These problems are eliminated by the *digital phase comparator* in Fig. 7.3–1b, where hard limiters convert input sinusoids to square waves applied to a switching circuit. The resulting plot of y versus ϵ has a triangular or sawtooth shape, depending on the switching circuit details. However, all three phase-comparison curves are essentially the same when $|\epsilon(t)| \ll 90°$—the intended operating condition in a PLL.

Hereafter, we'll work with the analog PLL structure in Fig. 7.3–2. We assume for convenience that the external input signal has constant amplitude $A_c = 2$ so that $x_c(t) = 2 \cos \theta_c(t)$ where, as usual,

$$\theta_c(t) = \omega_c t + \phi(t) \qquad \omega_c = 2\pi f_c \qquad [2]$$

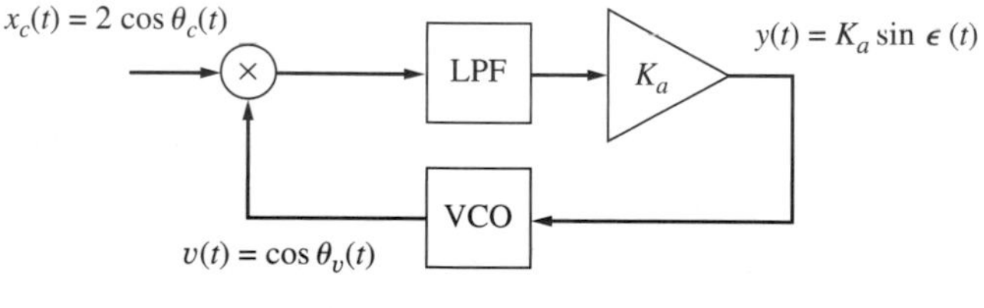

Figure 7.3–2   Phase-lock loop.

We also assume a unit-amplitude VCO output $v(t) = \cos \theta_v(t)$ and a loop amplifier with gain K_a. Hence,

$$y(t) = K_a \sin \epsilon(t) \qquad [3]$$

which is fed back for the control voltage to the VCO.

Since the VCO's *free-running frequency* with $y(t) = 0$ may not necessarily equal f_c, we'll write it as $f_v = f_c - \Delta f$ where Δf stands for the *frequency error*. Application of the control voltage produces the instantaneous angle

$$\theta_v(t) = 2\pi(f_c - \Delta f)t + \phi_v(t) + 90° \qquad [4a]$$

with

$$\phi_v(t) = 2\pi K_v \int^t y(\lambda) \, d\lambda \qquad [4b]$$

when K_v equals the frequency-deviation constant. The angular error is then

$$\epsilon(t) = \theta_c(t) - \theta_v(t) + 90°$$
$$= 2\pi\Delta ft + \phi(t) - \phi_v(t)$$

and differentiation with respect to t gives

$$\dot\epsilon(t) = 2\pi\Delta f + \dot\phi(t) - 2\pi K_v \, y(t)$$

Upon combining this expression with Eq. (3) we obtain the *nonlinear differential equation.*

$$\dot\epsilon(t) + 2\pi K \sin \epsilon(t) = 2\pi\Delta f + \dot\phi(t) \qquad [5]$$

in which we've introduced the *loop gain*

$$K \triangleq K_v K_a$$

This gain is measured in *hertz* and turns out to be a critical parameter.

Equation (5) governs the dynamic operation of a PLL, but it does not yield a closed-form solution with an arbitrary $\phi(t)$. To get a sense of PLL behavior and lock-in conditions, consider the case of a constant input phase $\phi(t) = \phi_0$ starting at $t = 0$. Then $\dot\phi(t) = 0$ and we rewrite Eq. (5) as

$$\frac{1}{2\pi K} \dot\epsilon(t) + \sin \epsilon(t) = \frac{\Delta f}{K} \qquad t \geq 0 \qquad [6]$$

Lock-in with a constant phase implies that the loop attains a *steady state* with $\dot\epsilon(t) = 0$ and $\epsilon(t) = \epsilon_{ss}$. Hence, $\sin \epsilon_{ss} = \Delta f/K$ at lock-in, and it follows that

$$\epsilon_{ss} = \arcsin \frac{\Delta f}{K} \qquad [7a]$$

$$y_{ss} = K_a \sin \epsilon_{ss} = \frac{\Delta f}{K_v} \qquad [7b]$$

$$v_{ss}(t) = \cos{(\omega_c t + \phi_0 - \epsilon_{ss} + 90°)} \qquad \text{[7c]}$$

Note that the nonzero value of y_{ss} cancels out the VCO frequency error, and $v_{ss}(t)$ is locked to the frequency of the input signal $x_c(t)$. The phase error ϵ_{ss} will be negligible if $|\Delta f/K| \ll 1$.

However, Eq. (6) has *no steady-state solution* and ϵ_{ss} in Eq. (7a) is undefined when $|\Delta f/K| > 1$. Therefore, lock-in requires the condition

$$K \ge |\Delta f| \qquad \text{[8]}$$

Stated another way, a PLL will lock to any constant input frequency within the range $\pm K$ hertz of the VCO's free-running frequency f_v.

Additional information regarding PLL behavior comes from Eq. (6) when we require sufficient loop gain that $\epsilon_{ss} \approx 0$. Then, after some instant $t_0 > 0$, $\epsilon(t)$ will be small enough to justify the approximation $\sin \epsilon(t) \approx \epsilon(t)$ and

$$\frac{1}{2\pi K}\dot{\epsilon}(t) + \epsilon(t) = 0 \qquad t \ge t_0 \qquad \text{[9a]}$$

This linear equation yields the well-known solution

$$\epsilon(t) = \epsilon(t_0)e^{-2\pi K(t-t_0)} \qquad t \ge t_0 \qquad \text{[9b]}$$

a *transient* error that virtually disappears after five time constants have elapsed, that is, $\epsilon(t) \approx 0$ for $t > t_0 + 5/(2\pi K)$. We thus infer that if the input $x_c(t)$ has a *time-varying* phase $\phi(t)$ whose variations are slow compared to $1/(2\pi K)$, and if the instantaneous frequency $f_c + \dot{\phi}(t)/2\pi$ does not exceed the range of $f_v \pm K$, then the PLL will stay in lock and track $\phi(t)$ with negligible error—provided that the LPF in the phase comparator passes the variations of $\phi(t)$ on to the VCO.

The *phase-plane plot* of $\dot{\epsilon}$ versus ϵ is defined by rewriting Eq. (6) in the form

EXERCISE 7.3–1

$$\dot{\epsilon} = 2\pi(\Delta f - K \sin \epsilon)$$

(a) Sketch $\dot{\epsilon}$ versus ϵ for $K = 2\,\Delta f$ and show that an arbitrary initial value $\epsilon(0)$ must go to $\epsilon_{ss} = 30° \pm m\,360°$ where m is an integer. *Hint:* $\epsilon(t)$ increases when $\dot{\epsilon}(t) > 0$ and decreases when $\dot{\epsilon}(t) < 0$. (b) Now sketch the phase-plane plot for $K < \Delta f$ to show that $|\dot{\epsilon}(t)| > 0$ for any $\epsilon(t)$ and, consequently, ϵ_{ss} does not exist.

Synchronous Detection and Frequency Synthesizers

The lock-in ability of a PLL makes it ideally suited to systems that have a pilot carrier for synchronous detection. Rather than attempting to filter the pilot out of the accompanying modulated waveform, the augmented PLL circuit in Fig. 7.3–3 can be used to generate a sinusoid synchronized with the pilot. To minimize clutter here, we've lumped the phase comparator, lowpass filter, and amplifier into a phase discriminator (PD) and we've assumed unity sinusoidal amplitudes throughout.

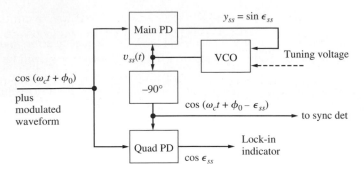

Figure 7.3–3   PLL pilot filter with two phase discriminators (PD).

Initial adjustment of the tuning voltage brings the VCO frequency close to f_c and $\epsilon_{ss} \approx 0$, a condition sensed by the quadrature phase discriminator and displayed by the lock-in indicator. Thereafter, the PLL automatically tracks any phase or frequency drift in the pilot, and the phase-shifted VCO output provides the LO signal needed for the synchronous detector. Thus, the whole unit acts as a *narrowband pilot filter* with a virtually noiseless output.

Incidentally, a setup like Fig. 7.3–3 can be used to *search* for a signal at some unknown frequency. You disconnect the VCO control voltage and apply a ramp generator to sweep the VCO frequency until the lock-in indicator shows that a signal has been found. Some radio scanners employ an automated version of this procedure.

For *synchronous detection* of DSB without a transmitted pilot, Costas invented the PLL system in Fig. 7.3–4. The modulated DSB waveform $x(t) \cos \omega_c t$ with bandwidth $2W$ is applied to a pair of phase discriminators whose outputs are proportional to $x(t) \sin \epsilon_{ss}$ and $x(t) \cos \epsilon_{ss}$. Multiplication and integration over $T \gg 1/W$ produces the VCO control voltage

$$y_{ss} \approx T\langle x^2(t)\rangle \, \sin \epsilon_{ss} \cos \epsilon_{ss} = \frac{T}{2} S_x \sin 2\epsilon_{ss}$$

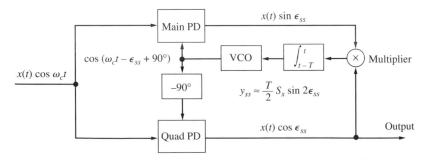

Figure 7.3–4   Costas PLL system for synchronous detection.

If $\Delta f \approx 0$, the PLL locks with $\epsilon_{ss} \approx 0$ and the output of the quadrature discriminator is proportional to the demodulated message $x(t)$. Of course the loop loses lock if $x(t) = 0$ for an extended interval.

The *frequency-offset loop* in Fig. 7.3–5 translates the input frequency (and phase) by an amount equal to that of an auxiliary oscillator. The intended output frequency is now $f_c + f_1$, so the free-running frequency of the VCO must be

$$f_v = (f_c + f_1) - \Delta f \approx f_c + f_1$$

The oscillator and VCO outputs are mixed and filtered to obtain the difference-frequency signal $\cos[\theta_v(t) - (\omega_1 t + \phi_1)]$ applied to the phase discriminator. Under locked conditions with $\epsilon_{ss} \approx 0$, the instantaneous angles at the input to the discriminator will differ by 90°. Hence, $\theta_v(t) - (\omega_1 t + \phi_1) = \omega_c t + \phi_0 + 90°$, and the VCO produces $\cos[(\omega_c + \omega_1)t + \phi_0 + \phi_1 + 90°]$.

By likewise equating instantaneous angles, you can confirm that Fig. 7.3–6 performs *frequency multiplication*. Like the frequency multiplier discussed in Sect. 5.2, this unit multiplies the instantaneous angle of the input by a factor of n. However, it does so with the help of a *frequency divider* which is easily implemented using a *digital counter*. Commercially available divide-by-n counters allow you to select any integer value for n from 1 to 10 or even higher. When such a counter is inserted in a PLL, you have an *adjustable* frequency multiplier.

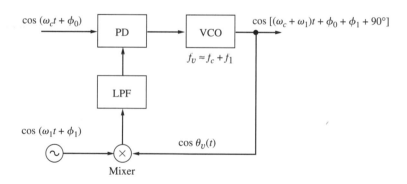

Figure 7.3–5   Frequency-offset loop.

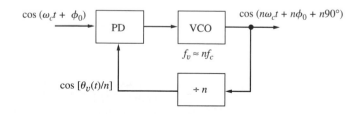

Figure 7.3–6   PLL frequency multiplier.

A *frequency synthesizer* starts with the output of one crystal-controlled master oscillator; various other frequencies are synthesized therefrom by combinations of frequency division, multiplication, and translation. Thus, all resulting frequencies are stabilized by and synchronized with the master oscillator. General-purpose laboratory synthesizers incorporate additional refinements and have rather complicated diagrams. So we'll illustrate the principles of frequency synthesis by an example.

EXAMPLE 7.3–1   Suppose a double-conversion SSB receiver needs fixed LO frequencies at 100 kHz (for synchronous detection) and 1.6 MHz (for the second mixer), and an adjustable LO that covers 9.90–9.99 MHz in steps of 0.01 MHz (for RF tuning). The custom-tailored synthesizer in Fig. 7.3–7 provides all the required frequencies by dividing down, multiplying up, and mixing with the output of a 10-MHz oscillator. You can quickly check out the system by putting a frequency-multiplication block in place of each PLL with a divider.

Observe here that all output frequencies are *less* than the master-oscillator frequency. This ensures that any frequency drift will be reduced rather than increased by the synthesis operations.

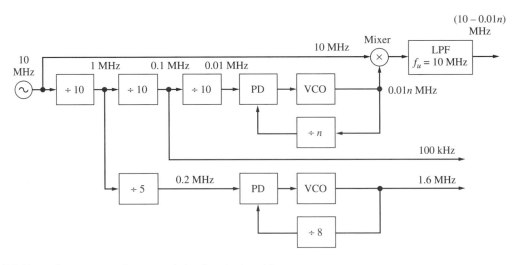

Figure 7.3–7   Frequency synthesizer with fixed and adjustable outputs.

EXERCISE 7.3–2   Draw the block diagram of a PLL system that synthesizes the output frequency nf_c/m from a master-oscillator frequency f_c. State the condition for locked operation in terms of the loop gain K and the VCO free-running frequency f_v.

Linearized PLL Models and FM Detection

Suppose that a PLL has been tuned to lock with the input frequency f_c, so $\Delta f = 0$. Suppose further that the PLL has sufficient loop gain to track the input phase $\phi(t)$ within a small error $\epsilon(t)$, so $\sin \epsilon(t) \approx \epsilon(t) = \phi(t) - \phi_v(t)$. These suppositions constitute the basis for the *linearized PLL model* in Fig. 7.3–8a, where the LPF has been represented by its impulse response $h(t)$.

Since we'll now focus on the phase variations, we view $\phi(t)$ as the input "signal" which is compared with the feedback "signal"

$$\phi_v(t) = 2\pi K_v \int^t y(\lambda)\, d\lambda$$

to produce the output $y(t)$. We emphasize that viewpoint by redrawing the linearized model as a *negative feedback system*, Fig. 7.3–8b. Note that the VCO becomes an integrator with gain $2\pi K_v$, while phase comparison becomes subtraction.

Fourier transformation finally takes us to the *frequency-domain model* in Fig. 7.3–8c, where $\Phi(f) = \mathcal{F}[\phi(t)]$, $H(f) = \mathcal{F}[h(t)]$, and so forth. Routine analysis yields

$$Y(f) = \frac{K_a H(f)}{1 + K_a H(f)(K_v/jf)}\, \Phi(f) = \frac{1}{K_v}\, \frac{jfKH(f)}{jf + KH(f)}\, \Phi(f) \qquad [10]$$

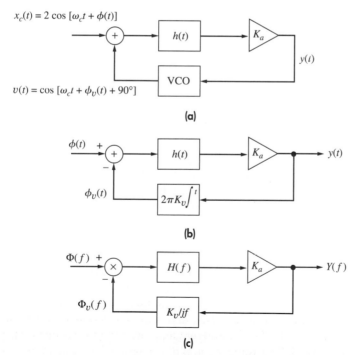

$$x_c(t) = 2 \cos [\omega_c t + \phi(t)]$$

$$v(t) = \cos [\omega_c t + \phi_v(t) + 90°]$$

(a)

$$\phi(t)$$
$$\phi_v(t)$$

(b)

$$\Phi(f)$$
$$\Phi_v(f)$$

(c)

Figure 7.3–8   Linearized PLL models. (a) Time domain; (b) phase; (c) frequency domain.

which expresses the frequency-domain relationship between the input phase and output voltage.

Now let $x_c(t)$ be an *FM wave* with $\dot{\phi}(t) = 2\pi f_\Delta \, x(t)$ and, accordingly,

$$\Phi(f) = 2\pi f_\Delta \, X(f)/(j2\pi f) = (f_\Delta/jf)X(f)$$

Substituting for $\Phi(f)$ in Eq. (10) gives

$$Y(f) = \frac{f_\Delta}{K_v} \, H_L(f)X(f) \qquad\qquad\qquad \text{[11a]}$$

where

$$H_L(f) = \frac{H(f)}{H(f) + j(f/K)} \qquad\qquad\qquad \text{[11b]}$$

which we interpret as the equivalent *loop transfer function*. If $X(f)$ has message bandwidth W and if

$$H(f) = 1 \qquad |f| \leq W \qquad\qquad\qquad \text{[12a]}$$

then $H_L(f)$ takes the form of a *first-order lowpass filter* with 3-dB bandwidth K, namely

$$H_L(f) = \frac{1}{1 + j(f/K)} \qquad |f| \leq W \qquad\qquad \text{[12b]}$$

Thus, $Y(f) \approx (f_\Delta/K_v)X(f)$ when $K \geq W$ so

$$y(t) \approx \frac{f_\Delta}{K_v} \, x(t) \qquad\qquad\qquad\qquad \text{[13]}$$

Under these conditions, the PLL recovers the message $x(t)$ from $x_c(t)$ and thereby serves as an *FM detector*.

A disadvantage of the first-order PLL with $H(f) = 1$ is that the loop gain K determines both the bandwidth of $H_L(f)$ and the lock-in frequency range. In order to track the instantaneous input frequency $f(t) = f_c + f_\Delta x(t)$ we must have $K \geq f_\Delta$. The large bandwidth of $H_L(f)$ may then result in excessive interference and noise at the demodulated output. For this reason, and other considerations, $H_L(f)$ is usually a more sophisticated *second-order* function in practical PLL frequency detectors.

7.4   TELEVISION SYSTEMS

The message transmitted by a television is a two-dimensional *image* with motion, and therefore a function of two spatial variables as well as time. This section introduces the theory and practice of image transmission via an electrical signal. Our initial discussion of monochrome (black and white) video signals and bandwidth requirements also applies to facsimile systems which transmit only still pictures. Then we'll describe TV transmitters, in block-diagram form, and the modifications needed for color television.

There are several types of television systems with numerous variations found in different countries. We'll concentrate on the NTSC (National Television System Committee) system used in North America, South America, and Japan and its digital replacement, the HDTV (high-definition television). More details about HDTV are given by Whitaker (1999), and ATSC (1995).

Video Signals, Resolution, and Bandwidth

To start with the simplest case, consider a motion-free monochrome intensity pattern $I(h, v)$, where h and v are the horizontal and vertical coordinates. Converting $I(h, v)$ to a signal $x(t)$—and vice versa—requires a discontinuous mapping process such as the *scanning raster* diagrammed in Fig. 7.4–1. The scanning device, which produces a voltage or current proportional to intensity, starts at point A and moves with constant but unequal rates in the horizontal and vertical directions, following the path AB. Thus, if s_h and s_v are the horizontal and vertical scanning speeds, the output of the scanner is the *video signal*

$$x(t) = I(s_h t, s_v t) \qquad\qquad [1]$$

since $h = s_h t$, and so forth. Upon reaching point B, the scanning spot quickly flies back to C (the horizontal retrace) and proceeds similarly to point D, where facsimile scanning would end.

In TV, however, image motion must be accommodated, so the spot retraces vertically to E and follows an interlaced pattern ending at F. The process is then repeated starting again at A. The two sets of lines are called the first and second *fields*; together they constitute one complete picture or *frame*. The frame rate is just rapid enough (25 to 30 per second) to create the illusion of continuous motion, while the field rate (twice the frame rate) makes the flickering imperceptible to the human eye. Hence, interlaced scanning allows the lowest possible picture repetition rate without visible flicker.

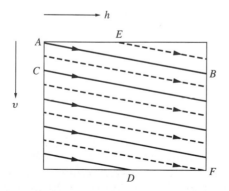

Figure 7.4–1   Scanning raster with two fields (line spacing grossly exaggerated).

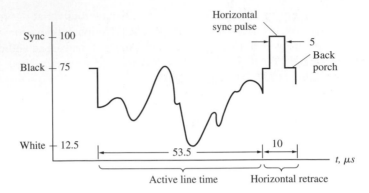

Figure 7.4–2   Video waveform for one full line (NTSC standards).

Two modifications are made to the video signal after scanning: *blanking pulses* are inserted during the retrace intervals to blank out retrace lines on the receiving picture tube; and *synchronizing pulses* are added on top of the blanking pulses to synchronize the receiver's horizontal and vertical sweep circuits. Figure 7.4–2 shows the waveform for one complete line, with amplitude levels and durations corresponding to NTSC standards. Other parameters are listed in Table 7.4–1 along with some comparable values for the European CCIR (International Radio Consultative Committee) system and the *high-definition* (HDTV) system.

Table 7.4–1   Television system parameters

	NTSC	CCIR	HDTV/USA
Aspect ratio, horizontal/vertical	4/3	4/3	16/9
Total of lines per frame	525	625	1125
Field frequency, Hz	60	50	60
Line frequency, kHz	15.75	15.625	33.75
Line time, μs	63.5	64	29.63
Video bandwidth, MHz	4.2	5.0	24.9
Optimal viewing distance	7H	7H	3H
Sound	Mono/Stereo output	Mono/Stereo output	6 channel Dolby Digital Surround
Horizontal retrace time, μS	10		3.7
Vertical retrace, lines/field	21		45

Analyzing the spectrum of the video signal in absence of motion is relatively easy with the aid of Fig. 7.4–3 where, instead of retraced scanning, the image has been periodically repeated in both directions so the equivalent scanning path is unbroken. Now any periodic function of two variables may be expanded as a *two-*

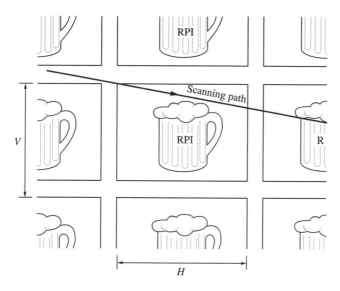

Figure 7.4–3   Periodically repeated image with unbroken scanning path.

dimensional Fourier series by straightforward extension of the one-dimensional series. For the case at hand with H and V the horizontal and vertical periods (including retrace allowance), the image intensity is

$$I(h, v) = \sum_{m=-\infty}^{\infty} \sum_{n=-\infty}^{\infty} c_{mn} \exp\left[j2\pi\left(\frac{mh}{H} + \frac{nv}{V} \right) \right] \qquad [2]$$

where

$$c_{mn} = \frac{1}{HV} \int_{0}^{H} \int_{0}^{V} I(h, v) \exp\left[-j2\pi\left(\frac{mh}{H} + \frac{nv}{V} \right) \right] dh \, dv \qquad [3]$$

Therefore, letting

$$f_h = \frac{s_h}{H} \qquad f_v = \frac{s_v}{V}$$

and using Eqs. (1) and (2), we obtain

$$x(t) = \sum_{m=-\infty}^{\infty} \sum_{n=-\infty}^{\infty} c_{mn} e^{j2\pi(mf_h + nf_v)t} \qquad [4]$$

This expression represents a doubly periodic signal containing all harmonics of the *line frequency* f_h and the *field frequency* f_v, plus their sums and differences. Since $f_h \gg f_v$ and since $|c_{mn}|$ generally decreases as the product mn increases, the amplitude spectrum has the form shown in Fig. 7.4–4, where the spectral lines cluster around the harmonics of f_h and there are large gaps between clusters.

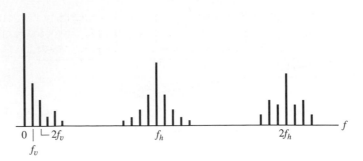

Figure 7.4–4   Video spectrum for still image.

Equation (4) and Fig. 7.4–4 are exact for a still picture, as in facsimile systems. When the image has motion, the spectral lines merge into continuous clumps around the harmonics of f_h. Even so, the spectrum remains mostly "empty" everywhere else, a property used to advantage in the subsequent development of color TV. Despite the gaps in Fig. 7.4–4, the video spectrum theoretically extends indefinitely—similar to an FM line spectrum. Determining the bandwidth required for a video signal thus involves additional considerations.

Two basic facts stand in the way of perfect image reproduction: (1) there can be only a finite number of lines in the scanning raster, which limits the image clarity or *resolution* in the vertical direction; and (2) the video signal must be transmitted with a finite bandwidth, which limits horizontal resolution. Quantitatively, we measure resolution in terms of the maximum number of discrete image lines that can be distinguished in each direction, say n_h and n_v. In other words, the most detailed image that can be resolved is taken to be a checkerboard pattern having n_h columns and n_v rows. We usually desire equal horizontal and vertical resolution in lines per unit distance, so $n_h/H = n_v/V$ and

$$\frac{n_h}{n_v} = \frac{H}{V} \tag{5}$$

which is called the *aspect ratio*.

Clearly, vertical resolution is related to the total number of raster lines N; indeed, n_v equals N if all scanning lines are active in image formation (as in facsimile but not TV) and the raster aligns perfectly with the rows of the image. Experimental studies show that arbitrary raster alignment reduces the effective resolution by a factor of about 70 percent, called the *Kerr factor*, so

$$n_v = 0.7(N - N_{vr}) \tag{6}$$

where N_{vr} is the number of raster lines lost during vertical retrace.

Horizontal resolution is determined by the baseband bandwidth B allotted to the video signal. If the video signal is a sinusoid at frequency $f_{max} = B$, the resulting picture will be a sequence of alternating dark and light spots spaced by one-half cycle in the horizontal direction. It then follows that

$$n_h = 2B(T_{line} - T_{hr}) \qquad [7]$$

Where T_{line} is the total duration of one line and T_{hr} is the horizontal retrace time. Solving Eq. (7) for B and using Eqs. (5) and (6) yields

$$B = \frac{(H/V)n_v}{2(T_{line} - T_{hr})} = 0.35(H/V) \frac{N - N_{vr}}{T_{line} - T_{hr}} \qquad [8]$$

Another, more versatile bandwidth expression is obtained by multiplying both sides of Eq. (8) by the frame time $T_{frame} = NT_{line}$ and explicitly showing the desired resolution. Since $N = n_v/0.7(1 - N_{vr}/N)$, this results in

$$BT_{frame} = \frac{0.714n_p}{\left(1 - \dfrac{N_{vr}}{N}\right)\left(1 - \dfrac{T_{hr}}{T_{line}}\right)} \qquad [9a]$$

where

$$n_p = \frac{H}{V}n_v^2 = n_h n_v \qquad [9b]$$

The parameter n_p represents the number of picture elements or *pixels*. Equation (9) brings out the fact that the bandwidth (or frame time) requirement increases in proportion to the number of pixels or as the square of the vertical resolution.

The NTSC system has $N = 525$ and $N_{vr} = 2 \times 21 = 42$ so there are 483 active lines. The line time is $T_{line} = 1/f_h = 63.5 \ \mu s$ and $T_{vr} = 10 \ \mu s$, leaving an active line time of $53.5 \ \mu s$. Therefore, using Eq. (8) with $H/V = 4/3$, we get the video bandwidth

EXAMPLE 7.4–1

$$B = 0.35 \times \frac{4}{3} \times \frac{483}{53.5 \times 10^{-6}} \approx 4.2 \text{ MHz}$$

This bandwidth is sufficiently large to reproduce the $5\text{-}\mu s$ sync pulses with reasonably square corners.

Facsimile systems require no vertical retrace and the horizontal retrace time is negligible. Calculate the time T_{frame} needed for facsimile transmission of a newspaper page, 37 by 59 cm, with a resolution of 40 lines/cm using a voice telephone channel with $B \approx 3.2$ kHz.

EXERCISE 7.4–1

Monochrome Transmitters and Receivers

The large bandwidth and significant low-frequency content of the video signal, together with the desired simplicity of envelope detection, have led to the selection of VSB + C (as described in Sect. 4.4) for TV broadcasting in the United States. However, since precise vestigial sideband shaping is more easily carried out at the receiver where the power levels are small, the actual modulated-signal spectrum is as indicated in Fig. 7.4–5a. The half-power frequency of the upper sideband is about 4.2 MHz above the video carrier f_{cv} while the lower sideband has a 1-MHz bandwidth. Figure 7.4–5b shows the frequency shaping at the receiver.

The audio signal is frequency-modulated on a separate carrier $f_{ca} = f_{cv} + f_a$, with $f_a = 4.5$ MHz and frequency deviation $f_\Delta = 25$ kHz. Thus, assuming an audio bandwidth of 10 kHz, $D = 2.5$ and the modulated audio occupies about 80 kHz. TV channels are spaced by 6 MHz, leaving a 250-kHz guard band. Carrier frequencies are assigned in the VHF ranges 54–72, 76–88, and 174–216 MHz, and in the UHF range 470–806 MHz.

The essential parts of a TV transmitter are block-diagrammed in Fig. 7.4–6. The synchronizing generator controls the scanning raster and supplies blanking and sync pulses for the video signal. The dc restorer and white clipper working together ensure that the amplified video signal levels are in proportion. The video modulator is of the high-level AM type with $\mu = 0.875$, and the power amplifier removes the lower portion of the lower sideband.

The antenna has a balance-bridge configuration such that the outputs of the audio and video transmitters are radiated by the same antenna without interfering with each other. The transmitted audio power is 50 to 70 percent of the video power.

As indicated in Fig. 7.4–7, a TV receiver is of the superheterodyne type. The main IF amplifier has f_{IF} in the 41- to 46-MHz range and provides the vestigial shaping per Fig. 7.4–5b. Note that the modulated audio signal is also passed by this

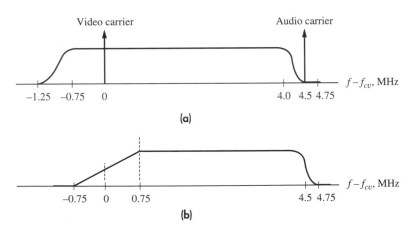

Figure 7.4–5   (a) Transmitted TV spectrum; (b) VSB shaping at receiver.

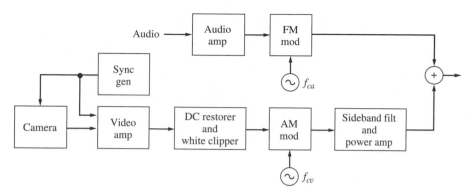

Figure 7.4–6   Monochrome TV transmitter.

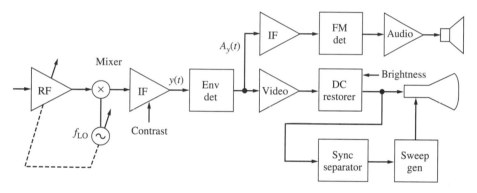

Figure 7.4–7   Monochrome TV receiver.

amplifier, but with substantially less gain. Thus, drawing upon Eq. (11), Sect. 4.4, the total signal at the input to the envelope detector is

$$y(t) = A_{cv}[1 + \mu x(t)] \cos \omega_{cv} t - A_{cv} \mu x_q(t) \sin \omega_{cv} t \tag{10}$$
$$+ A_{ca} \cos [(\omega_{cv} + \omega_a)t + \phi(t)]$$

where $x(t)$ is the video signal, $\phi(t)$ is the FM audio, and $\omega_a = 2\pi f_a$. Since $|\mu x_q(t)| \ll 1$ and $A_{ca} \ll A_{cv}$, the resulting envelope is approximately

$$A_y(t) = A_{cv}[1 + \mu x(t)] + A_{ca} \cos [\omega_a t + \phi(t)] \tag{11}$$

which gives the signal at the output of the envelope detector.

The video amplifier has a lowpass filter that removes the audio component from $A_y(t)$ as well as a dc restorer that electronically clamps the blanking pulses and thereby restores the correct dc level to the video signal. The amplified and dc-restored video signal is applied to the picture tube and to a sync-pulse separator that provides synchronization for the sweep generators. The "brightness" control permits manual adjustment of the DC level while the "contrast" control adjusts the gain of the IF amplifier.

Equation (11) shows that the envelope detector output also includes the modulated audio. This component is picked out and amplified by another IF amplifier tuned to 4.5 MHz. FM detection and amplification then yields the audio signal.

Observe that, although the transmitted composite audio and video signal is a type of frequency-division multiplexing, separate frequency conversion is not required for the audio. This is because the video carrier acts like a local oscillator for the audio in the envelope-detection process, an arrangement called the *intercarrier-sound system* having the advantageous feature that the audio and video are always tuned in together. Successful operation depends on the fact that the video component is large compared to the audio at the envelope detector input, as made possible by the white clipper at the transmitter (which prevents the modulated video signal from becoming too small) and the relative attenuation of the audio by the receiver's IF response.

Some additional features not shown on our transmitter and receiver diagrams relate to the *vertical retrace interval*. The NTSC system allots 21 lines per field to vertical retracing, or about 1.3 ms every 1/60 sec. The first 9 lines carry control pulses, but the remaining 12 may be utilized for other purposes while the retrace goes on. Applications of these available lines include: the vertical-interval test signal (VITS) for checking transmission quality; the vertical-interval reference (VIR) for receiver servicing and/or automatic adjustments; and digital signals that generate the closed-captioning characters on special receivers for the hearing impaired.

EXERCISE 7.4–2   Use a phasor diagram to derive Eq. (11) from Eq. (10).

Color Television

Any color can be synthesized from a mixture of the three additive primary colors, red, green, and blue. Accordingly, a brute-force approach to color TV would involve direct transmission of three video signals, say $x_R(t)$, $x_G(t)$, and $x_B(t)$—one for each primary. But, aside from the increased bandwidth requirement, this method would not be compatible with existing monochrome systems. A fully compatible color TV signal that fits into the monochrome channel was developed in 1954, drawing upon certain characteristics of human color perception. The salient features of that system are outlined here.

To begin with, the three primary color signals can be uniquely represented by any three other signals that are independent linear combinations of $x_R(t)$, $x_G(t)$, and $x_B(t)$. And, by proper choice of coefficients, one of the linear combinations can be made the same as the intensity or *luminance* signal of monochrome TV. In particular, it turns out that if

$$x_Y(t) = 0.30x_R(t) + 0.59x_G(t) + 0.11x_B(t) \qquad [12a]$$

then $x_Y(t)$ is virtually identical to the conventional video signal previously symbolized by $x(t)$. The remaining two signals, called the *chrominance* signals, are taken as

$$x_I(t) = 0.60x_R(t) - 0.28x_G(t) - 0.32x_B(t) \qquad [12b]$$

$$x_Q(t) = 0.21x_R(t) - 0.52x_G(t) + 0.31x_B(t) \qquad [12c]$$

Here, the color signals are normalized such that $0 \le x_R(t) \le 1$, and so forth, so the luminance signal is never negative while the chrominance signals are bipolar.

Understanding the chrominance signals is enhanced by introducing the *color vector*

$$x_C(t) = x_I(t) + jx_Q(t) \qquad [13]$$

whose magnitude $|x_c(t)|$ is the color intensity or *saturation* and whose angle $\arg x_c(t)$ is the *hue*. Figure 7.4–8 shows the vector positions of the saturated primary colors in the *IQ* plane. A partially saturated (pastel) blue-green, for instance, might have $x_R = 0$ and $x_B = x_G = 0.5$, so $x_C = -0.300 - j0.105$, $|x_C| = 0.318$, and $\arg x_C = -160°$. Since the origin of the *IQ* plane represents the absence of color, the luminance signal may be viewed as a vector perpendicular to this plane.

Because $x_Y(t)$ serves as the monochrome signal, it must be alloted the entire 4.2-MHz baseband bandwidth to provide adequate horizontal resolution. Consequently, there would seem to be no room for the chrominance signals. Recall, however, that the spectrum of $x_Y(t)$ has periodic gaps between the harmonics of the line frequency f_h—and the same holds for the chrominance signals. Moreover, subjective tests have shown that the human eye is less perceptive of chrominance resolution than luminance resolution, so that $x_I(t)$ and $x_Q(t)$ can be restricted to about 1.5 MHz and 0.5 MHz, respectively, without significant visible degradation of the color picture.

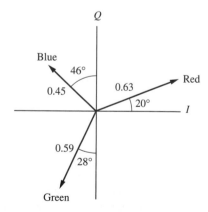

Figure 7.4–8   Saturated primary color vectors in the IQ plane.

Combining these factors permits multiplexing the chrominance signals in an interleaved fashion in the baseband spectrum of the luminance signal.

The chrominance signals are multiplexed on a *color subcarrier* whose frequency falls exactly halfway between the 227th and 228th harmonic of f_h, namely,

$$f_{cc} = \frac{455}{2} f_h \approx 3.6 \text{ MHz} \qquad [14]$$

Therefore, by extension of Fig. 7.4–3, the luminance and chrominance frequency components are interleaved as indicated in Fig. 7.4–9a, and there is 0.6 MHz between f_{cc} and the upper end of the baseband channel. The subcarrier modulation will be described shortly, after we examine frequency interleaving and compatibility.

What happens when a color signal is applied to a monochrome picture tube? Nothing, surprisingly, as far as the viewer sees. True, the color subcarrier and its sidebands produce sinusoidal variations on top of the luminance signal. But because all of these sinusoids are exactly an odd multiple of one-half the line frequency, they reverse in phase from line to line and from field to field—illustrated by Fig. 7.4–9b. This produces flickering in small areas that averages out over time and space to the correct luminance value and goes essentially unnoticed by the viewer.

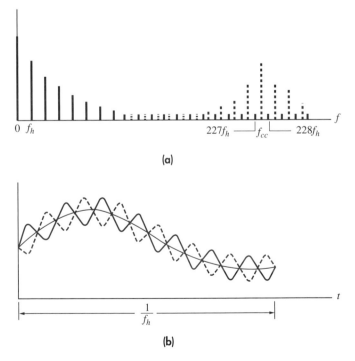

(a)

(b)

Figure 7.4–9   (a) Chrominance spectral lines (dashed) interleaved between luminance lines; (b) line-to-line phase reversal of chrominance variations on luminance.

By means of this averaging effect, frequency interleaving renders the color signal compatible with an unmodified monochrome receiver. It also simplifies the
design of color receivers since, reversing the above argument, the luminance signal
does not visibly interfere with the chrominance signals. There is a minor interference problem caused by the difference frequency $f_a - f_{cc}$ between the audio and
color subcarriers. That problem was solved by slightly changing the line frequency
to $f_h = f_a/286 = 15.73426$ kHz giving $f_a - f_{cc} = 4500 - 3,579.545 = 920.455$ kHz
$= (107/2)f_h$ which is an "invisible" frequency. (As a result of this change, the field
rate is actually 59.94 Hz rather than 60 Hz!)

A modified version of *quadrature-carrier multiplexing* puts both chrominance
signals on the color subcarrier. Figure 7.4–10 shows how the luminance and chrominance signals are combined to form the baseband signal $x_b(t)$ in a color transmitter.
Not shown is the nonlinear *gamma correction* introduced at the camera output to
compensate for the brightness distortion of color picture tubes.

The gamma-corrected color signals are first matrixed to obtain $x_Y(t)$, $x_I(t)$, and
$x_Q(t)$ in accordance with Eq. (12). Next, the chrominance signals are lowpass filtered (with different bandwidths) and applied to the subcarrier modulators. Subsequent bandpass filtering produces conventional DSB modulation for the Q channel
and modified VSB for the I channel—for example, DSB for baseband frequencies
of $x_I(t)$ below 0.5 MHz and LSSB for $0.5 < |f| < 1.5$ MHz. The latter keeps the

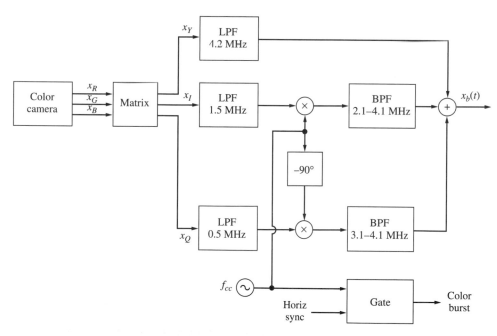

Figure 7.4–10   Color subcarrier modulation system.

modulated chrominance signals as high as possible in the baseband spectrum, thereby confining the flicker to small areas, while still allowing enough bandwidth for proper resolution of $x_I(t)$. Total sideband suppression cannot be used owing to the significant low-frequency content in $x_I(t)$ and $x_Q(t)$.

Including $x_Y(t)$, the entire baseband signal becomes

$$x_b(t) = x_Y(t) + x_Q(t) \sin \omega_{cc} t + x_I(t) \cos \omega_{cc} t + \hat{x}_{IH}(t) \sin \omega_{cc} t \qquad [15]$$

where $\hat{x}_{IH}(t)$ is the Hilbert transform of the high-frequency portion of $x_I(t)$ and accounts for the asymmetric sidebands. This baseband signal takes the place of the monochrome video signal in Fig. 7.4–6. Additionally, an 8-cycle piece of the color subcarrier known as the *color burst* is put on the trailing portion or "back porch" of the blanking pulses for purposes of synchronization.

Demultiplexing is accomplished in a color TV receiver after the envelope detector, as laid out in Fig. 7.4–11. Since the luminance signal is at baseband here, it requires no further processing save for amplification and a 3.6-MHz trap or rejection filter to eliminate the major flicker component; the chrominance sidebands need not be removed, thanks to frequency interleaving. The chrominance signals pass through a bandpass amplifier and are applied to a pair of synchronous detectors whose local oscillator is the VCO in a PLL synchronized by phase comparison with the received color burst. Manual controls usually labeled "color level" (i.e., satura-

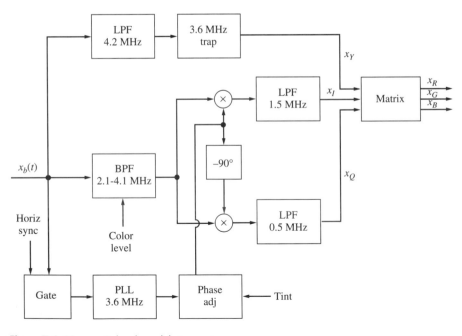

Figure 7.4–11   Color demodulation system.

tion) and "tint" (i.e., hue) are provided to adjust the gain of the chrominance amplifier and the phase of the VCO; their effect on the picture is readily explained in terms of the color vector and Fig. 7.4–8.

Assuming good synchronization, it follows from Eq. (15) that the detected but unfiltered I- and Q-channel signals are proportional to

$$v_I(t) = x_I(t) + 2x_{YH}(t) \cos \omega_{cc}t + x_I(t) \cos 2\omega_{cc}t \qquad [16a]$$
$$+ [x_Q(t) + \hat{x}_{IH}(t)] \sin 2\omega_{cc}t$$

$$v_Q(t) = x_Q(t) + \hat{x}_{IH}(t) + 2x_{YH}(t) \sin \omega_{cc}t + x_I(t) \sin 2\omega_{cc}t \qquad [16b]$$
$$- [x_Q(t) + \hat{x}_{IH}(t)] \cos 2\omega_{cc}t$$

where $x_{YH}(t)$ represents the luminance frequency components in the 2.1- to 4.1-MHz range. Clearly, lowpass filtering will remove the double-frequency terms, while the terms involving $x_{YH}(t)$ are "invisible" frequencies. Furthermore, $\hat{x}_{IH}(t)$ in Eq. (16b) has no components less than 0.5 MHz, so it is rejected by the LPF in the Q channel. (Imperfect filtering here results in a bothersome effect called quadrature color cross talk). Therefore, ignoring the invisible-frequency terms, $x_I(t)$ and $x_Q(t)$ have been recovered and can then be matrixed with $x_Y(t)$ to generate the color signals for the picture tube. Specifically, by inversion of Eq. (12),

$$x_R(t) = x_Y(t) + 0.95x_I(t) + 0.62x_Q(t) \qquad [17]$$
$$x_G(t) = x_Y(t) - 0.28x_I(t) - 0.64x_Q(t)$$
$$x_B(t) = x_Y(t) - 1.10x_I(t) + 1.70x_Q(t)$$

If the received signal happens to be monochrome, then the three color signals will be equal and the reproduced picture will be black-and-white. This is termed *reverse compatibility*.

The NTSC color system described here certainly ranks high as an extraordinary engineering achievement! It solved the problems of color reproduction with direct and reverse monochrome compatibility while staying within the confines of the 6-MHz channel allocation.

HDTV[†]

The tremendous advances in digital technology combined with consumer demand for better picture and sound quality, plus computer compatibility, has motivated television manufacturers to develop a new US color TV standard: high-definition television (HDTV). A digital standard provides multimedia options such as special effects, editing, and so forth, and better computer interfacing. The HDTV standard supports at least 18 different formats and is a significant advancement over NTSC with respect to TV quality. One of the HDTV standards is shown in Table 7.4–1. First, with

[†]João O. P. Pinto drafted this section.

respect to the NTSC system, the number of vertical and horizontal lines has doubled, and thus the *picture resolution* is four times greater. Second, the *aspect ratio* has been changed from 4/3 to 16/9. Third, as Figs. 7.4–12 and 7.4–13 indicate, HDTV has improved *scene capture* and *viewing angle* features. For example, with H equal to the TV screen height and with a viewing distance of 10 feet (7H) in the NTSC system, the viewing angle is approximately 10 degrees. Whereas with HDTV, the same 10 foot viewing distance (3H) yields a viewing angle of approximately 20 degrees.

HDTV has also adopted the *AC-3 surround sound* system instead of mono-phonic or stereo sound. This system has six channels: right, right surround, left, left surround, center, and *low-frequency effect* (*LFE*). The LFE channel has only a band-width of 120 Hz, effectively providing only 5.1 channels.

HDTV can achieve a given signal-to-noise ratio with 12 dB less radiated power than NTSC-TV. Thus, for the same transmitter power, reception that was marginal with NTSC broadcasts will be greatly improved with HDTV.

Although there was no attempt to make HDTV broadcast signals compatible with existing NTSC TV receivers, by 2006 the FCC will require that only digital signals be

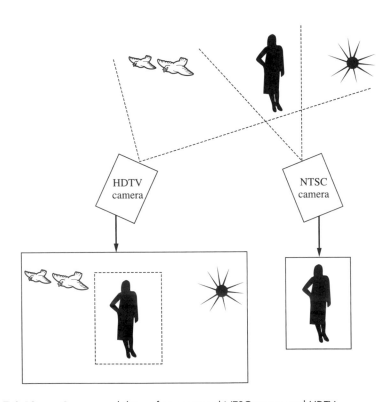

Figure 7.4–12   Scene capabilities of conventional NTSC system and HDTV.

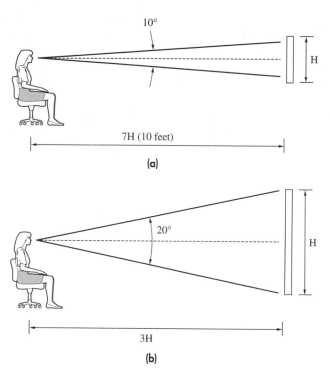

Figure 7.4–13   Viewing angles as a function of distance. (a) Conventional NTSC; (b) HDTV.

broadcast, of which HDTV is one kind. Thus, in order to receive TV broadcasts, existing TV sets will have to be replaced or modified by some type of converter.

The system for encoding and transmitting HDTV signals is shown in Figure 7.4–14. The transmitter consists of several stages. First, the 24.9-MHz video signal and corresponding audio signals are *compressed*, so they will fit into the allocated 6-MHz channel bandwidth. The compressed audio and video data is then combined with *ancillary data* that includes control data, closed captioning, and so forth, using a multiplexer. The multiplexer then formats the data into packets. Next, the packetized data is scrambled to remove any undesirable frequency discrete components, and is *channel encoded*. During channel coding the data is encoded with check or parity symbols using *Reed-Solomon coding* to enable error correction at the receiver. The symbols are interleaved to minimize the effects of *burst-type errors* where noise in the channel can cause successive symbols to be corrupted. Finally, the symbols are Trellis-Code Modulated (TCM). TCM, which will be discussed in Chapter 14, combines coding and modulation and makes it possible to increase the symbol transmission rate without an increase in error probability. The encoded data is combined with synchronization signals and is then

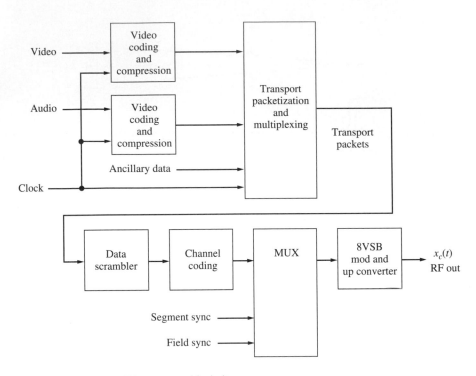

Figure 7.4–14   HDTV transmitter block diagram.

8VSB modulated. 8VSB is a VSB technique where an 8-level baseband code is VSB modulated onto a given carrier frequency.

The HDTV receiver shown in Figure 7.4–15 reverses the above process. As broadcasters and listeners make the transition from NTSC-TV to HDTV, they will be allowed to transmit both signals simultaneously. To overcome potential interference, the HDTV receiver uses the *NTSC rejection filter* to reject NTSC signals. A channel equalizer/ghost canceller stage, not shown, performs ghost cancellation and channel equalization. The *phase tracker* minimizes the effects of phase noise caused by the system's PLL.

When digitized, the 24.9-MHz video signal has a bit rate of 1 Gbps, whereas a 6-MHz television channel can only accommodate 20 Mbps. Therefore a compression ratio of more than 50:1 is required. The raw video signal obtained by the scanning process contains significant temporal and spatial redundancies. These are taken advantage of during the compression process. During the transmission of each frame, only those parts in the scene that move or change are actually transmitted. The specific compression process is the *MPEG-2* (Motion Picture Expert Group-2), which uses the *Discrete Cosine Transform* (DCT). See Gonzalez and Woods (1992) for more information on the DCT. The MPEG-2 signals are readily interfaced to computers for multimedia capability.

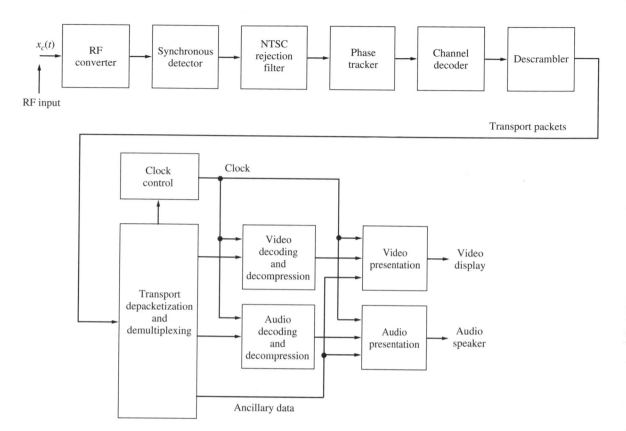

Figure 7.4–15   HDTV receiver block diagram.

7.5   PROBLEMS

7.1–1*   Suppose a commercial AM superhet has been designed such that the image frequency always falls above the broadcast band. Find the minimum value of f_{IF}, the corresponding range of f_{LO}, and the bounds on B_{RF}.

7.1–2   Suppose a commercial FM superhet has been designed such that the image frequency always falls below the broadcast band. Find the minimum value of f_{IF}, the corresponding range of f_{LO}, and the bounds on B_{RF}.

7.1–3*   Suppose a commercial AM superhet has $f_{IF} = 455$ kHz and $f_{LO} = 1/2\pi\sqrt{LC}$, where $L = 1\ \mu$H and C is a variable capacitor for tuning. Find the range of C when $f_{LO} = f_c + f_{IF}$ and when $f_{LO} = f_c - f_{IF}$.

7.1–4   Suppose the RF stage of a commercial AM superhet is a tuned circuit like Fig. 4.1–7 with $L = 1\ \mu$H and variable C for tuning. Find the range of C and the corresponding bounds on R.

7.1–5   Consider a superhet intended for USSB modulation with $W = 4$ kHz and $f_c = 3.57$–3.63 MHz. Take $f_{LO} = f_c + f_{IF}$ and choose the receiver parameters so that all bandpass stages have $B/f_0 \approx 0.02$. Then sketch $|H_{RF}(f)|$ to show that the RF stage can be fixed-tuned. Also sketch $|H_{IF}(f)|$, accounting for sideband reversal.

7.1–6   Do Prob. 7.1–5 for LSSB modulation with $W = 6$ kHz and $f_c = 7.14$–7.26 MHz.

7.1–7   Sketch the spectrum of $x_c(t) \times \cos 2\pi f_{LO} t$ to demonstrate the sideband-reversal effect in an SSB superhet when $f_{LO} = f_c + f_{IF}$.

7.1–8   For *automatic frequency control* in an FM superhet, the LO is replaced by a VCO that generates $A_{LO} \cos \theta(t)$ with $\dot{\theta}(t) = 2\pi[f_c - f_{IF} + Kv(t) + \epsilon(t)]$ where $\epsilon(t)$ is a slow and random frequency drift. The control voltage $v(t)$ is derived by applying the demodulated signal to an LPF with $B \ll W$. The demodulated signal is $y_D(t) = K_D \dot{\phi}_{IF}(t)/2\pi$ where $\phi_{IF}(t)$ is the instantaneous phase at the IF output. Analyze this AFC system by finding $y_D(t)$ in terms of $x(t)$ and $\epsilon(t)$.

7.1–9*   Consider a superhet that receives signals in the 50–54 MHz range with $f_{LO} = f_c + f_{IF}$. Assuming there is little filtering prior to the mixer, what range of input signals will be received if the f_{IF} is: (*a*) 455 kHz, (*b*) 7 MHz?

7.1–10   Design a receiver that will receive USSB signals in the 50–54 MHz range where $f_{IF} = 100$ MHz and does not exhibit sideband-reversal. Assume there is little filtering prior to the mixer. Specify f_{LO}, the product detector oscillator frequency, the center frequency of the IF bandpass filter, and any image frequencies that will be received.

7.1–11   Consider a superhet with $f_{LO} = f_c + f_{IF}$, $f_{IF} = 455$ kHz, and $f_c = 2$ MHz. The RF amplifier is preceded by a first-order RLC bandpass filter with $f_0 = 2$ MHz and $B = 0.5$ MHz. Assume the IF-BPF is nearly ideal and that the mixer has unity gain. What is the minimum spurious frequency input rejection ratio in dB?

7.1–12*   Suppose the receiver of Prob. 7.1–11 has a LO with a second harmonic whose voltage level is half that of the fundamental component. (*a*) What input frequencies will be accepted, and at what power level in dB as compared to the correct input? (*b*) Discuss all ways to minimize these interfering inputs.

7.1–13   Consider a superhet that receives signals in the 7.0 to 8.0 MHz range with $f_{LO} = f_c + f_{IF}$, and $f_{IF} = 455$ kHz. The receiver's RF amplifier has a passband of 2 MHz, and its IF-BPF is nearly ideal and has a bandwidth of 3 kHz. Design a frequency converter that has a fixed LO frequency that will enable the reception of 50.0- to 51.0-MHz signals. Assume the converter's RF amplifier is relatively wideband. (*a*) If the incoming frequency is supposed to be $f_c = 50$ MHz, what other spurious frequencies will this receiver respond to? (*b*) Describe how to minimize these spurious responses.

7.1–14*   What is the image rejection performance of a single conversion superhet receiver that receives signals in the 50–54 MHz range, $f_{LO} > f_c$, and has an RF amplifier that includes a fixed frequency RLC-BPF with $B = 4$ MHz with (*a*) $f_{IF} = 20$ MHz, (*b*) $f_{IF} = 100$ MHz?

7.1–15   Design a superhet receiver for a dual-mode cellular phone system that will accept either 850 MHz analog cellular signals or 1900 MHz digital personal communications systems (PCS) signals. Specify the F_{LO}, F_{IF}, and image frequencies.

7.1–16   Find suitable parameters of a double-conversion receiver having $RR = 60$ dB and intended for DSB modulation with $W = 10$ kHz and $f_c = 4$ MHz.

7.1–17   A double conversion receiver designed for $f_c = 300$ MHz has $f_{IF-1} = 30$ MHz and $f_{IF-2} = 3$ MHz, and each LO frequency is set at the higher of the two possible values. Insufficient filtering by the RF and first IF stages results in interference from *three* image frequencies. What are they?

7.1–18   Do Prob. 7.1–17 with each LO frequency set at the lower of the two possible values.

7.1–19   Specify the settings on a scanning spectrum analyzer to display the spectrum up to the 10th harmonic of a signal with a 50-ms period.

7.1–20   Specify the settings on a scanning spectrum analyzer to display the spectrum of a tone-modulated FM signal with $f_c = 100$ kHz, $f_m = 1$ kHz, and $\beta = 5$.

7.1–21‡   The magnitude spectrum of an *energy* signal $v(t)$ can be displayed by multiplying $v(t)$ with the swept-frequency wave $\cos(\omega_c t - \alpha t^2)$ and applying the product to a bandpass filter having $h_{bp}(t) = \cos(\omega_c t + \alpha t^2)$. Use equivalent lowpass time-domain analysis to show that $h_{\ell p}(t) = \frac{1}{2}e^{j\alpha t^2}$ and that the envelope of the bandpass output is proportional to $|V(f)|$ with $f = \alpha t/\pi$.

7.2–1   Four signals, each having $W = 3$ kHz, are to be multiplexed with 1-kHz guard bands between channels. The subcarrier modulation is USSB, except for the lowest channel which is unmodulated, and the carrier modulation is AM. Sketch the spectrum of the baseband and transmitted signal, and calculate the transmission bandwidth.

7.2–2   Do Prob. 7.2–1 with AM subcarrier modulation.

7.2–3   Let f_i be an arbitrary carrier in an FDM signal. Use frequency-translation sketches to show that the BPFs in Fig. 7.2–2 are not necessary if the subcarrier modulation is DSB and the detector includes an LPF. Then show that the BPFs are needed, in general, for SSB subcarrier modulation.

7.2–4*   Ten signals with bandwidth W are to be multiplexed using SSB subcarrier modulation and a guard band B_g between channels. The BPFs at the receiver have $|H(f)| = \exp\{-[1.2(f - f_0)/W]^2\}$, where f_0 equals the center frequency for each subcarrier signal. Find B_g so that the adjacent-channel response satisfies $|H(f)| \leq 0.1$. Then calculate the resulting transmission bandwidth of the FDM signal.

7.2–5   Suppose the voice channels in a group signal have $B_g = 1$ kHz and are separated at the receiver using BPFs with $|H(f)| = \{1 + [2(f - f_0)/B]^{2n}\}^{-1/2}$. Make a careful sketch of three adjacent channels in the group spectrum, taking account of the fact that a baseband voice signal has negligible content outside $200 < |f| < 3200$ Hz. Use your sketch to determine values for B, f_0, and n so that $|H(f)| \leq 0.1$ outside the desired passband.

7.2–6   Some FDM telemetry systems employ *proportional bandwidth* FM subcarrier modulation when the signals to be multiplexed have different bandwidths. All subcarrier signals have the same deviation ratio but the ith subcarrier frequency and message bandwidth are related by $f_i = W_i/\alpha$ where α is a constant. (a) Show that the subcarrier signal bandwidth B_i is proportional to f_i, and obtain an expression for f_{i+1} in terms of f_i to provide a guard band B_g between channels. (b) Calculate the next three subcarrier frequencies when $f_1 = 2$ kHz, $B_1 = 800$ Hz, and $B_g = 400$ Hz.

7.2–7   Find the output signals of the quadrature-carrier system in Fig. 7.2–6 when the receiver local oscillator has a phase error ϕ'.

7.2–8   In one proposed system for FM *quadraphonic multiplexing*, the baseband signal in Fig. 7.2–4 is modified as follows: The unmodulated signal is $x_0(t) = L_F + L_R + R_F + R_R$ (for monophonic compatibility), the 38-kHz subcarrier has quadrature-carrier multiplexing with modulating signals $x_1(t)$ and $x_2(t)$, and the SCA signal is replaced by a 76-kHz subcarrier with DSB modulation by $x_3(t) = L_F - L_R + R_F - R_R$. What should be the components of $x_1(t)$ for stereophonic compatibility? Now consider $x_0(t) \pm x_1(t) \pm x_3(t)$ to determine the components of $x_2(t)$. Draw a block diagram of the corresponding transmitter and quadraphonic receiver.

7.2–9‡   Suppose the transmission channel in Fig. 7.2–6 has linear distortion represented by the transfer function $H_C(f)$. Find the resulting spectrum at the lower output and show that the condition for no cross talk is $H_C(f - f_c) = H_C(f + f_c)$ for $|f| \leq W$. If this condition holds, what must be done to recover $x_1(t)$?

7.2–10*   Twenty-four voice signals are to be transmitted via multiplexed PAM with a marker pulse for frame synchronization. The sampling frequency is 8 kHz and the TDM signal has a 50 percent duty cycle. Calculate the signaling rate, pulse duration, and minimum transmission bandwidth.

7.2–11   Do Prob. 7.2–10 with a 6-kHz sampling frequency and 30 percent duty cycle.

7.2–12   Twenty signals, each with $W = 4$ kHz, are sampled at a rate that allows a 2-kHz guard band for reconstruction filtering. The multiplexed samples are transmitted on a CW carrier. Calculate the required transmission bandwidth when the modulation is: (a) PAM/AM with 25 percent duty cycle; (b) PAM/SSB with baseband filtering.

7.2–13*   Ten signals, each with $W = 2$ kHz, are sampled at a rate that allows a 1-kHz guard band for reconstruction filtering. The multiplexed samples are transmitted on a CW carrier. Calculate the required transmission bandwidth when the modulation is: (a) PPM/AM with 20 percent duty cycle; (b) PAM/FM with baseband filtering and $f_\Delta = 75$ kHz.

7.2–14   Given a 6-channel main multiplexer with $f_s = 8$ kHz, devise a telemetry system similar to Fig. 7.2–11 (including a marker) that accommodates six input signals having the following bandwidths: 8.0, 3.5, 2.0, 1.8, 1.5, and 1.2 kHz. Make sure that successive samples of each input signal are equispaced in time. Calculate the resulting baseband bandwidth and compare with the minimum transmission bandwidth for an FDM-SSB system.

7.2–15   Do Prob. 7.2–14 for seven input signals having the following bandwidths: 12.0, 4.0, 1.0, 0.9, 0.8, 0.5, and 0.3 kHz.

7.2–16   Do Prob. 7.2–14 for eight input signals having the following bandwidths: 12.0, 3.5, 2.0, 0.5, 0.4, 0.3, 0.2, and 0.1 kHz.

7.2–17   Calculate the bandwidth required so the cross talk does not exceed -40 dB when 25 voice signals are transmitted via PPM-TDM with $f_s = 8$ kHz and $t_0 = \tau = 0.2(T_s/M)$.

7.2–18*   Find the maximum number of voice signals that can be transmitted via TDM-PPM with $f_s = 8$ kHz and $t_0 = \tau = 0.25(T_s/M)$ when the channel has $B = 500$ kHz and the cross talk is to be kept below -30 dB.

7.2–19   Cross talk also occurs when a transmission system has inadequate *low-frequency response*, usually as a result of transformer coupling or blocking capacitors. Demonstrate this effect by sketching the pulse response of a high-pass filter whose step response is $g(t) = \exp(-2\pi f_\ell t)\, u(t)$. Consider the extreme cases $f_\ell \tau \ll 1$ and $f_\ell \tau \gg 1$.

7.3–1   For one implementation of digital phase comparison, the switching circuit in Fig. 7.3–1b has a set–reset flip-flop whose output becomes $s(t) = +A$ after a positive-going zero-crossing of $x_c(t)$ and $s(t) = -A$ after a positive-going zero-crossing of $v(t)$. (a) Take $x_c(t) = \cos \omega_c t$ and $v(t) = \cos(\omega_c t - \phi_v)$ and sketch one period of $s(t)$ for $\phi_v = 45, 135, 180, 225$, and $315°$. (b) Now plot y versus $\epsilon = \phi_v - 180°$ assuming that $y(t) = <s(t)>$. Note that this implementation requires $\pm 180°$ phase difference between the inputs for $y = 0$.

7.3–2   Do part (a) of Prob. 7.3–1 for a digital phase comparator with a switch controlled by $v(t)$ so its output is $s(t) = A \operatorname{sgn} x_c(t)$ when $v(t) > 0$ and $s(t) = 0$ when $v(t) < 0$. Now plot y versus $\epsilon = \phi_v - 90°$ assuming that $y(t) = <s(t)>$.

7.3–3   Consider a PLL in the steady state with $\epsilon_{ss} \ll 1$ for $t < 0$. The input frequency has a step change at $t = 0$, so $\phi(t) = 2\pi f_1 t$ for $t > 0$. Solve Eq. (5) to find and sketch $\epsilon(t)$, assuming that $K \gg |\Delta f + f_1|$.

7.3–4   Explain why the Costas PLL system in Fig. 7.3–4 cannot be used for synchronous detection of SSB or VSB.

7.3–5*   Consider a PLL in steady-state locked conditions. If the external input is $x_c(t) = A_c \cos(\omega_c t + \phi_0)$, then the feedback signal to the phase comparator must be proportional to $\cos(\omega_c t + \phi_0 + 90° - \epsilon_{ss})$. Use this property to find the VCO output in Fig. 7.3–5 when $\epsilon_{ss} \neq 0$.

7.3–6   Use the property stated in Prob. 7.3–5 to find the VCO output in Fig. 7.3–6 when $\epsilon_{ss} \neq 0$.

7.3–7   Modify the FM stereo receiver in Fig. 7.2–5 to incorporate a PLL with $f_v \approx 38$ kHz for the subcarrier. Also include a dc stereo indicator.

7.3–8*   Given a 100-kHz master oscillator and two adjustable divide-by-n counters with $n = 1$ to 10, devise a system that synthesizes any frequency from 1 kHz to 99 kHz in steps of 1 kHz. Specify the nominal free-running frequency of each VCO.

7.3-9 Referring to Table 7.1–1, devise a frequency synthesizer to generate $f_{LO} = f_c + f_{IF}$ for an FM radio. Assume you have available a master oscillator at 120.0 MHz and adjustable divide-by-n counters with $n = 1$ to 1000.

7.3-10 Referring to Table 7.1–1, devise a frequency synthesizer to generate $f_{LO} = f_c + f_{IF}$ for an AM radio. Assume you have available a master oscillator at 2105 kHz and adjustable divide-by-n counters with $n = 1$ to 1000.

7.3-11 The linearized PLL in Fig. 7.3–8 becomes a *phase* demodulator if we add an ideal integrator to get

$$z(t) = \int^t y(\lambda)\, d\lambda$$

Find $Z(f)/X(f)$ when the input is a PM signal. Compare with Eq. (11).

7.3-12* Consider the PLL model in Fig. 7.3–8c, where $E(f) = \Phi(f) - \Phi_v(f)$. (a) Find $E(f)/\Phi(f)$ and derive Eq. (10) therefrom. (b) Show that if the input is an FM signal, then $E(f) = (f_\Delta/K)H_\epsilon(f)X(f)$ with $H_\epsilon(f) = 1/[H(f) + j(f/K)]$.

7.3-13 Suppose an FM detector is a linearized first-order PLL with $H(f) = 1$. Let the input signal be modulated by $x(t) = A_m \cos 2\pi f_m t$ where $A_m \le 1$ and $0 \le f_m \le W$. (a) Use the relationship in Prob. 7.3–12b to find the steady-state amplitude of $\epsilon(t)$. (b) Since linear operation requires $|\epsilon(t)| \le 0.5$ rad, so $\sin \epsilon \approx \epsilon$, show that the minimum loop gain is $K = 2f_\Delta$.

7.3-14‡ Suppose an FM detector is a *second-order* PLL with loop gain K and $H(f) = 1 + K/j2f$. Let the input signal be modulated by $x(t) = A_m \cos 2\pi f_m t$ where $A_m \le 1$ and $0 \le f_m \le W$. (a) Use the relationship in Prob. 7.3–12b to show that the steady-state amplitude of $\epsilon(t)$ is maximum when $f_m = K/\sqrt{2}$ if $K/\sqrt{2} \le W$. (b) Now assume that $K/\sqrt{2} > W$ and $f_\Delta > W$. Since linear operation requires $|\epsilon(t)| < 0.5$ rad, so $\sin \epsilon \approx \epsilon$, show that the minimum loop gain is $K \approx 2\sqrt{f_\Delta W}$.

7.3-15‡ Consider the second-order PLL in Prob. 7.3–14. (a) Show that $H_L(f)$ becomes a second-order LPF with $|H_L|$ maximum at $f \approx 0.556K$ and 3-dB bandwidth $B \approx 1.14K$. (b) Use the loop-gain conditions in Probs. 7.3–13 and 7.3–14 to compare the minimum 3-dB bandwidths of a first-order and second-order PLL FM detector when $f_\Delta/W = 2$, 5, and 10.

7.4-1 Explain the following statements: (a) A TV frame should have an *odd* number of lines. (b) The waveform that drives the scanning path should be a *sawtooth*, rather than a sinusoid or triangle.

7.4-2 Consider a scanning raster with very small slope and retrace time. Sketch the video signal and its spectrum, without using Eq. (4), when the image consists of: (a) alternating black and white vertical bars of width $H/4$; (b) alternating black and white horizontal bars of height $V/4$.

7.4–3   Consider an image that's entirely black ($I = 0$) except for a centered white rectangle ($I = 1.0$) of width αH and height βV. (a) Show that $|c_{mn}| = \alpha\beta|$ sinc αm sinc $\beta n|$. (b) Sketch the resulting line spectrum when $\alpha = 1/2$, $\beta = 1/4$, and $f_v = f_h/100$.

7.4–4*   Calculate the number of pixels and the video bandwidth requirement for a low-resolution TV system with a square image, 230 active lines, and 100-μs active line time.

7.4–5   Calculate the number of pixels and the video bandwidth requirement for the HDTV system in Table 7.4–1 if $N_{vr} \ll N$ and $T_{hr} = 0.2T_{\text{line}}$.

7.4–6*   Calculate the number of pixels and the video bandwidth requirement for the CCIR system in Table 7.4–1 if $N_{vr} = 48$ and $T_{hr} = 10\ \mu$s.

7.4–7   Horizontal *aperture effect* arises when the scanning process in a TV camera produces the output.

$$\tilde{x}(t) = \int_{t-\tau}^{t} x(\lambda)\, d\lambda$$

where $x(t)$ is the desired video signal and $\tau \ll T_{\text{line}}$. (a) Describe the resulting TV picture. (b) Find an equalizer that will improve the picture quality.

7.4–8   Describe what happens to a color TV picture when: (a) the gain of the chrominance amplifier is too high or too low; (b) the phase adjustment of the color subcarrier is in error by ± 90 or $180°$.

7.4–9   Carry out the details leading from Eq. (15) to Eq. (16).

7.4–10   Obtain expressions equivalent to Eqs. (15) and (16) when all the filters in the x_Q channel (at transmitter and receiver) are the same as the x_I channel. Discuss your results.

Probability and Random Variables

CHAPTER OUTLINE

Chapters 2 through 7 dealt entirely with *deterministic signals*, for when we write an explicit time function $v(t)$ we presume that the behavior of the signal is known or determined for all time. In Chapter 9 we'll deal with *random signals* whose exact behavior cannot be described in advance. Random signals occur in communication both as unwanted noise and as desired information-bearing waveforms. Lacking detailed knowledge of the time variation of a random signal, we must speak instead in terms of probabilities and statistical properties. This chapter therefore presents the groundwork for the description of random signals.

The major topics include probabilities, random variables, statistical averages, and important probability models. We direct our coverage specifically toward those aspects used in later chapters and rely heavily on intuitive reasoning rather than mathematical rigor.

If you've previously studied probability and statistics, then you can skim over this chapter and go to Chapter 9. (However, be alert for possible differences of notation and emphasis.) If you want to pursue the subject in greater detail, you'll find a wealth of material in texts devoted to the subject.

OBJECTIVES

After studying this chapter and working the exercises, you should be able to do each of the following:

1. Calculate event probabilities using frequency of occurrence and the relationships for mutually exclusive, joint, conditional, and statistically independent events (Sect. 8.1).
2. Define and state the properties of the probability functions of discrete and continuous random variables (Sect. 8.2).
3. Write an expression for the probability of a numerical-valued event, given a frequency function, CDF, or PDF (Sect. 8.2).
4. Find the mean, mean-square, and variance of a random variable, given its frequency function or PDF (Sect. 8.3).
5. Define and manipulate the expectation operation (Sect. 8.3).
6. Describe applications of the binomial, Poisson, gaussian, and Rayleigh probability models (Sect. 8.4).
7. Write probabilities for a gaussian random variable in terms of the Q function (Sect. 8.4).

8.1   PROBABILITY AND SAMPLE SPACE

Probability theory establishes a mathematical framework for the study of *random phenomena*. The theory does not deal with the nature of random processes per se, but rather with their experimentally observable manifestations. Accordingly, we'll discuss probability here in terms of events associated with the outcomes of experiments. Then we'll introduce sample space to develop probability theory and to obtain the probabilities of various types of events.

Probabilities and Events

Consider an experiment involving some element of chance, so the outcome varies unpredictably from trial to trial. Tossing a coin is such an experiment, since a trial toss could result in the coin landing heads up or tails up. Although we cannot predict the outcome of a single trial, we may be able to draw useful conclusions about the results of a large number of trials.

For this purpose, let's identify a specific **event** A as something that might be observed on any trial of a chance experiment. We repeat the experiment N times and record N_A, the number of times A occurs. The ratio N_A/N then equals the **relative frequency of occurrence** of the event A for that sequence of trials.

The experiment obeys the **empirical law of large numbers** if N_A/N approaches a definite limit as N becomes very large and if every sequence of trials yields the same limiting value. Under these conditions we take the **probability** of A to be

$$P(A) = N_A/N \qquad N \to \infty \qquad\qquad [1]$$

The functional notation $P(A)$ emphasizes that the value of the probability depends upon the event in question. Nonetheless, every probability is a nonnegative number bounded by

$$0 \le P(A) \le 1$$

since $0 \le N_A \le N$ for any event A.

Our interpretation of probability as frequency of occurrence agrees with intuition and common experience in the following sense: You can't predict the specific result of a single trial of a chance experiment, but you expect that the number of times A occurs in $N \gg 1$ trials will be $N_A \approx NP(A)$. Probability therefore has meaning only in relation to a *large number of trials*.

By the same token, Eq. (1) implies the need for an *infinite* number of trials to measure an *exact* probability value. Fortunately, many experiments of interest possess inherent symmetry that allows us to deduce probabilities by logical reasoning, without resorting to actual experimentation. We feel certain, for instance, that an honest coin would come up heads half the time in a large number of trial tosses, so the probability of heads equals 1/2.

Suppose, however, that you seek the probability of getting two heads in three tosses of an honest coin. Or perhaps you know that there were two heads in three tosses and you want the probability that the first two tosses match. Although such problems could be tackled using relative frequencies, formal probability theory provides a more satisfactory mathematical approach, discussed next.

Sample Space and Probability Theory

A typical experiment may have several possible outcomes, and there may be various ways of characterizing the associated events. To construct a systematic model of a chance experiment let the **sample space** S denote the set of outcomes, and let S be partitioned into **sample points** $s_1, s_2, \ldots,$ corresponding to the specific outcomes. Thus, in set notation,

$$S = \{s_1, s_2, \ldots\}$$

Although the partitioning of S is not unique, the sample points are subject to two requirements:

1.   The set $\{s_1, s_2, \ldots\}$ must be **exhaustive,** so that S consists of *all* possible outcomes of the experiment in question.

2.   The outcomes $s_1, s_2, \ldots$ must be **mutually exclusive,** so that one and only one of them occurs on a given trial.

Consequently, any events of interest can be described by *subsets* of S containing zero, one, or more than one sample points.

By way of example, consider the experiment of tossing a coin three times and observing the sequence of heads (H) and tails (T). The sample space then contains $2 \times 2 \times 2 = 8$ distinct sequences, namely,

$$S = \{HHH, HTH, HHT, THH, THT, TTH, HTT, TTT\}$$

where the order of the listing is unimportant. What is important is that the eight sample-point sequences are exhaustive and mutually exclusive. The event $A = $ "two heads" can therefore be expressed as the subset

$$A = \{HTH, HHT, THH\}$$

Likewise, the events $B = $ "second toss differs from the other two" and $C = $ "first two tosses match" are expressed as

$$B = \{HTH, THT\} \qquad C = \{HHH, HHT, TTH, TTT\}$$

Figure 8.1–1 depicts the sample space and the relationships between A, B, and C in the form of a **Venn diagram,** with curves enclosing the sample points for each event. This diagram brings out the fact that B and C happen to be mutually exclusive events, having no common sample points, whereas A contains one point in common with B and another point in common with C.

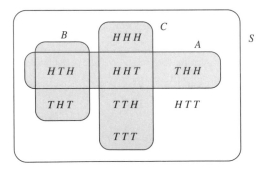

Figure 8.1–1   Sample space and Venn diagram of three events.

Other events may be described by particular combinations of event subsets, as follows:

- The **union event** $A + B$ (also symbolized by $A \cup B$) stands for the occurrence of *A or B or both,* so its subset consists of all s_i in either A or B.
- The **intersection event** AB (also symbolized by $A \cap B$) stands for the occurrence of *A and B,* so its subset consists only of those s_i in both A and B.

For instance, in Fig. 8.1–1 we see that

$$A + B = \{HTH, HHT, THH, THT\} \qquad AB = \{HTH\}$$

But since B and C are mutually exclusive and have no common sample points,

$$BC = \varnothing$$

where $\varnothing$ denotes the **empty set.**

Probability theory starts with the assumption that a probability $P(s_i)$ has been assigned to each point s_i in the sample space S for a given experiment. The theory says nothing about those probabilities except that they must be chosen to satisfy **three fundamental axioms:**

$$P(A) \geq 0 \text{ for any event } A \text{ in } S \qquad \text{[2a]}$$

$$P(S) = 1 \qquad \text{[2b]}$$

$$P(A_1 + A_2) = P(A_1) + P(A_2) \text{ if } A_1A_2 = \varnothing \qquad \text{[2c]}$$

These axioms form the basis of probability theory, even though they make no mention of frequency of occurrence. Nonetheless, axiom (2a) clearly agrees with Eq. (1), and so does axiom (2b) because one of the outcomes in S must occur on every trial. To interpret axiom (2c) we note that if A_1 occurs N_1 times in N trials and A_2 occurs N_2 times, then the event "A_1 or A_2" occurs $N_1 + N_2$ times since the stipulation $A_1A_2 = \varnothing$ means that they are mutually exclusive. Hence, as N becomes large, $P(A_1 + A_2) = (N_1 + N_2)/N = (N_1/N) + (N_2/N) = P(A_1) + P(A_2)$.

Now suppose that we somehow know all the sample-point probabilities $P(s_i)$ for a particular experiment. We can then use the three axioms to obtain relationships for the probability of any event of interest. To this end, we'll next state several important general relations that stem from the axioms. The omitted derivations are exercises in elementary set theory, and the relations themselves are consistent with our interpretation of probability as relative frequency of occurrence.

Axiom (2c) immediately generalizes for **three or more mutually exclusive events.** For if

$$A_1A_2A_3 \ldots = \varnothing$$

then

$$P(A_1 + A_2 + A_3 + \ldots) = P(A_1) + P(A_2) + P(A_3) + \ldots \qquad \text{[3]}$$

Furthermore, if M mutually exclusive events have the exhaustive property

$$A_1 + A_2 + \ldots + A_M = S$$

then, from axioms (2c) and (2b),

$$P(A_1 + A_2 + \cdots + A_M) = \sum_{i=1}^{M} P(A_i) = 1 \qquad [4]$$

Note also that Eq. (4) applies to the sample-point probabilities $P(s_i)$.

Equation (4) takes on special importance when the M events happen to be **equally likely,** meaning that they have *equal probabilities*. The sum of the probabilities in this case reduces to $M \times P(A_i) = 1$, and hence

$$P(A_i) = 1/M \qquad i = 1, 2, \ldots, M \qquad [5]$$

This result allows you to calculate probabilities when you can identify all possible outcomes of an experiment in terms of mutually exclusive, equally likely events. The hypothesis of equal likelihood might be based on experimental data or symmetry considerations—as in coin tossing and other honest games of chance.

Sometimes we'll be concerned with the *nonoccurrence* of an event. The event **"not A"** is called the **complement** of A, symbolized by A^C (also written $\overline{A}$). The probability of A^C is

$$P(A^C) = 1 - P(A) \qquad [6]$$

since $A + A^C = S$ and $AA^C = \varnothing$.

Finally, consider events A and B that are *not* mutually exclusive, so axiom (2c) does not apply. The probability of the union event $A + B$ is then given by

$$P(A + B) = P(A) + P(B) - P(AB) \qquad [7]$$

in which $P(AB)$ is the probability of the intersection or joint event AB. We call $P(AB)$ the **joint probability** and interpret it as

$$P(AB) = N_{AB}/N \qquad N \to \infty$$

where N_{AB} stands for the number of times A and B occur together in N trials. Equation (7) reduces to the form of axiom (2c) when $AB = \varnothing$, so A and B cannot occur together and $P(AB) = 0$.

EXAMPLE 8.1–1   As an application of our probability relationships, we'll calculate some event probabilities for the experiment of tossing an honest coin three times. Since H and T are equally likely to occur on each toss, the eight sample-point sequences back in Fig. 8.1–1 must also be equally likely. We therefore use Eq. (5) with $M = 8$ to get

$$P(s_i) = 1/8 \qquad i = 1, 2, \ldots, 8$$

The probabilities of the events A, B, and C are now calculated by noting that A contains three sample points, B contains two, and C contains four, so Eq. (3) yields

$$P(A) = \tfrac{1}{8} + \tfrac{1}{8} + \tfrac{1}{8} = \tfrac{3}{8} \qquad P(B) = \tfrac{2}{8} \qquad P(C) = \tfrac{4}{8}$$

Similarly, the joint-event subsets AB and AC each contain just one sample point, so

$$P(AB) = P(AC) = 1/8$$

whereas $P(BC) = 0$ since B and C are mutually exclusive.

The probability of the complementary event A^C is found from Eq. (6) to be

$$P(A^C) = 1 - \tfrac{3}{8} = \tfrac{5}{8}$$

The probability of the union event $A + B$ is given by Eq. (7) as

$$P(A + B) = \tfrac{3}{8} + \tfrac{2}{8} - \tfrac{1}{8} = \tfrac{4}{8}$$

Our results for $P(A^C)$ and $P(A + B)$ agree with the facts that the subset A^C contains five sample points and $A + B$ contains four.

A certain honest wheel of chance is divided into three equal segments colored green (G), red (R), and yellow (Y), respectively. You spin the wheel twice and take the outcome to be the resulting color sequence—GR, RG, and so forth. Let A = "neither color is yellow" and let B = "matching colors." Draw the Venn diagram and calculate $P(A)$, $P(B)$, $P(AB)$, and $P(A + B)$.

EXERCISE 8.1–1

Conditional Probability and Statistical Independence

Sometimes an event B depends in some way on another event A having $P(A) \neq 0$. Accordingly, the probability of B should be adjusted when you know that A has occurred. Mutually exclusive events are an extreme example of dependence, for if you know that A has occurred, then you can be sure that B did not occur on the same trial. Conditional probabilities are introduced here to account for event dependence and also to define statistical independence.

We measure the dependence of B on A in terms of the **conditional probability**

$$P(B|A) \triangleq P(AB)/P(A) \qquad \text{[8]}$$

The notation $B|A$ stands for the event **B given A,** and $P(B|A)$ represents the probability of B conditioned by the knowledge that A has occurred. If the events happen to be mutually exclusive, then $P(AB) = 0$ and Eq. (8) confirms that $P(B|A) = 0$ as expected. With $P(AB) \neq 0$, we interpret Eq. (8) in terms of relative frequency by inserting $P(AB) = N_{AB}/N$ and $P(A) = N_A/N$ as $N \rightarrow \infty$. Thus,

$$P(B|A) = \frac{N_{AB}/N}{N_A/N} = \frac{N_{AB}}{N_A}$$

which says that $P(B|A)$ equals the relative frequency of A and B together in the N_A trials where A occurred with or without B.

Interchanging B and A in Eq. (8) yields $P(A|B) = P(AB)/P(B)$, and we thereby obtain two relations for the joint probability, namely,

$$P(AB) = P(A|B)P(B) = P(B|A)P(A) \tag{9}$$

Or we could eliminate $P(AB)$ to get **Bayes' theorem**

$$P(B|A) = \frac{P(B)P(A|B)}{P(A)} \tag{10}$$

This theorem plays an important role in statistical decision theory because it allows us to *reverse* the conditioning event. Another useful expression is the **total probability**

$$P(B) = \sum_{i=1}^{M} P(B|A_i)P(A_i) \tag{11}$$

where the conditioning events $A_1, A_2, \ldots, A_M$ must be *mutually exclusive* and *exhaustive*.

Events A and B are said to be **statistically independent** when they do *not* depend on each other, as indicated by

$$P(B|A) = P(B) \qquad P(A|B) = P(A) \tag{12}$$

Inserting Eq. (12) into Eq. (9) then gives

$$P(AB) = P(A)P(B)$$

so the joint probability of statistically independent events equals the *product* of the individual event probabilities. Furthermore, if three or more events are all independent of each other, then

$$P(ABC \ldots) = P(A)P(B)P(C) \ldots \tag{13}$$

in addition to pairwise independence.

As a rule of thumb, *physical* independence is a sufficient condition for statistical independence. We may thus apply Eq. (12) to situations in which events have no physical connection. For instance, successive coin tosses are physically independent, and a sequence such as *TTH* may be viewed as a joint event. Invoking the equally likely argument for each toss alone, we have $P(H) = P(T) = 1/2$ and $P(TTH) = P(T)P(T)P(H) = (1/2)^3 = 1/8$—in agreement with our conclusion in Example 8.1–1 that $P(s_i) = 1/8$ for any three-toss sequence.

In Example 8.1–1 we calculated the probabilities $P(A) = 3/8$, $P(B) = 2/8$, and $P(AB) = 1/8$. We'll now use these values to investigate the dependence of events A and B.

EXAMPLE 8.1–2

Since $P(A)P(B) = 6/64 \neq P(AB)$, we immediately conclude that A and B are not statistically independent. The dependence is reflected in the conditional probabilities

$$P(B|A) = \frac{P(AB)}{P(A)} = \frac{1/8}{3/8} = \frac{1}{3} \qquad P(A|B) = \frac{P(AB)}{P(B)} = \frac{1/8}{2/8} = \frac{1}{2}$$

so $P(B|A) \neq P(B)$ and $P(A|B) \neq P(A)$.

Reexamination of Fig. 8.1–1 reveals why $P(B|A) > P(B)$. Event A corresponds to any one of three equally likely outcomes, and one of those outcomes also corresponds to event B. Hence, B occurs with frequency $N_{AB}/N_A = 1/3$ of the N_A trials in which A occurs—as contrasted with $P(B) = N_B/N = 2/8$ for all N trials. Like reasoning justifies the value of $P(A|B)$.

The resistance R of a resistor drawn randomly from a large batch has five possible values, all in the range 40–60 Ω. Table 8.1–1 gives the specific values and their probabilities.

EXAMPLE 8.1–3

Table 8.1–1

R:	40	45	50	55	60
$P_R(R)$:	0.1	0.2	0.4	0.2	0.1

Let the event A be "$R \leq 50\ \Omega$" so

$$P(A) = P(R = 40 \text{ or } R = 45 \text{ or } R = 50) = P_R(40) + P_R(45) + P_R(50) = 0.7$$

Similarly, the event $B = 45\ \Omega \leq R \leq 55\ \Omega$ has

$$P(B) = P_R(45) + P_R(50) + P_R(55) = 0.8$$

The events A and B are not independent since

$$P(AB) = P_R(45) + P_R(50) = 0.6$$

which does not equal the product $P(A)P(B)$. Then, using Eqs. (7) and (9),

$$P(A + B) = 0.7 + 0.8 - 0.6 = 0.9 \quad P(B|A) = \frac{0.6}{0.7} = 0.857 \quad P(A|B) = \frac{0.6}{0.8} = 0.75$$

The value of $P(A + B)$ is easily confirmed from Table 8.1–1, but the conditional probabilities are most easily calculated from Eq. (9).

EXERCISE 8.1–2   Referring to Fig. 8.1–1 let

$$D = \{THT, TTH, HTT, TTT\}$$

which expresses the event "two or three tails." Confirm that B and D are statistically independent by showing that $P(B|D) = P(B)$, $P(D|B) = P(D)$, and $P(B)P(D) = P(BD)$.

8.2   RANDOM VARIABLES AND PROBABILITY FUNCTIONS

Coin tossing and other games of chance are natural and fascinating subjects for probability calculations. But communication engineers are more concerned with random processes that produce *numerical* outcomes—the instantaneous value of a noise voltage, the number of errors in a digital message, and so on. We handle such problems by defining an appropriate *random variable,* or RV for short.

Despite the name, a random variable is neither random nor a variable. Instead, it's a *function* that generates numbers from the outcomes of a chance experiment. Specifically,

> A **random variable** is a rule or relationship, denoted by X, that assigns a real number $X(s)$ to every point in the sample space S.

Almost any relationship may serve as an RV, provided that X is real and single-valued and that

$$P(X = -\infty) = P(X = \infty) = 0$$

The essential property is that X maps the outcomes in S into numbers along the real line $-\infty < x < \infty$. (More advanced presentations deal with *complex* numbers.)

We'll distinguish between *discrete* and *continuous* RVs, and we'll develop *probability functions* for the analysis of numerical-valued random events.

Discrete Random Variables and CDFs

If S contains a countable number of sample points, then X will be a **discrete RV** having a *countable number of distinct values.* Figure 8.2–1 depicts the corresponding mapping processes and introduces the notation $x_1 < x_2 < \ldots$ for the values of $X(s)$ in ascending order. Each outcome produces a single number, but two or more outcomes may map into the same number.

Although a mapping relationship underlies every RV, we usually care only about the resulting numbers. We'll therefore adopt a more direct viewpoint and treat X itself as the general symbol for the experimental outcomes. This viewpoint allows us to deal with *numerical-valued events* such as $X = a$ or $X \le a$, where a is some point on the real line. Furthermore, if we replace the constant a with the independent variable x, then we get *probability functions* that help us calculate probabilities of numerical-valued events.

The probability function $P(X \le x)$ is known as the **cumulative distribution function** (or **CDF**), symbolized by

$$F_X(x) \triangleq P(X \le x) \qquad \text{[1]}$$

Pay careful attention to the notation here: The subscript X identifies the RV whose characteristics determine the function $F_X(x)$, whereas the argument x defines the event $X \le x$ so x is *not* an RV. Since the CDF represents a probability, it must be bounded by

$$0 \le F_X(x) \le 1 \qquad \text{[2a]}$$

with extreme values

$$F_X(-\infty) = 0 \qquad F_X(\infty) = 1 \qquad \text{[2b]}$$

The lower limit reflects our condition that $P(X = -\infty) = 0$, whereas the upper limit says that X always falls somewhere along the real line. The complementary events $X \le x$ and $X > x$ encompass the entire real line, so

$$P(X > x) = 1 - F_X(x) \qquad \text{[3]}$$

Other CDF properties will emerge as we go along.

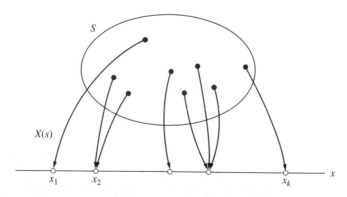

Figure 8.2–1   Sample points mapped by the discrete RV $X(s)$ into numbers on the real line.

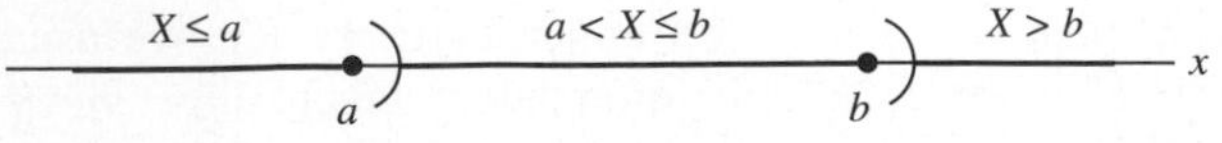

Figure 8.2–2   Numerical-valued events along the real line.

Suppose we know $F_X(x)$ and we want to find the probability of observing $a < X \le b$. Figure 8.2–2 illustrates the relationship of this event to the events $X \le a$ and $X > b$. The figure also brings out the difference between *open* and *closed* inequalities for specifying numerical events. Clearly, the three events here are mutually exclusive when $b > a$, and

$$P(X \le a) + P(a < X \le b) + P(X > b) = P(X \le \infty) = 1$$

Substituting $P(X \le a) = F_X(a)$ and $P(X > b) = 1 - F_X(b)$ yields the desired result

$$P(a < X \le b) = F_X(b) - F_X(a) \qquad b > a \tag{4}$$

Besides being an important relationship in its own right, Eq. (4) shows that $F_X(x)$ has the *nondecreasing property* $F_X(b) \le F_X(a)$ for any $b > a$. Furthermore, $F_X(x)$ is *continuous from the right* in the sense that if $\epsilon > 0$ then $F_X(x + \epsilon) \to F_X(x)$ as $\epsilon \to 0$.

Now let's take account of the fact that a discrete RV is restricted to distinct values $x_1, x_2, \ldots$. This restriction means that the possible outcomes $X = x_i$ constitute a set of mutually exclusive events. The corresponding set of probabilities will be written as

$$P_X(x_i) \triangleq P(X = x_i) \qquad i = 1, 2, \ldots \tag{5}$$

which we call the **frequency function.** Since the x_i are mutually exclusive, the probability of the event $X \le x_k$ equals the sum

$$P(X \le x_k) = P_X(x_1) + P_X(x_2) + \cdots + P_X(x_k)$$

Thus, the CDF can be obtained from the frequency function $P_X(x_i)$ via

$$F_X(x_k) = \sum_{i=1}^{k} P_X(x_i) \tag{6}$$

This expression indicates that $F_X(x)$ looks like a *staircase* with upward steps of height $P_X(x_i)$ at each $x = x_i$. The staircase starts at $F_X(x) = 0$ for $x < x_1$ and reaches $F_X(x) = 1$ at the last step. Between steps, where $x_k < x < x_{k+1}$, the CDF remains constant at $F_X(x_k)$.

EXAMPLE 8.2–1

Consider the experiment of transmitting a three-digit message over a noisy channel. The channel has *error probability* $P(E) = 2/5 = 0.4$ per digit, and errors are statistically independent from digit to digit, so the probability of receiving a correct digit is

$P(C) = 1 - 2/5 = 3/5 = 0.6$. We'll take X to be the *number of errors* in a received message, and we'll find the corresponding frequency function and CDF.

The sample space for this experiment consists of eight distinct error patterns, like the head-tail sequences back in Fig. 8.1–1. But now the sample points are not equally likely since the error-free pattern has $P(CCC) = P(C)P(C)P(C) = (3/5)^3 = 0.216$, whereas the all-error pattern has $P(EEE) = (2/5)^3 = 0.064$. Similarly, each of the three patterns with one error has probability $(2/5) \times (3/5)^2$ and each of the three patterns with two errors has probability $(2/5)^2 \times (3/5)$. Furthermore, although there are eight points in S, the RV X has only four possible values, namely, $x_i = 0, 1, 2,$ and 3 errors.

Figure 8.2–3*a* shows the sample space, the mapping for X, and the resulting values of $P_X(x_i)$. The values of $F_X(x_i)$ are then calculated via

$$F_X(0) = P_X(0) \qquad F_X(1) = P_X(0) + P_X(1)$$

and so forth in accordance with Eq. (6). The frequency function and CDF are plotted in Fig. 8.2–3*b*. We see from the CDF plot that the probability of less than two errors is $F_X(2 - \epsilon) = F_X(1) = 81/125 = 0.648$ and the probability of more than one error is $1 - F_X(1) = 44/125 = 0.352$.

Let a random variable be defined for the experiment in Exercise 8.1–1 (p. 317) by the following rule: The colors are assigned the numerical weights $G = 2, R = -1,$ and $Y = 0$, and X is taken as the *average* of the weights observed on a given trial of two spins. For instance, the outcome RY maps into the value $X(RY) = (-1 + 0)/2 = -0.5$. Find and plot $P_X(x_i)$ and $F_X(x_i)$. Then calculate $P(-1.0 < X \leq 1.0)$.

EXERCISE 8.2–1

Continuous Random Variables and PDFs

A **continuous RV** may take on *any* value within a certain range of the real line, rather than being restricted to a countable number of distinct points. For instance, you might spin a pointer and measure the final angle θ. If you take $X(\theta) = \tan^2 \theta$, as shown in Fig. 8.2–4, then every value in the range $0 \leq x < \infty$ is a possible outcome of this experiment. Or you could take $X(\theta) = \cos \theta$, whose values fall in the range $-1.0 \leq x \leq 1.0$.

Since a continuous RV has an *uncountable* number of possible values, the chance of observing $X = a$ must be vanishingly small in the sense that $P(X = a) = 0$ for any specific a. Consequently, frequency functions have no meaning for continuous RVs. However, events such as $X \leq a$ and $a < X \leq b$ may have nonzero probabilities, and $F_X(x)$ still provides useful information. Indeed, the properties stated before in Eqs. (1)–(4) remain valid for the CDF of a continuous RV.

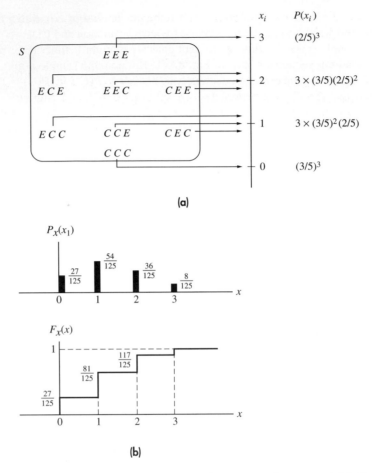

(a)

(b)

Figure 8.2-3 (a) Mapping for Example 8.2–1. (b) Frequency function and CDF for the discrete RV in Example 8.2–1.

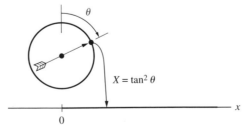

Figure 8.2-4 Mapping by a continuous RV.

But a more common description of a continuous RV is its **probability density function** (or PDF), defined by

$$p_X(x) \triangleq dF_X(x)/dx \qquad\qquad [7]$$

provided that the derivative exists. We don't lose information when differentiating $F_X(x)$ because we know that $F_X(-\infty) = 0$. We can therefore write

$$P(X \le x) = F_X(x) = \int_{-\infty}^{x} p_X(\lambda)\, d\lambda \qquad\qquad [8]$$

where we've used the dummy integration variable λ for clarity. Other important PDF properties are

$$p_X(x) \ge 0 \qquad \int_{-\infty}^{\infty} p_X(x)\, dx = 1 \qquad\qquad [9]$$

$$P(a < X \le b) = F_X(b) - F_X(a) = \int_{a}^{b} p_X(x)\, dx \qquad\qquad [10]$$

Thus,

A **PDF** is a nonnegative function whose total area equals unity and whose area in the range $a < x \le b$ equals the probability of observing X in that range.

As a special case of Eq. (10), let $a = x - dx$ and $b = x$. The integral then reduces to the *differential area* $p_X(x)\, dx$ and we see that

$$p_X(x)\, dx = P(x - dx < X < x) \qquad\qquad [11]$$

This relation serves as another interpretation of the PDF, emphasizing its nature as a probability *density*. Figure 8.2–5 shows a typical PDF for a continuous RV and the areas involved in Eqs. (10) and (11).

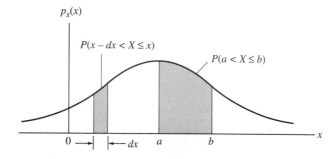

Figure 8.2–5   A typical PDF and the area interpretation of probabilities.

Occasionally we'll encounter **mixed random variables** having both continuous and discrete values. We treat such cases using *impulses* in the PDF, similar to our spectrum of a signal containing both nonperiodic and periodic components. Specifically, for any discrete value x_0 with nonzero probability $P_X(x_0) = P(X = x_0) \neq 0$, the PDF must include an impulsive term $P_X(x_0)\delta(x - x_0)$ so that $F_X(x)$ has an appropriate jump at $x = x_0$. Taking this approach to the extreme, the frequency function of a discrete RV can be converted into a PDF consisting entirely of impulses.

But when a PDF includes impulses, you need to be particularly careful with events specified by open and closed inequalities. For if $p_X(x)$ has an impulse at x_0, then the probability that $X \geq x_0$ should be written out as $P(X \geq x_0) = P(X > x_0) + P(X = x_0)$. In contrast, there's no difference between $P(X \geq x_0)$ and $P(X > x_0)$ for a strictly continuous RV having $P(X = x_0) = 0$.

EXAMPLE 8.2–2   **Uniform PDF**

To illustrate some of the concepts of a continuous RV, let's take $X = \theta$ (radians) for the angle of the pointer back in Fig. 8.2–4. Presumably all angles between 0 and 2π are equally likely, so $p_X(x)$ has some constant value C for $0 < x \leq 2\pi$ and $p_X(x) = 0$ outside this range. We then say that X has a **uniform PDF.**

The unit-area property requires that

$$\int_{-\infty}^{\infty} p_X(x)\,dx = \int_0^{2\pi} C\,dx = 1 \Rightarrow C = 1/2\pi$$

so

$$p_X(x) = \frac{1}{2\pi}[u(x) - u(x - 2\pi)] = \begin{cases} 1/2\pi & 0 < x \leq 2\pi \\ 0 & \text{otherwise} \end{cases}$$

which is plotted in Fig. 8.2–6a. Integrating $p_X(x)$ per Eq. (8) yields the CDF in Fig. 8.2–6b, where

$$F_X(x) = x/2\pi \qquad 0 < x \leq 2\pi$$

so, for example, $P(X \leq \pi) = F_X(\pi) = 1/2$. These functions describe a continuous RV **uniformly distributed** over the range $0 < x \leq 2\pi$.

But we might also define another random variable Z such that

$$Z = \begin{cases} \pi & X \leq \pi \\ X & X > \pi \end{cases}$$

Then $P(Z < \pi) = 0$, $P(Z = \pi) = P(X \leq \pi) = 1/2$, and $P(Z \leq z) = P(X \leq z)$ for $z > \pi$. Hence, using z as the independent variable for the real line, the PDF of Z is

$$p_Z(z) = \frac{1}{2}\delta(z - \pi) + \frac{1}{2\pi}[u(z - \pi) - u(z - 2\pi)]$$

The impulse here accounts for the discrete value $Z = \pi$.

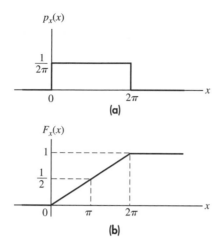

Figure 8.2–6   PDF and CDF of a uniformly distributed RV.

Use the PDFs in Example 8.2–2 to calculate the probabilities of the following   **EXERCISE 8.2–2**
events: (a) $\pi < X \leq 3\pi/2$, (b) $X > 3\pi/2$, (c) $\pi < Z \leq 3\pi/2$, and (d) $\pi \leq Z \leq 3\pi/2$.

Transformations of Random Variables

The preceding example touched upon a *transformation* that defines one RV in terms of another. Here, we'll develop a general expression for the resulting PDF when the new RV is a continuous function of a continuous RV with a known PDF.

Suppose we know $p_X(x)$ and we want to find $p_Z(z)$ for the RV related to X by the transformation function

$$Z = g(X)$$

We initially assume that $g(X)$ *increases monotonically,* so the probability of observing Z in the differential range $z - dz < Z \leq z$ equals the probability that X occurs in the corresponding range $x - dx < X \leq x$, as illustrated in Fig. 8.2–7. Equation (10) then yields $p_Z(z)\ dz = p_X(x)\ dx$, from which $p_Z(z) = p_X(x)\ dx/dz$. But if $g(X)$ *decreases monotonically,* then Z increases when X decreases and $p_Z(z) = p_X(x)(-dx/dz)$. We combine both of these cases by writing

$$p_Z(z) = p_X(x)\left|\frac{dx}{dz}\right|$$

Finally, since x transforms to $z = g(x)$, we insert the inverse transformation $x = g^{-1}(z)$ to obtain

$$p_Z(z) = p_X[g^{-1}(z)]\left|\frac{dg^{-1}(z)}{dz}\right| \qquad \text{[12]}$$

which holds for *any monotonic function.*

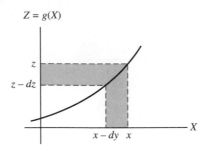

Figure 8.2–7   Transformation of an RV.

A simple but important monotonic function is the **linear transformation**

$$Z = \alpha X + \beta \qquad [13a]$$

where α and β are constants. Noting that $z = g(x) = \alpha x + \beta$, $x = g^{-1}(z) = (z - \beta)/\alpha$, and $dg^{-1}(z)/dz = 1/\alpha$, Eq. (12) becomes

$$p_Z(z) = \frac{1}{|\alpha|} p_X\left(\frac{z - \beta}{\alpha}\right) \qquad [13b]$$

Hence, $p_Z(z)$ has the same shape as $p_X(x)$ shifted by β and expanded or compressed by α.

If $g(X)$ is *not* monotonic, then two or more values of X produce the *same* value of Z. We handle such cases by subdividing $g(x)$ into a set of monotonic functions, $g_1(x), g_2(x), \ldots$, defined over different ranges of x. Since these ranges correspond to mutually exclusive events involving X, Eq. (12) generalizes as

$$p_Z(z) = p_X[g_1^{-1}(z)]\left|\frac{dg_1^{-1}(z)}{dz}\right| + p_X[g_2^{-1}(z)]\left|\frac{dg_2^{-1}(z)}{dz}\right| + \cdots \qquad [14]$$

The following example illustrates this method.

EXAMPLE 8.2–3   Consider the transformation $Z = \cos X$ with X being the uniformly distributed angle from Example 8.2–2. The plot of Z versus X in Fig. 8.2–8a brings out the fact that Z goes *twice* over the range -1 to 1 as X goes from 0 to 2π, so the transformation is not monotonic.

To calculate $p_Z(z)$, we first subdivide $g(x)$ into the two monotonic functions

$$g_1(x) = \cos x \qquad 0 < x \leq \pi$$

$$g_2(x) = \cos x \qquad \pi < x \leq 2\pi$$

which happen to be identical except for the defining ranges. For the range $0 < x \leq \pi$, we have $p_X(x) = 1/2\pi$ with $x = g_1^{-1}(z) = \cos^{-1} z$, so $dg_1^{-1}(z)/dz = -(1 - z^2)^{-1/2}$ and

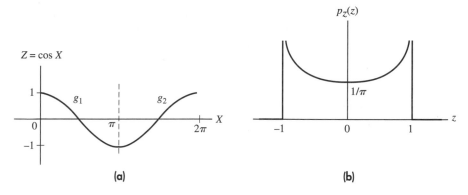

Figure 8.2–8

$p_X[g_1^{-1}(z)] = 1/2\pi$ over $-1 \le z \le 1$. The same results hold for $\pi < x \le 2\pi$ because $p_X(x)$ still equals $1/2\pi$ and $g_2(x) = g_1(x)$. Thus, from Eq. (14),

$$p_Z(z) = 2 \times \frac{1}{2\pi} \left| -(1-z)^{-1/2} \right| = \frac{1}{\pi\sqrt{1-z^2}} \qquad -1 \le z \le 1$$

As illustrated in Fig. 8.2–8b, this PDF has peaks at $z = \pm 1$ because $\cos X$ occurs more often near ± 1 than any other value when X is uniformly distributed over 2π radians.

Find $p_Z(z)$ when $Z = \sqrt{X}$ and $p_X(x)$ is uniform over $0 < x \le 4$.   **EXERCISE 8.2–3**

Joint and Conditional PDFs

Concluding our introduction to random variables, we briefly discuss the case of two continuous RVs that can be observed simultaneously. A classic example of this situation is the dart-throwing experiment, with X and Y taken as the rectangular coordinates of the dart's position relative to the center of the target.

The **joint probability density** function of X and Y will be denoted by $p_{XY}(x,y)$, where the comma stands for *and*. This function represents a surface over the x-y plane such that

$$p_{XY}(x, y)\, dx\, dy = P(x - dx < X \le x, y - dy < Y \le y)$$

which is the two-dimensional version of Eq. (11). In words, the *differential volume* $p_{XY}(x,y)\, dx\, dy$ equals the probability of observing the joint event $x - dx < X \le x$ and $y - dy < Y \le y$. Hence, the probability of the event $a < X \le b$ and $c < Y \le d$ is

$$P(a < X \le b, c < Y \le d) = \int_c^d \int_a^b p_{XY}(x, y)\, dx\, dy \qquad [15]$$

Equation (15) corresponds to the volume between the x-y plane and the surface $p_{XY}(x,y)$ for the stated ranges of x and y.

If X and Y happen to be *statistically independent,* then their joint PDF reduces to the product

$$p_{XY}(x, y) = p_X(x)p_Y(y) \tag{16}$$

Otherwise, the dependence of Y on X is expressed by the **conditional PDF**

$$p_Y(y|x) = \frac{p_{XY}(x, y)}{p_X(x)} \tag{17}$$

which corresponds to the PDF of Y given that $X = x$.

The PDF for X alone may be obtained from the joint PDF by noting that $P(X \leq x) = P(X \leq x, -\infty < Y < \infty)$ since the value of Y doesn't matter when we're only concerned with X. Thus,

$$p_X(x) = \int_{-\infty}^{\infty} p_{XY}(x, y)\, dy \tag{18}$$

We call $p_X(x)$ a **marginal PDF** when it's derived from a joint PDF per Eq. (18).

EXAMPLE 8.2–4

The joint PDF of two noise voltages is known to be

$$p_{XY}(x, y) = \frac{1}{2\pi} e^{-(y^2 - xy + x^2/2)} \qquad -\infty < x < \infty, -\infty < y < \infty$$

From Eq. (18), the marginal PDF for X alone is

$$p_X(x) = \int_{-\infty}^{\infty} \frac{1}{2\pi} e^{-(y^2 - xy + x^2/2)}\, dy = \frac{1}{\pi} e^{-x^2/4} \int_0^{\infty} e^{-\lambda^2}\, d\lambda = \frac{1}{2\sqrt{\pi}} e^{-x^2/4}$$

where we have made the change of variable $\lambda = y - x/2$. In like manner,

$$p_Y(y) = \int_{-\infty}^{\infty} \frac{1}{2\pi} e^{-(y^2 - xy + x^2/2)}\, dx = \frac{1}{\sqrt{2\pi}} e^{-y^2/2}$$

Thus, X and Y are not independent since $p_X(x)p_Y(y) \neq p_{XY}(x,y)$. But Eq. (17) yields the conditional PDFs

$$p_Y(y|x) = \frac{1}{\sqrt{\pi}} e^{-\left(y^2 - xy + \frac{x^2}{4}\right)} \qquad p_X(x|y) = \frac{1}{\sqrt{2\pi}} e^{-\left(\frac{y^2}{2} - xy + \frac{x^2}{2}\right)}$$

8.3   STATISTICAL AVERAGES

For some purposes, a probability function provides more information about an RV than actually needed. Indeed, the complete description of an RV may prove to be an embarrassment of riches, more confusing than illuminating. Thus, we often find it more convenient to describe an RV by a few characteristic *numbers.* These numbers are the various *statistical averages* presented here.

Means, Moments, and Expectation

The **mean** of the random variable X is a constant m_X that equals the sum of the values of X weighted by their probabilities. This statistical average corresponds to an ordinary experimental average in the sense that the sum of the values observed over $N \gg 1$ trials is expected to be about Nm_X. For that reason, we also call m_X the *expected value* of X, and we write $E[X]$ or $\overline{X}$ to stand for the **expectation operation** that yields m_X.

To formulate an expression for the statistical average or expectation, we begin by considering N independent observations of a *discrete* RV. If the event $X = x_i$ occurs N_i times, then the sum of the observed values is

$$N_1 x_1 + N_2 x_{2+\cdots} = \sum_i N_i x_i$$

Upon dividing by N and letting $N \to \infty$, the relative frequency N_i/N becomes $P(X = x_i) = P_X(x_i)$. Thus, the statistical average value is

$$m_X = \sum_i x_i P_X(x_i) \qquad\qquad [1]$$

which expresses the mean of a discrete RV in terms of its frequency function $P_X(x_i)$.

For the mean of a *continuous* RV, we replace $P_X(x_i)$ with $P(x - dx < X < x) = p_X(x)\,dx$ and pass from summation to integration so that

$$m_X = \int_{-\infty}^{\infty} x\, p_X(x)\, dx \qquad\qquad [2]$$

This expression actually includes Eq. (1) as a special case obtained by writing the discrete PDF as

$$p_X(x) = \sum_i P_X(x_i)\delta(x - x_i)$$

Hence, when we allow impulses in the PDF, Eq. (2) applies to *any* RV—continuous, discrete, or mixed. Hereafter, then, statistical averages will be written mostly in integral form with PDFs. The corresponding expressions for a strictly discrete RV are readily obtained by substituting the frequency function in place of the PDF or, more directly, by replacing the integration with the analogous summation.

When a function $g(X)$ transforms X into another random variable Z, its expected value can be found from $p_X(x)$ by noting that the event $X = x$ transforms to $Z = g(x)$, so

$$E[g(X)] = \int_{-\infty}^{\infty} g(x)p_X(x)\, dx \qquad\qquad [3]$$

If $g(X) = X^n$, then $E[X^n]$ is known as the **nth moment** of X. The first moment, of course, is just the mean value $E[X] = m_X$. The second moment $E[X^2]$ or $\overline{X^2}$ is called the **mean-square** value, as distinguished from the *mean squared* $m_X^2 = \overline{X}^2$. Writing out Eq. (3) with $g(X) = X^2$, we have

$$\overline{X^2} = \int_{-\infty}^{\infty} x^2 p_X(x)\, dx$$

or, for a discrete RV,

$$\overline{X^2} = \sum_i x_i^2 P_X(x_i)$$

The mean-square value will be particularly significant when we get to random signals and noise.

Like time averaging, the expectation in Eq. (3) is a *linear* operation. Thus, if α and β are constants and if $g(X) = \alpha X + \beta$, then

$$E[\alpha X + \beta] = \alpha \overline{X} + \beta \qquad \text{[4]}$$

Although this result seems rather trivial, it leads to the not-so-obvious relation

$$E[\overline{X}X] = \overline{X}E[X] = \overline{X}^2$$

since $\overline{X}$ is a constant inside $E[\overline{X}X]$.

Standard Deviation and Chebyshev's Inequality

The **standard deviation** of X, denoted by σ_X, provides a measure of the *spread* of observed values of X relative to m_X. The square of the standard deviation is called the **variance,** or **second central moment,** defined by

$$\sigma_X^2 \triangleq E[(X - m_X)^2] \qquad \text{[5]}$$

But a more convenient expression for the standard deviation emerges when we expand $(X - m_X)^2$ and invoke Eq. (4), so

$$E[(X - m_X)^2] = E[X^2 - 2m_X X + m_X^2] = \overline{X^2} - 2m_X\overline{X} + m_X^2 = \overline{X^2} - m_X^2$$

and

$$\sigma_X = \sqrt{\overline{X^2} - m_X^2} \qquad \text{[6]}$$

Hence, the standard deviation equals the square root of the mean-square value minus the mean value squared.

For an interpretation of σ_X, let k be any positive number and consider the event $|X - m_X| \geq k\sigma_X$. **Chebyshev's inequality** (also spelled Tchebycheff) states that

$$P(|X - m_X| \geq k\sigma_X) \leq 1/k^2 \qquad \text{[7a]}$$

regardless of $p_X(x)$. Thus, the probability of observing any RV outside $\pm k$ standard deviations of its mean is no larger than $1/k^2$. By the same token,

$$P(|X - m_X| < k\sigma_X) > 1 - 1/k^2 \tag{7b}$$

With $k = 2$, for instance, we expect X to occur within the range $m_X \pm 2\sigma_X$ for more than $3/4$ of the observations. A *small standard deviation* therefore implies a *small spread* of likely values, and vice versa.

The proof of Eq. (7a) starts by taking $Z = X - m_X$ and $a = k\sigma_X > 0$. We then let ϵ be a small positive quantity and note that

$$E[Z^2] = \int_{-\infty}^{\infty} z^2 p_Z(z) \, dz \geq \int_{-\infty}^{-a} z^2 p_Z(z) \, dz + \int_{a-\epsilon}^{\infty} z^2 p_Z(z) \, dz$$

But $z^2 \geq a^2$ over the range $a \leq |z| < \infty$, so

$$E[Z^2] \geq a^2 \left[\int_{-\infty}^{-a} p_Z(z) \, dz + \int_{a-\epsilon}^{\infty} p_Z(z) \, dz \right]$$

where the first integral inside the brackets represents $P(Z \leq -a)$ whereas the second represents $P(Z \geq a)$. Therefore,

$$P(|Z| \geq a) = P(Z \leq -a) + P(Z \geq a) \leq E[Z^2]/a^2$$

and Eq. (7a) follows by inserting $Z = X - m_X$, $E[Z^2] = \sigma_X^2$, and $a = k\sigma_X$.

To illustrate the calculation of statistical averages, let's take the case where

EXAMPLE 8.3–1

$$p_X(x) = \frac{a}{2} e^{-a|x|} \qquad -\infty < x < \infty$$

with a being a positive constant. This PDF describes a continuous RV with a **Laplace distribution.**

Drawing upon the even symmetry of $p_X(x)$, Eqs. (2) and (3) yield

$$m_X = \int_{-\infty}^{\infty} x \frac{a}{2} e^{-a|x|} \, dx = 0 \qquad E[X^2] = 2 \int_{0}^{\infty} x^2 \frac{a}{2} e^{-ax} \, dx = \frac{2}{a^2}$$

Hence, from Eq. (6), $\sigma_X = \sqrt{E[X^2] - m_X^2} = \sqrt{2}/a$.

The probability that an observed value of a Laplacian RV falls within $\pm 2\sigma_X$ of the mean is given by

$$P(|X - 0| < 2\sqrt{2}/a) = \int_{-2\sqrt{2}/a}^{2\sqrt{2}/a} \frac{a}{2} e^{-a|x|} \, dx = 0.94$$

as compared with the lower bound of 0.75 from Chebyshev's inequality.

EXERCISE 8.3–1

Let X have a uniform distribution over $0 < X \leq 2\pi$, as in Example 8.2–2 (p. 326). Calculate m_X, $\overline{X^2}$, and σ_X. What's the probability of $|X - m_X| < 2\sigma_X$?

Multivariate Expectations

Multivariate expectations involve two or more RVs, and they are calculated using multiple integration over joint PDFs. Specifically, when $g(X,Y)$ defines a function of X and Y, its expected value is

$$E[g(X, Y)] \triangleq \iint\limits_{-\infty}^{\infty} g(x, y)p_{XY}(x, y) \, dx \, dy \qquad [8]$$

However, we'll restrict our attention to those cases in which the multiple integration reduces to separate integrals.

First, suppose that X and Y are *independent*, so $p_{XY}(x,y) = p_X(x)p_Y(y)$. Assume further that we can write $g(X,Y)$ as a *product* in the form $g(X,Y) = g_X(X)g_Y(Y)$. Equation (8) thereby becomes

$$E[g(X, Y)] = \iint\limits_{-\infty}^{\infty} g_X(x)g_Y(y)p_X(x)p_Y(y) \, dx \, dy \qquad [9]$$

$$= \int_{-\infty}^{\infty} g_X(x)p_X(x) \, dx \int_{-\infty}^{\infty} g_Y(y)p_Y(y) \, dy$$

$$= E[g_X(X)]E[g_Y(Y)]$$

If we take $g(X,Y) = XY$, for instance, then $g_X(X) = X$ and $g_Y(Y) = Y$ so $E[XY] = E[X]E[Y]$ or

$$\overline{XY} = \overline{X}\,\overline{Y} = m_X m_Y \qquad [10]$$

Hence, the mean of the product of independent RVs equals the product of their means.

Next, consider the sum $g(X,Y) = X + Y$, where X and Y are not necessarily independent. Routine manipulation of Eq. (8) now leads to $E[X + Y] = E[X] + E[Y]$ or

$$\overline{X + Y} = \overline{X} + \overline{Y} = m_X + m_Y \qquad [11]$$

Hence, the mean of the sum equals the sum of the means, irrespective of statistical independence.

Finally, let $Z = X + Y$ so we know that $m_Z = m_X + m_Y$. But what's the variance σ_Z^2? To answer that question we calculate the mean-square value $E[Z^2]$ via

$$\overline{Z^2} = E[X^2 + 2XY + Y^2] = \overline{X^2} + 2\overline{XY} + \overline{Y^2}$$

Thus,

$$\sigma_Z^2 = \overline{Z^2} - (m_X + m_Y)^2 = (\overline{X^2} - m_X^2) + (\overline{Y^2} - m_Y^2) + 2(\overline{XY} - m_X m_Y)$$

The last term of this result vanishes when Eq. (10) holds, so the variance of the sum of independent RVs is

$$\sigma_Z^2 = \sigma_X^2 + \sigma_Y^2 \tag{12}$$

Equations (9)–(12) readily generalize to include three or more RVs. Keep in mind, however, that only Eq. (11) remains valid when the RVs are not independent.

Sample Mean and Frequency of Occurrence   EXAMPLE 8.3–2

Let $X_1, X_2, \ldots, X_N$ be *sample values* obtained from N independent observations of a random variable X having mean m_X and variance σ_X^2. Each sample value is an RV, and so is the sum

$$Z = X_1 + X_2 + \ldots + X_N$$

and the **sample mean**

$$\mu = Z/N$$

We'll investigate the statistical properties of μ, and we'll use them to reexamine the meaning of probability.

From Eqs. (11) and (12) we have $\bar{Z} = \bar{X}_1 + \bar{X}_2 + \cdots + \bar{X}_N = Nm_X$ and $\sigma_Z^2 = N\sigma_X^2$. Thus, $\bar{\mu} = \bar{Z}/N = m_X$, whereas

$$\sigma_\mu^2 = E[(\mu - \bar{\mu})^2] = \frac{1}{N^2} E[(Z - \bar{Z})^2] = \frac{1}{N^2} \sigma_Z^2 = \frac{1}{N} \sigma_X^2$$

Since $\sigma_\mu = \sigma_X/\sqrt{N}$, the spread of the sample mean decreases with increasing N, and μ approaches m_X as $N \to \infty$. Furthermore, from Chebyshev's inequality, the probability that μ differs from m_X by more than some positive amount ϵ is upper-bounded by

$$P(|\mu - m_X| \geq \epsilon) \leq \sigma_X^2/N\epsilon^2$$

Although not immediately obvious, this result provides further justification for the relative-frequency interpretation of probability.

To develop that point, let A be a chance-experiment event and let X be a discrete RV defined such that $X = 1$ when A occurs on a given trial and $X = 0$ when A does not occur. If A occurs N_A times in N independent trials, then $Z = N_A$ and $\mu = N_A/N$. Thus, our definition of X makes the sample mean μ equal to the *frequency of occurrence* of A for that set of trials. Furthermore, since $P(A) = P(X = 1) = P_X(1)$ and $P(A^c) = P_X(0)$, the statistical averages of X are

$$m_X = 0 \times P_X(0) + 1 \times P_X(1) = P(A)$$
$$E[X^2] = 0^2 \times P_X(0) + 1^2 \times P_X(1) = P(A)$$
$$\sigma_X^2 = P(A) - P^2(A)$$

so

$$P(|\mu - m_X| \geq \epsilon) = P\left[\left|\frac{N_A}{N} - P(A)\right| \geq \epsilon\right] \leq \frac{P(A) - P^2(A)}{N\epsilon^2}$$

We therefore conclude that, as $N \to \infty$, N_A/N must approach $P(A)$ in the sense that the probability of a significant difference between N_A/N and $P(A)$ becomes negligibly small.

EXERCISE 8.3–2   Prove Eq. (11) using marginal PDFs as defined by Eq. (18), Sect. 8.2 (p. 330), for $p_X(x)$ and $p_Y(y)$.

Characteristic Functions ★

Having found the mean and variance of a sum $Z = X + Y$, we'll now investigate the PDF of Z and its relation to $p_X(x)$ and $p_Y(y)$ when X and Y are independent. This investigation is appropriate here because the best approach is an indirect one using a special type of expectation.

The **characteristic function** of an RV X is an expectation involving an auxiliary variable ν defined by

$$\Phi_X(\nu) \triangleq E[e^{j\nu X}] = \int_{-\infty}^{\infty} e^{j\nu x} p_X(x)\, dx \qquad [13]$$

Upon closer inspection of Eq. (13), the presence of the complex exponential $e^{j\nu X}$ suggests similarity to a *Fourier integral*. We bring this out explicitly by letting $\nu = 2\pi t$ and $x = f$ so that

$$\Phi_X(2\pi t) = \int_{-\infty}^{\infty} p_X(f) e^{j2\pi ft}\, df = \mathscr{F}^{-1}[p_X(f)] \qquad [14a]$$

Consequently, by the Fourier integral theorem,

$$p_X(f) = \mathscr{F}[\Phi_X(2\pi t)] = \int_{-\infty}^{\infty} \Phi_X(2\pi t) e^{-j2\pi ft}\, dt \qquad [14b]$$

Hence, the characteristic function and PDF of a random variable constitute the *Fourier-transform pair* $\Phi_X(2\pi t) \leftrightarrow p_X(f)$.

Now, for the PDF of a sum of independent RVs, we let $Z = X + Y$ and we use Eqs. (13) and (9) to write

$$\Phi_Z(\nu) = E[e^{j\nu(X+Y)}] = E[e^{j\nu X} e^{j\nu Y}] = E[e^{j\nu X}]E[e^{j\nu Y}] = \Phi_X(\nu)\Phi_Y(\nu)$$

Then, from the convolution theorem,

$$p_Z(f) = \mathscr{F}[\Phi_X(2\pi t)\Phi_Y(2\pi t)] = p_X(f) * p_Y(f)$$

Appropriate change of variables yields the final result

$$p_Z(z) = \int_{-\infty}^{\infty} p_X(z - \lambda)p_Y(\lambda)\, d\lambda = \int_{-\infty}^{\infty} p_X(\lambda)p_Y(z - \lambda)\, d\lambda \qquad \text{[15]}$$

Thus, the PDF of $X + Y$ equals the *convolution* of $p_X(x)$ and $p_Y(y)$ when X and Y are independent.

Other applications of characteristic functions are explored in problems at the end of the chapter.

Use Eq. (14a) to find $\Phi_X(\nu)$ for the uniform PDF $p_X(x) = a^{-1}\,\Pi(x/a)$.   **EXERCISE 8.3–3**

8.4   PROBABILITY MODELS

Many probability functions have been devised and studied as models for various random phenomena. Here we discuss the properties of two discrete functions (binomial and Poisson) and two continuous functions (gaussian and Rayleigh). These models, together with the uniform and Laplace distributions, cover most of the cases encountered in our latter work. Table T.5 at the back of the book summarizes our results and includes a few other probability functions for reference purposes.

Binomial Distribution

The binomial model describes an integer-valued discrete RV associated with repeated trials. Specifically,

> A **binomial random variable** I corresponds to the number of times an event with probability α occurs in n independent trials of a chance experiment.

This model thus applies to repeated coin tossing when I stands for the number of heads in n tosses and $P(H) = \alpha$. But, more significantly for us, it also applies to digital transmission when I stands for the number of errors in an n-digit message with per-digit error probability α.

To formulate the binomial frequency function $P_I(i) = P(I = i)$, consider any sequence of n independent trials in which event A occurs i times. If $P(A) = \alpha$, then $P(A^C) = 1 - \alpha$ and the sequence probability equals $\alpha^i(1 - \alpha)^{n-i}$. The number of different sequences with i occurrences is given by the binomial coefficient, denoted $\binom{n}{i}$, so we have

$$P_I(i) = \binom{n}{i} \alpha^i (1 - \alpha)^{n-i} \qquad i = 0, 1 \ldots, n \qquad \text{[1]}$$

The corresponding CDF is

$$F_I(k) = \sum_{i=0}^{k} P_I(i) \qquad k = 0, 1 \ldots, n$$

These functions were previously evaluated in Fig. 8.2–3 for the case of $n = 3$ and $\alpha = 2/5$.

The **binomial coefficient** in Eq. (1) equals the coefficient of the $(i + 1)$ term in the expansion of $(a + b)^n$, defined in general by the factorial expression

$$\binom{n}{i} \triangleq \frac{n!}{i!(n - i)!} \qquad \text{[2]}$$

where it's understood that $0! = 1$ when $i = 0$ or $i = n$. This quantity has the symmetry property

$$\binom{n}{i} = \binom{n}{n - i}$$

We thus see that

$$\binom{n}{0} = \binom{n}{n} = 1 \qquad \binom{n}{1} = \binom{n}{n-1} = n \qquad \binom{n}{2} = \binom{n}{n-2} = \frac{n(n - 1)}{2}$$

and so on. Other values can be found using *Pascal's triangle,* tables, or a calculator or computer with provision for factorials.

The statistical averages of a binomial RV are obtained by inserting Eq. (1) into the appropriate discrete expectation formulas. Some rather laborious algebra then yields simple results for the mean and variance, namely,

$$m = n\alpha \qquad \sigma^2 = n\alpha(1 - \alpha) = m(1 - \alpha) \qquad \text{[3]}$$

where we've omitted the subscript I for simplicity. The *relative spread* σ/m decreases as $1/\sqrt{n}$, meaning that the likely values of I cluster around m when n is large.

EXAMPLE 8.4–1   Suppose 10,000 digits are transmitted over a noisy channel with per-digit error probability $\alpha = 0.01$. Equation (3) then gives

$$m = 10,000 \times 0.01 = 100 \qquad \sigma^2 = 100(1 - 0.01) = 99$$

Hence, the likely range $m \pm 2\sigma$ tells us to expect about 80 to 120 errors.

Poisson Distribution

The Poisson model describes another integer-valued RV associated with repeated trials, in that

A **Poisson random variable** I corresponds to the number of times an event occurs in an interval T when the probability of a single occurrence in the small interval ΔT is $\mu \Delta T$.

The resulting Poisson frequency function is

$$P_I(i) = e^{-\mu T} \frac{(\mu T)^i}{i!} \qquad [4]$$

from which

$$m = \mu T \qquad \sigma^2 = m$$

These expressions describe random phenomena such as *radioactive decay* and *shot noise* in electronic devices.

The Poisson model also approximates the *binomial* model when n is very large, α is very small, and the product $n\alpha$ remains finite. Equation (1) becomes awkward to handle in this case, but we can let $\mu T = m$ in Eq. (4) to obtain the more convenient approximation

$$P_I(i) \approx e^{-m} \frac{m^i}{i!} \qquad [5]$$

Neither n nor α appears here since they have been absorbed in the mean value $m = n\alpha$.

Use Eq. (5) to estimate the probability of $I \leq 2$ errors when 10,000 digits are transmitted over a noisy channel having error probability $\alpha = 5 \times 10^{-5}$.

EXERCISE 8.4–1

Gaussian PDF

The gaussian model describes a continuous RV having the *normal distribution* encountered in many different applications in engineering, physics, and statistics. The remarkable versatility of this model stems from the famous **central limit theorem,** which states in essence that

If X represents the sum of N independent random components, and if each component makes only a small contribution to the sum, then the CDF of X approaches a gaussian CDF as N becomes large, regardless of the distribution of the individual components.

A more complete statement of the theorem and its proof involves sophisticated mathematical arguments and won't be pursued here. However, we'll draw upon the implication that the gaussian model often holds when the quantity of interest results from the summation of many small fluctuating components. Thus, for instance, random *measurement errors* usually cause experimental values to have a gaussian distribution around the true value. Similarly, the random motion of thermally agitated electrons produces the gaussian random variable known as *thermal noise.*

A **gaussian RV** is a continuous random variable X with mean m, variance σ^2, and PDF

$$p_X(x) = \frac{1}{\sqrt{2\pi\sigma^2}} e^{-(x-m)^2/2\sigma^2} \qquad -\infty < x < \infty \qquad [6]$$

This function defines the bell-shaped curve plotted in Fig. 8.4–1. The even symmetry about the peak at $x = m$ indicates that

$$P(X \le m) = P(X > m) = \tfrac{1}{2}$$

so observed values of X are just as likely to fall above or below the mean.

Now assume that you know the mean m and variance σ^2 of a gaussian RV and you want to find the probability of the event $X > m + k\sigma$. Since the integral in question cannot be evaluated in closed form, numerical methods have been used to generate extensive tables of the normalized integral

$$Q(k) \triangleq \frac{1}{\sqrt{2\pi}} \int_k^\infty e^{-\lambda^2/2} \, d\lambda \qquad [7]$$

The change of variable $\lambda = (x - m)/\sigma$ then shows that

$$P(X > m + k\sigma) = Q(k)$$

We therefore call $Q(k)$ the **area under the gaussian tail,** as illustrated by Fig. 8.4–2. This figure also brings out the fact that $P(m < X \le m + k\sigma) = \tfrac{1}{2} - Q(k)$, which follows from the symmetry and unit-area properties of $p_X(x)$.

You can calculate any desired gaussian probability in terms of $Q(k)$ using Fig. 8.4–2 and the symmetry of the PDF. In particular,

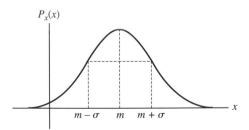

Figure 8.4–1   Gaussian PDF.

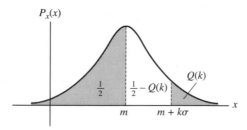

Figure 8.4-2   Area interpretation of $Q(k)$.

$$P(X > m + k\sigma) = P(X \le m - k\sigma) = Q(k) \qquad \text{[8a]}$$

$$P(m < X \le m + k\sigma) = P(m - k\sigma < X \le m) = \tfrac{1}{2} - Q(k) \qquad \text{[8b]}$$

$$P(|X - m| > k\sigma) = 2Q(k) \qquad \text{[8c]}$$

$$P(|X - m| \le k\sigma) = 1 - 2Q(k) \qquad \text{[8d]}$$

Table 8.4–1 compares some values of this last quantity with the lower bound $1 - 1/k^2$ from Chebyshev's inequality. The lower bound is clearly conservative, and the likely range of observed values is somewhat less than $m \pm 2\sigma$. We usually take the likely range of a gaussian RV to be $m \pm \sigma$ since $P(|X - m| \le \sigma) \approx 0.68$.

Table 8.4-1

k	$1 - 2Q(k)$	$1 - 1/k^2$
0.5	0.38	
1.0	0.68	0.00
1.5	0.87	0.56
2.0	0.95	0.75
2.5	0.99	0.84

For larger values of k, the area under the gaussian tail becomes too small for numerical integration. But we can then use the analytical approximation

$$Q(k) \approx \frac{1}{\sqrt{2\pi k^2}} e^{-k^2/2} \qquad k > 3 \qquad \text{[9]}$$

This approximation follows from Eq. (7) by integration by parts.

Table T.6 at the back of the book gives a detailed plot of $Q(k)$ for $0 \le k \le 7$. Also given are relationships between $Q(k)$ and other gaussian probability functions found in the literature.

EXAMPLE 8.4–2	Suppose you want to evaluate the probability of $|X - m| \leq 3$ when X is gaussian with $\sigma^2 = 4$. You can use Eq. (8d) and Table 8.4–1 by noting that $\sigma = 2$ and $k\sigma = 3$ when $k = 3/\sigma = 1.5$. Hence,

$$P(|X - m| \leq 3) = P(|X - m| \leq 1.5\sigma) = 1 - 2Q(1.5) = 0.87$$ |

EXERCISE 8.4–2	Given that X is gaussian with $m = 5$ and $\sigma = 8$, sketch the PDF and mark the boundaries of the area in question to show that $P(9 < X \leq 25) = Q(0.5) - Q(2.5) \approx 0.3$.

Rayleigh PDF

The Rayleigh model describes a continuous RV obtained from *two* gaussian RVs as follows:

> If X and Y are independent gaussian RVs with zero mean and the same variance σ^2, then the random variable defined by $R = +\sqrt{X^2 + Y^2}$ has a **Rayleigh distribution.**

Thus, as shown in Fig. 8.4–3, the Rayleigh model applies to any *rectangular-to-polar conversion* when the rectangular coordinates are identical but independent gaussians with zero mean.

To derive the corresponding Rayleigh PDF, we introduce the random angle Φ from Fig. 8.4–3 and start with the joint PDF relationship

$$p_{R\Phi}(r,\varphi)|dr \, d\varphi| = p_{XY}(x, y)|dx \, dy|$$

where

$$r^2 = x^2 + y^2 \qquad \varphi = \arctan(y/x) \qquad dx \, dy = r \, dr \, d\varphi$$

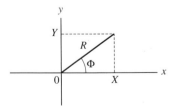

Figure 8.4–3   Rectangular to polar conversion.

Since X and Y are independent gaussians with $m = 0$ and variance σ^2,

$$p_{XY}(x, y) = p_X(x)p_Y(y) = \frac{1}{2\pi\sigma^2} e^{-(x^2+y^2)/2\sigma^2}$$

Hence, including $u(r)$ to reflect the fact that $r \geq 0$, we have

$$p_{R\Phi}(r, \varphi) = \frac{r}{2\pi\sigma^2} e^{-r^2/2\sigma^2} u(r) \qquad [10]$$

The angle φ does not appear explicitly here, but its range is clearly limited to 2π radians.

We now obtain the PDF for R alone by integrating Eq. (10) with respect to φ. Taking either $0 < \varphi \leq 2\pi$ or $-\pi < \varphi \leq \pi$ yields the **Rayleigh PDF**

$$p_R(r) = \frac{r}{\sigma^2} e^{-r^2/2\sigma^2} u(r) \qquad [11]$$

which is plotted in Fig. 8.4–4. The resulting mean and second moment of R are

$$\bar{R} = \sqrt{\tfrac{\pi}{2}}\, \sigma \qquad \overline{R^2} = 2\sigma^2 \qquad [12]$$

For probability calculations, the Rayleigh CDF takes the simple form

$$F_R(r) = P(R \leq r) = \left(1 - e^{-r^2/2\sigma^2}\right)u(r) \qquad [13]$$

derived by integrating $p_R(\lambda)$ over $0 \leq \lambda \leq r$.

Returning to Eq. (10), we get the marginal PDF for the random angle Φ via

$$p_\Phi(\varphi) = \int_0^\infty p_{R\Phi}(r, \varphi)\, dr = \frac{1}{2\pi}$$

so Φ has a *uniform distribution* over 2π radians. We also note that $p_{R\Phi}(r,\varphi) = p_R(r)p_\Phi(\varphi)$, which means that the polar coordinates R and Φ are *statistically independent*. These results will be of use in Chapter 10 for the representation of *bandpass noise*.

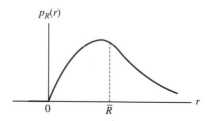

Figure 8.4–4   Rayleigh PDF.

EXAMPLE 8.4–3   Suppose you throw darts at a target whose bull's eye has a radius of 3 cm. If the rectangular coordinates of the impact points are gaussianly distributed about the origin with spread $\sigma = 4$ cm in either direction, then your probability of hitting the bull's eye is given by Eq. (13) as

$$P(R \leq 3) = 1 - e^{-9/32} \approx 25\%$$

EXERCISE 8.4–3   Derive Eq. (13) from Eq. (11).

Bivariate Gaussian Distribution ★

Lastly, we want to investigate the joint PDF of two gaussian RVs that are neither identically distributed nor independent. As preparation, we first introduce a general measure of interdependence between any two random variables.

Let X and Y be arbitrarily distributed with respective means and variances m_X, m_Y, σ_X^2, and σ_Y^2. The degree of dependence between them is expressed by the **correlation coefficient**

$$\rho \triangleq \frac{1}{\sigma_X \sigma_Y} \; E[(X - m_X)(Y - m_Y)] \qquad [14]$$

where the expectation $E[(X - m_X)(Y - m_Y)]$ is called the **covariance** of X and Y. At one extreme, if X and Y are statistically independent, then $E[(X - m_X)(Y - m_Y)] = E[(X - m_X)]E[(Y - m_Y)] = 0$ so the covariance equals zero and $\rho = 0$. At the other extreme, if Y depends entirely upon X in the sense that $Y = \pm \alpha X$, then $\sigma_Y^2 = (\alpha \sigma_X)^2$ and the covariance equals $\pm \alpha \sigma_X^2$ so $\rho = \pm 1$. Thus, the correlation coefficient ranges over $-1 \leq \rho \leq 1$, and $|\rho|$ reflects the degree of interdependence.

When X and Y are interdependent *gaussian* RVs, their joint PDF is given by the **bivariate gaussian model**

$$p_{XY}(x, y) = \frac{1}{2\pi \sigma_X \sigma_Y \sqrt{1 - \rho^2}} \; e^{-f(x,\, y)/(1 - \rho^2)} \qquad [15a]$$

with

$$f(x, y) = \frac{(x - m_X)^2}{2\sigma_X^2} + \frac{(y - m_Y)^2}{2\sigma_Y^2} - \frac{(x - m_X)(y - m_Y)\rho}{\sigma_X \sigma_Y} \qquad [15b]$$

This formidable-looking expression corresponds to a bell-shaped surface over the x-y plane, the peak being at $x = m_X$ and $y = m_Y$.

If $\rho = 0$, then the last term of Eq. (15b) disappears and $p_{XY}(x,y) = p_X(x)p_Y(y)$. We thus conclude that

●   Uncorrelated gaussian RVs are statistically independent.

Further study leads to the additional conclusions that

- The marginal and conditional PDFs derived from the bivariate gaussian PDF with any ρ are all gaussian functions.
- Any linear combination $Z = \alpha X + \beta Y$ will also be a gaussian RV.

These three properties are unique characteristics of the bivariate gaussian case, and they do not necessarily hold for other distributions.

All of the foregoing analysis can be extended to the case of three or more jointly gaussian RVs. In such cases, matrix notation is usually employed for compact PDF expressions.

8.5   PROBLEMS

8.1–1*   The outcome of an experiment is an integer I whose value is equally likely to be any integer in the range $1 \le I \le 12$. Let A be the event that I is odd, let B be the event that I is exactly divisible by 3, and let C be the event that I is exactly divisible by 4. Draw the Venn diagram and find the probabilities of the events A, B, C, AB, AC, BC, and $A^C B$.

8.1–2   The outcome of an experiment is an integer I whose value is equally likely to be any integer in the range $1 \le I \le 4$. The experiment is performed twice, yielding the outcomes I_1 and I_2. Let A be the event that $I_1 = I_2$, let B be the event that $I_1 > I_2$, and let C be the event that $I_1 + I_2 \ge 6$. Draw the Venn diagram and find the probabilities of the events A, B, C, AB, AC, BC, and $A^C B$.

8.1–3   If A and B are not mutually exclusive events, then the number of times A occurs in N trials can be written as $N_A = N_{AB} + N_{AB^C}$, where N_{AB^C} stands for the number of times A occurs without B. Use this notation to show that $P(AB^C) = P(A) - P(AB)$.

8.1–4   Use the notation in Prob. 8.1–3 to justify Eq. (7), (p. 316).

8.1–5*   Let C stand for the event that either A or B occurs but not both. Use the notation in Prob. 8.1–3 to express $P(C)$ in terms of $P(A)$, $P(B)$, and $P(AB)$.

8.1–6   A biased coin is loaded such that $P(H) = (1 + \epsilon)/2$ with $0 < |\epsilon| < 1$. Show that the probability of a match in two independent tosses will be greater than 1/2.

8.1–7   A certain computer becomes inoperable if two components C_A and C_B both fail. The probability that C_A fails is 0.01 and the probability that C_B fails is 0.005. However, the probability that C_B fails is increased by a factor of 4 if C_A has failed. Calculate the probability that the computer becomes inoperable. Also find the probability that C_A fails if C_B has failed.

8.1–8   An honest coin is tossed twice and you are given partial information about the outcome. (*a*) Use Eq. (8) (p. 317) to find the probability of a match when you are told that the first toss came up heads. (*b*) Use Eq. (8) (p. 317) to find the probability of a match when you are told that heads came up on at least one toss. (*c*) Use Eq. (10)

(p. 318) to find the probability of heads on at least one toss when you are told that a match has occurred.

8.1–9   Do Prob. 8.1–8 for a loaded coin having $P(H) = 1/4$.

8.1–10   Derive from Eq. (9) (p. 318) the **chain rule**

$$P(XYZ) = P(X)P(Y|X)P(Z|XY)$$

8.1–11* A box contains 10 fair coins with $P(H) = 1/2$ and 20 loaded coins with $P(H) = 1/4$. A coin is drawn at random from the box and tossed twice. (a) Use Eq. (11) (p. 318) to find the probability of the event "all tails." Let the conditioning events be the honesty of the coin. (b) If the event "all tails" occurs, what's the probability that the coin was loaded?

8.1–12   Do Prob. 8.1–11 for the case when the withdrawn coin is tossed three times.

8.1–13   Two marbles are randomly withdrawn without replacement from a bag initially containing 5 red marbles, 3 white marbles, and 2 green marbles. (a) Use Eq. (11) (p. 318) to find the probability that the withdrawn marbles have matching colors. Let the conditioning events be the color of the first marble withdrawn. (b) If the withdrawn marbles have matching colors, what's the probability that they are white?

8.1–14   Do Prob. 8.1–13 for the case when three marbles are withdrawn from the bag.

8.2–1*   Let $X = \frac{1}{2}N^2$, where N is a random integer whose value is equally likely to be any integer in the range $-1 \leq N \leq 3$. Plot the CDF of X and use it to evaluate the probabilities of: $X \leq 0$, $2 < X \leq 3$, $X < 2$, and $X \geq 2$.

8.2–2   Do Prob. 8.2–1 with $X = 4 \cos \pi N/3$.

8.2–3   Let $p_X(x) = xe^{-x}u(x)$. Find $F_X(x)$ and use it to evaluate $P(X \leq 1)$, $P(1 < X \leq 2)$, and $P(X > 2)$.

8.2–4   Let $p_X(x) = \frac{1}{2} e^{-|x|}$. Find $F_X(x)$ and use it to evaluate $P(X \leq 0)$, $P(0 < X \leq 1)$, and $P(X > 1)$.

8.2–5*   Suppose a certain random variable has the CDF

$$F_X(x) = \begin{cases} 0 & x \leq 0 \\ Kx^2 & 0 < x \leq 10 \\ 100K & x > 10 \end{cases}$$

Evaluate K, write the corresponding PDF, and find the values of $P(X \leq 5)$ and $P(5 < X \leq 7)$.

8.2–6   Do Prob. 8.2–5 with

$$F_X(x) = \begin{cases} 0 & x \leq 0 \\ K \sin \pi x/40 & 0 < x \leq 10 \\ K \sin \pi/4 & x > 10 \end{cases}$$

8.2–7   Given that $F_X(x) = (\pi + 2 \tan^{-1} x)/2\pi$, find the CDF and PDF of the random variable Z defined by $Z = 0$ for $X \le 0$ and $Z = X$ for $X > 0$.

8.2–8‡   Do Prob. 8.2–7 with $Z = -1$ for $X \le 0$ and $Z = X$ for $X > 0$.

8.2–9*   Let $p_X(x) = 2e^{-2x}u(x)$. Find the PDF of the random variable defined by the transformation $Z = 2X - 5$. Then sketch both PDFs on the same set of axes.

8.2–10   Do Prob. 8.2–9 with $Z = -2X + 1$.

8.2–11   Let X have a uniform PDF over $-1 \le x \le 3$. Find and sketch the PDF of Z defined by the transformation $Z = \sqrt{X} + 1$.

8.2–12   Do Prob. 8.2–11 with $Z = |X|$.

8.2–13‡   Do Prob. 8.2–11 with $Z = \sqrt{|X|}$.

8.2–14   Consider the **square-law transformation** $Z = X^2$. Show that

$$p_Z(z) = \frac{1}{2\sqrt{z}} [p_X(\sqrt{z}) + p_X(-\sqrt{z})]u(z)$$

8.2–15*   Find $p_Y(y)$ when $p_{XY}(x,y) = ye^{-y(x+1)}u(x)u(y)$. Then show that X and Y are not statistically independent, and find $p_X(x|y)$.

8.2–16   Do Prob. 8.2–15 with $p_{XY}(x,y) = [(x + y)^2/40]\Pi(x/2)\Pi(y/6)$.

8.2–17   Show that $\int_{-\infty}^{\infty} p_X(x|y)\, dx = 1$. Explain why this must be true.

8.2–18   Obtain an expression for $p_Y(y|x)$ in terms of $p_X(x|y)$ and $p_Y(y)$.

8.3–1*   Find the mean, second moment, and standard deviation of X when $p_X(x) = ae^{-ax}u(x)$ with $a > 0$.

8.3–2   Find the mean, second moment, and standard deviation of X when $p_X(x) = a^2xe^{-ax}u(x)$ with $a > 0$.

8.3–3   Find the mean, second moment, and standard deviation of X when

$$p_X(x) = \frac{\sqrt{2}}{\pi[1 + (x - a)^4]}$$

8.3–4   A discrete RV has two possible values, a and b. Find the mean, second moment, and standard deviation in terms of $p = P(X = a)$.

8.3–5*   A discrete RV has K equally likely possible values, $0, a, 2a, \ldots, (K - 1)a$. Find the mean, second moment, and standard deviation.

8.3–6   Find the mean, second moment, and standard deviation of $Y = a \cos X$, where a is a constant and X has a uniform PDF over $\theta \le x \le \theta + 2\pi$.

8.3–7   Do Prob. 8.3–6 for the case when X has a uniform PDF over $\theta \le x \le \theta + \pi$.

8.3–8   Let $Y = \alpha X + \beta$. Show that $\sigma_Y = |\alpha|\sigma_X$.

8.3–9*   Let $Y = X + \beta$. What value of β minimizes $E[Y^2]$?

8.3–10‡ Let X be a nonnegative continuous RV and let a be any positive constant. By considering $E[X]$, derive **Markov's inequality** $P(X \geq a) \leq m_X/a$.

8.3–11 Use $E[(X \leq Y)^2]$ to obtain upper and lower bounds on $E[XY]$ when X and Y are not statistically independent.

8.3–12 The **covariance** of X and Y is defined as $C_{XY} = E[(X - m_X)(Y - m_Y)]$. Expand this joint expectation and simplify it for the case when: (*a*) X and Y are statistically independent; (*b*) Y is related to X by $Y = \alpha X + \beta$.

8.3–13 In **linear estimation** we estimate Y from X by writing $\tilde{Y} = \alpha X + \beta$. Obtain expressions for α and β to minimize the mean square error $\epsilon^2 = E[(Y - \tilde{Y})^2]$.

8.3–14 Show that the nth moment of X can be found from its characteristic function via

$$E[X^n] = j^{-n} \left. \frac{d^n \Phi_X(\nu)}{d\nu^n} \right|_{\nu = 0}$$

8.3–15* Obtain the characteristic function of X when $p_X(x) = ae^{-ax}u(x)$ with $a > 0$. Then use the relation in Prob. 8.3–14 to find the first three moments.

8.3–16 Let X have a known PDF and let $Y = g(X)$, so

$$\Phi_Y(\nu) = E[e^{j\nu g(X)}] = \int_{-\infty}^{\infty} e^{j\nu g(x)} p_X(x) \, dx$$

If this integral can be rewritten in the form

$$\Phi_Y(\nu) = \int_{-\infty}^{\infty} e^{j\nu\lambda} h(\lambda) \, d\lambda$$

then $p_Y(y) = h(y)$. Use this method to obtain the PDF of $Y = X^2$ when $p_X(x) = 2axe^{-ax^2}u(x)$.

8.3–17‡ Use the method in Prob. 8.3–16 to obtain the PDF of $Y = \sin X$ when X has a uniform PDF over $|x| \leq \pi/2$.

8.4–1* Ten honest coins are tossed. What's the likely range of the number of heads? What's the probability that there will be fewer than three heads?

8.4–2 Do Prob. 8.4–1 with biased coins having $P(H) = 3/5$.

8.4–3 The **one-dimensional random walk** can be described as follows. An inebriated man walking in a narrow hallway takes steps of equal length l. He steps forward with probability $\alpha = \frac{3}{4}$ or backwards with probability $1 - \alpha = \frac{1}{4}$. Let X be his distance from the starting point after 100 steps. Find the mean and standard deviation of X.

8.4–4 A noisy transmission channel has per-digit error probability $\alpha = 0.01$. Calculate the probability of more than one error in 10 received digits. Repeat this calculation using the Poisson approximation.

8.4–5* A radioactive source emits particles at the average rate of 0.5 particles per second. Use the Poisson model to find the probability that: (*a*) exactly one particle is emitted in two seconds; (*b*) more than one particle is emitted in two seconds.

8.4–6‡ Show that the Poisson distribution in Eq. (5), (p. 339), yields $E[I] = m$ and $E[I^2] = m^2 + m$. The summations can be evaluated by writing the series expansion for e^m and differentiating it twice.

8.4–7 Observations of a noise voltage X are found to have a gaussian distribution with $m = 100$ and $\sigma = 2$. Evaluate $\overline{X^2}$ and the probability that X falls outside the range $m \pm \sigma$.

8.4–8 A gaussian RV has $\overline{X} = 2$ and $\overline{X^2} = 13$. Evaluate the probabilities of the events $X > 5$ and $2 < X \le 5$.

8.4–9* A gaussian RV has $E[X] = 10$ and $E[X^2] = 500$. Find $P(X > 20)$, $P(10 < X \le 20)$, $P(0 < X \le 20)$, and $P(X > 0)$.

8.4–10 When a binomial CDF has $n \gg 1$, it can be approximated by a gaussian CDF with the same mean and variance. Suppose an honest coin is tossed 100 times. Use the gaussian approximation to find the probability that: (*a*) heads occurs more than 70 times; (*b*) the number of heads is between 40 and 60.

8.4–11 Let X be a gaussian RV with mean m and variance σ^2. Write an expression in terms of the Q function for $P(a < X \le b)$ with $a < m < b$.

8.4–12 A random noise voltage X is known to be gaussian with $E[X] = 0$ and $E[X^2] = 9$. Find the value of c such that $|X| < c$ for: (*a*) 90 percent of the time; (*b*) 99 percent of the time.

8.4–13 Write $e^{-\lambda^2/2}\, d\lambda = -(1/\lambda)d(e^{-\lambda^2/2})$ to show that the approximation in Eq. (9), (p. 341), is an upper bound on $Q(k)$. Then justify the approximation for $k \gg 1$.

8.4–14‡ Let X be a gaussian RV with mean m and variance σ^2. (*a*) Show that $E[(X - m)^n] = 0$ for odd n. (*b*) For even n, use integration by parts to obtain the recursion relation

$$E[(X - m)^n] = (n - 1)\sigma^2\, E[(X - m)^{n-2}]$$

Then show for $n = 2, 4, 6, \ldots$ that

$$E[(X - m)^n] = 1 \cdot 3 \cdot 5 \cdots (n - 1)\sigma^n$$

8.4–15 Let X be a gaussian RV with mean m and variance σ^2. Show that its characteristic function is $\Phi_X(\nu) = e^{-\sigma^2\nu^2/2}e^{jm\nu}$.

8.4–16 Let $Z = X + Y$, where X and Y are independent gaussian RVs with different means and variances. Use the characteristic function in Prob. 8.4–15 to show that Z has a gaussian PDF. Then extrapolate your results for

$$Z = \frac{1}{n}\sum_{i=1}^{n} X_i$$

where the X_i are mutually independent gaussian RVs.

8.4–17*   A random variable Y is said to be **log-normal** if the transformation $X = \ln Y$ yields a gaussian RV. Use the gaussian characteristic function in Prob. 8.4–15 to obtain $E[Y]$ and $E[Y^2]$ in terms of m_X and σ_X^2. Do not find the PDF or characteristic function of Y.

8.4–18   Let $Z = X^2$, where X is a gaussian RV with zero mean and variance σ^2. (*a*) Use Eqs. (3) and (13), Sect. 8.3, to show that

$$\Phi_Z(\nu) = (1 - j2\sigma^2\nu)^{-1/2}$$

(*b*) Apply the method in Prob. 8.3–16 to find the first three moments of Z. What statistical properties of X are obtained from these results?

8.4–19   The resistance R of a resistor drawn randomly from a large batch is found to have a Rayleigh distribution with $\overline{R^2} = 32$. Write the PDF $p_R(r)$ and evaluate the probabilities of the events $R > 6$ and $4.5 < R \le 5.5$.

8.4–20   The noise voltage X at the output of a rectifier is found to have a Rayleigh distribution with $\overline{X^2} = 18$. Write the PDF $p_X(x)$ and evaluate $P(X < 3)$, $P(X > 4)$, and $P(3 < X \le 4)$.

8.4–21   Certain radio channels suffer from **Rayleigh fading** such that the received signal power is a random variable $Z = X^2$ and X has a Rayleigh distribution. Use Eq. (12), Sect. 8.2, to obtain the PDF

$$p_Z(z) = \frac{1}{m} e^{-z/m} u(z)$$

where $m = E[Z]$. Evaluate the probability $P(Z \le km)$ for $k = 1$ and $k = 0.1$.

8.4–22‡   Let R_1 and R_2 be independent Rayleigh RVs with $E[R_1^2] = E[R_2^2] = 2\sigma^2$. (*a*) Use the characteristic function from Prob. 8.4–18 to obtain the PDF of $A = R_1^2$. (*b*) Now apply Eq. (15), Sect. 8.3, to find the PDF of $W = R_1^2 + R_2^2$.

8.4–23   Let the bivariate gaussian PDF in Eq. (15), (p. 344), have $m_X = m_Y = 0$ and $\sigma_X = \sigma_Y = \sigma$. Show that $p_Y(y)$ and $p_X(x|y)$ are gaussian functions.

8.4–24*   Find the PDF of $Z = X + 3Y$ when X and Y are gaussian RVs with $m_X = 6$, $m_Y = -2$, $\sigma_X = \sigma_Y = 4$, and $E[XY] = -22$.

8.4–25   Let $X = Y^2$, so X and Y are not independent. Nevertheless, show that they are uncorrelated if the PDF of X has even symmetry.

Random Signals and Noise

CHAPTER OUTLINE

A ll meaningful communication signals are unpredictable or random as viewed from the receiving end. Otherwise, there would be little value in transmitting a signal whose behavior was completely known beforehand. Furthermore, all communication systems suffer to some degree from the adverse effects of electrical noise. The study of random signals and noise undertaken here is therefore essential for evaluating the performance of communication systems.

Sections 9.1 and 9.2 of this chapter combine concepts of signal analysis and probability to construct mathematical models of random electrical processes, notably random signals and noise. Don't be discouraged if the material seems rather theoretical and abstract, for we'll put our models to use in Sects. 9.3 through 9.5. Specifically, Sect. 9.3 is devoted to the descriptions of noise per se, while Sects. 9.4 and 9.5 examine signal transmission in the presence of noise. Most of the topics introduced here will be further developed and extended in later chapters of the text.

OBJECTIVES

After studying this chapter and working the exercises, you should be able to do each of the following:

1. Define the mean and autocorrelation function of a random process, and state the properties of a stationary or gaussian process (Sect. 9.1)
2. Relate the time and ensemble averages of a random signal from an ergodic process (Sect. 9.1).
3. Obtain the mean-square value, variance, and power spectrum of a stationary random signal, given its autocorrelation function (Sect. 9.2).
4. Find the power spectrum of a random signal produced by superposition, modulation, or filtering (Sect. 9.2).
5. Write the autocorrelation and spectral density of white noise, given the noise temperature (Sect. 9.3).
6. Calculate the noise bandwidth of a filter, and find the power spectrum and total output power with white noise at the input (Sect. 9.3).
7. State the conditions under which signal-to-noise ratio is meaningful (Sect. 9.4).
8. Analyze the performance of an analog baseband transmission system with noise (Sect. 9.4).
9. Find the optimum filter for pulse detection in white noise (Sect. 9.5).
10. Analyze the performance of a pulse transmission system with noise (Sect. 9.5).

9.1   RANDOM PROCESSES

A random signal is the manifestation of a *random electrical process* that takes place over time. Such processes are also called *stochastic processes*. When time enters the picture, the complete description of a random process becomes quite complicated—especially if the statistical properties change with time. But many of the random processes encountered in communication systems have the property of *stationarity* or even *ergodicity*, which leads to rather simple and intuitively meaningful relationships between statistical properties, time averages, and spectral analysis.

This section introduces the concepts and description of random process and briefly sets forth the conditions implied by stationarity and ergodicity.

Ensemble Averages and Correlation Functions

Previously we said that a random variable maps the outcomes of a chance experiment into numbers $X(s)$ along the real line. We now include time variation by saying that

> A **random process** maps experimental outcomes into real functions of time. The collection of time functions is known as an **ensemble,** and each member is called a **sample function.**

We'll represent ensembles formally by $v(t,s)$. When the process in question is electrical, the sample functions are **random signals.**

Consider, for example, the set of voltage waveforms generated by thermal electron motion in a large number of identical resistors. The underlying experiment might be: Pick a resistor at random and observe the waveform across its terminals. Figure 9.1–1 depicts some of the random signals from the ensemble $v(t,s)$ associated with this experiment. A particular outcome (or choice of resistor) corresponds to the sample function $v_i(t) = v(t,s_i)$ having the value $v_i(t_1) = v(t_1,s_i)$ at time t_1. If you know the experimental outcome then, in principle, you know the entire behavior of the sample function and all randomness disappears.

But the basic premise regarding random processes is that you don't know which sample function you're observing. So at time t_1, you could expect *any* value from the ensemble of possible values $v(t_1,s)$. In other words, $v(t_1,s)$ constitutes a *random variable,* say V_1, defined by a "vertical slice" through the ensemble at $t = t_1$, as illustrated in Fig. 9.1–1. Likewise, the vertical slice at t_2 defines another random variable V_2. Viewed in this light, *a random process boils down to a family of RVs.*

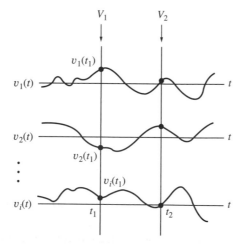

Figure 9.1–1   Waveforms in an ensemble $v(t,s)$.

Now let's omit s and represent the random process by $v(t)$, just as we did when we used X for a single random variable. The context will always make it clear when we're talking about random processes rather than nonrandom signals, so we'll not need a more formal notation. (Some authors employ boldface letters or use $V(t)$ for the random process and V_{t_i} for the random variables.) Our streamlined symbol $v(t)$ also agrees with the fact that we hardly ever know nor care about the details of the underlying experiment. What we do care about are the statistical properties of $v(t)$.

For a given random process, the **mean value** of $v(t)$ at arbitrary time t is defined as

$$\overline{v(t)} \triangleq E[v(t)] \tag{1}$$

Here, $E[v(t)]$ denotes an **ensemble average** obtained by averaging over all the sample functions with time t held fixed at an arbitrary value. Setting $t = t_1$ then yields $E[v(t_1)] = \overline{V}_1$, which may differ from $\overline{V}_2$.

To investigate the relationship between the RVs V_1 and V_2 we define the **autocorrelation function**

$$R_v(t_1, t_2) \triangleq E[v(t_1)v(t_2)] \tag{2}$$

where the lowercase subscript has been used to be consistent with our previous work in Chapter 3. This function measures the *relatedness* or dependence between V_1 and V_2. If they happen to be statistically independent, then $R_v(t_1, t_2) = \overline{V}_1\overline{V}_2$. However, if $t_2 = t_1$, then $V_2 = V_1$ and $R_v(t_1, t_2) = \overline{V_1^2}$. More generally, setting $t_2 = t_1 = t$ yields

$$R_v(t, t) = E[v^2(t)] = \overline{v^2(t)}$$

which is the mean-square value of $v(t)$ as a function of time.

Equations (1) and (2) can be written out explicitly when the process in question involves an ordinary random variable X in the functional form

$$v(t) = g(X, t)$$

Thus, at any time t_i, we have the RV transformation $V_i = g(X, t_i)$. Consequently, knowledge of the PDF of X allows you to calculate the ensemble average and the autocorrelation function via

$$\overline{v(t)} = E[g(X, t)] = \int_{-\infty}^{\infty} g(x, t)p_X(x)\, dx \tag{3a}$$

$$R_v(t_1, t_2) = E[g(X, t_1)g(X, t_2)] = \int_{-\infty}^{\infty} g(x, t_1)g(x, t_2)p_X(x)\, dx \tag{3b}$$

Equations (3a) and (3b) also generalize to the case of a random process defined in terms of two or more RVs. If $v(t) = g(X, Y, t)$, for instance, then Eq. (3b) becomes $R_v(t_1, t_2) = E[g(X, Y, t_1)g(X, Y, t_2)]$.

Occasionally we need to examine the joint statistics of two random processes, say $v(t)$ and $w(t)$. As an extension of Eq. (2), their relatedness is measured by the **crosscorrelation function**

$$R_{vw}(t_1, t_2) \triangleq E[v(t_1)w(t_2)] \qquad [4]$$

The processes are said to be **uncorrelated** if, for all t_1 and t_2,

$$R_{vw}(t_1, t_2) = \overline{v(t_1)} \times \overline{w(t_2)} \qquad [5]$$

Physically independent random processes are usually statistically independent and, hence, uncorrelated. However, uncorrelated processes are not necessarily independent.

Consider the random processes $v(t)$ and $w(t)$ defined by

EXAMPLE 9.1–1

$$v(t) = t + X \qquad w(t) = tY$$

where X and Y are random variables. Although the PDFs of X and Y are not given, we can still obtain expressions for the ensemble averages from the corresponding expectation operations.

The mean and autocorrelation of $v(t)$ are found using Eqs. (3a) and (3b), keeping in mind that time is *not* a random quantity. Thus,

$$\overline{v(t)} = E[t + X] = t + E[X] = t + \overline{X}$$

$$R_v(t_1, t_2) = E[(t_1 + X)(t_2 + X)]$$

$$= E[t_1t_2 + (t_1 + t_2)X + X^2] = t_1t_2 + (t_1 + t_2)\overline{X} + \overline{X^2}$$

Likewise, for $w(t)$,

$$\overline{w(t)} = E[tY] = t\overline{Y} \qquad R_w(t_1, t_2) = E[t_1Yt_2Y] = t_1t_2\overline{Y^2}$$

Taking the crosscorrelation of $v(t)$ with $w(t)$, we get

$$R_{vw}(t_1, t_2) = E[(t_1 + X)t_2Y] = E[t_1t_2Y + t_2XY] = t_1t_2\overline{Y} + t_2\overline{XY}$$

If X and Y happen to be independent, then $\overline{XY} = \overline{X} \times \overline{Y}$ and

$$R_{vw}(t_1, t_2) = t_1t_2\overline{Y} + t_2\overline{X}\,\overline{Y} = (t_1 + \overline{X})(t_2\overline{Y}) = \overline{v(t_1)} \times \overline{w(t_2)}$$

so the processes are uncorrelated.

Randomly Phased Sinusoid

EXAMPLE 9.1–2

Suppose you have an oscillator set at some nonrandom amplitude A and frequency ω_0, but you don't know the phase angle until you turn the oscillator on and observe the waveform. This situation can be viewed as an experiment in which you pick an oscillator at random from a large collection with the same amplitude and frequency but no phase synchronization. A particular oscillator having phase angle φ_i generates the sinusoidal sample function $v_i(t) = A \cos(\omega_0 t + \varphi_i)$, and the ensemble of sinusoids constitutes a random process defined by

$$v(t) = A \cos(\omega_0 t + \Phi)$$

where Φ is a random angle presumably with a uniform PDF over 2π radians. We'll find the mean value and autocorrelation function of this randomly phased sinusoid.

Since $v(t)$ is defined by transformation of Φ, we can apply Eqs. (3a) and (3b) with $g(\Phi, t) = A \cos(\omega_0 t + \Phi)$ and $p_\Phi(\varphi) = 1/2\pi$ for $0 < \varphi \le 2\pi$. As a preliminary step, let n be a nonzero integer and consider the expected value of $\cos(\alpha + n\Phi)$. Treating α as a constant with respect to the integration over φ,

$$E[\cos(\alpha + n\Phi)] = \int_{-\infty}^{\infty} \cos(\alpha + n\Phi)\, p_\Phi(\varphi)\, d\varphi = \int_{0}^{2\pi} \cos(\alpha + n\Phi)\frac{1}{2\pi}\, d\varphi$$

$$= [\sin(\alpha + 2\pi n) - \sin\alpha]/2\pi n = 0 \qquad n \ne 0$$

But, with $n = 0$, $E[\cos\alpha] = \cos\alpha$ because $\cos\alpha$ does not involve the random variable Φ.

Now we find the mean value of $v(t)$ by inserting $g(\Phi,t)$ with $\alpha = \omega_0 t$ into Eq. (3a), so

$$\overline{v(t)} = E[g(\Phi, t)] = AE[\cos(\omega_0 t + \Phi)] = 0$$

which shows that the mean value equals zero at any time. Next, for the autocorrelation function, we use Eq. (3b) with $\alpha_1 = \omega_0 t_1$ and $\alpha_2 = \omega_0 t_2$. Trigonometric expansion then yields

$$R_v(t_1, t_2) = E[A \cos(\alpha_1 + \Phi) \times A \cos(\alpha_2 + \Phi)]$$

$$= (A^2/2)\, E[\cos(\alpha_1 - \alpha_2) + \cos(\alpha_1 + \alpha_2 + 2\Phi)]$$

$$= (A^2/2)\{E[\cos(\alpha_1 - \alpha_2)] + E[\cos(\alpha_1 + \alpha_2 + 2\Phi)]\}$$

$$= (A^2/2)\,\{\cos(\alpha_1 - \alpha_2) + 0\}$$

and hence

$$R_v(t_1, t_2) = (A^2/2)\cos\omega_0(t_1 - t_2)$$

Finally, setting $t_2 = t_1 = t$ gives

$$\overline{v^2(t)} = R_v(t, t) = A^2/2$$

so the mean-square value stays constant over time.

EXERCISE 9.1–1   Let $v(t) = X + 3t$ where X is an RV with $\overline{X} = 0$ and $\overline{X^2} = 5$. Show that $\overline{v(t)} = 3t$ and $R_v(t_1,t_2) = 5 + 9t_1 t_2$.

Ergodic and Stationary Processes

The randomly phased sinusoid in Example 9.1–2 illustrates the property that some *ensemble* averages may equal the corresponding *time* averages of an arbitrary sample function. To elaborate, recall that if $g[v_i(t)]$ is any function of $v_i(t)$, then its time average is given by

$$<g[v_i(t)]> = \lim_{T \to \infty} \frac{1}{T} \int_{-T/2}^{T/2} g[v_i(t)]\, dt$$

With $v_i(t) = a \cos(\omega_0 t + \varphi_i)$, for instance, time averaging yields $<v_i(t)> = 0 = E[v(t)]$ and $<v_i^2(t)> = \frac{1}{2}a^2 = E[v^2(t)]$.

Using a time average instead of an ensemble average has strong practical appeal when valid, because an ensemble average involves every sample function rather than just one. We therefore say that

A random process is **ergodic** if all time averages of sample functions equal the corresponding ensemble averages.

This means that we can take time averages of one sample function to determine or at least estimate ensemble averages.

The definition of ergodicity requires that an ergodic process has $<g[v_i(t)]> = E\{g[v(t)]\}$ for any $v_i(t)$ and any function $g[v_i(t)]$. But the value of $<g[v_i(t)]>$ must be independent of t, so we conclude that

All ensemble averages of an ergodic process are independent of time.

The randomly phased sinusoid happens to ergodic, whereas the process in Exercise 9.1–1 is not since $E[v(t)]$ varies with time.

When a random signal comes from an ergodic source, the mean and mean-square values will be constants. Accordingly, we write

$$E[v(t)] = \bar{v} = m_V \qquad E[v^2(t)] = \overline{v^2} = \sigma_V^2 + m_V^2 \qquad \text{[6]}$$

where m_V and σ_V^2 denote the mean and variance of $v(t)$. Then, observing that any sample function has $<v_i(t)> = E[v(t)]$ and $<v_i^2(t)> = E[v^2(t)]$, we can interpret certain ensemble averages in more familiar terms as follows:

1. The mean value m_V equals the *dc component* $<v_i(t)>$.
2. The mean squared m_V^2 equals the *dc power* $<v_i(t)>^2$.

3. The mean-square value $\overline{v^2}$ equals the *total average power* $<v_i^2(t)>$.

4. The variance σ_V^2 equals the *ac power,* meaning the power in the time-varying component.

5. The standard deviation σ_V equals the *rms value* of the time-varying component.

These relations help make an electrical engineer feel more at home in the world of random processes.

Regrettably, testing a given process for ergodicity generally proves to be a daunting task because we must show that $<g[v_i(t)]> = E\{g[v(t)]\}$ for any and all $g[v(t)]$. Instead, we introduce a useful but less stringent condition by saying that

> A random process is **wide-sense stationary** (WSS) when the mean $E[v(t)]$ is independent of time and the autocorrelation function $R_v(t_1, t_2)$ depends only on the time difference $t_1 - t_2$.

Expressed in mathematical form, wide-sense stationarity requires that

$$E[v(t)] = m_V \qquad R_v(t_1, t_2) = R_v(t_1 - t_2) \qquad \text{[7]}$$

Any ergodic process satisfies Eq. (7) and thus is wide-sense stationary. However, *stationarity does not guarantee ergodicity* because any sample function of an ergodic process must be representative of the entire ensemble. Furthermore, an ergodic process is *strictly* stationary in that *all* ensemble averages are independent of time. Hereafter, unless otherwise indicated, the term *stationary* will mean *wide-sense stationary* per Eq. (7).

Although a stationary process is not necessarily ergodic, its autocorrelation function is directly analogous to the autocorrelation function of a deterministic signal. We emphasize this fact by letting $\tau = t_1 - t_2$ and taking either $t_1 = t$ or $t_2 = t$ to rewrite $R_v(t_1 - t_2)$ as

$$R_v(\tau) = E[v(t)v(t - \tau)] = E[v(t + \tau)v(t)] \qquad \text{[8]}$$

Equation (8) then leads to the properties

$$R_v(-\tau) = R_v(\tau) \qquad \text{[9a]}$$

$$R_v(0) = \overline{v^2} = \sigma_V^2 + m_V^2 \qquad \text{[9b]}$$

$$|R_v(\tau)| \le R_v(0) \qquad \text{[9c]}$$

so the autocorrelation function $R_v(\tau)$ of a stationary process has even symmetry about a maximum at $\tau = 0$, which equals the mean-square value.

For $\tau \ne 0$, $R_v(\tau)$ measures the statistical similarity of $v(t)$ and $v(t \pm \tau)$. On the one hand, if $v(t)$ and $v(t \pm \tau)$ become *independent* as $\tau \to \infty$, then

$$R_v(\pm\infty) = \overline{v}^2 = m_V^2 \qquad \text{[10]}$$

On the other hand, if the sample functions are *periodic* with period T_0, then

$$R_v(\tau \pm nT_0) = R_v(\tau) \qquad n = 1, 2, \ldots \qquad \text{[11]}$$

and $R_v(\tau)$ does not have a unique limit as $|\tau| \to \infty$.

Returning to the randomly phased sinusoid, we now see that the stationarity conditions in Eq. (7) are satisfied by $E[v(t)] = 0$ and $R_v(t_1,t_2) = (A^2/2) \cos \omega_0(t_1 - t_2) = R_v(t_1 - t_2)$. We therefore write $R_v(\tau) = (A^2/2) \cos \omega_0\tau$, which illustrates the properties in Eqs. (9a)–(9c). Additionally, each sample function $v_i(t) = A \cos (\omega_0 t + \varphi_i)$ has period $T_0 = 2\pi/\omega_0$ and so does $R_v(\tau)$, in agreement with Eq. (11).

Finally, we define the **average power** of a random process $v(t)$ to be the ensemble average of $<v^2(t)>$, so

$$P \triangleq E[<v^2(t)>] = <E[v^2(t)]> \qquad \text{[12]}$$

This definition agrees with our prior observation that the average power of an *ergodic* process is $<v_i^2(t)> = \overline{v^2}$, since an ergodic process has $E[v^2(t)] = <v_i^2(t)>$ and $<E[v^2(t)]> = E[v^2(t)]$ when $E[v^2(t)]$ is independent of time. If the process is *stationary* but not necessarily ergodic, then $E[v^2(t)] = R_v(0)$ and Eq. (12) reduces to

$$P = R_v(0) \qquad \text{[13]}$$

All stationary processes of practical interest have $R_v(0) > 0$, so most of the sample functions are *power signals* rather than finite-energy signals.

Random Digital Wave   **EXAMPLE 9.1–3**

The random digital wave comes from an ensemble of rectangular pulse trains like the sample function in Fig. 9.1–2a. All pulses have fixed nonrandom duration D, but the ensemble involves two random variables, as follows:

1. The delay T_d is a continuous RV uniformly distributed over $0 < t_d \leq D$, indicating that the ensemble consists of unsynchronized waveforms.

2. The amplitude a_k of the kth pulse is a discrete RV with mean $E[a_k] = 0$ and variance σ^2, and the amplitudes in different intervals are statistically independent so $E[a_j a_k] = E[a_j]E[a_k] = 0$ for $j \neq k$.

Note that we're using the lowercase symbol a_k here for the random amplitude, and that the subscript k denotes the sequence position rather than the amplitude value. We'll investigate the stationarity of this process, and we'll find its autocorrelation function.

Consider the kth pulse interval defined by $kD + T_d < t < (k + 1)D + T_d$ and shown in Fig. 9.1–2b. Since $v(t_i) = a_k$ when t_i falls in this interval, and since all such intervals have the same statistics, we conclude that

$$E[v(t)] = E[a_k] = 0 \qquad E[v^2(t)] = E[a_k^2] = \sigma^2$$

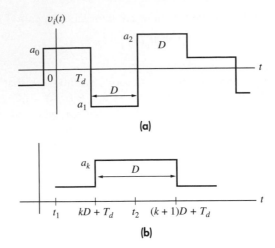

(a)

(b)

Figure 9.1–2   Random digital wave. (a) Sample function; (b) kth pulse interval.

Being independent of time, these results suggest a stationary process. To complete the test for wide-sense stationarity, we must find $R_v(t_1,t_2)$. However, since the probability function for the pulse amplitudes is not known, our approach will be based on the expectation interpretation of the ensemble average $E[v(t_1)v(t_2)]$ when t_1 and t_2 fall in the same or different pulse intervals.

Clearly, t_1 and t_2 must be in *different* intervals when $|t_2 - t_1| > D$, in which case $v(t_1) = a_j$ and $v(t_2) = a_k$ with $j \neq k$ so

$$E[v(t_1)v(t_2)] = E[a_j a_k] = 0 \qquad |t_2 - t_1| > D$$

But if $|t_2 - t_1| < D$, then either t_1 and t_2 are in *adjacent* intervals and $E[v(t_1)v(t_2)] = 0$, or else t_1 and t_2 are in the same interval and $E[v(t_1)v(t_2)] = E[a_k^2] = \sigma^2$.

We therefore let A stand for the event "t_1 and t_2 in adjacent intervals" and write

$$E[v(t_1)v(t_2)] = E[a_j a_k]P(A) + E[a_k^2][1 - P(A)]$$
$$= \sigma^2[1 - P(A)] \qquad |t_2 - t_1| < D$$

From Fig. 9.1–2b, the probability $P(A)$ involves the random delay T_d as well as t_1 and t_2. For $t_1 < t_2$ as shown, t_1 and t_2 are in adjacent intervals if $t_1 < kD + T_d < t_2$, and

$$P(t_1 - kD < T_d < t_2 - kD) = \int_{t_1 - kD}^{t_2 - kD} \frac{1}{D} \, dt_d = \frac{t_2 - t_1}{D}$$

Including the other case when $t_2 < t_1$, the probability of t_1 and t_2 being in adjacent intervals is

$$P(A) = |t_2 - t_1|/D$$

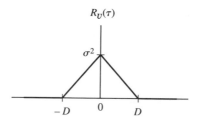

Figure 9.1–3   Autocorrelation of the random digital wave.

and hence,

$$E[v(t_1)v(t_2)] = \sigma^2[1 - |t_2 - t_1|/D] \qquad |t_2 - t_1| < D$$

Combining this result with our previous result for $|t_2 - t_1| > D$, we have

$$R_v(t_1,t_2) = \sigma^2\left(1 - \frac{|t_2 - t_1|}{D}\right) \qquad |t_2 - t_1| < D$$

$$= 0 \qquad\qquad |t_2 - t_1| > D$$

Since $R_v(t_1,t_2)$ depends only on the time difference $t_1 - t_2$, the random digital wave is wide-sense stationary.

Accordingly, we now let $\tau = t_1 - t_2$ and express the correlation function in the compact form

$$R_v(\tau) = \sigma^2\Lambda(\tau/D)$$

where $\Lambda(\tau/D)$ is the triangle function. The corresponding plot of $R_v(\tau)$ in Fig. 9.1–3 deserves careful study because it further illustrates the autocorrelation properties stated in Eqs. (9a)–(9c), with $m_V = 0$ and $\overline{v^2} = \sigma_V^2 + m_V^2 = \sigma^2$.

The average power of this process is then given by Eq. (13) as

$$P = R_v(0) = \sigma^2$$

However, the process is *not ergodic* and the average power of a particular sample function could differ from P. By way of example, if $v_i(t)$ happens to have $a_k = a_0$ for all k, then $<v_i^2(t)> = <a_0^2> = a_0^2 \neq P$. We use P as the "best" prediction for the value of $<v_i^2(t)>$ because we don't know the behavior of $v_i(t)$ in advance.

Let $v(t)$ be a stationary process and let $z(t_1,t_2) = v(t_1) \pm v(t_2)$. Use the fact that $E[z^2(t_1,t_2)] \geq 0$ to prove Eq. (9c).   **EXERCISE 9.1–2**

Gaussian Processes

A random process is called a **gaussian process** if all marginal, joint, and conditional PDFs for the random variables $V_i = v(t_i)$ are gaussian functions. But instead of finding all these PDFs, we usually invoke the central-limit to determine if a given process is gaussian. Gaussian processes play a major role in the study of communication systems because the gaussian model applies to so many random electrical phenomena—at least as a first approximation.

Having determined or assumed that $v(t)$ is gaussian, several important and convenient properties flow therefrom. Specifically, more advanced investigations show that:

1. The process is completely described by $E[v(t)]$ and $R_v(t_1,t_2)$.
2. If $R_v(t_1,t_2) = E[v(t_1)]E[v(t_2)]$, then $v(t_1)$ and $v(t_2)$ are uncorrelated and statistically independent.
3. If $v(t)$ satisfies the conditions for wide-sense stationarity, then the process is also strictly stationary and ergodic.
4. Any linear operation on $v(t)$ produces another gaussian process.

These properties greatly simplify the analysis of random signals, and they will be drawn upon frequently hereafter. Keep in mind, however, that they hold in general only for gaussian processes.

EXERCISE 9.1–3   By considering $R_w(t_1,t_2)$, determine the properties of $w(t) = 2v(t) - 8$ when $v(t)$ is a gaussian process with $E[v(t)] = 0$ and $R_v(t_1,t_2) = 9e^{-5|t_1-t_2|}$.

9.2   RANDOM SIGNALS

This section focuses on random signals from ergodic or at least stationary sources. We'll apply the Wiener-Kinchine theorem to obtain the power spectrum, and we'll use correlation and spectral analysis to investigate filtering and other operations on random signals.

Power Spectrum

When a random signal $v(t)$ is stationary, then we can meaningfully speak of its power spectrum $G_v(f)$ as the distribution of the average power P over the frequency domain. According to the **Wiener-Kinchine theorem,** $G_v(f)$ is related to the autocorrelation $R_v(\tau)$ by the Fourier transform

$$G_v(f) = \mathcal{F}_\tau[R_v(\tau)] \triangleq \int_{-\infty}^{\infty} R_v(\tau)e^{-j2\pi f\tau}\, d\tau \qquad \text{[1a]}$$

Conversely,

$$R_v(\tau) = \mathcal{F}_\tau^{-1}[G_v(f)] \triangleq \int_{-\infty}^{\infty} G_v(f)e^{j2\pi f\tau}\,df \qquad [1b]$$

Thus, the autocorrelation and spectral density constitute a Fourier transform pair, just as in the case of deterministic power signals. Properties of $G_v(f)$ include

$$\int_{-\infty}^{\infty} G_v(f)\,df = R_v(0) = \overline{v^2} = P \qquad [2]$$

$$G_v(f) \geq 0 \qquad G_v(-f) = G_v(f) \qquad [3]$$

The even-symmetry property comes from the fact that $R_v(\tau)$ is real and even, since $v(t)$ is real.

The power spectrum of a random process may be continuous, impulsive, or mixed, depending upon the nature of the source. By way of illustration, the randomly phased sinusoid back in Example 9.1–2 has

$$R_v(\tau) = \frac{A^2}{2}\,\cos 2\pi f_0\tau \longleftrightarrow G_v(f) = \frac{A^2}{4}\,\delta(f-f_0) + \frac{A^2}{4}\,\delta(f+f_0) \qquad [4]$$

The resulting impulsive spectrum, plotted in Fig. 9.2–1a, is identical to that of a deterministic sinusoid because the randomly phased sinusoid comes from an ergodic process whose sinusoidal sample functions differ only in phase angle. As contrast, the random digital wave in Example 9.1–3 has

$$R_v(\tau) = \sigma^2 \Lambda(\tau/D) \longleftrightarrow G_v(f) = \sigma^2 D \operatorname{sinc}^2 fD \qquad [5]$$

Figure 9.2–1b shows this continuous power spectrum.

Since the autocorrelation of a random signal has the same mathematical properties as those of a deterministic power signal, justification of the Wiener-Kinchine theorem for random signals could rest on our prior proof for deterministic signals. However, an independent derivation based on physical reasoning provides additional insight and a useful alternative definition of power spectrum.

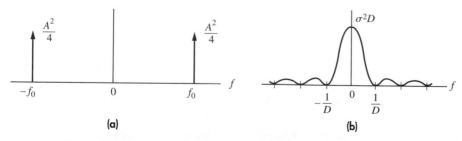

(a)   (b)

Figure 9.2–1   Power spectra. (a) Randomly phased sinusoid; (b) random digital wave.

Consider the finite-duration or *truncated* random signal

$$v_T(t) \triangleq \begin{cases} v(t) & |t| < T/2 \\ 0 & |t| > T/2 \end{cases}$$

Since each truncated sample function has finite energy, we can introduce the Fourier transform

$$V_T(f, s) = \int_{-\infty}^{\infty} v_T(t)e^{-j\omega t}\, dt = \int_{-T/2}^{T/2} v(t)e^{-j\omega t}\, dt \qquad [6]$$

Then $|V_T(f,s_i)|^2$ is the *energy spectral density* of the truncated sample function $v_T(t,s_i)$. Furthermore, drawing upon Rayleigh's energy theorem in the form

$$\int_{-T/2}^{T/2} v^2(t)\, dt = \int_{-\infty}^{\infty} v_T^2(t)\, dt = \int_{-\infty}^{\infty} |V_T(f, s)|^2\, df$$

the average power of $v(t)$ becomes

$$P = \lim_{T\to\infty} \int_{-T/2}^{T/2} E[v^2(t)]\, dt$$

$$= \lim_{T\to\infty} E\left[\frac{1}{T}\int_{-\infty}^{\infty} |V_T(f, s)|^2\, df\right] = \int_{-\infty}^{\infty} \lim_{T\to\infty} \frac{1}{T}\, E\left[|V_T(f, s)|^2\right] df$$

Accordingly, we now *define* the power spectrum of $v(t)$ as

$$G_v(f) \triangleq \lim_{T\to\infty} \frac{1}{T}\, E\left[|V_T(f, s)|^2\right] \qquad [7]$$

which agrees with the properties in Eqs.(2) and (3).

Conceptually, Eq. (7) corresponds to the following steps: (1) calculate the energy spectral density of the truncated sample functions, (2) average over the ensemble, (3) divide by T to obtain power, and (4) take the limit $T \to \infty$. Equation (7) provides the basis for experimental *spectra estimation.* For if you observe a sample function $v(t,s_i)$ for a long time T, then you can estimate $G_v(f)$ from

$$\tilde{G}_v(f) = \frac{1}{T} |V_T(f,s_i)|^2$$

The spectral estimate $\tilde{G}_v(f)$ is called the **periodogram** because it originated in the search for periodicities in seismic records and similar experimental data.

Now, to complete our derivation of the Wiener-Kinchine theorem, we outline the proof that $G_v(f)$ in Eq. (7) equals $\mathcal{F}_T[R_v(\tau)]$. First, we substitute Eq. (6) into $E[|V_T(f,s)|^2] = E[V_T(f,s)V_T^*(f,s)]$ and interchange integration and expectation to get

$$E\left[|V_T(f, s)|^2\right] = \iint_{-T/2}^{T/2} E[v(t)v(\lambda)]e^{-j\omega(t-\lambda)}\, dt\, d\lambda$$

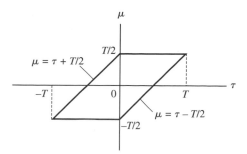

Figure 9.2–2   Integration region in the $\tau - \mu$ plane.

in which

$$E[v(t)v(\lambda)] = R_v(t, \lambda) = R_v(t - \lambda)$$

Next, let $\tau = t - \lambda$ and $\mu = t$ so the double integration is performed over the region of the $\tau - \mu$ plane shown in Fig. 9.2–2. Integrating with respect to μ for the two cases $\tau < 0$ and $\tau > 0$ then yields

$$E\big[|V_T(f, s)|^2\big] = \int_{-T}^{0} R_v(\tau)e^{-j\omega\tau}\left(\int_{-T/2}^{\tau+T/2} d\mu \right)d\tau + \int_{0}^{T} R_v(\tau)e^{-j\omega\tau}\left(\int_{\tau-T/2}^{T/2} d\mu \right)d\tau$$

$$= \int_{-T}^{0} R_v(\tau)e^{-j\omega\tau}(T + \tau)d\tau + \int_{0}^{T} R_v(\tau)e^{-j\omega\tau}(T - \tau)d\tau$$

Finally, since $\tau = -|\tau|$ for $\tau < 0$, we have

$$E\big[|V_T(f, s)|^2\big] = T\int_{-T}^{T}\left(1 - \frac{|\tau|}{T} \right)R_v(\tau)e^{-j\omega\tau}\, d\tau \qquad \text{[8]}$$

Therefore,

$$\lim_{T\to\infty} \frac{1}{T}\, E\big[|V_T(f, s)|^2\big] = \int_{-\infty}^{\infty} R_v(\tau)e^{-j\omega\tau}\, d\tau$$

which confirms that $G_v(f) = \mathcal{F}_\tau[R_v(\tau)]$.

Random Telegraph Wave

EXAMPLE 9.2–1

Figure 9.2–3a represents a sample function of a *random telegraph wave*. This signal makes independent random shifts between two equally likely values, A and 0. The number of shifts per unit time is governed by a Poisson distribution, with μ being the average shift rate.

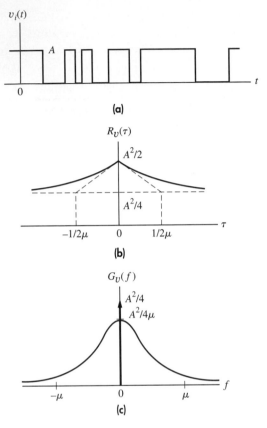

9.2–3   Random telegraph wave. (a) Sample function; (b) autocorrelation; (c) power spectrum.

We'll find the power spectrum given the autocorrelation function

$$R_v(\tau) = \frac{A^2}{4}\left(e^{-2\mu|\tau|} + 1\right)$$

which is sketched in Fig. 9.2–3b. From $R_v(\tau)$ we see that

$$P = \overline{v^2} = R_v(0) = \frac{A^2}{2} \qquad m_V^2 = R_v(\pm\infty) = \frac{A^2}{4}$$

so the rms value is $\sigma_V = \sqrt{\overline{v^2} - m_V^2} = A/2$.

Taking the Fourier transform of $R_v(\tau)$ gives the power spectrum

$$G_v(f) = \frac{A^2}{4\mu\left[1 + (\pi f/\mu)^2\right]} + \frac{A^2}{4}\,\delta(f)$$

which includes an impulse at the origin representing the dc power $m_v^2 = A^2/4$. This mixed spectrum is plotted in Fig. 9.2–3c. Although μ equals the average shift rate, about 20 percent of the ac power (measured in terms of σ_v^2) is contained in the higher frequencies $|f| > \mu$.

To confirm in general that $G_v(f)$ includes an impulse when $m_V \neq 0$, let $z(t) = v(t) - m_V$ and show from $R_z(\tau)$ that $G_v(f) = G_z(f) + m_v^2\delta(f)$.   **EXERCISE 9.2–1**

Superposition and Modulation

Some random signals may be viewed as a combination of other random signals. In particular, let $v(t)$ and $w(t)$ be *jointly stationary* so that

$$R_{vw}(t_1, t_2) = R_{vw}(t_1 - t_2)$$

and let

$$z(t) = v(t) \pm w(t) \qquad [9]$$

Then

$$R_z(\tau) = R_v(\tau) + R_w(\tau) \pm [R_{vw}(\tau) + R_{wv}(\tau)]$$

and

$$G_z(f) = G_v(f) + G_w(f) \pm [G_{vw}(f) + G_{wv}(f)]$$

where we have introduced the **cross-spectral density**

$$G_{vw}(f) \triangleq \mathcal{F}_\tau[R_{vw}(\tau)] \qquad [10]$$

The cross-spectral density vanishes when $v(t)$ and $w(t)$ are uncorrelated and $m_V m_W = 0$, so

$$R_{vw}(\tau) = R_{wv}(\tau) = 0 \qquad [11a]$$

Under this condition

$$R_z(\tau) = R_v(\tau) + R_w(\tau) \qquad [11b]$$

$$G_z(f) = G_v(f) + G_w(f) \qquad [11c]$$

$$\overline{z^2} = \overline{v^2} + \overline{w^2} \qquad [11d]$$

Thus, we have **superposition** of autocorrelation, power spectra, and average power.

When Eq. (11a) holds, the random signals are said to be **incoherent.** Signals from independent sources are usually incoherent, and superposition of average power is a common physical phenomenon. For example, if two musicians play in unison but without perfect synchronization, then the total acoustical power simply equals the sum of the individual powers.

Now consider the **modulation** operation defined by the product

$$z(t) = v(t) \cos(\omega_c t + \Phi) \qquad [12]$$

where $v(t)$ is stationary random signal and Φ is a random angle independent of $v(t)$ and uniformly distributed over 2π radians. If we didn't include Φ, then $z(t)$ would be nonstationary; including Φ merely recognizes the arbitrary choice of the time origin when $v(t)$ and $\cos \omega_c t$ come from independent sources.

Since modulation is a time-varying process, we must determine $R_z(t)$ from $R_z(t_1,t_2) = E[z(t_1)z(t_2)]$ by taking the joint expectation with respect to both v and Φ. After trigonometric expansion we get

$$R_z(t_1, t_2) = \tfrac{1}{2}E[v(t_1)v(t_2)]\{\cos \omega_c(t_1 - t_2) + E[\cos(\omega_c t_1 - \omega_c t_2 + 2\Phi)]\}$$

$$= \tfrac{1}{2}R_v(t_1, t_2) \cos \omega_c(t_1 - t_2)$$

Thus, with $\tau = t_1 - t_2$,

$$R_z(\tau) = \tfrac{1}{2}R_v(\tau) \cos 2\pi f_c \tau \qquad [13a]$$

and Fourier transformation yields

$$G_z(f) = \tfrac{1}{4}\left[G_v(f - f_c) + G_v(f + f_c)\right] \qquad [13b]$$

Not surprisingly, modulation translates the power spectrum of $v(t)$ up and down by f_c units.

Modulation is a special case of the product operation

$$z(t) = v(t)w(t) \qquad [14]$$

If $v(t)$ and $w(t)$ are independent and jointly stationary, then

$$R_z(\tau) = R_v(\tau)R_w(\tau) \qquad [15a]$$

and

$$G_z(f) = G_v(f)*G_w(f) \qquad [15b]$$

which follows from the convolution theorem.

EXERCISE 9.2–2   Derive Eq. (13b) from Eq. (15b) by making a judicious choice for $w(t)$.

Filtered Random Signals

Figure 9.2–4 represents a random signal $x(t)$ applied to the input of a filter (or any LTI system) having transfer function $H(f)$ and impulse response $h(t)$. The resulting output signal is given by the convolution

$$y(t) = \int_{-\infty}^{\infty} h(\lambda)x(t - \lambda)d\lambda \qquad [16]$$

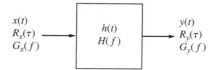

Figure 9.2–4   Random signal applied to a filter.

Since convolution is a linear operation, a *gaussian input* produces a *gaussian output* whose properties are completely described by m_Y and $R_y(\tau)$. These output statistics can be found from $H(f)$ or $h(t)$, and they will be useful even in the nongaussian case.

We'll assume that $h(t)$ is *real,* so both $x(t)$ and $y(t)$ are real. We'll also assume that $x(t)$ is a stationary power signal and that the system is *stable.* Under these conditions, $y(t)$ will be a stationary power signal and the output-input crosscorrelation is a convolution with the impulse response, namely,

$$R_{yx}(\tau) = h(\tau)*R_x(\tau) \qquad\qquad [17a]$$

The proof of this relation starts with $R_{yx}(t_1, t_2) = E[y(t_1)x(t_2)]$. Inserting $y(t_1)$ from Eq. (16) and exchanging the order of operations yields

$$R_{yx}(t_1, t_2) = \int_{-\infty}^{\infty} h(\lambda)E[x(t_1 - \lambda)x(t_2)]d\lambda$$

Then, since $x(t)$ is stationary,

$$E[x(t_1 - \lambda)x(t_2)] = R_x(t_1 - \lambda, t_2) = R_x(t_1 - \lambda - t_2)$$

Finally, letting $t_2 = t_1 - \tau$,

$$R_{yx}(t_1, t_1 - \tau) = R_{yx}(\tau) = \int_{-\infty}^{\infty} h(\lambda)R_x(\tau - \lambda)\, d\lambda$$

so $R_{yx}(\tau) = h(\tau)*R_x(\tau)$. Proceeding in the same fashion, the output autocorrelation is found to be

$$R_y(\tau) = h(-\tau)*R_{yx}(\tau) = h(-\tau)*h(\tau)*R_x(\tau) \qquad\qquad [17b]$$

which also establishes the fact that $y(t)$ is at least wide-sense stationary.

From Eq. (17b), it follows that the power spectra are related by

$$G_y(f) = |H(f)|^2 G_x(f) \qquad\qquad [18]$$

Consequently,

$$R_y(\tau) = \mathcal{F}_\tau^{-1}[|H(f)|^2\, G_x(f)] \qquad\qquad [19a]$$

$$\overline{y^2} = R_y(0) = \int_{-\infty}^{\infty} |H(f)|^2\, G_x(f)\, df \qquad\qquad [19b]$$

Furthermore, the mean value of the output is

$$m_Y = \left[\int_{-\infty}^{\infty} h(\lambda)d\lambda \right] m_X = H(0)m_X \qquad \text{[20]}$$

where $H(0)$ equals the system's dc gain.

The power spectrum relation in Eq. (18) has additional value for the study of *linear operations* on random signals, whether or not filtering is actually involved. In particular, suppose that you know $G_x(f)$ and you want to find $G_y(f)$ when $y(t) = dx(t)/dt$. Conceptually, $y(t)$ could be obtained by applying $x(t)$ to an ideal differentiator which we know has $H(f) = j2\pi f$. We thus see from Eq. (18) that if

$$y(t) = dx(t)/dt \qquad \text{[21a]}$$

then

$$G_y(f) = (2\pi f)^2 G_x(f) \qquad \text{[21b]}$$

Conversely, if

$$y(t) = \int_{-\infty}^{t} x(\lambda)d\lambda \qquad m_X = 0 \qquad \text{[22a]}$$

then

$$G_y(f) = (2\pi f)^{-2} G_x(f) \qquad \text{[22b]}$$

These relations parallel the differentiation and integration theorems for Fourier transforms.

EXAMPLE 9.2–2   Let the random telegraph wave from Example 9.2–1 be applied to an ideal bandpass filter with unit gain and narrow bandwidth B centered at $f_c = \mu/\pi$. Figure 9.2–5 shows the resulting output power spectrum $G_y(f) = |H(f)|^2 G_x(f)$.

With $G_x(\pm f_c) = A^2/8\mu$ and $B \ll f_c$, we have

$$G_y(f) \approx \begin{cases} A^2/8\mu & -B/2 < |f - f_c| < B/2 \\ 0 & \text{otherwise} \end{cases}$$

and the output power equals the area under $G_y(f)$, namely

$$\overline{y^2} \approx 2B \times (A^2/8\mu) = A^2 B/4\mu \ll A^2/4\pi$$

whereas $\overline{x^2} = A^2/2$. Moreover, since $H(0) = 0$, we know from Eq. (20) that $m_Y = 0$ even though $m_X = A/2$. Note that we obtained these results without the added labor of finding $R_y(\tau)$.

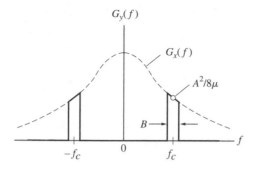

Figure 9.2–5   Filtered power spectrum in Example 9.2–2.

Hilbert Transform of a Random Signal   **EXAMPLE 9.2–3**

The Hilbert transform $\hat{x}(t)$ was previously defined as the output of a quadrature phase-shifting filter having

$$h_Q(t) = \frac{1}{\pi t} \qquad H_Q(f) = -j \operatorname{sgn} f$$

Since $|H_Q(f)|^2 = 1$, we now conclude that if $x(t)$ is a random signal and $y(t) = \hat{x}(t)$, then

$$G_{\hat{x}}(f) = G_x(f) \qquad R_{\hat{x}}(\tau) = \mathcal{F}_\tau^{-1}[G_{\hat{x}}(f)] = R_x(\tau)$$

Thus, Hilbert transformation does not alter the values of m_X or $\overline{x^2}$.

However, from Eq. (17a),

$$R_{\hat{x}x}(\tau) = h_Q(\tau) * R_x(\tau) = \hat{R}_x(\tau)$$

where $\hat{R}_x(\tau)$ stand for the Hilbert transform of $R_x(\tau)$. It can also be shown that

$$R_{x\hat{x}}(\tau) = -\hat{R}_x(\tau)$$

We'll apply these results in the next chapter.

Let the random digital wave described by Eq. (5) (p. 363) be applied to a first-order   **EXERCISE 9.2–3**
LPF with $H(f) = [1 + j(f/B)]^{-1}$ and $B \ll 1/D$. Obtain approximate expressions for
$G_y(f)$ and $R_y(\tau)$.

9.3   NOISE

Unwanted electric signals come from a variety of sources, generally classified as either human interference or naturally occurring noise. Human interference is produced by other communication systems, ignition and commutator sparking, ac hum, and so forth. Natural noise-generating phenomena include atmospheric disturbances, extraterrestrial radiation, and random electron motion. By careful engineering, the effects of many unwanted signals can be reduced or virtually eliminated. But there always remains certain inescapable random signals that present a fundamental limit to system performance.

One unavoidable cause of electrical noise is the thermal motion of electrons in conducting media—wires, resistors, and so on. Accordingly, this section begins with a brief discussion of *thermal noise* that, in turn, leads to the convenient abstraction of *white noise*—a useful model in communication. We then consider filtered white noise and input–output relations. Other aspects of noise analysis are developed in subsequent chapters and the appendix. In particular, the special properties of *bandpass* noise will be discussed in Sect. 10.1.

Thermal Noise and Available Power

For our purposes,

Thermal noise is the noise produced by the random motion of charged particles (usually electrons) in conducting media.

From kinetic theory, the average energy of a particle at absolute temperature $\mathcal{T}$ is proportional to $k\mathcal{T}$, k being the Boltzmann constant. We thus expect thermal-noise values to involve the product $k\mathcal{T}$. In fact, we'll develop a measure of noise power in terms of temperature. Historically, Johnson (1928) and Nyquist (1928b) first studied noise in metallic resistors—hence, the designation *Johnson noise* or *resistance noise*. There now exists an extensive body of theoretical and experimental studies pertaining to noise, from which we'll freely draw.

When a metallic resistance R is at temperature $\mathcal{T}$, random electron motion produces a noise voltage $v(t)$ at the open terminals. Consistent with the central-limit theorem, $v(t)$ has a *gaussian distribution* with zero mean and variance

$$\overline{v^2} = \sigma_V^2 = \frac{2(\pi k\mathcal{T})^2}{3h} R \qquad \text{V}^2 \qquad [1]$$

where $\mathcal{T}$ is measured in kelvins (K) and

$$k = \text{Boltzmann constant} = 1.38 \times 10^{-23} \text{ J/K}$$

$$h = \text{Planck constant} = 6.62 \times 10^{-34} \text{ J-s}$$

The presence of the Planck constant in Eq. (1) indicates a result from quantum theory. The theory further shows that the mean square spectral density of thermal noise is

$$G_v(f) = \frac{2Rh|f|}{e^{h|f|/k\mathcal{T}} - 1} \quad \text{V}^2/\text{Hz} \quad \text{[2a]}$$

which is plotted in Fig. 9.3–1 for $f \geq 0$. This expression reduces at "low" frequencies to

$$G_v(f) \approx 2Rk\mathcal{T}\left(1 - \frac{h|f|}{2k\mathcal{T}}\right) \quad |f| \ll \frac{k\mathcal{T}}{h} \quad \text{[2b]}$$

Both Eq. (2a) and (2b) omit a term corresponding to the zero-point energy, which is independent of temperature and plays no part in thermal noise transferred to a load.

Fortunately, communication engineers almost never have need for Eqs. (2a) and (2b). To see why, let room temperature or the **standard temperature** be

$$\mathcal{T}_0 \triangleq 290 \text{ K } (63°\text{F}) \quad \text{[3a]}$$

which is rather on the chilly side but simplifies numerical work since

$$k\mathcal{T}_0 \approx 4 \times 10^{-21} \quad \text{W-s} \quad \text{[3b]}$$

If the resistance in question is at $\mathcal{T}_0$, then $G_v(f)$ is essentially *constant* for $|f| < 0.1k\mathcal{T}_0/h \approx 10^{12}$ Hz. But this upper limit falls in the infrared part of the electromagnetic spectrum, far above the point where conventional electrical components have ceased to respond. And this conclusion holds even at cryogenic temperatures ($\mathcal{T} \approx 0.001\mathcal{T}_0$).

Therefore, for almost all purposes we can say that the mean square voltage spectral density of thermal noise is constant at

$$G_v(f) = 2Rk\mathcal{T} \quad \text{V}^2/\text{Hz} \quad \text{[4]}$$

obtained from Eq. (2b) with $h|f|/k\mathcal{T} \ll 1$. The one trouble with Eq. (4) is that it erroneously predicts $\overline{v^2} = \infty$ when $G_v(f)$ is integrated over all f. However, you seldom have to deal directly with $\overline{v^2}$ because $v(t)$ is always subject to the filtering effects of other circuitry. That topic will be examined shortly. Meanwhile, we'll use Eq. (4) to construct the Thévenin equivalent model of a resistance, as shown in Fig. 9.3–2a. Here the resistance is replaced by a *noiseless resistance* of the same value, and the

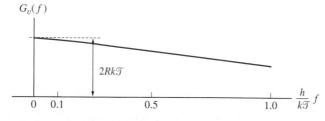

Figure 9.3–1 Thermal noise spectra density, V²/Hz.

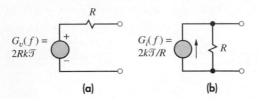

Figure 9.3–2   Thermal resistance noise. (a) Thévenin equivalent; (b) Norton equivalent.

noise is represented by a mean square voltage generator. Similarly, Fig. 9.3–2b is the Norton equivalent with a mean square current generator having $G_i(f) = G_v(f)/R^2 = 2k\mathcal{T}/R$. Both generators are shaded to indicate their special nature.

Instead of dealing with mean square voltage or current, describing thermal noise by its *available power* cleans up the notation and speeds calculations. Recall that available power is the *maximum* power that can be delivered to a load from a source having fixed nonzero source resistance. The familiar theorem for maximum power transfer states that this power is delivered only when the load impedance is the complex conjugate of the source impedance. The load is then said to be **matched** to the source, a condition usually desired in communication systems.

Let the sinusoidal source in Fig. 9.3–3a have impedance $Z_s = R_s + jX_s$, and let the open-circuit voltage be v_s. If the load impedance is matched, so that $Z_L = Z_s^* = R_s - jX_s$, then the terminal voltage is $v_s/2$ and the available power is

$$P_a = \frac{<[v_s(t)/2]^2>}{R_s} = \frac{<v_s^2(t)>}{4R_s}$$

Using our Thévenin model, we extend this concept to a thermal resistor viewed as a noise source in Fig. 9.3–3b. By comparison, the **available spectral density** at the load resistance is

$$G_a(f) = \frac{G_v(f)}{4R} = \frac{1}{2}k\mathcal{T} \quad \text{W/Hz} \tag{5}$$

A thermal resistor therefore delivers a maximum power density of $k\mathcal{T}/2$ W/Hz to a matched load, regardless of the value of R!

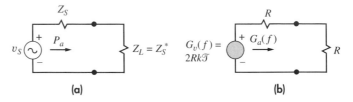

Figure 9.3–3   Available power. (a) AC source with matched load; (b) thermal resistance with matched load.

Calculate from Eq. (1) the rms noise voltage across the terminals of a $1-\text{k}\Omega$ resistance at $\mathcal{T} = 29$ K. Then use Eq. (2b) to find the percentage of the mean square voltage that comes from frequency components in the range $|f| < 1$ GHz.

EXERCISE 9.3–1

White Noise and Filtered Noise

Besides thermal resistors, many other types of noise sources are gaussian and have a flat spectral density over a wide frequency range. Such a spectrum has all frequency components in equal proportion, and is therefore called **white noise** by analogy to white light. White noise is a convenient model (and often an accurate one) in communications, and the assumption of a gaussian process allows us to invoke all the aforementioned properties—but some applications (beyond our scope) may need a more advanced model for the noise.

We'll write the spectral density of white noise in general as

$$G(f) = N_0/2 \qquad \text{[6a]}$$

where N_0 represents a *density* constant in the standard notation. The seemingly extraneous factor of 1/2 is included to indicate that half the power is associated with positive frequency and half with negative frequency, as shown in Fig. 9.3–4a. Alternatively, N_0 stands for the *one-sided* spectral density. The autocorrelation function for white noise follows immediately by Fourier transformation of $G(f)$, so

$$R(\tau) = \frac{N_0}{2}\delta(\tau) \qquad \text{[6b]}$$

as in Fig. 9.3–4b.

We thus see that $R(\tau) = 0$ for $\tau \neq 0$, so any two different samples of a gaussian white noise signal are *uncorrelated* and hence *statistically independent*. This observation, coupled with the constant power spectrum, leads to an interesting conclusion, to wit: When white noise is displayed on an oscilloscope, successive sweeps are always different from each other; but the waveform always looks the same, no

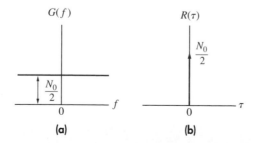

Figure 9.3–4   White noise. (a) Power spectrum; (b) autocorrelation.

matter what sweep speed is used, since all rates of time variation (frequency components) are contained in equal proportion. Similarly, when white noise drives a loudspeaker, it always sounds the same, somewhat like a waterfall.

The value of N_0 in Eqs. (6a) and (6b) depends upon two factors: the type of noise, and the type of spectral density. If the source is a *thermal resistor*, then the mean square voltage and mean square current densities are

$$N_{0v} = 4Rk\mathcal{T} \qquad N_{0i} = 4k\mathcal{T}/R \qquad \text{[7]}$$

where the added subscripts v and i identify the type of spectral density. Moreover, any *thermal* noise source by definition has the available one-sided noise spectral density $2G_a(f) = k\mathcal{T}$. Other white noise sources are *nonthermal* in the sense that the available power is unrelated to a physical temperature. Nonetheless, we can speak of the **noise temperature** $\mathcal{T}_N$ of almost any white noise source, thermal or nonthermal, by defining

$$\mathcal{T}_N \triangleq \frac{2G_a(f)}{k} = \frac{N_0}{k} \qquad \text{[8a]}$$

Then, given a source's noise temperature,

$$N_0 = k\mathcal{T}_N \qquad \text{[8b]}$$

It must be emphasized that $\mathcal{T}_N$ is not necessarily a physical temperature. For instance, some noise generators have $\mathcal{T}_N \approx 10\mathcal{T}_0 \approx 3000$ K (5000° F), but the devices surely aren't that hot.

Now consider gaussian white noise $x(t)$ with spectral density $G_x(f) = N_0/2$ applied to an LTI filter having transfer function $H(f)$. The resulting output $y(t)$ will be gaussian noise described by

$$G_y(f) = \frac{N_0}{2}|H(f)|^2 \qquad \text{[9a]}$$

$$R_y(\tau) = \frac{N_0}{2}\mathcal{F}_\tau^{-1}\left[|H(f)|^2\right] \qquad \text{[9b]}$$

$$\overline{y^2} = \frac{N_0}{2}\int_{-\infty}^{\infty}|H(f)|^2\,df \qquad \text{[9c]}$$

Pay careful attention to Eq. (9a) which shows that the spectral density of filtered white noise takes the *shape* of $|H(f)|^2$. We therefore say that filtering white noise produces *colored* noise with frequency content primarily in the range passed by the filter.

As an illustration of filtered noise, let $H(f)$ be an ideal lowpass function with unit gain and bandwidth B. Then

$$G_y(f) = \frac{N_0}{2}\Pi\left(\frac{f}{2B}\right) \qquad R_y(\tau) = N_0B \text{ sinc } 2B\tau \qquad \text{[10a]}$$

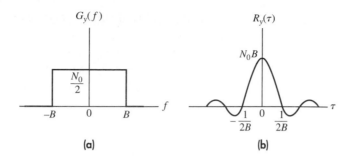

Figure 9.3–5 White noise passed by an ideal LPF. (a) Power spectrum; (b) autocorrelation.

which are plotted in Fig. 9.3–5. Besides the spectral shaping, we see that lowpass filtering causes the output noise to be correlated over time intervals of about $1/2B$. We also see that

$$\overline{y^2} = N_0 B \qquad [10b]$$

so the output power is directly proportional to the filter's bandwidth.

Thermal Noise in an *RC* Network **EXAMPLE 9.3–1**

To pull together several of the topics so far, consider the *RC* network in Fig. 9.3–6a with the resistor at temperature $\mathcal{T}$. Replacing this thermal resistor with its Thévenin model leads to Fig. 9.3–6b, a white noise mean square voltage source with $G_x(f) = 2Rk\mathcal{T}$ V²/Hz applied to a *noiseless RC* LPF. Since $|H(f)|^2 = [1 + (f/B)^2]^{-1}$, the output spectral density is

$$G_y(f) = |H(f)|^2 G_x(f) = \frac{2Rk\mathcal{T}}{1 + (f/B)^2} \qquad B = \frac{1}{2\pi RC} \qquad [11a]$$

The inverse transform then yields

$$R_y(\tau) = 2Rk\mathcal{T}\pi B e^{-2\pi B|\tau|} = \frac{k\mathcal{T}}{C} e^{-|\tau|/RC} \qquad [11b]$$

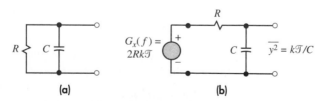

Figure 9.3–6 (a) *RC* network with resistance noise; (b) noise equivalent model.

which shows that the interval over which the filtered noise voltage has appreciable correlation approximately equals the network's time constant RC, as might have been suspected.

We can further say that $y(t)$ is a gaussian random signal with no dc component—since $x(t)$ is a zero-mean gaussian—and

$$\overline{y^2} = R_y(0) = \frac{k\mathcal{T}}{C} \qquad [12]$$

Surprisingly, $\overline{y^2}$ depends on C but not on R, even though the noise source is the thermal resistor! This paradox will be explained shortly; here we conclude our example with a numerical calculation.

Suppose the resistor is at room temperature $\mathcal{T}_0$ and $C = 0.1~\mu\text{F}$; then

$$\overline{y^2} = 4 \times 10^{-21}/10^{-7} = 4 \times 10^{-14}~~\text{V}^2$$

and the rms output voltage is $\sigma_Y = 2 \times 10^{-7} = 0.2~\mu\text{V}$. Such exceedingly small values are characteristic of thermal noise, which is why thermal noise goes unnoticed in ordinary situations. However, the received signal in a long-distance communication system may be of the same order of magnitude or even smaller.

Noise Equivalent Bandwidth

Filtered white noise usually has finite power. To emphasize this property, we designate **average noise power** by N and write Eq. (9c) in the form

$$N = \frac{N_0}{2} \int_{-\infty}^{\infty} |H(f)|^2\, df = N_0 \int_{0}^{\infty} |H(f)|^2\, df$$

Noting that the integral depends only on the filter's transfer function, we can simplify discussion of noise power by defining a **noise equivalent bandwidth** B_N (or just the **noise bandwidth**) as

$$B_N \triangleq \frac{1}{g} \int_{0}^{\infty} |H(f)|^2\, df \qquad [13]$$

where

$$g = |H(f)|^2_{\max}$$

which stands for the center-frequency power ratio (assuming that the filter has a meaningful center frequency). Hence the filtered noise power is

$$N = gN_0 B_N \qquad [14]$$

This expression becomes more meaningful if you remember that N_0 represents *density.*

Examining Eq. (14) shows that the effect of the filter has been separated into two parts: the *relative frequency selectivity* as described by B_N, and the *power gain* (or attenuation) represented by g. Thus, B_N equals the bandwidth of an ideal rectangular filter with the same gain that would pass as much white noise power as the filter in question, as illustrated in Fig. 9.3–7 for a bandpass filter.

Let's apply Eqs. (13) and (14) to the RC LPF in Example 9.3–1. The filter's center frequency is $f = 0$ so $g = |H(0)|^2 = 1$, and

$$B_N = \int_0^\infty \frac{df}{1 + (f/B)^2} = \frac{\pi}{2} B = \frac{1}{4RC} \tag{15}$$

The reason why $\overline{y^2}$ in Eq. (12) is independent of R now becomes apparent if we write $\overline{y^2} = N = N_0 B_N = (4Rk\mathcal{T}) \times (1/4RC)$. Thus, increasing R increases the noise density N_0 (as it should) but decreases the noise bandwidth B_N. These two effects precisely cancel each other and $\overline{y^2} = k\mathcal{T}/C$.

By definition, the noise bandwidth of an ideal filter is its actual bandwidth. For practical filters, B_N is somewhat greater than the 3-dB bandwidth. However, as the filter becomes more selective (sharper cutoff characteristics), its noise bandwidth approaches the 3-dB bandwidth, and for most applications we are not too far off in taking them to be equal.

In summary, if $y(t)$ is filtered white noise of zero mean, then

$$\overline{y^2} = \sigma_Y^2 = N = g N_0 B_N \qquad \sigma_Y = \sqrt{N} = \sqrt{g N_0 B_N} \tag{16}$$

This means that given a source of white noise, an average power meter (or mean square voltage meter) will read $\overline{y^2} = N = N_0 B_N$, where B_N is the noise bandwidth of the meter itself. Working backward, the source power density can be inferred via $N_0 = N/B_N$, provided you're sure that the noise is white over the frequency-response range of the meter.

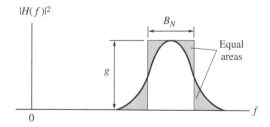

Figure 9.3–7   Noise equivalent bandwidth of a bandpass filter.

EXERCISE 9.3–2   Consider an nth-order Butterworth LPF defined by Eq. (4), p. 113. Show that the noise bandwidth B_N is related to the 3-dB bandwidth B by

$$B_N = \frac{\pi B}{2n \sin(\pi/2n)} \qquad \text{[17]}$$

Hence, $B_N \to B$ as $n \to \infty$.

System Measurements Using White Noise ★

Since white noise contains all frequencies in equal proportion, it's a convenient signal for system measurements and experimental design work. Consequently, white noise sources with calibrated power density have become standard laboratory instruments. A few of the measurements that can be made with these sources are outlined here.

Noise Equivalent Bandwidth   Suppose the gain of an amplifier is known, and we wish to find its noise equivalent bandwidth. To do so, we can apply white noise to the input and measure the average output power with a meter whose frequency response is essentially constant over the amplifier's passband. The noise bandwidth in question is then, from Eq. (14), $B_N = N/gN_0$.

Amplitude Response   To find the amplitude response (or amplitude ratio) of a given system, we apply white noise to the input so the output power spectrum is proportional to $|H(f)|^2$. Then we scan the output with a tunable bandpass filter whose bandwidth is constant and small compared to the variations of $|H(f)|^2$. Figure 9.3–8a diagrams the experimental setup. If the scanning filter is centered at f_c, the rms noise voltage at its output is proportional to $|H(f_c)|$. By varying f_c, a point-by-point plot of $|H(f)|$ is obtained.

Impulse Response   Figure 9.3–8b shows a method for measuring the impulse response $h(t)$ of a given system. The instrumentation required is a white noise source, a variable time delay, a multiplier, and an averaging device. Denoting the input noise as $x(t)$, the system output is $h*x(t)$, and the delayed signal is $x(t - t_d)$. Thus, the output of the multiplier is

$$z(t) = x(t - t_d)[h*x(t)] = x(t - t_d) \int_{-\infty}^{\infty} h(\lambda)x(t - \lambda)d\lambda$$

Now suppose that $z(t)$ is averaged over a long enough time to obtain $<z(t)>$. If the noise source is ergodic and the system is linear and time-invariant, then the average output approximates the ensemble average

$$E[z(t)] = \int_{-\infty}^{\infty} h(\lambda)E[x(t - t_d)x(t - \lambda)]d\lambda = \int_{-\infty}^{\infty} h(\lambda)R_x(\lambda - t_d)d\lambda$$

But with $x(t)$ being white noise, Eq. (6b) says that $R_x(\lambda - t_d) = (N_0/2)\delta(\lambda - t_d)$. Hence,

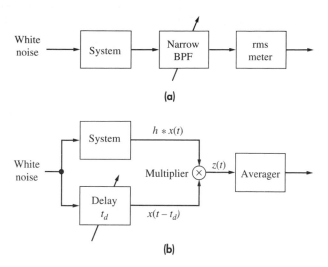

Figure 9.3–8   System measurements using white noise. (a) Amplitude response; (b) impulse response.

$$<z(t)> \approx \frac{N_0}{2} \int_{-\infty}^{\infty} h(\lambda)\delta(\lambda - t_d)d\lambda = \frac{N_0}{2} \ h(t_d)$$

and $h(t)$ can be measured by varying the time delay t_d.

The measurement techniques in Fig. 9.3–8 have special value for industrial processing or control systems. A conventional sinusoidal or pulsed input could drive such a system out of its linear operating region and, possibly, cause damage to the system. Low-level white noise then provides an attractive alternative for the test input signal.

9.4   BASEBAND SIGNAL TRANSMISSION WITH NOISE

At last we're ready to investigate the effects of noise on electrical communication. We begin here by studying baseband signal transmission systems with additive noise, and we'll introduce the signal-to-noise ratio as a measure of system performance in regard to analog communication. Section 9.5 focuses on pulse transmission.

Throughout this section and Section 9.5, we'll restrict our consideration to a linear system that does not include carrier modulation. This elementary type of signal transmission will be called **baseband communication.** The results obtained for baseband systems serve as a benchmark for comparison when we discuss carrier modulation systems with noise in subsequent chapters.

Additive Noise and Signal-to-Noise Ratios

Contaminating noise in signal transmission usually has an **additive** effect in the sense that

> Noise often adds to the information-bearing signal at various points between the source and the destination.

For purposes of analysis, all the noise will be lumped into one source added to the signal $x_R(t)$ at the input of the receiver. Figure 9.4–1 diagrams our model of additive noise.

This model emphasizes the fact that the most vulnerable point in a transmission system is the receiver input where the signal level has its weakest value. Furthermore, noise sources at other points can be "referred" to the receiver input using techniques covered in the appendix.

Since the receiver is linear, its combined input produces output *signal plus noise* at the destination. Accordingly, we write the output waveform as

$$y_D(t) = x_D(t) + n_D(t) \qquad [1]$$

where $x_D(t)$ and $n_D(t)$ stand for the signal and noise waveforms at the destination. The total output power is then found by averaging $y_D^2(t) = x_D^2(t) + 2x_D(t)n_D(t) + n_D^2(t)$.

To calculate this average, we'll treat the signal as a sample function of an ergodic process and we'll make two reasonable assumptions about additive noise:

1. The noise comes from an ergodic source with zero mean and power spectral density $G_n(f)$;

2. The noise is physically independent of the signal and therefore uncorrelated with it.

Under these conditions the statistical average of the crossproduct $x_D(t)n_D(t)$ equals zero because $x_D(t)$ and $n_D(t)$ are *incoherent*. Thus, the statistical average of $y_D^2(t)$ yields

$$\overline{y_D^2} = \overline{x_D^2} + \overline{n_D^2} \qquad [2]$$

which states that we have *superposition of signal and noise power* at the destination.

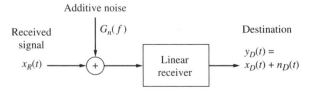

Figure 9.4–1   Model of received signal with additive noise.

Let's underscore the distinction between desired signal power and unwanted noise power by introducing the notation

$$S_D \triangleq \overline{x_D^2} \qquad N_D \triangleq \overline{n_D^2} \qquad \text{[3a]}$$

so that

$$\overline{y_D^2} = S_D + N_D \qquad \text{[3b]}$$

The **signal-to-noise ratio** will now be defined as the ratio of signal power to noise power, symbolized by

$$(S/N)_D \triangleq S_D/N_D = \overline{x_D^2}/\overline{n_D^2} \qquad \text{[4]}$$

and often expressed in decibels.

This ratio provides an important and handy indication of the degree to which the signal has been contaminated with additive noise. But note that the interpretation of signal-to-noise ratio is meaningful only when Eq. (2) holds. Otherwise, $\overline{y_D^2}$ would include additional terms involving the crossproduct of signal *times* noise.

Superposition of signal and noise power is a helpful condition in experimental work because you can't turn off the noise to determine S_D alone. Instead, you must measure N_D alone (with the signal off) and measure $\overline{y_D^2} = S_D + N_D$ (with the signal on). Given these measured values, you can calculate $(S/N)_D$ from the relationship $\overline{y_D^2}/N_D = (S_D + N_D)/N_D = (S/N)_D + 1$.

For analytic work, we generally take the case of *white noise* with $G_n(f) = N_0/2$. If the receiver has gain g_R and noise bandwidth B_N, the destination noise power becomes

$$N_D = g_R N_0 B_N \qquad \text{[5]}$$

When the noise has a gaussian distribution, this case is called **additive white gaussian noise** (AWGN), which is often the assumed model.

In any white-noise case, the noise density may also be expressed in terms of the noise temperature $\mathcal{T}_N$ referred to the receiver input, so that

$$N_0 = k\mathcal{T}_N = k\mathcal{T}_0(\mathcal{T}_N/\mathcal{T}_0) \approx 4 \times 10^{-21}(\mathcal{T}_N/\mathcal{T}_0) \quad \text{W/Hz} \qquad \text{[6]}$$

where we've introduced the standard temperature $\mathcal{T}_0$ for numerical convenience. Typical values of $\mathcal{T}_N$ range from about $0.2\mathcal{T}_0$ (60 K) for a carefully designed low-noise system up to $10\mathcal{T}_0$ or more for a "noisy" system.

Analog Signal Transmission

Figure 9.4–2 represents a simple analog signal baseband transmission system. The information generates an *analog message waveform x(t)*, which is to be reproduced at the destination. We'll model the source as an ergodic process characterized by a *message bandwidth W* such that any sample function $x(t)$ has negligible spectral content for $|f| > W$. The channel is assumed to be distortionless over the message bandwidth

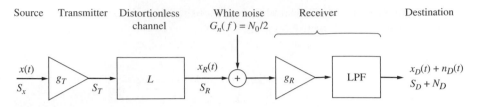

Figure 9.4–2   Analog baseband transmission system with noise.

so that $x_D(t) = Kx(t - t_d)$, where K and t_d account for the total amplification and time delay of the system. We'll concentrate on the contaminating effects of additive white noise, as measured by the system's signal-to-noise ratio at the destination.

The average signal power generated at the source can be represented as $S_x \triangleq \overline{x^2}$. Since the channel does not require equalization, the transmitter and receiver merely act as amplifiers over the message band with power gains g_T and g_R compensating for the transmission loss L. Thus, the transmitted signal power, the received signal power, and the destination signal power are related by

$$S_T = g_T\overline{x^2} = g_T S_x \quad S_R = \overline{x_R^2} = S_T/L \quad S_D = \overline{x_D^2} = g_R S_R \qquad [7]$$

These three parameters are labeled at the corresponding locations in Fig. 9.4–2.

The figure also shows a *lowpass filter* as part of the receiver. This filter has the crucial task of passing the message while reducing the noise at the destination. Obviously, the filter should reject all noise frequency components that fall outside the message band—which calls for an *ideal* LPF with bandwidth $B = W$. The resulting destination noise power will be $N_D = g_R N_0 W$, obtained from Eq. (5) with $B_N = B = W$.

We now see that the receiver gain g_R amplifies signal and noise equally. Therefore, g_R cancels out when we divide S_D by N_D, and

$$(S/N)_D = S_R/N_0 W \qquad [8]$$

This simple result gives the destination signal-to-noise ratio in terms of three fundamental system parameters: the signal power S_R and noise density N_0 at the receiver input, and the message bandwidth W. We can also interpret the denominator $N_0 W$ as the *noise power in the message band* before amplification by g_R. Consequently, a wideband signal suffers more from noise contamination than a narrowband signal.

For decibel calculations of $(S/N)_D$, we'll express the signal power in milliwatts (or dBm) and write the noise power in terms of the noise temperature $\mathcal{T}_N$. Thus,

$$\left(\frac{S}{N}\right)_{D_{dB}} = 10 \log_{10} \frac{S_R}{k\mathcal{T}_N W} \approx S_{R_{dBm}} + 174 - 10 \log_{10}\left(\frac{\mathcal{T}_N}{\mathcal{T}_0} W\right) \qquad [9]$$

where the constant 174 dB comes from Eq. (6) converted to milliwatts.

Table 9.4–1 lists typical dB values of $(S/N)_D$ along with the frequency range needed for various types of analog communication systems. The upper limit of the frequency range equals the nominal message bandwidth W. The lower limit also has design significance because many transmission systems include transformers or coupling capacitors that degrade the low-frequency response.

Table 9.4–1   Typical transmission requirements for selected analog signals

Signal Type	Frequency Range	Signal-to-Noise Ratio, dB
Barely intelligible voice	500 Hz to 2 kHz	5–10
Telephone-quality voice	200 Hz to 3.2 KHz	25–35
AM broadcast quality audio	100 Hz to 5 kHz	40–50
High-fidelity audio	20 Hz to 20 kHz	55–65
Video	60 Hz to 4.2 MHz	45–55

The destination signal-to-noise ratio doesn't depend on the receiver gain, which only serves to produce the desired signal level at the output. However, $(S/N)_D$ will be affected by any gains or losses that enter the picture *before* the noise has been added. Specifically, substituting $S_R = S_T/L$ in Eq. (8) yields

$$(S/N)_D = S_T/LN_0W \qquad [10]$$

so $(S/N)_D$ is directly proportional to the transmitted power S_T and inversely proportional to the transmission loss L—a rather obvious but nonetheless significant conclusion.

When all the parameters in Eq. (10) are fixed and $(S/N)_D$ turns out to be too small, consideration should be given to the use of *repeaters* to improve system performance. In particular, suppose that the transmission path is divided into equal sections, each having loss L_1. If a repeater amplifier with noise density N_0 is inserted at the end of the first section, its output signal-to-noise ratio will be

$$(S/N)_1 = S_T/L_1N_0W$$

which follows immediately from Eq. (10). Repeaters are often designed so the amplifier gain just equals the section loss. The analysis in the appendix then shows that if the system consists of m identical repeater sections (including the receiver), the overall signal-to-noise ratio becomes

$$\left(\frac{S}{N}\right)_D \approx \frac{1}{m}\left(\frac{S}{N}\right)_1 = \frac{L}{mL_1}\left(\frac{S_T}{LN_0W}\right) \qquad [11]$$

Compared to direct transmission, this result represents potential improvement by a factor of L/mL_1.

It should be stressed that all of our results have been based on distortionless transmission, additive white noise, and ideal filtering. Consequently, Eqs. (8)–(11) represent *upper bounds* on $(S/N)_D$ for analog communication. If the noise bandwidth of the lowpass filter in an actual system is appreciably greater than the message bandwidth, the signal-to-noise ratio will be reduced by the factor W/B_N. System nonlinearities that cause the output to include signal-times-noise terms also reduce $(S/N)_D$. However, nonlinear companding may yield a net improvement.

EXAMPLE 9.4–1

Consider a cable system having $L = 140$ dB $= 10^{14}$ and $\mathcal{T}_N = 5\mathcal{T}_0$. If you want high-fidelity audio transmission with $W = 20$ kHz and $(S/N)_D \geq 60$ dB, the necessary signal power at the receiver can be found from Eq. (9) written as

$$S_{R_{\text{dBm}}} + 174 - 10\log_{10}(5 \times 20 \times 10^3) \geq 60 \text{ dB}$$

Hence, $S_R \geq -64$ dBm $\approx 4 \times 10^{-7}$ mW and the transmitted power must be $S_T = LS_R \geq 4 \times 10^7$ mW $= 40,000$ W.

Needless to say, you wouldn't even try to put 40 kW on a signal transmission cable! Instead, you might insert a repeater at the midpoint so that $L_1 = 70$ dB $= 10^7$. (Recall that cable loss in dB is directly proportional to length.) The resulting improvement factor in Eq. (11) is

$$\frac{L}{mL_1} = \frac{10^{14}}{2 \times 10^7} = 5 \times 10^6$$

which reduces the transmitted-power requirement to $S_T \geq (4 \times 10^7 \text{ mW})/(5 \times 10^6) = 8$ mW—a much more realistic value. You would probably take S_T in the range of 10–20 mW to provide a margin of safety.

EXERCISE 9.4–1

Repeat the calculations in Example 9.4–1 for the case of video transmission with $W = 4.2$ MHz and $(S/N)_D \geq 50$ dB.

9.5   BASEBAND PULSE TRANSMISSION WITH NOISE

This section looks at baseband pulse transmission with noise, which differs from analog signal transmission in two major respects. First, rather than reproducing a waveform, we're usually concerned with measuring the pulse amplitude or arrival time or determining the presence or absence of a pulse. Second, we may know the pulse shape in advance, but not its amplitude or arrival time. Thus, the concept of signal-to-noise ratio introduced in Sect. 9.4 has little meaning here.

Pulse Measurements in Noise

Consider initially the problem of measuring some parameter of a single received pulse $p(t)$ contaminated by additive noise, as represented by the receiver diagrammed in Fig. 9.5–1a. Let the pulse be more-or-less rectangular with received amplitude A, duration τ, and energy $E_p = A^2\tau$. Let the noise be white with power

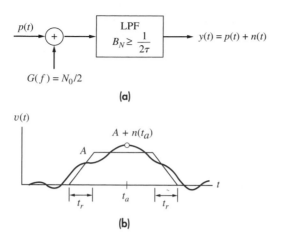

Figure 9.5–1   Pulse measurement in noise. (a) Model; (b) waveform.

spectral density $G(f) = N_0/2$ and zero mean value. The pulse will be passed and the excess noise will be removed by a reasonably selective lowpass filter having unit gain and bandwidth $B \approx B_N \geq 1/2\tau$. Thus, the output $y(t) = p(t) + n(t)$ sketched in Fig. 9.5–1b consists of noise variations superimposed on a trapezoidal pulse shape with risetime $t_r \approx 1/2B_N$.

If you want to measure the pulse *amplitude*, you should do so at some instant t_a near the center of the output pulse. A single measurement then yields the random quantity

$$y(t_a) = A + n(t_a) = A + \epsilon_A$$

where $\epsilon_A = n(t_a)$ represents the **amplitude error.** Thus, the error variance is

$$\sigma_A^2 = \overline{n^2} = N_0 B_N \qquad [1]$$

which should be small compared to A^2 for an accurate measurement. Since $A^2 = E_p/\tau$ and $B_N \geq 1/2\tau$, we can write the lower bound

$$\sigma_A^2 \geq \frac{N_0}{2\tau} = \frac{N_0 A^2}{2E_p} \qquad [2]$$

Any filter bandwidth less than about $1/2\tau$ would reduce the output pulse amplitude as well as the noise. Achieving the lower bound requires a *matched* filter as discussed later.

Measurements of pulse *arrival time* or *duration* are usually carried out by detecting the instant t_b when $y(t)$ crosses some fixed level such as $A/2$. The noise perturbation $n(t_b)$ shown in the enlarged view of Fig. 9.5–2 causes a **time-position error** ϵ_t. From the similar triangles here we see that $\epsilon_t/n(t_b) = t_r/A$, so $\epsilon_t = (t_r/A)n(t_b)$ and

$$\sigma_t^2 = (t_r/A)^2 \overline{n^2} = (t_r/A)^2 N_0 B_N$$

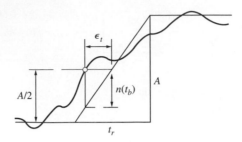

Figure 9.5–2   Time-position error caused by noise.

Substituting $t_r \approx 1/2B_N$ and $A^2 = E_p/\tau$ yields

$$\sigma_t^2 \approx \frac{N_0}{4B_N A^2} = \frac{N_0 \tau}{4B_N E_p}$$ [3]

which implies that we can make σ_t arbitrarily small by letting $B_N \to \infty$ so that $t_r \to 0$. But the received pulse actually has a nonzero risetime determined by the *transmission bandwidth* B_T. Hence,

$$\sigma_t^2 \geq \frac{N_0}{4B_T A^2} = \frac{N_0 \tau}{4B_T E_p}$$ [4]

and the lower bound is obtained with filter bandwidth $B_N \approx B_T$—in contrast to the lower bound on σ_A obtained with $B_N \approx 1/2\tau$.

EXERCISE 9.5–1   Suppose a 10-μs pulse is transmitted on a channel having $B_T = 800$ kHz and $N_0 = E_p/50$. Calculate σ_A/A and σ_t/τ when: (a) $B_N = 1/2\tau$; (b) $B_N = B_T$.

Pulse Detection and Matched Filters

When we know the pulse shape, we can design *optimum* receiving filters for detecting pulses buried in additive noise of almost any known spectral density $G_n(f)$. Such optimum filters, called **matched filters,** have extensive applications in digital communication, radar systems, and the like.

Figure 9.5–3a will be our model for pulse detection. The received pulse has known shape $p(t)$ but unknown amplitude A_p and arrival time t_0, so the received signal is

$$x_R(t) = A_p p(t - t_0)$$ [5a]

Thus, the Fourier transform of $x_R(t)$ will be

$$X_R(f) = A_p P(f) e^{-j\omega t_0}$$ [5b]

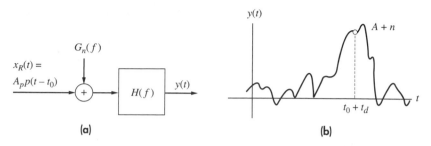

Figure 9.5-3   Pulse detection in noise. (a) Model; (b) filtered output.

where $P(f) = \mathcal{F}[p(t)]$. The total *received pulse energy* is

$$E_p = \int_{-\infty}^{\infty} |X_R(f)|^2 \, df = A_p^2 \int_{-\infty}^{\infty} |P(f)|^2 \, df \qquad \text{[5c]}$$

To detect the pulse in the face of additive noise, the receiving filter should compress the pulse energy into an output pulse with *peak amplitude A* at some specific instant, say $t = t_0 + t_d$, and the filter should also minimize the rms output noise. The output waveform would thus look something like Fig. 9.5–3b. We seek the filter transfer function $H(f)$ that achieves this goal, given $p(t)$ and $G_n(f)$.

First, we write the peak pulse amplitude in terms of $H(f)$ and $P(f)$, namely,

$$A = \mathcal{F}^{-1}[H(f)X_R(f)]\Big|_{t=t_0+t_d} = A_p \int_{-\infty}^{\infty} H(f)P(f)e^{+j\omega t_d} \, df \qquad \text{[6]}$$

Next, assuming that the noise has zero mean, its output variance is

$$\sigma^2 = \int_{-\infty}^{\infty} |H(f)|^2 \, G_n(f) \, df$$

We want to maximize the ratio of peak output amplitude A to rms noise σ or, equivalently,

$$\left(\frac{A}{\sigma}\right)^2 = A_p^2 \frac{\left|\int_{-\infty}^{\infty} H(f)P(f)e^{j\omega t_d} \, df\right|^2}{\int_{-\infty}^{\infty} |H(f)|^2 \, G_n(f) \, df} \qquad \text{[7]}$$

where $H(f)$ is the only function at our disposal. Normally, optimization requires the methods of variational calculus. But this particular problem (and a few others) can be solved by adroit application of *Schwarz's inequality*.

For our present purposes we draw upon the inequality from p. 128 in the form

$$\frac{\left|\int_{-\infty}^{\infty} V(f)W^*(f)df\right|^2}{\int_{-\infty}^{\infty} |V(f)|^2\, df} \leq \int_{-\infty}^{\infty} |W(f)|^2\, df$$

where $V(f)$ and $W(f)$ are arbitrary functions of f. The left-hand side of this inequality is identical to Eq. (7) with

$$V(f) = H(f)\sqrt{G_n(f)}$$

$$W^*(f) = \frac{A_p H(f)P(f)e^{j\omega t_d}}{V(f)} = \frac{A_p P(f)e^{j\omega t_d}}{\sqrt{G_n(f)}}$$

and the inequality becomes an equality when $V(f)$ is proportional to $W(f)$. Therefore, if $V(f) = KW(f)/A_R$, then we obtain the maximum value

$$\left(\frac{A}{\sigma}\right)^2_{\max} = A_p^2 \int_{-\infty}^{\infty} \frac{|P(f)|^2}{G_n(f)}\, df \qquad\qquad [8]$$

The corresponding optimum filter must have

$$H_{\mathrm{opt}}(f) = K\frac{P^*(f)e^{-j\omega t_d}}{G_n(f)} \qquad\qquad [9]$$

where K is an arbitrary gain constant. Observe that $|H_{\mathrm{opt}}(f)|$ is proportional to $|P(f)|$ and inversely proportional to $G_n(f)$. Hence, the optimum filter *emphasizes* those frequencies where the pulse spectrum is large and *deemphasizes* those frequencies where the noise spectrum is large—a very sensible thing to do.

In the special but important case of *white noise* with $G_n(f) = N_0/2$, Eq. (8) reduces to

$$\left(\frac{A}{\sigma}\right)^2_{\max} = \frac{2A_p^2}{N_0} \int_{-\infty}^{\infty} |P(f)|^2\, df = \frac{2E_p}{N_0} \qquad\qquad [10]$$

which brings out the significance of the energy E_p for pulse detection. The impulse response of the optimum filter is then

$$h_{\mathrm{opt}}(t) = \mathscr{F}^{-1}\left[\frac{2K}{N_0}\, P^*(f)e^{-j\omega t_d}\right] = \frac{2K}{N_0} p(t_d - t) \qquad\qquad [11]$$

The name *matched filter* comes from the fact that $h_{\mathrm{opt}}(t)$ has the same shape as the pulse $p(t)$ reversed in time and shifted by t_d. The value of t_d equals the delay between the pulse arrival and the output peak, and it may be chosen by the designer.

Sometimes the matched filter turns out to be physically *unrealizable* because $h_{opt}(t)$ has precursors for $t < 0$. However, a reasonable approximation may be obtained by taking t_d large enough so that the precursors are negligible.

Consider the rectangular pulse shape $p(t) = u(t) - u(t - \tau)$. Find $h_{opt}(t)$ from Eq. (11) when $K = N_0/2$, and determine the condition on t_d for realizability of the matched filter. Then use the convolution $h_{opt}(t)^* x_R(t)$ to obtain the peak value A of the output pulse.

EXERCISE 9.5–2

9.6 PROBLEMS

9.1–1* The random variable X has a uniform distribution over $0 \le x \le 2$. Find $\overline{v(t)}$, $R_v(t_1,t_2)$, and $\overline{v^2(t)}$ for the random process $v(t) = 6e^{Xt}$.

9.1–2 Do Prob. 9.1–1 with $v(t) = 6 \cos Xt$.

9.1–3 Let X and Y be independent RVs. Given that X has a uniform distribution over $-1 \le x \le 1$ and that $\overline{Y} = 2$ and $\overline{Y^2} = 6$, find $\overline{v(t)}$, $R_v(t_1,t_2)$, and $\overline{v^2(t)}$ for the random process $v(t) = (Y + 3Xt)t$.

9.1–4 Do Prob. 9.1–3 with $v(t) = Ye^{Xt}$.

9.1–5 Do Prob. 9.1–3 with $v(t) = Y \cos Xt$.

9.1–6‡ Let $v(t) = A \cos (2\pi Ft + \Phi)$ where A is a constant and F and Φ are RVs. If Φ has a uniform distribution over 2π radians and F has an arbitrary PDF $p_F(f)$, show that

$$R_v(t_1, t_2) = \frac{A^2}{2} \int_{-\infty}^{\infty} \cos 2\pi\lambda(t_1 - t_2)p_F(\lambda) \, d\lambda$$

Also find $\overline{v(t)}$ and $\overline{v^2(t)}$.

9.1–7* Let X and Y be independent RVs, both having zero mean and variance σ^2. Find the crosscorrelation function of the random processes

$$v(t) = X \cos \omega_0 t + Y \sin \omega_0 t$$
$$w(t) = Y \cos \omega_0 t - X \sin \omega_0 t$$

9.1–8 Consider the process $v(t)$ defined in Prob. 9.1–7. (a) Find $\overline{v(t)}$ and $R_v(t_1,t_2)$ to confirm that this process is wide-sense stationary. (b) Show, from $E[v^2(t)]$ and $<v_i^2(t)>$, that the process is not ergodic.

9.1–9 Let $v(t) = A \cos (\omega_0 t + \Phi)$, where A and Φ are independent RVs and Φ has a uniform distribution over 2π radians. (a) Find $\overline{v(t)}$ and $R_v(t_1,t_2)$ to confirm that this process is wide-sense stationary. (b) Show, from $E[v^2(t)]$ and $<v_i^2(t)>$, that the process is not ergodic.

9.1-10* Let $z(t) = v(t) - v(t + T)$, where $v(t)$ is a stationary nonperiodic process and T is a constant. Find the mean and variance of $z(t)$ in terms of $R_v(\tau)$.

9.1-11 Do Prob. 9.1–10 with $z(t) = v(t) + v(t - T)$.

9.2-1* A certain random signal has $R_v(\tau) = 16e^{-(8\tau)^2} + 9$. Find the power spectrum and determine the signal's dc value, average power, and rms value.

9.2-2 Do Prob. 9.2–1 with $R_v(\tau) = 32 \text{ sinc}^2 8\tau + 4 \cos 8\tau$.

9.2-3 Consider the signal defined in Prob. 9.1–6. Show that $G_v(f) = (A^2/4)[p_F(f) + p_F(-f)]$. Then simplify this expression for the case when $F = f_0$, where f_0 is a constant.

9.2-4 Consider the spectral estimate $\tilde{G}_v(f) = |V_T(f, s)|^2/T$. Use Eq. (8), p. 365, to show that $E[\tilde{G}_v(f)] = (T \text{ sinc}^2 fT) * G_v(f)$ What happens in the limit as $T \to \infty$?

9.2-5‡ Let $v(t)$ be a randomly phased sinusoid. Show that Eq. (7), p. 364, yields $G_v(f)$ in Eq. (4), p. 363.

9.2-6 Modify Eqs. (11b)–(11d), p. 367, for the case when $v(t)$ and $w(t)$ are independent stationary signals with $m_V m_W \neq 0$. Show from your results that $R_z(\pm\infty) = (m_V \pm m_W)^2$ and that $\overline{z^2} > 0$.

9.2-7 Let $v(t)$ and $w(t)$ be jointly stationary, so that $R_{vw}(t_1, t_2) = R_{vw}(t_1 - t_2)$. Show that

$$R_{wv}(\tau) = R_{vw}(-\tau)$$

What's the corresponding relationship between the cross-spectral densities?

9.2-8 Let $z(t) = v(t) - v(t + T)$, where $v(t)$ is a stationary random signal and T is a constant. Start with $R_z(t_1, t_2)$ to find $R_z(\tau)$ and $G_z(f)$ in terms of $R_v(\tau)$ and $G_v(f)$.

9.2-9 Do Prob. 9.2–8 with $z(t) = v(t) + v(t - T)$.

9.2-10 Let $z(t) = A \cos (2\pi f_1 t + \Phi_1) \cos (2\pi f_2 t + \Phi_2)$, where A, f_1, and f_2 are constants, and Φ_1 and Φ_2 are independent RVs, both uniformly distributed over 2π radians. Find $G_z(f)$ and simplify it for the case $f_1 = f_2$.

9.2-11 Confirm that $R_y(\tau) = h(-\tau) * R_{yx}(\tau)$.

9.2-12* Let $y(t) = dx(t)/dt$. Find $R_y(\tau)$ and $R_{yx}(\tau)$ in terms of $R_x(\tau)$ by taking the inverse transforms of $G_y(f)$ and $G_{yx}(f)$.

9.2-13 Let $y(t) = x(t) - \alpha x(t - T)$, where α and T are constants. Obtain expressions for $G_y(f)$ and $R_y(\tau)$.

9.2-14 Use the relation in Prob. 9.2–7 to show that

$$R_{x\hat{x}}(\tau) = -\hat{R}_x(\tau)$$

9.2-15‡ The moving average of a random signal $x(t)$ is defined as

$$y(t) = \frac{1}{T} \int_{t - T/2}^{t + T/2} x(\lambda) \, d\lambda$$

Find $H(f)$ such that $y(t) = h(t)^*x(t)$ and show that

$$R_y(\tau) = \frac{1}{T}\int_{-T}^{T}\left(1 - \frac{|\lambda|}{T}\right)R_x(\tau - \lambda)d\lambda$$

9.3-1   Derive Eq. (2b), p. 373, by using series approximations in Eq. (2a), p. 373.

9.3-2   Use Eq. (17b), p. 369, to show that Eq. (9b), p. 376, can be written in the alternate form

$$R_y(\tau) = \frac{N_0}{2}\int_{-\infty}^{\infty} h(t)h(t + \tau)dt$$

9.3-3*   Find $G_y(f)$, $R_y(\tau)$, and $\overline{y^2}$ when white noise is filtered by the zero-order hold on p. 241.

9.3-4   Do Prob. 9.3–3 with a gaussian filter having $H(f) = Ke^{-(af)^2}$.

9.3-5   Do Prob. 9.3–3 with an ideal BPF having gain K and delay t_0 over the frequency range $f_0 - B/2 \le |f| \le f_0 + B/2$.

9.3-6   Do Prob. 9.3–3 with an ideal HPF having gain K and delay t_0 over the frequency range $|f| \ge f_0$.

9.3-7*   Figure P9.3–7 represents a white-noise voltage source connected to a noiseless RL network. Find $G_y(f)$, $R_y(\tau)$, and $\overline{y^2}$ taking $y(t)$ as the voltage across R.

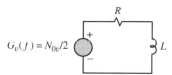

Figure P9.3–7

9.3-8‡   Do Prob. 9.3–7 taking $y(t)$ as the voltage across L.

9.3-9   The spectral density of the current $i(t)$ in Fig. P9.3–7 is $G_i(f) = N_{0v}/(2|R + j\omega L|^2)$. If the source represents thermal noise from the resistance, then the **equipartition theorem** of statistical mechanics requires that $\frac{1}{2}L\overline{i^2} = \frac{1}{2}k\mathcal{T}$. As a consequence, show that $N_{0v} = 4Rk\mathcal{T}$, in agreement with Eq. (7), p. 376.

9.3-10*   Thermal noise from a 10 kΩ resistance at room temperature is applied to an ideal LPF with $B = 2.5$ MHz and unit gain. The filtered noise is applied to a full-wave rectifier, producing $z(t) = |y(t)|$. Calculate the mean and rms value of $z(t)$.

9.3-11‡ Do Prob. 9.3–10 with a half-wave rectifier, so that $z(t) = 0$ for $y(t) < 0$.

9.3-12   Thermal noise from a 10 kΩ resistance at room temperature is applied to an ideal LPF with $B = 2.5$ MHz and unit gain. The filtered noise voltage is then applied to a delay line producing $z(t) = y(t - T)$. Use Eqs. (14) and (15), p. 344, to find the joint PDF of the random variables $Y = y(t)$ and $Z = z(t)$ when $T = 1\ \mu s$.

9.3–13‡   Do Prob. 9.3–12 with $T = 0.1\ \mu s$.

9.3–14*   Find B_N for the gaussian LPF in Prob. 9.3–4. Compare B_N with the 3-dB bandwidth B.

9.3–15   **Impulse noise,** which occurs in some communication systems, can be modeled as the random signal

$$x(t) = \sum_{k=-\infty}^{\infty} A_k \delta(t - T_k)$$

where the A_k and T_k are independent sets of random variables. The impulse weights A_k are independent and have zero mean and variance σ^2. The delay times T_k are governed by a Poisson process such that the expected number of impulses in time T equals μT. Use Eq. (7), p. 364, to show that impulse noise has a constant spectral density given by $G_x(f) = \mu \sigma^2$.

9.4–1*   Calculate $(S/N)_D$ in dB for a baseband system with $\mathcal{T}_N = \mathcal{T}_0$, $W = 4$ MHz, and $S_R = 0.02\ \mu W$.

9.4–2   Calculate $(S/N)_D$ in dB for a baseband system with $\mathcal{T}_N = 5\mathcal{T}_0$, $W = 2$ MHz, and $S_R = 0.004\ \mu W$.

9.4–3   A baseband analog transmission system with $W = 5$ kHz has $(S/N)_D = 46$ dB when $S_T = 100$ mW. If the receiver bandwidth is changed accordingly, what value of S_T is appropriate to: (*a*) upgrade the system for high-fidelity audio; (*b*) downgrade the system for telephone-quality voice?

9.4–4   Consider an AWGN baseband transmission system with $W = 10$ kHz. Express $(S/N)_D$ in a form like Eq. (8), p. 384, when the receiving filter is: (*a*) a first-order LPF with $B = 15$ kHz; (*b*) a second-order Butterworth LPF with $B = 12$ kHz.

9.4–5*   Consider a baseband transmission system with additive white noise and a distorting channel having $|H_C(f)|^2 = 1/\{L[1 + (f/W)^2]\}$. The distortion is equalized by a receiving filter with $|H_R(f)|^2 = [K/|H_C(f)|]^2 \Pi(f/2W)$. Obtain an expression for $(S/N)_D$ in the form like Eq. (10), p. 385.

9.4–6   Do Prob. 9.4–5 with $|H_C(f)|^2 = 1/\{L[1 + (2f/W)^4]\}$.

9.4–7   A baseband signal with $W = 5$ kHz is transmitted 40 km via a cable whose loss is $\alpha = 3$ dB/km. The receiver has $\mathcal{T}_N = 10\mathcal{T}_0$. (*a*) Find S_T needed to get $(S/N)_D = 60$ dB. (*b*) Repeat the calculation assuming a repeater at the midpoint.

9.4–8   A cable transmission system with $L = 240$ dB has $m = 6$ equal-length repeater sections and $(S/N)_D = 30$ dB. Find the new value of $(S/N)_D$ if: (*a*) m is increased to 12; (*b*) m is decreased to 4.

9.4–9*   The cable for a 400-km repeater system has $\alpha = 0.5$ dB/km. Find the minimum number of equal-length repeater sections needed to get $(S/N)_D \geq 30$ dB if $S_T/N_0 W = 80$ dB.

9.4–10   If all other parameters of a cable transmission system are fixed, show that the number of equal-length repeater sections that maximizes $(S/N)_D$ is $m = 0.23 L_{dB}$.

9.4–11‡ The open-circuit signal voltage output of an oscillator is $A \cos 2\pi f_0 t$. The oscillator has a noiseless source resistance R and internally generated thermal noise at temperature $\mathcal{T}_N$. A capacitance C is connected across the output terminals to improve the signal-to-noise ratio. Obtain an expression for S/N in terms of C, and find the value of C that maximizes S/N.

9.5–1* A baseband pulse transmission system has $B_N = 1/\tau$, $\mathcal{T}_N = \mathcal{T}_0$, and $E_p = 10^{-20}$ J. Find $(\sigma_A/A)^2$.

9.5–2 Show that $\sigma_t/t_r = \sigma_A/A = \sqrt{N_0 B_N \tau / E_p}$ when $B_N \le B_T$.

9.5–3 A baseband pulse transmission system has $\tau = 5$ μs, $B_T = 1$ MHz, and $N_0 = 10^{-12}$ W/Hz. Find the minimum value of E_p so that $\sigma_A \le A/100$, and calculate the corresponding value of σ_t/τ.

9.5–4 A baseband pulse transmission system has $\tau = 50$ μs, $B_T = 1$ MHz, and $N_0 = 10^{-12}$ W/Hz. Find the minimum value of E_p so that $\sigma_t \le \tau/100$, and calculate the corresponding value of σ_A/A.

9.5–5* A baseband pulse transmission system has $\tau = 1$ ms, $B_T = 1$ MHz, and $N_0 = 10^{-12}$. Find the minimum value of E_p and the corresponding value of B_N so that $\sigma_A \le A/100$ and $\sigma_t \le \tau/1000$.

9.5–6 A rectangular pulse with duration τ has been contaminated by white noise. The receiving filter is a first-order LPF with 3-dB bandwidth $B \gg 1/\tau$, rather than a matched filter. Obtain an expression for $(A/\sigma)^2$ in terms of E_p/N_0.

9.5–7 Do Prob. 9.5–6 for arbitrary B.

9.5–8* Let the shape of a received pulse be $p(t) = \Lambda(t/\tau)$. Find the characteristics of the matched filter assuming white noise. Then find the condition on t_d so that this filter can be realized.

9.5–9 Let the shape of a received pulse be $p(t) = e^{-bt}u(t)$. Find the characteristics of the matched filter assuming white noise. Then find the condition on t_d so that a good approximation of this filter can be realized.

chapter
10

Noise in Analog Modulation Systems

CHAPTER OUTLINE

This chapter rounds out our study of analog modulation with an examination of system performance in the presence of contaminating noise. Our emphasis will be on CW modulation systems, but we'll also look at analog pulse modulation.

We'll begin with the properties of the *bandpass* noise that appears in CW modulation systems. The assumption of bandpass-filtered white noise from a stationary gaussian process leads to mathematical descriptions that we employ to investigate additive noise effects in linear and exponential modulation systems. Our work culminates in a relative comparison of the several types of CW modulation. Then we'll examine the effects of additive noise on phase-lock loops and analog pulse modulation.

Further comparisons including both CW and pulse modulation will be made in Chap. 16, based upon absolute standards derived from information theory.

OBJECTIVES

After studying this chapter and working the exercises, you should be able to do each of the following:

1. Sketch the noise power spectrum at the input of a CW demodulator (Sect. 10.1).
2. Write the quadrature and envelope-and-phase expressions for bandpass noise, and relate the components to the power spectrum (Sect. 10.1).
3. Write expressions for the predetection and postdetection signal-plus-noise when a CW modulation system has additive noise (Sects. 10.2 and 10.3).
4. Sketch the noise power spectrum and calculate the noise power at the output of a CW demodulator with input bandpass noise (Sects. 10.2 and 10.3).
5. Explain the meaning and significance of threshold effect, deemphasis improvement, and wideband noise reduction (Sect. 10.3).
6. Calculate $(S/N)_D$ and the threshold level, if any, for a CW modulation system with specified parameters (Sects. 10.2 and 10.3).
7. Select a suitable analog CW modulation type, given the desired system performance and constraints (Sect. 10.4).
8. Describe the effects of additive noise on phase-lock loop performance (Sect. 10.5).
9. Determine suitable parameters for a PAM, PDM, or PPM system given the desired $(S/N)_D$ (Sect. 10.6).
10. Explain the meaning and significance of false-pulse threshold effect in PDM or PPM (Sect. 10.6).

10.1   BANDPASS NOISE

Although we must give individual attention to the effects of noise on specific types of analog modulation, all analog CW communication systems have the same general structure and suffer from bandpass noise. This section summarizes the system models and describes bandpass noise, assuming that the noise comes from an additive white gaussian noise (AWGN) process. In particular, we'll state the statistical properties of the quadrature components and the envelope and phase of bandpass noise. The stated properties are subsequently justified by studying the correlation functions.

System Models

Figure 10.1–1 depicts our generalized model of an analog CW communication system. The message $x(t)$ is a lowpass signal coming from an ergodic process with bandwidth W, normalized such that

$$|x(t)| \leq 1 \qquad S_x = \overline{x^2} = <x^2(t)> \leq 1$$

All message signal averages will be taken in the statistical sense over the ensemble.

The channel has transmission loss L but otherwise provides nearly distortionless transmission and negligible time delay. (A distorting channel would require equalization by a bandpass device located before the detector because demodulation is a nonlinear operation—except for synchronous detection which permits baseband equalization.) To minimize notational changes, the modulated signal at the output of the channel will be represented by $x_c(t)$ with carrier amplitude A_c. The **received signal power** is then

$$S_R = \frac{S_T}{L} = \overline{x_c^2} \tag{1}$$

The corresponding transmitted waveform is $\sqrt{L}x_c(t)$, so our previous expressions for S_T still apply but with A_c replaced by $\sqrt{L}A_c$.

We'll model the entire predetection portion of the receiver as a *bandpass filter* with transfer function $H_R(f)$ having unit gain over the transmission bandwidth B_T. We can ignore any predetection frequency translation since it has the same effect on signal and noise—which is also true for any predetection gain. Thus, in the usual case of superheterodyne receiver, $H_R(f)$ is the frequency response of the IF amplifier with $f_{\mathrm{IF}} = f_c$.

Under the foregoing conditions with the assumption of additive noise at the receiver's input, the total signal-plus-noise at the detector becomes

$$v(t) = x_c(t) + n(t) \tag{2}$$

where $n(t)$ represents the **predetection noise.** Eventually, we'll recast $v(t)$ in envelope-and-phase or quadrature carrier form as

$$v(t) = A_v(t) \cos[\omega_c t + \phi_v(t)] = v_i(t) \cos \omega_c t - v_q(t) \sin \omega_c t$$

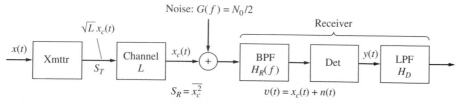

Figure 10.1–1   Model of a CW communication system with noise.

which facilitates analysis of the demodulated signal-plus-noise $y(t)$. Lastly, the demodulated waveform goes through a *lowpass filter* with $H_D(f)$ to yield the final output $y_D(t)$ at the destination. This postdetection filtering may include deemphasis or other processing operations.

The additive form of Eq. (2) together with the reasonable assumption of statistically independent signal and noise allows us to write

$$\overline{v^2} = \overline{x_c^2} + \overline{n^2} = S_R + N_R \qquad [3]$$

where $N_R = \overline{n^2}$ is the **predetection noise power.** Before discussing the signal-to-noise ratio, we'll turn off the signal and examine N_R by itself.

Figure 10.1–2a isolates that part of the system diagram relevant to N_R. Here, we treat the channel noise plus any noise generated in the predetection portion of the receiver as being equivalent to *white noise.* Hence, the filtered output $n(t)$ has spectral density

$$G_n(f) = \frac{N_0}{2} |H_R(f)|^2$$

as sketched in Fig. 10.1–2b. The density parameter N_0 includes all noise referred to the input of the receiver. We then say that

Bandpass noise results when white noise passes through a bandpass filter.

Figure 10.1–2b is based upon a predetection filter with nearly square frequency response, so its noise bandwidth equals B_T and

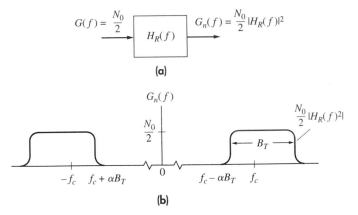

(a)

(b)

Figure 10.1–2   Bandpass filtered white noise. (a) Block diagram; (b) power spectrum.

$$N_R = \int_{-\infty}^{\infty} G_n(f)df = N_0 B_T \qquad [4]$$

A less selective filter would, of course, pass more noise power.

Note that the carrier frequency f_c does not necessarily fall in the center of the passband in Fig. 10.1–2b. We've written the lower cutoff as $f_c - \alpha B_T$ to include the symmetric-sideband case ($\alpha = 1/2$) and the suppressed-sideband case ($\alpha = 0$ or 1). Needless to say, the value of B_T depends upon the message bandwidth W and the type of modulation.

From Eqs. (3) and (4), we now define the **predetection signal-to-noise ratio**

$$\left(\frac{S}{N}\right)_R \triangleq \frac{S_R}{N_R} = \frac{S_R}{N_0 B_T} \qquad [5]$$

which looks similar to the *destination* signal-to-noise ratio $S_R/N_0 W$ defined in Sect. 9.4 relative to baseband transmission. But $(S/N)_R$ should not be confused with $S_R/N_0 W$. To bring out the distinction, we'll introduce the system parameter

$$\gamma \triangleq S_R/N_0 W \qquad [6]$$

such that

$$\left(\frac{S}{N}\right)_R = \frac{W}{B_T}\gamma \qquad [7]$$

and hence $(S/N)_R \le \gamma$ since $B_T \ge W$. You should keep in mind the interpretation that γ equals the *maximum* destination S/N for analog baseband transmission with identical values of S_R and N_0 at the receiver. By the same token, Eqs. (5) and (7) are actually *upper bounds* on $(S/N)_R$ since the various imperfections in a practical system inevitably degrade the signal-to-noise ratio to some extent.

Quadrature Components

Now let $n(t)$ be a sample function of an AWGN process. Then

$$\bar{n} = 0 \qquad \overline{n^2} = \sigma_N^2 = N_R$$

which follows from the absence of a dc component in $G_n(f)$. And the shape of $G_n(f)$ in Fig. 10.1–2b suggests expressing the noise in the usual *bandpass* form

$$n(t) = n_i(t) \cos \omega_c t - n_q(t) \sin \omega_c t \qquad [8]$$

with in-phase component $n_i(t)$ and quadrature component $n_q(t)$. These components are jointly *stationary* and *gaussian,* like $n(t)$, and have the following properties:

$$\overline{n_i} = \overline{n_q} = 0 \qquad \overline{n_i(t)n_q(t)} = 0 \qquad [9a]$$

$$\overline{n_i^2} = \overline{n_q^2} = \overline{n^2} = N_R \qquad [9b]$$

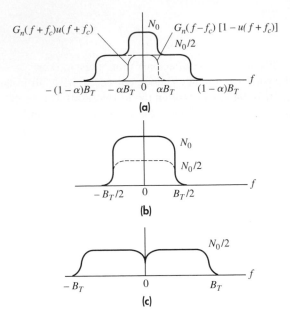

Figure 10.1–3   Lowpass power spectra of the quadrature components of bandpass noise. (a) General case; (b) symmetric-sideband case; (c) suppressed-sideband case.

Equation (9a) means that the random variables $n_i(t)$ and $n_q(t)$ are independent at any instant t, so their joint probability density function is the product of identical gaussian PDF's.

The **power spectral densities** of the quadrature components are identical *lowpass* functions related to $G_n(f)$ by

$$G_{n_i}(f) = G_{n_q}(f) = G_n(f + f_c)u(f + f_c) + G_n(f - f_c)[1 - u(f - f_c)] \quad [10]$$

where the term $G_n(f + f_c)u(f + f_c)$ simply stands for the positive frequency portion of $G_n(f)$ translated *downward* and $G_n(f - f_c)[1 - u(f - f_c)]$ stands for the negative-frequency portion translated *upward*. These terms then overlap and add for $|f| \leq \alpha B_T$, as illustrated by Fig. 10.1–3a. Figure 10.1–3b shows the complete overlap in the symmetric-sideband case ($\alpha = 1/2$), and Fig. 10.1–3c shows the lack of overlap in the suppressed-sideband case ($\alpha = 0$ or 1). When $G_n(f)$ has *local symmetry* around $\pm f_c$, as reflected in Fig. 10.1–3b, the quadrature components are *uncorrelated* processes.

EXERCISE 10.1–1   Sketch $G_n(f)$ and $G_{n_i}(f)$ when $|H_R(f)|$ has the VSB shaping shown in Fig. 10.1–4 for $f > 0$.

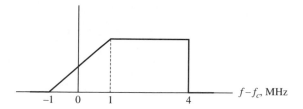

Figure 10.1–4

Envelope and Phase

As an alternative to Eq. (8), we also want to express bandpass noise in the form

$$n(t) = A_n(t) \cos \left[\omega_c t + \phi_n(t) \right] \qquad [11]$$

with envelope $A_n(t)$ and phase $\phi_n(t)$. The standard phasor diagram in Fig. 10.1–5 relates our two sets of components. Clearly, at any instant of time,

$$A_n^2 = n_i^2 + n_q^2 \qquad \phi_n = \tan^{-1} \frac{n_q}{n_i} \qquad [12a]$$

and conversely

$$n_i = A_n \cos \phi_n \qquad n_q = A_n \sin \phi_n \qquad [12b]$$

These nonlinear relationships make spectral analysis of A_n and ϕ_n difficult, even though we know $G_{n_i}(f)$ and $G_{n_q}(f)$. However, the lowpass spectrum of the quadrature components suggests that the time variations of $A_n(t)$ and $\phi_n(t)$ will be slow compared to f_c, in agreement with the bandpass nature of $n(t)$.

Furthermore, Eq. (12a) constitutes a rectangular-to-polar conversion of independent gaussian RVs, just like the one that led to the Rayleigh distribution. We thus conclude that the PDF of the envelope is a *Rayleigh* function given by

$$p_{A_n}(A_n) = \frac{A_n}{N_R} e^{-A_n^2/2N_R} u(A_n) \qquad [13]$$

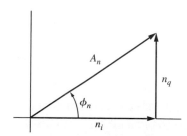

Figure 10.1–5   Phasor diagram for bandpass noise components.

with mean and second moment

$$\overline{A_n} = \sqrt{\pi N_R/2} \qquad \overline{A_n^2} = 2N_R$$

The probability that A_n exceeds some specified positive value a is then

$$P(A_n > a) = e^{-a^2/2N_R} \qquad\qquad [14]$$

These results follow from Eqs. (11)–(13) in Sect. 8.4.

The phase ϕ_n has a *uniform* PDF over 2π radians, independent of A_n. Hence,

$$\overline{n^2} = \overline{A_n^2 \cos^2(\omega_c t + \phi_n)} = \overline{A_n^2} \times 1/2 = N_R$$

which explains the factor of 2 in $\overline{A_n^2} = 2N_R$.

EXERCISE 10.1–2   Suppose bandpass noise with $N_R = 1~\mu W$ is applied to a one-ohm resistor. Calculate the mean and rms value of the envelope voltage. Also evaluate $P(A_n > 2\overline{A_n})$.

Correlation Functions★

The properties of the quadrature components of bandpass noise were presented without proof in order to put the important results up front. Now we'll outline the derivation of those results by drawing upon various correlation functions. This analysis brings together concepts and relations from several previous chapters to shed further light on bandpass noise.

We begin with the fictitious **lowpass equivalent** noise waveform defined by

$$n_{\ell p}(t) \triangleq \tfrac{1}{2}[n(t) + j\hat{n}(t)]e^{-j\omega_c t}$$

in which $\hat{n}(t)$ is the Hilbert transform of the bandpass noise $n(t)$. The lowpass nature of $n_{\ell p}(t)$ is easily confirmed by deterministic Fourier transformation. But the quadrature components of $n(t)$ should be such that

$$\tfrac{1}{2}[n_i(t) + jn_q(t)] = n_{\ell p}(t)$$

Thus, equating the real and imaginary parts of $n_{\ell p}(t)$ yields

$$n_i(t) = n(t) \cos \omega_c t + \hat{n}(t) \sin \omega_c t \qquad\qquad [15a]$$

$$n_q(t) = \hat{n}(t) \cos \omega_c t - n(t) \sin \omega_c t \qquad\qquad [15b]$$

which establishes explicit relationships between the quadrature components and $n(t)$.

This expression contains much valuable information, as follows:

1. It states the physically obvious fact that $n_i(t)$ and $n_q(t)$ depend entirely on $n(t)$— remember that $\hat{n}(t)$ represents a linear operation on $n(t)$.

2. If $n(t)$ is gaussian then $\hat{n}(t)$ is gaussian, and since Eq. (15) shows that the quadrature components are linear combinations of gaussian RVs at any instant, they must also be gaussian.

3. Equation (15) provides the starting point for correlation analysis.
4. Equation (15) brings out the importance of Hilbert transforms in the study of bandpass noise.

The Hilbert transform of a random signal was previously considered in Example 9.2–3. Applying those results to the case at hand, we have

$$G_{\hat{n}}(f) = G_n(f) \qquad R_{\hat{n}}(\tau) = R_n(\tau) \qquad [16a]$$

and

$$R_{\hat{n}n}(\tau) = \hat{R}_n(\tau) \qquad R_{n\hat{n}}(\tau) = -\hat{R}_n(\tau) \qquad [16b]$$

Here, $\hat{R}_n(\tau)$ stands for the Hilbert transform of $R_n(\tau)$, defined by $\hat{R}_n(\tau) = h_Q(\tau)*R_n(\tau)$ with $h_Q(\tau) = 1/\pi\tau$.

Having completed the necessary groundwork, we proceed to the autocorrelation function of the in-phase component $n_i(t)$. Into the basic definition

$$R_{n_i}(t, t - \tau) = E[n_i(t)n_i(t - \tau)]$$

we insert Eq. (15a) and perform some manipulations to get

$$R_{n_i}(t, t - \tau) = \frac{1}{2}\Big\{\big[R_n(\tau) + R_{\hat{n}}(\tau)\big] \cos \omega_c t + \big[R_{\hat{n}n}(\tau) - R_{n\hat{n}}(\tau)\big] \sin \omega_c t$$

$$+ \big[R_n(\tau) - R_{\hat{n}}(\tau)\big] \cos \omega_c(2t - \tau) + \big[R_{\hat{n}n}(\tau) + R_{n\hat{n}}(\tau)\big] \sin \omega_c(2t - \tau)\Big\}$$

This cumbersome expression then simplifies with the help of Eq. (16) to

$$R_{n_i}(t, t - \tau) = R_n(\tau) \cos \omega_c\tau + \hat{R}_n(\tau) \sin \omega_c\tau$$

which is independent of t. The same result holds for the autocorrelation of $n_q(t)$. Thus

$$R_{n_i}(\tau) = R_{n_q}(\tau) = R_n(\tau) \cos \omega_c\tau + \hat{R}_n(\tau) \sin \omega_c\tau \qquad [17]$$

so the quadrature components are *stationary* and have identical autocorrelation and spectral density functions.

To obtain the power spectral density via Fourier transformation of Eq. (17), we note that

$$\mathcal{F}_\tau[R_n(\tau) \cos \omega_c\tau] = \frac{1}{2}[G_n(f - f_c) + G_n(f + f_c)]$$

Then, using the convolution and modulation theorems,

$$\mathcal{F}_\tau[\hat{R}_n(\tau)] = \mathcal{F}_\tau[h_Q(\tau)]\mathcal{F}_\tau[R_n(\tau)] = (-j \operatorname{sgn} f)G_n(f)$$

and

$$\mathcal{F}_\tau[\hat{R}_n(\tau) \sin \omega_c t] = \mathcal{F}_\tau[\hat{R}_n(\tau) \cos (\omega_c\tau - \pi/2)]$$

$$= -\frac{j}{2}[-j \operatorname{sgn} (f - f_c)G_n(f - f_c)] + \frac{j}{2}[-j \operatorname{sgn} (f + f_c)G_n(f + f_c)]$$

Therefore,

$$G_{n_i}(f) = G_{n_q}(f) = \tfrac{1}{2}[1 + \text{sgn}\,(f + f_c)]G_n(f + f_c)$$

$$+ \tfrac{1}{2}[1 - \text{sgn}\,(f - f_c)]G_n(f - f_c)$$

which reduces to Eq. (10) because the first term vanishes for $f < -f_c$ whereas the second term vanishes for $f > f_c$.

Finally, a similar analysis for the crosscorrelation of the quadrature components produces

$$R_{n_i n_q}(\tau) = R_n(\tau)\,\sin\,\omega_c\tau - \hat{R}_n(\tau)\,\cos\,\omega_c\tau \qquad [18]$$

and

$$\mathscr{F}_\tau\!\left[R_{n_i n_q}(\tau)\right] = j\{G_n(f + f_c)u(f + f_c) - G_n(f - f_c)[1 - u(f - f_c)]\} \qquad [19]$$

If $G_n(f)$ has local symmetry around $f = \pm f_c$, then the right-hand side of Eq. (19) equals zero for all f. This means that $R_{n_i n_q}(\tau) = 0$ for all τ, so the quadrature components are uncorrelated and statistically independent processes.

10.2   LINEAR CW MODULATION WITH NOISE

Now we're prepared to deal with the situation in Fig. 10.2–1. The linearly modulated signal $x_c(t)$ is contaminated by AWGN at the input to the receiver. Predetection bandpass filtering produces $v(t) = x_c(t) + n(t)$ with $\overline{x_c^2} = S_R$ and $\overline{n^2} = N_R$ so

$$\left(\frac{S}{N}\right)_R = \frac{S_R}{N_R} = \frac{S_R}{N_0 B_T} = \frac{W}{B_T}\gamma$$

The bandpass noise can be expressed in quadrature form as

$$n(t) = n_i(t)\,\cos\,\omega_c t - n_q(t)\,\sin\,\omega_c t$$

where $\overline{n_i^2} = \overline{n_q^2} = N_R = N_0 B_T$. The demodulation operation will be represented by one of the following idealized mathematical models:

$$y(t) = \begin{cases} v_i(t) & \text{Synchronous detector} \\ A_v(t) - \overline{A_v} & \text{Envelope detector} \end{cases}$$

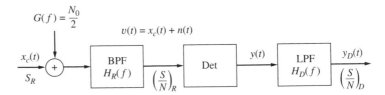

Figure 10.2–1   Model of receiver for CW modulation with noise.

These models presuppose perfect synchronization, and so forth, as appropriate. The term $\overline{A_v} = <A_v(t)>$ reflects the dc block normally found in an envelope detector. A detection constant could be included in our models but it adds nothing in the way of generality.

The questions at hand are these: Given $x_c(t)$ and the type of detector, what's the final signal-plus-noise waveform $y_D(t)$ at the destination? And if the signal and noise are additive at the output, what's the destination signal-to-noise ratio $(S/N)_D$?

Synchronous Detection

An ideal synchronous detector simply extracts the in-phase component of $v(t)$. If the modulation is DSB, then $x_c(t) = A_c\, x(t) \cos \omega_c t$ so

$$v(t) = [A_c\, x(t) + n_i(t)] \cos \omega_c t - n_q(t) \sin \omega_c t \tag{1}$$

and $y(t) = v_i(t) = A_c\, x(t) + n_i(t)$. Thus, if the postdetection filter approximates an ideal LPF with bandwidth W,

$$y_D(t) = A_c x(t) + n_i(t) \tag{2}$$

We see that the output signal and noise are, indeed, additive, and that the quadrature noise component $n_q(t)$ has been rejected by the detector.

Furthermore, if the predetection filter has a relatively square response with bandwidth $B_T = 2W$ centered at f_c, then the output noise power will take the shape of Fig. 10.1–3b. Hence,

$$G_{n_i}(f) \approx N_0 \Pi(f/2W) \tag{3}$$

which looks like lowpass-filtered white noise. Under these conditions, we don't need any postdetection filter beyond the LPF within the synchronous detector.

Next we obtain the postdetection S/N from Eq. (2) by taking the mean square values of the signal and noise terms. Upon noting that $N_D = \overline{n_i^2}$ and $S_D = A_c^2 \overline{x^2} = A_c^2 S_x$, whereas $S_R = \overline{x_c^2} = A_c^2 S_x/2$, we get

$$\left(\frac{S}{N}\right)_D = \frac{S_D}{N_D} = \frac{2S_R}{N_0 B_T} = 2\left(\frac{S}{N}\right)_R \tag{4a}$$

or, since $B_T = 2W$,

$$\left(\frac{S}{N}\right)_D = \frac{S_R}{N_0 W} = \gamma \qquad \text{DSB} \tag{4b}$$

Therefore, insofar as noise is concerned, DSB with ideal synchronous detection has the same performance as analog baseband transmission.

You might have suspected a different result in view of the predetection noise power $N_R = N_0 B_T = 2N_0 W$. However, the signal sidebands add in a *coherent* fashion, when translated to baseband, whereas the noise sidebands add *incoherently*. The sideband coherence in synchronous detection of DSB exactly counterbalances the double-sideband noise power passed by the predetection filter.

The preceding analysis is readily adapted to the case of an AM signal $x_c(t) = A_c[1 + x(t)] \cos \omega_c t$, in which we've taken $\mu = 1$ for simplicity. If the synchronous detector includes an ideal dc block, then $y_D(t)$ will be as given in Eq. (2) so $S_D = A_c^2 S_x$ and $N_D = \overline{n_i^2}$. But when we account for the unmodulated carrier power in $S_R = A_c^2(1 + S_x)/2$ we find that $S_D = 2S_x S_R/(1 + S_x)$ and

$$\left(\frac{S}{N}\right)_D = \frac{2S_x}{1 + S_x}\left(\frac{S}{N}\right)_R = \frac{S_x}{1 + S_x}\gamma \qquad \text{AM} \qquad [5]$$

This ratio is bounded by $(S/N)_D \leq \gamma/2$ since $S_x \leq 1$.

Full-load tone modulation corresponds to $S_x = 1/2$ and $(S/N)_D = \gamma/3$, which is about 5 dB below that of DSB with the same parameters. More typically, however, $S_x \approx 0.1$ and AM would be some 10 dB inferior to DSB. AM broadcasting stations usually combat this effect with special techniques such as volume compression and peak limiting of the modulating signal to keep the carrier fully modulated most of the time. These techniques actually distort $x(t)$.

For SSB modulation (or VSB with a small vestige) we have $x_c(t) = (A_c/2)[x(t) \cos \omega_c t \pm \hat{x}(t) \sin \omega_c t]$ with $B_T = W$ and $S_R = A_c^2 S_x/4$. Synchronous detection rejects the quadrature component of both the signal and noise, leaving

$$y_D(t) = \tfrac{1}{2}A_c x(t) + n_i(t) \qquad [6]$$

so $S_D = A_c^2 S_x/4 = S_R$. Since f_c falls at either edge of an ideal predetection filter, $G_{n_i}(f)$ has the shape of Fig. 10.1–3c. Hence,

$$G_{n_i}(f) \approx \frac{N_0}{2}\Pi(f/2W) \qquad [7]$$

and $N_D = \overline{n_i^2} = N_0 W$. Therefore,

$$\left(\frac{S}{N}\right)_D = \left(\frac{S}{N}\right)_R = \gamma \qquad \text{SSB} \qquad [8]$$

which shows that SSB yields the same noise performance as analog baseband or DSB transmission.

Finally, consider VSB plus carrier. If the vestigial band is small compared to W, then the predetection and postdetection noise will be essentially the same as SSB. But the signal will be essentially the same as AM with all the information-bearing power in one sideband. Hence,

$$\left(\frac{S}{N}\right)_D \approx \frac{S_x}{1 + S_x}\left(\frac{S}{N}\right)_R \approx \frac{S_x}{1 + S_x}\gamma \qquad \text{VSB + C} \qquad [9]$$

assuming that $B_T \approx W$ and $\mu \approx 1$.

To summarize the results in Eqs. (2)–(9), we state the following general properties of synchronously detected linear modulation with noise:

1. The message and noise are additive at the output if they are additive at the detector input.

2.   If the predetection noise spectrum is reasonably flat over the transmission band, then the destination noise spectrum is essentially constant over the message band.

3.   Relative to $(S/N)_D$, suppressed-sideband modulation has no particular advantage over double-sideband modulation because the coherence property of double sideband compensates for the reduced predetection noise power of single sideband.

4.   Making due allowance for the "wasted" power in unsuppressed-carrier systems, all types of linear modulation have the same performance as baseband transmission on the basis of average transmitted power and fixed noise density.

These statements presume nearly ideal systems with fixed *average* power.

Comparisons based on *peak envelope power* indicate that SSB yields a postdetection S/N about 3 dB better than DSB and 9 dB better than AM, assuming a reasonably smooth modulating signal. But SSB is inferior to DSB if the message has pronounced discontinuities causing envelope horns.

Suppose the predetection filter for a USSB signal actually passes $f_c - W/4 \leq |f| \leq f_c + W$. Use Fig. 10.1–3a to sketch the postdetection noise power spectrum. Then show that $(S/N)_D$ will be about 1 dB less than the value predicted by Eq. (8).   **EXERCISE 10.2–1**

Envelope Detection and Threshold Effect

Inasmuch as AM is normally demodulated by an envelope detector, we should examine how this differs from synchronous detection when noise is present. At the detector input we have

$$v(t) = A_c[1 + x(t)] \cos \omega_c t + [n_i(t) \cos \omega_c t - n_q(t) \sin \omega_c t] \qquad \text{[10]}$$

where we're still taking $\mu = 1$. The phasor construction of Fig. 10.2–2 shows that the resultant envelope and phase are

$$A_v(t) = \sqrt{\{A_c[1 + x(t)] + n_i(t)\}^2 + [n_q(t)]^2} \qquad \text{[11]}$$

$$\phi_v(t) = \tan^{-1} \frac{n_q(t)}{A_c[1 + x(t)] + n_i(t)}$$

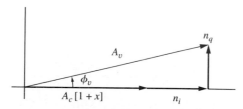

Figure 10.2–2   Phasor diagram for AM plus noise with $(S/N)_R \gg 1$.

Clearly, further progress calls for some simplifications, so let's assume that the signal is either very large or very small compared to the noise.

Taking the signal to dominate, say $A_c^2 \gg \overline{n^2}$, then $A_c[1 + x(t)]$ will be large compared to $n_i(t)$ and $n_q(t)$, at least most of the time. The envelope can then be approximated by

$$A_v(t) \approx A_c[1 + x(t)] + n_i(t) \tag{12}$$

which shows the envelope modulation due to noise, similar to interference modulation. An ideal envelope detector reproduces the envelope minus its dc component, so

$$y_D(t) = A_v(t) - \overline{A_v} = A_c x(t) + n_i(t)$$

which is identical to that of a synchronous detector. The postdetection S/N is then as previously given in Eq. (5). Likewise, Eq. (9) will hold for envelope detection of VSB + C.

But bear in mind that these results hold only when $A_c^2 \gg \overline{n^2}$. Since $A_c^2/\overline{n^2}$ is proportional to $S_R/N_0 B_T$, an equivalent requirement is $(S/N)_R \gg 1$. (There is no such condition with synchronous detection.) Thus, providing that the predetection signal-to-noise ratio is large, envelope demodulation in the presence of noise has the same performance quality as synchronous demodulation.

At the other extreme, with $(S/N)_R \ll 1$, the situation is quite different. For if $A_c^2 \ll \overline{n^2}$, the noise dominates in a fashion similar to strong interference, and we can think of $x_c(t)$ as modulating $n(t)$ rather than the reverse. To expedite the analysis of this case, $n(t)$ is represented in *envelope-and-phase* form $n(t) = A_n(t) \cos [\omega_c t + \phi_n(t)]$, leading to the phasor diagram of Fig. 10.2–3. In this figure the noise phasor is the reference because we are taking $n(t)$ to be dominant. The envelope is then approximated by the horizontal component, so

$$A_v(t) \approx A_n(t) + A_c[1 + x(t)] \cos \phi_n(t) \tag{13}$$

from which

$$y(t) = A_n(t) + A_c x(t) \cos \phi_n(t) - \overline{A_n} \tag{14}$$

where $\overline{A_n} = \sqrt{\pi N_R/2}$.

The principal output component is obviously the noise envelope $A_n(t)$, as expected. Furthermore, there is no term in Eq. (14) strictly proportional to the mes-

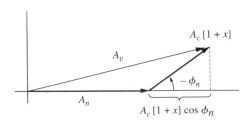

Figure 10.2–3   Phasor diagram for AM plus noise with $(S/N)_R \ll 1$.

sage $x(t)$. Though signal and noise were *additive* at the input, the detected message term is *multiplied* by noise in the form of cos $\phi_n(t)$, which is random. The message is therefore hopelessly **mutilated,** and its information has been lost. Under these circumstances, an output signal-to-noise ratio is difficult to define, if not meaningless.

The mutilation or loss of message at low predetection signal-to-noise ratios is called **threshold effect.** The name comes about because

> There is some value of $(S/N)_R$ above which mutilation is negligible and below which system performance rapidly deteriorates.

With synchronous detection, the output signal and noise are always additive. True, the message is buried in noise if $(S/N)_R \ll 1$, but the identity of $x(t)$ is preserved.

Actually, the threshold is not a unique point unless some convention is established for its definition. Generally speaking, threshold effects are minimal if $A_c \gg A_n$ most of the time. To be more specific we define the **threshold level** as that value of $(S/N)_R$ for which $A_c \geq A_n$ with probability 0.99. Then

$$\left(\frac{S}{N}\right)_{Rth} = 4 \ln 10 \approx 10 \qquad [15a]$$

or, since $(S/N)_R = \gamma/2$,

$$\gamma_{th} = 8 \ln 10 \approx 20 \qquad [15b]$$

If $(S/N)_R < (S/N)_{Rth}$ (or $\gamma < \gamma_{th}$), message mutilation must be expected, along with the consequent loss of information.

Looking at the value of $(S/N)_{Rth}$ and recalling that $(S/N)_D < (S/N)_R$ leads to a significant conclusion:

> Threshold effect is usually not a serious limitation for AM broadcasting.

For audio transmission demands a postdetection signal-to-noise ratio of 30 dB or more, so $(S/N)_R$ is well above the threshold level. In other words, additive noise obscures the signal long before multiplicative noise mutilates it. On the other hand, sophisticated processing techniques exist for recovering digital signals buried in additive noise. Hence, if AM is used for digital transmission, synchronous detection may be necessary to avoid threshold effects.

Lastly, let's consider how an envelope detector can act in a synchronous fashion and why this requires large $(S/N)_R$. Assuming the input noise is negligible, the diode in an envelope detector functions as a switch, closing briefly on the carrier peaks of the proper polarity; therefore the switching is perfectly synchronized with the carrier.

But when noise dominates, the switching is controlled by the noise peaks and synchronism is lost. The latter effect never occurs in true synchronous detectors, where the locally generated carrier can always be much greater than the noise.

EXERCISE 10.2–2   Use Eq. (14), p. 404, to derive the threshold level in Eq. (15a). Specifically, show that if $S_x = 1$, then $P(A_c \geq A_n) = 0.99$ requires $(S/N)_R = 4 \ln 10 \approx 10$.

10.3   EXPONENTIAL CW MODULATION WITH NOISE

This section deals with noise in analog PM and FM systems. The demodulation operation will be represented by

$$y(t) = \begin{cases} \phi_v(t) & \text{Phase detector} \\ \frac{1}{2\pi}\dot{\phi}_v(t) & \text{Frequency detector} \end{cases}$$

As we saw in Chap. 5, the inherent nonlinear nature of exponential modulation leads to analytic difficulties—all the more so when noise must be considered. We'll therefore begin with the large signal condition $(S/N)_R \gg 1$ to determine the postdetection noise characteristics and signal-to-noise ratios for PM and FM. Our efforts here pay off in results that quantify the valuable *wideband noise reduction* property, a property further enhanced by postdetection FM deemphasis filtering.

But wideband noise reduction involves a *threshold effect* that, unlike the AM case, may pose a significant performance limitation. We'll qualitatively discuss operation near threshold, and take a brief look at the FM feedback receiver as one technique for threshold extension.

Postdetection Noise

The predetection portion of an exponential modulation receiver has the structure previously diagrammed in Fig. 10.2–1 (p. 406). The received signal is

$$x_c(t) = A_c \cos\left[\omega_c t + \phi(t)\right]$$

where $\phi(t) = \phi_\Delta x(t)$ for a PM wave or $\dot{\phi}(t) = 2\pi f_\Delta x(t)$ for an FM wave. In either case, the carrier amplitude remains *constant* so

$$S_R = \tfrac{1}{2}A_c^2 \qquad \left(\frac{S}{N}\right)_R = \frac{A_c^2}{2N_0 B_T} \tag{1}$$

and $(S/N)_R$ is often called the **carrier-to-noise ratio** (CNR). The predetection BPF is assumed to have a nearly ideal response with bandwidth B_T centered at f_c.

Figure 10.3–1 portrays our model for the remaining portion of the receiver, with the detector input $v(t) = x_c(t) + n(t) = A_v(t) \cos\left[\omega_c t + \phi_v(t)\right]$. The limiter sup-

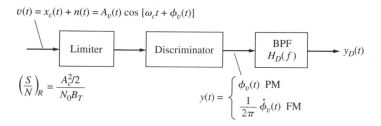

$$v(t) = x_c(t) + n(t) = A_v(t) \cos \left[\omega_c t + \phi_v(t) \right]$$

Limiter → Discriminator → BPF $H_D(f)$ → $y_D(t)$

$$\left(\frac{S}{N} \right)_R = \frac{A_c^2/2}{N_0 B_T}$$

$$y(t) = \begin{cases} \phi_v(t) \ \text{PM} \\ \dfrac{1}{2\pi}\, \dot{\phi}_v(t) \ \text{FM} \end{cases}$$

Figure 10.3–1   Model for detection of exponential modulation plus noise.

presses any amplitude variation represented by $A_v(t)$. To find the signal and noise contained in $\phi_v(t)$, we express $n(t)$ in envelope-and-phase form and write

$$v(t) = A_c \cos \left[\omega_c t + \phi(t) \right] + A_n(t) \cos \left[\omega_c t + \phi_n(t) \right] \qquad [2]$$

The phasor construction of Fig. 10.3–2 then shows that

$$\phi_v(t) = \phi(t) + \tan^{-1} \frac{A_n(t) \sin \left[\phi_n(t - \phi(t) \right]}{A_c + A_n(t) \cos \left[\phi_n(t) - \phi(t) \right]} \qquad [3]$$

The first term of $\phi_v(t)$ is the signal phase by itself, but the contaminating second term involves both noise *and* signal. Clearly, this expression is very unclear and we can't go much further without some simplifications.

A logical simplification comes from the large-signal condition $(S/N)_R \gg 1$, so $A_c \gg A_n(t)$ most of the time and we can use the small-argument approximation for the inverse tangent function. A less obvious simplification ignores $\phi(t)$ in Eq. (3), replacing $\phi_n(t) - \phi(t)$ with $\phi_n(t)$ alone. We justify this step for purposes of noise analysis by recalling that ϕ_n has a *uniform* distribution over 2π radians; hence, in the sense of ensemble averages, $\phi_n - \phi$ differs from ϕ_n only by a shift of the mean value. With these two simplifications Eq. (3) becomes

$$\phi_v(t) \approx \phi(t) + \psi(t) \qquad [4]$$

where

$$\psi(t) \triangleq \frac{A_n \sin \phi_n(t)}{A_c} = \frac{1}{\sqrt{2S_R}} n_q(t) \qquad [5]$$

in which we've substituted $n_q = A_n \sin \phi_n$ and $S_R = A_c^2/2$.

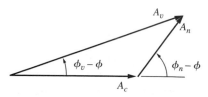

Figure 10.3–2   Phasor diagram of exponential modulation plus noise.

Equation (4) says that the signal phase $\phi(t)$ and the equivalent phase noise $\psi(t)$ are *additive* under the large-signal condition. Equation (5) brings out the fact that $\psi(t)$ depends on the *quadrature* component of $n(t)$ and decreases with increasing signal power.

Now let $\phi(t) = 0$ and consider the resulting noise $\psi(t)$ at the output of a *phase detector*. The PM postdetection noise power spectrum has the shape of $G_{n_q}(f)$ in Fig. 10.1–3b (p. 402), but multiplied by $1/2S_R$ because $\overline{\psi^2} = \overline{n_q^2}/2S_R$. Hence,

$$G_\psi(f) \approx \frac{N_0}{2S_R} \Pi\left(\frac{f}{B_T}\right) \qquad [6]$$

which is essentially flat over $|f| \leq B_T/2$, as sketched in Fig. 10.3–3.

Since $B_T/2$ exceeds the message bandwidth W, save for the special case of NBPM, the receiver should include a *postdetection filter* with transfer function $H_D(f)$ to remove out-of-band noise. If $H_D(f)$ approximates the response of an ideal LPF with unit gain and bandwidth W, then the output noise power at the destination will be

$$N_D = \int_{-W}^{W} G_\psi(f)\,df = \frac{N_0 W}{S_R} \qquad \text{PM} \qquad [7]$$

The shaded area in Fig. 10.3–3 equals N_D.

Next consider a *frequency detector* with input $\phi_v(t) = \psi(t)$, so the output is the *instantaneous frequency noise*

$$\xi(t) \triangleq \frac{1}{2\pi} \dot{\psi}(t) = \frac{1}{2\pi\sqrt{2S_R}} \dot{n}_q(t) \qquad [8]$$

Thus, from Eq. (21), p. 370, we get the FM postdetection noise power spectrum

$$G_\xi(f) = (2\pi f)^2 \frac{1}{8\pi^2 S_R} G_{n_q}(f) = \frac{N_0 f^2}{2S_R} \Pi\left(\frac{f}{B_T}\right) \qquad [9]$$

This *parabolic* function sketched in Fig. 10.3–4 has components beyond $W < B_T/2$, like PM, but increases as f^2.

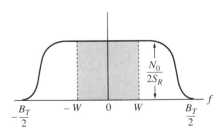

Figure 10.3–3   PM postdetection noise spectrum.

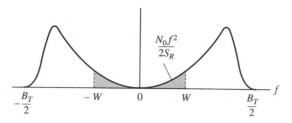

Figure 10.3–4   FM postdetection noise spectrum.

If we again take the postdetection filter to be an essentially ideal LPF that passes the shaded region in Fig. 10.3–4, the destination noise power will be

$$N_D = \int_{-W}^{W} G_\xi(f)\,df = \frac{N_0 W^3}{3S_R} \qquad \text{FM} \qquad \text{[10]}$$

However, if we also incorporate *deemphasis* filtering such that $|H_D(f)| = |H_{\text{de}}(f)|\Pi(f/2W)$ with $|H_{\text{de}}(f)| = [1 + (f/B_{\text{de}})^2]^{-1/2}$ then

$$N_D = \int_{-W}^{W} |H_{\text{de}}(f)|^2 G_\xi(f)\,df = \frac{N_0 B_{\text{de}}^3}{S_R}\left[\left(\frac{W}{B_{\text{de}}}\right) - \tan^{-1}\left(\frac{W}{B_{\text{de}}}\right)\right] \qquad \text{[11a]}$$

In the usual case where $W/B_{\text{de}} \gg 1$, Eq. (11a) simplifies to

$$N_D \approx N_0 B_{\text{de}}^2 W/S_R \qquad \text{Deemphasized FM} \qquad \text{[11b]}$$

since $\tan^{-1}(W/B_{\text{de}}) \approx \pi/2 \ll W/B_{\text{de}}$.

Let's summarize and comment on our results up to this point as follows:

1. The postdetection noise spectral densities in PM and FM have out-of-band components that call for postdetection filtering.

2. The PM noise spectrum is flat, like linear modulation except for the out-of-band components.

3. The FM noise spectrum increases parabolically, so higher baseband signal frequencies suffer more noise contamination than lower frequencies. Deemphasis filtering compensates for this effect, provided that the message has been preemphasized at the transmitter.

4. The destination noise power N_D in PM and FM decreases as S_R increases, a phenomenon known as **noise quieting.** You may hear noise quieting on an FM radio with the volume turned up when you tune between stations.

Find the deemphasized noise spectrum $|H_{\text{de}}(f)|^2 G_\xi(f)$ *without* a lowpass cutoff for $|f| > W$. Then estimate the total area N_D assuming $B_T \gg W \gg B_{\text{de}}$. Compare your result with Eq. (11b).   **EXERCISE 10.3–1**

Destination S/N

Now we can calculate the signal-to-noise ratios for PM, FM, and deemphasized FM. We continue the large-signal condition $(S/N)_R \gg 1$ so our previous results for N_D still hold. The presence of a signal phase $\phi(t)$ does not void those results even though we replaced $\phi_n(t) - \phi(t)$ with $\phi_n(t)$ in Eq. (3). Had we included $\phi(t)$ in the phase noise, a more complicated analysis would have revealed that the postdetection noise spectrum includes additional components that fall outside the message band and are rejected by $H_D(f)$.

The demodulated signal plus noise in a PM system with $\phi(t) = \phi_\Delta x(t)$ is

$$y(t) = \phi_v(t) = \phi_\Delta x(t) + \psi(t)$$

The postdetection filter passes the signal term $\phi_\Delta x(t)$ so $S_D = \phi_\Delta^2 \overline{x^2} = \phi_\Delta^2 S_x$, and the output noise power N_D is given by Eq. (7). Hence,

$$\left(\frac{S}{N}\right)_D = \frac{\phi_\Delta^2 S_x}{(N_0 W / S_R)} = \phi_\Delta^2 S_x \frac{S_R}{N_0 W} = \phi_\Delta^2 S_x \gamma \qquad \text{PM} \qquad [12]$$

Since γ equals the output S/N for analog baseband transmission (or suppressed-carrier linear modulation) with received power S_R, bandwidth W, and noise density N_0, we see that PM gives an improvement over baseband of exactly $\phi_\Delta^2 S_x$. But in view of the ambiguity constraint $\phi_\Delta \le \pi$, the PM improvement is no greater than $\phi_\Delta^2 S_x|_{\max} = \pi^2$, or about 10 dB at best. In fact if $\phi_\Delta^2 S_x < 1$, then PM performance is inferior to baseband but the transmission bandwidth is still $B_T \ge 2W$.

The demodulated signal plus noise in an FM system with $\dot{\phi}(t) = 2\pi f_\Delta x(t)$ is

$$y(t) = \frac{1}{2\pi} \dot{\phi}_v(t) = f_\Delta x(t) + \xi(t)$$

The postdetection filter passes the signal term $f_\Delta x(t)$ so $S_D = f_\Delta^2 S_x$, and N_D is given by Eq. (10). Hence,

$$\left(\frac{S}{N}\right)_D = \frac{f_\Delta^2 S_x}{(N_0 W^3 / 3 S_R)} = 3\left(\frac{f_\Delta}{W}\right)^2 S_x \frac{S_R}{N_0 W}$$

in which we spot the *deviation ratio* $D = f_\Delta / W$. We therefore write

$$\left(\frac{S}{N}\right)_D = 3 D^2 S_x \gamma \qquad \text{FM} \qquad [13]$$

and it now appears that $(S/N)_D$ can be made *arbitrarily large* by increasing D without increasing the signal power S_R—a conclusion that requires further qualification and will be reexamined shortly.

Meanwhile, recall that the transmission bandwidth requirement B_T increases with the deviation ratio. Therefore, Eq. (13) represents **wideband noise reduction** in that we have the latitude to

Exchange increased bandwidth for reduced transmitter power while keeping $(S/N)_D$ fixed.

To emphasize this property, take the case of wideband FM with $D \gg 1$ and $B_T = 2f_\Delta \gg W$. Then $D = B_T/2W$ and Eq. (13) becomes

$$\left(\frac{S}{N}\right)_D = \frac{3}{4}\left(\frac{B_T}{W}\right)^2 S_x \gamma \qquad \text{WBFM} \qquad [14]$$

which shows that $(S/N)_D$ increases as the square of the bandwidth ratio B_T/W. With smaller deviation ratios, the break-even point compared to baseband transmission occurs when $3D^2 S_x = 1$ or $D = 1/\sqrt{3S_x} \geq 0.6$. The dividing line between NBFM and WBFM is sometimes designated to be $D \approx 0.6$ for this reason.

Finally, if the receiver includes deemphasis filtering and $B_{de} \ll W$, the output noise is further reduced in accordance with Eq. (11b). Thus,

$$\left(\frac{S}{N}\right)_D \approx \left(\frac{f_\Delta}{B_{de}}\right)^2 S_x \gamma \qquad \text{Deemphasized FM} \qquad [15]$$

and we have a deemphasis improvement factor of about $(W/B_{de})^2/3$. This improvement requires preemphasis filtering at the transmitter and may carry a hidden penalty. For if the message amplitude spectrum does not roll off at least as fast as $1/f$, like an audio signal does, then preemphasis increases the deviation ratio and the transmission bandwidth requirement.

Just how much can be gained from wideband noise reduction is well illustrated with the broadcast FM parameters $f_\Delta = 75$ kHz, $W = 15$ kHz, and $D = 5$. Taking $S_x = 1/2$ for a representative value, Eq. (13) gives

EXAMPLE 10.3–1

$$(S/N)_D = (3 \times 5^2 \times 1/2)\gamma \approx 38\gamma$$

or about 16 dB better than analog baseband transmission. Deemphasis filtering with $B_{de} = 2.1$ kHz increases $(S/N)_D$ to about 640γ. Thus, other factors being equal, a 1-W FM system with deemphasis could replace a 640-W baseband system. The cost of this transmitted power reduction is increased bandwidth, since FM with $D = 5$ requires $B_T \approx 14W$.

But several practical factors work against full realization of increased bandwidth in exchange for reduced transmitter power S_T. And indeed the goal of FM broadcasting is maximum $(S/N)_D$ rather than minimum S_T. However, other applications involve minimizing S_T or squeezing as much as possible from every available transmitted watt. The large-signal condition $(S/N)_R \gg 1$ then poses a serious limitation for such applications, and the FM threshold effect becomes a matter of grave concern.

EXERCISE 10.3–2   Calculate the minimum transmitted power needed when a PM system replaces the 1-W FM system in Example 10.3–1 and the value of $(S/N)_D$ remains the same.

FM Threshold Effect

The small-signal condition $(S/N)_R \ll 1$ can be represented by a phasor diagram like Fig. 10.3–2 with signal and noise phasors interchanged. Then, since $A_n(t) \gg A_c$ most of the time, the resultant phase at the detector input is

$$\phi_v(t) \approx \phi_n(t) + \frac{A_c \sin\left[\phi(t) - \phi_n(t)\right]}{A_n(t)} \tag{16}$$

The noise now dominates and the message, contained in $\phi(t)$, has been *mutilated* beyond all hope of recovery.

Actually, significant mutilation begins to occur when $(S/N)_R \approx 1$ and $\overline{A_n} \approx A_c$. With phasors of nearly equal length, we have a situation similar to cochannel interference when $\rho = A_i/A_c \approx 1$. Small noise variations may then produce large **spikes** in the demodulated FM output. The phasor diagram in Fig. 10.3–5a illustrates this

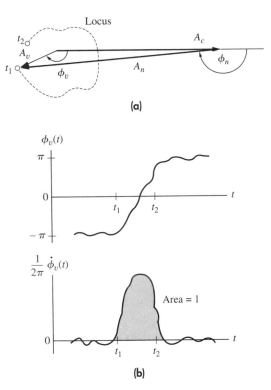

(a)

(b)

Figure 10.3–5   FM near threshold. (a) Phasor diagram; (b) instantaneous phase and frequency.

point taking $\phi(t) = 0$ and $\phi_n(t_1) \approx -\pi$ so $\phi_v(t_1) \approx -\pi$. If the variations of $A_n(t)$ and $\phi_n(t)$ follow the dashed locus from t_1 to t_2 then $\phi_v(t_2) \approx +\pi$. Correspondingly, the waveform $\phi_v(t)$ in Fig. 10.3–5b has a step of height 2π and the output $y(t) = \dot{\phi}_v(t)/2\pi$ has a unit-area spike. These spikes would be heard on an FM radio as a crackling or clicking sound that masks the signal.

We infer from this qualitative picture that the output noise spectrum is no longer parabolic but tends to fill in at dc, the output spikes producing appreciable low-frequency content. This conclusion has been verified through detailed analysis using the "click" approach as refined by Rice. The analysis is complicated (and placed beyond our scope) by the fact that the spike characteristics change when the carrier is modulated—called the *modulation-suppression effect*. Thus, quantitative results are obtained only for specific modulating signals. In the case of tone modulation, the total output noise becomes

$$N_D = \frac{N_0 W^3}{3S_R} \left[1 + \frac{12D}{\pi} \gamma e^{-(W/B_T)\gamma} \right] \qquad [17]$$

where the second term is the contribution of the spikes. See Rice (1948) and Stumpers (1948) for the original work.

Figure 10.3–6 shows $(S/N)_D$ in decibels plotted versus γ in decibels for two values of the deviation ratio D, taking tone modulation and N_D given by Eq. (17). The rather sudden drop-off of these curves is the FM threshold effect, traced to the exponential factor in Eq. (17). We see that

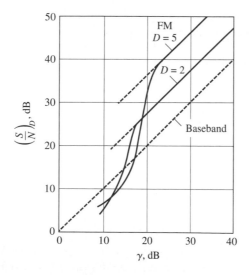

Figure 10.3–6   FM noise performance (without deemphasis).

When the system is operating near the "knee" of a curve, small variations of received signal power cause sizable changes in the output signal—one moment it's there, the next moment it's gone.

Below threshold, noise *captures* the output just like strong cochannel interference.

Experimental studies indicate that noise mutilation is negligible in most cases of interest if $(S/N)_R \geq 10$ or thereabouts. Hence, we define the *threshold point* to be at

$$(S/N)_{Rth} = 10 \qquad [18]$$

Equivalently, since $(S/N)_R = (W/B_T)\gamma$,

$$\gamma_{th} = 10\frac{B_T}{W} = 20M(D) \qquad [19a]$$

$$\approx 20(D + 2) \qquad D > 2 \qquad [19b]$$

where use has been made of the FM bandwidth equation $B_T = 2M(D)W \approx 2(D + 2)W$. Equations (18) and (19) also apply to PM with D replaced by ϕ_Δ.

Figure 10.3–6 correctly demonstrates that FM performance above threshold is quite impressive—after all, baseband transmission at best gives $(S/N)_D = \gamma$. And these curves do not include the additional improvement afforded by deemphasis filtering. But observe what happens if we attempt to make $(S/N)_D$ arbitrarily large by increasing only the deviation ratio while holding γ fixed, say at 20 dB. With $D = 2$ $(B_T \approx 7W)$ we are just above threshold and $(S/N)_D \approx 28$ dB. But with $D = 5$ $(B_T \approx 14W)$ we are below threshold, and the output signal is useless because of mutilation. We therefore cannot achieve an unlimited exchange of bandwidth for signal-to-noise ratio, and system performance may actually deteriorate with increased deviation.

Swapping bandwidth in favor of reduced signal power is likewise restricted. Suppose, for example, that a 30-dB signal-to-noise ratio is desired with a minimum of transmitted power but the transmission bandwidth can be as large as $B_T = 14W$. Were it not for threshold effect, we could use FM with $D = 5$ and $\gamma = 14$ dB, a power saving of 16 dB compared to baseband. But the threshold point for $D = 5$ is at $\gamma_{th} \approx 22$ dB, for which $(S/N)_D \approx 37$ dB. Thus,

The design may be dictated by the threshold point rather than the desired signal-to-noise ratio.

Correspondingly, the potential power reduction may not be fully realized.

In view of these considerations, it's useful to calculate $(S/N)_D$ at the threshold point. Thus, again omitting deemphasis, we substitute Eq. (19) into Eq. (13) to get

$$\left(\frac{S}{N}\right)_{D\text{th}} = 3D^2 S_x \gamma_{\text{th}} \qquad \text{[20a]}$$

$$\approx 60D^2(D+2)S_x \qquad D>2 \qquad \text{[20b]}$$

which equals the minimum value of $(S/N)_D$ as a function of D. Given a specified value for $(S/N)_D$ and no bandwidth constraint, you can solve Eq. (20) for the deviation ratio D that yields the *most efficient performance* in terms of signal power. Of course, some allowance must be made for possible signal fading since it is unadvisable to operate with no margin relative to the threshold point.

EXAMPLE 10.3–2

Suppose a minimum-power FM system is to be designed such that $(S/N)_D \approx 50$ dB, given $S_x = 1/2$, $W = 10$ kHz, $N_0 = 10^{-8}$ W/Hz, and no constraint on B_T. Temporarily ignoring threshold, we might use Eq. (13) to get $10^5 = 1.5D^2\gamma$ so $\gamma \approx 296$ when $D = 15$. But taking threshold into account with the stated values and the assumption that $D > 2$, Eq. (20) becomes $10^5 \approx 60D^2(D+2)/2$, and trial-and-error solution yields $D \approx 15$ so $B_T \approx 2(D+2)W = 340$ kHz. Then, from Eq. (19a), $S_R/N_0W \geq \gamma_{\text{th}} = 10 \times 34 = 340$, which requires $S_R \geq 340N_0W = 34$ mW.

EXERCISE 10.3–3

Find the minimum useful value of $(S/N)_D$ for a deemphasized FM system with $B_T = 5W$, $f_\Delta = 10B_{\text{de}}$, and $S_x = 1/2$.

Threshold Extension by FM Feedback★

Since the threshold limitation yields a constraint on the design of minimum-power analog FM systems, there has been interest in *threshold extension* techniques. Long ago Chaffee (1939) proposed a means for extending the FM threshold point using a frequency-following or frequency-compressive feedback loop in the receiver, called an FM feedback (FMFB) receiver.

The FMFB receiver diagrammed in Fig. 10.3–7 embodies features of a phase-lock loop within the superheterodyne structure. Specifically, the superhet's LO has been replaced by a VCO whose free-running frequency equals $f_c - f_{IF}$. The control voltage for the VCO comes from the demodulated output $y_D(t)$. If the loop has sufficient gain K and $(S/N)_D$ is reasonably large, then the VCO tracks the instantaneous phase of $x_c(t)$. This tracking action reduces the frequency deviation from f_Δ to $f_\Delta/(1+K)$, as well as translating the signal down to the IF band. Thus, if K is such that $f_\Delta/(1+K)W < 1$, then the IF input becomes a *narrowband* FM signal and the IF bandwidth need be no larger than $B_{IF} \approx 2W$.

VCO tracking likewise reduces the noise frequency deviation by the same factor, so $(S/N)_D$ equals that of a conventional receiver when $(S/N)_R \gg 1$. But note that the

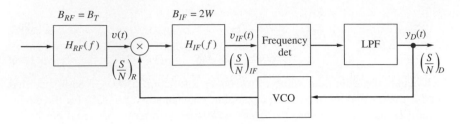

Figure 10.3–7   FMFB receiver for threshold extension.

IF has a larger predetection signal-to-noise ratio, namely $(S/N)_{IF} = S_R/N_{0BIF} \approx (B_T/2W)(S/N)_R$. Since threshold effects now depend primarily on the value of $(S/N)_{IF}$, the threshold level has been extended down to a lower value. Experimental studies confirm a threshold extension of 5–7 dB for FMFB receivers—a signficant factor for minimum-power designs. A conventional receiver with a PLL demodulator also provides threshold extension and has the advantage of simpler implementation.

10.4   COMPARISON OF CW MODULATION SYSTEMS

At last we're in a position to make a meaningful comparison of the various types of analog CW modulation. Table 10.4–1 summarizes the points to be compared: normalized transmission bandwidth $b = B_T/W$, destination signal-to-noise ratio $(S/N)_D$ normalized by γ, threshold point if any, dc response (or low-frequency response), and instrumentation complexity. The table also includes baseband transmission for reference purposes. As before, we have used $\gamma = S_R/N_0W$ where S_R is the received signal power, W is the message bandwidth, and $N_0 = k\mathcal{T}_N$ is the noise density referred to the receiver input. We have also used $S_x = \overline{x^2} = <x^2(t)>$, where $x(t)$ is the message. Nearly ideal systems are assumed, so the values of $(S/N)_D$ are *upper bounds*.

Of the several types of linear modulation, suppressed carrier methods are superior to conventional AM on at least two counts: signal-to-noise ratios are better, and there is no threshold effect. When **bandwidth conservation** is important, single sideband and vestigial sideband are particularly attractive. But you seldom get something for nothing in this world, and the price of efficient linear modulation is the increased complexity of instrumentation, especially at the receiver. Synchronous detection, no matter how it's accomplished, requires sophisticated circuitry compared to the envelope detector. For **point-to-point communication** (one transmitter, one receiver) the price might be worthwhile. But for **broadcast communication** (one transmitter, *many* receivers) economic considerations tip the balance toward the simplest possible receiver, and hence envelope detection.

From an instrumentation viewpoint AM is the least complex linear modulation, while suppressed-carrier VSB, with its special sideband filter and synchronization requirements, is the most complex. Of DSB and SSB (in their proper applications) the latter is less difficult to instrument because synchronization is not so critical. In addition, improved filter technology has made the required sideband filters more readily available. Similarly, VSB + C is classed as of "moderate" complexity, despite the vestigial filter, since envelope detection is sufficient.

Table 10.4–1   Comparison of CW modulation systems

Type	$b = B_T/W$	$(S/N)_D \div \gamma$	γ_{th}	DC	Complexity	Comments
Baseband	1	1	. . .	No[1]	Minor	No modulation
AM	2	$\dfrac{\mu^2 S_x}{1 + \mu^2 S_x}$	20	No	Minor	Envelope detection $\mu \leq 1$
DSB	2	1	. . .	Yes	Major	Synchronous detection
SSB	1	1	. . .	No	Moderate	Synchronous detection
VSB	1+	1	. . .	Yes	Major	Synchronous detection
VSB + C	1+	$\dfrac{\mu^2 S_x}{1 + \mu^2 S_x}$	20	Yes[2]	Moderate	Envelope detection $\mu < 1$
PM[3]	$2M(\phi_\Delta)$	$\phi_\Delta^2 S_x$	$10b$	Yes	Moderate	Phase detection, constant amplitude $\phi_\Delta \leq \pi$
FM[3,4]	$2M(D)$	$3D^2 S_x$	$10b$	Yes	Moderate	Frequency detection, constant amplitude

[1]Unless direct coupled.
[2]With electronic dc restoration.
[3]$b \geq 2$.
[4]Deemphasis not included.

Compared to baseband or linear modulation, exponential modulation can provide substantially increased values of $(S/N)_D$—especially FM with deemphasis—with only moderately complex instrumentation. Figure 10.4–1 illustrates this in a form similar to Fig. 10.3–6 (again taking $S_x = 1/2$) except that a representative 12-dB deemphasis improvement has been added to the FM curves and performance below threshold is omitted. All curves are labeled with the bandwidth ratio b.

Clearly, for equal values of b, FM is markedly superior to PM insofar as noise performance is concerned. And as long as the system is above threshold, the improvement can be made arbitrarily large by increasing b, whereas PM is limited to $b \leq 10$ since $\phi_\Delta \leq \pi$. The penalty for the FM improvement is excessive transmission bandwidth. Therefore, wideband exponential modulation is most appropriate when extremely clean output signals are desired and bandwidth conservation is a secondary factor. At microwave frequencies, both the noise-reduction and constant-amplitude

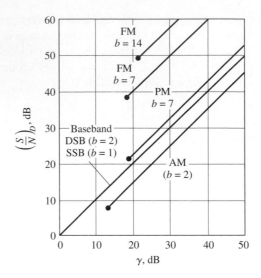

Figure 10.4–1   Performance of CW modulation systems, including 12-dB deemphasis improvement for FM.

properties are advantageous. As to power conservation, FM with moderate values of *b* does offer a saving over linear modulation, threshold limitations notwithstanding.

Regarding transmission of modulating signals having significant low-frequency components, we have already argued the superiority of DSB and VSB. For facsimile and video, electronic dc restoration makes envelope-detected VSB possible and desirable. (AM could be used in this way, but the bandwidth is prohibitive. Suppressed-carrier single sideband is virtually out of the question.) Also we noted previously that a balanced discriminator has excellent low-frequency response; hence, the low-frequency performance of FM can equal that of DSB or VSB, and without troublesome synchronization. For similar reasons, high-quality magnetic-tape recorders are often equipped with an FM mode in which the input is recorded as a frequency-modulated wave.

Not shown in the table is relative system performance in the face of time-varying transmission characteristics, frequency-selective fading, multiple-path propagation, and so forth. An unstable transmission medium has a *multiplicative* effect, which is particularly disastrous for envelope detection. (Late-night listeners to distant AM stations are familiar with the garbled result.) Similarly, transmission instabilities often preclude wideband modulation.

To summarize briefly is impossible. There is no universal solution to all communication problems. The communication engineer must therefore approach each new task with an open mind and a careful review of all available information.

10.5   PHASE-LOCK LOOP NOISE PERFORMANCE★

In Chapter 7, we assumed the PLL input was noise free. This is usually not the case, so we now want to consider the effects of additive white noise on PLL performance.

Let's augment the linearized models of Figs. 7.3–8b and 7.3–8c with a noise source as shown in Figure 10.5–1. The output of the second summer is the phase error $e(t)$. Since the system is linear, we can also consider the signal and noise inputs separately, and so with just the noise input, the power spectral density (PSD) of the VCO output phase is

$$G_{\phi_v}(f) = |H_L(f)|^2 \, G_n(f) \qquad\qquad [1]$$

where $G_n(f)$ is the PSD of the noise and $H_L(f)$ is the PLL closed loop transfer function previously derived in Sect. 7.3. If the noise source is white, then $G_n(f) = N_0/2$ and the VCO output variance (or noise) becomes

$$\sigma_{\phi_v}^2 = N_0 \int_0^\infty |H_L(f)|^2 \, df \qquad\qquad [2]$$

Since we have assumed the input $\phi(t) = 0$ then the *phase error noise* becomes $\sigma_e^2 = \sigma_{\phi_v}^2$. Given that the noise equivalent bandwidth is

$$B_N = \frac{1}{g} \int_0^\infty |H_L(f)|^2 \, df \qquad\qquad [3]$$

and assuming that $g = 1$, the phase error noise becomes

$$\sigma_e^2 = N_0 B_N \qquad\qquad [4]$$

This phase noise represents the amount of variation, or jitter, contained in the VCO output and is called **phase-jitter.**

When Eqs. (3) and (4) are examined closely, we see that phase jitter decreases with narrower loop bandwidths. Therefore loop bandwidth versus phase jitter becomes another trade-off we have to deal with. Decreasing the loop bandwidth will

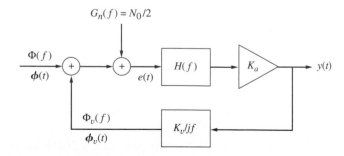

Figure 10.5–1   Linearized PLL model with additive noise

decrease phase jitter, but adversely affects the ability of the system to track the input signal's phase variations.

If we assume the input signal is sinuosidal with amplitude A_c, then we can express the loop's signal-to-noise ratio by

$$\gamma = \frac{A_c^2}{\sigma_e^2} \qquad [5]$$

Equations (4) and (5) assume a linear PLL model with a relatively high signal-to-noise ratio. If this is not the case, a better approximation for phase-jitter for a first-order PLL is given by Vitterbi (1966) as

$$\sigma_e^2 = \int_{-\pi}^{\pi} \phi^2 \frac{\exp(\gamma \cos \phi)}{2\pi J_0(\gamma)} \, d\phi \qquad [6]$$

where $J_0(\gamma)$ is the zeroth-order modified Bessel function of the first kind.

10.6 ANALOG PULSE MODULATION WITH NOISE

Finally, we take a brief look at analog pulse modulation with noise. For simplicity, we'll assume that the pulse-modulated signal has been sent at baseband, so there will be no CW modulation and no bandpass noise. Instead, the predetection noise will have a *lowpass* spectrum.

Signal-to-Noise Ratios

Regardless of the particular circuitry employed in the receiver, demodulating an analog pulse-modulated wave boils down to message reconstruction from sample values. A generalized demodulator would therefore consist of a pulse converter that transforms the pulse-modulated wave into a train of weighted impulses from which an ideal LPF reconstructs the message. Figure 10.6–1 diagrams this demodulation model including additive noise.

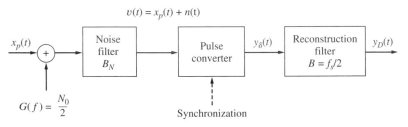

Figure 10.6–1   Model for demodulation of analog pulse modulation with noise.

A noise-limiting filter at the input passes the received pulsed signal $x_p(t)$ plus lowpass noise $n(t)$, so $v(t) = x_p(t) + n(t)$. The converter measures the amplitude, duration, or position of each pulse in $v(t)$ and generates

$$y_\delta(t) = \sum_k [\mu_p x(kT_s) + \epsilon_k]\delta(t - kT_s) \tag{1}$$

where μ_p is the modulation constant relating $x(kT_s)$ to $x_p(t)$ and ϵ_k is the measurement error induced by the noise. We'll assume for convenience that the reconstruction filter has bandwidth $B = f_s/2$, gain $K = T_s = 1/2B$, and zero time delay. Hence, the impulse-train input $y_\delta(t)$ produces the final output

$$y_D(t) = \sum_k [\mu_p x(kT_s) + \epsilon_k]\, \text{sinc}\,(f_s t - k) = \mu_p x(t) + n_D(t) \tag{2a}$$

with

$$n_D(t) = \sum_k \epsilon_k \,\text{sinc}\,(f_s t - k) \tag{2b}$$

which represents the noise at the destination.

Since the errors ϵ_k are proportional to sample values of the lowpass noise $n(t)$ spaced by T_s, and since the noise-limiting filter has $B_N > 1/T_s$, the values of ϵ_k will be essentially uncorrelated and will have zero mean. We can therefore write the destination noise power as $N_D = \overline{n_D^2} = \overline{\epsilon_k^2} = \sigma^2$ and the signal power is $S_D = \mu_p^2 \overline{x^2} = \mu_p^2 S_x$. Hence,

$$\left(\frac{S}{N}\right)_D = \frac{S_D}{N_D} = \frac{\mu_p^2}{\sigma^2} S_x \tag{3}$$

which expresses the destination signal-to-noise ratio in terms of the error variance σ^2 caused by reconstruction from noisy samples. Our next task is to determine μ_p^2 and σ^2 for specific types of analog pulse modulation.

A PAM signal contains the message samples in the modulated pulse amplitude $A_0[1 + \mu x(kT_s)]$, so the modulation constant is $\mu_p = A_0\mu \le A_0$. From Sect. 9.5, the amplitude error variance is $\sigma^2 = \sigma_A^2 = N_0 B_N$. Thus, under the best conditions of maximum modulation ($\mu = 1$) and minimum noise bandwidth ($B_N \approx 1/2\tau$), we have

$$\left(\frac{S}{N}\right)_D = \frac{2A_0^2 \tau}{N_0} S_x$$

where τ is the pulse duration.

When $\mu = 1$ and $x(t)$ has no dc component, the *average energy* per modulated pulse is $A_0^2[1 + x(kT_s)]^2\tau = A_0^2(1 + S_x)\tau$. Multiplying this average energy by the pulse rate f_s gives the *received signal power* $S_R = f_s A_0^2(1 + S_x)\tau$. We thus obtain our final result in the form

$$\left(\frac{S}{N}\right)_D = \frac{S_x}{1 + S_x} \frac{2S_R}{N_0 f_s} = \frac{S_x}{1 + S_x}\left(\frac{2W}{f_s}\right)\gamma \qquad \text{PAM} \qquad\qquad \text{[4]}$$

This result shows that $(S/N)_D \leq \gamma/2$, so that PAM performance is at least 3 dB below unmodulated baseband transmission—just like AM CW modulation. The maximum value is seldom achieved in practice, nor is it sought after. The merit of PAM resides in its simplicity for multiplexing, not in its noise performance.

However, PPM and PDM do offer some improvement by virtue of *wideband noise reduction*. For if $B_N \approx B_T$, the time-position error variance is $\sigma^2 = \sigma_t^2 = N_0/(4B_T A^2)$. Since the pulse amplitude A is a constant, the received power can be written as $S_R = f_s A^2 \tau_0$, where τ_0 denotes the *average* pulse duration in PDM or the fixed pulse duration in PPM. Equation (3) then becomes

$$\left(\frac{S}{N}\right)_D = \frac{4\mu_p^2 B_T A^2}{N_0} S_x = 4\mu_p^2 B_T \frac{S_R}{N_0 f_s \tau_0} S_x \qquad\qquad \text{[5]}$$

$$= 4\mu_p^2 B_T \left(\frac{W}{f_s \tau_0}\right) S_x \gamma \qquad \text{PDM or PPM}$$

This expression reveals that $(S/N)_D$ increases with increasing transmission bandwidth B_T. The underlying physical reason should be evident from Fig. 9.5–2, p. 388, with $t_r \approx 1/2B_T$.

The PPM modulation constant is the maximum pulse displacement, so $\mu_p = t_0$. The parameters t_0 and τ_0 are constrained by

$$t_0 \leq T_s/2 \qquad \tau_0 = \tau \geq 2t_r \approx 1/B_T$$

and $f_s = 1/T_s \geq 2W$. Taking all values to be optimum with respect to noise reduction, we obtain the upper bound

$$\left(\frac{S}{N}\right)_D \leq \frac{1}{8}\left(\frac{B_T}{W}\right)^2 S_x \gamma \qquad \text{PPM} \qquad\qquad \text{[6]}$$

Hence, PPM performance improves as the *square* of the bandwidth ratio B_T/W. A similar optimization for PDM with $\mu_p = \mu\tau_0 \leq \tau_0$ yields the less-impressive result

$$\left(\frac{S}{N}\right)_D \leq \frac{1}{2}\frac{B_T}{W} S_x \gamma \qquad \text{PDM} \qquad\qquad \text{[7]}$$

To approach the upper bound, a PDM wave must have a 50 percent duty cycle so that $\tau_0 \approx T_s/2$.

Practical PPM and PDM systems may fall short of the maximum values predicted here by 10 dB or more. Consequently, the noise reduction does not measure up to that of wideband FM. But remember that the average power S_R comes from short-duration high-power pulses rather than being continuously delivered as in CW modulation. Power-supply considerations may therefore favor pulsed operation in some circumstances.

EXERCISE 10.6–1

Explain why a single-channel PDM system must have $\mu\tau_0 \le 1/4W$. Then derive Eq. (7) from Eq. (5) with $\mu_p = \mu\tau_0$.

False-Pulse Threshold Effect

Suppose you try to increase the value of $(S/N)_D$ in a PDM or PPM system by making B_T very large. Since $\overline{n^2}$ increases with B_T, the noise variations in $v(t) = x_p(t) + n(t)$ will eventually dominate and be mistaken for signal pulses. If these **false pulses** occur often, the reconstructed waveform has no relation to $x(t)$ and the message will have been completely lost. Hence, pulse-time modulation involves a false-pulse *threshold effect,* analogous to the threshold effect in wideband FM. This effect does not exist in PAM with synchronization because we always know when to measure the amplitude.

To determine the threshold level, we'll say that false pulses are sufficiently infrequent if $P(n \ge A) \le 0.01$. For gaussian noise with $\sigma_N^2 = \overline{n^2} = N_0 B_T$, the corresponding threshold condition is approximately $A \ge 2\sigma_N$, so the pulse must be strong enough to "lift" the noise by at least twice its rms value. (This is the same condition as the **tangential sensitivity** in pulsed radar systems.) Using the fact that $A^2 = S_R/\tau_0 f_s$, we have $S_R/\tau_0 f_s \ge 4N_0 B_T$ or

$$\gamma_{th} = \left(\frac{S_R}{N_0 W}\right)_{min} = 4\tau_0 f_s \frac{B_T}{W} \ge 8 \qquad [8]$$

This threshold level is appreciably less than that of FM, so PPM could be advantageous for those situations where FM would be below its threshold point.

10.7   PROBLEMS

10.1–1   White noise with $N_0 = 10$ is applied to a BPF having $|H_R(f)|^2$ plotted in Fig. P10.1–1. Sketch $G_{n_i}(f)$ taking $f_c = f_1$, and show therefrom that $n_i^2 = \overline{n^2}$.

Figure P10.1–1

10.1–2   Do Prob. 10.1–1 taking $f_c = f_2$.

10.1–3*   White noise is applied to a tuned circuit whose transfer function is

$$H(f) = \left[1 + j\frac{4}{3}\left(\frac{f}{f_c} - \frac{f_c}{f}\right)\right]^{-1}$$

Evaluate $G_n(f) \div N_0$ at $f/f_c = 0, \pm 0.5, \pm 1, \pm 1.5,$ and ± 2. Then plot $G_{n_i}(f)$.

10.1–4   The BPF in Fig. 10.1–2 (p. 400) will have **local symmetry** around $\pm f_c$ if its lowpass equivalent function $H_{\ell p}(f) \triangleq H_R(f + f_c)u(f + f_c)$ has the even-symmetry property $|H_{\ell p}(-f)| = |H_{\ell p}(f)|$. (a) Let $G_{\ell p}(f) \triangleq (N_0/2)|H_{\ell p}(f)|^2$ and show that we can write

$$G_n(f) = G_{\ell p}(f - f_c) + G_{\ell p}(f + f_c) = \begin{cases} G_{\ell p}(f - f_c) & f > 0 \\ G_{\ell p}(f + f_c) & f < 0 \end{cases}$$

(b) Use this result to show that $G_{n_i}(f) = 2G_{\ell p}(f)$.

10.1–5   A tuned circuit with $Q = f_c/B_T \gg 1$ approximates the local-symmetry property in Prob. 10.1–4, and has $H_R(f) \approx 1/[1 + j2(f - f_c)/B_T]$ for $f > 0$. (a) Find $G_{\ell p}(f)$. (b) Evaluate $\overline{n^2}$ by calculating $\overline{n_i^2}$.

10.1–6   Let $y(t) = 2n(t)\cos(\omega_c t + \theta)$, where $n(t)$ is bandpass noise centered at f_c. Show that $y(t)$ consists of a lowpass component and a bandpass component. Find the mean value and variance of each component in terms of the properties of $n(t)$.

10.1–7*   Bandpass gaussian noise with $\sigma_N = 2$ is applied to an ideal envelope detector, including a dc block. Find the PDF of the output $y(t)$ and calculate σ_Y.

10.1–8   Bandpass gaussian noise with variance σ_N^2 is applied to an ideal square-law device, producing $y(t) = A_n^2(t)$. Find $\overline{y}$, $\overline{y^2}$, and the PDF of $y(t)$.

10.1–9   Let $v_{\ell p}(t) = \frac{1}{2}[v(t) + j\hat{v}(t)]e^{-j\omega_c t}$. Assuming that $v(t)$ is Fourier transformable, show that $V_{\ell p}(f) = \mathcal{F}[v_{\ell p}(t)] = V(f + f_c)u(f + f_c)$. Then sketch $V_{\ell p}(f)$ taking $V(f)$ as in Fig. P10.1–1 with $f_c = f_1 + b$.

10.1–10*   Let $H_R(f)$ in Fig. 10.1–2 (p. 400) be an ideal BPF with unit gain and let $\alpha = 1/2$ so f_c is the center frequency. Find $R_n(\tau)$ and $\hat{R}_n(\tau)$ by taking the inverse transform of $G_n(f)$ and $(-j \, \text{sgn} \, f)G_n(f)$. Then obtain $R_{n_i}(\tau)$ from Eq. (17) (p. 405) and confirm that it equals the inverse transform of $G_{n_i}(f)$.

10.1–11   Do Prob. 10.1–10 with $\alpha = 0$, so f_c is the lower cutoff frequency.

10.1–12   Suppose $G_n(f)$ has local symmetry around $\pm f_c$, as detailed in Prob. 10.1–4. Write the inverse transforms of $G_{n_i}(f)$ and $G_n(f)$ to show that $R_n(\tau) = R_{n_i}(\tau)\cos\omega_c\tau$. Then show from Eq. (17) that $\hat{R}_n(\tau) = R_{n_i}(\tau)\sin\omega_c\tau$.

10.1–13   Derive Eq. (17) starting from $E[n_q(t)n_q(t - \tau)]$.

10.1–14   Derive Eq. (18) starting from $E[n_q(t)n_q(t - \tau)]$.

10.1–15   Let $G_n(f)$ have local symmetry around $\pm f_c$. Prove that $R_{n_i n_q}(\tau) = 0$ for all τ using: (a) the correlation relations in Prob. 10.1–12; (b) the spectral relations in Prob. 10.1–4.

10.1–16*   Let $H_R(f)$ in Fig. 10.1–2 (p. 400) be an ideal BPF with unit gain and let $\alpha = 0$, so f_c is the lower cutoff frequency. Use Eq. (19) to find $R_{n_i n_q}(\tau)$.

10.1–17   Do Prob. 10.1–16 with $\alpha = 1/4$, so the lower cutoff frequency is $f_c - B_T/4$.

10.2–1*   A DSB signal plus noise is demodulated by synchronous detection. Find $(S/N)_D$ in dB given that $S_R = 20$ nW, $W = 5$ MHz, and $\mathcal{T}_N = 10\mathcal{T}_0$.

10.2–2   An AM signal plus noise is demodulated by synchronous detection. Find $(S/N)_D$ in dB given that $S_x = 0.4$, $S_R = 20$ nW, $W = 5$ MHz, $\mu = 1$, and $\mathcal{T}_N = 10\mathcal{T}_0$.

10.2–3   A DSB signal plus noise is demodulated by a product detector with phase error ϕ'. Take the local oscillator signal to be $2\cos(\omega_c t + \phi')$ and show that $(S/N)_D = \gamma \cos^2 \phi'$.

10.2–4   Rewrite Eqs. (4b) and (5) (p. 408) in terms of $\gamma_p = S_p/N_0 W$, where S_p is the peak envelope power of the DSB or AM signal.

10.2–5*   Let $x_c(t)$ have quadrature multiplexing as on p. 271, where $x_1(t)$ and $x_2(t)$ are independent signals and $\overline{x_1^2} = \overline{x_2^2}$. Assume an ideal receiver with AWGN and two synchronous detectors. Find the output signal plus noise for each channel, and express $(S/N)_D$ in terms of γ.

10.2–6   Explain why an SSB receiver should have a nearly rectangular BPF with bandwidth $B_T = W$, whereas predetection filtering is not critical for DSB.

10.2–7   Modify Eq. (8) (p. 408) for LSSB when $|H_R(f)|^2$ has the shape in Fig. P10.1–1 with $f_2 = f_c$ and $2b = W$.

10.2–8   Some receivers have additive "one-over-f" noise with power spectral density $G(f) = N_0 f_c/2|f|$ for $f > 0$. Obtain the resulting expressions for $(S/N)_D$ in terms of γ and W/f_c for USSB and DSB modulation. Compare your results when $W/f_c = 1/5$ and $1/50$.

10.2–9‡   When a demodulated signal includes **multiplicative noise** or related effects, the postdetection S/N cannot be defined unambiguously. An alternative performance measure is then the **normalized mean-square error** $\epsilon^2 \triangleq E\{[x(t) - Ky_D(t)]^2\}/S_x$ where K is chosen such that $Ky_D(t) = x(t)$ in absence of multiplicative effects. Find $y_D(t)$ and show that $\epsilon^2 = 2[1 - \overline{\cos \phi}] + 1/\gamma$ when a USSB signal with AWGN is demodulated by a product detector whose local-oscillator signal is $2\cos[\omega_c t + \phi(t)]$, where $\phi(t)$ is a slowly drifting random phase. *Hint*: $\overline{\hat{x}^2} = \overline{x^2}$ and $\overline{x\hat{x}} = -\hat{R}_x(0) = 0$ since $\hat{R}_x(\tau)$ is an odd function.

10.2–10   Explain why an AM receiver should have a nearly rectangular BPF with bandwidth $B_T = 2W$ for envelope detection, whereas predetection filtering is not critical for synchronous detection.

10.2–11*   An AM system with envelope detection is operating at the threshold point. Find the power gain in dB needed at the transmitter to get up to $(S/N)_D = 40$ dB with full-load tone modulation.

10.2–12   An AM system with envelope detection has $(S/N)_D = 30$ dB under full-load tone-modulation conditions with $W = 8$ kHz. If all bandwidths are increased accordingly, while other parameters are held fixed, what is the largest useable value of W?

10.2–13‡   Consider an AM system with envelope detection operating below threshold. Find ϵ^2 defined in Prob. 10.2–9 assuming that $y_D(t) = y(t)$, $\overline{x^2} = 1$, and $\bar{x} = 0$. Express your answer in terms of γ.

10.3–1*   An exponentially modulated signal plus noise has $S_R = 10$ nW, $W = 500$ kHz, and $\mathcal{T}_N = 10\mathcal{T}_0$. Find the value of N_D for PM detection, FM detection, and deemphasized FM detection with $B_{de} = 5$ kHz.

10.3–2   Suppose an nth-order Butterworth LPF is used for the postdetection filter in an FM receiver. Obtain an upper bound on N_D and simplify for $n \gg 1$.

10.3–3   Find $G_\xi(f)$ when the predetection BPF in an FM receiver has $H_R(f)$ as given in Prob. 10.1–5. Then calculate N_D and simplify taking $B_T \gg W$.

10.3–4   An FM signal plus noise has $S_R = 1$ nW, $W = 500$ kHz, $S_x = 0.1$, $f_\Delta = 2$ MHz, and $\mathcal{T}_N = 10\mathcal{T}_0$. Find $(S/N)_D$ in dB for FM detection and for deemphasized FM detection with $B_{de} = 5$ kHz.

10.3–5   Obtain an expression for $(S/N)_D$ for PM with deemphasis filtering (p. 416). Simplify your result taking $B_{de} \ll W$.

10.3–6*   The signal $x(t) = \cos 2\pi 200t$ is sent via FM without preemphasis. Calculate $(S/N)_D$ when $f_\Delta = 1$ kHz, $S_R = 500N_0$, and the postdetection filter is an ideal BPF passing $100 \leq |f| \leq 300$ Hz.

10.3–7   Obtain an expression for $(S/N)_D$ for FM with a gaussian deemphasis filter having $|H_{de}(f)|^2 = e^{-(f/B_{de})^2}$. Calculate the resulting deemphasis improvement factor when $B_{de} = W/7$.

10.3–8   A certain PM system has $(S/N)_D = 30$ dB. Find the new value of $(S/N)_D$ when the modulation is changed to preemphasized FM with $B_{de} = W/10$, while B_T and all other parameters are held fixed.

10.3–9   Modify Eq. (20) (p. 421) to include deemphasis filtering and rework the calculations in Example 10.3–2.

10.3–10*   Obtain an expression like Eq. (20) (p. 421) for PM, and determine the upper bound on $(S/N)_D$ at threshold.

10.3–11   Consider the FMFB receiver in Fig. 10.3–7 (p. 422). Let $v(t)$ be a noise-free FM signal with deviation ratio D and take the VCO output to be $2 \cos [(\omega_c - \omega_{IF})t + K\phi_D(t)]$ where $\dot{\phi}_D(t) = 2\pi y_D(t)$. Show that the deviation ratio of the IF signal is $D_{IF} = D/(1 + K)$.

10.4–1*   An analog communication system has $\overline{x^2} = 1/2$, $W = 10$ kHz, $N_0 = 10^{-15}$ W/Hz, and transmission loss $L = 100$ dB. Calculate S_T needed to get $(S/N)_D = 40$ dB when the modulation is: (*a*) SSB; (*b*) AM with $\mu = 1$ and $\mu = 0.5$; (*c*) PM with $\phi_\Delta = \pi$; (*d*) FM with $D = 1$, 5, and 10. Omit deemphasis in the FM case, but check for threshold limitations.

10.4–2   Do Prob. 10.4–1 with $\overline{x^2} = 1$ and $W = 20$ kHz.

10.4–3 An analog communication system has $\overline{x^2} = 1/2$, $W = 10$ kHz, $S_T = 10$ W, and $N_0 = 10^{-13}$ W/Hz. Calculate the path length corresponding to $(S/N)_D = 40$ dB for a transmission cable with loss factor $\alpha = 1$ dB/km when the modulation is: (a) SSB; (b) AM with $\mu = 1$; (c) FM with $D = 2$ and 8.

10.4–4 Do Prob. 10.4–3 for line-of-sight radio transmission at $f_c = 300$ MHz with antenna gains of 26 dB at transmitter and receiver.

10.4–5* A signal with $\overline{x^2} = 1/2$ is transmitted via AM with $\mu = 1$ and $(S/N)_D = 13$ dB. If the modulation is changed to FM (without deemphasis) and the bandwidths are increased accordingly while other parameters remain fixed, what's the largest usable deviation ratio and the resulting value of $(S/N)_D$?

10.4–6‡ A frequency-division multiplexing system has USSB subcarrier modulation and FM carrier modulation without preemphasis. There are K independent input signals, each having bandwidth W_0, and the subcarrier frequencies are $f_k = (k - 1)W_0$, $k = 1, 2, \ldots, K$. The baseband signal at the FM modulator is $x_b(t) = \Sigma\, \alpha_k x_k(t)$, where the α_k are constants and $x_k(t)$ is the kth subcarrier signal, each having $\overline{x_k^2} = 1$. The discriminator at the receiver is followed by a bank of BPFs and synchronous detectors. Use the property that the predetection and postdetection S/N are equal for SSB to show that the output of the kth channel has $(S/N)_k = f_\Delta^2 \alpha_k^2 / N_k$ where $N_k = (3k^2 - 3k + 1)N_0 W_0^3 / 3S_R$.

10.4–7‡ In the system described in Prob. 10.4–6, the α_k are chosen such that $\overline{x_b^2} = 1$ and all channels have the same output S/N. Obtain expressions for α_k and $(S/N)_k$ under these conditions.

10.4–8‡ Consider an FM stereo multiplexing system (p. 270) where, for purposes of analysis, the preemphasis filtering may be done after matrixing and the deemphasis filtering before matrixing. Let the input signals before preemphasis be $x_1(t) = x_L(t) + x_R(t)$ and $x_2(t) = x_L(t) - x_R(t)$, and let the deemphasized outputs before matrixing be $y_1(t) = f_\Delta x_1(t) + n_1(t)$ and $y_2(t) = f_\Delta x_2(t) + n_2(t)$ so the final outputs are $y_1(t) \pm y_2(t)$. (a) Show that the power spectral density of $n_2(t)$ is $|H_{de}(f)|^2 (N_0/S_R)(f^2 + f_0^2)\Pi(f/2W)$ where $f_0 = 38$ kHz, and that $\overline{n_2^2} \gg \overline{n_1^2}$ when $W = 15$ kHz and $B_{de} = 2.1$ kHz. (b) Taking $\overline{x_L x_R} = 0$ and $\overline{x_L^2} = \overline{x_R^2} \approx 1/3$—so that $\overline{x_b^2} \approx 1$ if the pilot is small—show that $(S/N)_D$ for each channel in stereo transmission is about 20 dB less than $(S/N)_D$ for monaural transmission.

10.6–1* A single-channel PPM signal plus noise has $S_x = 0.4$, $S_R = 10$ nW, $W = 500$ kHz, $f_s = 1.2$ MHz, $t_0 = \tau_0 = 0.1T_s$, $B_T = 10$ MHz, and $\mathcal{T}_N = 10\mathcal{T}_0$. Find the value of $(S/N)_D$ in decibels.

10.6–2 A single-channel PDM signal plus noise has $S_x = 0.1$, $W = 100$ Hz, $f_s = 250$ Hz, $\mu = 0.2$, $\tau_0 = T_s/50$, $B_T = 3$ kHz, and $\mathcal{T}_N = 50\mathcal{T}_0$. Find the value of S_R so that $(S/N)_D \geq 40$ dB.

10.6–3 Calculate the upper bound on $(S/N)_D$ for PPM baseband transmission with $B_T = 20W$ and compare with the actual value when $t_0 = 0.3T_s$, $\tau = 0.2T_s$, and $f_s = 2.5W$.

10.6–4   Equation (5), p. 428, also holds for each output of a TDM system if we write A^2 in terms of average power in the multiplexed waveform. Use this method to find $(S/N)_D$ for each channel of a PPM-TDM system having nine voice signals plus a marker, $S_x = 1$, $W = 3.2$ kHz, $f_s = 8$ kHz, $\tau = T_s/50$, $B_T = 400$ kHz, and the marker has duration 3τ. Express your result in terms of $\gamma/9$ since there are nine transmitted signals.

10.6–5   Consider an M-channel TDM-PPM system with guard time $T_g = \tau \geq 1/B_T$ and no synchronization marker, so $S_R = Mf_sA^2\tau$. Start with Eq. (5), p. 428, to show that $(S/N)_D < (1/8)(B_T/MW)^2 S_x(\gamma/M)$.

chapter

11

Baseband Digital Transmission

CHAPTER OUTLINE

This chapter launches our study of digital communication systems. We'll first describe the advantages of digital systems over analog systems. We then focus on baseband transmission to emphasize the generic concepts and problems of digital communication, with or without carrier modulation, and to bring out the differences between digital and analog transmission. Following an overview of digital signals and systems, we'll analyze the limitations imposed by additive noise and transmission bandwidth. We'll also look at practical design considerations such as regenerative repeaters, equalization, and synchronization.

The obvious question we should ask is: what do we gain with digital systems over analog ones? After all, an analog system can be implemented with relatively few components, whereas a digital system requires significantly more hardware. For example, a simple analog LPF is implemented with a resistor and capacitor. The equivalent digital implementation requires an analog-to-digital converter (ADC), digital signal processor (DSP), digital-to-analog converter (DAC), and an LPF for antialiasing. Nonetheless, despite the apparent increase in hardware complexity, we gain the following advantages:

1. **Stability.** Digital systems are inherently time invariant. Key system parameters are imbedded in algorithms that change only if reprogrammed, making for greater accuracy in signal reproduction. With analog hardware, the signal and its parameters are subject to change with component aging, external temperatures, and other environmental factors.

2. **Flexibility.** Once digital hardware is in place, we have a great deal of flexibility in changing the system. This enables us to employ a multitude of signal processing algorithms to more efficiently (a) improve signal fidelity, (b) do error correction/detection for data accuracy, (c) perform encryption for privacy and security, (d) employ compression algorithms to remove redundancies, and (e) allow for multiplexing of various types of signals such as voice, picture, video, text, and so on. Furthermore, an algorithm can be easily and remotely modified.

3. **Reliable reproduction.** An analog message traveling through a channel becomes degraded by distortion and noise. While we can employ amplifiers (repeaters) to boost the signal, they amplify the noise as well as the signal and can only increase distortion. Thus distortion becomes cumulative. Making a photocopy of another photocopy is an example of this phenomenon. As we will see, with digital communication signal reproduction is extremely reliable whether we employ regenerative repeaters for a long haul digital channel or make copies of digital audio recordings.

It should be noted, however, that we can implement much of an analog communication system using digital hardware and the appropriate ADC and DAC steps, and thereby secure for an analog system many of the advantages of a digital system.

OBJECTIVES

After studying this chapter and working the exercises, you should be able to do each of the following:

1. State the advantages of digital over analog communication.
2. Identify the format of a digital PAM waveform and calculate the mean and variance of the amplitude sequence (Sect. 11.1).
3. Find and sketch the power spectrum of a digital waveform with uncorrelated message symbols (Sect. 11.1).
4. Sketch the conditional PDFs at a digital regenerator, given the noise PDF, and formulate an expression for the error probability (Sect. 11.2).

5. Calculate the equivalent bit error probability for an optimum *M*-ary system with a distortionless channel and white noise (Sect. 11.2).

6. Relate the transmission bandwidth, signaling rate, and bit rate for an *M*-ary system with Nyquist pulse shaping (Sect. 11.3).

7. Determine appropriate parameters for a digital baseband system to satisfy stated specifications (Sects. 11.2 and 11.3).

8. Understand when data scrambling and synchronization are appropriate and how they are accomplished (Sect. 11.4).

9. Identify the properties of a maximal-length sequence produced from an *n*-bit shift register with feedback (Sect. 11.4).

10. Determine the output sequence from an *n*-bit shift register with given set of feedback connections and initial conditions (Sect. 11.4).

11.1   DIGITAL SIGNALS AND SYSTEMS

Fundamentally, a digital message is nothing more than an **ordered sequence of symbols** produced by a discrete information source. The source draws from an **alphabet** of $M \geq 2$ different symbols, and produces output symbols at some average **rate** r. For instance, a typical computer terminal has an alphabet of $M \approx 90$ symbols, equal to the number of character keys multiplied by two to account for the shift key. When you operate the terminal as fast as you can, you become a discrete information source producing a digital message at a rate of perhaps $r \approx 5$ symbols per second. The computer itself works with just $M = 2$ internal symbols, represented by LOW and HIGH electrical states. We usually associate these two symbols with the **binary digits** 0 and 1, known as **bits** for short. Data transfer rates within a computer may exceed $r = 10^8$.

The task of a digital communication system is to transfer a digital message from the source to the destination. But finite transmission bandwidth sets an upper limit to the symbol rate, and noise causes errors to appear in the output message. Thus, **signaling rate** and **error probability** play roles in digital communication similar to those of bandwidth and signal-to-noise ratio in analog communication. As preparation for the analysis of signaling rate and error probability, we must first develop the description and properties of digital signals.

Digital PAM Signals

Digital message representation at baseband commonly takes the form of an **amplitude-modulated pulse train.** We express such signals by writing

$$x(t) = \sum_k a_k p(t - kD) \qquad [1]$$

where the modulating amplitude a_k represents the *k*th symbol in the message sequence, so the amplitudes belong to a set of *M* discrete values. The index *k* ranges from $-\infty$ to $+\infty$ unless otherwise stated. Equation (1) defines a digital PAM signal, as distinguished from those rare cases when pulse-duration or pulse-position modulation is used for digital transmission.

The unmodulated pulse $p(t)$ may be rectangular or some other shape, subject to the conditions

$$p(t) = \begin{cases} 1 & t = 0 \\ 0 & t = \pm D, \pm 2D, \ldots \end{cases} \qquad \text{[2]}$$

This condition ensures that we can recover the message by sampling $x(t)$ periodically at $t = KD$, $K = 0, \pm 1, \pm 2, \ldots$, since

$$x(KD) = \sum_k a_k p(KD - kD) = a_K$$

The rectangular pulse $p(t) = \Pi(t/\tau)$ satisfies Eq. (2) if $\tau \leq D$, as does any time-limited pulse with $p(t) = 0$ for $|t| \geq D/2$.

Note that D does not necessarily equal the pulse duration but rather the pulse-to-pulse interval or the time allotted to one symbol. Thus, the *signaling rate* is

$$r \triangleq 1/D \qquad \text{[3a]}$$

measured in symbols per second or **baud.** In the special but important case of *binary* signaling ($M = 2$), we write $D = T_b$ for the bit duration and the bit rate is[†]

$$r_b = 1/T_b \qquad \text{[3b]}$$

measured in bits per second, abbreviated bps or b/s. The notation T_b and r_b will be used to identify results that apply only for binary signaling.

Figure 11.1–1 depicts various PAM **formats** or line codes for the binary message 10110100, taking rectangular pulses for clarity. The simple **on-off** waveform in part *a* represents each 0 by an "off" pulse ($a_k = 0$) and each 1 by an "on" pulse with amplitude $a_k = A$ and duration $T_b/2$ followed by a return to the zero level. We therefore call this a **return-to-zero** (RZ) format. A **nonreturn-to-zero** (NRZ) format has "on" pulses for full bit duration T_b, as indicated by the dashed lines. Internal computer waveforms are usually of this type. The NRZ format puts more energy into each pulse, but requires synchronization at the receiver because there's no separation between adjacent pulses.

The unipolar nature of an on-off signal results in a *dc component* that carries no information and wastes power. The **polar** signal in part *b* has opposite polarity pulses, either RZ or NRZ, so its dc component will be zero if the message contains 1s and 0s in equal proportion. This property also applies to the **bipolar** signal in part *c*, where successive 1s are represented by pulses with alternating polarity. The bipolar format, also known as **pseudo-trinary** or **alternate mark inversion** (AMI),

---

[†]The more common notation R for bit rate risks confusion with autocorrelation functions and with information rate defined in Chap. 16.

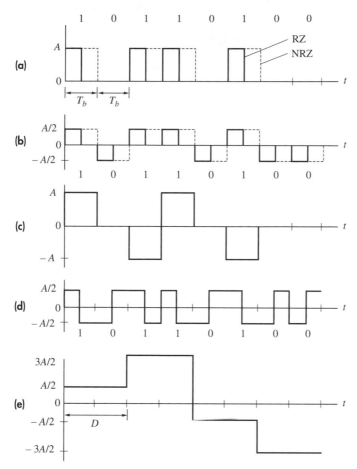

Figure 11.1–1 Binary PAM formats with rectangular pulses. (a) Unipolar RZ and NRZ; (b) polar RZ and NRZ; (c) bipolar NRZ; (d) split-phase Manchester; (e) polar quaternary NRZ.

eliminates ambiguities that might be caused by transmission *sign inversions*—a problem characteristic of switched telephone links.

The **split-phase Manchester** format in part *d* represents 1s with a positive half-interval pulse followed by a negative half-interval pulse, and vice versa for the representation of 0s. This format is also called **twinned binary.** It guarantees zero dc component regardless of the message sequence. However, it requires an absolute sense of polarity at the receiver.

Finally, Fig. 11.1–1*e* shows a **quaternary** signal derived by grouping the message bits in blocks of two and using four amplitude levels to prepresent the four possible combinations 00, 01, 10, and 11. Thus, $D = 2T_b$ and $r = r_b/2$. Different

Table 11.1–1

a_k	Natural Code	Gray Code
$3A/2$	11	10
$A/2$	10	11
$-A/2$	01	01
$-3A/2$	00	00

assignment rules or *codes* may relate the a_k to the grouped message bits. Two such codes are listed in Table 11.1–1. The Gray code has advantages relative to noise-induced errors because only one bit changes going from level to level.

Quaternary coding generalizes to **M-ary** coding in which blocks of n message bits are represented by an M-level waveform with

$$M = 2^n \qquad [4a]$$

Since each pulse now corresponds to $n = \log_2 M$ bits, the M-ary signaling rate has been decreased to

$$r = \frac{r_b}{\log_2 M} \qquad [4b]$$

But increased signal power would be required to maintain the same spacing between amplitude levels.

Transmission Limitations

Now consider the linear baseband transmission system diagrammed in Fig. 11.1–2a. We'll assume for convenience that the transmitting amplifier compensates for the transmission loss, and we'll lump any interference together with the additive noise. After lowpass filtering to remove out-of-band contaminations, we have the signal-plus-noise waveform

$$y(t) = \sum_k a_k \tilde{p}(t - t_d - kD) + n(t)$$

where t_d is the transmission delay and $\tilde{p}(t)$ stands for the pulse shape with transmission distortion. Figure 11.1–2b illustrates what $y(t)$ might look like when $x(t)$ is the unipolar NRZ signal in Fig. 11.1–1a.

Recovering the digital message from $y(t)$ is the task of the **regenerator.** An auxiliary synchronization signal may help the regeneration process by identifying the optimum sampling times

$$t_K = KD + t_d$$

If $\tilde{p}(0) = 1$ then

$$y(t_K) = a_K + \sum_{k \neq K} a_k \tilde{p}(KD - kD) + n(t_K) \qquad [5]$$

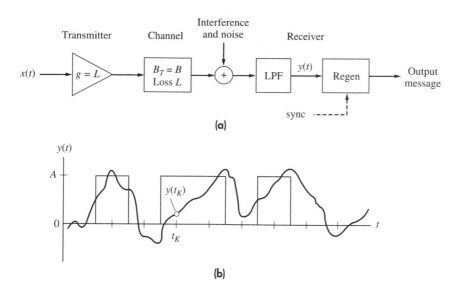

Figure 11.1–2 (a) Baseband transmission system; (b) signal-plus-noise waveform.

whose first term is the desired message information. The last term of Eq. (5) is the noise contamination at t_K, while the middle term represents cross talk or spillover from other signal pulses—a phenomenon given the descriptive name **inter-symbol interference** (ISI). The combined effects of noise and ISI may result in *errors* in the regenerated message. For instance, at the sample time t_K indicated in Fig. 11.1–2b, $y(t_K)$ is closer to 0 even though $a_K = A$.

We know that if $n(t)$ comes from a white-noise source, then its mean square value can be reduced by reducing the bandwidth of the LPF. We also know that low-pass filtering causes pulses to spread out, which would increase the ISI. Consequently, a fundamental limitation of digital transmission is the relationship between ISI, bandwidth, and signaling rate.

This problem emerged in the early days of telegraphy, and Harry Nyquist (1924, 1928a) first stated the relationship as follows:

> Given an ideal lowpass channel of bandwidth B, it is possible to transmit independent symbols at a rate $r \leq 2B$ baud without intersymbol interference. It is not possible to transmit independent symbols at $r > 2B$.

The condition $r \leq 2B$ agrees with our pulse-resolution rule $B \geq 1/2\tau_{min}$ in Sect. 3.4 if we require $p(t)$ to have duration $\tau \leq D = 1/r$.

The second part of Nyquist's statement is easily proved by assuming that $r = 2(B + \epsilon) > 2B$. Now suppose that the message sequence happens to consist of two symbols alternating forever, such as $101010\ldots$. The resulting waveform $x(t)$ then is *periodic* with period $2D = 2/r$ and contains only the fundamental frequency $f_0 = B + \epsilon$ and its harmonics. Since no frequency greater than B gets through the channel, the output signal will be zero—aside from a possible but useless dc component.

Signaling at the maximum rate $r = 2B$ requires a special pulse shape, the *sinc pulse*

$$p(t) = \text{sinc } rt = \text{sinc } t/D \qquad \text{[6a]}$$

having the *bandlimited* spectrum

$$P(f) = \mathcal{F}[p(t)] = \frac{1}{r}\,\Pi\!\left(\frac{f}{r}\right) \qquad \text{[6b]}$$

Since $P(f) = 0$ for $|f| > r/2$, this pulse suffers no distortion from an ideal lowpass frequency response with $B \geq r/2$ and we can take $r = 2B$. Although $p(t)$ is not time-limited, it does have *periodic zero crossings* at $t = \pm D, \pm 2D, \ldots$, which satisfies Eq. (2). (See Fig. 6.1–6 for an illustration of this property.) Nyquist also derived other bandlimited pulses with periodic zero crossings spaced by $D > 1/2B$ so $r < 2B$, a topic we set aside to pick up again in Sect. 11.3 after discussing noise and errors in Sect. 11.2.

Meanwhile, note that any real channel needs *equalization* to approach an ideal frequency response. Such equalizers often require experimental adjustment in the field because we don't know the channel characteristics exactly. An important experimental display is the so-called **eye pattern,** which further clarifies digital transmission limitations.

Consider the distorted but noise-free polar binary signal in Fig. 11.1–3a. When displayed on a long-persistence oscilloscope with appropriate synchronization and sweep time, we get the superposition of successive symbol intervals shown in Fig. 11.1–3b. The shape of this display accounts for the name "eye pattern." A distorted M-ary signal would result in $M - 1$ "eyes" stacked vertically.

Figure 11.1–4 represents a generalized binary eye pattern with labels identifying significant features. The optimum sampling time corresponds to the maximum eye opening. ISI at this time partially closes the eye and thereby reduces the **noise**

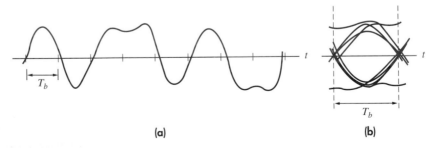

(a)   (b)

Figure 11.1–3   (a) Distorted polar binary signal; (b) eye pattern.

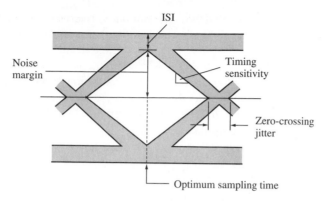

Figure 11.1–4   Generalized binary eye pattern.

margin. If synchronization is derived from the zero crossings, as it usually is, zero-crossing distortion produces **jitter** and results in nonoptimum sampling times. The slope of the eye pattern in the vicinity of the zero crossings indicates the sensitivity to timing error. Finally, any nonlinear transmission distortion would reveal itself in an asymmetric or "squinted" eye.

Determine the relation between r and B when $p(t) = \text{sinc}^2 at$.   **EXERCISE 11.1–1**

Power Spectra of Digital PAM

The pulse spectrum $P(f) = \mathcal{F}[p(t)]$ provides some hints about the *power spectrum* of a digital PAM signal $x(t)$. If $p(t) = \text{sinc } rt$, as a case in point, then $P(f)$ in Eq. (6b) implies that $G_x(f) = 0$ for $|f| > r/2$. However, detailed knowledge of $G_x(f)$ provides additional and valuable information relative to digital transmission.

A simplified random digital wave with $p(t) = \Pi(t/D)$ was tackled in Chap. 9. Under the conditions

$$E[a_k a_i] = \begin{cases} \sigma_a^2 & i = k \\ 0 & i \neq k \end{cases}$$

we found that $G_x(f) = \sigma_a^2 D \, \text{sinc}^2 fD$. Now, substituting $P(f) = D \, \text{sinc } fD$, we write

$$G_x(f) = \frac{\sigma_a^2}{D} |P(f)|^2 \qquad [7]$$

This expression holds for *any* digital PAM signal with pulse spectrum $P(f)$ when the a_k are uncorrelated and have zero mean value.

But unipolar signal formats have $\overline{a_k} \neq 0$ and, in general, we can't be sure that the message source produces uncorrelated symbols. A more realistic approach therefore models the source as a *discrete stationary random process*. Ensemble averages of the a_k are then given by the autocorrelation function

$$R_a(n) = E[a_k a_{k-n}]$$ [8]

analogous to writing $R_v(\tau) = E[v(t)v(t - \tau)]$ for a stationary random signal $v(t)$. The integers n and k in Eq. (8) reflect the time-discrete nature of a digital sequence.

If a digital PAM signal $x(t)$ has the pulse spectrum $P(f)$ and amplitude autocorrelation $R_a(n)$, its power spectrum is

$$G_x(f) = \frac{1}{D}|P(f)|^2 \sum_{n=-\infty}^{\infty} R_a(n)e^{-j2\pi nfD}$$ [9]

Despite the formidable appearance of Eq. (9), it easily reduces to Eq. (7) when $R_a(0) = \sigma_a^2$ and $R_a(n) = 0$ for $n \neq 0$. In the case of uncorrelated message symbols but $\overline{a_k} = m_a \neq 0$,

$$R_a(n) = \begin{cases} \sigma_a^2 + m_a^2 & n = 0 \\ m_a^2 & n \neq 0 \end{cases}$$ [10]

and

$$\sum_{n=-\infty}^{\infty} R_a(n)e^{-j2\pi nfD} = \sigma_a^2 + m_a^2 \sum_{n=-\infty}^{\infty} e^{-j2\pi nfD}$$

Then, drawing upon Poisson's sum formula,

$$\sum_{n=-\infty}^{\infty} e^{-j2\pi nfD} = \frac{1}{D} \sum_{n=-\infty}^{\infty} \delta\left(f - \frac{n}{D}\right)$$

and therefore

$$G_x(f) = \sigma_a^2 r|P(f)|^2 + (m_a r)^2 \sum_{n=-\infty}^{\infty} |P(nr)|^2 \delta(f - nr)$$ [11]

Here we have inserted $r = 1/D$ and used the sampling property of impulse multiplication.

The important result in Eq. (11) reveals that the power spectrum of $x(t)$ contains impulses at *harmonics* of the signaling rate r, unless $m_a = 0$ or $P(f) = 0$ at all $f = nr$. Hence, a synchronization signal can be obtained by applying $x(t)$ to a narrow BPF (or PLL filter) centered at one of these harmonics. We can also calculate the total *average power* $\overline{x^2}$ by integrating $G_x(f)$ over all f. Thus,

$$\overline{x^2} = \sigma_a^2 r E_p + (m_a r)^2 \sum_{n=-\infty}^{\infty} |P(nr)|^2$$ [12]

where E_p equals the energy in $p(t)$. For Eq. (12), and hereafter, we presume the conditions in Eq. (10) barring information to the contrary.

The derivation of Eq. (9) starts with the definition of power spectrum from Eq. (7), Sect. 9.2. Specifically, we write

$$G_x(f) \triangleq \lim_{T \to \infty} \frac{1}{T} E[|X_T(f)|^2]$$

in which $X_T(f)$ is the Fourier transform of a truncated sample function $x_T(t) = x(t)$ for $|t| < T/2$. Next, let $T = (2K + 1)D$ so the limit $T \to \infty$ corresponds to $K \to \infty$. Then, for $K \gg 1$,

$$x_T(t) = \sum_{k=-K}^{K} a_k p(t - kD)$$

$$X_T(f) = \sum_{k=-K}^{K} a_k P(f) e^{-j\omega kD}$$

and

$$|X_T(f)|^2 = |P(f)|^2 \left(\sum_{k=-K}^{K} a_k e^{-j\omega kD} \right) \left(\sum_{i=-K}^{K} a_i e^{+j\omega iD} \right)$$

After interchanging the order of expectation and summation we have

$$E[|X_T(f)|^2] = |P(f)|^2 \rho_K(f)$$

with

$$\rho_K(f) = \sum_{k=-K}^{K} \sum_{i=-K}^{K} E[a_k a_i] e^{-j\omega(k-i)D}$$

where $E[a_k a_i] = R_a(k - i)$.

The double summation for $\rho_K(f)$ can be manipulated into the single sum

$$\rho_K(f) = (2K + 1) \sum_{n=-2K}^{2K} \left(1 - \frac{|n|}{2K + 1} \right) R_a(n) e^{-j\omega nD}$$

Substituting these expressions in the definition of $G_x(f)$ finally gives

$$G_x(f) = \lim_{K \to \infty} \frac{1}{(2K + 1)D} |P(f)|^2 \rho_K(f)$$

$$= \frac{1}{D} |P(f)|^2 \sum_{n=-\infty}^{\infty} R_a(n) e^{-j\omega nD}$$

as stated in Eq. (9).

Consider the unipolar binary RZ signal in Fig. 11.1–1a, where $p(t) = \Pi(2r_b t)$ so

EXAMPLE 11.1–1

$$P(f) = \frac{1}{2r_b} \operatorname{sinc} \frac{f}{2r_b}$$

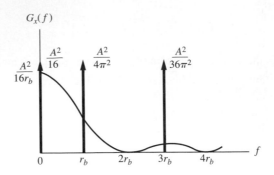

Figure 11.1–5   Power spectrum of unipolar binary RZ signal.

If the source bits are equally likely and statistically independent, then $\overline{a_k} = A/2$, $\overline{a_k^2} = A^2/2$, and Eq. (10) applies with

$$m_a^2 = \sigma_a^2 = \frac{A^2}{4}$$

Using Eq. (11) we find the power spectrum to be

$$G_x(f) = \frac{A^2}{16r_b} \operatorname{sinc}^2 \frac{f}{2r_b} + \frac{A^2}{16} \sum_{n=-\infty}^{\infty} \left(\operatorname{sinc}^2 \frac{n}{2} \right) \delta(f - nr_b)$$

which is sketched in Fig. 11.1–5 for $f \geq 0$. Notice the absence of impulses at the even harmonics because $P(nr_b) = 0$ when $n = \pm 2, \pm 4, \ldots$.

We could also, in principle, use Eq. (12) to calculate $\overline{x^2}$. However, it should be evident from the waveform that $\overline{x^2} = A^2/4$ when 1s and 0s are equally likely.

EXERCISE 11.1–2   Modify the results of Example 11.1–1 for $p(t) = \Pi(r_b t)$, corresponding to an NRZ waveform. In particular, show that the only impulse in $G_x(f)$ is at $f = 0$.

Spectral Shaping by Precoding★

Precoding refers to operations that cause the statistics of the amplitude sequence a_k to differ from the statistics of the message sequence. Usually, the purpose of precoding is to *shape* the power spectrum via $R_a(n)$, as distinguished from $P(f)$. To bring out the potential for statistical spectral shaping, we rewrite Eq. (9) in the form

$$G_x(f) = r|P(f)|^2 \left[R_a(0) + 2 \sum_{n=1}^{\infty} R_a(n) \cos (2\pi nf/r) \right] \qquad [13]$$

having drawn upon the property $R_a(-n) = R_a(n)$.

Now suppose that $x(t)$ is to be transmitted over a channel having poor low-frequency response—a voice telephone channel perhaps. With appropriate precoding, we can force the bracketed term in Eq. (13) to equal zero at $f = 0$ and thereby eliminate any dc component in $G_x(f)$, irrespective of the pulse spectrum $P(f)$. The *bipolar* signal format back in Fig. 11.1–1c is, in fact, a precoding technique that removes dc content.

The bipolar signal has three amplitude values, $a_k = +A, 0,$ and $-A$. If 1s and 0s are equally likely in the message, then the amplitude probabilities are $P(a_k = 0) = 1/2$ and $P(a_k = +A) = P(a_k = -A) = 1/4$, so the amplitude statistics differ from the message statistics. Furthermore, the assumption of uncorrelated message bits leads to the amplitude correlation

$$R_a(n) = \begin{cases} A^2/2 & n = 0 \\ -A^2/4 & n = 1 \\ 0 & n \geq 2 \end{cases} \qquad \text{[14a]}$$

Therefore,

$$G_x(f) = r_b|P(f)|^2 \frac{A^2}{2}(1 - \cos 2\pi f/r_b) \qquad \text{[14b]}$$

$$= r_b|P(f)|^2 A^2 \sin^2 \pi f/r_b$$

which is sketched in Fig. 11.1–6 taking $p(t) = \Pi(r_b t)$.

Two other precoding techniques that remove dc content are the split-phase Manchester format (Fig. 11.1–1d) and the family of **high density bipolar codes** denoted as HDBn. The HDBn scheme is a bipolar code that also eliminates long signal "gaps" by substituting a special pulse sequence whenever the source produces n successive 0s.

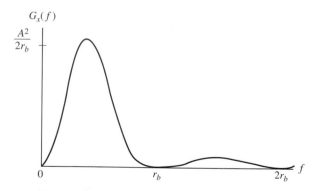

Figure 11.1–6   Power spectrum of bipolar signal.

Sketch $G_x(f)$ for a bipolar signal with $p(t) = \text{sinc } r_b t$. Then use your sketch to show that $\overline{x^2} = A^2/2$.

11.2   NOISE AND ERRORS

Here we investigate noise, errors, and error probabilities in baseband digital transmission, starting with the binary case and generalizing to M-ary signaling.

We assume throughout a distortionless channel, so the received signal is free of ISI. We also assume additive white noise with zero mean value, independent of the signal. (Some of these restrictions will be lifted in the next section.)

Binary Error Probabilities

Figure 11.2–1 portrays the operations of a baseband binary receiver. The received signal plus noise is applied to a lowpass filter whose transfer function has been designed to remove excess noise without introducing ISI. A sample-and-hold (S/H) device triggered at the optimum times extracts from $y(t)$ the sample values

$$y(t_k) = a_k + n(t_k)$$

Comparing successive values of $y(t_k)$ with a fixed *threshold level* V completes the regeneration process. If $y(t_k) > V$, the comparator goes HIGH to indicate a 1; if $y(t_k) < V$, the comparator goes LOW to indicate 0. The regenerator thereby acts as an *analog-to-digital converter*, converting the noisy analog waveform $y(t)$ into a noiseless digital signal $x_e(t)$ with occasional *errors*.

We begin our analysis taking $x(t)$ to be a *unipolar* signal in which $a_k = A$ represents the message bit 1 and $a_k = 0$ represents the message bit 0. Intuitively, then, the threshold should be set at some intermediate level, $0 < V < A$. The regeneration process is illustrated by the waveforms in Fig. 11.2–2. Errors occur when $a_k = 0$ but a positive noise excursion results in $y(t_k) > V$, or when $a_k = A$ but a negative noise excursion results in $y(t_k) < V$.

To formulate the error probabilities, let the *random variable Y* represent $y(t_k)$ at an arbitrary sampling time and let n represent $n(t_k)$. The probability density function

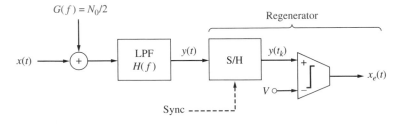

Figure 11.2–1   Baseband binary receiver.

of Y obviously involves the noise PDF, but it also depends upon the presence or absence of the signal pulse. We therefore need to work with *conditional* probabilities. In particular, if H_0 denotes the *hypothesis* or assumption that $a_k = 0$ and $Y = n$, we can write the conditional PDF

$$p_Y(y|H_0) = p_N(y) \qquad \text{[1a]}$$

where $p_N(n)$ is the PDF of the noise alone. Similarly, if H_1 denotes the hypothesis that $a_k = A$ and $Y = A + n$, then

$$p_Y(y|H_1) = p_N(y - A) \qquad \text{[1b]}$$

obtained from the linear transformation of $p_N(n)$ with $n = y - A$.

Figure 11.2–3 shows typical curves of $p_Y(y|H_0)$ and $p_Y(y|H_1)$ along with a proposed threshold V. The comparator implements the following **decision rule:**

Choose hypothesis H_0 ($a_k = 0$) if $Y < V$

Choose hypothesis H_1 ($a_k = A$) if $Y > V$

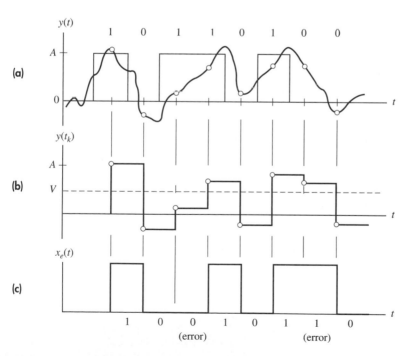

Figure 11.2–2   Regeneration of a unipolar signal. (*a*) Signal plus noise; (*b*) S/H output; (*c*) comparator output.

(We ignore the borderline case $Y = V$ whose probability of occurrence will be vanishingly small.) The corresponding regeneration error probabilities are then given by

$$P_{e0} \triangleq P(Y > V|H_0) = \int_V^\infty p_Y(y|H_0)\, dy \qquad [2a]$$

$$P_{e1} \triangleq P(Y < V|H_1) = \int_{-\infty}^V p_Y(y|H_1)\, dy \qquad [2b]$$

equivalent to the shaded areas indicated in Fig. 11.2–3. The area interpretation helps bring out the significance of the threshold level when all other factors remain fixed. Clearly, lowering the threshold reduces P_{e1} and simultaneously increases P_{e0}. Raising the threshold has the opposite effect.

But an error in digital transmission is an error, regardless of type. Hence, the threshold level should be adjusted to minimize the **average error probability**

$$P_e = P_0 P_{e0} + P_1 P_{e1} \qquad [3a]$$

where

$$P_0 = P(H_0) \qquad\qquad P_1 = P(H_1) \qquad [3b]$$

which stand for the *source digit probabilities*. The *optimum* threshold level V_{opt} must therefore satisfy $dP_e/dV = 0$, and Leibniz's rule for differentiating the integrals in Eq. (2) leads to the general relation

$$P_0 p_Y(V_{\text{opt}}|H_0) = P_1 p_Y(V_{\text{opt}}|H_1) \qquad [4]$$

But we normally expect 1s and 0s to be *equally likely* in a long string of message bits, so

$$P_0 = P_1 = \tfrac{1}{2} \qquad\qquad P_e = \tfrac{1}{2}(P_{e0} + P_{e1}) \qquad [5a]$$

and

$$p_Y(V_{\text{opt}}|H_0) = p_Y(V_{\text{opt}}|H_1) \qquad [5b]$$

We'll work hereafter with the equally likely condition, unless stated otherwise.

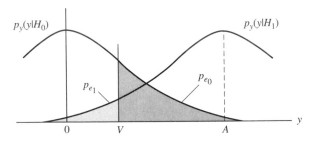

Figure 11.2–3   Conditional PDFs with decision threshold and error probabilities.

Closer examination of Eq. (5b) reveals that V_{opt} corresponds to the point where the conditional PDF curves *intersect*. Direct confirmation of this conclusion is provided by the graphical construction in Fig. 11.2–4 labeled with four relevant areas, α_1 through α_4. The optimum threshold yields $P_{e1} = \alpha_1 + \alpha_2$, $P_{e0} = \alpha_3$, and $P_{e_{\min}} = \frac{1}{2}(\alpha_1 + \alpha_2 + \alpha_3)$. A nonoptimum threshold such as $V < V_{\text{opt}}$ yields $P_{e1} = \alpha_1$ and $P_{e0} = \alpha_2 + \alpha_3 + \alpha_4$; thus, $P_e = \frac{1}{2}(\alpha_1 + \alpha_2 + \alpha_3 + \alpha_4) = P_{e_{\min}} + \frac{1}{2}\alpha_4 > P_{e_{\min}}$.

Next we make the usual assumption that the noise is *gaussian* with zero mean and variance σ^2, so

$$p_N(n) = \frac{1}{\sqrt{2\pi\sigma^2}}\, e^{-n^2/2\sigma^2}$$

Substituting this gaussian function into Eqs. (1) and (2) gives

$$P_{e0} = \int_V^\infty p_N(y)\,dy = Q\left(\frac{V}{\sigma}\right) \tag{6a}$$

$$P_{e1} = \int_{-\infty}^V p_N(y - A)\,dy = Q\left(\frac{A - V}{\sigma}\right) \tag{6b}$$

where Q is the area under the gaussian tail as previously defined in Fig. 8.4-2. Since $p_N(n)$ has even symmetry, the conditional PDFs $p_Y(y|H_0)$ and $p_Y(y|H_1)$ intersect at the *midpoint* and $V_{\text{opt}} = A/2$ when $P_0 = P_1 = \frac{1}{2}$. Furthermore, with $V = V_{\text{opt}}$ in Eq. (6), $P_{e0} = P_{e1} = Q(A/2\sigma)$ so the optimum threshold yields *equal digit error probabilities* as well as minimizing the net error probability. Thus, $P_e = \frac{1}{2}(P_{e0} + P_{e1}) = P_{e0} = P_{e1}$ and

$$P_e = Q\left(\frac{A}{2\sigma}\right) \tag{7}$$

which is the minimum net error probability for binary signaling in gaussian noise when the source digits are equally likely.

Based on Eq. (7), the plot of the Q function in Table T.6 can be interpreted now as a plot of P_e versus $A/2\sigma$. This plot reveals that P_e drops off dramatically when

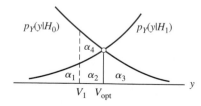

Figure 11.2–4

$A/2\sigma$ increases. For instance, $P_e \approx 2 \times 10^{-2}$ at $A/2\sigma = 2.0$ whereas $P_e \approx 10^{-9}$ at $A/2\sigma = 6.0$.

Although we derived Eq. (7) for the case of a unipolar signal, it also holds in the *polar* case if $a_k = \pm A/2$ so the level spacing remains unchanged. The only difference at the receiver is that $V_{opt} = 0$, midway between the two levels. However, the transmitter needs less signal power to achieve a specified level spacing in the polar signal.

Let's bring out that advantage of polar signaling by expressing A in terms of the average received signal power S_R. If we assume equal digit probabilities and more-or-less rectangular pulses with full-interval duration T_b, then $S_R = A^2/2$ for unipolar signaling while $S_R = A^2/4$ for polar signaling. (These relations should be obvious from the NRZ waveforms in Fig. 11.1–1.) Hence,

$$A = \begin{cases} \sqrt{2S_R} & \text{Unipolar} \\ \sqrt{4S_R} & \text{Polar} \end{cases} \qquad [8]$$

and the factor of $\sqrt{2}$ can make a considerable difference in the value of P_e.

Since the noise has zero mean, the variance σ^2 equals the noise power N_R at the output of the filter. Therefore, we can write $A/2\sigma$ in terms of the signal-to-noise power ratio $(S/N)_R = S_R/N_R$, namely

$$\left(\frac{A}{2\sigma}\right)^2 = \frac{A^2}{4N_R} = \begin{cases} \frac{1}{2}(S/N)_R & \text{Unipolar} \\ (S/N)_R & \text{Polar} \end{cases} \qquad [9]$$

But Eq. (9) conceals the effect of the signaling rate r_b. In order to pass pulses of duration $T_b = 1/r_b$, the noise-limiting filter must have $B_N \geq r_b/2$, so

$$N_R = N_0 B_N \geq N_0 r_b/2 \qquad [10]$$

Rapid signaling thus requires more signal power to maintain a given error probability P_e.

EXAMPLE 11.2–1   Suppose a computer produces unipolar pulses at the rate $r_b = 10^6$ bps $= 1$ Mbps for transmission over a noisy system with $N_0 = 4 \times 10^{-20}$ W/Hz. The error rate is specified to be no greater than one bit per hour, or $P_e \leq 1/3600 r_b \approx 3 \times 10^{-10}$. Table T.6 indicates that we need $A/2\sigma \geq 6.2$, and Eqs. (9) and (10) give the corresponding signal-power requirement

$$S_R = 2\left(\frac{A}{2\sigma}\right)^2 N_R \geq 1.5 \times 10^{-12} = 1.5 \text{ pW}$$

Clearly, any reasonable signal power ensures almost *errorless* transmission insofar as additive noise is concerned. Hardware glitches and other effects would be the limiting factors on system performance.

EXERCISE 11.2–1

Consider a unipolar system with equally likely digits and $(S/N)_R = 50$. Calculate P_{e_0}, P_{e_1}, and P_e when the threshold is set at the nonoptimum value $V = 0.4A$. Compare P_e with the minimum value from Eq. (7).

Regenerative Repeaters

Long-haul transmission requires repeaters, be it for analog or digital communication. But unlike analog-message repeaters, digital repeaters can be *regenerative*. If the error probability per repeater is reasonably low and the number of hops m is large, the regeneration advantage turns out to be rather spectacular. This will be demonstrated for the case of polar binary transmission.

When analog repeaters are used and Eq. 11, Sect. 9.4 applies, the final signal-to-noise ratio is $(S/N)_R = (1/m)(S/N)_1$ and

$$P_e = Q\left[\sqrt{\frac{1}{m}\left(\frac{S}{N}\right)_1}\right] \qquad [11]$$

where $(S/N)_1$ is the signal-to-noise ratio after one hop. Therefore, the transmitted power *per repeater* must be increased linearly with m just to stay even, a factor not to be sneezed at since, for example, it takes 100 or more repeaters to cross the continent. The $1/m$ term in Eq. (11) stems from the fact that the contaminating noise progressively builds up from repeater to repeater.

In contrast, a regenerative repeater station consists of a complete receiver and transmitter back to back in one package. The receiving portion converts incoming signals to message digits, making a few errors in the process; the digits are then delivered to the transmitting portion, which in turn generates a new signal for transmission to the next station. The regenerated signal is thereby completely stripped of random noise but does contain some errors.

To analyze the performance, let α be the error probability at each repeater, namely,

$$\alpha = Q\left[\sqrt{\left(\frac{S}{N}\right)_1}\right]$$

assuming identical units. As a given digit passes from station to station, it may suffer cumulative conversion errors. If the number of erroneous conversions is *even*, they cancel out, and a correct digit is delivered to the destination. (Note that this is true only for *binary* digits.) The probability of i errors in m successive conversions is given by the *binomial frequency function*,

$$P_I(i) = \binom{m}{i}\alpha^i(1 - \alpha)^{m-i}$$

Since we have a destination error only when i is *odd*,

$$P_e = \sum_{i \text{ odd}} P_I(i) = \binom{m}{1}\alpha(1-\alpha)^{m-1} + \binom{m}{3}\alpha^3(1-\alpha)^{m-3} + \cdots \approx m\alpha$$

where the approximation applies for $\alpha \ll 1$ and m not too large. Hence,

$$P_e \approx mQ\left[\sqrt{\left(\frac{S}{N}\right)_1}\right] \tag{12}$$

so P_e increases linearly with m, which generally requires a much smaller power increase to counteract than Eq. (11).

Figure 11.2–5 illustrates the power saving provided by regeneration as a function of m, the error probability being fixed at $P_e = 10^{-5}$. Thus, for example, a 10-station nonregenerative baseband system requires about 8.5 dB more transmitted power (per repeater) than a regenerative system.

Matched Filtering

Every baseband digital receiver—whether at the destination or part of a regenerative repeater—should include a lowpass filter designed to remove excess noise without introducing ISI. But what's the *optimum* filter design for this purpose? For the case of timelimited pulses in white noise, the answer is a **matched filter.** We'll pursue that case here and develop the resulting minimum error probability for binary signaling in white noise.

Let the received signal be a single timelimited pulse of duration τ centered at time $t = kD$, so

$$x(t) = a_k p(t - kD)$$

where $p(0) = 1$, $p(t) = 0$ for $|t| > \tau/2$, and $\tau \le D$. Maximizing the output ratio $(a_k/\sigma)^2$ at time $t_k = kD + t_d$ will minimize the error probability. As we learned in

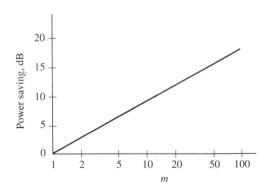

Figure 11.2–5   Power saving gained by m regenerative repeaters when $P_e = 10^{-5}$.

Sect. 9.5, this maximization calls for a matched filter whose impulse response is proportional to $p(t_d - t)$. In particular, we take

$$h(t) = \frac{1}{\tau_{eq}} p(t_d - t) \qquad\qquad [13a]$$

with

$$\tau_{eq} = \int_{-\infty}^{\infty} p^2(t)dt \qquad\qquad t_d = \tau/2 \qquad\qquad [13b]$$

The delay $t_d = \tau/2$ is the minimum value that yields a causal impulse response, and the proportionality constant $1/\tau_{eq}$ has been chosen so that the peak output amplitude equals a_k. The parameter τ_{eq} can be interpreted from the property that $a_k^2 \tau_{eq}$ equals the *energy* of the pulse $x(t)$.

In absence of noise, the resulting output pulse is $y(t) = h(t) * x(t)$, with peak value $y(t_k) = a_k$ as desired. This peak occurs $\tau/2$ seconds after the peak of $x(t)$. Thus, matched filtering introduces an unavoidable *delay*. However, it does not introduce ISI at the sampling times for adjacent pulses since $y(t) = 0$ outside of $t_k \pm \tau$. Figure 11.2–6 illustrates these points taking a rectangular shape for $p(t)$, in which case $\tau_{eq} = \tau$ and $y(t)$ has a triangular shape.

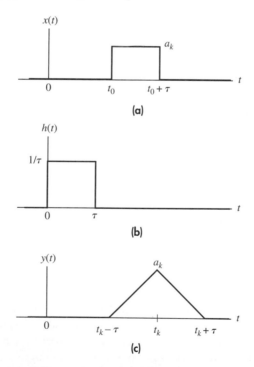

(a)

(b)

(c)

Figure 11.2–6   Matched filtering with rectangular pulses. (a) Received pulse; (b) impulse response; (c) output pulse.

When $x(t)$ is accompanied by white noise, the output noise power from the matched filter will be

$$N_R = \sigma^2 = \frac{N_0}{2} \int_{-\infty}^{\infty} |H(f)|^2 \, df$$

$$= \frac{N_0}{2} \int_{-\infty}^{\infty} |h(t)|^2 \, dt = \frac{N_0}{2\tau_{eq}} \qquad [14]$$

This result agrees with the lower bound in Eq. (10) since, for binary signaling, $\tau_{eq} \le T_b = 1/r_b$. We'll use this result to evaluate the maximum value of $(A/2\sigma)^2$ and the corresponding minimum binary error probability when the noise is white and gaussian and the receiver has an optimum filter matched to the timelimited pulse shape.

Consider a binary transmission system with rate r_b, average received power S_R, and noise density N_0. We can characterize this system in terms of two new parameters E_b and γ_b defined by

$$E_b \triangleq S_R/r_b \qquad [15a]$$

$$\gamma_b \triangleq S_R/N_0 r_b = E_b/N_0 \qquad [15b]$$

The quantity E_b corresponds to the *average energy per bit*, while γ_b represents the ratio of bit energy to noise density. If the signal consists of timelimited pulses $p(t)$ with amplitude sequence a_k, then

$$E_b = \overline{a_k^2} \int_{-\infty}^{\infty} p^2(t) \, dt = \overline{a_k^2} \, \tau_{eq}$$

where $\overline{a_k^2} = A^2/2$ for a unipolar signal or $\overline{a_k^2} = A^2/4$ for a polar signal. Thus, since the output noise power from a matched filter is $\sigma^2 = N_0/2\tau_{eq}$, we have

$$(A/2\sigma)^2 = \begin{cases} E_b/N_0 = \gamma_b & \text{Unipolar} \\ 2E_b/N_0 = 2\gamma_b & \text{Polar} \end{cases}$$

and Eq. (7) becomes

$$P_e = \begin{cases} Q(\sqrt{\gamma_b}) & \text{Unipolar} \\ Q(\sqrt{2\gamma_b}) & \text{Polar} \end{cases} \qquad [16]$$

This is the minimum possible error probability, attainable only with matched filtering.

Finally, we should give some attention to the implementation of a matched filter described by Eq. (13). The impulse response for an arbitrary $p(t)$ can be approximated with passive circuit elements, but considerable design effort must be expended to get $h(t) \approx 0$ for $t > \tau$. Otherwise, the filter may produce significant ISI. For a rectangular pulse shape, you can use an active circuit such as the one diagrammed in Fig. 11.2–7a, called an **integrate-and-dump** filter. The op-amp integrator integrates each incoming pulse so that $y(t_k) = a_k$ at the end of the pulse, after

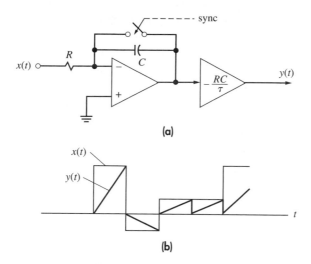

Figure 11.2–7   Integrate-and-dump filter. (a) Op-amp circuit; (b) polar M-ary waveforms.

which the dump switch resets the integrator to zero—thereby ensuring no ISI at subsequent sampling times. The integrate-and-dump filter is probably the best practical implementation of matched filtering. Figure 11.2–7b illustrates its operation with a polar M-ary waveform.

Let $x(t)$ be the unipolar RZ signal in Fig. 11.1–1a. (a) Sketch the corresponding output waveform from a matched filter and from an integrate-and-dump filter. (b) Confirm that matched filtering yields $(A/2\sigma)^2 = \gamma_b$ even though $\sigma^2 = N_0 r_b$ so $N_R > N_0 r_b/2$.

EXERCISE 11.2–2

M-ary Error Probabilities

Binary signaling provides the greatest immunity to noise for a given S/N because it has only two amplitude levels—and you can't send information with fewer than two levels. Multilevel M-ary signaling requires more signal power but less transmission bandwidth because the signaling rate will be smaller than the bit rate of an equivalent binary signal. Consequently, M-ary signaling suits applications such as digital transmission over voice channels where the available bandwidth is limited and the signal-to-noise ratio is relatively large.

   Here we calculate M-ary error probabilities in zero-mean gaussian noise. We'll take the most common case of polar signaling with an even number of equispaced levels at

$$a_k = \pm A/2, \pm 3A/2, \ldots, \pm(M-1)A/2 \qquad \textbf{[17]}$$

We'll also assume equally likely M-ary symbols, so that

$$P_e = \frac{1}{M}\left(P_{e_0} + P_{e_1} + \cdots + P_{e_{M-1}}\right) \qquad [18]$$

which is the M-ary version of Eq. (5a).

Figure 11.2–8 shows the conditional PDFs for a quaternary $(M = 4)$ polar signal plus gaussian noise. The decision rule for regeneration now involves *three* threshold levels, indicated in the figure at $y = -A$, 0, and $+A$. These are the optimum thresholds for minimizing P_e, but they do not result in equal error probabilities for all symbols. For the two extreme levels at $a_k = \pm 3A/2$ we get

$$P_{e_0} = P_{e_3} = Q(A/2\sigma)$$

whereas

$$P_{e_1} = P_{e_2} = 2Q(A/2\sigma)$$

because both positive and negative noise excursions produce errors for the inner levels at $a_k = \pm A/2$. The resulting average error probability is

$$P_e = \frac{1}{4} \times 6Q\left(\frac{A}{2\sigma}\right) = \frac{3}{2}Q\left(\frac{A}{2\sigma}\right)$$

or 50 percent greater than binary signaling with the same level spacing.

The foregoing analysis readily generalizes to an arbitrary even value of M with $M - 1$ decision thresholds at

$$y = 0, \pm A, \pm 2A, \ldots, \pm \frac{M-2}{2}A \qquad [19]$$

Then $P_{e_0} = P_{e_{M-1}} = Q(A/2\sigma)$ while the $M - 2$ inner levels have doubled error probability, yielding the average error probability

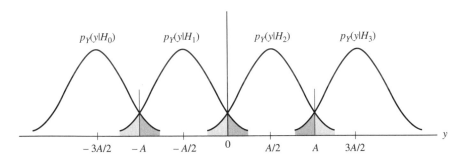

Figure 11.2–8   Conditional PDFs for a quaternary polar signal with gaussian noise.

$$P_e = \frac{1}{M}\left[2 \times Q\left(\frac{A}{2\sigma}\right) + (M-2) \times 2Q\left(\frac{A}{2\sigma}\right)\right] \qquad [20]$$

$$= \frac{2M-2}{M}Q\left(\frac{A}{2\sigma}\right) = 2\left(1 - \frac{1}{M}\right)Q\left(\frac{A}{2\sigma}\right)$$

Equation (20) clearly reduces to Eq. (7) when $M = 2$, whereas $P_e \approx 2Q(A/2\sigma)$ when $M \gg 2$.

Next, we relate $A/2\sigma$ to the signal power and noise density assuming a time-limited pulse shape $p(t)$ so the average energy per M-ary digit is $E_M = \overline{a_k^2}\,\tau_{eq}$ where

$$\tau_{eq} = \int_{-\infty}^{\infty} p^2(t)\,dt$$

as before. If the M amplitude levels are equally likely and given by Eq. (17), then

$$\overline{a_k^2} = 2 \times \frac{1}{M}\sum_{i=1}^{M/2}(2i-1)^2\left(\frac{A}{2}\right)^2 = \frac{M^2-1}{12}A^2 \qquad [21]$$

Hence, since $S_R = rE_M$,

$$\left(\frac{A}{2\sigma}\right)^2 = \frac{3}{M^2-1}\frac{E_M}{\tau_{eq}\sigma^2} \qquad [22]$$

$$= \frac{3}{M^2-1}\frac{1}{r\tau_{eq}}\frac{S_R}{N_R} \leq \frac{6}{M^2-1}\frac{S_R}{N_0 r}$$

where the upper bound corresponds to $N_R = N_0/2\tau_{eq}$ obtained with matched filtering. Equations (20) and (22) constitute our final result for error probability in a polar M-ary system with gaussian white noise.

More often than not, M-ary signaling is used to transmit *binary* messages and the value of M is selected by the system engineer to best fit the available channel. We should therefore investigate the design considerations in the selection of M, especially the impact on error probability. But Eqs. (20) and (22) fail to tell the full story for two reasons: first, the M-ary signaling rate differs from the bit rate r_b; second, the M-ary error probability differs from the bit error probability.

We can easily account for the signaling-rate difference when the message bits are encoded in blocks of length $\log_2 M$. Then r_b and r are related by

$$r_b = r\log_2 M \qquad [23]$$

from Eq. (4), Sect. 11.1. To relate the M-ary symbol error probability P_e to the resulting error probability per bit, we'll assume a *Gray code* and a reasonably large signal-to-noise ratio. Under these conditions a noise excursion seldom goes beyond one amplitude level in the M-ary waveform, which corresponds to just one erroneous bit in the block of $\log_2 M$ bits. Therefore,

$$P_{be} \approx P_e/\log_2 M \qquad [24]$$

where P_{be} stands for the equivalent **bit error probability,** also called the bit error rate (BER).

Combining Eqs. (23) and (24) with our previous *M*-ary expressions, we finally have

$$P_{be} \approx 2\,\frac{M-1}{M\log_2 M}\,Q\!\left(\frac{A}{2\sigma}\right) \qquad \text{[25a]}$$

in which

$$\left(\frac{A}{2\sigma}\right)^2 \le \frac{6}{M^2-1}\,\frac{S_R}{N_0 r} = \frac{6\log_2 M}{M^2-1}\,\gamma_b \qquad \text{[25b]}$$

Notice that the upper bound with matched filtering has been written in terms of $\gamma_b = S_R/N_0 r_b = S_R/(N_0 r\log_2 M)$. This facilitates the study of *M*-ary signaling as a function of energy per message bit.

EXAMPLE 11.2–2   **COMPARISON OF BINARY AND M-ARY SIGNALING**

Suppose the channel in question has a fixed signaling rate $r = 3000$ baud $= 3$ kbaud and a fixed signal-to-noise ratio $(S/N)_R = 400 \approx 26$ dB. (These values would be typical of a voice telephone channel, for instance.) We'll assume matched filtering of NRZ rectangular pulses, so $r\tau_{\text{eq}} = 1$ and

$$\left(\frac{A}{2\sigma}\right)^2 = \frac{3}{M^2-1}\,(S/N)_R = \frac{6\log_2 M}{M^2-1}\,\gamma_b$$

which follow from Eqs. (22) and (25b).

Binary signaling yields a vanishingly small error probability when $(S/N)_R = 400$, but at the rather slow rate $r_b = r = 3$ kbps. *M*-ary signaling increases the bit rate, per Eq. (23), but the error probability also increases because the spacing between amplitude levels gets smaller when you increase *M* with the signal power held fixed. Table 11.2–1 brings out the trade-off between bit rate and error probability for this channel.

Table 11.2–1   *M*-ary signaling with $r = 3$ kilobaud and $(S/N)_R = 400$

M	r_b (kbps)	$A/2\sigma$	P_{be}
2	3	20.0	3×10^{-89}
4	6	8.9	1×10^{-19}
8	9	4.4	4×10^{-6}
16	12	2.2	7×10^{-3}
32	15	1.1	6×10^{-2}

Table 11.2–2 M-ary signaling with $r_b = 9$ kbps and $P_{be} = 4 \times 10^{-6}$

M	r (kbaud)	γ_b
2	9.00	10
4	4.50	24
8	3.00	67
16	2.25	200
32	1.80	620

Another type of trade-off is illustrated by Table 11.2–2, where the bit rate and error probability are both held fixed. Increasing M then yields a lower signaling rate r—implying a smaller transmission bandwidth requirement. However, the energy per bit must now be increased to keep the error probability unchanged. Observe that going from $M = 2$ to $M = 32$ reduces r by 1/5 but increases γ_b by more than a factor of 60. This type of trade-off will be reconsidered from the broader viewpoint of information theory in Chap. 16.

EXERCISE 11.2–3

Consider the three-level bipolar binary format in Fig. 11.1–1c with amplitude probabilities $P(a_k = 0) = 1/2$ and $P(a_k = +A) = P(a_k = -A) = 1/4$. Make a sketch similar to Fig. 11.2–8 and find P_e in terms of A and σ when the decision thresholds are at $y = \pm A/2$. Then calculate S_R and express P_e in a form like Eq. (16).

11.3   BANDLIMITED DIGITAL PAM SYSTEMS

This section develops design procedures for baseband digital systems when the transmission channel imposes a bandwidth limitation. By this we mean that the available transmission bandwidth is not large compared to the desired signaling rate and, consequently, rectangular signaling pulses would be severely distorted. Instead, we must use *bandlimited* pulses specially shaped to avoid ISI.

Accordingly, we begin with Nyquist's strategy for bandlimited pulse shaping. Then we consider the optimum terminal filters needed to minimize error probability. The assumption that the noise has a gaussian distribution with zero mean value will be continued, but we'll allow an arbitrary noise power spectrum. We'll also make allowance for linear transmission distortion, which leads to the subject of equalization for digital systems. The section closes with an introductory discussion of correlative coding techniques that increase the signaling rate on a bandlimited channel.

Nyquist Pulse Shaping

Our presentation of Nyquist pulse shaping will be couched in general terms of M-ary signaling with $M \geq 2$ and symbol interval $D = 1/r$. In order to focus on potential ISI problems at the receiver, we'll let $p(t)$ be the pulse shape at the output of the receiving filter. Again assuming that the transmitter gain compensates for transmission loss, the output waveform in absence of noise is

$$y(t) = \sum_k a_k p(t - t_d - kD)$$

As before, we want $p(t)$ to have the property

$$p(t) = \begin{cases} 1 & t = 0 \\ 0 & t = \pm D, \pm 2D, \dots \end{cases} \qquad [1a]$$

which eliminates ISI. Now we impose the additional requirement that the pulse spectrum be *bandlimited*, such that

$$P(f) = 0 \qquad |f| \geq B \qquad [1b]$$

where

$$B = \frac{r}{2} + \beta \qquad 0 \leq \beta \leq \frac{r}{2}$$

This spectral requirement permits signaling at the rate

$$r = 2(B - \beta) \qquad B \leq r \leq 2B \qquad [2]$$

in which B may be interpreted as the minimum required transmission bandwidth, so that $B_T \geq B$.

Nyquist's *vestigial-symmetry theorem* states that Eq. (1) is satisfied if $p(t)$ has the form

$$p(t) = p_\beta(t)\,\text{sinc}\,rt \qquad [3a]$$

with

$$\mathcal{F}[p_\beta(t)] = P_\beta(f) = 0 \qquad |f| > \beta \qquad [3b]$$

$$p_\beta(0) = \int_{-\infty}^{\infty} P_\beta(f)\,df = 1$$

Clearly, $p(t)$ has the time-domain properties of Eq. (1a). It also has the frequency-domain properties of Eq. (1b) since

$$P(f) = P_\beta(f) * [(1/r)\Pi(f/r)]$$

and the convolution of two bandlimited spectra produces a new bandlimited spectrum whose bandwidth equals the sum of the bandwidths, namely, $B = \beta + r/2$. Usually we take $p_\beta(t)$ to be an even function so $P_\beta(f)$ is real and even; then $P(f)$ has vestigial symmetry around $f = \pm r/2$, like the symmetry of a vestigial sideband filter.

Infinitely many functions satisfy Nyquist's conditions, including the case when $p_\beta(t) = 1$ so $\beta = 0$ and $p(t) = \text{sinc}\,rt$, as in Eq. (6), Sect. 11.1. We know that this pulse shape allows bandlimited signaling at the maximum rate $r = 2B$. However, synchronization turns out to be a very touchy matter because the pulse shape falls off no faster than $1/|t|$ as $|t| \to \infty$. Consequently, a small timing error ϵ results in the sample value

$$y(t_K) = a_K \operatorname{sinc} r\epsilon + \sum_{k \neq K} a_k \operatorname{sinc} (KD - kD + r\epsilon)$$

and the ISI in the second term can be quite large.

Synchronization problems are eased by reducing the signaling rate and using pulses with a *cosine rolloff spectrum*. Specifically, if

$$P_\beta(f) = \frac{\pi}{4\beta} \cos \frac{\pi f}{2\beta} \, \Pi\!\left(\frac{f}{2\beta}\right) \tag{4a}$$

then

$$P(f) = \begin{cases} \dfrac{1}{r} & |f| < \dfrac{r}{2} - \beta \\[2ex] \dfrac{1}{r} \cos^2 \dfrac{\pi}{4\beta}(|f| - \dfrac{r}{2} + \beta) & \dfrac{r}{2} - \beta < |f| < \dfrac{r}{2} + \beta \\[2ex] 0 & |f| > \dfrac{r}{2} + \beta \end{cases} \tag{4b}$$

and the corresponding pulse shape is

$$p(t) = \frac{\cos 2\pi\beta t}{1 - (4\beta t)^2} \operatorname{sinc} rt \tag{5}$$

Plots of $P(f)$ and $p(t)$ are shown in Fig. 11.3–1 for two values of β along with $\beta = 0$. When $\beta > 0$, the spectrum has a smooth rolloff and the leading and trailing oscillations of $p(t)$ decay more rapidly than those of sinc rt.

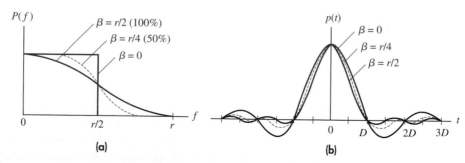

(a)   (b)

Figure 11.3–1   Nyquist pulse shaping. (a) Spectra; (b) waveforms.

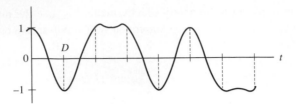

Figure 11.3–2   Baseband waveform for 10110100 using Nyquist pulses with $\beta = r/2$.

Further consideration of Eqs. (4) and (5) reveals two other helpful properties of $p(t)$ in the special case when $\beta = r/2$, known as 100 percent rolloff. The spectrum then reduces to the *raised cosine* shape

$$P(f) = \frac{1}{r} \cos^2 \frac{\pi f}{2r} = \frac{1}{2r}\left[1 + \cos\left(\frac{\pi f}{r}\right) \right] \qquad |f| \le r \qquad [6a]$$

and

$$p(t) = \frac{\operatorname{sinc} 2rt}{1 - (2rt)^2} \qquad [6b]$$

The half-amplitude width of this pulse exactly equals the symbol interval D, that is, $p(\pm 0.5\,D) = 1/2$, and there are additional zero crossings at $t = \pm 1.5D, \pm 2.5D, \ldots$. A polar signal constructed with this pulse shape will therefore have zero crossings precisely halfway between the pulse centers whenever the amplitude changes polarity. Figure 11.3–2 illustrates this feature with the binary message 10110100. These zero crossings make it a simple task to extract a synchronization signal for timing purposes at the receiver. However, the penalty is a 50 percent reduction of signaling speed since $r = B$ rather than $2B$. Nyquist proved that the pulse shape defined by Eq. (6) is the *only* one possessing all of the aforementioned properties.

EXERCISE 11.3–1   Sketch $P(f)$ and find $p(t)$ for the Nyquist pulse generated by taking $P_\beta(f) = (2/r)\Lambda(2f/r)$. Compare your results with Eq. (6).

Optimum Terminal Filters

Having abandoned rectangular pulses, we must likewise abandon the conventional matched filter and reconsider the design of the optimum receiving filter that minimizes error probability. This turns out to be a relatively straightforward problem under the following reasonable conditions:

1. The signal format is polar, and the amplitudes a_k are uncorrelated and equally likely.

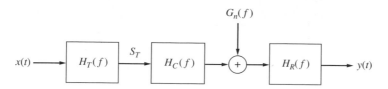

Figure 11.3–3

2. The transmission channel is linear but not necessarily distortionless.
3. The filtered output pulse $p(t)$ is to be Nyquist shaped.
4. The noise is additive and has a zero-mean gaussian distribution but may have a nonwhite power spectrum.

To allow for possible channel distortion and/or nonwhite noise, our optimization must involve filters at both the transmitter and receiver. As a bonus, the source waveform $x(t)$ may have a more-or-less arbitrary pulse shape $p_x(t)$.

Figure 11.3–3 lays out the system diagram, including a transmitting filter function $H_T(f)$, a channel function $H_C(f)$, and a receiving filter function $H_R(f)$. The input signal has the form

$$x(t) = \sum_k a_k p_x(t - kD) \qquad \text{[7a]}$$

and its power spectrum is

$$G_x(f) = \sigma_a^2 r |P_x(f)|^2 \qquad \text{[7b]}$$

where $P_x(f) = \mathscr{F}[p_x(t)]$ and

$$\sigma_a^2 = \overline{a_k^2} = \frac{M^2 - 1}{12} A^2 \qquad \text{[7c]}$$

These relations follow from Eq. (12), Sect. 11.1, and Eq. (21), Sect. 11.2, with our stated conditions on a_k. Thus, the transmitted signal power will be

$$S_T = \int_{-\infty}^{\infty} |H_T(f)|^2 G_x(f)\, df = \frac{M^2 - 1}{12} A^2 r \int_{-\infty}^{\infty} |H_T(f)P_x(f)|^2\, df \qquad \text{[8]}$$

a result we'll need shortly.

At the output of the receiving filter we want the input pulse $p_x(t)$ to produce a Nyquist-shaped pulse $p(t - t_d)$, where t_d represents any transmission time delay. The transfer functions in Fig. 11.3–3 must therefore obey the relationship

$$P_x(f)H_T(f)H_C(f)H_R(f) = P(f)e^{-j\omega t_d} \qquad \text{[9]}$$

so both terminal filters help shape $p(t)$. But only the receiving filter controls the output noise power

$$N_R = \sigma^2 = \int_{-\infty}^{\infty} |H_R(f)|^2 G_n(f)\, df \qquad [10]$$

where $G_n(f)$ is the noise power spectrum at the input to the receiver.

Equations (7)–(10) constitute the information relevant to our design problem. Specifically, since the error probability decreases as $A/2\sigma$ increases, we seek the terminal filters that maximize $(A/2\sigma)^2$ subject to two constraints: (1) the transmitted power must be held fixed at some specified value S_T, and (2) the filter transfer functions must satisfy Eq. (9).

We incorporate Eq. (9) and temporarily eliminate $H_T(f)$ by writing

$$|H_T(f)| = \frac{|P(f)|}{|P_x(f)H_C(f)H_R(f)|} \qquad [11]$$

Then we use Eqs. (8) and (10) to express $(A/2\sigma)^2$ as

$$\left(\frac{A}{2\sigma}\right)^2 = \frac{3S_T}{(M^2-1)r}\frac{1}{I_{HR}} \qquad [12a]$$

where

$$I_{HR} = \int_{-\infty}^{\infty} |H_R(f)|^2 G_n(f)\, df \int_{-\infty}^{\infty} \frac{|P(f)|^2}{|H_C(f)H_R(f)|^2}\, df \qquad [12b]$$

Maximizing $(A/2\sigma)^2$ thus boils down to minimizing the product of integrals I_{HR}, in which $H_R(f)$ is the only function under our control.

Now observe that Eq. (12b) has the form of the right-hand side of Schwarz's inequality as stated in Eq. 17, Sect. 3.6. The minimum value of I_{HR} therefore occurs when the two integrands are proportional. Consequently, the optimum receiving filter has

$$|H_R(f)|^2 = \frac{g|P(f)|}{\sqrt{G_n(f)}|H_C(f)|} \qquad [13a]$$

where g is an arbitrary gain constant. Equation (11) then gives the optimum transmitting filter characteristic

$$|H_T(f)|^2 = \frac{|P(f)|\sqrt{G_n(f)}}{g|P_x(f)|^2|H_C(f)|} \qquad [13b]$$

These expressions specify the optimum amplitude ratios for the terminal filters. Note that the receiving filter deemphasizes those frequencies where $G_n(f)$ is large, and the transmitting filter supplies the corresponding preemphasis. The phase shifts are arbitrary, providing that they satisfy Eq. (9).

Substituting Eq. (13) into Eq. (12) yields our final result

$$\left(\frac{A}{2\sigma}\right)^2_{\max} = \frac{3S_T}{(M^2 - 1)r}\left[\int_{-\infty}^{\infty} \frac{|P(f)|\sqrt{G_n(f)}}{|H_C(f)|}\, df\right]^{-2} \qquad \text{[14]}$$

from which the error probability can be calculated using Eq. (20), Sect. 11.2. As a check of Eq. (14), take the case of white noise with $G_n(f) = N_0/2$ and a distortionless channel with transmission loss L so $|H_C(f)|^2 = 1/L$; then

$$\left(\frac{A}{2\sigma}\right)^2_{\max} = \frac{6S_T/L}{(M^2 - 1)rN_0}\left[\int_{-\infty}^{\infty} |P(f)|\, df\right]^{-2}$$

But since $S_T/L = S_R$ and Nyquist-shaped pulses have

$$\int_{-\infty}^{\infty} |P(f)|\, df = 1$$

we thus obtain

$$\left(\frac{A}{2\sigma}\right)^2_{\max} = \frac{6}{M^2 - 1}\frac{S_R}{N_0 r} = \frac{6\log_2 M}{M^2 - 1}\gamma_b$$

which confirms that the optimum terminal filters yield the same upper bound as matched filtering—see Eqs. (22) and (25), Sect. 11.2.

EXAMPLE 11.3–1

Consider a system with white noise, transmission loss L, and a distortionless channel response over $|f| \leq B_T$ where $B_T \geq r$. This transmission bandwidth allows us to use the pulse shape $p(t)$ in Eq. (6), thereby simplifying synchronization. Simplicity also suggests using the rectangular input pulse $p_x(t) = \Pi(t/\tau)$ with $\tau \leq 1/r$, so $P_x(f) = \tau\operatorname{sinc} f\tau$. Taking the gain constant g in Eq. (13) such that $|H_R(0)| = 1$, we have

$$|H_R(f)| = \cos\frac{\pi f}{2r} \qquad |H_T(f)| = \sqrt{L}\,\frac{\cos(\pi f/2r)}{r\tau\operatorname{sinc} f\tau} \qquad |f| \leq r$$

as plotted in Fig. 11.3–4. Notice the slight high-frequency rise in $|H_T(f)|$ compared to $|H_R(f)|$. If the input pulses have a small duration $\tau \ll 1/r$, then this rise becomes negligible and $|H_T(f)| \approx |H_R(f)|$, so one circuit design serves for both filters.

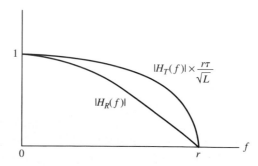

Figure 11.3–4
Amplitude ratio of optimum filters in Example 11.3–1.

EXERCISE 11.3–2   Carry out the details going from Eq. (12) to Eqs. (13) and (14).

Equalization

Regardless of which particular pulse shape has been chosen, some amount of residual ISI inevitably remains in the output signal as a result of imperfect filter design, incomplete knowledge of the channel characteristics, and so on. Hence, an adjustable equalizing filter is often inserted between the receiving filter and the regenerator. These "mop-up" equalizers usually have the structure of a transversal filter previously discussed in Sect. 3.2 relative to linear distortion of analog signals. However, mop-up equalization of digital signals involves different design strategies that deserve attention here.

Figure 11.3–5 shows a transversal equalizer with $2N + 1$ taps and total delay $2ND$. The distorted pulse shape $\tilde{p}(t)$ at the input to the equalizer is assumed to have its peak at $t = 0$ and ISI on both sides. The equalized output pulse will be

$$p_{eq}(t) = \sum_{n=-N}^{N} c_n \tilde{p}(t - nD - ND)$$

and sampling at $t_k = kD + ND$ yields

$$p_{eq}(t_k) = \sum_{n=-N}^{N} c_n \tilde{p}(kD - nD) = \sum_{n=-N}^{N} c_n \tilde{p}_{k-n} \qquad [15]$$

where we've introduced the shorthand notation $\tilde{p}_{k-n} = \tilde{p}[(k - n)D]$. Equation (15) thus takes the form of a *discrete convolution*.

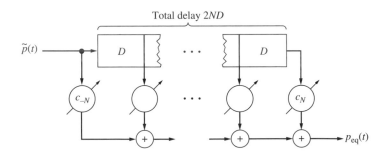

Total delay $2ND$

Figure 11.3–5   Transversal equalizer with $2N + 1$ taps.

Ideally, we would like the equalizer to eliminate all ISI, resulting in

$$p_{eq}(t_k) = \begin{cases} 1 & k = 0 \\ 0 & k \neq 0 \end{cases}$$

But this cannot be achieved, in general, because the $2N + 1$ tap gains are the only variables at our disposal. We might settle instead for choosing the tap gains such that

$$p_{eq}(t_k) = \begin{cases} 1 & k = 0 \\ 0 & k = \pm 1, \pm 2, \ldots, \pm N \end{cases} \qquad \text{[16]}$$

thereby forcing N zero values on each side of the peak of $p_{eq}(t)$. The corresponding tap gains are computed from Eqs. (15) and (16) combined in the matrix equation

$$
\begin{bmatrix}
\widetilde{p}_0 & \cdots & \widetilde{p}_{-2N} \\
\vdots & & \vdots \\
\widetilde{p}_{N-1} & \cdots & \widetilde{p}_{-N-1} \\
\widetilde{p}_N & \cdots & \widetilde{p}_{-N} \\
\widetilde{p}_{N+1} & \cdots & \widetilde{p}_{-N+1} \\
\vdots & & \vdots \\
\widetilde{p}_{2N} & \cdots & \widetilde{p}_0
\end{bmatrix}
\begin{bmatrix}
c_{-N} \\
\vdots \\
c_{-1} \\
c_0 \\
c_1 \\
\vdots \\
c_N
\end{bmatrix}
=
\begin{bmatrix}
0 \\
\vdots \\
0 \\
1 \\
0 \\
\vdots \\
0
\end{bmatrix}
\qquad \text{[17]}
$$

Equation (17) describes a **zero-forcing equalizer.** This equalization strategy is optimum in the sense that it minimizes the peak intersymbol interference, and it has the added advantage of simplicity. Other optimization criteria lead to different strategies with more complicated tap-gain relationships.

When the transmission channel is a switched telephone link or a radio link with slowly changing conditions, the values of $\widetilde{p}_k$ will not be known in advance. The tap gains must then be adjusted on-line using a training sequence transmitted before the actual message sequence. An **adaptive equalizer** incorporates a microprocessor for automatic rather than manual tap-gain adjustment. More sophisticated versions adjust themselves continuously using error measures derived from the message sequence. Further details regarding adaptive equalization are given by Haykin (2001, Sect. 4.1) and Proakis (2001, Chap. 11).

Mop-up equalization, whether fixed or adaptive, does have one hidden catch in that the equalizer somewhat increases the noise power at the input to the regenerator. This increase is usually more than compensated for by the reduction of ISI.

Suppose a three-tap zero forcing equalizer is to be designed for the distorted pulse plotted in Fig. 11.3–6a. Inserting the values of $\widetilde{p}_k$ into Eq. (17) with $N = 1$, we have

EXAMPLE 11.3–2

$$\begin{bmatrix} 1.0 & 0.1 & 0.0 \\ -0.2 & 1.0 & 0.1 \\ 0.1 & -0.2 & 1.0 \end{bmatrix} \begin{bmatrix} c_{-1} \\ c_0 \\ c_1 \end{bmatrix} = \begin{bmatrix} 0 \\ 1 \\ 0 \end{bmatrix}$$

Therefore,

$$c_{-1} = -0.096 \qquad c_0 = 0.96 \qquad c_1 = 0.2$$

and the corresponding sample values of $p_{eq}(t)$ are plotted in Fig. 11.3–6b with an interpolated curve. As expected, there is one zero on each side of the peak. However, zero forcing has produced some small ISI at points further out where the unequalized pulse was zero.

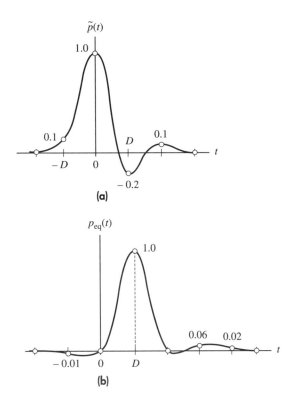

Figure 11.3–6   (a) Distorted pulse; (b) equalized pulse.

Correlative Coding★

Correlative coding, also known as **partial-response signaling,** is a strategy for band-limited transmission at $r = 2B$ that avoids the problems associated with $p(t) = \text{sinc } rt$.

The strategy involves two key functional operations, correlative filtering and digital precoding. *Correlative filtering* purposely introduces controlled intersymbol interference, resulting in a pulse train with an increased number of amplitude levels and a correlated amplitude sequence. Nyquist's signaling-rate limitation no longer applies because the correlated symbols are not independent, and therefore higher signaling rates are possible. *Digital precoding* of the message sequence before waveform generation facilitates message recovery from the correlative-filtered pulse train.

Figure 11.3–7a shows the general model of a transmission system with correlative coding, omitting noise. The digital precoder takes the sequence of message symbols m_k and produces a coded sequence m_k' applied to an impulse generator. (In practice, the impulses would be replaced by short rectangular pulses with duration $\tau \ll D$.) The resulting input signal is an impulse train

$$x(t) = \sum_k a_k \delta(t - kD)$$

whose weights a_k represent the m_k'. The terminal filters and channel have the overall transfer function $H(f)$, producing the output waveform

$$y(t) = \sum_k a_k h(t - kD) \tag{18}$$

where $h(t) = \mathcal{F}^{-1}[H(f)]$.

Although a correlative filter does not appear as a distinct block in Fig. 11.3–7a, the transfer function $H(f)$ must be equivalent to that of Fig. 11.3–7b—which consists of a transversal filter and an ideal LPF. The transversal filter has total delay ND

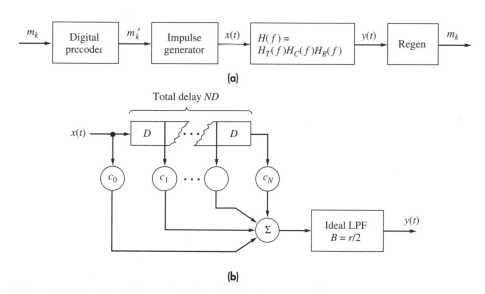

(a)

(b)

Figure 11.3–7   (a) Transmission system with correlative coding; (b) equivalent correlative filter.

and $N + 1$ tap gains. Since the impulse response of the LPF is sinc rt and $r = 1/D$, the cascade combination yields the overall impulse response

$$h(t) = \sum_{n=0}^{N} c_n \operatorname{sinc} (rt - n) \qquad [19]$$

Hence, Eq. (18) becomes

$$y(t) = \sum_{k} a_k \left[\sum_{n=0}^{N} c_n \operatorname{sinc} (rt - n - k) \right] \qquad [20a]$$

$$= \sum_{k} a_k' \operatorname{sinc} (rt - k)$$

where

$$a_k' \triangleq c_0 a_k + c_1 a_{k-1} + \cdots + c_N a_{k-N} = \sum_{n=0}^{N} c_n a_{k-n} \qquad [20b]$$

Message regeneration must then be based on the sample values $y(t_k) = a_k'$.

Equation (20) brings out the fact that correlative filtering changes the amplitude sequence a_k into the modified sequence a_k'. We say that this sequence has a **correlation span** of N symbols, since each a_k' depends on the previous N values of a_k. Furthermore, when the a_k sequence has M levels, the a_k' sequence has $M' > M$ levels. To demonstrate that these properties of correlative filtering lead to practical bandlimited transmission at $r = 2B$, we must look at a specific case.

Duobinary signaling is the simplest type of correlative coding, having $M = 2$, $N = 1$, and $c_0 = c_1 = 1$. The equivalent correlative filter is diagrammed in Fig. 11.3–8 along with its impulse response

$$h(t) = \operatorname{sinc} r_b t + \operatorname{sinc} (r_b t - 1) \qquad [21a]$$

and the magnitude of the transfer function

$$H(f) = \frac{2}{r_b} \cos \frac{\pi f}{r_b} e^{-j\pi f/r_b} \qquad |f| \leq r_b/2 \qquad [21b]$$

The smooth rolloff of $H(f)$ is similar to the spectrum of a Nyquist-shaped pulse and can be synthesized to a good approximation. But, unlike Nyquist pulse shaping, duobinary signaling achieves this rolloff without increasing the bandwidth requirement above $B = r_b/2$. In exchange for the signaling rate advantage, a duobinary waveform has intentional ISI and $M' = 3$ levels—attributed to the property of the impulse response that $h(t) = 1$ at both $t = 0$ and $t = T_b$.

To bring out the ISI effect, let the amplitude sequence a_k be related to the precoded binary message sequence m_k' by

$$a_k = (m_k' - 1/2)A = \begin{cases} + A/2 & m_k' = 1 \\ - A/2 & m_k' = 0 \end{cases} \qquad [22]$$

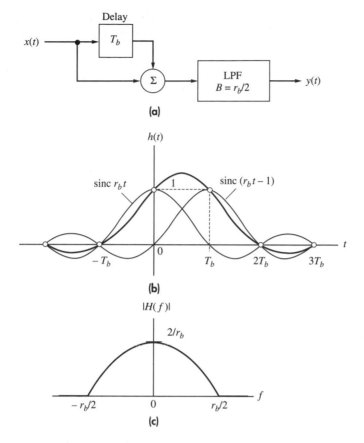

Figure 11.3–8   Duobinary signaling. (a) Equivalent correlative filter; (b) impulse response; (c) amplitude ratio.

which is equivalent to a polar binary format with level spacing A. Equation (20) then gives the corresponding output levels

$$y(t_k) = a'_k = a_k + a_{k-1} = (m'_k + m'_{k-1} - 1)A \qquad \text{[23a]}$$

$$= \begin{cases} +A & m'_k = m'_{k-1} = 1 \\ 0 & m'_k \neq m'_{k-1} \\ -A & m'_k = m'_{k-1} = 0 \end{cases} \qquad \text{[23b]}$$

In principle, you could use Eq. (23b) to recover m'_k from $y(t_k)$ if you have previously recovered m'_{k-1}. However, when noise causes an erroneous value of m'_{k-1}, all subsequent message bits will be in error until the next noise-induced error—a phenomenon called **error propagation.**

The digital precoder for duobinary signaling shown in Fig. 11.3–9a prevents error propagation and makes it possible to recover the input message sequence m_k from $y(t_k)$. The precoder consists of an exclusive-OR gate with feedback through a D-type flip-flop. Figure 11.3–9b lists the coding truth table along with the algebraic sum $m_k' + m_{k-1}'$ that appears in Eq. (23a). Substitution now yields

$$y(t_k) = \begin{cases} \pm A & m_k = 0 \\ 0 & m_k = 1 \end{cases} \qquad \text{[24]}$$

which does not involve m_{k-1} thanks to the precoder. When $y(t)$ includes additive gaussian white noise, the appropriate decision rule for message regeneration is:

Choose $m_k = 0$ if $|y| > A/2$

Choose $m_k = 1$ if $|y| < A/2$

This rule is easily implemented with a rectifier and a single decision threshold set at $A/2$. Optimum terminal filter design gives the minimum error probability

$$P_e = \frac{3}{2} Q\left(\frac{\pi}{4} \sqrt{2\gamma_b} \right) \qquad \text{[25]}$$

which is somewhat higher than that of a polar binary system.

When the transmission channel has poor dc response, **modified duobinary signaling** may be employed. The correlative filter has $N = 2$, $c_0 = 1$, $c_1 = 0$, and $c_2 = -1$, so that

$$H(f) = \frac{2j}{r_b} \sin \frac{2\pi f}{r_b} e^{-j2\pi f/r_b} \qquad |f| \leq r_b/2 \qquad \text{[26]}$$

Figure 11.3–10 shows $|H(f)|$ and the block diagram of the correlative filter. The precoder takes the form of Fig. 11.3–9 with two flip-flops in series to feed back

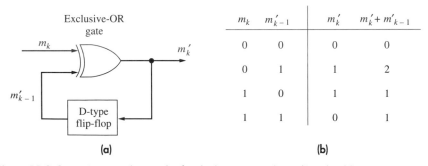

m_k	m_{k-1}'	m_k'	$m_k' + m_{k-1}'$
0	0	0	0
0	1	1	2
1	0	1	1
1	1	0	1

(a)   (b)

Figure 11.3–9   (a) Digital precoder for duobinary signaling; (b) truth table.

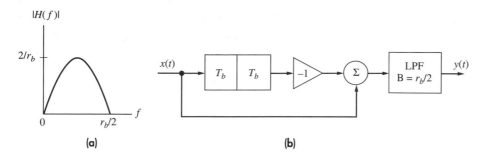

Figure 11.3-10   Correlative filter for modified duobinary signaling. (a) Amplitude ratio of correlative filter; (b) block diagram.

m'_{k-2}. If the pulse generation is in accordance with Eq. (22), then

$$y(t_k) = a_k - a_{k-2} = (m'_k - m'_{k-2})A \qquad [27]$$

$$= \begin{cases} 0 & m_k = 0 \\ \pm A & m_k = 1 \end{cases}$$

which can be compared with Eqs. (23) and (24).

Consider a duobinary system that uses the precoding of Fig. 11.3–9 and has binary message $m_k = 000010011000$. For the purposes of initial conditions, we set the initial values of m'_{k-1} and a_{k-1} to 0 and $-A$, respectively. Table 11.3–1 shows the message input, m_k; the precoded output, m'_k; the input to the correlative filter, a_k, with $A = 2$; and finally the correlative filter output, $y(t_k)$. Note that if we apply the rule of Eq. (24), we will recover the original message from $y(t_k) \rightarrow \hat{m}_k$. Now let's consider the modified duobinary system of Fig. 11.3–10 that includes precoding with two flip-flops. Here we set the initial values of m'_{k-2} and a_{k-2} to 0 and $-A$, respectively. The results are shown in Table 11.3–2.

EXAMPLE 11.3-3

Table 11.3–1   Duobinary signaling example with precoding

m_k	0	0	0	0	1	0	0	1	1	0	0	0
m'_{k-1}	0	0	0	0	0	1	1	1	0	1	1	1
m'_k	0	0	0	0	1	1	1	0	1	1	1	1
a_k	-1	-1	-1	-1	1	1	1	-1	1	1	1	1
a_{k-1}	-1	-1	-1	-1	-1	1	1	1	-1	1	1	1
$y(t_k)$	-2	-2	-2	-2	0	2	2	0	0	2	2	2
$\hat{m}_k$	0	0	0	0	1	0	0	1	1	0	0	0

Table 11.3–2   Modified duobinary signaling example with precoding

m_k	0	0	0	0	1	0	0	1	1	0	0	0
m'_{k-2}	0	0	0	0	0	0	1	0	1	1	0	1
m'_k	0	0	0	0	1	0	1	1	0	1	0	1
a_k	-1	-1	-1	-1	1	-1	1	1	-1	1	-1	1
a_{k-2}	-1	-1	-1	-1	-1	-1	1	-1	1	1	1	1
$y(t_k)$	-2	-2	-2	-2	0	-2	2	0	0	2	-2	2
$\hat{m}_k$	0	0	0	0	1	0	0	1	1	0	0	0

EXERCISE 11.3–3   Construct a truth table like Fig. 11.3–9b for the case of modified duobinary signaling and use it to obtain Eq. (27).

11.4   SYNCHRONIZATION TECHNIQUES

Synchronization is the art of making clocks tick together. The clocks in a digital communication system are at the transmitter and receiver, and allowance must be made for the transmission time delay between them. Besides *symbol* synchronization, most systems also require *frame* synchronization to identify the start of a message or various subdivisions within the message sequence. Additionally, carrier synchronization is essential for digital transmission with coherent carrier modulation—a topic to be discussed in Chap. 14.

Here we'll consider symbol and frame synchronization in baseband binary systems. Our attention will be focused on extracting synchronization from the received signal itself, rather than using an auxiliary sync signal. By way of an overview, Fig. 11.4–1 illustrates the position of the **bit synchronizer** relative to the clock and regenerator. Framing information is usually derived from the regenerated message and the clock, as indicated by the location of the **frame synchronizer.** We'll look at typical synchronization techniques, along with the related topics of shift-register operations for message scrambling and pseudonoise (PN) sequences for framing purposes.

Our coverage of synchronization will be primarily descriptive and illustrative. Detailed treatments of digital synchronization with additive noise are given by Mengali and D'Andrea (1997).

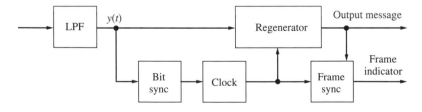

Figure 11.4–1   Synchronization in a binary receiver.

Bit Synchronization

Bit-sync generation becomes almost trivial when $y(t)$ has unipolar RZ format so its power spectrum $G_y(f)$ includes $\delta(f \pm r_b)$, like Fig. 11.1–5. A PLL or narrow BPF tuned to $f = r_b$ will then extract a sinusoid proportional to $\cos(2\pi r_b t + \phi)$, and phase adjustment yields a sync signal for the clock. The same technique works with a *polar* format if $y(t)$ is first processed by a square-law device, as diagrammed in Fig. 11.4–2a. The resulting unipolar waveform $y^2(t)$ shown in Fig. 11.4–2b now has the desired sinusoidal component at $f = r_b$. Various other nonlinear polar-to-unipolar operations on $y(t)$ achieve like results in open-loop bit synchronizers.

But a closed-loop configuration that incorporates the clock in a feedback loop provides more reliable synchronization. Figure 11.4–3 gives the diagram and explanatory waveforms for a representative closed-loop bit synchronizer. Here a zero-crossing detector generates a rectangular pulse with half-bit duration $T_b/2$ starting at each zero-crossing in $y(t)$. The pulsed waveform $z(t)$ then multiplies the square-wave clock signal $c(t)$ coming back from the *voltage-controlled clock* (VCC). The control voltage $v(t)$ is obtained by integrating and lowpass-filtering the product $z(t)c(t)$. The loop reaches steady-state conditions when the edges of $c(t)$ and $z(t)$ are synchronized and offset by $T_b/4$, so the product has zero area and $v(t)$ remains constant. Practical implementations of this system usually feature digital components in place of the analog multiplier and integrator.

Both of the foregoing techniques work best when the zero-crossings of $y(t)$ are spaced by integer multiples of T_b. Otherwise, the synchronization will suffer from

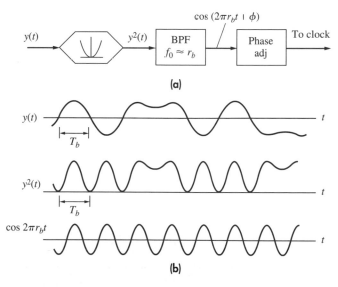

(a)

(b)

Figure 11.4–2   Bit synchronization by polar to unipolar conversion. (a) Block diagram; (b) waveforms.

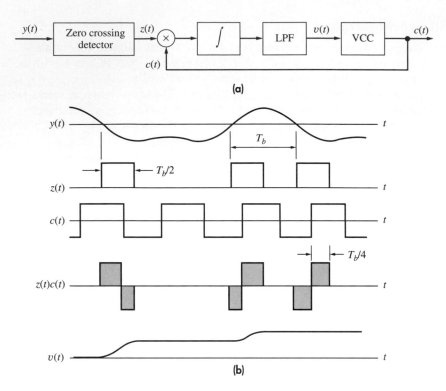

(a)

(b)

Figure 11.4–3   Closed-loop bit synchronization with a voltage-controlled clock. (a) Block diagram; (b) waveforms.

timing jitter. An additional problem arises if the message includes a long string of 1s or 0s, so $y(t)$ has no zero-crossings, and synchronism may be lost. Message scramblers discussed shortly help alleviate this problem.

A different approach to synchronization, independent of zero-crossings, relies on the fact that a properly filtered digital signal has peaks at the optimum sampling times and is reasonably symmetric on either side. Thus, if t_k is synchronized and $\delta < T_b/2$, then

$$|y(t_k - \delta)| \approx |y(t_k + \delta)| < |y(t_k)|$$

However, a late sync signal produces the situation shown in Fig. 11.4–4a where $|y(t_k - \delta)| > |y(t_k + \delta)|$, while an early sync signal would result in $|y(t_k - \delta)| < |y(t_k + \delta)|$. The **early-late synchronizer** in Fig. 11.4–4b uses these properties to develop the control voltage for a VCC in a feedback loop. A late sync signal results in $v(t) = |y(t_k - \delta)| - |y(t_k + \delta)| > 0$, which speeds up the clock, and conversely for an early sync signal.

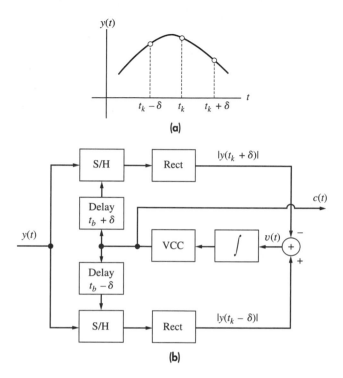

Figure 11.4–4 Early-late bit synchronization. (a) Waveform; (b) block diagram.

Scramblers and PN Sequence Generators

Scrambling is a coding operation applied to the message at the transmitter that "randomizes" the bit stream, eliminating long strings of like bits that might impair receiver synchronization. Scrambling also eliminates most periodic bit patterns that could produce undesirable discrete-frequency components (including dc) in the power spectrum. Needless to say, the scrambled sequence must be unscrambled at the receiver so as to preserve overall *bit sequence transparency*.

Simple but effective scramblers and unscramblers are built from tapped *shift registers* having the generic form of Fig. 11.4–5, the digital counterpart of a tapped delay line. Successive bits from the binary input sequence b_k enter the register and shift from one stage to the next at each tick of the clock. The output b'_k is formed by combining the bits in the register through a set of tap gains and mod-2 adders, yielding

$$b'_k = \alpha_1 b_{k-1} \oplus \alpha_2 b_{k-2} \oplus \cdots \oplus \alpha_n b_{k-n} \qquad [1]$$

The tap gains themselves are binary digits, so $\alpha_1 = 1$ simply means a direct connection while $\alpha_1 = 0$ means no connection. The symbol $\oplus$ stands for *modulo-2 addition*, defined by the properties

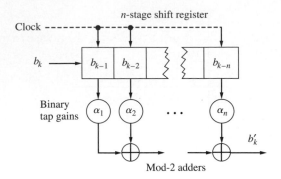

Figure 11.4–5   Tapped shift register.

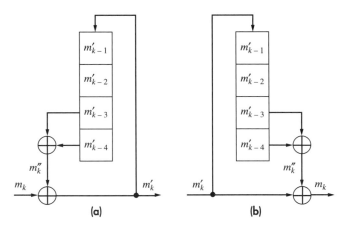

Figure 11.4–6   (a) Binary scrambler; (b) unscrambler.

$$b_1 \oplus b_2 = \begin{cases} 0 & b_1 = b_2 \\ 1 & b_1 \neq b_2 \end{cases} \qquad \text{[2a]}$$

and

$$b_1 \oplus b_2 \oplus b_3 = (b_1 \oplus b_2) \oplus b_3 = b_1 \oplus (b_2 \oplus b_3) \qquad \text{[2b]}$$

where b_1, b_2, and b_3 are arbitrary binary digits. Mod-2 addition is implemented with exclusive-OR gates, and obeys the rules of ordinary addition except that $1 \oplus 1 = 0$.

Figure 11.4–6 shows an illustrative scrambler and unscrambler, each employing a 4-stage shift register with tap gains $\alpha_1 = \alpha_2 = 0$ and $\alpha_3 = \alpha_4 = 1$. (For clarity, we omit the clock line here and henceforth.) The binary message sequence m_k at the input to the scrambler is mod-2 added to the register output m_k'' to form the scrambled message m_k' which is also fed back to the register input. Thus, $m_k'' = m_{k-3}' \oplus m_{k-4}'$ and

$$m'_k = m_k \oplus m''_k \qquad [3a]$$

The unscrambler has essentially the reverse structure of the scrambler and reproduces the original message sequence, since

$$m'_k \oplus m''_k = (m_k \oplus m''_k) \oplus m''_k \qquad [3b]$$

$$= m_k \oplus (m''_k \oplus m''_k) = m_k \oplus 0 = m_k$$

Equations (3a) and (3b) hold for any shift-register configuration as long as the scrambler and unscrambler have identical registers.

The scrambling action does, of course, depend on the shift-register configuration. Table 11.4–1 portrays the scrambling produced by our illustrative scrambler when the initial state of the register is all 0s. Note that the string of nine 0s in m_k has been eliminated in m'_k. Nonetheless, there may be some specific message sequence that will result in a long string of like bits in m'_k. Of more serious concern is *error propagation* at the unscrambler, since one erroneous bit in m'_k will cause several output bit errors. Error propagation stops when the unscrambler register is full of correct bits.

Next, in preparation for the subsequent discussion of frame synchronization, we consider shift register *sequence generation*. When a shift register has a nonzero initial state and the output is fed back to the input, the unit acts as a periodic sequence generator. Figure 11.4–7 shows an illustrative sequence generator using a five-stage shift register where the second and fifth cells are tapped and mod-2 added, and the result is fed back to the first stage so

$$m_1 = m_2 \oplus m_5 \qquad [4]$$

and the sequence output is m_5. For shorthand purposes this can also be referred to as a [5, 2] configuration. If the initial state of the register is 11111, it then produces a

Table 11.4–1

	m'_{k-1}	0	1	0	1	0	1	1	1	1	0	0	0	1	0
Registers	m'_{k-2}	0	0	1	0	1	0	1	1	1	1	0	0	0	1
Contents	m'_{k-3}	0	0	0	1	0	1	0	1	1	1	1	0	0	0
	m'_{k-4}	0	0	0	0	1	0	1	0	1	1	1	1	0	0

Register Output	m''_k	0	0	0	1	1	1	1	1	0	0	0	1	0	0

Input Sequence	m_k	1	0	1	1	0	0	0	0	0	0	0	0	0	1

Output Sequence	m'_k	1	0	1	0	1	1	1	1	0	0	0	1	0	1

31-bit sequence 1111100110100100001010111011000, which repeats periodically thereafter. In general

$$N = 2^n - 1 \qquad [5]$$

If the appropriate feedback tap connections are made, then an n-bit register can produce a **maximal-length** (ml) sequence per Eq. (5). (Figure 11.4–7 is, in fact, an ml sequence generator with $n = 5$ and $N = 31$.) An ml sequence has the following properties:

1. **Balance.** The number of ones generated is one more than the number of zeros generated.

2. **Run.** A *run* is a sequence of a single type of digit. An ml sequence will have one-half of its runs of length 1, one-quarter of its runs of length 2, one-eighth of its runs of length 3, and so on.

3. **Autocorrelation.** Its autocorrelation function has properties similar to the correlation properties of random noise, in that there is a single autocorrelation peak.

4. The mod-2 addition of an ml sequence and any shifted version of it results in a shifted version of the original sequence.

5. Except for the zero state, all of the 2^n possible states will exist during the sequence generation.

Maximal-length sequences are all also called *pseudonoise* (PN) sequences. The name *pseudonoise* comes from the *correlation* properties of PN sequences. To develop this point, let a PN sequence s_k be used to form a binary polar NRZ signal

$$s(t) = \sum_k (2s_k - 1)p(t - kT_c) \qquad [6]$$

where $p(t)$ is a rectangular pulse and the amplitude of the kth pulse is

$$c(t) = 2s_k - 1 = \begin{cases} +1 & s_k = 1 \\ -1 & s_k = 0 \end{cases} \qquad [7]$$

The signal $s(t)$ is deterministic and periodic, with period NT_c, and has a periodic autocorrelation function given by

$$R_c(\tau) = [(N + 1)\Lambda(\tau/T_c) - 1]/N \qquad |\tau| \le NT_c/2 \qquad [8]$$

which is plotted in Fig 11.4–8. If N is very large and T_c very small, then

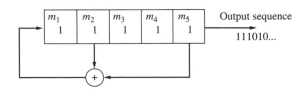

Figure 11.4–7
Shift register sequence generator with [5, 2] configuration.

$$R_s(\tau) \approx T_c\delta(\tau) - 1/N \qquad |\tau| \le NT_c/2 \qquad [9]$$

so the PN signal acts essentially like *white noise* with a small dc component. This noiselike correlation property leads to practical applications in test instruments and radar ranging, spread spectrum communication (described in Chap. 15) and digital framing. When we do use PN generators to operate on a data stream, to avoid significant dc content in the PN sequence and minimize information loss to the signal we are operating on, we use the function $c(t)$ instead of the polar s_k for our scrambling sequence.

A 3-bit, [3, 1] shift register configuration with values of 111 produces a periodic ml sequence of 1110100 with $N = 7$. A simple way used by Dixon (1994) to calculate the autocorrelation function for the comparable $c(t)$ sequence is to compare the original s_k sequence to each of its $N - 1$ shifted versions ($\tau = 0$ to $6T_c$). Then, for each pair, let $v(\tau)$ be the difference between the number of bit matches and mismatches, so the autocorrelation function is $R_{[3,1]}(\tau) = v(\tau)/N$ and hence the results shown in Table 11.4–2.

EXAMPLE 11.4–1

Table 11.4–2

τ	Original/shifted	$v(\tau)$	$R_{[3,1]}(\tau) = v(\tau)/N$
0	1110100 1110100	7.00	1.00
1	1110100 0111010	−1.00	−0.14
2	1110100 0011101	−1.00	−0.14
3	1110100 1001110	−1.00	−0.14
4	1110100 0100111	−1.00	−0.14
5	1110100 1010011	−1.00	−0.14
6	1110100 1101001	−1.00	−0.14
0	1110100 1110100	7.00	1.00

If plotted, the autocorrelation function would look like the one in Figure 11.4–8 with $N = 7$ and $T_c = 1$. From this, we observe our [3, 1] shift register configuration produces an ml sequence.

Show that the output sequence from a 5-stage, [5, 4, 3, 2] register with initial conditions 11111 is 1111100100110000101101010001110 and has period 31.

EXERCISE 11.4–1

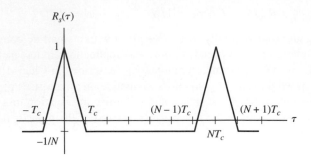

Figure 11.4–8   Autocorrelation of a PN sequence.

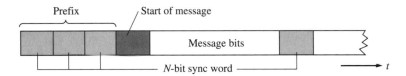

Figure 11.4–9

Frame Synchronization

A digital receiver needs to know when a signal is present. Otherwise, the input noise alone may produce random output bits that could be mistaken for a message. Therefore, identifying the start of a message is one aspect of frame synchronization. Another aspect is identifying subdivisions or frames within the message. To facilitate frame synchronization, binary transmission usually includes special N-bit *sync words* as represented in Fig. 11.4–9. The initial **prefix** consists of several repetitions of the sync word, which marks the beginning of transmission and allows time for bit-sync acquisition. The prefix is followed by a different codeword labeling the start of the message itself. Frames are labeled by sync words inserted periodically in the bit stream.

The elementary frame synchronizer in Fig. 11.4–10 is designed to detect a sync word $s_1 s_2 \ldots s_N$ whenever it appears in the regenerated sequence m_k. Output bits with the polar format

$$a_k = 2m_k - 1 = \pm 1$$

are loaded into an N-stage polar shift register having polar tap gains given by

$$c_i = 2s_{N+1-i} - 1 \qquad \text{[10]}$$

This awkward-looking expression simply states that the gains equal the sync-word bits in polar form and reverse order, that is, $c_1 = 2s_N - 1$ while $c_N = 2s_1 - 1$. The tap-gain outputs are summed *algebraically* to form

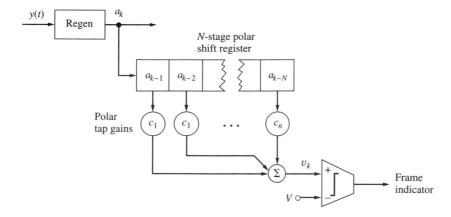

Figure 11.4–10   Frame synchronizer.

$$v_k = \sum_{i=1}^{N} c_i a_{k-i} \qquad [11]$$

This voltage is compared with a threshold voltage V, and the frame-sync indicator goes HIGH when $v_k > V$.

If the register word is identical to the sync word, then $a_{k-i} = c_i$ so $c_i a_{k-i} = c_i^2 = 1$ and $v_k = N$. If the register word differs from the sync word in just one bit, then $v_k = N - 2$ (why?). Setting the threshold voltage V slightly below $N - 2$ thus allows detection of error-free sync words and sync words with one bit error. Sync words with two or more errors go undetected, but that should be an unlikely event with any reasonable value of P_e. *False* frame indication occurs when N or $N - 1$ successive message bits match the sync-word bits. The probability of this event is

$$P_{ff} = \left(\frac{1}{2}\right)^{N} + \left(\frac{1}{2}\right)^{N-1} = 3 \times 2^{-N} \qquad [12]$$

assuming equally likely 1s and 0s in the bit stream.

Further examination of Eqs. (10) and (11) reveals that the frame synchronizer calculates the *cross-correlation* between the bit stream passing through the register and the sync word, represented by the tap gains. The correlation properties of a PN sequence therefore make it an ideal choice for the sync word. In particular, suppose the prefix consists of several periods of a PN sequence. As the prefix passes through the frame-sync register, the values of v_k will trace out the shape of $R_s(\tau)$ in Fig. 11.4–8 with peaks $v_k = N$ occurring each time the initial bit s_1 reaches the end of the register. An added advantage is the ease of PN sequence generation at the transmitter, even with large values of N. For instance, getting $P_{ff} < 10^{-4}$ in Eq. (12) requires $N > 14.8$, which can be accomplished with a 4-stage PN generator.

11.5　PROBLEMS

11.1–1　Sketch $x(t)$ and construct the corresponding eye pattern (without transmission distortion) for binary PAM with the data sequence 1011100010 when the signal has a unipolar format and pulse shape $p(t) = \cos^2(\pi t/2T_b)\Pi(t/2T_b)$.

11.1–2　Do Prob. 11.1–1 with $p(t) = \cos(\pi t/2T_b)\Pi(t/2T_b)$.

11.1–3　Do Prob. 11.1–1 with a polar format and $p(t) = \cos(\pi t/2T_b)\Pi(t/2T_b)$.

11.1–4　Do Prob. 11.1–1 with a bipolar format and $p(t) = \Lambda(t/T_b)$.

11.1–5　Do Prob. 11.1–1 with a bipolar format.

11.1–6　Modify Table 11.1–1 for an *octal* signal ($M = 8$).

11.1–7*　A certain computer generates binary words, each consisting of 16 bits, at the rate 20,000 words per second. (*a*) Find the bandwidth required to transmit the output as a binary PAM signal. (*b*) Find M so that the output could be transmitted as an M-ary signal on a channel having $B = 60$ kHz.

11.1–8　A certain digital tape reader produces 3000 symbols per second, and there are 128 different symbols. (*a*) Find the bandwidth required to transmit the output as a binary PAM signal. (*b*) Find M so that the output could be transmitted as an M-ary signal on a telephone link having $B = 3$ kHz.

11.1–9　Binary data is transmitted as a unipolar signal with $A = 1$ and $p(t) = u(t + T_b) - u(t)$. The transmission system's step response is $g(t) = K_0(1 - e^{-bt})u(t)$, where $b = 2/T_b$. (*a*) Sketch $\tilde{p}(t)$ and find K_0 such that $\tilde{p}(0) = 1$. (*b*) Sketch $y(t)$ for the data sequence 10110 and evaluate $y(t)$ and the ISI at the optimum sampling times.

11.1–10　Do Prob. 11.1–9 for a polar signal with $A/2 = 1$ and a transmission system having $b = 1/T_b$.

11.1–11*　Consider digital transmission with a gaussian pulse shape $p(t) = \exp[-\pi(bt)^2]$, which is neither timelimited nor bandlimited and does not have periodic zero crossings. Let $p(kD) \leq 0.01$ for $k \neq 0$, to limit the ISI, and let the bandwidth B be such that $P(f) \leq 0.01P(0)$ for $|f| > B$. Find the resulting relationship between r and B.

11.1–12　Find and sketch the power spectrum of a binary PAM signal with polar RZ format and rectangular pulses, assuming independent and equiprobable message bits. Then show that the time-domain and frequency-domain calculations of $\overline{x^2}$ are in agreement.

11.1–13　Consider a unipolar binary PAM signal whose pulse shape is the *raised cosine* in Example 2.5–2 with total duration $2\tau = T_b$. (*a*) Sketch $x(t)$ for the data sequence 10110100. Then find and sketch the power spectrum, assuming independent equiprobable message bits. (*b*) Repeat part *a* with $2\tau = 2T_b$.

11.1–14　Find and sketch the power spectrum of a polar binary PAM signal with $p(t) = \text{sinc } r_b t$ when the message bits are independent but have unequal probabilities, say α and $1 - \alpha$. Use your sketch to show that $\overline{x^2}$ is independent of α.

11.1–15 Find and plot the power spectrum of a binary signal with the split-phase Manchester format in Fig. 11.1–1c assuming independent equiprobable message bits. Compare your result with Fig. 11.1–6.

11.1–16 Verify the correlation properties of bipolar precoding given in Eq. (14a) by tabulating the possible sequences $a_{k-n}a_k$ and their probabilities for $n = 0, n = 1$, and $n \geq 2$. Assume independent equiprobable message bits.

11.1–17‡ Let $g(n)$ be a function of the discrete variable n. By expanding and rearranging the summation, show that

$$\sum_{k=-K}^{K} \sum_{i=-K}^{K} g(k - i) = (2K + 1) \sum_{n=-2K}^{2K} (1 - \frac{|n|}{2K + 1})g(n)$$

Apply this result to carry out the manipulation of the sum $\rho_K(f)$ used to derive $G_x(f)$ from $E[|X_T(f)|^2]$.

11.1–18‡ Consider a binary signal constructed using *different pulse shapes* $p_1(t)$ and $p_0(t)$ to represent 1s and 0s. For purposes of analysis, we can write

$$x(t) = \sum_{k} [a_k p_1(t - kT_b) + b_k p_0(t - kT_b)]$$

where a_k equals the kth bit and $b_k = 1 - a_k$. (a) Assuming independent and equiprobable bits, show that $\overline{a_k} = \overline{b_k} = \overline{a_k^2} = \overline{b_k^2} = 1/2$, $\overline{a_k b_k} = 0$, and $\overline{a_k a_i} = \overline{b_k b_i} = \overline{a_k b_i} = 1/4$ for $i \neq k$. (b) Now form the truncated signal $x_T(t)$ with $T = (2K + 1)T_b \gg T_b$ and show that

$$E[|X_T(f)|^2] = \frac{2K + 1}{4}|P_1(f) - P_0(f)|^2$$

$$+ \frac{1}{4}|P_1(f) + P_0(f)|^2 \sum_{k=-K}^{K} \sum_{i=-K}^{K} e^{-j\omega(k-i)T_b}$$

where $P_1(f) = \mathcal{F}[p_1(t)]$ and $P_0(f) = \mathcal{F}[p_0(t)]$. (c) Finally, use the relationship in Prob. 11.1–17 and Poisson's sum formula to obtain

$$G_x(f) = \frac{r_b}{4}|P_1(f) - P_0(f)|^2$$

$$+ \frac{r_b^2}{4} \sum_{n=-\infty}^{\infty} |P_1(nr_b) - P_0(nr_b)|^2 \delta(f - nr_b)$$

11.2–1* Find $(S/N)_R$ such that a unipolar binary system with AWGN has $P_e = 0.001$. What would be the error probability of a polar system with the same $(S/N)_R$?

11.2–2 A binary system has AWGN with $N_0 = 10^{-8}/r_b$. Find S_R for polar and unipolar signaling so that $P_e \leq 10^{-6}$.

11.2–3   Some switching circuits generate *impulse noise*, which can be modeled as filtered white noise with average power σ^2 and an exponential PDF

$$p_n(n) = \frac{1}{\sqrt{2\sigma^2}} e^{-\sqrt{2}|n|/\sigma}$$

(a) Develop an expression for P_e in terms of A and σ for polar binary signaling contaminated by impulse noise. (b) Compare the effect of impulse noise with that of gaussian noise by considering the condition $P_e \leq 0.001$.

11.2–4*   Consider a polar binary system with ISI and AWGN such that $y(t_k) = a_k + \epsilon_k + n(t_k)$, where the ISI ϵ_k is equally likely to be $+\alpha$ or $-\alpha$. (a) Develop an expression for P_e in terms of A, α, and σ. (b) Evaluate P_e when $A = 8\sigma$ and $\alpha = 0.1A$. Compare your result with the error probability without ISI.

11.2–5   Do Prob. 11.2–4 taking $\epsilon_k = +\alpha$, 0, and $-\alpha$ with probabilities 0.25, 0.5, and 0.25, respectively.

11.2–6   Use Eq. (4) to obtain an expression for the optimum threshold in a polar binary system with AWGN when $P_0 \neq P_1$.

11.2–7   Derive Eq. (4) using Leibniz's rule, which states that

$$\frac{d}{dz}\int_{a(z)}^{b(z)} g(z, \lambda)\, d\lambda = g[z, b(z)]\frac{db(z)}{dz}$$

$$- g[z, a(z)]\frac{da(z)}{dz} + \int_{a(z)}^{b(z)} \frac{\partial}{\partial z}[g(z, \lambda)]\, d\lambda$$

where z is an independent variable and $a(z)$, $b(z)$, and $g(z, \lambda)$ are arbitrary functions.

11.2–8*   A polar binary system has 20 repeaters, each with $(S/N)_1 = 20$ dB. Find P_e when the repeaters are regenerative and nonregenerative.

11.2–9   A polar binary system with 50 repeaters is to have $P_e = 10^{-4}$. Find $(S/N)_1$ in dB when the repeaters are regenerative and nonregenerative.

11.2–10   Consider the split-phase Manchester format in Fig. 11.1–1(d), where

$$p(t) = \begin{cases} 1 & -T_b/2 < t < 0 \\ -1 & 0 < t < T_b/2 \end{cases}$$

Plot the matched filter's impulse response. Then use superposition to plot the pulse response $a_k\, p(t - t_0) * h(t)$ and compare with Fig. 11.2–6(c).

11.2–11   Consider a unipolar RZ binary system with $p(t) = u(t) - u(t - T_b/2)$. Instead of a matched filter, the receiver has a first-order LPF with impulse response $h(t) = K_0 e^{-bt}u(t)$ where $K_0 = b/(1 - e^{-bT_b/2})$. (a) Find and sketch $Ap(t) * h(t)$ and obtain the condition on b such that the ISI at any subsequent sampling time does not exceed $0.1A$. (b) Show that $(A/2\sigma)^2 = (4br_b/K_0^2)\gamma_b \leq 0.812\gamma_b$, where the upper bound comes from the ISI condition.

11.2–12   A binary data transmission system is to have $r_b = 500$ kbps and $P_{be} \leq 10^{-4}$. The noise is white gaussian with $N_0 = 10^{-17}$ W/Hz. Find the minimum value of S_R when: (a) $M = 2$; (b) $M = 8$ with Gray coding.

11.2–13*   Suppose the transmission bandwidth of the system in Prob. 11.2–12 is $B = 80$ kHz. Find the smallest allowed value of M and the corresponding minimum value of S_R, assuming a Gray code.

11.2–14   A binary data transmission system has AWGN, $\gamma_b = 100$, and Gray coding. What's the largest value of M that yields $P_{be} \leq 10^{-5}$?

11.2–15   Derive the result stated in Eq. (21).

11.2–16   Suppose binary data is converted to $M = 4$ levels via the Gray code in Table 11.1–1. Use Fig. 11.2–8 to derive an expression for P_{be} in terms of $k = A/2\sigma$. Simplify your result by assuming $k > 1$.

11.2–17   Do Prob. 11.2–16 with the natural code in Table 11.1–1.

11.3–1   Use Eq. (3a) to find and sketch $p(t)$ and $P(f)$ when $P_\beta(f) = (1/2\beta)\Pi(f/2\beta)$ with $\beta = r/4$.

11.3–2   Use Eq. (3a) to find and sketch $p(t)$ and $P(f)$ when $P_\beta(f) = (1/2\beta)\Pi(f/2\beta)$ with $\beta = r/3$.

11.3–3*   Suppose $B = 3$ kHz. Using pulses with a cosine rolloff spectrum, what is the maximum baud rate if the rolloffs are: (a) 100 percent, (b) 50 percent, (c) 25 percent?

11.3–4   Given a binary message of 10110100, sketch the baseband waveform using 50 percent rolloff raised cosine pulses. How does this compare with Figure 11.3–2?

11.3–5   We want to transmit binary data at 56 kbps using pulses with a cosine rolloff spectrum. What is B for rolloffs of: (a) 100 percent, (b) 50 percent, (c) 25 percent?

11.3–6   Obtain Eq. (6b) from Eq. (5) with $\beta = r/2$. Then show that $p(\pm D/2) = 1/2$.

11.3–7   Carry out the details leading to Eqs. (4b) and (5) starting from Eq. (4a).

11.3–8   A more general form of Nyquist's signaling theorem states that if $P(f) = \mathcal{F}[p(t)]$ and

$$\sum_{n=-\infty}^{\infty} P(f - nr) = 1/r \qquad -\infty < f < \infty$$

then $p(t)$ has the property in Eq. (1a) with $D = 1/r$. (a) Prove this theorem by taking the Fourier transform of both sides of

$$p(t) \sum_{k=-\infty}^{\infty} \delta(t - kD) = \sum_{k=-\infty}^{\infty} p(kD)\delta(t - kD)$$

Then use Poisson's sum formula. (b) Use a sketch to show that $P(f)$ in Eq. (4b) satisfies the foregoing condition.

11.3–9 Consider an arbitrary bandlimited pulse spectrum with $P(f) = 0$ for $|f| \geq B$. Use simple sketches to obtain the additional requirements on $P(f)$ needed to satisfy Nyquist's signaling theorem as stated in Prob. 11.3–5 when: (a) $r/2 < B < r$; (b) $B = r/2$; (c) $B < r/2$.

11.3–10 A binary data system is to be designed for $r_b = 600$ kbps and $P_{be} \leq 10^{-5}$. The waveform will have $M = 2^n$, Gray coding, and Nyquist pulse shaping. The noise is white gaussian with $N_0 = 1$pW/Hz. The transmission channel has loss $L = 50$ dB and is distortionless over the allocated bandwidth $B = 200$ kHz. Choose M to minimize the transmitted power, and find the resulting values of r, β, and S_T.

11.3–11 Do Prob. 11.3–10 with $B = 120$ kHz.

11.3–12 Do Prob. 11.3–10 with $B = 80$ kHz.

11.3–13 Consider a data transmission system with $M = 2$, $r = 20,000$, $p_x(t) = \Pi(2rt)$, $|H_C(f)| = 0.01$, $G_n(f) = 10^{-10}(1 + 3 \times 10^{-4}|f|)^2$, and $p(t)$ per Eq. (6b). (a) Find and sketch the amplitude ratio for the optimum terminal filters. (b) Calculate S_T needed to get $P_e = 10^{-6}$, assuming gaussian noise.

11.3–14 Consider a data transmission system with $M = 4$, $r = 100$, $p_x(t) = \Pi(10rt)$, $G_n(f) = 10^{-10}$, $|H_C(f)|^2 = 10^{-6}/(1 + 32 \times 10^{-4}f^2)$, and $p(t) = $ sinc rt. (a) Find and sketch the amplitude ratio for the optimum terminal filters. (b) Calculate S_T needed to get $P_e = 10^{-6}$, assuming gaussian noise.

11.3–15‡ Consider a polar system in the form of Fig. 11.3–3 with $G_n(f) = N_0/2$ and a *time-limited* input pulse shape $p_x(t)$. Let $H_T(f) = K$ and $H_R(f) = [P_x(f)H_C(f)]^*e^{-j\omega t_d}$, so the receiving filter is *matched* to the received pulse shape. Since this scheme does not shape the output pulse $p(t - t_d)$ for zero ISI, it can be used only when the duration of $p_x(t)$ is small compared to $1/r$. (a) Obtain an expression for K such that $p(0) = 1$. Then develop expressions for S_T and σ^2 to show that $(A/2\sigma)^2$ is given by Eq. (12a) with

$$I_{HR} = \frac{N_0 \int_{-\infty}^{\infty} |P_x(f)|^2 \, df}{2 \int_{-\infty}^{\infty} |P_x(f)H_C(f)|^2 \, df}$$

(b) Show that this result is equivalent to the maximum value $(A/2\sigma)^2 = 6S_R/(M^2 - 1)N_0 r$.

11.3–16‡ A certain system has been built with optimum terminal filters for Nyquist pulse shaping, assuming white noise and a distortionless channel with loss L. However, it turns out that the channel actually introduces some linear distortion, so an equalizer with $H_{eq}(f) = [\sqrt{L}H_C(f)]^{-1}$ has been added at the output. Since the terminal filters were not modified to account for the channel distortion, the system has less than optimum performance. (a) Obtain expressions for S_T and σ^2 to show that

$$\left(\frac{A}{2\sigma}\right)^2 = \frac{6S_T/L}{K(M^2-1)N_0r} \qquad \text{where} \qquad K = \frac{1}{L}\int_{-\infty}^{\infty} \frac{|P(f)|}{|H_C(f)|^2}\, df$$

(b) Evaluate K in dB when $P(f)$ is as given by Eq. (6a) and $H_C(f) = \{\sqrt{L}[1 + j2(f/r)]\}^{-1}$.

11.3–17   Find the tap gains for a three-tap zero-forcing equalizer when $\tilde{p}_{-1} = 0.4$, $\tilde{p}_0 = 1.0$, $\tilde{p}_1 = 0.2$, and $\tilde{p}_k = 0$ for $|k| > 1$. Then find and plot $p_{eq}(t_k)$.

11.3–18   Obtain expressions for the tap gains of a three-tap zero-forcing equalizer when $\tilde{p}_{-1} = \epsilon$, $\tilde{p}_0 = 1$, $\tilde{p}_1 = \delta$, and $\tilde{p}_k = 0$ for $|k| > 1$.

11.3–19‡   Find the tap gains for a five-tap zero-forcing equalizer for $\tilde{p}(t)$ in Fig. 11.3–6a. (You can solve the simultaneous equations by successive substitution.) Then find the resulting values of $p_{eq}(t_k)$ and compare with Fig. 11.3–6b.

11.3–20   Consider binary correlative coding with $N = 2$, $c_0 = 1$, $c_1 = 2$, and $c_2 = 1$. (a) Find and sketch $h(t)$ and $|H(f)|$. (b) Use Eqs. (20b) and (22) to develop an expression for $y(t_k)$ like Eq. (23b).

11.3–21   Do Prob. 11.3–20 with $N = 4$, $c_0 = 1$, $c_1 = 0$, $c_2 = -2$, $c_3 = 0$, and $c_4 = 1$.

11.3–22‡   Consider a duobinary system with $H(f)$ as given by Eq. (21b), $p_x(t) = \delta(t)$, $H_C(f) = 1/\sqrt{L}$, and gaussian white noise. Obtain an expression for $(A/2\sigma)^2$ and apply Schwarz's inequality to show that the optimum filters have $|H_T(f)|^2 = gL|H(f)|$ and $|H_R(f)|^2 = |H(f)|/g$. Then derive Eq. (25).

11.3–23   Consider a duobinary system with an input data sequence of 101011101 and $A = 2$. (a) Find the coder output and then verify the receiver will have the correct output. (b) Calculate the dc value of the coder's output. (c) What is the receiver's output if the third bit has a value of zero?

11.3–24   Do Prob. 11.3–23 with a precoder.

11.3–25   Do Prob. 11.3–23 with modified duobinary signaling and precoding.

11.4–1†   Given a five-stage shift register scrambler/unscrambler system with $m_k'' = m_{k-1}' \oplus m_{k-4}' \oplus m_{k-2}'$ and zero initial shift register conditions, compute the scrambled output and the output from the corresponding unscrambler for an input sequence of $m_k = 011111101110111$. What are the dc levels for the unscrambled and scrambled bit streams?

11.4–2   Suppose we have a five-stage shift register sequence generator with a [5, 4, 3, 2] configuration, and its contents are all initially ones. Determine the output sequence, its length, and plot the corresponding autocorrelation function. Is the output an ml sequence?

11.4–3   Do Prob. 11.4–2 using a shift register with a [4, 2] configuration.

chapter

12

Digitization Techniques for Analog Messages and Computer Networks

CHAPTER OUTLINE

aving studied the basic concepts of digital transmission in Chap. 11, we return once more to analog communication. But now we consider *digital* transmission of analog messages via *coded pulse modulation*. Coded pulse modulation systems employ sampling, quantizing, and coding to convert analog waveforms into digital signals.

Digital coding of analog information produces a *rugged* signal with a high degree of immunity to transmission distortion, interference, and noise. Digital coding also allows the use of regenerative repeaters for long-distance analog communication. However, the quantizing process essential for digital coding results in *quantization noise* which becomes the fundamental limitation on waveform reconstruction. To keep the quantization noise small enough for suitable fidelity, a coded pulse modulation system generally requires a much larger bandwidth than a comparable analog transmission system.

We'll develop these properties first in conjunction with **pulse-code modulation** (PCM). Next, we describe **delta modulation** (DM) and other schemes that involve predictive coding. To better appreciate the advantages of digital over analog systems we will look at digital audio recording using the audio **compact disk** (CD). We then consider **digital multiplexing,** a valuable technique that makes it possible to combine analog and digital information for transmission in the form of a multiplexed digital signal. Finally, the chapter closes with a brief discussion of computer networks by considering the **Open Systems Interconnection** (OSI) and the **Transmission Control Protocol/Internet Protocol** (TCP/IP) systems.

OBJECTIVES

After studying this chapter and working the exercises, you should be able to do each of the following:

1. Define and relate the parameters of a PCM system, and distinguish between quantization noise and random noise (Sect. 12.1).

2. Find the conditions for PCM transmission above threshold, and calculate the value of $(S/N)_D$ (Sects. 12.1 and 12.2).

3. Identify and compare the distinctive features and relative advantages of PCM (with and without companding), delta modulation, and differential PCM (Sects. 12.2 and 12.3).

4. Describe the operation of a compact disk digital audio system, how it achieves error control, and its advantages over analog systems (Sect. 12.4).

5. Diagram a digital multiplexing system that accommodates both analog and digital signals in a standard multiplexing hierarchy including the North American and CCIT (Sect. 12.5).

6. Explain the concepts of the Integrated Services Digital Network (ISDN) and Synchronous Optical Network (SONET) hardware multiplexing schemes (Sect. 12.5).

7. Describe the concepts of packet switching, frame relay, and asynchronous transfer mode (ATM) data switching schemes (Sect. 12.5).

8. Explain the OSI and TCP/IP computer network architectures (Sect. 12.6).

12.1   PULSE-CODE MODULATION

This section describes the functional operation of pulse-code modulation (PCM)

> PCM is a digital transmission system with an *analog-to-digital converter* (ADC) at the input and a *digital-to-analog converter* (DAC) at the output.

When the digital error probability is sufficiently small, PCM performance as an *analog* communication system depends primarily on the quantization noise introduced by the ADC. Here we'll analyze analog message reconstruction with quantization noise, temporarily deferring to the next section the effects of random noise and digital errors.

PCM Generation and Reconstruction

Figure 12.1–1a diagrams the functional blocks of a PCM generation system. The analog input waveform $x(t)$ is lowpass filtered and sampled to obtain $x(kT_s)$. A **quantizer** rounds off the sample values to the nearest discrete value in a set of q **quantum levels.** The resulting quantized samples $x_q(kT_s)$ are discrete in time (by virtue of sampling) and discrete in amplitude (by virtue of quantizing).

To display the relationship between $x(kT_s)$ and $x_q(kT_s)$, let the analog message be a voltage waveform normalized such that $|x(t)| \leq 1$ V. Uniform quantization subdivides the 2-V peak-to-peak range into q equal steps of height $2/q$ V, as shown in Fig. 12.1–1b. The quantum levels are then taken to be at $\pm 1/q, \pm 3/q, \ldots, \pm(q-1)/q$ in the usual case when q is an even integer. A quantized value such as $x_q(kT_s) = 5/q$ corresponds to any sample value in the range $4/q < x(kT_s) < 6/q$.

Next, an **encoder** translates the quantized samples into **digital code words.** The encoder works with M-ary digits and produces for each sample a codeword consisting of ν digits in parallel. Since there are M^ν possible M-ary codewords with ν digits per word, unique encoding of the q different quantum levels requires that $M^\nu \geq q$. The parameters M, ν, and q should be chosen to satisfy the equality, so that

$$q = M^\nu \qquad \nu = \log_M q \qquad\qquad [1]$$

Thus, the number of quantum levels for *binary* PCM equals some power of 2, namely $q = 2^\nu$.

Finally, successive codewords are read out serially to constitute the PCM waveform, an M-ary digital signal. The PCM generator thereby acts as an ADC, performing analog-to-digital conversions at the sampling rate $f_s = 1/T_s$. A timing circuit coordinates the sampling and parallel-to-serial readout.

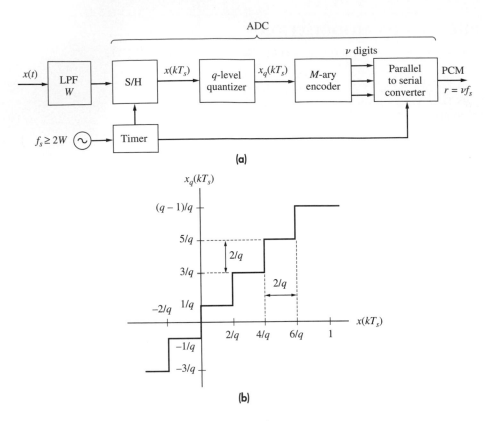

Figure 12.1–1   (a) PCM generation system; (b) quantization characteristic.

Each encoded sample is represented by a ν-digit output word, so the signaling rate becomes $r = \nu f_s$ with $f_s \geq 2W$. Therefore, the bandwidth needed for PCM baseband transmission is

$$B_T \geq \tfrac{1}{2}r = \tfrac{1}{2}\nu f_s \geq \nu W \qquad [2]$$

Fine-grain quantization for accurate reconstruction of the message waveform requires $q \gg 1$, which increases the transmission bandwidth by the factor $\nu = \log_M q$ times the message bandwidth W.

Now consider a PCM receiver with the reconstruction system in Fig. 12.1–2a. The received signal may be contaminated by noise, but regeneration yields a clean and nearly errorless waveform if $(S/N)_R$ is sufficiently large. The DAC operations of serial-to-parallel conversion, M-ary decoding, and sample-and-hold generate the analog waveform $x_q(t)$ drawn in Fig. 12.1–2b. This waveform is a "staircase" approximation of $x(t)$, similar to flat-top sampling except that the sample values have been quantized. Lowpass filtering then produces the smoothed output signal

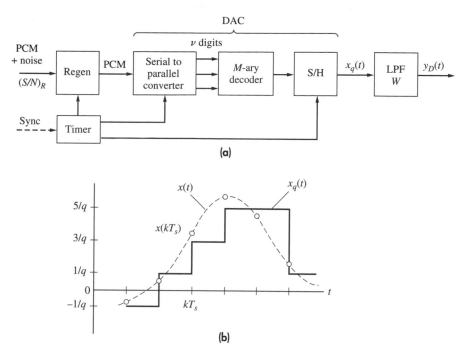

Figure 12.1–2   (a) PCM receiver; (b) reconstructed waveform.

$y_D(t)$, which differs from the message $x(t)$ to the extent that the quantized samples differ from the exact sample values $x(kT_s)$.

Perfect message reconstruction is therefore impossible in PCM, even when random noise has no effect. The ADC operation at the transmitter introduces permanent errors that appear at the receiver as quantization noise in the reconstructed signal. We'll study this quantization noise after an example of PCM hardware implementation.

EXAMPLE 12.1-1

Suppose you want to build a binary PCM system with $q = 8$ so $\nu = \log_2 8 = 3$ bits per codeword. Figure 12.1–3a lists the 8 quantum levels and two types of binary codes. The "natural" code assigns the word 000 to the lowest level and progresses upward to 111 in the natural order of binary counting. The sign/magnitude code uses the leading bit b_2 for the algebraic sign of x_q while the remaining bits $b_1 b_0$ represent the magnitude. Other encoding algorithms are possible, and may include additional bits for error protection—the topic of Chap. 13.

A direct-conversion ADC circuit for the sign/magnitude code is shown in Fig. 12.1–3b. This circuit consists of one comparator for the sign bit and three parallel comparators plus combinational logic to generate the magnitude bits. Direct-conversion ADCs have the advantage of high operating speed and are called **flash**

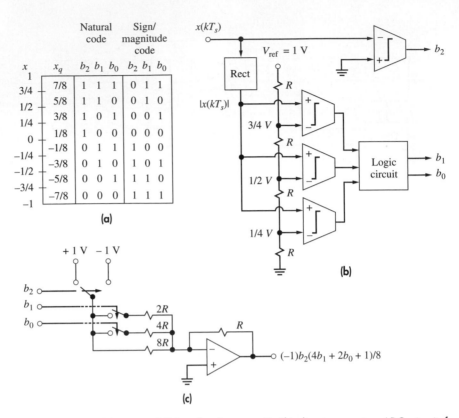

x	x_q	Natural code $b_2\ b_1\ b_0$	Sign/ magnitude code $b_2\ b_1\ b_0$
1	7/8	1 1 1	0 1 1
3/4	5/8	1 1 0	0 1 0
1/2	3/8	1 0 1	0 0 1
1/4	1/8	1 0 0	0 0 0
0	−1/8	0 1 1	1 0 0
−1/4	−3/8	0 1 0	1 0 1
−1/2	−5/8	0 0 1	1 1 0
−3/4	−7/8	0 0 0	1 1 1
−1			

(a)

(b)

(c)

Figure 12.1–3   (a) Binary PCM codes for $q = 8$; (b) direct-conversion ADC circuit for sign/magnitude code; (c) weighted-resistor decoder circuit.

encoders, but they require a total of $q/2$ comparators. At lower speeds you can get by with one comparator and a feedback loop, a configuration found in the dual-slope, counter-comparison, and successive-approximation encoders.

Figure 12.1–3c shows the circuit for a **weighted-resistor decoder** that goes with a 3-bit sign/magnitude code. The sign bit operates a polarity selector switch, while the magnitude bits control the resistors to be connected to the reference voltage. The overall circuit acts as an inverting op-amp summer with output voltage $(-1)^{b_2}(4b_1 + 2b_0 + 1)/8$.

EXAMPLE 12.1–2   **Direct Digital Synthesis of Analog Waveforms**

An alternative to the PLL frequency synthesizer is the **Direct Digital Synthesis** (DDS) as shown in Fig. 12.1–4. Here a waveform is quantized with the samples stored in computer memory. The memory contents are repeatedly sent to a DAC

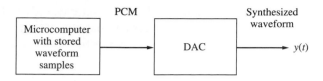

Figure 12.1–4   DDS for waveform generation.

which converts these to an equivalent analog signal. The rate of the output samples determines the waveform's frequency.

EXERCISE 12.1–1

A binary channel with $r_b = 36,000$ bits/sec is available for PCM voice transmission. Find appropriate values of ν, q, and f_s assuming $W \approx 3.2$ kHz.

Quantization Noise

Although PCM reconstruction most often takes the form of staircase filtering, as in Fig. 12.1–2, we'll find the *impulse reconstruction* model in Fig. 12.1–5 more convenient for the analysis of quantization noise. Here, a pulse converter in place of the sample-and-hold circuit generates the weighted impulse train

$$y_\delta(t) = \sum_k [x(kT_s) + \epsilon_k]\, \delta(t - kT_s) \qquad \text{[3a]}$$

where ϵ_k represents the **quantization error,** namely

$$\epsilon_k = x_q(kT_s) - x(kT_s) \qquad \text{[3b]}$$

Lowpass filtering with $B = f_s/2$ yields the final output

$$y_D(t) = x(t) + \sum_k \epsilon_k \operatorname{sinc}(f_s t - k) \qquad \text{[4]}$$

This expression has the same form as reconstruction of analog pulse modulation with noisy samples; see Eq. (2) Sect. 10.6. Furthermore, when q is large enough for reasonable signal approximation, the ϵ_k will be uncorrelated and independent of $x(t)$. Accordingly, we identify $\overline{\epsilon_k^2}$ as the mean-square **quantization noise.**

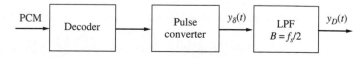

Figure 12.1–5   Impulse reconstruction model.

Round-off quantization with equispaced levels ensures that $|\epsilon_k| \leq 1/q$. Lacking additional information to the contrary, we assume that the quantization error has zero mean value and a uniform probability density function over $-1/q \leq \epsilon_k \leq 1/q$. Thus, the quantization noise power is

$$\sigma_q^2 = \overline{\epsilon_k^2} = \frac{1}{(2/q)} \int_{-1/q}^{1/q} \epsilon^2 \, d\epsilon = \frac{1}{3q^2} \qquad [5]$$

which reflects the intuitive observation that the quantization noise decreases when the number of quantum levels increases.

Now we measure PCM performance in terms of the destination signal power $S_D = \overline{x^2} = S_x \leq 1$ and the quantization noise power σ_q^2. The destination signal-to-noise ratio then becomes

$$\left(\frac{S}{N}\right)_D = \frac{S_x}{\sigma_q^2} = 3q^2 S_x \qquad [6]$$

A more informative relation for binary PCM is obtained by setting $q = 2^\nu$ and expressing $(S/N)_D$ in decibels. Thus,

$$\left(\frac{S}{N}\right)_D = 10 \log_{10}(3 \times 2^{2\nu} S_x) \leq 4.8 + 6.0\nu \text{ dB} \qquad [7]$$

where the upper bound holds when $S_x = 1$. Voice telephone PCM systems typically have $\nu = 8$ so $(S/N)_D \leq 52.8$ dB.

But many analog signals—especially voice and music—are characterized by a large **crest factor**, defined as the ratio of peak to rms value, $|x(t)|_{\text{max}}/\sigma_x$. Our signal normalization establishes $|x(t)|_{\text{max}} \leq 1$, and a large crest factor then implies that $S_x = \sigma_x^2 \ll 1$. Consequently, the actual signal-to-noise ratio will be significantly less than the theoretical upper bound. For instance, some digital audio recording systems take $\nu = 14$ to get high-fidelity quality with $(S/N)_D \approx 60$ dB, compared to $(S/N)_D \leq 88.8$ dB predicted by Eq. (7). As a bandwidth-conserving alternative to increasing ν, PCM performance may be improved through the use of *companding*, equivalent to *nonuniform quantization*.

EXAMPLE 12.1–3

Consider a maternal-fetal electrocardiograph system (ECG) where the ECG signal strengths of the maternal and fetal components are on the order of 1 volt and 100 μV respectively. Based on these signal levels, let's assume that the fetal signal has $S_x = 5 \times 10^{-9}$. If we employ a 1-volt bipolar 12-bit ADC, then using Eq. (7) we get a $(S/N)_D = 4.8 + 6 \times 12 + 10 \log(5 \times 10^{-9}) = -6.2$ dB. On the other hand, if we use a 16-bit ADC, we get a $(S/N)_D = 4.8 + 6 \times 16 + 10 \log(5 \times 10^{-9}) = 17.8$ dB. Furthermore, if we compare these two cases with respect to voltage levels, the 12-bit ADC gives us a peak-to-peak quantization noise of $2/2^{12} = 488$ μV which exceeds the 100 μV input signal level. However, with the 16-bit ADC, we only get a peak-to-peak quantization noise of 31 μV.

EXERCISE 12.1-2

Consider binary PCM transmission of a video signal with $f_s = 10$ MHz. (a) Calculate the signaling rate needed to get $(S/N)_D \geq 50$ dB when $S_x = 1$. (b) Repeat (a) with $S_x = 0.1$.

Nonuniform Quantizing and Companding★

A normalized signal $x(t)$ with a large crest factor can be represented by the typical probability density function $p_x(x)$ sketched in Fig. 12.1–6. The even symmetry and absence of an impulse at $x = 0$ correspond to $\bar{x} = 0$, so $\sigma_x^2 = \overline{x^2} = S_x$ and

$$S_x = \int_{-1}^{1} x^2 p_X(x)\, dx = 2\int_{0}^{1} x^2 p_X(x)\, dx \qquad [8]$$

This integration yields $S_x \ll 1$ because the PDF has a dominant peak at $x = 0$.

The shape of $p_x(x)$ also means that $|x(t)| \ll 1$ most of the time. It would therefore make good sense to use **nonuniform** quantization as indicated by the dashed lines. The quantum levels $\pm x_1, \ldots, \pm x_{q/2}$ are closely spaced near $x = 0$, but more widely spaced for the large values of $|x(t)|$ which occur infrequently. We calculate the resulting quantization noise as follows.

Consider a sample value $x = x(kT_s)$ in the band $a_i < x < b_i$ around the quantum level x_i. The quantizing error $\epsilon_i = x_i - x$ then has the mean square value

$$\overline{\epsilon_i^2} = \int_{a_i}^{b_i} (x_i - x)^2 p_X(x)\, dx \qquad [9a]$$

Summing $\overline{\epsilon_i^2}$ over all q levels gives the quantization noise

$$\sigma_q^2 = 2\sum_{i=1}^{q/2} \overline{\epsilon_i^2} \qquad [9b]$$

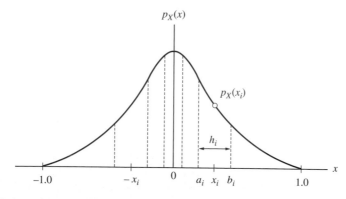

Figure 12.1–6   Message PDF with nonuniform quantization bands.

where we've taken advantage of the even symmetry. In the usual case of $q \gg 1$, the step height $h_i = b_i - a_i$ will be small enough that $p_X(x) \approx p_X(x_i)$ over each integration band and x_i will fall roughly in the middle of the step. Under these conditions Eq. (9a) simplifies to

$$\overline{\epsilon_i^2} \approx p_X(x_i) \int_{x_i - h_i/2}^{x_i + h_i/2} (x_i - x)^2 \, dx = p_X(x_i) \frac{h_i^3}{12}$$

and thus

$$\sigma_q^2 \approx \frac{1}{6} \sum_{i=1}^{q/2} p_X(x_i) h_i^3 \qquad [10]$$

As a check on this expression, we note that if the signal has the uniform PDF $p_X(x) = 1/2$ and if the steps have equal height $h_i = 2/q$, then $\sigma_q^2 = (1/6)(q/2)(1/2)(2/q)^3 = 1/3q^2$ which agrees with our earlier result in Eq. (5).

Theoretically, you could *optimize* PCM performance by finding the values of x_i, a_i, and b_i that result in *minimum quantization noise*. Such optimization is a difficult procedure that requires knowledge of the signal's PDF. Additionally, the custom-tailored hardware needed for nonlinear quantizing costs far more than standard uniform quantizers. Therefore, the approach taken in practice is to use *uniform* quantizing after **nonlinear signal compression,** the compression characteristics being determined from experimental studies with representative signals.

Figure 12.1–7 plots an illustrative compressor curve $z(x)$ versus x for $0 \leq x \leq 1$; the complete curve must have odd symmetry such that $z(x) = -z(|x|)$ for $-1 \leq x \leq 0$. Uniform quantization of $z(x)$ then corresponds to nonuniform quantization of x, as shown in the figure. The nonlinear distortion introduced by the com-

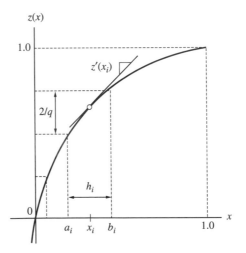

Figure 12.1–7   Compressor characteristic curve.

pressor is corrected after reconstruction by a complementary expander, identical to the *companding* strategy discussed in Sect. 3.2. Hence, the postdetection signal-to-noise ratio for companded PCM is $(S/N)_D = S_x/\sigma_q^2$, with σ_q^2 given by Eq. (10).

Our next task is to obtain σ_q^2 in terms of the compressor curve. For that purpose let

$$z'(x) \triangleq \frac{dz(x)}{dx}$$

so $z'(x_i)$ equals the slope of $z(x)$ at $x = x_i$. The conditions $q \gg 1$ and $h_i \ll 1$ justify the approximation $z'(x_i) \approx (2/q)/h_i$, so

$$h_i^2 \approx \left[\frac{2/q}{z'(x_i)}\right]^2 = \frac{4}{q^2[z'(x_i)]^2}$$

Equation (10) then becomes

$$\sigma_q^2 \approx \frac{4}{6q^2} \sum_{i=1}^{q/2} \frac{p_X(x_i)}{[z'(x_i)]^2} h_i \approx \frac{2}{3q^2} \int_0^1 \frac{p_X(x)}{[z'(x)]^2} dx$$

where we've passed from summation to integration via $h_i \to dx$. Therefore, our final result is

$$\left(\frac{S}{N}\right)_D = \frac{S_x}{\sigma_q^2} \approx \frac{3q^2 S_x}{K_z} \qquad\qquad \textbf{[11a]}$$

with

$$K_z \triangleq 2 \int_0^1 \frac{p_X(x)}{[z'(x)]^2} dx \qquad\qquad \textbf{[11b]}$$

which takes account of the compression. If $K_z < 1$ then $(S/N)_D > 3q^2 S_x$ and companding improves PCM performance by reducing the quantization noise.

μ-Law Companding for Voice PCM   **EXAMPLE 12.1–4**

The popular μ-law companding for voice telephone PCM employs the following compressor characteristic

$$z(x) = x_{\max}\frac{\ln(1 + \mu|x|/x_{\max})}{\ln(1 + \mu)} \, \text{sgn}\,(x/x_{\max}) \qquad |x|/x_{\max} \le 1 \qquad \textbf{[12]}$$

where $x_{\max}$ is the maximum input level that can be quantized without overload, and in the United States, the standard value for μ is 255. The function $z(x)$ then feeds the input of an n-bit uniform quantizer to generate x_q'. To get back an estimate of the original voice signal, we apply the quantized signal to an expander which is the inverse of Eq. (12) or

$$z^{-1}(x_q') = \frac{x_{\max}}{\mu}\left[(1 + \mu)^{|x_q'|/x_{\max}} - 1\right] \text{sgn}\,(x_q'/x_{\max}) \to \hat{x} \qquad \textbf{[13]}$$

If we assume normalized inputs, such that $|x| \leq 1$ then from Eq. (12) we get

$$z'(x) = \frac{\mu}{\ln (1 + \mu)} \frac{1}{1 + \mu|x|} \qquad [14]$$

The parameter μ is a large number so $z'(x) \gg 1$ for $|x| \ll 1$ whereas $z'(x) \ll 1$ for $|x| \approx 1$. Substituting $z'(x)$ into Eq. (11b) and performing the integration, we obtain

$$K_z = \frac{\ln^2 (1 + \mu)}{\mu^2}(1 + 2\mu\overline{|x|} + \mu^2 S_x) \qquad [15]$$

Now we need values for $\overline{|x|}$ and S_x to test the efficacy of μ-law companding.

Laboratory investigations have shown that the PDF of a voice signal can be modeled by a *Laplace* distribution in the form

$$p_X(x) = \frac{\alpha}{2} e^{-\alpha|x|} \qquad [16a]$$

with

$$S_x = \sigma_x^2 = \frac{2}{\alpha^2} \qquad \overline{|x|} = \frac{1}{\alpha} = \sqrt{\frac{S_x}{2}} \qquad [16b]$$

This distribution cannot be normalized for $|x(t)|_{\max} \leq 1$, but the probability of $|x(t)| > 1$ will be less than 1 percent if $S_x < 0.1$. Other voice PDF models yield about the same relationship between $\overline{|x|}$ and S_x, which is the critical factor for evaluating K_z.

Taking the standard value $\mu = 255$ and putting $\overline{|x|} = \sqrt{S_x/2}$ in Eq. (15), we obtain

$$K_z = 4.73 \times 10^{-4}(1 + 361\sqrt{S_x} + 65,025S_x)$$

Numerical calculations then show that $K_z < 1$ for $S_x < 0.03$. But more significant is the fact that S_x/K_z stays nearly constant over a wide range of S_x. Consequently, μ-law companding for voice PCM provides an essentially *fixed* value of $(S/N)_D$, despite wide variations of S_x among individual talkers. Figure 12.1–8 brings out this desirable feature by plotting $(S/N)_D$ in dB versus S_x in dB, with and without companding, when $q = 2^8$. Notice the companding improvement for $S_x < -20$ dB.

EXERCISE 12.1–3   Companding will reduce the quantization errors. Consider a $\mu = 255$ compandor to be used with a $\nu = 3$ bit quantizer where the output varies over ± 8.75 V. For an input of 0.6 V, what is the ϵ_k with and without companding?

12.2   PCM WITH NOISE

In this section we account for the effects of random noise in PCM transmission. The resulting *digital errors* produce *decoding noise*. After defining the error threshold

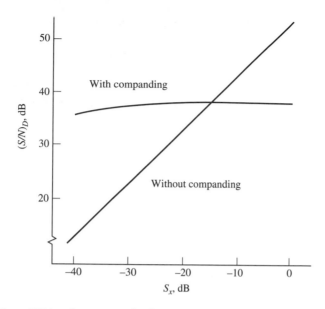

Figure 12.1–8   PCM performance with μ-law companding.

level, we'll be in a position to make a meaningful comparison of PCM with analog modulation methods.

Decoding Noise

Random noise added to the PCM signal at the receiver causes regeneration errors that appear as erroneous digits in the codewords. The decoder then puts out a different quantum level than the one intended for a particular sample.

Hence,

The reconstructed message waveform becomes contaminated with decoding noise as well as quantization noise.

The analysis of decoding noise is not too difficult if we restrict our attention to binary PCM with uniform quantization and a relatively small bit error probability P_e. The number of bit errors in a ν-digit codeword is a random variable governed by the binomial distribution. However, when $P_e \ll 1$, the probability of one error in a given word approximately equals νP_e and the probability of two or more errors is small enough for us to ignore that event. Of course, the effect of a single error depends on where it falls in the word, because the different bit positions have different decoding interpretations.

Consider a "natural" binary codeword of the form $b_{v-1}b_{v-2} \cdots b_1 b_0$ in which the mth bit distinguishes between quantum levels spaced by 2^m times the step height $2/q$. (The sign/magnitude code illustrated back in Fig. 12.1–3 follows this pattern for the magnitude bits, but the sign bit has a variable level meaning that makes analysis more complicated.) An error in the mth bit then shifts the decoded level by the amount $\epsilon_m = \pm(2/q)2^m$, and the average of ϵ_m^2 over the v bit positions equals the mean-square decoding error for a random bit-error location. Thus,

$$\overline{\epsilon_m^2} = \frac{1}{v} \sum_{m=0}^{v-1} \left(\frac{2}{q} 2^m \right)^2 = \frac{4}{vq^2} \sum_{m=0}^{v-1} 4^m \tag{1}$$

$$= \frac{4}{vq^2} \frac{4^v - 1}{3} = \frac{4}{3v} \frac{q^2 - 1}{q^2} \approx \frac{4}{3v}$$

where we've used the formula for summing a geometric progression and substituted $4^v = 2^{2v} = q^2 \gg 1$. The **decoding noise power** is therefore

$$\sigma_d^2 = v P_e \overline{\epsilon_m^2} \approx \tfrac{4}{3} P_e \tag{2}$$

since erroneous words occur with probability $v P_e$.

The *total* destination noise power consists of decoding noise σ_d^2 and quantization noise $\sigma_q^2 = 1/3q^2$, which come from essentially independent processes. Therefore,

$$N_D = \sigma_q^2 + \sigma_d^2 = \frac{1 + 4q^2 P_e}{3q^2}$$

and

$$\left(\frac{S}{N} \right)_D = \frac{3q^2}{1 + 4q^2 P_e} S_x \tag{3}$$

so the effect of decoding noise depends upon the relative value of the quantity $4q^2 P_e$. Indeed, we see in Eq. (3) the two extreme conditions

$$\left(\frac{S}{N} \right)_D \approx \begin{cases} 3q^2 S_x & P_e \ll 1/4q^2 \\ \dfrac{3}{4 P_e} S_x & P_e \gg 1/4q^2 \end{cases} \tag{4}$$

Quantization noise dominates when P_e is small, but decoding noise dominates and reduces $(S/N)_D$ when P_e is large compared to $1/4q^2$.

Now recall that the value of P_e is determined by the received signal-to-noise ratio $(S/N)_R$ at the input to the digital regenerator. Specifically, for polar binary signaling in gaussian white noise we know that $P_e = Q[\sqrt{(S/N)_R}]$. Figure 12.2–1 plots $(S/N)_D$ versus $(S/N)_R$ for this case, with $S_x = 1/2$ and two values of q. The precipitous decline of $(S/N)_D$ as $(S/N)_R$ decreases constitutes a *threshold effect* caused by increasing errors. Below the error threshold, when $P_e \gg 1/4q^2$, the errors occur so often that the reconstructed waveform bears little resemblance to the original signal and the message has been mutilated beyond recognition.

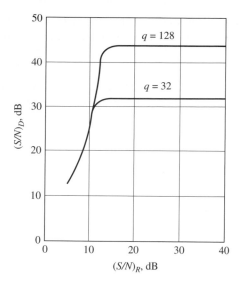

Figure 12.2–1   Noise performance of PCM.

Error Threshold

The PCM error threshold level is usually defined at the point where decoding noise reduces $(S/N)_D$ by 1 dB. Unfortunately, this definition does not lend itself to analytical investigations. As a more convenient alternative, we'll say that decoding errors have negligible effect if $P_e \leq 10^{-5}$. Then we obtain the corresponding condition on $(S/N)_R$ for polar M-ary signaling using Eq. (20) Sect. 11.2, namely

$$P_e = 2\left(1 - \frac{1}{M}\right)Q\left[\sqrt{\frac{3}{M^2 - 1}\left(\frac{S}{N}\right)_R}\right] \leq 10^{-5}$$

Solving for the minimum value of $(S/N)_R$ yields the threshold level

$$\left(\frac{S}{N}\right)_{R_{th}} \approx 6(M^2 - 1) \qquad \text{[5]}$$

If $(S/N)_R < 6(M^2 - 1)$, the PCM output will be hopelessly mutilated by decoding noise.

A subtle but important implication of Eq. (5) relates to the digital signaling rate and transmission bandwidth. We'll bring out that relationship with the help of the analog transmission parameter $\gamma = S_R/N_0 W = (B_T/W)(S/N)_R$. The PCM transmission bandwidth is $B_T \geq r/2 \geq \nu W$, so

$$\gamma_{th} = (B_T/W)(S/N)_{R_{th}} \qquad \text{[6]}$$

$$\approx 6\frac{B_T}{W}(M^2 - 1) \geq 6\nu(M^2 - 1)$$

Given ν and M, this equation tells you the minimum value of γ needed for PCM operation above threshold. It also facilitates the comparison of PCM with other transmission schemes.

EXAMPLE 12.2–1

Example 12.1–4 showed that a voice PCM system with $M = 2$, $\nu = 8$, and μ-law companding has $(S/N)_D \approx 37$ dB. Equation (6) gives the corresponding threshold level $\gamma_{\text{th}} \geq 144 \approx 22$ dB. Hence, the PCM system has a potential 15-dB advantage over direct analog baseband transmission in which $(S/N)_D = \gamma$. The full advantage would not be realized in practice where allowance must be made for $B_T > \nu W$ and $\gamma > \gamma_{\text{th}}$.

EXERCISE 12.2–1

The 1-dB definition for error threshold is equivalent to $10 \log_{10} (1 + 4q^2 P_e) = 1$. Calculate P_e from this definition when $q = 2^8$. Then find the corresponding value of $(S/N)_R$ for polar binary signaling, and compare your result with Eq.(5).

PCM versus Analog Modulation

The PCM threshold effect reminds us of analog modulation methods such as FM and PPM that have the property of *wideband noise reduction* above their threshold levels. As our initial point of comparison, let's demonstrate that PCM also provides wideband noise reduction when operated above threshold so $(S/N)_D = 3q^2 S_x$. For this purpose, we'll assume that the sampling frequency is close to the Nyquist rate and $B_T \approx \nu W$. Then $q = M^\nu \approx M^b$ where $b = B_T/W$ is the bandwidth ratio. Hence,

$$\left(\frac{S}{N}\right)_D \approx 3M^{2b} S_x \tag{7}$$

which exhibits noise reduction as an *exponential* exchange of bandwidth for signal-to-noise ratio. Granted that the noise being reduced is quantization noise, but random noise has no effect on PCM above threshold. The exponential factor in Eq. (7) is far more dramatic than that of wideband analog modulation, where $(S/N)_D$ increases proportionally to b or b^2.

For further comparison including threshold limitations, Fig. 12.2–2 illustrates the performance of several modulation types as a function of γ. All curves are calculated with $S_x = 1/2$, and the heavy dots indicate the threshold points. The PCM curves are based on Eqs. (6) and (7) with $M = 2$ and $\nu = b$.

Clearly, in the name of power efficiency, PCM should be operated just above threshold, since any power increase beyond $\gamma \approx \gamma_{\text{th}}$ yields no improvement of $(S/N)_D$. Near threshold, PCM does offer some advantage over FM or PPM with the same value of b and $(S/N)_D$. And even a 3-dB power advantage, being a factor of 2, may spell the difference between success and failure for some applications. But that

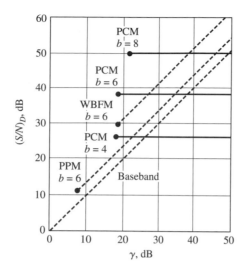

Figure 12.2–2   Performance comparison of PCM and analog modulation.

advantage is gained at the price of more complicated and costly hardware. In fact, PCM was deemed totally impractical prior to the development of high-speed digital electronics in the late 1950s—two decades after the invention of PCM.

Two other benefits don't appear in Fig. 12.2–2.

> PCM allows the advantage of *regenerative repeaters* when the transmission link requires many repeater stations.

Another benefit comes from the fact that

> PCM allows analog message transmission as a digital signal.

Digital multiplexing then makes it possible to combine PCM and digital data signals for flexible and efficient utilization of a communication channel. Taken together, these two benefits account for PCM's preeminence in the design of new systems for long-distance telephony.

But PCM is not suited to all applications. In radio broadcasting, for instance, we want a relatively large signal-to-noise ratio, say $(S/N)_D \approx 60$ dB. Figure 12.2–2 reveals that this would require binary PCM with $b > 8$, or FM with a smaller bandwidth ratio $b = 6$ and much simpler hardware at the transmitter and receivers.

Likewise, bandwidth and hardware considerations would reject PCM for most single-channel systems.

EXERCISE 12.2–2   Starting with Eqs. (6) and (7), show that a PCM system operated at the threshold point has

$$\left(\frac{S}{N}\right)_{D_{th}} = 3\left(1 + \frac{\gamma_{th}}{6b}\right)^b S_x \qquad \qquad \text{[8]}$$

Compare this expression with that of WBFM by setting $D = b/2 \gg 1$ in Eq. (20), Sect. 10.3.

12.3   DELTA MODULATION AND PREDICTIVE CODING

Sample values of analog waveforms derived from physical processes often exhibit *predictability* in the sense that the average change from sample to sample is small. Hence, you can make a reasonable guess of the next sample value based on previous values. The predicted value has some error, of course, but the range of the error should be much less than the peak-to-peak signal range. Predictive coded modulation schemes exploit this property by transmitting just the *prediction errors*. An identical prediction circuit at the destination combines the incoming errors with its own predicted values to reconstruct the waveform.

Predictive methods work especially well with audio and video signals, and much effort has been devoted to prediction strategies for efficient voice and image transmission. **Delta modulation** (DM) employs prediction to simplify hardware in exchange for increased signaling rate compared to PCM. **Differential pulse-code modulation** (DPCM) reduces signaling rate but involves more elaborate hardware. We'll discuss both DM and DPCM in this section, along with the related and fascinating topic of **speech synthesis** using prediction.

Delta Modulation

Let an analog message waveform $x(t)$ be lowpass filtered and sampled every T_s seconds. We'll find it convenient here to use **discrete-time** notation, with the integer independent variable k representing the sampling instant $t = kT_s$. We thus write $x(k)$ as a shorthand for $x(kT_s)$, and so on.

When the sampling frequency is greater than the Nyquist rate, we expect that $x(k)$ roughly equals the previous sample value $x(k-1)$. Therefore, given the quantized sample value $x_q(k-1)$, a reasonable guess for the next value would be

$$\widetilde{x}_q(k) = x_q(k-1) \qquad \qquad \text{[1]}$$

where $\widetilde{x}_q(k)$ denotes our **prediction** of $x_q(k)$. A delay line with time delay T_s then serves as the prediction circuit. The difference between the predicted and actual value can be expressed as

$$x_q(k) = \tilde{x}_q(k) + \epsilon_q(k) \qquad\qquad\qquad [2]$$

in which $\epsilon_q(k)$ is the **prediction error.**

If we transmit $\epsilon_q(k)$, we can use the system in Fig. 12.3–1 to generate $x_q(k)$ by delaying the current output and adding it to the input. This system implements Eqs. (1) and (2), thereby acting as an **accumulator.** The accumulation effect is brought out by writing $x_q(k) = \epsilon_q(k) + x_q(k-1)$ with $x_q(k-1) = \epsilon_q(k-1) + x_q(k-2)$, and so forth; hence

$$\begin{aligned} x_q(k) &= \epsilon_q(k) + \epsilon_q(k-1) + x_q(k-2) \\ &= \epsilon_q(k) + \epsilon_q(k-1) + \epsilon_q(k-2) + \cdots \end{aligned}$$

An **integrator** accomplishes the same accumulation when $\epsilon_q(k)$ takes the form of brief rectangular pulses.

At the transmitting end, prediction errors are generated by the simple delta modulation system diagrammed in Fig. 12.3–2. The comparator serves as a *binary quantizer* with output values $\pm\Delta$, depending on the difference between the predicted value $\tilde{x}_q(k)$ and the unquantized sample $x(k)$. Thus, the resulting DM signal is

$$\epsilon_q(k) = [\operatorname{sgn} \epsilon(k)]\Delta \qquad\qquad\qquad [3a]$$

where

$$\epsilon(k) = x(k) - \tilde{x}_q(k) \qquad\qquad\qquad [3b]$$

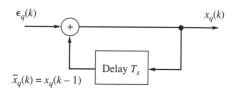

Figure 12.3–1   Accumulator for delta modulation.

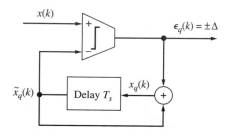

Figure 12.3–2   DM transmitter.

which represents the unquantized error. An accumulator (or integrator) in a feedback loop produces $\widetilde{x}_q(k)$ from $\epsilon_q(k)$, similar to Fig. 12.3–1 except that the feedback signal comes from the delayed output.

Observe that this DM transmitter requires no analog-to-digital conversion other than the comparator. Also observe that an accumulator like Fig. 12.3–1 performs the digital-to-analog conversion at the receiver, reconstructing $x_q(k)$ from $\epsilon_q(k)$. Thus,

> A DM system achieves digital transmission of analog signals with very simple hardware compared to a PCM system.

The name delta modulation reflects the fact that each input sample $x(k)$ has been encoded as a single pulse of height $+\Delta$ or $-\Delta$. But we can also view $\epsilon_q(k)$ as a *binary waveform* with signaling rate $r_b = f_s$, or *one bit per sample*. For this reason DM is sometimes called "one-bit PCM." The corresponding transmission bandwidth requirement is

$$B_T \geq r_b/2 = f_s/2 \qquad\qquad \textbf{[4]}$$

We get by with just one bit per sample because we're transmitting prediction errors, not sample values. Nonetheless, successful operation requires rather high sampling rates, as we'll soon see.

Figure 12.3–3 depicts illustrative continuous-time waveforms $x(t)$, $\widetilde{x}_q(t)$, and $\epsilon_q(t)$ involved in DM. The staircase waveform $x_q(t)$ at the receiver differs from $\widetilde{x}_q(t)$ only by a time shift of T_s seconds. The transmitter starts with an arbitrary initial prediction such as $\widetilde{x}_q(0) < x(0)$ so $\epsilon_q(0) = +\Delta$. Then $\epsilon_q(0)$ is fed back through the accumulator to form the updated prediction $\widetilde{x}_q(T_s) = x_q(0) + \epsilon_q(0)$. Continual updating at each sampling instant causes $\widetilde{x}_q(t)$ to increase by steps of Δ until the start-up interval ends when $\widetilde{x}_q(kT_s) > x(kT_s)$ and $\epsilon_q(kT_s) = -\Delta$. If $x(t)$ remains constant, $\widetilde{x}_q(t)$ takes on a hunting behavior. When $x(t)$ varies with time, $\widetilde{x}_q(t)$ follows it in stepwise fashion as long as the rate of change does not exceed the DM

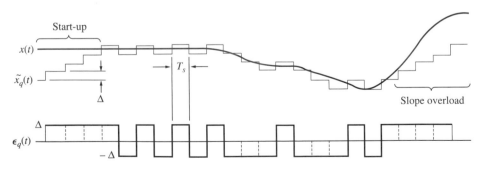

Figure 12.3–3   DM waveforms.

tracking capability. The difference between $\tilde{x}_q(t)$ and $x(t)$ is called **granular noise,** analogous to quantization noise in PCM. The reconstructed and smoothed waveform at the receiver will be a reasonable approximation for $x(t)$ if Δ and T_s are sufficiently small.

But when $x(t)$ increases or decreases too rapidly, $\tilde{x}_q(t)$ lags behind and we have the phenomenon known as **slope overload,** a fundamental limitation of DM. Since $\tilde{x}_q(t)$ changes by $\pm\Delta$ every $T_s = 1/f_s$ seconds, the maximum DM slope is $\pm f_s\Delta$ and a sufficient condition for slope tracking is

$$f_s\Delta \geq |\dot{x}(t)|_{\max} \qquad [5]$$

where $\dot{x}(t) = dx/dt$. Consider, for instance, the modulating tone $x(t) = A_m \cos 2\pi f_m t$ so $\dot{x}(t) = -2\pi f_m A_m \sin 2\pi f_m t$ and $|\dot{x}(t)|_{\max} = 2\pi f_m A_m \leq 2\pi W$, where the upper bound incorporates our message conventions $A_m \leq 1$ and $f_m \leq W$. Equation (5) therefore calls for a high sampling frequency $f_s \geq 2\pi W/\Delta \gg 2W$, since we want $\Delta \ll 2$ to make the steps of $\tilde{x}_q(t)$ small compared to the peak-to-peak signal range $-1 \leq x(t) \leq 1$.

DM performance quality depends on the granular noise, slope-overload noise, and regeneration errors. However, only granular noise has significant effect under normal operating conditions, which we assume hereafter. Even so, the analysis of granular noise is a difficult problem best tackled by computer simulation for accurate results or by approximations for rough results.

We'll estimate DM performance using the receiver modeled by Fig. 12.3–4a with $\epsilon_q(t - T_s) = \epsilon_q(k - 1)$ at the input to the accumulator. Equations (1) and (3b) then give the accumulator output as $x_q(k - 1) = \tilde{x}_q(k) = x(k) - \epsilon(k)$ or

$$\tilde{x}_q(t) = x(t) - \epsilon(t) \qquad [6]$$

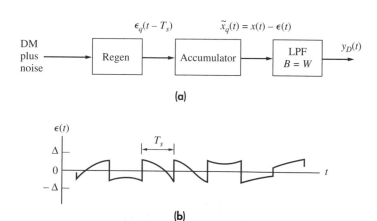

(a)

(b)

Figure 12.3–4   (a) DM receiver; (b) granular noise waveform.

where $\epsilon(t)$ is the granular noise waveform sketched in Fig. 12.3–4b. The shape of $\epsilon(t)$ and the fact that $|\epsilon(t)| \leq \Delta$ suggest a uniform amplitude distribution with

$$\overline{\epsilon^2} = \Delta^2/3$$

Furthermore, experimental studies confirm that the power spectrum of $\epsilon(t)$ is essentially flat over $|f| \leq 1/T_s = f_s$. Thus,

$$G_\epsilon(f) \approx \overline{\epsilon^2}/2f_s \qquad\qquad |f| \leq f_s$$

and lowpass filtering yields

$$N_g = \int_{-W}^{W} G_\epsilon(f)\, df = \frac{W}{f_s}\,\overline{\epsilon^2} = \frac{W}{f_s}\frac{\Delta^2}{3} \qquad\qquad [7]$$

which is the average power of the granular noise component of $x_q(t)$.

When granular noise is the only contamination in the filtered output, we obtain the signal-to-noise ratio

$$\left(\frac{S}{N}\right)_D = \frac{S_x}{N_g} = \frac{3f_s}{\Delta^2 W}\,S_x \qquad\qquad [8]$$

This result is almost identical to the PCM expression $(S/N)_D = 3q^2 S_x$ if $f_s = 2W$ and $\Delta = 1/q$. But f_s and Δ must satisfy Eq. (5). We therefore need a more general relationship for the slope-tracking condition.

Recall from Eq. (21), Sect. 9.2, that if $G_x(f)$ is the power spectrum of $x(t)$, then the power spectrum of the derivative $\dot{x}(t)$ is $(2\pi f)^2 G_x(f)$. Hence, the *mean square signal slope* can be put in the form

$$\overline{|\dot{x}(t)|^2} = \int_{-\infty}^{\infty} (2\pi f)^2 G_x(f)\, df = (2\pi\sigma\, W_{\mathrm{rms}})^2 \qquad\qquad [9]$$

where $\sigma = \sqrt{S_x}$ is the signal's rms value and W_{rms} is its **rms bandwidth,** defined by

$$W_{\mathrm{rms}} \triangleq \frac{1}{\sigma}\left[\int_{-\infty}^{\infty} f^2 G_x(f)\, df\right]^{1/2} \qquad\qquad [10]$$

Now we introduce the so-called **slope loading factor**

$$s \triangleq \frac{f_s \Delta}{2\pi\sigma\, W_{\mathrm{rms}}} \qquad\qquad [11]$$

which is the ratio of the maximum DM slope to the rms signal slope.

A reasonably large value of s ensures negligible slope overload. Hence, we'll incorporate this factor explicitly in Eq. (8) by writing Δ in terms of s from Eq. (11). Thus,

$$\left(\frac{S}{N}\right)_D = \frac{3}{4\pi^2}\frac{f_s^3}{s^2 W_{\mathrm{rms}}^2 W}$$

$$= \frac{6}{\pi^2} \left(\frac{W}{W_{\mathrm{rms}}} \right)^2 \frac{b^3}{s^2} \qquad \textbf{[12]}$$

where

$$b \triangleq f_s/2W \qquad \textbf{[13]}$$

The parameter b equals our usual bandwidth ratio B_T/W when B_T has the minimum value $f_s/2$. Equation (12) brings out the fact that DM performance falls between PCM and PPM since the wideband noise reduction goes as b^3 rather than b^2 or exponentially. DM generally requires a larger transmission bandwidth than PCM to achieve the same signal-to-noise ratio, so its applications are limited to those cases where ease of implementation takes precedence over bandwidth considerations.

Computer simulations by Abate (1967) indicate that Eq. (12) holds for $\ln 2b \leq s < 8$. If $s < \ln 2b$, then **slope-overload noise** dominates and $(S/N)_D$ drops off quite rapidly. Figure 12.3–5 illustrates how $(S/N)_D$ varies with s. For a specified bandwidth ratio, DM performance is maximized by taking the empirically determined optimum slope loading factor

$$s_{\mathrm{opt}} \approx \ln 2b \qquad \textbf{[14]}$$

The maximum value of $(S/N)_D$ is then given by Eq. (12) with $s = s_{\mathrm{opt}}$.

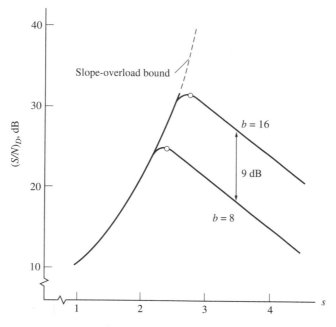

Figure 12.3–5   DM performance versus slope loading factor.

EXAMPLE 12.3–1   **DM Voice Transmission**

When a typical voice signal has been prefiltered so that $W \approx 4$ kHz, its rms bandwidth will be $W_{\mathrm{rms}} \approx 1.3$ kHz. Substituting these values and $s = s_{\mathrm{opt}} = \ln 2b$ in Eq. (12) yields the maximum DM signal-to-noise ratio $(S/N)_D \approx 5.8b^3/(\ln 2b)^2$. If $b = 16$, then $(S/N)_D \approx 33$ dB which is comparable to binary PCM with $\nu = 7$ and μ-law companding. But the DM signal requires $r_b = f_s = 2bW = 128$ kbps and $B_T \geq 64$ kHz, whereas the PCM signal would have $r_b = \nu f_s \geq 56$ kbps and $B_T \geq 28$ kHz.

EXERCISE 12.3–1   Consider a signal with a uniform power spectrum $G_x(f) = (S_x/2W)\Pi(f/2W)$. Show that $W_{\mathrm{rms}} = W/\sqrt{3}$. Then calculate the optimum value of Δ in terms of S_x when $b = 16$.

Delta-Sigma Modulation

The preceding DM scheme takes the derivative of the input. In cases where the data are noisy, the noise can cause cumulative errors in the demodulated signal. There is further difficulty if the signal has a significant dc component. An alternative to the DM is a **Delta-Sigma Modulator** (DSM), also called sigma-delta modulation.

We first consider an equivalent version of the conventional delta modulator transmitter/receiver system as shown in Fig. 12.3–6a. Now let's add an integrator (or accumulator) to the input which has the effect of *preemphasizing* the low frequencies. This can also integrate the signal in the feedback path and thereby eliminate the feedback integrator. To compensate for the additional integrator, we then add a differentiator at the receiver. However, since the DM receiver already has an integrator, which is the inverse of a differentiator, we can eliminate both of these at the receiver and thus simplify its design, giving us the system shown in Fig. 12.3–6b.

Adaptive Delta Modulation

Adaptive delta modulation (ADM) involves additional hardware designed to provide *variable step size*, thereby reducing slope-overload effects without increasing the granular noise. A reexamination of Fig. 12.3–3 reveals that slope overload appears in $\epsilon_q(t)$ as a sequence of pulses having the same polarity, whereas the polarity tends to alternate when $x_q(t)$ tracks $x(t)$. This sequence information can be utilized to adapt the step size in accordance with the signal's characteristics.

Figure 12.3–7 portrays the action of an ADM transmitter in which the step size in the feedback loop is adjusted by a variable gain $g(k)$ such that

$$\tilde{x}_q(k) = \tilde{x}_q(k-1) + g(k-1)\epsilon_q(k-1)$$

The step-size controller carries out the adjustment algorithm

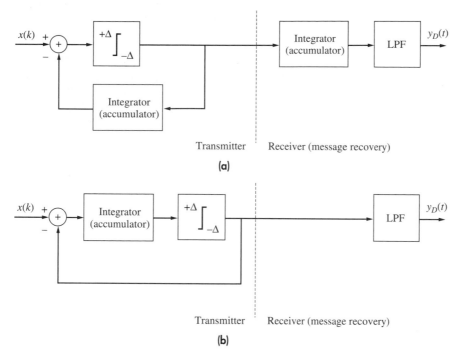

Figure 12.3–6   (a) DM system; (b) DSM system.

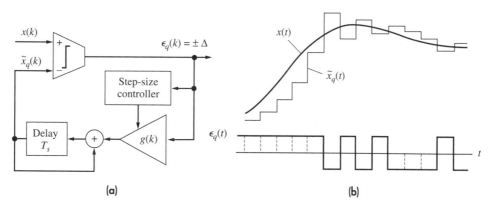

Figure 12.3–7   (a) Adaptive DM transmitter; (b) waveforms.

$$g(k) = \begin{cases} g(k-1) \times K & \epsilon_q(k) = \epsilon_q(k-1) \\ g(k-1)/K & \epsilon_q(k) \neq \epsilon_q(k-1) \end{cases}$$

where K is a constant taken to be in the range $1 < K < 2$. Thus, the effective step size increases by successive powers of K during slope-overload conditions, as signified by $\epsilon_q(k) = \epsilon_q(k-1)$, but decreases when $\epsilon_q(k) \neq \epsilon_q(k-1)$. Another adaptive scheme called **continuously variable slope delta modulation** (CVSDM) provides a continuous range of step-size adjustment instead of a set of discrete values.

The signal-to-noise ratio of ADM is typically 8–14 dB better than ordinary DM. Furthermore, the variable step size yields a wider *dynamic range* for changing values of S_x, similar to the effect of μ-law companding in PCM. As a net result, ADM voice transmission gets by with a bandwidth ratio of $b = 6$–8 or $B_T = 24$–32 kHz.

Differential PCM

Differential pulse-code modulation (DPCM) combines prediction with multilevel quantizing and coding. The transmitter diagrammed in Fig. 12.3–8a has a *q*-level

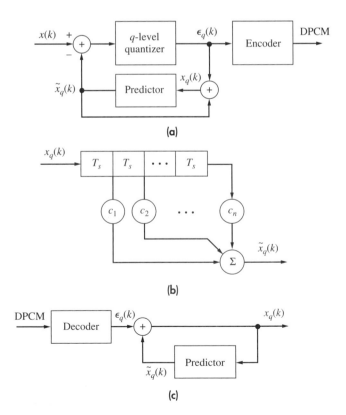

(a)

(b)

(c)

Figure 12.3–8   Differential PCM. (*a*) Transmitter; (*b*) prediction circuit; (*c*) receiver.

quantizer with quantum levels at $\pm\Delta, \pm 3\Delta, \ldots, \pm(q-1)\Delta$. The unquantized error $x(k) - \tilde{x}_q(k)$ is applied to the quantizer to produce the prediction error $\epsilon_q(k)$ which is then encoded as a binary word with $\nu = \log_2 q$ bits, just like binary PCM. DPCM transmission therefore requires

$$r_b = \nu f_s \qquad\qquad B_T \geq \nu f_s/2$$

However, since $\tilde{x}_q(k)$ changes as much as $\pm(q-1)\Delta$ from sample to sample, the slope tracking condition becomes

$$f_s(q-1)\Delta \geq |\dot{x}(t)|_{\max}$$

If $q \gg 1$ and $W_{\mathrm{rms}} \ll W$, the sampling frequency can be nearly as low as the Nyquist rate.

Multilevel quantization of the prediction error obviously provides better information for message reconstruction at the receiver. To gain full advantage of this potential, the DPCM prediction circuit usually takes the form of a transversal filter shown in Fig. 12.3–8b, where

$$\tilde{x}_q(k) = \sum_{i=1}^{n} c_i x_q(k-i)$$

so the predictor draws upon the previous n samples. The tap gains c_i are chosen to minimize the mean square value of the error $x(k) - \tilde{x}_q(k)$. The receiver in Fig. 12.3–8c includes an identical prediction filter after the decoder.

Assuming $q \gg 1$ and no slope overload, DPCM performs essentially like PCM enhanced by a **prediction gain** G_p such that

$$\left(\frac{S}{N}\right)_D = G_p \, 3q^2 \, S_x \tag{15}$$

The gain of an optimum predictor is given by

$$G_p = \left[1 - \sum_{i=1}^{n} c_i \rho_i \right]^{-1} \tag{16a}$$

where $\rho_i = R_x(iT_s)/S_x$ is the normalized signal correlation and the tap gains satisfy the matrix relationship

$$
\begin{bmatrix}
\rho_0 & \rho_1 & \cdots & \rho_{n-1} \\
\rho_1 & \rho_0 & \cdots & \rho_{n-2} \\
\vdots & & & \vdots \\
\rho_{n-1} & \rho_{n-2} & \cdots & \rho_0
\end{bmatrix}
\begin{bmatrix}
c_1 \\
c_2 \\
\vdots \\
c_n
\end{bmatrix}
=
\begin{bmatrix}
\rho_1 \\
\rho_2 \\
\vdots \\
\rho_n
\end{bmatrix}
\tag{16b}
$$

Jayant and Noll (1984) outline the derivation of Eqs. (15) and (16) and present experimental data showing that $G_p \approx 5$–10 dB for voice signals. The higher correlation of a TV video signal results in prediction gains of about 12 dB.

In contrast to delta modulation, DPCM employs more elaborate hardware than PCM for the purpose of improving performance quality or reducing the signaling rate and transmission bandwidth. *Adaptive* DPCM (ADPCM) achieves even greater improvement by adapting the quantizer or predictor or both to the signal characteristics.

EXERCISE 12.3–2   Suppose a DPCM predictor yields $G_p = 6$ dB. Show that the DPCM word needs one less bit than that of binary PCM, all other factors being equal. *Hint:* See Eq. (7), Sect. 12.1.

LPC Speech Synthesis

Linear predictive coding (LPC) is a novel approach to digital representation of analog signals. The method uses a transversal filter (or its digital-circuit equivalent) plus some auxiliary components to *synthesize* the waveform in question. The parameters of the waveform synthesizer are then encoded for transmission, instead of the actual signal. Considerable efficiency results if the synthesizer accurately mimics the analog process. Since there already exists extensive knowledge about speech processes, LPC is particularly well suited to speech synthesis and transmission.

Figure 12.3–9 diagrams a speech synthesizer consisting of two input generators, a variable-gain amplifier, and a transversal filter in a feedback loop. The amplifier gain and filter tap gains are adjusted to model the acoustical properties of the vocal tract. Unvoiced speech (such as hissing sound) is produced by connecting the white-noise generator. Voiced speech is produced by connecting the impulse-train generator set at an appropriate pitch frequency.

If the filter has about 10 tap gains, and all parameter values are updated every 10 to 25 ms, the synthesized speech is quite intelligible, although it may sound

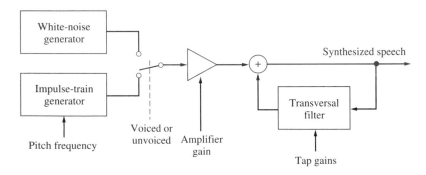

Figure 12.3–9   Speech synthesizer.

rather artificial, like a robot. Some talking toys and recorded-message systems generate speech sounds by the synthesis method, drawing upon parameter values stored in a digital memory. Systems that generate speech in this manner are called **vocoders.** An alternative to vocoders are **waveform encoders** where a given algorithm such as PCM is employed to replicate a given speech signal.

Now consider the LPC transmitter of Fig. 12.3–10a. Sample values of voice input are analyzed to determine the parameters for the synthesizer, whose output is compared with the input. The resulting error is encoded along with the parameter values to form the transmitted digital signal. The receiver in Fig. 12.3–10b uses the parameter values and quantized error to reconstruct the voice waveform.

A complete LPC codeword consists of 80 bits—1 bit for the voiced/unvoiced switch, 6 bits for the pitch frequency, 5 for the amplifier gain, 6 for each of the 10 tap gains, and a few bits for the error. Updating the parameters every 10–25 ms is equivalent to sampling at 40–100 Hz, so LPC requires a very modest bit rate in the vicinity of 3000 to 8000 bps. Table 12.3–1 compares LPC with other voice encoding methods. The substantial bit-rate reduction made possible by LPC has stimulated efforts to improve the quality of speech synthesis for voice communication. In fact, as shown in Table 12.3–1, using LPC we can compress speech from a 56-kbps to 3-kbps rate. See Rabiner and Schafer (1978) for a general introduction to digital processing applied to speech signals.

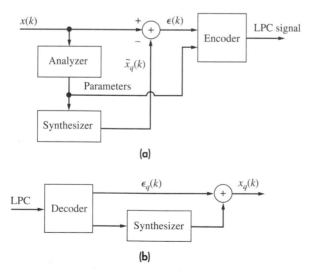

(a)

(b)

Figure 12.3–10   LPC transmission system. (a) Transmitter; (b) receiver.

Table 12.3–1   Comparison of voice encoding methods

Encoding method	Sampling rate, kHz	Bits per sample	Bit rate, kbps
DM	64–128	1	64–128
PCM	8	7–8	56–64
ADM	48–64	1	48–64
DPCM	8	4–6	32–48
ADPCM	8	3–4	24–32
LPC	0.04–0.1	≈ 80	3–8

12.4   DIGITAL AUDIO RECORDING

Much of our communication infrastructure deals with the transmission of signals whose content is music, so we are interested in preserving fidelity, dynamic range, and so on. The study of digital audio recording provides us with a good lesson in the advantages of digital over analog systems and some methods we can employ to maintain signal quality.

Analog audio storage media such as magnetic tape have several quality limitations. The media itself wears out with constant use or degenerates over time: tapes stretch out and exhibit wow and flutter. Their dynamic range is usually limited to 70 dB, whereas a live orchestra can have a dynamic range of 100 to 120 dB. Therefore, to record the music and accommodate the wide range of amplitudes, the recording level has to be adjusted so that soft music isn't lost, and loud music doesn't saturate the amplifier.

The recent development of *compact disk* (CD) technology for digital recording of audio signals is a significant advancement in the art of music recording. A CD consists of a plastic disk about 120 mm in diameter. It has about 20,000 tracks, each having a width of 0.5 μm and spaced 1.6 μm apart. Each track has a series of microscopic **pits** that stick out from the plastic. The regions between pits are called **lands.** The presence or absence of a pit is detected and converted to a binary electrical signal by the laser diode and photodiode system of Fig. 12.4–1. The laser light reflects off the pit and is transmitted into the photodiode via the beam splitter. Because the laser beam is focused at the specific depth, fingerprints, dust, scratches and other surface defects are blurred and therefore usually not detected by the system. Unlike magnetic media, there is no mechanical contact with the surface and thus no wear on the surface. Figure 12.4–2 shows the equivalent electrical signal of the programmed tracks. Note that a binary 1 occurs when there is a transition from *pit* to *land* or vice versa. The CD has a total track length of 5300 m and a scan velocity of 1.2 m/s, thus giving a total playing time of 74 minutes.

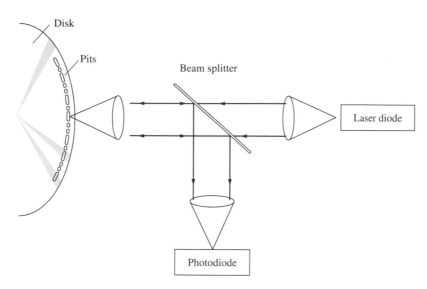

Figure 12.4–1   Optical readout system for a CD.

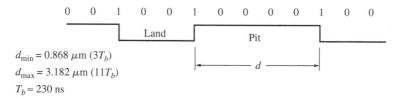

$d_{min} = 0.868\ \mu m\ (3T_b)$
$d_{max} = 3.182\ \mu m\ (11T_b)$
$T_b \approx 230$ ns

Figure 12.4–2   CD NRZI signal corresponding to programmed pits.

CD Recording

The system for recording *left* and *right channel* audio information to a CD is shown in Fig. 12.4–3. With high-fidelity music having a bandwidth of 20 kHz, the signal is sampled at 44.1 kHz to meet the Nyquist criterion and provide a guard band to minimize aliasing. The system has an analog antialiasing LPF which may also incorporate *preemphasis*. If preemphasis is included, a control word is also recorded to activate deemphasis during playback. Prior to PCM, a small amount of *dither*, or white noise, is added to mask any granular noise. The signal is then digitized using PCM with 16-bit uniform quantization. For an ADC, the dynamic range is defined by $20 \log(2^v)$ dB, and thus a 16-bit system has a dynamic range of $20 \log(2^{16}) = 96$ dB. With a sample rate of 44.1 kHz and 16 bits of quantization, the PCM has an output bit rate of 705.6 kpbs.

The two PCM outputs are sent to the *Cross Interleave Reed-Solomon Error Control Code* (CIRC) stage giving an effective bit rate of 1.4112 Mbps. The purpose

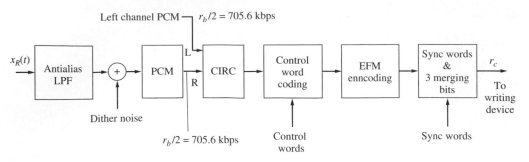

Figure 12.4–3   CD audio recording system.

of the CIRC is to encode the data with parity or check symbols to enable error correction during the playback process. Reed-Solomon codes are used for this purpose. Encoding occurs as follows: (*a*) Each 16-bit PCM signal is split into two 8-bit symbols. (*b*) Twelve symbols from each channel (i.e., twenty-four 8-bit symbols) are *scrambled* and then applied to the first CIRC encoder. (*c*) This encoder inserts four 8-bit parity symbols into the 24-symbol sequence to generate a 28-symbol sequence. These extra symbols are called *Q words* and enable error correction. (*d*) The symbol sequence is then dispersed according to a specified system of *interleaving*. Interleaving symbols minimizes the effects of **burst**-type errors where noise and media defects cause successive symbols to be corrupted. Interleaving also better enables the Reed-Solomon codes to correct errors. (*e*) The interleaved words undergo another encoding process in which an additional four 8-bit parity symbols called *P words* are added to the 28-symbol sequence. The end result is that the CIRC module converts a 24-symbol sequence into a 32-symbol sequence.

The *control-word module* adds an 8-bit symbol to each 32-symbol data block from the CIRC. This extra word is used to provide information about track separation, end of disk flags, track number, and so forth.

The *eight-to-fourteen* (EFM) module takes each 8-bit symbol from the control word module and converts it to a 14-bit symbol. Adding these additional bits reduces the signal bandwidth and dc content, and adds additional synchronization information, the purpose being to improve the recording and playback process and to allow for easier manufacturing tolerances for the lands/pits. Recall that an 8-bit code can represent 256 possible bit combinations, and a 14-bit code represents 16,384 combinations. Thus with EFM encoding, we chose 256 out of the 16,384 possibilities to represent the data from the control word module. To further reduce the signal's dc content, the EFM stage adds three additional bits we call *merging bits* to each 14-bit word. That is, each binary 1 is separated by at least two but not more than ten 0s, and the signal has a length that varies between three and eleven clock periods. The bit stream is then converted into a nonreturn-to-zero inverse (NRZI) format as shown in Fig. 12.4–2. Thus we have added an additional 9 bits to the incoming 8-bit symbol.

The data is then organized into blocks called *frames*, each one consisting of thirty-three 17-bit symbols (i.e., 561 bits). In order to indicate the start of each frame, and to specify the player's motor frequency, an additional 17-bit sync word and three merging bits are added, resulting in a frame length of 588 bits. The data output r_c is then written into the CD. Therefore, starting with the 1.4112 Mbs data rate at the input of the CIRC, the output rate is calculated as follows:

$$r_c = 1.4112 \text{ Mbps} \times \left(\frac{32}{24}\right)_{\text{CIRC}} \times \left(\frac{33}{32}\right)_{\substack{\text{control} \\ \text{word} \\ \text{coding}}} \times \left(\frac{17}{8}\right)_{\text{EFM}}$$

$$\times \left(\frac{588}{561}\right)_{\substack{\text{sync word} \\ \text{\& 3 merging} \\ \text{bits}}} = 4.3218 \text{ Mbps} = 1/T_b$$

Given that a 1 is separated from two to ten 0s, a bit rate of 4.3218 Mbps, and a scan velocity of 1.2 m/s means the CD land/pit size varies from 0.833 μm $(3T_b)$ to 3.054 μm $(11T_b)$.

For a CD, calculate the number of music information bits in each frame and the frame rate.

EXERCISE 12.4–1

CD Playback

The CD playback process is shown in Fig. 12.4–4. The *decoder* decodes the EFM signal, removes the merging bits, and extracts sync and control word information. The signal then goes to the error detector.

CD technology does a superior job of error control. CDs contain two types of errors. *Random errors* are caused by air bubbles or pit inaccuracies in the CD material, and *burst errors* are caused by scratches, fingerprints, and so forth. When errors do occur, they manifest themselves by "click" sounds at playback. Error control is

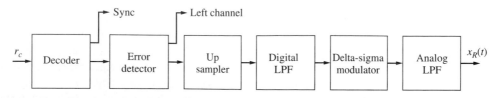

Figure 12.4–4   CD audio playback.

done by correction or concealment in the following order. First we employ check digits to correct errors that occur. If a specific bit is incorrect, we simply change it to the opposite state. On the other hand, if the bit was received with ambiguity, and there is enough redundancy in the code, we can use the redundancy to determine the correct data value. Otherwise we *conceal* the incorrect value by interpolating between neighboring samples. If that isn't satisfactory, the remaining choice is to mute the unreliable data value. As long as muting does not exceed a few milliseconds, it is not noticed by the listener.

At this point, we would expect to use a 16-bit DAC to reconstruct the signal. However, 16-bit DACs are relatively expensive, and a more economical and clever alternative exists. We first **up-sample** the data from the error detector by inserting $N - 1$ all zero words between the PCM samples, thus effectively increasing the sampling rate to $f_s' = Nf_s$. A *digital* LPF interpolates between the original samples. We usually use $N = 256$, so the sample rate increases from $f_s = 44.1$ kHz to $f_s = 11.3$ MHz. This massive increase in sampling rate causes the amplitude differences between successive samples to be relatively small and therefore allows us to use a delta-sigma modulator as a 1-bit DAC. A Butterworth analog LPF with a bandwidth of 20 kHz completes the reconstruction process. See Baert et al. (1998) for more information on audio recording.

12.5 DIGITAL MULTIPLEXING

Analog signal multiplexing was previously discussed under the heading frequency-division and time-division multiplexing. While those same techniques could be applied to waveforms representing digital signals, we gain greater efficiency and flexibility by taking advantage of the inherent nature of a digital signal as a sequence of symbols.

> Thus, digital multiplexing is based on the principle of **interleaving symbols** from two or more digital signals.

It is similar to time-division multiplexing but free from the rigid constraints of periodic sampling and waveform preservation.

More profoundly, however, the digital revolution has also eliminated the distinctions between the various telecommunications services such as telephone, television, and the Internet. Instead, voice, video, graphic, or text information is encoded, or digitized, and simply becomes data to be multiplexed with other data and then transmitted over an available channel. It no longer matters whether or not the data originated from a voice or picture; it's all just data.

In this section we want to first consider the general concepts and problems of digital multiplexing. The signals to be multiplexed may have come from digital data sources or analog sources that have been digitally encoded. We'll consider specific cases including telephone system hierarchies, the Integrated Services Digital Network (ISDN), and the Synchronous Optical NETwork (SONET). We will also briefly consider data multiplexing schemes such as packet switching, frame relay, and asynchronous transfer mode (ATM).

Multiplexers and Hierarchies

A binary multiplexer (MUX) merges input bits from different sources into one signal for transmission via a digital communication system. In other words, a MUX divides the capacity of the system between several pairs of input and output terminals. The multiplexed signal consists of source digits interleaved bit-by-bit or in clusters of bits (words or characters).

Successful demultiplexing at the destination requires a carefully constructed multiplexed signal with a constant bit rate. Towards this end, a MUX usually must perform the four functional operations:

1. Establish a frame as the smallest time interval containing at least one bit from every input.
2. Assign to each input a number of unique bit slots within a frame.
3. Insert control bits for frame identification and synchronization.
4. Make allowance for any variations of the input bit rates.

Bit rate variation poses the most vexing design problem in practice, and leads to three broad categories of multiplexers.

Synchronous multiplexers are used when a master clock governs all sources, thereby eliminating bit-rate variations. Synchronous multiplexing systems attain the highest throughput efficiency, but they require elaborate provision for distributing the master-clock signal.

Asynchronous multiplexers are used for digital data sources that operate in a start/stop mode, producing bursts of characters with variable spacing between bursts. Buffering and character interleaving make it possible to merge these sources into a synchronous multiplexed bit stream, as discussed later in conjunction with computer networks.

Quasi-synchronous multiplexers are used when the input bit rates have the same nominal value but vary within specified bounds. These multiplexers, arranged in a hierarchy of increasing bit rates, constitute the building blocks of interconnected digital telecommunication systems.

Two slightly different multiplexing patterns have been adopted for digital telecommunication: the AT&T hierarchy in North America and Japan and the CCIT hierarchy in Europe. (CCIT stands for International Telegraph and Telephone Consultive Committee of the International Telecommunications Union.) Both hierarchies are based on a 64-kbps voice PCM unit, and have the same structural layout shown in Fig. 12.5–1. The third level is intended only for multiplexing purposes,

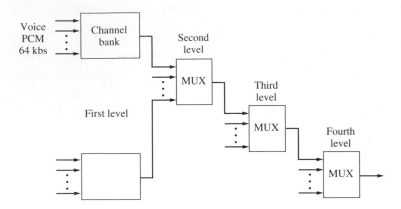

Figure 12.5–1   Multiplexing hierarchy for digital telecommunications.

whereas the other three levels are designed for point-to-point transmission as well as multiplexing. The parameters of the AT&T and CCIT hierarchies are listed in Table 12.5–1.

Observe in all cases that the output bit rate at a given level exceeds the sum of the input bit rates. This surplus allows for control bits and additional *stuff bits* needed to yield a steady output rate. Consequently, when we include the PCM bandwidth expansion, a digital network has very low *bandwidth efficiency* if devoted entirely to voice transmission. For instance, the fourth level of the AT&T multiplexing scheme requires $B_T \geq r_b/2 \approx 137$ MHz to transmit $24 \times 4 \times 7 \times 7 = 4032$ voice PCM signals, so the bandwidth efficiency is $(4032 \times 4 \text{ kHz})/137$ MHz $\approx 12\%$. The old AT&T FDM hierarchy had a bandwidth efficiency of approximately 85 percent. Thus digital multiplexing sacrifices analog bandwidth efficiency in exchange for the advantages of digital transmission.

Previously noted advantages of digital transmission include hardware cost reduction made possible by digital integrated circuits and power cost reduction made possible by regenerative repeaters. Now we can begin to appreciate the flexibility made

Table 12.5–1   Multiplexing hierarchies

	AT&T		CCIT	
	Number of inputs	**Output rate, Mbps**	**Number of inputs**	**Output rate, Mbps**
First Level	24	1.544	30	2.048
Second Level	4	6.312	4	8.448
Third Level	7	44.736	4	34.368
Fourth Level	6	274.176	4	139.264

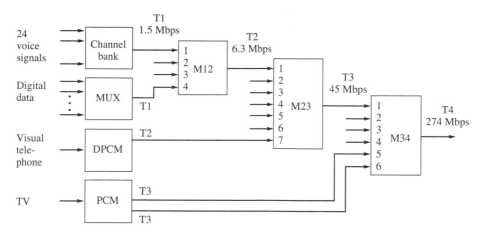

Figure 12.5–2   Illustrative configuration of the AT&T hierarchy.

possible by digital multiplexing, since the input bit streams at any level in Fig. 12.5–1 can be any desired mix of digital data and digitally encoded analog signals.

By way of example, Fig. 12.5–2 shows an illustrative configuration of the AT&T hierarchy with voice, digital data, visual telephone, and color TV signals for transmission on the fourth-level T4 line. Each of the twenty-four 64-kbps encoded voice signals is referred to as a **digital signal level zero** (DS0). The first-level T1 signals include PCM voice and multiplexed digital data. The T1 line is often referred to as a digital signal level one (DS1), the T2 line as DS2, and so on. The second-level T2 signals are multiplexed T1 signals along with visual telephone signals encoded as binary differential PCM (DPCM) with $f_s \approx 2$ MHz and $\nu = 2$ bits per word. PCM encoding of color TV requires a 90-Mbps bit rate ($f_s \approx 10$ MHz, $\nu = 9$), so two third-level T3 lines are allocated to this signal. The higher-level multiplexers labeled M12, M23, and M34 belong to the quasi-synchronous class. Let's consider the first-level synchronous multiplexer called the *channel bank*.

T1 Voice PCM Channel Bank   EXAMPLE 12.5–1

Synchronous multiplexing of voice PCM requires that the signals be delivered in *analog* form to the channel bank. Then, as diagrammed in Fig. 12.5–3*a*, sequential sampling under the control of a local clock generates an analog TDM PAM signal (see Figs. 7.2–7 and 7.2–8). This signal is transformed by the encoder into TDM PCM with interleaved words. Finally, the processor appends framing and signaling information to produce the output T1 signal.

The T1 frame structure is represented by Fig. 12.5–3*b*. Each frame contains one 8-bit word from each of the 24 input channels plus one bit for framing, giving a total of 193 bits. The 8-kHz sampling frequency corresponds to 125-μs frame duration,

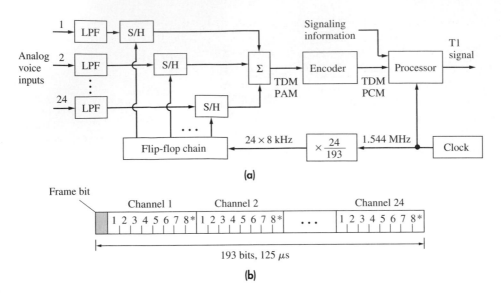

Figure 12.5–3   (a) T1 channel bank; (b) frame structure.

so the T1 bit rate is r_b = 193 bits ÷ 125 μs = 1.544 Mbps. Signaling information (dial pulses, "busy" signals, etc.) is incorporated by a method aptly known as *bit robbing*. Every sixth frame, a signaling bit replaces the least-significant bit of each channel word—denoted by the starred bit locations in the figure. Bit-robbing reduces the effective voice-PCM word length to $\nu = 7\frac{5}{6}$ and has inconsequential effect on reproduction quality. Yet it allows 24 signaling bits every 6 × 125 μs, or an equivalent signaling rate of 32 kbps.

T1 signals may be either combined at an M12 multiplexer or transmitted directly over short-haul links for local service up to 80 km. The T1 transmission line is a twisted-pair cable with regenerative repeaters every 2 km. A bipolar signal format eliminates the problems of absolute polarity and dc transmission.

EXERCISE 12.5–1   Assume that the first-level multiplexer in the CCIT hierarchy is a synchronous voice-PCM channel bank with 30 input signals, output bit rate r_b = 2.048 Mbps, and no bit-robbing. Find the number of framing plus signaling bits per frame.

Digital Subscriber Lines

The *plain old telephone* or POT is no longer used for just analog voice signals. It has become the channel for computer, fax, video and other digital data. With standard voice-only telephone lines, the "last-mile" connection between the telephone customer and the central office (CO) consists of a twisted-wire pair that feeds to a POT line card located at the CO. The POT card interfaces the voice line to a DS0 signal connected to the rest of the telephone network. While the twisted pair of wires may be capable of carrying signals of up to 30 MHz, the POT card is designed for voice signals, and thus the bandwidth is limited to 3.2 kHz. If the computer interface is made via a modem (modulator/demodulator), the data rate is limited to 30 kbps or, in some cases, up to 56 kbps. This is not acceptable for video, interactive video, and other high-speed services. Instead, we want to consider a **digital subscriber line** (DSL) that can handle increased data rates. The term *DSL* is somewhat of a misnomer since DSL is primarily a set of standards that defines the CO interface, and may or may not affect the existing twisted pair cable arrangement.

There are several DSL standards; some of these include:

1. **Asymmetric Digital Subscriber Line (ADSL).**   This is an FDM system where the existing twisted pair cable supports three services: (*a*) POTs, (*b*) 640 kbps digital data from subscriber to CO (upstream), and (*c*) 6.144 Mbps digital data from CO to subscriber (downstream). The ADSL system is shown in Fig. 12.5–4 with its corresponding spectrum shown in Fig. 12.5–5.

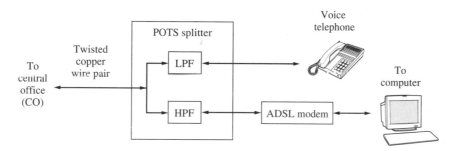

Figure 12.5–4   ADSL telephone twisted-wire pair interface.

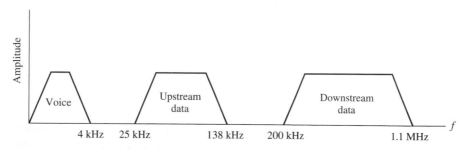

Figure 12.5–5   Spectrum of an ADSL signal.

2. **High Bit Rate Digital Subscriber Line (HDSL).** Consists of one to three twisted-wire pairs to transmit and receive digital data from 1.544 to 2.048 Mbps.

3. **Symmetrical Digital Subscriber Line (SDSL).** This is a single twisted pair version of HDSL and carries 0.192 to 2.23 Mbps digital data in both directions.

4. **Very High Bit Rate Digital Subscriber Line (VDSL).** A single twisted pair line that can carry as much as 26 or 52 Mbps in either direction.

5. **Integrated Services Digital Network (ISDN).** This may be a single twisted cable pair that allows for a data rate of at least 128 kbps and enables the integration of voice, video, and other data sources. The specifics of ISDN will be considered in the next section.

See Dutta-Roy (2000) for more information on DSLs.

Integrated Services Digital Network

Integrated Services Digital Network (ISDN) is a TDM digital telephone network that integrates voice, video, computer, and other data sources. There are two ISDN structures: **Basic Rate Interface** (BRI) and **Primary Rate Interface** (PRI). The BRI consists of two **B (bearer)** 64-kbps channels and a 16-kbps **D (delta)** channel. It is also referred to as 2B + D. BRI is used primarily by residential and small business customers. In North America, Japan, and Korea, PRI has twenty-three 64-kbps B channels and one 64-kbps D channel, or 23B + D. The B and D channels are full duplex (i.e., both directions) and carry voice, computer data, and so on, while the D channel is generally used to transmit control, signaling, telemetry, or other connection information. With BRI, the two B channels could carry two 64-kbps PCM voice messages or one 64-kbps voice message and up to six multiplexed 9600 bps data streams. Because of faster data rates and increased data channels, ISDN is an alternative to the **modem** (modulator/demodulator) for interfacing home computers to the telephone lines.

With BRI, the two B 64-kbps channels, the 16-kbps D channel, and some additional overhead bits are multiplexed together for a 192-kbps output. Specifically, 16 bits from each B channel and 4 bits from the D channel are combined with an additional 12 bits of overhead to make a 48-bit frame. Each frame is transmitted with a period of 250 μs to get an output bit stream rate of 192 kbps. PRI has a frame consisting of 8 bits from each of the 24 channels plus a framing bit used for synchronization and control yielding 193 bits (i.e., $24 \times 8 + 1 = 193$). The frame rate is 8000 frames per second yielding an output bit stream rate of 1.544 Mbps. The European and rest of the world standard for PRI consists of 30 B + 1 D 64-kbps channels where each frame is 256 bits long. Thus, with overhead, the output rate is 2.048 Mbps.

The ISDN also has high-capacity H channels for information in excess of 64 kbps. These include video, high-resolution graphics, high-fidelity audio, HDTV, and so on. The channels and their respective capacities are shown in Table 12.5–2. Note that Channels H0 and H11 are North American IDSNs, whereas H12 is used in Europe.

Table 12.5–2   ISDN H channel capacities

Channel	Capacity, kbps
H0	384
H11	1,536
H12	1,920
H21	32,768
H22	44,160
H4	135,168

So far we have discussed what is now referred to as **narrowband** ISDN (N-ISDN). However ISDN specifications have been increased to handle data rates of 2.5 Gbps and beyond, creating what we now call **broadband**-ISDN (BISDN). BISDN enables such services as interactive video, HDTV, other multimedia, and of course allows the system to carry more services.

Synchronous Optical Network

Synchronous Optical NETwork (SONET) is a fiber-optic standard that enables multiple broadband signals and even relatively low-rate signals to be multiplexed onto a fiber-optic channel. The standard was developed by BellCore (Bell Communications Research) and standardized by ANSI (American National Standards Institute). The European equivalent is referred to as **Synchronous Digital Hierarchy** (SDH). A complete tutorial on SONET is available on the World Wide Web at location *http://www.webproforum.com/sonet*. We will present a brief description of its operation.

Figure 12.5–6 shows a block diagram of a SONET system, whereas the details of the **path-terminating element** (PTE) and its associated multiplexing are shown in

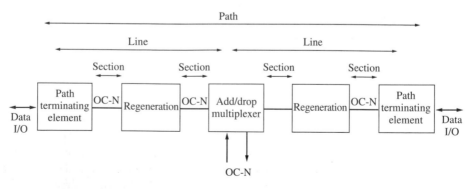

Figure 12.5–6   SONET system.

Fig. 12.5–7. SONET communication works as follows. A communication's link, which we will refer to as a **path,** is established between the left- and right-most PTE. The PTE multiplexes various input electrical signals to generate a single optical signal and vice versa. First the PTE multiplexes and formats the input data to generate a single 51,840 Mbps **synchronous transport signal** (STS-1) at the output. It also demultiplexes the STS signal at the destination. The inputs can be a single broadband signal such as an HDTV or multiple low-rate signals that are mapped into a **virtual tributary** (VT) which, in turn, is mapped to the STS-1 signal. The STS-1 output is an electrical signal with a frame structure such that each frame has 90 bytes $\times$ 8 bits/byte $\times$ 9 rows/frame $\times$ 8000 frames/sec and yields an STS-1 signal rate of 51.840 Mbps. In the case of SONETs, bytes are often referred to as **octets.**

The N STS-1 signals are then byte multiplexed to generate an STS-N signal. It is scrambled and fed to an electrical-to-optical converter where it is converted to an equivalent optical **OC-N** signal at some specified wavelength. The signal then travels over a fiber-optic channel to its destination. It is possible to multiplex additional optical wavelengths over the channel. The optical channel may contain regeneration to overcome path losses. The channel may also contain **add/drop multiplexers** (ADM) that enable additional STS signals to be added or dropped from the channel. For example, when a signal arrives at the ADM, the header information is examined to determine if a section of data is to proceed to the same PTE as the rest of the STS data or is to be routed to another PTE. The input/output lines of the ADM are OC-Ns, but some ADMs may have electrical input/outputs as well. A **section** defines a

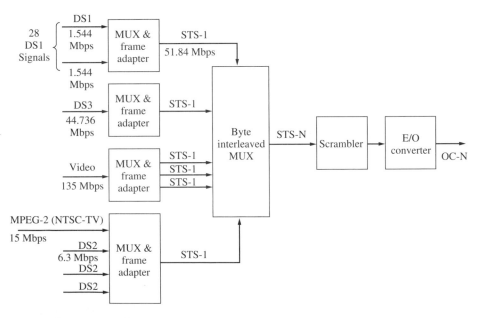

Figure 12.5–7   Path-terminating elements.

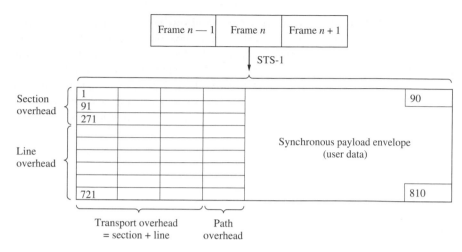

Figure 12.5–8   STS-1 SONET frame, 9 rows × 90 bytes per row.

link between a PTE and regeneration or regeneration and the ADM. A **line** defines the link between the PTE and the ADM.

The frame structure is shown in Fig. 12.5–8. The data in the first three columns are **transport overhead** (TOH). In the TOH, the first three rows are **section overhead** (SOH), and the next six rows are **line overhead** (LOH). The data in the fourth column is **path overhead** (POH). The remaining 86 columns × 9 rows are for user data. The user data and path overhead constitute the **synchronous payload envelope** (SPE). Each of the overhead sections defines the addressing and multiplexing functions at their particular level. The SOH, LOH, and POH also contain information relating to framing, coding, synchronization, performance monitoring, and status. See *http://www.webproforum.com/sonet* for more specific information.

Data Multiplexers

Multiplexers for computer communication depart in two significant respects from general-purpose telecommunications service. On the one hand, complications arise from the fact that each computer has its own independent clock and operates in an asynchronous start/stop mode. On the other hand, simplifications come from the fact that computers don't require the nearly instantaneous response needed for two-way voice communication. Consequently buffering is essential and the associated time delay is tolerable.

Conventional communication links such as a telephone T1 line employ **circuit switching** in which a dedicated line is assigned to connect the source and destination. This is the case whether the line is available at all times, or is shared by other users in a TDM scheme shown in Figs. 7.2–7 or 12.5–2. Consequently, circuit switching is not very efficient if the data are bursty or intermittent.

Message switching, or **statistical time division multiplexing,** on the other hand, only uses a channel when we want to send a specific message. After the message is sent the channel is relinquished for another message from the same or another source. Message switching is also called **store and forward switching** because when a particular message reaches a node, it may have to be stored until a connection to the next node is available. While store and forward switching is inherently more efficient than circuit switching, it may involve significant time delays—delays that would be unacceptable in two-way voice communication. But such delays are not a problem for applications like the Internet since Internet users can tolerate brief turnaround lags. Having said this, however, improvements in router technology (i.e., the device that determines the data route) have greatly reduced the latency time in store and forward switching, which is thus becoming an alternative to circuit switching for telephone systems.

The first message switching system we'll consider is **packet switching.** Here the message length is limited to short blocks of data called **packets.** The packets may be sent using a dedicated line we call a **virtual circuit** such that when all the packets are sent, the line is given up. Virtual and switched circuits may be the same except that the virtual circuit is only used for the duration of the message. A packet has a header that contains the router information that specifies the location to be routed to. The packet also includes extra bits for error control purposes. Packet switching poses some interesting problems since, at any given time, the network contains numerous packets attempting to reach their respective destinations. With packet switching, multiple packets from the same source may take different paths depending on availability, may arrive in scrambled order, or may in fact get lost. If too many packets have entered a switch and compete for available lines and buffer space, we have a problem of *contention.* If the input rate to the switch exceeds the output rate, the excess packets are stored in a buffer. However, if the buffer is not large enough to store these excess packets, some may get discarded. If this occurs relatively frequently, the switch is said to be *congested.*

Packet switching is capable of working at data rates up to 64 kbps. The rate is limited because much of the packet contains error control information. **Frame relay** switching is similar to packet switching, but uses variable length packets called **frames.** Much of the error bits have been removed and thus frame relay can achieve significantly faster data rates, up to 2 Mbps. **Asynchronous transfer mode** (ATM), or cell-relay, has even lower overhead than frame relay and uses fixed-length packets called **cells.** An extremely broadband switching scheme with rates in the 10s and 100s of Mbps, it allows for multiple virtual channels with the rate of each one dynamically set on demand. ATM is used in conjunction with BISDN and SONET to carry widely different services (voice, video, business information, etc.).

EXAMPLE 12.5–2   Consider a voice TDM system with 24 channels, each one having a data rate of 64 kbps but, because of the intermittent nature of speech, is utilized only 10 percent of the time. We want to transmit a fax page consisting of 8 million bits. If we employ message switching using all the channels that are available 90 percent of the time, our waiting time is $1/(24 \times 64 \text{ kbps} \times 0.9) = 0.72 \ \mu\text{s/bit} \times 8$ million bits = 5.8 seconds. On the other hand, if we use circuit switching and only one of the TDM channels, our waiting time is $1/(64 \text{ kbps}) = 16.6 \ \mu\text{s/bit} \times 8$ million bits = 125 seconds. Therefore, if we can tolerate the intermittent delays associated with store-and-forward switching, we have a relatively efficient method of quickly transferring data.

12.6   COMPUTER NETWORKS

In 1969, the Advanced Research Projects Agency (ARPA) of the US Department of Defense established the **ARPANET,** the acronym for **A**dvanced **R**esearch **P**rojects **A**gency **NET**work, for the interconnection of computers and terminals. It was a packet-switched network. A model is shown in Fig. 12.6–1. Here the network consists of five computers, symbolized by circles, and four terminal interfaces, symbolized by squares. Each circle or square represents a communication **node** equipped with a concentrator having the ability to transmit, receive, and store data. The ARPANET evolved and has become what is commonly known as the **Internet.**

The Internet has enabled the interconnection of millions of computers worldwide via telephone lines, satellite links, cable TV networks, RF (wireless), and other media. Its rapid growth in the last several years has been fueled by low-cost hardware connections, low-cost personal computers, and easy-to-use software interfaces in the form of email and web browsers. The available technology has brought the Internet into the communication mainstream just like the telephone 100 years ago. With increasing bandwidth on existing and new channels, the Internet will be able to deliver interactive video and other multimedia information (i.e., streaming audio and video), and thus may rival broadcast and cable TV for video/sound entertainment.

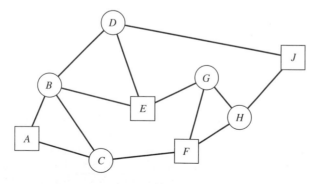

Figure 12.6–1   Computer network.

There are several types of computer networks that can be interconnected via the Internet. **Local area networks** (LAN) allow computer systems in a building or within an enterprise to be connected together and share common databases, programs, and other sources of information. Similarly, other networks cover a large area of the country and are called **wide area networks** (WAN). An example would be banking systems, in which branch computers and automatic teller machines are interconnected. Finally, there are **global area networks** (GAN), which enable worldwide network interconnection. In the remainder of this section we will look at two network models: the Open Systems Interconnection and the Transmission Control Protocol/Internet Protocol.

Open Systems Interconnection

Open Systems Interconnection (OSI) was developed in the late 1970s by the International Standards Organization (ISO) in association with the International Telecommunications Union (ITU) to be a model for the standardization of computer networks. "Open" means that any two systems can communicate if they adhere to the OSI model.

Figure 12.6–2 shows the OSI model, which consists of seven layers. Layers 1 to 3 are part of the network connection, while layers 4 to 7 are part of the data originator (source) or recipient (destination). This architecture allows any one layer to be modified or improved without affecting the other layers. The layers are described as follows.

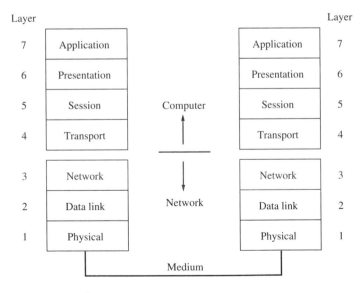

Figure 12.6–2   OSI architecture.

1. **Physical.** This is where bit stream information is inserted into the network. Here we are directed to topics covered in Chapter 11 and the earlier part of this chapter. The physical layer includes electrical and mechanical standards (i.e., electrical signals, connectors, etc.). Examples include the Ethernet, CCITT 1.430 BISDN, and the ISO-8877 ISDN connector.

2. **Data link.** Structures the data into frames so that each one includes bits for synchronization, error detection/correction, and the beginning and end of a frame, plus the source's and destination's addresses. It also sends and receives acknowledgements if the frame is received correctly. For example, if a burst error occurred and a frame was corrupted, the software in the data link layer would cause the source to retransmit the frame. If an acknowledgment signal was not received by the source, the source would then attempt to send a duplicate frame. Therefore, the data link layer would minimize the problem of duplicate frames. The data link layer along with the other layers manages the data flow so a source with a high data rate does not overwhelm a slow destination. A data link layer example is the CCIT 1.441/Q.921 ISDN Data Link Protocol (LAPD).

3. **Network.** Decides how the data is to be routed from node to node, and is used to manage data flow if there is network congestion. This layer may also be involved in sending and receiving acknowledgments.

4. **Transport.** It either breaks the message data from the session layer into packets (or some other unit) or takes the packets from the network layer and connects them to form a message.

5. **Session.** Establishes connections between computers or reestablishes the connection in the event of a failure. It also links applications such as audio and video services and authenticates network users.

6. **Presentation.** Performs data formatting function such as code conversion (e.g., ASCII to EBCDIC), encryption/decryption, and data compression/decompression.

7. **Applications.** This is the user interface to the OSI system and includes such services as file transfer programs, airline reservation data, email or web browsers.

In practical systems, some or all of these layers can be integrated at various degrees into a single hardware device. For example, some personal computers have a plug-in modem card (i.e., stand-alone modem) that performs the functions of layers 1–3 while the rest of the computer performs the functions of layers 4–7. On the other hand, some personal computers have a built-in modem that shares the computer's CPU at a lower priority to perform the functions of layers 1–3, using a process known as **cycle stealing.** Of course, with this system, if there is a great deal of interaction with the network, the system slows down.

Transmission Control Protocol/Internet Protocol

The Transmission Control Protocol/Internet Protocol (TCP/IP) was developed in conjunction with the original ARPANET and preceded the ISO model. While the ISO model has served as a guideline for network architectures and can certainly help us understand its functions, it has never become *the* standard its creators hoped it would. The TCP/IP system is not a formal standard like the OSI one, but was developed with the "implement as you go, experiment some more" philosophy that has become part of the Internet culture, versus the "specify first, implement later" approach taken by the ISO committee (Peterson and Davie, 2000). The TCP/IP is part of the Internet and preceded the OSI system. However, the TCP/IP can be implemented using the OSI structure, and some of the OSI layers are part of the TCP/IP system. A TCP/IP model is shown in Fig. 12.6–3. The application and physical layers are similar to the OSI model with the other ones described as follows.

1. **Network.** Manages the exchange of data between the computer and the attached network.

2. **Internet.** Causes the data to be transferred across the network or, using a router, sends the data to a different network. A **router** is a processor that transfers data from one network to another as it moves from source to destination. A router also determines the best path for data to travel. As Fig. 12.6–3 shows, the Internet layer enables the linking of multiple networks where each one may have a different technology and data format. As an aside, a **bridge** is a device that performs a similar function as a router. However, a bridge links like networks together and thus is much simpler than a router. For example, a bridge could link systems that conform to IEEE 802.3 such as an Ethernet and token ring LANs. A **repeater** can also be used to link networks, but only if the network and data for-

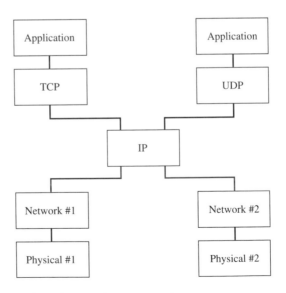

Figure 12.6–3   TCP/IP architecture for two networks.

mat are completely identical, as in the linking of two Ethernets, for example. A repeater makes two networks look like one and therefore is usually found in systems where, due to cable length restrictions, a network has to be split into two separate pieces. A **firewall** is a router that is programmed to prevent unauthorized access to a given site or to a section of a given site. This is in addition to any encryption that individual users may already employ. Firewalls are a relatively centralized security system. For example, a bank may have a firewall preventing unauthorized access to certain parts of its site, but allowing individual depositors access to their accounts using personalized encryption.

3. **TCP.** This is also called the **host-to-host** or **transport** layer. It numbers each data segment, so it can be put in the correct sequence at the source; at the destination, it puts the data in the correct sequence, so it is reliably delivered to the application. If reliability is not a major concern, an alternative logical channel to data delivery between the IP and network layers is the **user datagram protocol** (UDP) for the transfer of single packet messages we call **datagrams.** The UDP has the advantage of lower overhead than the TCP channel.

You will note that in contrast to the OSI system, except for the physical and application layers, there is not a great deal of similarity between the OSI and TCP/IP layers. OSI defines a rigid set of requirements that specifies what function is done in each layer, whereas TCP/IP primarily defines a method for exchanging data between different networks. Generally, there is no rigid requirement that a given function such as error correction and detection reside in any one particular layer.

The subject of computer networks is an entire study in itself. More detailed information with respect to analysis and design of computer networks can be found in Tannenbaum (1989), Stallings (2000), Miller (2000), and Peterson and Davie (2000).

12.7 PROBLEMS

12.1–1* An analog waveform with $W = 15$ kHz is to be quantized to $q \geq 200$ levels and transmitted via an M-ary PCM signal having $M = 2^n$. Find the maximum allowed values of ν and f_s and the corresponding value of n when the available transmission bandwidth is $B_T = 50$ kHz.

12.1–2 Do Prob. 12.1–1 with $B_T = 80$ kHz.

12.1–3 *Hyperquantization* is the process whereby N successive quantized sample values are represented by a single pulse with q^N possible values. Describe how PCM with hyperquantization can achieve *bandwidth compression*, so $B_T < W$.

12.1–4 Suppose the PCM quantization error ϵ_k is specified to be no greater than $\pm P$ percent of the peak-to-peak signal range. Obtain the corresponding condition on ν in terms of M and P.

12.1–5*   A voice signal having $W = 3$ kHz and $S_x = 1/4$ is to be transmitted via M-ary PCM. Determine values for M, ν, and f_s such that $(S/N)_D \geq 40$ dB if $B_T = 16$ kHz.

12.1–6   Do Prob. 12.1–5 with $(S/N)_D \geq 36$ dB and $B_T = 20$ kHz.

12.1–7   An audio signal with $S_x = 0.3$ is to be transmitted via a PCM system whose parameters must satisfy the standards for broadcast-quality audio transmission listed in Table 9.4–1. (*a*) If $M = 2$, then what are the required values of ν and B_T? (*b*) If $B_T = 4W$, then what's the minimum value of M?

12.1–8   Do Prob. 12.1–7 for high-fidelity audio transmission standards.

12.1–9   What is the minimum number of bits required for an ADC to quantize a signal that varies from 5 μV to 200 mV so that $(S/N)_D \geq 40$ dB?

12.1–10*   What is the q, Δ and $(S/N)_D$ for a PCM system with a $\nu = 12$-bit ADC and a bipolar ± 10 sinusoidal input?

12.1–11   Do Prob. 12.1–10 with $\nu = 16$ bits.

12.1–12   How many bits are required for an ADC to encode music where the dynamic range is 120 dB?

12.1–13*   What is the minimum size memory required to store 10 minutes of a sampled and quantized voice assuming $S_x/\sigma_q^2 = 35$ dB and $f_s = 8$ kHz?

12.1–14   Consider a uniform quantizer with $\nu = 12$ bits and inputs between $0+/-10$ V. What is ϵ_k for an input of (*a*) $+0.02$ volts, (*b*) $+0.2$ volts?

12.1–15‡   Consider the signal PDF in Fig. P12.1–15. Use Eq. (9) to show that uniform quantization with $q = 2^\nu$ levels and $\nu \geq 3$ yields $\sigma_q^2 = 1/3q^2$. Then calculate S_x and find $(S/N)_D$ in terms of q.

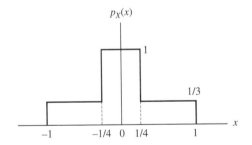

Figure P12.1–15

12.1–16‡   Consider a signal whose PDF is reasonably smooth and continuous and satisfies the condition $P[|x(t)| > 1] \ll 1$. From Eq. (9), show that uniform quantization with $q \gg 1$ yields $\sigma_q^2 \approx 1/3q^2$.

12.1–17   Let $x(t)$ have a zero-mean gaussian PDF with variance σ_x^2 such that $P[|x(t)| > 1] \leq 0.01$. Use the result from Prob. 12.1–16 to obtain $(S/N)_D$ in a form like Eq. (7), assuming uniform quantization with $q \gg 1$.

12.1–18   Let $x(t)$ have a Laplace PDF, as given by Eq. (13), with variance σ_x^2 such that $P[|x(t)| > 1] \leq 0.01$. Use the result from Prob. 12.1–16 to obtain $(S/N)_D$ in a form like Eq. (7), assuming uniform quantization with $q \gg 1$.

12.1–19   Consider the compressor characteristic $z(x) = (\text{sgn } x)\sqrt{|x|}$ for $|x| \leq 1$. (a) Find and sketch the complementary expander characteristic $x(z)$. (b) Evaluate K_z when $x(t)$ has the PDF in Fig. P12.1–15.

12.1–20   Consider the μ-law compressor in Example 12.1–4. (a) Derive the complementary expander characteristic $x(z)$. (b) Carry out the details leading to K_z as given in Eq. (15).

12.1–21   Repeat Prob. 12.1–15 if the quantizer is preceded by a $\mu = 255$ logarithmic compandor.

12.1–22   A voice signal with a Laplace distribution is applied to a compressor having $z(x) = (\text{sgn } x)(1 - e^{-3|x|})$ for $|x| \leq 1$. (a) Obtain an expression for K_z in terms of α. Then show that $K_z \approx 1/9$ for $\alpha \gg 1$ and express $(S/N)_D$ in dB in terms of ν and S_x for binary voice PCM with this compressor. (b) Now take $q = 2^8$ and evaluate S_x, K_z, and $(S/N)_D$ for $\alpha = 4, 8$, and 16. Use these results to make a plot like Fig. 12.1–8. Your plot should demonstrate that this compressor yields a higher $(S/N)_D$ but less dynamic range than the standard μ-law compressor.

12.1–23‡   The *A-law companding system* employs a compressor with

$$z(x) = \begin{cases} Ax/(1 + \ln A) & 0 \leq x \leq 1/A \\ (1 + \ln Ax)/(1 + \ln A) & 1/A < x \leq 1 \end{cases}$$

and $z(x) = -z(x)$ for $-1 \leq x \leq 0$. (a) Assume $p_X(x)$ has even symmetry and negligible area outside $|x| \leq 1$ to show that

$$K_z = (1 + \ln A)^2 \left[S_x + 2 \int_0^{1/A} \left(\frac{1}{A^2} - x^2 \right) p_X(x)\, dx \right]$$

(b) Obtain the expression for K_z in terms of A and α when $x(t)$ has a Laplace distribution, and show that $K_z \approx (1 + \ln A)^2/A^2$ when $\alpha \gg A$. (c) Let $A = 100$ and $q = 2^8$. Evaluate S_x, K_z, and $(S/N)_D$ for $\alpha = 4, 16, 64$ and $\alpha \gg 100$. Use these results to make a plot like Fig. 12.1–8. Your plot should demonstrate that A-law companding yields a lower $(S/N)_D$ but greater dynamic range than standard μ-law companding.

12.2–1*   A signal with $S_x = 1/2$ and $W = 6$ kHz is to be transmitted via M-ary PCM on a channel having $N_0 = 0.01$ μW/Hz and $B_T = 15$ kHz. Find the smallest value of M and the corresponding smallest value of ν that yields $(S/N)_D \geq 36$ dB. Then calculate the minimum value of S_R for operation above threshold, assuming the PCM signal occupies the full transmission bandwidth.

12.2–2   Do Prob. 12.2–1 with $B_T = 20$ kHz.

12.2–3   Do Prob. 12.2–1 with $B_T = 50$ kHz.

12.2–4*   Consider a voice PCM system with $M = 2, \nu = 8$, and μ-law companding, so $(S/N)_D = 37$ dB. The PCM signal is transmitted via a regenerative repeater system with 20 identical sections. Find the minimum value of γ at the input to the first repeater for operation above threshold. Then determine the PCM advantage in dB by calculating the value of γ at the input to the first repeater of an analog repeater system that yields $(S/N)_D = 37$ dB after 20 sections.

12.2–5   Do Prob. 12.2–4 with 100 sections.

12.2–6   Show from Eq. (3) that decoding noise decreases $(S/N)_D$ by 1 dB when $P_\epsilon \approx 1/15q^2$. Calculate the corresponding threshold value of γ for binary PCM in gaussian noise when $\nu = 4, 8, 12$. Then take $S_x = 1$ and plot $(S/N)_D$ in dB versus γ_{th} in dB.

12.2–7‡   Consider binary PCM with sign/magnitude codewords. Figure 12.1–1b reveals that an error in the sign bit shifts the decoded level by an amount $\pm 2(2i - 1)/q$ where $i = 1, 2, \ldots, q/2$. Show that $\overline{\epsilon_m^2} = (5q^2 - 8)/3\nu q^2$, so $\overline{\epsilon_m^2} \approx 5/3\nu$ if $q \gg 1$.

12.2–8   Suppose a PCM system has fixed values of γ and $b = B_T/W$. Taking threshold into account, obtain an expression for the maximum value of q.

12.2–9   Find and plot γ_{th} in dB versus $b = B_T/W$ for a PCM system with $q = 256$ and $\nu = b$.

12.3–1   What Δ minimizes slope overload for a DM system where the input is a normalized sinusoid, given $f_s = 30$ kHz and $W = 3$ kHz?

12.3–2*   What is the maximum amplitude for a 1 kHz sinusoidal input for a DM system that prevents slope-overload that has been sampled at 10 times the Nyquist rate with $\Delta = 0.117$ V?

12.3–3   The signal $x(t) = 8 \cos 2\pi Wt$ is the input to a delta modulator with $T_s = 1/24W$. Plot the sample points $x(kT_s)$ for $0 \le k \le 30$. Then take $\tilde{x}_q(0) = 0$ and plot $\tilde{x}_q(t)$ with $\Delta = 1$ and $\Delta = 3$.

12.3–4   Use the results of Example 12.3–1 to plot $(S/N)_D$ in dB versus b for DM voice transmission with $b = 4, 8$, and 16. Compare with binary PCM transmission, taking $\nu = b$ and $S_x = 1/30$.

12.3–5   Use the results of Example 12.3–1 to tabulate f_z and the optimum value of Δ for DM voice transmission with $b = 4, 8$, and 16 when $S_x = 1/9$.

12.3–6   Suppose a signal to be transmitted via DM has a dominant spectral peak at $f = f_0 < W$, so that $|\dot{x}(t)|_{max} = 2\pi f_0$. Obtain an upper bound on $(S/N)_D$ in a form like Eq. (12).

12.3–7*   The power spectrum of a lowpass filtered voice signal can be approximated as $G_x(f) = [K/(f_0^2 + f^2)]\Pi(f/2W)$. Find K in terms of S_x, and obtain an expression for the rms bandwidth. Evaluate W_{rms} when $W = 4$ kHz and $f_0 = 0.8$ kHz.

12.3–8   An approximate expression for DM slope-overload noise is given by Abate (1967) as

$$N_{so} = \frac{8\pi^2}{27}\left(\frac{W_{rms}}{W}\right)^2(3s + 1)e^{-3s}S_x$$

Write Δ^2/f_s in Eq. (7) in terms of s and b, and show that the total quantization noise $N_q + N_{so}$ is minimized by taking $s = \ln 2b$.

12.3–9   Find the tap gains and evaluate the prediction gain in dB for DPCM with a one-tap and a two-tap transversal filter when the input is a voice signal having $\rho_1 = 0.8$ and $\rho_2 = 0.6$.

12.3–10*   Do Prob. 12.3–9 for a TV image signal having $\rho_1 = 0.95$ and $\rho_2 = 0.90$.

12.3–11   Consider a DPCM system with a two-tap transversal prediction filter. Assuming $q \gg 1$ so that $x_q(kT_s) \approx x(kT_s)$, find the tap gains to implement the prediction strategy

$$\tilde{x}_q(kT_s) \approx x[(k - 1)T_s] + T_s\frac{dx(t)}{dt}\bigg|_{i=(k-1)T_s}$$

12.3–12   Consider a DPCM system with a one-tap transversal prediction filter and $q \gg 1$, so $x_q(kT_s) \approx x(kT_s)$ and $\epsilon_q(kT_s) \approx x(kT_s) - \tilde{x}_q(kT_s)$. Obtain an expression for the mean square prediction error $\overline{\epsilon^2} = E[\epsilon_q^2(k)]$ in terms of S_x and the signal's autocorrelation $R_x(\tau) = E[x(t)x(t - \tau)]$. Then find the value of the tap gain that minimizes $\overline{\epsilon^2}$.

12.3–13   Use the method outlined in Prob. 12.3–12 to find the tap gains for a two-tap transversal prediction filter. Express your results in a matrix like Eq. (16b).

12.4–1*   How many bits can a CD store?

12.4–2   What percentage of a CD will it take to store the Bible if it consists of 981 pages, two columns per page, 57 lines per column, 45 characters per line, and each character has 7 bits?

12.4–3   How many minutes of music can be recorded on a 2-Gbyte hard drive with $f_s = 44.1$ kHz, $\nu = 16$ bits, and two recording channels?

12.5–1*   Several high-fidelity audio channels having $W = 15$ kHz are to be transmitted via binary PCM with $\nu = 12$. Determine how many of the PCM signals can be accommodated by the first level of the AT&T multiplexing hierarchy. Then calculate the corresponding bandwidth efficiency.

12.5–2   Do Prob. 12.5–1 for the first level of the CCIT hierarchy.

12.5–3   Determine the number of voice telephone signals that can be carried by a STS-1 SONET.

12.5–4*   How long does it take to transmit an 8×10 inch image with 600 dots per inch resolution and 1 bit per dot over a BRI channel?

12.5–5   Repeat Prob. 12.5–4 with a 56-kbps modem.

chapter

13

Channel Coding and Encryption

CHAPTER OUTLINE

A primary objective of transmitting digital information is to minimize errors and, in some cases, to maintain data security. Transmission errors in digital communication depend on the signal-to-noise ratio. If a particular system has a fixed value of *S/N* and the error rate is unacceptably high, then some other means of improving reliability must be sought. Error-control coding often provides the best solution.

The tremendous expansion of electronic commerce (e-commerce) has been made possible by improved and economical data security systems. Data security has three goals: **secrecy,** to prevent unauthorized eavesdropping; **authenticity,** to verify the sender's signature and prevent forgery; and finally **integrity,** to prevent message alteration by unauthorized means. All three of these goals require a system of coding called **encryption.**

Error-control coding involves the systematic addition of extra digits to the transmitted message. These extra check digits convey no information by themselves, but make it possible to detect or correct errors in the regenerated message digits. In principle, information theory holds out the promise of nearly errorless transmission, as will be discussed in Chap. 16. In practice, we seek some compromise between conflicting considerations of reliability, efficiency, and equipment complexity. A multitude of error-control codes have therefore been devised to suit various applications.

Encryption codes, on the other hand, are used to scramble a message by **diffusion** and/or **confusion** and thereby prevent unauthorized access. Diffusion is where the message is spread out, thus making it more difficult for the enemy to look for patterns and use other statistical methods to decode, or decipher, the message. Confusion is where the *cryptographer* uses a complex set of transformations to hide the message. Obviously the goal is to create a *ciphertext* that is *unconditionally secure* such that no amount of computing power can decipher the code. However, practical systems are *computationally secure,* meaning it would take a certain number of years to break the cipher.

This chapter starts with an overview of error-control coding, emphasizing the distinction between **error detection** and **error correction** and systems that employ these strategies. Subsequent sections describe the two major types of codes implementations, **block codes** and **convolutional codes.** We will then present the basic ideas behind data encryption and decryption and look at the widely used **data encryption standard (DES)** and the **Rivest-Shamir-Adleman (RSA) system.** We'll omit formal mathematical analysis. Detailed treatments of error-control coding and encryption are provided by the references cited in the supplementary reading list.

OBJECTIVES

After studying this chapter, and working the exercises, you should be able to do each of the following:

1. Explain how parity checking can be used for error detection or correction, and relate the error-control properties of a code to its minimum distance (Sect. 13.1).
2. Explain how interleaving codewords can be used to make error correction and detection methods more effective (Sect. 13.1).
3. Calculate the message bit rate and error probability for a forward error correction (FEC) system with a given block code (Sect. 13.1).
4. Analyze the performance of an ARQ system with a given block code (Sect. 13.1).
5. Describe the structure of a systematic linear block code or cyclic code (Sect. 13.2).
6. Use matrix or polynomial operations to perform encoding and decoding operations of a given code (Sect. 13.2).
7. Describe the operation of convolutional codes (Sect. 13.3).
8. Describe the basic concepts of data secret-key and public-key encryption systems, and encrypt data using the RSA encryption algorithm (Sect. 13.4).

13.1   ERROR DETECTION AND CORRECTION

Coding for error detection, without correction, is simpler than error-correction coding. When a two-way channel exists between source and destination, the receiver can request retransmission of information containing detected errors.

This error-control strategy, called **automatic repeat request** (ARQ), particularly suits data communication systems such as computer networks. However, when retransmission is impossible or impractical, error control must take the form of **forward error correction** (FEC) using an error-correcting code. Both strategies will be examined here, after an introduction to simple but illustrative coding techniques.

Repetition and Parity-Check Codes

When you try to talk to someone across a noisy room, you may need to repeat yourself to be understood. A brute-force approach to binary communication over a noisy channel likewise employs *repetition*, so each message bit is represented by a *codeword* consisting of n identical bits. Any transmission error in a received codeword alters the repetition pattern by changing a 1 to a 0 or vice versa.

If transmission errors occur randomly and independently with probability $P_e = \alpha$, then the binomial frequency function gives the probability of i errors in an n-bit codeword as

$$P(i, n) = \binom{n}{i} \alpha^i (1 - \alpha)^{n-i} \tag{1a}$$

$$\approx \binom{n}{i} \alpha^i \qquad \alpha \ll 1$$

where

$$\binom{n}{i} = \frac{n!}{i!(n - i)!} = \frac{n(n - 1) \cdots (n - i + 1)}{i!} \tag{1b}$$

We'll proceed on the assumption that $\alpha \ll 1$—which does not necessarily imply reliable transmission since $\alpha = 0.1$ satisfies our condition but would be an unacceptable error probability for digital communication. Repetition codes improve reliability when α is sufficiently small that $P(i + 1, n) \ll P(i, n)$ and, consequently, several errors per word are much less likely than a few errors per word.

Consider, for instance, a triple-repetition code with codewords 000 and 111. All other received words, such as 001 or 101, clearly indicate the presence of errors. Depending on the decoding scheme, this code can *detect* or *correct* erroneous words. For error detection without correction, we say that any word other than 000 or 111 is a detected error. Single and double errors in a word are thereby detected, but triple errors result in an undetected **word error** with probability

$$P_{we} = P(3, 3) = \alpha^3$$

For error correction, we use majority-rule decoding based on the assumption that at least two of the three bits are correct. Thus, 001 and 101 are decoded as 000 and 111, respectively. This rule corrects words with single errors, but double or triple errors result in a decoding error with probability

$$P_{we} = P(2, 3) + P(3, 3) = 3\alpha^2 - 2\alpha^3$$

Since $P_e = \alpha$ would be the error probability without coding, we see that either decoding scheme for the triple-repetition code greatly improves reliability if, say, $\alpha \le 0.01$. However, this improvement is gained at the cost of reducing the message bit rate by a factor of 1/3.

More efficient codes are based on the notion of **parity.** The parity of a binary word is said to be even when the word contains an even number of 1s, while odd parity means an odd number of 1s. The codewords for an **error-detecting parity-check code** are constructed with $n - 1$ message bits and one check bit chosen such that all codewords have the same parity. With $n = 3$ and even parity, the valid codewords are 000, 011, 101, and 110, the last bit in each word being the parity check. When a received word has odd parity, 001 for instance, we immediately know that it contains a transmission error—or three errors or, in general, an odd number of errors. Error correction is not possible because we don't know where the errors fall within the word. Furthermore, an even number of errors preserves valid parity and goes unnoticed.

Under the condition $\alpha \ll 1$, double errors occur far more often than four or more errors per word. Hence, the probability of an undetected error in an n-bit parity-check codeword is

$$P_{we} \approx P(2, n) \approx \frac{n(n - 1)}{2}\alpha^2 \qquad [2]$$

For comparison purposes, **uncoded** transmission of words containing $n - 1$ message bits would have

$$P_{uwe} = 1 - P(0, n - 1) \approx (n - 1)\alpha$$

Thus, if $n = 10$ and $\alpha = 10^{-3}$, then $P_{uwe} \approx 10^{-2}$ whereas coding yields $P_{we} \approx 5 \times 10^{-5}$ with a rate reduction of just 9/10. These numbers help explain the popularity of parity checking for error detection in computer systems.

As an example of parity checking for *error correction*, Fig. 13.1–1 illustrates an error-correcting scheme in which the codeword is formed by arranging k message bits in a square array whose rows and columns are checked by $2\sqrt{k}$ parity bits. A transmission error in one message bit causes a row and column parity failure with the error at the intersection, so single errors can be corrected. This code also detects double errors.

Interleaving

Throughout the foregoing discussion we've assumed that transmission errors appear randomly and independently in a codeword. This assumption hold for errors caused

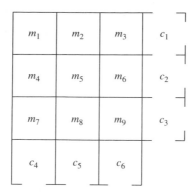

Figure 13.1–1   Square array for error correction by parity checking.

by white or filtered white noise. But *impulse noise,* produced by lightning and switching transients, causes errors to occur in **bursts** that span several successive bits. Burst errors also appear when radio-transmission systems suffer from rapid fading or if the channel has memory where a given data symbol is a function of present and past symbols. Such multiple errors wreak havoc on the performance of conventional codes and must be combated by special techniques. Conventional error control methods such as parity checking are designed for errors that are isolated or statistically independent events.

One solution is to spread out the transmitted codewords using a system of **interleaving** as represented in Fig. 13.1–2. Here the message bits are dispersed with the curved line connecting the original message and parity bit sequence in one parity word. These check bits enable us to check for single errors. Consider the case of a system that can only correct single errors. An error burst occurs such that several successive bits have been corrupted. If this happens to the original bit sequence, the system would be overwhelmed and unable to correct the problem. However, let's say the same error burst occurs in the interleaved transmission. In this case successive bits from *different* message words have been corrupted. When received, the bit sequence is reordered to its original form and then the FEC can correct the faulty bits. Therefore, our single error correction system is able to fix several errors.

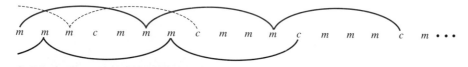

Figure 13.1–2   Interleaved check bits for error control with burst errors.

Code Vectors and Hamming Distance

Rather than continuing a piecemeal survey of particular codes, we now introduce a more general approach in terms of code **vectors.** An arbitrary n-bit codeword can be visualized in an n-dimensional space as a vector whose elements or coordinates equal the bits in the codeword. We thus write the codeword 101 in row-vector notation as $X = (1\ 0\ 1)$. Figure 13.1–3 portrays all possible 3-bit codewords as dots corresponding to the vector tips in a three-dimensional space. The solid dots in part (a) represent the triple-repetition code, while those in part (b) represent a parity-check code.

Notice that the triple-repetition code vectors have greater separation than the parity-check code vectors. This separation, measured in terms of the **Hamming distance,** has direct bearing on the error-control power of a code. The Hamming distance $d(X, Y)$ between two vectors X and Y is defined to equal the number of different elements. For instance, if $X = (1\ 0\ 1)$ and $Y = (1\ 1\ 0)$ then $d(X, Y) = 2$ because the second and third elements are different.

The **minimum distance** d_{min} of a particular code is the smallest Hamming distance between valid code vectors. Consequently, error detection is always possible when the number of transmission errors in a codeword is less than d_{min} so the erroneous word is not a valid vector. Conversely, when the number of errors equals or exceeds d_{min}, the erroneous word may correspond to another valid vector and the errors cannot be detected.

Further reasoning along this line leads to the following distance requirements for various degrees of error control capability:

Detect up to ℓ errors per word $\qquad\qquad\qquad\qquad\qquad d_{min} \geq \ell + 1$ **[3a]**

Correct up to t errors per word $\qquad\qquad\qquad\qquad\qquad d_{min} \geq 2t + 1$ **[3b]**

Correct up to t errors and detect $\ell > t$ errors per word $\qquad d_{min} \geq t + \ell + 1$ **[3c]**

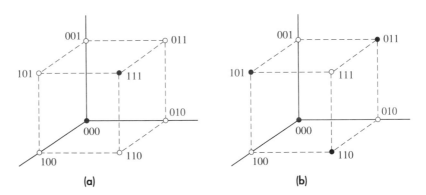

(a)   **(b)**

Figure 13.1–3   Vectors representing 3-bit codewords. (a) Triple-repetition code; (b) parity-check code.

By way of example, we see from Fig. 13.1–3 that the triple-repetition code has $d_{\min} = 3$. Hence, this code could be used to detect $\ell \leq 3 - 1 = 2$ errors per word or to correct $t \leq (3 - 1)/2 = 1$ error per word—in agreement with our previous observations. A more powerful code with $d_{\min} = 7$ could correct triple errors or it could correct double errors and detect quadruple errors.

The power of a code obviously depends on the number of bits added to each codeword for error-control purposes. In particular, suppose that the codewords consist of $k < n$ message bits and $n - k$ parity bits checking the message bits. This structure is known as an (n, k) **block code.** The minimum distance of an (n, k) block code is upper-bounded by

$$d_{\min} \leq n - k + 1 \qquad [4]$$

and the code's efficiency is measured by the **code rate**

$$R_c \triangleq k/n \qquad [5]$$

Regrettably, the upper bound in Eq. (4) is realized only by repetition codes, which have $k = 1$ and very inefficient code rate $R_c = 1/n$. Considerable effort has thus been devoted to the search for powerful and reasonably efficient codes, a topic we'll return to in the next section.

FEC Systems

Now we're prepared to examine the forward error correction system diagrammed in Fig. 13.1–4. Message bits come from an information source at rate r_b. The encoder takes blocks of k message bits and constructs an (n, k) block code with code rate $R_c = k/n < 1$. The bit rate on the channel therefore must be greater than r_b, namely

$$r = (n/k)r_b = r_b/R_c \qquad [6]$$

The code has $d_{\min} = 2t + 1 \leq n - k + 1$, and the decoder operates strictly in an error-correction mode.

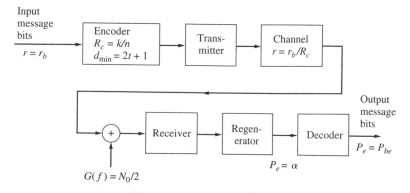

Figure 13.1–4   FEC system.

We'll investigate the performance of this FEC system when additive white noise causes random errors with probability $\alpha \ll 1$. The value of α depends, of course, on the signal energy and noise density at the receiver. If E_b represents the average energy per *message* bit, then the average energy per *code* bit is $R_c E_b$ and the ratio of bit energy to noise density is

$$\gamma_c \triangleq R_c E_b / N_0 = R_c \gamma_b \qquad \text{[7]}$$

where $\gamma_b = E_b / N_0$. Our performance criterion will be the probability of output message-bit errors, denoted by P_{be} to distinguish it from the word error probability P_{we}.

The code always corrects up to t errors per word and some patterns of more than t errors may also be correctable, depending upon the specific code vectors. Thus, the probability of a decoding word error is upper-bounded by

$$P_{we} \leq \sum_{i=t+1}^{n} P(i, n)$$

For a rough but reasonable performance estimate, we'll take the approximation

$$P_{we} \approx P(t + 1, n) \approx \binom{n}{t + 1} \alpha^{t+1} \qquad \text{[8]}$$

which means that an uncorrected word typically has $t + 1$ bit errors. On the average, there will be $(k/n)(t + 1)$ message-bit errors per uncorrected word, the remaining errors being in check bits. When Nk bits are transmitted in $N \gg 1$ words, the expected total number of erroneous message bits at the output is $(k/n)(t + 1)NP_{we}$. Hence,

$$P_{be} = \frac{t + 1}{n} P_{we} \approx \binom{n - 1}{t} \alpha^{t+1} \qquad \text{[9]}$$

in which we have used Eq. (1b) to combine $(t + 1)/n$ with the binomial coefficient.

If the noise has a gaussian distribution and the transmission system has been optimized (i.e., polar signaling and matched filtering), then the transmission error probability is given by

$$\alpha = Q\left(\sqrt{2\gamma_c}\right) = Q\left(\sqrt{2R_c \gamma_b}\right) \qquad \text{[10]}$$

$$\approx (4\pi R_c \gamma_b)^{-1/2} e^{-R_c \gamma_b} \qquad R_c \gamma_b \geq 5$$

The gaussian tail approximation invoked here follows from Eq. (9), Sect. 8.4, and is consistent with the assumption that $\alpha \ll 1$. Thus, our final result for the output error probability of the FEC system becomes

$$P_{be} = \binom{n - 1}{t} \left[Q\left(\sqrt{2R_c \gamma_b}\right)\right]^{t+1} \qquad \text{[11]}$$

$$\approx \binom{n - 1}{t} (4\pi R_c \gamma_b)^{-(t+1)/2} e^{-(t+1)R_c \gamma_b}$$

Uncoded transmission on the same channel would have

$$P_{ube} = Q\left(\sqrt{2\gamma_b}\right) \approx (4\pi\gamma_b)^{-1/2}e^{-\gamma_b} \qquad \text{[12]}$$

since the signaling rate can be decreased from r_b/R_c to r_b.

A comparison of Eqs. (11) and (12) brings out the importance of the code parameters $t = (d_{\min} - 1)/2$ and $R_c = k/n$. The added complexity of an FEC system is justified provided that t and R_c yield a value of P_{be} significantly less than P_{ube}. The exponential approximations show that this essentially requires $(t + 1)R_c > 1$. Hence, a code that only corrects single or double errors should have a relatively high code rate, while more powerful codes may succeed despite lower code rates. The channel parameter γ_b also enters into the comparison, as demonstrated by the following example.

Suppose we have a (15, 11) block code with $d_{\min} = 3$, so $t = 1$ and $R_c = 11/15$. An FEC system using this code would have $\alpha = Q\left[\sqrt{(22/15)\gamma_b}\,\right]$ and $P_{be} = 14\alpha^2$, whereas uncoded transmission on the same channel would yield $P_{ube} = Q(\sqrt{2\gamma_b})$. These three probabilities are plotted versus γ_b in dB in Fig. 13.1–5. If $\gamma_b > 8$ dB, we see that coding decreases the error probability by at least an order of magnitude compared to uncoded transmission. At $\gamma_b = 10$ dB, for instance, uncoded transmission yields $P_{ube} \approx 4 \times 10^{-6}$ whereas the FEC system has $P_{be} \approx 10^{-7}$ even though the higher channel bit rate increases the transmission error probability to $\alpha \approx 6 \times 10^{-5}$.

EXAMPLE 13.1–1

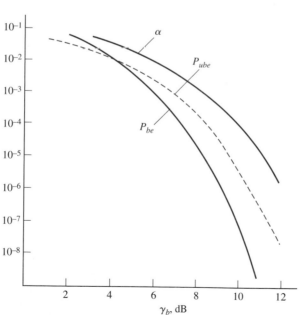

Figure 13.1–5   Curves of error probabilities in Example 13.1–1.

If $\gamma_b < 8$ dB, however, coding does not significantly improve reliability and actually makes matters worse when $\gamma_b < 4$ dB. Furthermore, an uncoded system could achieve better reliability than the FEC system simply by increasing the signal-to-noise ratio about 1.5 dB. Hence, this particular code doesn't save much signal power, but it would be effective if γ_b has a fixed value in the vicinity of 8–10 dB.

EXERCISE 13.1–1   Suppose the system in Example 13.1–1 is operated at $\gamma_b \approx 8$ dB so $\alpha = 0.001$. Evaluate $P(i, n)$ for $i = 0, 1, 2,$ and 3. Do your results support the approximation in Eq. (8)?

ARQ Systems

The automatic-repeat-request strategy for error control is based on error detection and retransmission rather than forward error correction. Consequently, ARQ systems differ from FEC systems in three important respects. First, an (n, k) block code designed for error detection generally requires fewer check bits and has a higher k/n ratio than a code designed for error correction. Second, an ARQ system needs a return transmission path and additional hardware in order to implement repeat transmission of codewords with detected errors. Third, the forward transmission bit rate must make allowance for repeated word transmissions. The net impact of these differences becomes clearer after we describe the operation of the ARQ system represented by Fig. 13.1–6.

Each codeword constructed by the encoder is stored temporarily and transmitted to the destination where the decoder looks for errors. The decoder issues a **positive acknowledgment** (ACK) if no errors are detected, or a **negative acknowledgment** (NAK) if errors are detected. A negative acknowledgment causes the input controller to retransmit the appropriate word from those stored by the input buffer. A particular word may be transmitted just once, or it may be transmitted two or more times, depending on the occurrence of transmission errors. The function of the output controller and buffer is to assemble the output bit stream from the codewords that have been accepted by the decoder.

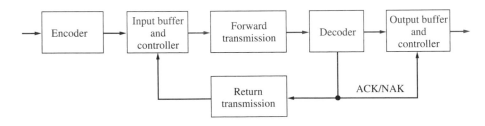

Figure 13.1–6   ARQ system.

Compared to forward transmission, return transmission of the ACK/NAK signal involves a low bit rate and we can reasonably assume a negligible error probability on the return path. Under this condition, all codewords with detected errors are retransmitted as many times as necessary, so the only output errors appear in words with undetected errors. For an (n, k) block code with $d_{\min} = \ell + 1$, the corresponding output error probabilities are

$$P_{we} = \sum_{i=\ell+1}^{n} P(i, n) \approx P(\ell + 1, n) \approx \binom{n}{\ell + 1} \alpha^{\ell+1} \qquad [13]$$

$$P_{be} = \frac{\ell + 1}{n} P_{we} \approx \binom{n - 1}{\ell} \alpha^{\ell+1} \qquad [14]$$

which are identical to the FEC expressions, Eqs. (8) and (9), with ℓ in place of t. Since the decoder accepts words that have either no errors or undetectable errors, the **word retransmission probability** is given by

$$p = 1 - [P(0, n) + P_{we}]$$

But a good error-detecting code should yield $P_{we} \ll P(0, n)$. Hence,

$$p \approx 1 - P(0, n) = 1 - (1 - \alpha)^n \approx n\alpha \qquad [15]$$

where we've used the approximation $(1 - \alpha)^n \approx 1 - n\alpha$ based on $n\alpha \ll 1$.

As for the retransmission process itself, there are three basic ARQ schemes illustrated by the timing diagrams in Fig. 13.1–7. The asterisk marks words received with detected errors which must be retransmitted. The **stop-and-wait** scheme in part *a* requires the transmitter to stop after every word and wait for acknowledgment from the receiver. Just one word needs to be stored by the input buffer, but the transmission time delay t_d in each direction results in an **idle time** of duration $D \geq 2t_d$ between words. Idle time is eliminated by the **go-back-N** scheme in part *b* where codewords are transmitted continuously. When the receiver sends a NAK signal, the transmitter goes back N words in the buffer and retransmits starting from that point. The receiver discards the $N - 1$ intervening words, correct or not, in order to preserve proper sequence. The **selective-repeat** scheme in part *c* puts the burden of sequencing on the output controller and buffer, so that only words with detected errors need to be retransmitted.

Clearly, a selective-repeat ARQ system has the highest *throughput efficiency*. To set this on a quantitative footing, we observe that the total number of transmissions of a given word is a discrete random variable m governed by the event probabilities $P(m = 1) = 1 - p$, $P(m = 2) = p(1 - p)$, and so on. The average number of transmitted words per accepted word is then

$$\bar{m} = 1(1 - p) + 2p(1 - p) + 3p^2(1 - p) + \cdots \qquad [16]$$

$$= (1 - p)(1 + 2p + 3p^2 + \cdots) = \frac{1}{1 - p}$$

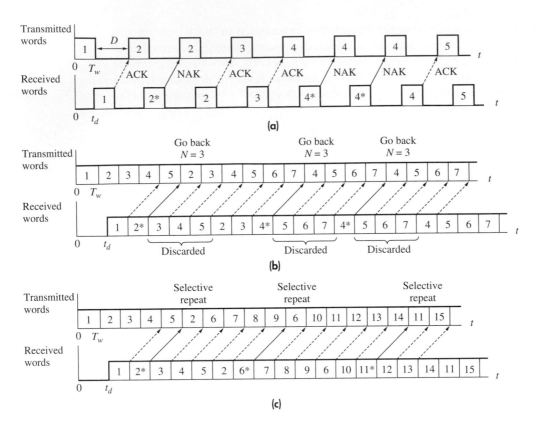

Figure 13.1–7   ARQ schemes. (a) Stop-and-wait; (b) go-back-N; (c) selective-repeat.

since $1 + 2p + 3p^2 + \cdots = (1 - p)^{-2}$. On the average, the system must transmit $n\bar{m}$ bits for every k message bits, so the throughput efficiency is

$$R'_c = \frac{k}{n\bar{m}} = \frac{k}{n}(1 - p) \qquad [17]$$

in which $p \approx n\alpha$, from Eq. (15).

We use the symbol R'_c here to reflect the fact that the forward-transmission bit rate r and the message bit rate r_b are related by

$$r = r_b / R'_c$$

comparable to the relationship $r = r_b/R_c$ in an FEC system. Thus, when the noise has a gaussian distribution, the transmission error probability α is calculated from Eq. (10) using R'_c instead of $R_c = k/n$. Furthermore, if $p \ll 1$, then $R'_c \approx k/n$. But an error-detecting code has a larger k/n ratio than an error-correcting code of equivalent error-control power. Under these conditions, the more elaborate hardware needed for selective-repeat ARQ may pay off in terms of better performance than an FEC system would yield on the same channel.

The expression for $\bar{m}$ in Eq. (16) also applies to a stop-and-wait ARQ system. However, the idle time reduces efficiency by the factor $T_w/(T_w + D)$ where $D \geq 2t_d$ is the round-trip delay and T_w is the word duration given by $T_w = n/r \leq k/r_b$. Hence,

$$R'_c = \frac{k}{n} \frac{1 - p}{1 + (D/T_w)} \leq \frac{k}{n} \frac{1 - p}{1 + (2t_d r_b/k)} \qquad [18]$$

in which the upper bound comes from writing $D/T_w \geq 2t_d r_b/k$.

A go-back-N ARQ system has no idle time, but N words must be retransmitted for each word with detected errors. Consequently, we find that

$$\bar{m} = 1 + \frac{Np}{1 - p} \qquad [19]$$

and

$$R'_c = \frac{k}{n} \frac{1 - p}{1 - p + Np} \leq \frac{k}{n} \frac{1 - p}{1 - p + (2t_d r_b/k)p} \qquad [20]$$

where the upper bound reflects the fact that $N \geq 2t_d/T_w$.

Unlike selective-repeat ARQ, the throughput efficiency of the stop-and-wait and go-back-N schemes depends on the round-trip delay. Equations (18) and (20) reveal that both of these schemes have reasonable efficiency if the delay and bit rate are such that $2t_d r_b \ll k$. However, stop-and-wait ARQ has very low efficiency when $2t_d r_b \geq k$, whereas the go back-N scheme may still be satisfactory provided that the retransmission probability p is small enough.

Finally, we should at least describe the concept of **hybrid** ARQ systems. These systems consist of an FEC subsystem within the ARQ framework, thereby combining desirable properties of both error-control strategies. For instance, a hybrid ARC system might employ a block code with $d_{\min} = t + \ell + 1$, so the decoder can correct up to t errors per word and detect but not correct words with $\ell > t$ errors. Error correction reduces the number of words that must be retransmitted, thereby increasing the throughput without sacrificing the higher reliability of ARQ.

Suppose a selective-repeat ARQ system uses a simple parity-check code with $k = 9$, $n = 10$, and $\ell = 1$. The transmission channel is corrupted by gaussian noise and we seek the value of γ_b needed to get $P_{be} \approx 10^{-5}$. Equation (14) yields the required transmission error probability $\alpha = (10^{-5}/9)^{1/2} \approx 1.1 \times 10^{-3}$, and the corresponding word retransmission probability in Eq. (15) is $p \approx 10\alpha \ll 1$. Hence, the throughput efficiency will be $R'_c \approx k/n = 0.9$, from Eq. (17). Since $\alpha = Q(\sqrt{2R'_c\gamma_b})$ we call upon the plot of Q in Table T.6 to obtain our final result $\gamma_b \approx 3.1^2/1.8 = 5.3$ or 7.3 dB. As a comparison, Fig. 13.1–5 shows that uncoded transmission would have

EXAMPLE 13.1–2

$P_{ube} \approx 6 \times 10^{-4}$ if $\gamma_b = 7.3$ dB and requires $\gamma_b \approx 9.6$ dB to get $P_{ube} = 10^{-5}$. The ARQ system thus achieves a power saving of about 2.3 dB.

EXERCISE 13.1–2   Assume that the system in Example 13.1–2 has $r_b = 50$ kbps and $t_d = 0.2$ ms. By calculating R'_c show that the go-back-N scheme would be acceptable but not the stop-and-wait scheme when channel limitations require $r \leq 100$ kbps.

13.2   LINEAR BLOCK CODES

This section describes the structure, properties, and implementation of block codes. We start with a matrix representation of the encoding process that generates the check bits for a given block of message bits. Then we use the matrix representation to investigate decoding methods for error detection and correction. The section closes with a brief introduction to the important class of cyclic block codes.

Matrix Representation of Block Codes

An (n, k) block code consists of n-bit vectors, each vector corresponding to a unique block of $k < n$ message bits. Since there are 2^k different k-bit message blocks and 2^n possible n-bit vectors, the fundamental strategy of block coding is to choose the 2^k code vectors such that the minimum distance is as large as possible. But the code should also have some structure that facilitates the encoding and decoding processes. We'll therefore focus on the class of **systematic linear** block codes.

Let an arbitrary code vector be represented by

$$X = (x_1 \, x_2 \quad \cdots \quad x_n)$$

where the elements $x_1, x_2, \ldots,$ are, of course, binary digits. A code is *linear* if it includes the all-zero vector and if the sum of any two code vectors produces another vector in the code. The *sum* of two vectors, say X and Z, is defined as

$$X + Z \triangleq (x_1 \oplus z_1 \; x_2 \oplus z_2 \quad \cdots \quad x_n \oplus z_n) \tag{1}$$

in which the elements are combined according to the rules of mod-2 addition given in Eq. (2), Sect. 11.4.

As a consequence of linearity, we can determine a code's minimum distance by the following argument. Let the number of nonzero elements of a vector X be symbolized by $w(X)$, called the vector *weight*. The Hamming distance between any two code vectors X and Z is then

$$d(X, Z) = w(X + Z)$$

Error correction necessarily entails more circuitry but it, too, can be based on the syndrome. We develop the decoding method by introducing an *n*-bit **error vector** *E* whose nonzero elements mark the positions of transmission errors in *Y*. For instance, if $X = (1\ 0\ 1\ 1\ 0)$ and $Y = (1\ 0\ 0\ 1\ 1)$ then $E = (0\ 0\ 1\ 0\ 1)$. In general,

$$Y = X + E \qquad\qquad [11a]$$

and, conversely,

$$X = Y + E \qquad\qquad [11b]$$

since a second error in the same bit location would cancel the original error. Substituting $Y = X + E$ into $S = YH^T$ and invoking Eq. (9), we obtain

$$S = (X + E)H^T = XH^T + EH^T = EH^T \qquad\qquad [12]$$

which reveals that the syndrome depends entirely on the error pattern, not the specific transmitted vector.

However, there are only 2^q different syndromes generated by the 2^n possible *n*-bit error vectors, including the no-error case. Consequently, a given syndrome does not uniquely determine *E*. Or, putting this another way, we can correct just $2^q - 1$ patterns with one or more errors, and the remaining patterns are *uncorrectable*. We should therefore design the decoder to correct the $2^q - 1$ most likely error patterns— namely those patterns with the *fewest* errors, since single errors are more probable than double errors, and so forth. This strategy, known as **maximum-likelihood decoding,** is optimum in the sense that it minimizes the word-error probability. Maximum-likelihood decoding corresponds to choosing the code vector that has the smallest Hamming distance from the received vector.

To carry out maximum-likelihood decoding, you must first compute the syndromes generated by the $2^q - 1$ most probable error vectors. The **table-lookup decoder** diagrammed in Fig. 13.2–2 then operates as follows. The decoder calculates *S* from the received vector *Y* and looks up the assumed error vector $\hat{E}$ stored in the table. The sum $Y + \hat{E}$ generated by exclusive-OR gates finally constitutes the decoded word. If there are no errors, or if the errors are uncorrectable, then $S = (0\ 0 \cdots 0)$ so $Y + \hat{E} = Y$. The check bits in the last *q* elements of $Y + \hat{E}$ may be omitted if they are of no further interest.

The relationship between syndromes and error patterns also sheds some light on the design of error-correcting codes, since each of the $2^q - 1$ nonzero syndromes must represent a specific error pattern. Now there are $\binom{n}{1} = n$ single-error patterns for an *n*-bit word, $\binom{n}{2}$ double-error patterns, and so forth. Hence, if a code is to correct up to *t* errors per word, *q* and *n* must satisfy

$$2^q - 1 \geq n + \binom{n}{2} + \cdots + \binom{n}{t} \qquad\qquad [13]$$

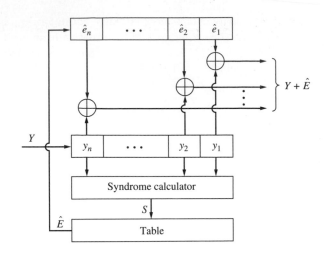

Figure 13.2–2   Table-lookup decoder.

In the particular case of a single-error-correcting code, Eq. (13) reduces to $2^q - 1 \geq n$. Furthermore, when E corresponds to a single error in the jth bit of a codeword, we find from Eq. (12) that S is identical to the jth row of H^T. Therefore, to provide a distinct syndrome for each single-error pattern and for the no-error pattern, the rows of H^T (or columns of H) must all be different and each must contain at least one nonzero element. The generator matrix of a Hamming code is designed to satisfy this requirement on H, while q and n satisfy $2^q - 1 = n$.

EXAMPLE 13.2–2

Let's apply table-lookup decoding to a $(7, 4)$ Hamming code used for single-error correction. From Eq. (8) and the P submatrix given in Example 13.2–1, we obtain the 3×7 parity-check matrix

$$H = [P^T \mid I_q] = \begin{bmatrix} 1 & 1 & 1 & 0 & 1 & 0 & 0 \\ 0 & 1 & 1 & 1 & 0 & 1 & 0 \\ 1 & 1 & 0 & 1 & 0 & 0 & 1 \end{bmatrix}$$

There are $2^3 - 1 = 7$ correctable single-error patterns, and the corresponding syndromes listed in Table 13.2–2 follow directly from the columns of H. To accommodate this table the decoder needs to store only $(q + n) \times 2^q = 80$ bits.

But suppose a received word happens to have two errors, such that $E = (1\ 0\ 0\ 0\ 0\ 1\ 0)$. The decoder calculates $S = YH^T = EH^T = (1\ 1\ 1)$ and the syndrome table gives the assumed single-error pattern $\hat{E} = (0\ 1\ 0\ 0\ 0\ 0\ 0)$. The decoded output word $Y + \hat{E}$ therefore contains *three* errors, the two transmission errors plus the erroneous correction added by the decoder.

If multiple transmission errors per word are sufficiently infrequent, we need not be concerned about the occasional extra errors committed by the decoder. If multi-

Table 13.2–2 Syndromes for the (7, 4) Hamming code

S			$\hat{E}$						
0	0	0	0	0	0	0	0	0	0
1	0	1	1	0	0	0	0	0	0
1	1	1	0	1	0	0	0	0	0
1	1	0	0	0	1	0	0	0	0
0	1	1	0	0	0	1	0	0	0
1	0	0	0	0	0	0	1	0	0
0	1	0	0	0	0	0	0	1	0
0	0	1	0	0	0	0	0	0	1

ple errors are frequent, a more powerful code would be required. For instance, an *extended* Hamming code has an additional check bit that provides double-error detection along with single-error correction.

Use Eqs. (8) and (10) to show that the *j*th bit of *S* is given by

$$s_j = y_1 p_{1j} \oplus y_2 p_{2j} \oplus \cdots \oplus y_k p_{kj} \oplus y_{k+j}$$

Then diagram the syndrome-calculation circuit for a (7, 4) Hamming code, and compare it with Fig. 13.2–1.

EXERCISE 13.2–2

Cyclic Codes

The code for a forward-error-correction system must be capable of correcting $t \geq 1$ errors per word. It should also have a reasonably efficient code rate $R_c = k/n$. These two parameters are related by the inequality

$$1 - R_c \geq \frac{1}{n} \log_2 \left[\sum_{i=0}^{t} \binom{n}{i} \right]$$ [14]

which follows from Eq. (13) with $q = n - k = n(1 - R_c)$. This inequality underscores the fact that if we want $R_c \approx 1$, we must use codewords with $n \gg 1$ and $k \gg 1$. However, the hardware requirements for encoding and decoding long codewords may be prohibitive unless we impose further structural conditions on the code. **Cyclic codes** are a subclass of linear block codes with a cyclic structure that leads to more practical implementation. Thus, block codes used in FEC systems are almost always cyclic codes.

To describe a cyclic code, we'll find it helpful to change our indexing scheme and express an arbitrary *n*-bit code vector in the form

$$X = (x_{n-1} \; x_{n-2} \; \cdots \; x_1 \; x_0) \tag{15}$$

Now suppose that X has been loaded into a shift register with feedback connection from the first to last stage. Shifting all bits one position to the left yields the **cyclic shift** of X, written as

$$X' \triangleq (x_{n-2} \;\; x_{n-3} \;\; \cdots \;\; x_1 \;\; x_0 \;\; x_{n-1}) \tag{16}$$

A second shift produces $X'' = (x_{n-3} \;\; \cdots \;\; x_1 \;\; x_0 \;\; x_{n-1} \;\; x_{n-2})$, and so forth. A linear code is *cyclic* if every cyclic shift of a code vector X is another vector in the code.

This cyclic property can be treated mathematically by associating a code vector X with the *polynomial*

$$X(p) = x_{n-1} p^{n-1} + x_{n-2} p^{n-2} + \cdots + x_1 p + x_0 \tag{17}$$

where p is an arbitrary real variable. The powers of p denote the positions of the codeword bits represented by the corresponding coefficients of p. Formally, binary code polynomials are defined in conjunction with **Galois fields,** a branch of modern algebra that provides the theory needed for a complete treatment of cyclic codes. For our informal overview of cyclic codes we'll manipulate code polynomials using ordinary algebra modified in two respects. First, to be in agreement with our earlier definition for the sum of two code vectors, the sum of two polynomials is obtained by *mod-2 addition* of their respective coefficients. Second, since all coefficients are either 0 or 1, and since $1 \oplus 1 = 0$, the *subtraction* operation is the same as mod-2 *addition*. Consequently, if $X(p) + Z(p) = 0$ then $X(p) = Z(p)$.

We develop the polynomial interpretation of cyclic shifting by comparing

$$pX(p) = x_{n-1} p^{n} + x_{n-2} p^{n-1} + \cdots + x_1 p^2 + x_0 p$$

with the shifted polynomial

$$X'(p) = x_{n-2} p^{n-1} + \cdots + x_1 p^2 + x_0 p + x_{n-1}$$

If we sum these polynomials, noting that $(x_1 \oplus x_1)p^2 = 0$, and so on, we get

$$pX(p) + X'(p) = x_{n-1} p^{n} + x_{n-1}$$

and hence

$$X'(p) = pX(p) + x_{n-1}(p^{n} + 1) \tag{18}$$

Iteration yields similar expressions for multiple shifts.

The polynomial $p^{n} + 1$ and its factors play major roles in cyclic codes. Specifically, an (n, k) cyclic code is defined by a **generator polynomial** of the form

$$G(p) = p^{q} + g_{q-1} p^{q-1} + \cdots + g_1 p + 1 \tag{19}$$

where $q = n - k$ and the coefficients are such that $G(p)$ is a factor of $p^{n} + 1$. Each codeword then corresponds to the polynomial product

$$X(p) = Q_M(p)G(p) \tag{20}$$

in which $Q_M(p)$ represents a block of k message bits. All such codewords satisfy the cyclic condition in Eq. (18) since $G(p)$ is a factor of both $X(p)$ and $p^n + 1$.

Any factor of $p^n + 1$ that has degree q may serve as the generator polynomial for a cyclic code, but it does not necessarily generate a good code. Table 13.2–3 lists the generator polynomials of selected cyclic codes that have been demonstrated to possess desirable parameters for FEC systems. The table includes some cyclic Hamming codes, the famous Golay code, and a few members of the important family of BCH codes discovered by Bose, Chaudhuri, and Hocquenghem. The entries under $G(p)$ denote the polynomial's coefficients; thus, for instance, 1 0 1 1 means that $G(p) = p^3 + 0 + p + 1$.

Cyclic codes may be systematic or nonsystematic, depending on the term $Q_M(p)$ in Eq. (20). For a systematic code, we define the message-bit and check-bit polynomials

$$M(p) = m_{k-1} p^{k-1} + \cdots + m_1 p + m_0$$

$$C(p) = c_{q-1} p^{q-1} + \cdots + c_1 p + c_0$$

and we want the codeword polynomials to be

$$X(p) = p^q M(p) + C(p) \tag{21}$$

Equations (20) and (21) therefore require $p^q M(p) + C(p) = Q_M(p)G(p)$, or

$$\frac{p^q M(p)}{G(p)} = Q_M(p) + \frac{C(p)}{G(p)} \tag{22a}$$

This expression says that $C(p)$ equals the *remainder* left over after dividing $p^q M(p)$ by $G(p)$, just as 14 divided by 3 leaves a remainder of 2 since $14/3 = 4 + 2/3$. Symbolically, we write

$$C(p) = \text{rem} \left[\frac{p^q M(p)}{G(p)} \right] \tag{22b}$$

where rem [ ] stands for the remainder of the division within the brackets.

Table 13.2–3   Selected cyclic codes

Type	n	k	R_c	$d_{\min}$			$G(p)$			
Hamming	7	4	0.57	3					1	011
Codes	15	11	0.73	3					10	011
	31	26	0.84	3					100	101
BCH	15	7	0.46	5				111	010	001
Codes	31	21	0.68	5			11	101	101	001
	63	45	0.71	7	1	111	000	001	011	001	111
Golay	23	12	0.52	7				101	011	100	011
Code										

The division operation needed to generate a systematic cyclic code is easily and efficiently performed by the shift-register encoder diagrammed in Fig. 13.2–3. Encoding starts with the feedback switch closed, the output switch in the message-bit position, and the register initialized to the all-zero state. The k message bits are shifted into the register and simultaneously delivered to the transmitter. After k shift cycles, the register contains the q check bits. The feedback switch is now opened and the output switch is moved to deliver the check bits to the transmitter.

Syndrome calculation at the receiver is equally simple. Given a received vector Y, the syndrome is determined from

$$S(p) = \text{rem} \left[\frac{Y(p)}{G(p)} \right] \qquad \text{[23]}$$

If $Y(p)$ is a valid code polynomial, then $G(p)$ will be a factor of $Y(p)$ and $Y(p)/G(p)$ has zero remainder. Otherwise we get a nonzero syndrome polynomial indicating detected errors.

Besides simplified encoding and syndrome calculation, cyclic codes have other advantages over noncyclic block codes. The foremost advantage comes from the ingenious error-correcting decoding methods that have been devised for specific cyclic codes. These methods eliminate the storage needed for table-lookup decoding and thus make it practical to use powerful and efficient codes with $n \gg 1$. Another advantage is the ability of cyclic codes to detect error bursts that span many successive bits.

EXAMPLE 13.2–3

Consider the cyclic $(7, 4)$ Hamming code generated by $G(p) = p^3 + 0 + p + 1$. We'll use long division to calculate the check-bit polynomial $C(p)$ when $M = (1\ 1\ 0\ 0)$. We first write the message-bit polynomial

$$M(p) = p^3 + p^2 + 0 + 0$$

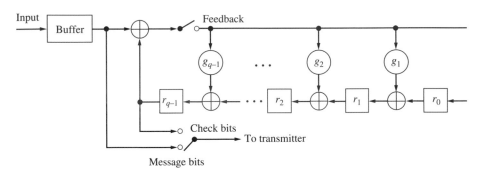

Figure 13.2–3   Shift-register encoder.

so $p^q M(p) = p^3 M(p) = p^6 + p^5 + 0 + 0 + 0 + 0 + 0$. Next, we divide $G(p)$ into $p^q M(p)$, keeping in mind that subtraction is the same as addition in mod-2 arithmetic. Thus,

$$
\begin{array}{r}
Q_M(p) = p^3 + p^2 + p + 0 \\
p^3 + 0 + p + 1 \, \overline{)\, p^6 + p^5 + 0 \ + 0 \ + 0 \ + 0 + 0} \\
\underline{p^6 + 0 \ + p^4 + p^3} \\
p^5 + p^4 + p^3 + 0 \\
\underline{p^5 + 0 \ + p^3 + p^2} \\
p^4 + 0 \ + p^2 + 0 \\
\underline{p^4 + 0 \ + p^2 + p} \\
0 + 0 + p + 0 \\
\underline{0 + 0 + 0 + 0} \\
C(p) = 0 + p + 0
\end{array}
$$

so the complete code polynomial is

$$
X(p) = p^3 M(p) + C(p) = p^6 + p^5 + 0 + 0 + 0 + p + 0
$$

which corresponds to the codeword

$$
X = (1 \quad 1 \quad 0 \quad 0 \mid 0 \quad 1 \quad 0)
$$

You'll find this codeword back in Table 13.2–1, where you'll also find the cyclic shift $X' = (1\,0\,0\,0 \mid 1\,0\,1)$ and all multiple shifts.

Finally, Fig. 13.2–4 shows the shift-register encoder and the register bits for each cycle of the encoding process when the input is $M = (1\,1\,0\,0)$. After four shift cycles, the register holds $C = (0\,1\,0)$—in agreement with our manual division.

Remember that an error is only detected when $Y(p)$ is not evenly divisible by $G(p)$. If $G(p)$ is not properly chosen, errors could occur in $Y(p)$ making it evenly divisible by $G(p)$ and, therefore, some errors could go undetected. A class of cyclic codes we call the **cyclic redundancy codes** (CRCs) have been designed to minimize the possibility of errors slipping through, particularly for burst-error detection. CRCs are also characterized by any end-around-cyclic shift of a codeword that produces another codeword. Their structure makes for efficient coding and decoding. Some of the CRCs are given in Table 13.2–4.

With CRCs, the following error types will be detected: (*a*) all single bit errors; (*b*) any odd number of errors, assuming $p + 1$ is a factor of $G(p)$; (*c*) burst errors of length not exceeding q, where q is the number of check bits; (*d*) double errors if $G(p)$ contains at least three 1s.

If all patterns of error bursts are equally likely, then for bursts of length $q + 1$ the probability of the error being undetected is $1/2^{q-1}$, and if the burst length is greater than $q + 1$, then the probability of it going undetected is $1/2^q$. See Peterson and Brown (1961) for a more extensive discussion on the error detection capabilities of CRCs.

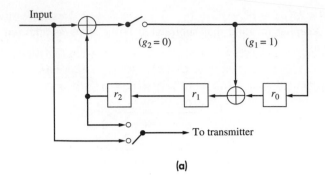

Input

$(g_2 = 0)$ $(g_1 = 1)$

r_2 r_1 r_0

To transmitter

(a)

Input bit	Register bits before shifts			Register bits after shifts		
				$r_2' =$	$r_1' =$	$r_0' =$
m	r_2	r_1	r_0	r_1	$r_0 \oplus r_2 \oplus m$	$r_2 \oplus m$
1	0	0	0	0	1	1
1	0	1	1	1	0	1
0	1	0	1	0	0	1
0	0	0	1	0	1	0

(b)

Figure 13.2–4   (a) Shift-register encoder for (7, 4) Hamming code; (b) register bits when $M = (1\ 1\ 0\ 0)$.

Table 13.2–4   Cyclic redundancy codes

Code	$G(p)$	$q = n - k$
CRC-8	100 000 111	8
CRC-12	1 100 000 001 111	12
CRC-16	11 000 000 000 000 101	16
CRC-CCIT	10 001 000 000 100 001	16

EXERCISE 13.2–3   We want to transmit an ASCII letter "J," which in binary is 1001010, and then be able to check for errors using CRC-8 code. Determine the transmitted sequence X and then show how the receiver can detect errors in the two left-most message bits using the CRC calculation of Eq. (23).

M-ary Codes

A subset of the BCH codes that perform well under burst error conditions and can be used with *M*-ary modulation systems are the **Reed-Solomon** (RS) codes. These are nonbinary codes that are members of an *M*-ary alphabet. If we use an *m*-bit digital encoder we will then have an alphabet of $M = 2^m$ symbols. An RS code's minimum distance is

$$d_{min} = n - k + 1 \qquad \text{[24]}$$

and, as with the binary codes, n is the total number of symbols in the code block and k is the number of message symbols. RS codes are capable of correcting t or fewer symbol errors with

$$t = \frac{d_{min} - 1}{2} = \frac{n - k}{2} \qquad \text{[25]}$$

With an *M* symbol alphabet, we have $n = 2^m - 1$ and thus $k = 2^m - 1 - 2t$.

Consider a binary BCH $(n, k) = (7, 4)$ system. From this $d_{min} = 3$ bits. Let's say we have an RS $(n, k) = (7, 4)$ system where each symbol has $m = 3$ bits. Thus $d_{min} = n - k = 7 - 4 = 3$ symbols. However, in terms of bits, the RS code gives us a distance of $(2^3)^3 = 512$ bits, whereas the binary BCH code only gave us a 3-bit distance. Thus an RS system has the potential for large code distances. In general an *M*-ary (n, k) RS code with *m* bits per symbol has $(2^m)^n$ code vectors out of which $(2^m)^k$ are message vectors. With more symbol bits, the greater the possible code distance.

EXAMPLE 13.2–4

How many symbols in error can a (63, 15) RS code correct?

EXERCISE 13.2–4

13.3   CONVOLUTIONAL CODES

Convolutional codes have a structure that effectively extends over the entire transmitted bit stream, rather than being limited to codeword blocks. The convolutional structure is especially well suited to space and satellite communication systems that require simple encoders and achieve high performance by sophisticated decoding methods. Our treatment of this important family of codes consists of selected examples that introduce the salient features of convolutional encoding and decoding.

Convolutional Encoding

The fundamental hardware unit for convolutional encoding is a tapped shift register with $L + 1$ stages, as diagrammed in Fig. 13.3–1. Each tap gain g is a binary digit

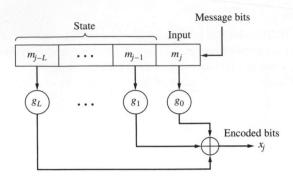

Figure 13.3–1   Tapped shift register for convolutional encoding.

representing a short-circuit connection or an open circuit. The message bits in the register are combined by mod-2 addition to form the encoded bit

$$x_j = m_{j-L}g_L \oplus \cdots \oplus m_{j-1}g_1 \oplus m_j g_0 \qquad \text{[1]}$$

$$= \sum_{i=0}^{L} m_{j-i} g_i \qquad (\text{mod-2})$$

The name *convolutional encoding* comes from the fact that Eq. (1) has the form of a *binary convolution*, analogous to the convolutional integral

$$x(t) = \int m(t - \lambda)g(\lambda) \, d\lambda$$

Notice in Eq. (1) that x_j depends on the current input m_j and on the *state* of the register defined by the previous L message bits. Also notice that a particular message bit influences a span of $L + 1$ successive encoded bits as it shifts through the register.

To provide the extra bits needed for error control, a complete convolutional encoder must generate output bits at a rate greater than the message bit rate r_b. This is achieved by connecting two or more mod-2 summers to the register and interleaving the encoded bits via a commutator switch. For example, the encoder in Fig. 13.3–2 generates $n = 2$ encoded bits

$$x_j' = m_{j-2} \oplus m_{j-1} \oplus m_j \quad x_j'' = m_{j-2} \oplus m_j \qquad \text{[2]}$$

which are interleaved by the switch to produce the output stream

$$X = x_1' x_1'' x_2' x_2'' x_3' x_3'' \cdots$$

The output bit rate is therefore $2r_b$ and the code rate is $R_c = 1/2$—like an (n, k) block code with $R_c = k/n = 1/2$.

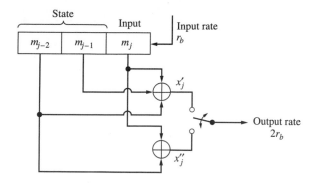

Figure 13.3–2   Convolutional encoder with $n = 2$, $k = 1$, and $L = 2$.

However, unlike a block code, the input bits have not been grouped into words. Instead, each message bit influences a span of $n(L + 1) = 6$ successive output bits. The quantity $n(L + 1)$ is called the **constraint length** measured in terms of encoded output bits, whereas L is the encoder's **memory** measured in terms of input message bits. We say that this encoder produces an (n, k, L) convolutional code† with $n = 2$, $k = 1$, and $L = 2$.

Three different but related graphical representations have been devised for the study of convolutional encoding: the code tree, the code trellis, and the state diagram. We'll present each of these for our $(2, 1, 2)$ encoder in Fig. 13.3–2, starting with the code tree. In accordance with normal operating procedure, we presume that the register has been cleared to contain all 0s when the first message bit m_1 arrives. Hence, the initial state is $m_{-1}m_0 = 00$ and Eq. (2) gives the output $x_1'x_1'' = 00$ if $m_1 = 0$ or $x_1'x_1'' = 11$ if $m_1 = 1$.

The **code tree** drawn in Fig. 13.3–3 begins at a branch point or *node* labeled a representing the initial state. If $m_1 = 0$, you take the upper branch from node a to find the output 00 and the next state, which is also labeled a since $m_0 m_1 = 00$ in this case. If $m_1 = 1$, you take the lower branch from a to find the output 11 and the next state $m_0 m_1 = 01$ signified by the label b. The code tree progressively evolves in this fashion for each new input bit. Nodes are labeled with letters denoting the current state $m_{j-2} m_{j-1}$; you go up or down from a node, depending on the value of m_j; each branch shows the resulting encoded output $x_j' x_j''$ calculated from Eq. (2), and it terminates at another node labeled with the next state. There are 2^j possible branches for the jth message bit, but the branch pattern begins to repeat at $j = 3$ since the register length is $L + 1 = 3$.

---

†Notation for convolutional codes has not been standardized and varies from author to author, as does the definition of constraint length.

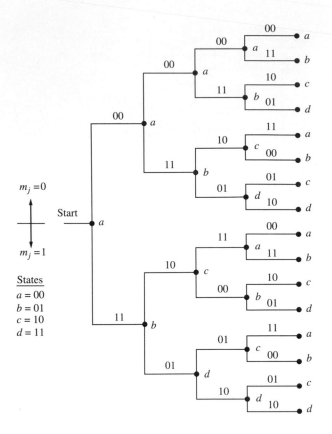

Figure 13.3–3   Code tree for (2, 1, 2) encoder.

Having observed repetition in the code tree, we can construct a more compact picture called the **code trellis** and shown in Fig. 13.3–4a. Here, the nodes on the left denote the four possible current states, while those on the right are the resulting next states. A solid line represents the state transition or branch for $m_j = 0$, and a broken line represents the branch for $m_j = 1$. Each branch is labeled with the resulting output bits $x'_j x''_j$. Going one step further, we coalesce the left and right sides of the trellis to obtain the **state diagram** in Fig. 13.3–4b. The self-loops at nodes a and d represent the state transitions a-a and d-d.

Given a sequence of message bits and the initial state, you can use either the code trellis or state diagram to find the resulting state sequence and output bits. The procedure is illustrated in Fig. 13.3–4c, starting at initial state a.

Numerous other convolutional codes are obtained by modifying the encoder in Fig. 13.3–2. If we just change the connections to the mod-2 summers, then the code tree, trellis, and state diagram retain the same structure since the states and branching pattern reflect only the register contents. The output bits would be different, of course, since they depend specifically on the summer connections.

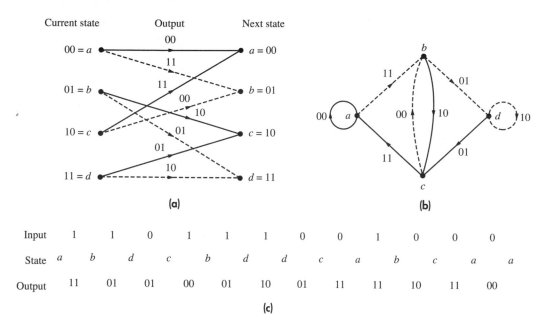

Current state   Output   Next state

(a)

(b)

Input	1	1	0	1	1	1	0	0	1	0	0	0
State	a	b	d	c	b	d	d	c	a	b	c	a	a
Output	11	01	01	00	01	10	01	11	11	10	11	00

(c)

Figure 13.3–4   (a) Code trellis; (b) state diagram for (2, 1, 2) encoder; (c) illustrative sequence.

If we extend the shift register to an arbitrary length $L + 1$ and connect it to $n \geq$ 2 mod-2 summers, we get an (n, k, L) convolutional code with $k = 1$ and code rate $R_c = 1/n \leq 1/2$. The state of the encoder is defined by L previous input bits, so the code trellis and state diagram have 2^L different states, and the code-tree pattern repeats after $j = L + 1$ branches. Connecting one commutator terminal directly to the first stage of the register yields the encoded bit stream

$$X = m_1 x_1'' x_1''' \cdots m_2 x_2'' x_2''' \cdots m_3 x_3'' x_3''' \cdots \qquad [3]$$

which defines a *systematic* convolutional code with $R_c = 1/n$.

Code rates higher than $1/n$ require $k \geq 2$ shift registers and an input distributor switch. This scheme is illustrated by the (3, 2, 1) encoder in Fig. 13.3–5. The message bits are distributed alternately between $k = 2$ registers, each of length $L + 1 = 2$. We regard the pair of bits $m_{j-1}m_j$ as the current input, while the pair $m_{j-3} m_{j-2}$ constitute the state of the encoder. For each input pair, the mod-2 summers generate $n = 3$ encoded output bits given by

$$x_j' = m_{j-3} \oplus m_{j-2} \oplus m_j \qquad \qquad x_j'' = m_{j-3} \oplus m_{j-1} \oplus m_j \qquad [4]$$

$$x_j''' = m_{j-2} \oplus m_j$$

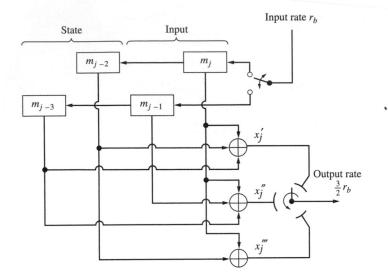

Figure 13.3–5   (3, 2, 1) encoder.

Thus, the output bit rate is $3r_b/2$, corresponding to the code rate $R_c = k/n = 2/3$. The constraint length is $n(L + 1) = 6$, since a particular input bit influences a span of $n = 3$ output bits from each of its $L + 1 = 2$ register positions.

Graphical representation becomes more cumbersome for convolutional codes with $k > 1$ because we must deal with input bits in groups of 2^k. Consequently, 2^k branches emanate and terminate at each node, and there are 2^{kL} different states. As an example, Fig. 13.3–6 shows the state diagram for the (3, 2, 1) encoder in Fig. 13.3–5. The branches are labeled with the $k = 2$ input bits followed by the resulting $n = 3$ output bits.

The convolutional codes employed for FEC systems usually have small values of n and k, while the constraint length typically falls in the range of 10 to 30. All convolutional encoders require a commutator switch at the output, as shown in Figs. 13.3–2 and 13.3–5.

For codes with $k > 1$, the input distributor switch can be eliminated by using a single register of length kL and shifting the bits in groups of k.

In any case, convolutional encoding hardware is simpler than the hardware for block encoding since message bits enter the register unit at a steady rate r_b and an input buffer is not needed.

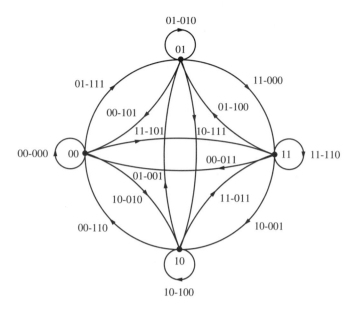

Figure 13.3–6   State diagram for (3, 2, 1) encoder.

The (2, 1, 2) convolutional encoder of Fig. 13.3–2 can be expressed as two generator polynomials with $G_1(D) = 1 + D^1 + D^2$ and $G_2(D) = 1 + D^2$. If we transform the message sequence into polynomial form, the message sequence of (1 1 0 1 1 1 0 0 1 0 0 0) is thus described as $M(D) = 1 + D + D^3 + D^4 + D^5 + D^8$. This transformation of the message and register gains into the D domain is analogous to the Fourier transform converting a convolution into a multiplication. The output from the upper branch of our encoder becomes $X'_j = M(D)G_1(D) = (1 + D + D^3 + D^4 + D^5 + D^8)(1 + D + D^2) = (1 + D^5 + D^7 + D^8 + D^9 + D^{10})$ and thus $x'_j = 100001011110$. Note when doing the multiplication operation, $D^i + D^i = 0$. Similarly for the lower branch we have $X''_j = M(D)G_2(D) = (1 + D + D^3 + D^4 + D^5 + D^8)(1 + D^2) = (1 + D + D^2 + D^4 + D^6 + D^7 + D^8 + D^{10})$ and hence $x''_j = 1 1 1 0 1 0 1 1 1 0 1 0$. With interleaving, the output becomes $x_j = $ 11 01 01 00 01 10 01 11 11 10 11 00, which is the same result as we obtained using convolution.

EXAMPLE 13.3–1

Consider a systematic (3, 1, 3) convolutional code. List the possible states and determine the state transitions produced by $m_j = 0$ and $m_j = 1$. Then construct and label the state diagram, taking the encoded output bits to be m_j, $m_{j-2} \oplus m_j$, and $m_{j-3} \oplus m_{j-1}$. (See Fig. P13.3–4 for a convenient eight-state pattern.)

EXERCISE 13.3–1

Free Distance and Coding Gain

We previously found that the error-control power of a block code depends upon its minimum distance, determined from the weights of the codewords. A convolutional code does not subdivide into codewords, so we consider instead the weight $w(X)$ of an entire transmitted sequence X generated by some message sequence. The **free distance** of a convolutional code is then defined to be

$$d_f \triangleq [w(X)]_{\min} \qquad X \neq 000 \cdots \tag{5}$$

The value of d_f serves as a measure of error-control power.

It would be an exceedingly dull and tiresome task to try to evaluate d_f by listing all possible transmitted sequences. Fortunately, there's a better way based on the normal operating procedure of appending a "tail" of 0s at the end of a message to clear the register unit and return the encoder to its initial state. This procedure eliminates certain branches from the code trellis for the last L transitions.

Take the code trellis in Fig. 13.3–4a, for example. To end up at state a, the next-to-last state must be either a or c so the last few branches of any transmitted sequence X must follow one of the paths shown in Fig. 13.3–7. Here the final state is denoted by e, and each branch has been labeled with the number of 1s in the encoded bits—which equals the *weight* associated with that branch. The total weight of a transmitted sequence X equals the sum of the branch weights along the path of X. In accordance with Eq. (5), we seek the path that has the smallest branch-weight sum, other than the trivial all-zero path.

Looking backwards $L + 1 = 3$ branches from e, we locate the last path that emanates from state a before terminating at e. Now suppose all earlier transitions followed the all-zero path along the top line, giving the state sequence $aa \cdots abce$. Since an a-a branch has weight 0, this state sequence corresponds to a **minimum-weight nontrivial path.** We therefore conclude that $d_f = 0 + 0 + \cdots + 0 + 2 + 1 + 2 = 5$. There are other minimum-weight paths, such as $aa \cdots abcae$ and $aa \cdots abcbce$, but no nontrivial path has less weight than $d_f = 5$.

Another approach to the calculation of free distance involves the **generating function** of a convolutional code. The generating function may be viewed as the

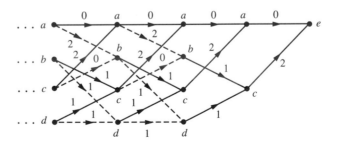

Figure 13.3–7   Termination of (2, 1, 2) code trellis.

transfer function of the encoder with respect to state transitions. Thus, instead of relating the input and output bits streams by *convolution*, the generating function relates the initial and final states by *multiplication*. Generating functions provide important information about code performance, including the free distance and decoding error probability.

We'll develop the generating function for our $(2, 1, 2)$ encoder using the modified state diagram in Fig. 13.3–8a. This diagram has been derived from Fig. 13.3–4b with four modifications. First, we've eliminated the a-a loop which contributes nothing to the weight of a sequence X. Second, we've drawn the c-a branch as the final c-e transition. Third, we've assigned a **state variable** W_a at node a, and likewise at all other nodes. Fourth, we've labeled each branch with two "gain" variables D and I such that the exponent of D equals the branch weight (as in Fig. 13.3–7), while the exponent of I equals the corresponding number of nonzero message bits (as signified by the solid or dashed branch line). For instance, since the c-e branch represents $x'_j x''_j = 11$ and $m_j = 0$, it is labeled with $D^2 I^0 = D^2$. This exponential trick allows us to perform *sums* by *multiplying* the D and I terms, which will become the independent variables of the generating function.

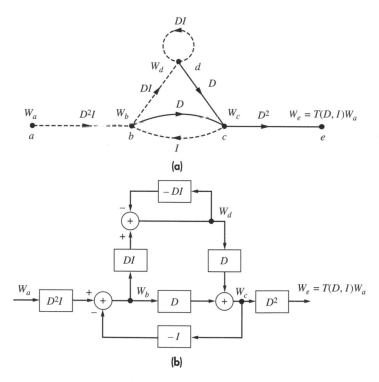

(a)

(b)

Figure 13.3–8   (a) Modified state diagram for $(2, 1, 2)$ encoder; (b) equivalent block diagram.

Our modified state diagram now looks like a *signal-flow graph* of the type sometimes used to analyze feedback systems. Specifically, if we treat the nodes as summing junctions and the *DI* terms as branch gains, then Fig. 13.3–8*a* represents the set of algebraic *state equations*

$$W_b = D^2 I W_a + I W_c \qquad W_c = D W_b + D W_d \qquad \text{[6a]}$$

$$W_d = D I W_b + D I W_d \qquad W_e = D^2 W_c$$

The encoder's generating function $T(D, I)$ can now be defined by the **input-output equation**

$$T(D, I) \triangleq W_e / W_a \qquad \text{[6b]}$$

These equations are also equivalent to the *block diagram* in Fig. 13.3–8*b*, which further emphasizes the relationships between the state variables, the branch gains, and the generating function. Note that minus signs have been introduced here so that the two feedback paths *c-b* and *d-d* correspond to negative feedback.

Next, the expression for $T(D, I)$ is obtained by algebraic solution of Eq. (6), or by block-diagram reduction of Fig. 13.3–8*b* using the transfer-function relations for parallel, cascade, and feedback connections in Fig. 3.1–8. (If you know *Mason's rule*, you could also apply it to Fig. 13.3–8*a*.) Any of these methods produces the final result

$$T(D, I) = \frac{D^5 I}{1 - 2DI} \qquad \text{[7a]}$$

$$= D^5 I + 2 D^6 I^2 + 4 D^7 I^3 + \cdots$$

$$= \sum_{d=5}^{\infty} 2^{d-5} D^d I^{d-4} \qquad \text{[7b]}$$

where we've expanded $(1 - 2DI)^{-1}$ to get the series in Eq. (7*b*). Keeping in mind that $T(D, I)$ represents all possible transmitted sequences that terminate with a *c-e* transition, Eq. (7*b*) has the following interpretation: for any $d \geq 5$, there are exactly 2^{d-5} valid paths with weight $w(X) = d$ that terminate with a *c-e* transition, and those paths are generated by messages containing $d - 4$ nonzero bits. The smallest value of $w(X)$ is the free distance, so we again conclude that $d_f = 5$.

As a generalization of Eq. (7), the generating function for an arbitrary convolutional code takes the form

$$T(D, I) = \sum_{d=d_f}^{\infty} \sum_{i=1}^{\infty} A(d, i) D^d I^i \qquad \text{[8]}$$

Here, $A(d, i)$ denotes the number of different input-output paths through the modified state diagram that have weight d and are generated by messages containing i nonzero bits.

Now consider a received sequence $Y = X + E$, where E represents transmission errors. The path of Y then diverges from the path of X and may or may not be a valid path for the code in question. When Y does not correspond to a valid path, a *maximum-likelihood decoder* should seek out the valid path that has the smallest Hamming distance from Y. Before describing how such a decoder might be implemented, we'll state the relationship between generating functions, free distance, and error probability in maximum-likelihood decoding of convolutional codes.

If transmission errors occur with equal and independent probability α per bit, then the probability of a decoded message-bit error is upper-bounded by

$$P_{be} \leq \frac{1}{k} \frac{\partial T(D, I)}{\partial I}\bigg|_{D=2\sqrt{\alpha(1-\alpha)}, I=1} \tag{9}$$

When α is sufficiently small, series expansion of $T(D, I)$ yields the approximation

$$P_{be} \approx \frac{M(d_f)}{k} 2^{d_f} \alpha^{d_f/2} \quad \sqrt{\alpha} \ll 1 \tag{10}$$

where

$$M(d_f) = \sum_{i=1}^{\infty} iA(d_f, i)$$

The quantity $M(d_f)$ simply equals the total number of nonzero message bits over all minimum-weight input-output paths in the modified state diagram.

Equation (10) supports our earlier assertion that the error-control power of a convolutional code depends upon its free distance. For a performance comparison with uncoded transmission, we'll make the usual assumption of gaussian white noise and $(S/N)_R = 2R_c\gamma_b \geq 10$ so Eq. (10), Sect. 13.1, gives the transmission error probability

$$\alpha \approx (4\pi R_c\gamma_b)^{-1/2}e^{-R_c\gamma_b}$$

The decoded error probability then becomes

$$P_{be} \approx \frac{M(d_f)2^{d_f}}{k(4\pi R_c\gamma_b)^{d_f/4}} e^{-(R_c d_f/2)\gamma_b} \tag{11}$$

whereas uncoded transmission would yield

$$P_{ube} \approx \frac{1}{(4\pi\gamma_b)^{1/2}} e^{-\gamma_b} \tag{12}$$

Since the exponential terms dominate in these expressions, we see that convolutional coding improves reliability when $R_c d_f/2 > 1$. Accordingly, the quantity $R_c d_f/2$ is known as the **coding gain,** usually expressed in dB.

Explicit design formulas for d_f do not exist, unfortunately, so good convolutional codes must be discovered by computer search and simulation. Table 13.3–1 lists the maximum free distance and coding gain of convolutional codes for selected values of n, k, and L. Observe that the free distance and coding gain increase with increasing memory L when the code rate R_c is held fixed. All listed codes are nonsystematic; a

Table 13.3–1   Maximum free distance and coding gain of selected convolutional codes

n	k	R_c	L	d_f	$R_c d_f/2$
4	1	1/4	3	13	1.63
3	1	1/3	3	10	1.68
2	1	1/2	3	6	1.50
			6	10	2.50
			9	12	3.00
3	2	2/3	3	7	2.33
4	3	3/4	3	8	3.00

systematic convolutional code has a smaller d_f than an optimum nonsystematic code with the same rate and memory.

We should also point out that some convolutional codes exhibit **catastrophic error propagation.** This occurs when a finite number of channel errors causes an infinite number of decoding errors, even if subsequent symbols are correct. Encoders that exhibit this behavior will show in their state diagram a state where a

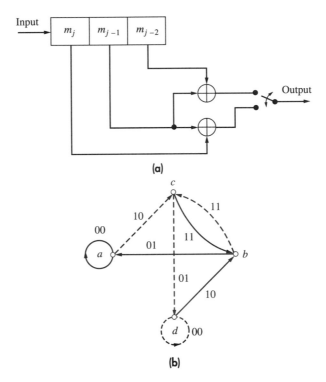

(a)

(b)

Figure 13.3–9   Encoder subject to catastrophic error propagation. (a) Encoder; (b) state diagram.

given nonzero input causes a transition back to that state itself producing a zero output. Catastrophic codes can also be identified if their generator polynomials have a common factor of degree at least one. For example, Fig. 13.3–9 shows a catastrophic encoder and its state diagram. Note at state d a nonzero input causes the encoder to branch back to itself with a zero output. The code is also found to be catastrophic because generator polynomials $G_1(D) = 1 + D$ and $G_2(D) = D + D^2$ have the common factor of $1 + D$.

EXAMPLE 13.3–2

The (2, 1, 2) encoder back in Fig. 13.3–2 has $T(D, I) = D^5I/(1 - 2DI)$, so $\partial T(D, I)/\partial I = D^5/(1 - 2DI)^2$. Equation (9) therefore gives

$$P_{be} \leq \frac{2^5[\alpha(1 - \alpha)]^{5/2}}{[1 - 4\sqrt{\alpha(1 - \alpha)}]^2} \approx 2^5\alpha^{5/2}$$

and the small-α approximation agrees with Eq. (10). Specifically, in Fig. 13.3–8a we find just one minimum-weight nontrivial path $abce$, which has $w(X) = 5 = d_f$ and is generated by a message containing one nonzero bit, so $M(d_f) = 1$.

If $\gamma_b = 10$, then $R_c\gamma_b = 5$, $\alpha \approx 8.5 \times 10^{-4}$, and maximum-likelihood decoding yields $P_{be} \approx 6.7 \times 10^{-7}$, as compared with $P_{ube} \approx 4.1 \times 10^{-6}$. This rather small reliability improvement agrees with the small coding gain $R_c d_f/2 = 5/4$.

EXERCISE 13.3–2

Let the connections to the mod-2 summers in Fig. 13.3–2 be changed such that $x_j' = m_j$ and $x_j'' = m_{j-2} \oplus m_{j-1} \oplus m_j$.

(a) Construct the code trellis and modified state diagram for this systematic code. Show that there are two minimum-weight paths in the state diagram, and that $d_f = 4$ and $M(d_f) = 3$. It is not necessary to find $T(D, I)$.

(b) Now assume $\gamma_b = 10$. Calculate α, P_{be}, and P_{ube}. What do you conclude about the performance of a convolutional code when $R_c d_f/2 = 1$?

Decoding Methods

There are three generic methods for decoding convolutional codes. At one extreme, the Viterbi algorithm executes *maximum-likelihood decoding* and achieves optimum performance but requires extensive hardware for computation and storage. At the other extreme, *feedback decoding* sacrifices performance in exchange for simplified hardware. Between these extremes, *sequential decoding* approaches optimum performance to a degree that depends upon the decoder's complexity. We'll describe how these methods work with a (2, 1, L) code. The extension to other codes is conceptually straightforward, but becomes messy to portray for $k > 1$.

Recall that a maximum-likelihood decoder must examine an entire received sequence Y and find a valid path that has the smallest Hamming distance from Y.

However, there are 2^N possible paths for an arbitrary message sequence of N bits (or Nn/k bits in Y), so an exhaustive comparison of Y with all valid paths would be an absurd task in the usual case of $N \gg 1$. The **Viterbi algorithm** applies maximum-likelihood principles to limit the comparison to 2^{kL} *surviving paths*, independent of N, thereby bringing maximum-likelihood decoding into the realm of feasibility.

A Viterbi decoder assigns to each branch of each surviving path a **metric** that equals its Hamming distance from the corresponding branch of Y. (We assume here that 0s and 1s have the same transmission-error probability; if not, the branch metric must be redefined to account for the differing probabilities.) Summing the branch metrics yields the path metric, and Y is finally decoded as the surviving path with the smallest metric. To illustrate the metric calculations and explain how surviving paths are selected, we'll walk through an example of Viterbi decoding.

Suppose that our $(2, 1, 2)$ encoder is used at the transmitter, and the received sequence starts with $Y = 11\ 01\ 11$. Figure 13.3–10 shows the first three branches of the valid paths emanating from the initial node a_0 in the code trellis. The number in parentheses beneath each branch is the branch metric, obtained by counting the differences between the encoded bits and the corresponding bits in Y. The circled number at the right-hand end of each branch is the running path metric, obtained by summing branch metrics from a_0. For instance, the metric of the path $a_0 b_1 c_2 b_3$ is $0 + 2 + 2 = 4$.

Now observe that another path $a_0 a_1 a_2 b_3$ also arrives at node b_3 and has a smaller metric $2 + 1 + 0 = 3$. Regardless of what happens subsequently, this path will have a smaller Hamming distance from Y than the other path arriving at b_3 and is therefore

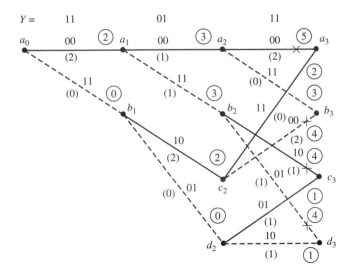

Figure 13.3–10

more likely to represent the actual transmitted sequence. Hence, we discard the larger-metric path, marked by an $\times$, and we declare the path with the smaller metric to be the survivor at this node. Likewise, we discard the larger-metric paths arriving at nodes a_3, c_3, and d_3, leaving a total of $2^{kL} = 4$ surviving paths. The fact that none of the surviving path metrics equals zero indicates the presence of *detectable errors* in Y.

Figure 13.3–11 depicts the continuation of Fig. 13.3–10 for a complete message of $N = 12$ bits, including tail 0s. All discarded branches and all labels except the running path metrics have been omitted for the sake of clarity. The letter T under a node indicates that the two arriving paths had equal running metrics, in which case we just flip a coin to choose the survivor (why?). The maximum-likelihood path follows the heavy line from a_0 to a_{12}, and the final value of the path metric signifies at least two transmission errors in Y. The decoder assumes the corresponding transmitted sequence $Y + \hat{E}$ and message sequence $\hat{M}$ written below the trellis.

A Viterbi decoder must calculate two metrics for each node and store 2^{kL} surviving paths, each consisting of N branches. Hence, decoding complexity increases exponentially with L and linearly with N. The exponential factor limits practical application of the Viterbi algorithm to codes with small values of L.

When $N \gg 1$, storage requirements can be reduced by a truncation process based on the following metric-divergence effect: if two surviving paths emanate from the same node at some point, then the running metric of the less likely path tends to increase more rapidly than the metric of the other survivor within about $5L$ branches from the common node. This effect appears several times in Fig. 13.3–11; consider, for instance, the two paths emanating from node b_1. Hence, decoding need not be delayed until the end of the transmitted sequence. Instead, the first k message bits can be decoded and the first set of branches can be deleted from memory after

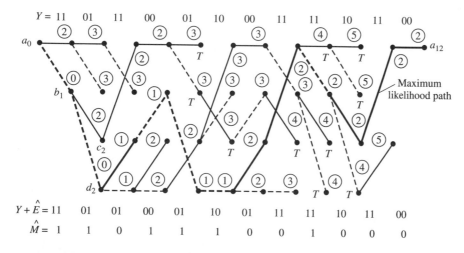

Figure 13.3–11   Illustration of the Viterbi algorithm for maximum-likelihood decoding.

the first $5Ln$ received bits have been processed. Successive groups of k message bits are then decoded for each additional n bits received thereafter.

Sequential decoding, which was invented before the Viterbi algorithm, also relies on the metric-divergence effect. A simplified version of the sequential algorithm is illustrated in Fig. 13.3–12a, using the same trellis, received sequence, and metrics as in Fig. 13.3–11. Starting at a_0, the sequential decoder pursues a single path by taking the branch with the smallest branch metric at each successive node. If two or more branches from one node have the same metric, such as at node d_2, the decoder selects one at random and continues on. Whenever the current path happens to be unlikely, the running metric rapidly increases and the decoder eventually decides to go back to a lower-metric node and try another path. There are three of these abandoned paths in our example. Even so, a comparison with Fig. 13.3–11 shows that sequential decoding involves less computation than Viterbi decoding.

The decision to backtrack and try again is based on the expected value of the running metric at a given node. Specifically, if α is the transmission error probability per bit, then the expected running metric at the jth node of the correct path equals

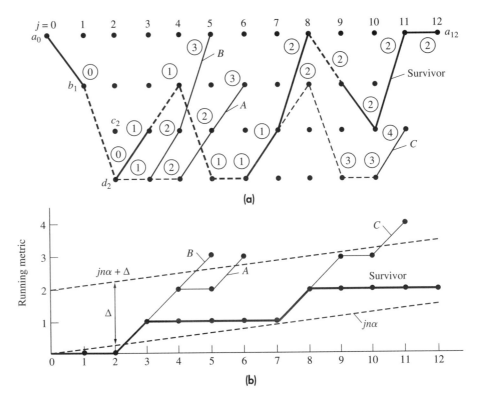

(a)

(b)

Figure 13.3–12   Illustration of sequential decoding.

$jn\alpha$, the expected number of bit errors in Y at that point. The sequential decoder abandons a path when its metric exceeds some specified *threshold* Δ above $jn\alpha$. If no path survives the threshold test, the value of Δ is increased and the decoder backtracks again. Figure 13.3–12b plots the running metrics versus j, along with $jn\alpha$ and the threshold line $jn\alpha + \Delta$ for $\alpha = 1/16$ and $\Delta = 2$.

Sequential decoding approaches the performance of maximum-likelihood decoding when the threshold is loose enough to permit exploration of all probable paths. However, the frequent backtracking requires more computations and results in a decoding delay significantly greater than Viterbi decoding. A tighter threshold reduces computations and decoding delay but may actually eliminate the most probable path, thereby increasing the output error probability compared to that of maximum-likelihood decoding with the same coding gain. As compensation, sequential decoding permits practical application of convolutional codes with large L and large coding gain since the decoder's complexity is essentially independent of L.

We've described sequential decoding and Viterbi decoding in terms of *algorithms* rather than block diagrams of hardware. Indeed, these methods are usually implemented as software that performs the metric calculations and stores the path data. When circumstances preclude algorithmic decoding, and a higher error probability is tolerable, *feedback decoding* may be the appropriate method. A feedback decoder acts in general like a "sliding block decoder" that decodes message bits one by one based on a block of L or more successive tree branches. We'll focus on the special class of feedback decoding that employs *majority logic* to achieve the simplest hardware realization of a convolutional decoder.

Consider a message sequence $M = m_1 m_2 \cdots$ and the *systematic* $(2, 1, L)$ encoded sequence

$$X = x_1' x_1'' x_2' x_2'' \cdots \qquad \text{[13a]}$$

where

$$x_j' = m_j \qquad x_j'' = \sum_{i=0}^{L} m_{j-i} g_i \qquad (\text{mod-2}) \qquad \text{[13b]}$$

We'll view the entire sequence X as a codeword of indefinite length. Then, borrowing from the matrix representation used for block codes, we'll define a *generator matrix G* and a *parity-check matrix H* such that

$$X = MG \qquad \qquad XH^T = 0\,0 \cdots$$

To represent Eq. (13), G must be a semi-infinite matrix with a diagonal structure given by

$$G = \begin{bmatrix} 1 & g_0 & 0 & g_1 & 0 & \cdots & 0 & g_L & & & \\ & & 1 & g_0 & 0 & g_1 & 0 & \cdots & 0 & g_L & \\ & & & \cdot & \cdot & \cdot & & & & & \cdot \\ & & & & \cdot & \cdot & \cdot & & & & & \cdot \\ & & & & & \cdot & \cdot & \cdot & & & & & \cdot \end{bmatrix} \qquad \text{[14a]}$$

This matrix extends indefinitely to the right and down, and the triangular blank spaces denote elements that equal zero. The parity-check matrix is

$$H = \begin{bmatrix} g_0 & 1 & & & & & & \\ g_1 & 0 & g_0 & 1 & & & & \\ \cdot & \cdot & g_1 & 0 & g_0 & 1 & & \\ \cdot & & \cdot & \cdot & \cdot & & \cdot & \cdot \\ \cdot & & & \cdot & & & \cdot & \cdot \\ g_L & 0 & & \cdot & \cdot & & & \cdot \\ & & g_L & 0 & & & & \\ & & & & \cdot & \cdot & & \\ & & & & & \cdot & \cdot & \end{bmatrix} \qquad [14b]$$

which also extends indefinitely to the right and down.

Next, let E be the transmission error pattern in a received sequence $Y = X + E$. We'll write these sequences as

$$Y = y_1' y_1'' y_2' y_2'' \cdots \qquad E = e_1' e_1'' e_2' e_2'' \cdots$$

so that $y_j' = m_j \oplus e_j'$. Hence, given the error bit e_j', the jth message bit is

$$m_j = y_j' \oplus e_j' \qquad [15]$$

A feedback decoder estimates errors from the *syndrome* sequence

$$S = YH^T = (X + E)H^T = EH^T$$

Using Eq. (14b) for H, the jth bit of S is

$$s_j = \sum_{i=0}^{L} y_{j-i}' g_i \oplus y_j'' = \sum_{i=0}^{L} e_{j-i}' g_i \oplus e_j'' \qquad [16]$$

where the sums are mod-2 and it's understood that $y_{j-i}' = e_{j-i}' = 0$ for $j - i \leq 0$.

As a specific example, take a (2, 1, 6) encoder with $g_0 = g_2 = g_5 = g_6 = 1$ and $g_1 = g_3 = g_4 = 0$, so

$$s_j = y_{j-6}' \oplus y_{j-5}' \oplus y_{j-2}' \oplus y_j' \oplus y_j'' \qquad [17a]$$

$$= e_{j-6}' \oplus e_{j-5}' \oplus e_{j-2}' \oplus e_j' \oplus e_j'' \qquad [17b]$$

Equation (17a) leads directly to the shift-register circuit for syndrome calculation diagrammed in Fig. 13.3–13. Equation (17b) is called a *parity-check sum* and will lead us eventually to the remaining portion of the feedback decoder.

To that end, consider the parity-check table in Fig. 13.3–14a where checks indicate which error bits appear in the sums s_{j-6}, s_{j-4}, s_{j-1}, and s_j. This table brings out the fact that e_{j-6}' is checked by all four of the listed sums, while no other error bit is checked by more than one. Accordingly, this set of check sums is said to be *orthogonal* on e_{j-6}'. The tap gains of the encoder were carefully chosen to obtain orthogonal check sums.

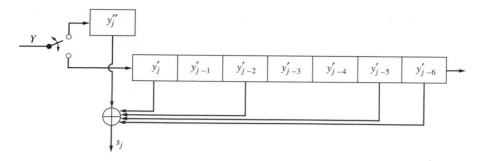

Figure 13.3–13   Shift-register circuit for syndrome calculation for a systematic (2, 1, 6) code.

	e'_{j-12}	e'_{j-11}	e'_{j-10}	e'_{j-9}	e'_{j-8}	e'_{j-7}	e'_{j-6}	e''_{j-6}	e'_{j-5}	e'_{j-4}	e''_{j-4}	e'_{j-3}	e'_{j-2}	e'_{j-1}	e''_{j-1}	e'_j	e''_j
s_{j-6}	✓	✓			✓		✓	✓									
s_{j-4}			✓	✓			✓			✓	✓						
s_{j-1}						✓	✓					✓		✓	✓		
s_j							✓		✓			✓				✓	✓

Figure 13.3–14   Parity-check table for a systematic (2, 1, 6) code.

When the transmission error probability is reasonably small, we expect to find at most one or two errors in the 17 transmitted bits represented by the parity check table. If one of the errors corresponds to $e'_{j-6} = 1$, then the four check sums will contain three or four 1s. Otherwise, the check sums contain less than three 1s. Hence, we can apply these four check sums to a *majority-logic gate* to generate the most likely estimate of e'_{j-6}.

Figure 13.3–15 diagrams a complete majority-logic feedback decoder for our systematic (2, 1, 6) code. The syndrome calculator from Fig. 13.3–13 has two outputs, y'_{j-6} and s_j. The syndrome bit goes into another shift register with taps that connect the check sums to the majority-logic gate, whose output equals the estimated error $\hat{e}'_{j-6}$. The mod-2 addition $y'_{j-6} \oplus \hat{e}'_{j-6} = \hat{m}_{j-6}$ carries out error correction based on Eq. (15). The error is also fed back to the syndrome register to improve the reliability of subsequent check sums. This feedback path accounts for the name *feedback decoding*.

Our example decoder can correct any single-error or double-error pattern in six consecutive message bits. However, more than two transmission errors produces erroneous corrections and error propagation via the feedback path. These effects result in a higher output error than that of maximum-likelihood decoding. See Lin and Costello (1983, Chap. 13) for the error analysis and further treatment of majority-logic decoding.

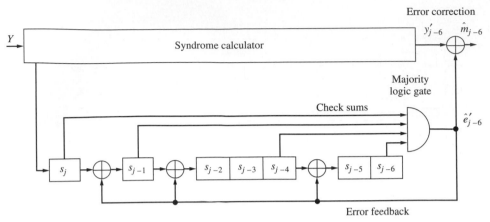

Figure 13.3–15   Majority-logic feedback decoder for a systematic (2, 1, 6) code.

Turbo Codes

Turbo codes, or **parallel concatenated codes** (PCC), are a relatively new class of convolutional codes first introduced in 1993. They have enabled channel capacities to nearly reach the Shannon limit. Our coverage of turbo codes will be primarily descriptive. Detailed treatments of turbo codes are given by Berrou et al. (1993), Berrou (1996), Hagenauer et al. (1996), and Johannesson and Zigangirov (1999).

Shannon's theorem for channel capacity assumes *random* coding with the bit error rate (BER) approaching zero as the code's block or constraint length approaches infinity. It is therefore not feasible to decode a completely truly random code. Increases in code complexity accomplished by longer block or constraint lengths require a corresponding exponential increase in decoding complexity. We can, however, achieve a given BER if the code is sufficiently unstructured, and we are willing to pay the associated cost in decoding complexity. This leads us to the following paradox by J. Wolfowitz:

> Almost all codes are good, except those we can think of.

Turbo codes overcome this paradox in that they can be made sufficiently random to achieve a given BER and, by using iterative methods, can be efficiently and feasibly decoded.

Figure 13.3–16 illustrates a turbo encoder. Here we have the parallel concatenation of two codes produced from two rate 1/2 *recursive systematic convolutional* (RSC) encoders. The second RSC is preceded by a pseudo-random *interleaver* whose length can vary from 100–10,000 bits or more to permute the symbol sequence. The RSCs are not necessarily identical, nor do they have to be convolu-

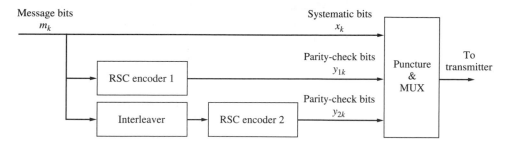

Figure 13.3–16   Turbo encoder.

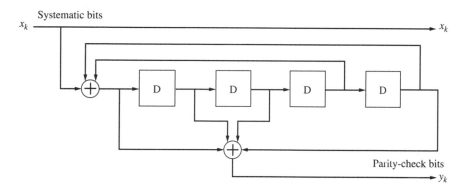

Figure 13.3–17   Recursive systematic convolutional (RSC) encoder with $R = 1/2$, $G_1 = 23$, $G_2 = 35$, and $L = 2$.

tional, and more than two can be used. Both encoders produce the parity-check bits. These parity bits and the original bit stream (called the *systematic bits*) are multiplexed and then transmitted. As given, the overall rate is 1/3. However, we may be able to increase this rate to $R = 1/2$ using a process of **puncturing,** whereby some bits are deleted. This could be done, for example, by eliminating the odd parity bits from the first RSC and the even parity bits from the second RSC.

Figure 13.3–17 shows a G_1, G_2-RSC encoder that has been used for turbo coding. Unlike the nonsystematic convolutional (NSC) encoders described earlier in this section, this encoder is *systematic* in that the message and parity bits are separate. The feedback connections from the state outputs make this encoder *recursive*, and thus single output errors produce a large quantity of parity errors. For this particular encoder, the polynomial describing the feedback connections is $1 + D^3 + D^4 = 10\,011_2 = 23_8$ and the polynomial for the output is $1 + D + D^2 + D^4 = 11\,101_2 = 35_8$. Hence, the literature often refers this to as a $G_1 = 23$, $G_2 = 35$, or simply a (23, 35) encoder.

The turbo decoder is shown in Fig. 13.3–18. It consists of two *maximum a posterior* (MAP) decoders and a feedback path that works in similar manner to that of an automobile turbo engine, hence the term *turbo code*. The first decoder takes the information from the received signal and calculates the *a posteriori probability*

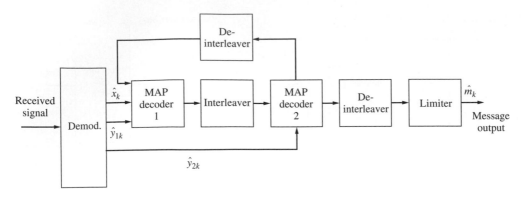

Figure 13.3–18   Turbo decoder.

(APP) value. This value is then used as the *a priori* probability value for the second decoder. The output is then fed back to the first decoder where the process repeats in an iterative fashion with each iteration producing more refined estimates.

Instead of using the Viterbi algorithm, which minimizes the error for the entire sequence, this MAP decoder uses a modified form of the BCJR (Bahl, Cocke, Jelinek, and Raviv, 1972) algorithm that takes into account the recursive character of the RSC codes and computes a log-likelihood ratio to estimate the APP for each bit.

The results by Berrou et al. are impressive. When encoding using rate $R = 1/2$, $G_1 = 37$ and $G_2 = 21$, 65,537 interleaving, and 18 iterations, they were able to achieve a BER of 10^{-5} at $E_b / N_0 = 0.7$ dB. The main disadvantage of turbo codes with their relatively large codewords and iterative decoding process is their long latency. A system with 65,537 interleaving and 18 iterations may have too long a latency for voice telephony. On the other hand, turbo codes have excellent performance in deep space applications.

13.4   DATA ENCRYPTION

Up to now, the purpose of channel coding was to reduce errors. Later, in Chapter 15, we will see how coding can be employed to spread out the information over the frequency band to increase a system's immunity to jamming. However, in this section we want to look at **encryption coding,** in which the goal is secure communication. We will first consider the basics of encryption, and then look at two popular encryption methods: the data encryption standard (DES) and Rivest-Shamir-Adleman (RSA) systems.

Encryption is the transformation of *plaintext* into *ciphertext; decryption* is the inverse of this process. The *cipher* is the set of transformations, and the *keys* are the transformation parameters. There are two types of ciphers: *Block encryption* is

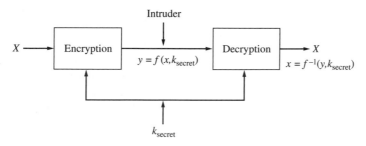

Figure 13.4–1   Secret key system.

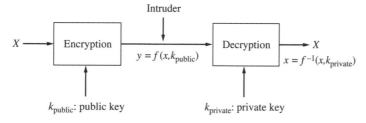

Figure 13.4–2   Public/private key system.

where fixed blocks of plaintext are operated on in a combinatorial fashion, and *stream encryption* is similar to convolution encoding in that small segments of the plaintext are processed.

A single *or* **secret key** encryption system is shown in Fig. 13.4–1. Here the plaintext sequence x is transformed into a ciphertext sequence y as

$$y = f(x, k_{secret}) \qquad [1]$$

where f is an encryption function consisting of a combination of substitutions, permutations, and/or mathematical operations. Parameter k_{secret} is a *secret key* that determines the coding and decoding parameters. The message recipient deciphers the message by applying an inverse function

$$x = f^{-1}(x, k_{secret}) \qquad [2]$$

With *secret key encryption*, the keys for the sender and recipient are identical and must be kept secret. Therefore, to ensure security, key distribution becomes a challenge. It must occur over a secure channel. That could be the postal system, another electronic communications channel, or perhaps carrier pigeon. The point is that the message is only as secure as the key. If an intruder is able to steal the key, the message is no longer secure. Periodically changing the key may also foil an intruder.

Now let's consider the public-key system of Fig. 13.4–2 where the sender and recipient have a different key so

$$y = f(x, k_{public}) \qquad [3]$$

and

$$x = f^{-1}(x, k_{\text{private}}) \qquad \textbf{[4]}$$

Let's say Alice wants to send a secret message to Nate. Nate's public directory lists both his address and *public key*, k_{public}. Anytime someone wants to send a secret message to Nate, they encrypt it using his public key and the transform of Eq. (3). In this case, Alice sends ciphertext y to Nate. Nate then decrypts the message using the transform of Eq. (4) and his *private key*, k_{private}, which is hidden from the world. The transforms in Eqs. (3) and (4) are designed such that the inverse of each function can only be computed if we know the other key. In other words, message x can only be extracted from sequence y if the recipient knows the private key; it is nearly impossible to extract x from y if we only know function f and the public key.

We can also use this system for message authentication, when we want to send a message with an *electronic signature*. Let's say Nate wants to send Alice a signed message. Nate encrypts his signature using his private key and Eq. (4) and then sends the sequence x to Alice. Alice then uses Nate's public key and Eq. (3) to read Nate's signature. In this scenario, Nate's signature can only be forged if the forger knows his private key.

Before we study the more widely used and sophisticated encryption systems, let's look at the two elementary operations, *substitution* and *permutation*, which are the basis for other systems.

Substitution is where plaintext symbols are converted into another set of symbols. A simple example is as follows.

Plaintext: A B C D E F G H I J K L M N O P Q R S T U V W X Y Z

Substitute text: Z V E C L M O F P T X B I K S D Q U G W N R Y J A H

Therefore, if we wanted to send the message "THE DOG BITES" we would perform the substitution and send

"WFL-CSO-VPWLG"

But we could also transpose or permute the text by some fixed rule where the characters are put in fixed block lengths and circularly shifted left or right. For example, we could put the characters in groups of four and then permute by performing circular shift to the right by two:

Plaintext: $X_1 X_2 X_3 X_4$

Permuted text: $X_3 X_4 X_1 X_2$

If we extend our previous example, our substituted text becomes,

Substituted and permuted text: "L-WFO-CSWLVP-G"

If the message is sufficiently long, the principle flaw of this system is that it can be broken by *frequency or statistical analysis*, where the intruder takes advantage of certain letter patterns that occur in the English language. For example, letters *e, a, th*, and *the* occur frequently, while letters *x* and *z* rarely occur; letter *q* is always followed by a *u*; a consonant is usually followed by an identical consonant or a vowel. If the text is transmitted via a data network, the words "login" and/or "userid" often occur. Code breaking is also facilitated if the intruder knows the message's context. For example, in ship-to-shore communication, the beginning of the message may be a weather report.

Permutation operations can be performed with the **P-box** structures shown in Fig. 13.4–3. The first is a *straight permutation*; the second is an *expanded permutation*, where the box maps more than one input to the output; and the third is a *compressed*, or *choice, permutation* where the box *chooses* only some of the inputs to be mapped to the output (i.e., some inputs are left out). The P-box by itself could also be used for substitution, but the key would be inordinately long. For example, if we wanted to do a substitution for any 8-bit word, we would need a key size of $2^8 = 256$. Instead, we can simplify our key requirements by using the S-box structure of Fig. 13.4–4. The P-box and S-box structures can be combined to form a **product cipher** and thereby increase the security of our cipher. The widely used data encryption standard is a product cipher.

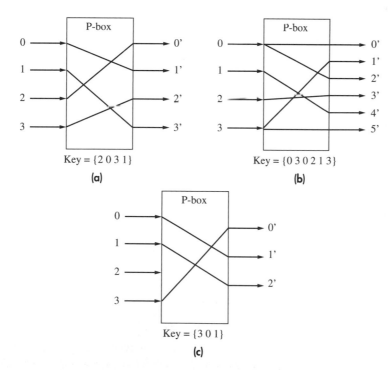

Figure 13.4–3   Permutation examples. (a) straight; (b) expanded; (c) compressed.

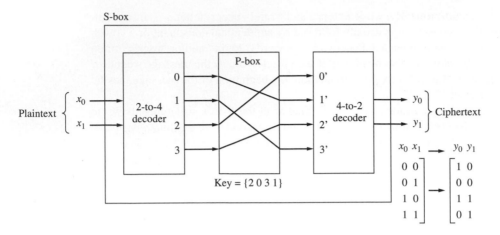

Figure 13.4–4   S-box (substitution) examples.

Data Encryption Standard

The **data encryption standard** (DES) is a single-key system developed by Diffie and Hellman (1979). It is one of the most widely known and used algorithms and has found wide acceptance in electronic commerce activities such as banking, credit card operations, electronic fund transfers, and so on. It is recognized by NIST and is fully described in FIPS Pub 46–2.

DES encrypts 64-bit blocks of data using a series of permutations and substitutions. Its immunity to breaking is based on its complexity. The key is 56 bits wide; thus employing brute force methods to break it would require an exhaustive test of 2^{56} possible keys. On the average, however, we would find the correct key after 2^{55} tries. If we did this on a processor that performs one test per microsecond, the correct key could be found in about 1200 years. On the other hand, if each of the approximately 300 million Americans worked in parallel, it would only take two minutes to determine the key. We may not, however, have to resort to such brute force to discover the key. If we knew it originated from some pseudo-random number generator such a telephone directory or computer's PN generator, we might be able to shortcut the process.

The overall DES process is described in the flowchart of Fig. 13.4–5. The system works as follows (Diffie et al., FIPS–46): (*a*) Sixty-four-bit plaintext undergoes an *initial permutation* stage using a P-Box and is divided into two 32-bit left-right segments, L_0 and R_0; (*b*) Sixteen *substitute and permute* operations (or *rounds*) are performed on each of these two blocks. The outcome of each round is a function of the subkey, k_i. (*c*) The two 32-bit outputs feed into the P-box that is an inverse of the *initial permutation* with the output producing 64-bit ciphertext.

The details of the *i*th *substitute and permute* block are given in Fig. 13.4–6. This block works as follows: (*a*) The right-most 32 bits from the previous stage are

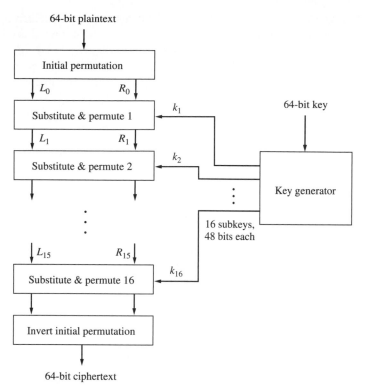

Figure 13.4–5   DES flowchart.

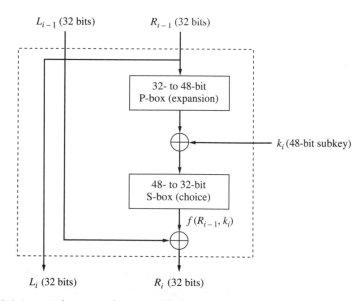

Figure 13.4–6   Substitute and permute block.

expanded to 48 bits using a P-box; (*b*) The 48-bit result is then exclusively ORed with a 48-bit subkey, k_i, producing a 48-bit word. (*c*) This 48-bit word is then compressed to 32 bits using an S-box whose output we call $f(R_{i-1}, k_i)$. (*d*) Function $f(R_{i-1}, k_i)$ is then exclusively ORed with the leftmost 32 bits, L_{i-1}, producing the right 32-bit word, R_i, for the next round. The new left output is equal to the former right input. Mathematically, a round consists of

$$L_i = R_{i-1} \qquad\qquad i = 1, 2, \ldots 16 \qquad\qquad \textbf{[5a]}$$

$$R_i = L_{i-1} \oplus f(R_{i-1}, k_i) \qquad i = 1, 2, \ldots 16 \qquad\qquad \textbf{[5b]}$$

The process to generate the key is shown in Fig. 13.4–7. We first start out with a 64-bit key that includes 8 parity bits. A 64- to 56-bit P-box strips off these parity bits leaving us with a 56-bit key. This key is then divided into two 28-bit words. Each 28-bit word undergoes one or two left shifts. The two shifted 28-bit words are then fed to a 56- to 48-bit P-box whose output is subkey k_1. This process is then repeated 15 more times to generate an additional 15 subkeys. In the NIST scheme of FIPS–46, stages 1, 2, 9, and 16 undergo only one left shift, whereas the other stages

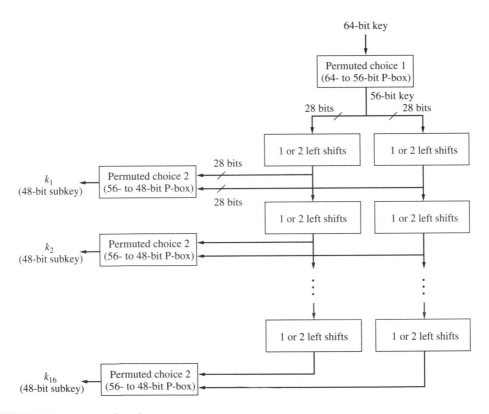

Figure 13.4–7   Key generator flowchart.

undergo two left shifts. To convert the ciphertext back into plaintext, we employ the reverse process of Fig. 13.4–5.

The DES structures and connections of the P- and S-boxes of Figs. 13.4–6 and 13.4–7 have been standardized and are fully described in paper FIPS–46. Therefore, a given key is the sole determining factor of how the data is encrypted. Let's now look at the details of a few of the more interesting blocks. The 32- to 48-bit expansion block used in Fig. 13.4–6 is shown in Fig. 13.4–8. Here, eight 4-bit words are expanded to make eight 6-bit words by repeating the edge bits. An expansion block causes intruder errors to be magnified, thereby making it more difficult to break the code. Fig. 13.4–9 illustrates the 48- to 32-bit compression block used in calculating $f(R_{i-1}, k_i)$. Again, the details of the S-block, S_i, and the permutation block are given in the FIPS–46 paper.

For increased security, there is also **Triple DES,** which consists of three stages of DES encryption, the first and last using the same key and the middle one using a different key. Three stages of encryption enable a triple DES to be compatible with older single DES systems whereby a Triple DES can decrypt single DES data.

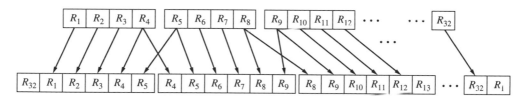

Figure 13.4–8   32- to 48-bit expansion.

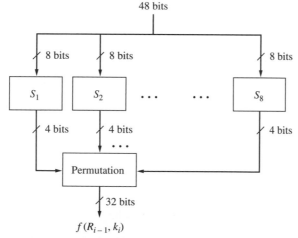

Figure 13.4–9   Details of 48- to 32-bit S-box.

Rivest-Shamir-Adleman System

As previously mentioned, the significant challenge of secret-key systems is secure but fast and efficient key distribution. Not only must the key channel be secure, but in today's e-commerce environment, the key must be readily available. This precludes such channels as the conventional mail system for key distribution. To overcome this challenge, the **Rivest-Shamir-Adleman** (RSA) (1978) public key system was developed. While DES's strength is based on the complexity of its algorithm, the RSA's security is based on the inherent difficulty of determining the prime factors of large numbers. The following elementary example illustrates RSA encryption in which we want to send an encrypted message consisting of letter "x," whose value in ASCII is 88, and then decrypt the received message.

1. Pick two prime numbers, p and q, and calculate their product or

$$n = pq \qquad\qquad [6]$$

Both p and q are kept secret. In this example, use $p = 13, q = 17$, and then $n = pq = 221$.

2. Compute the *Euler-quotient* function

$$\phi(n) = (p - 1)(q - 1) = 12 \times 16 = 192 \qquad\qquad [7]$$

3. Pick another prime number, e, that *is relatively prime* to $\phi(n)$ (i.e., the prime factors of $\phi(n)$ do not include e). In this case, pick $e = 5$, since 5 is not a prime factor of 192.

4. The *public keys* are then $e = 5$ and $n = 221$.

5. A message is then encrypted using the public keys so that

$$y = x^e \bmod n \qquad\qquad [8]$$

giving $y = 88^5 \bmod 221$ in this example. As an aside, the mod (modulo) operation returns the remainder portion of the quotient of the two numbers or $y = a \bmod b$ means $y = \mathrm{rem}\,[a \div b]$.

In typical cases calculating x^e is too large for most practical computers. Instead we factor out the base term in Eq. (6), and thus for this example obtain

$$y = 88^5 \bmod 221$$

$$= [88^2 \bmod 221][88^2 \bmod 221][88^1 \bmod 221] \bmod 221$$

$$= [9 \times 9 \times 88] \bmod 221 = 56$$

Our encrypted message is $y = 56$.

6. To decrypt the message, y, we apply the following function

$$x = y^d \bmod n \qquad\qquad [9]$$

where d is the *private key*.

7. To determine the private key needed to decrypt the message we use the algorithm:

$$d\,e = 1 \bmod [\phi(n)] \qquad \text{[10a]}$$

where d is less than $\phi(n)$. Equivalently, we can get d using Euclid's algorithm with

$$d\,e = \phi(n) \times Q + 1 \qquad \text{[10b]}$$

where Q is any integer that satisfies Eq. (10b).

In this example with $\phi(n) = 192$ and $e = 5$, we get $5d = 192Q + 1$, and if $Q = 2$ we get $d = 77$. Thus, one of our private keys is $d = 77$.

8. To decrypt y we have

$$x = 56^{77} \bmod 221$$

Again, to do this calculation, we have to factor $y^d \bmod n$ to get

$$\{[(56^4 \bmod 221)^{19} \bmod 221](56 \bmod 221)\} \bmod 221 = 88 = x$$

And we have recovered our original message.

9. In summary, the public keys are n and e, and the private keys are p, q, and d. Encryption and decryption are done using these and Eqs. (8) and (9).

Determining the private key, d, requires that the intruder determine the prime factors of public key, n. For large values of $n = pq$, calculating p and q from only knowing n takes a great deal of computation time. Thus, RSA can be made extremely secure. Whenever the state of the art of computation improves, making it easier for an intruder to factor n and thus break the code, we can simply go to a larger value of n. RSA isn't the only public key encryption system, but many believe it is the most robust. Similar to RSA, there is also the Diffie-Hellman system (1976) and the Merkle-Hellman scheme (1978).

13.5   PROBLEMS

13.1–1* Calculate the probabilities that a word has no errors, detected errors, and undetected errors when a parity-check code with $n = 4$ is used and $\alpha = 0.1$.

13.1–2 Do Prob. 13.1–1 with $n = 9$ and $\alpha = 0.05$.

13.1–3 Consider the square-array code in Fig. 13.1–1. (a) Confirm that if a word has two errors, then they can be detected but not corrected. (b) Discuss what happens when a word contains three errors.

13.1–4   An FEC system contaminated by gaussian white noise must achieve $P_{be} \leq 10^{-4}$ with minimum transmitted power. Three block codes under consideration have the following parameters:

n	k	d_{min}
31	26	3
31	21	5
31	16	7

Determine which code should be used, and calculate the power saving in dB compared to uncoded transmission.

13.1–5*   Do Prob. 13.1–4 with $P_{be} \leq 10^{-6}$.

13.1–6   Calculate α, P_{be}, and P_{ube} at $\gamma_b = 2$, 5, and 10 for an FEC system with gaussian white noise using a (31, 26) block code having $d_{min} = 3$. Plot your results in a form like Fig. 13.1–5.

13.1–7   Do Prob. 13.1–6 for a (31, 21) code having $d_{min} = 5$.

13.1–8   A selective-repeat ARQ system with gaussian white noise is to have $P_{be} = 10^{-5}$ using one of the following block codes for error detection:

n	k	d_{min}
12	11	2
15	11	3
16	11	4

Calculate r_b/r and γ_b for each code and for uncoded transmission. Then plot γ_b in dB versus r_b/r.

13.1–9   Do Prob. 13.1–8 for $P_{be} = 10^{-6}$.

13.1–10*   A go-back-N ARQ system has gaussian white noise, $\gamma_c = 6$ dB, $r = 500$ kbps, and a one-way path length of 45 km. Find P_{be}, the minimum value of N, and the maximum value of r_b using a (15, 11) block code with $d_{min} = 3$ for error detection.

13.1–11   Do Prob. 13.1–10 using a (16, 11) block code with $d_{min} = 4$.

13.1–12   A stop-and-wait ARQ system uses simple parity checking with $n = k + 1$ for error detection. The system has gaussian white noise, $r = 10$ kbps, and a one-way path length of 18 km. Find the smallest value of k such that $P_{be} \leq 10^{-6}$ and $r_b \geq 7200$ bps. Then calculate γ_b in dB.

13.1–13   Do Prob. 13.1–12 with a 60-km path length.

13.1–14‡   Derive m as given in Eq. (19) for a go-back-N ARQ system. *Hint:* If a given word has detected errors in i successive transmissions, then the total number of transmitted words equals $1 + Ni$.

13.1–15 Consider a hybrid ARQ system using a code that corrects t errors and detects $\ell > t$ errors per n-bit word. Obtain an expression for the retransmission probability p when $\alpha \ll 1$. Then take $d_{\min} = 4$ and compare your result with Eq. (15).

13.1–16 Suppose a hybrid selective-repeat ARQ system uses an (n, k) block code with $d_{\min} = 2t + 2$ to correct up to t errors and detect $t + 1$ errors per word. (a) Assume $\alpha \ll 1$ to obtain an approximate expression for the retransmission probability p, and show that

$$P_{be} \approx \binom{n-1}{t+1} \alpha^{t+2}$$

(b) Evaluate α and p for a $(24, 12)$ code with $d_{\min} = 8$ when $P_{be} = 10^{-5}$. Then assume gaussian white noise and find P_{be} for uncoded transmission with the same value of γ_b.

13.2–1 Let U and V be n-bit vectors. (a) By considering the number of 1's in U, V, and $U + V$, confirm that $d(U, V) \le w(U) + w(V)$. (b) Now let $U = X + Y$ and $V = Y + Z$. Show that $U + V = X + Z$ and derive the *triangle inequality*

$$d(X, Z) \le d(X, Y) + d(Y, Z)$$

13.2–2 Let X be a code vector, let Z be any other vector in the code, and let Y be the vector that results when X is received with i bit errors. Use the triangle inequality in Prob. 13.2–1 to show that if the code has $d_{\min} \ge \ell + 1$ and if $i \le \ell$, then the errors in Y are detectable.

13.2–3 Let X be a code vector, let Z be any other vector in the code, and let Y be the vector that results when X is received with i bit errors. Use the triangle inequality in Prob. 13.2–1 to show that if the code has $d_{\min} \ge 2t + 1$ and if $i \le t$, then the errors in Y are correctable.

13.2–4 A triple-repetition code is a systematic $(3, 1)$ block code generated using the submatrix $P = \begin{bmatrix} 1 & 1 \end{bmatrix}$. Tabulate all possible received vectors Y and $S = YH^T$. Then determine the corresponding maximum-likelihood errors patterns and corrected vectors $Y + \hat{E}$.

13.2–5 Construct the lookup table for the $(6, 3)$ block code in Exercise 13.2–1.

13.2–6 Consider a $(5, 3)$ block code obtained by deleting the last column of the P submatrix in Exercise 13.2–1. Construct the lookup table, and show that this code could be used for error detection but not correction.

13.2–7 Let the P submatrix for a $(15, 11)$ Hamming code be arranged such that the row words increase in numerical value from top to bottom. Construct the lookup table and write the check-bit equations.

13.2–8 Suppose a block code with $t = 1$ is required to have $k = 6$ message bits per word. (a) Find the minimum value of n and the number of bits stored in the lookup table. (b) Construct an appropriate P submatrix.

13.2–9   Do Prob. 13.2–8 with $k = 8$.

13.2–10   It follows from Eq. (4) that $XH^T = MA$ with $A = GH^T$. Prove Eq. (9) by showing that any element of A has the property $a_{ij} = 0$.

13.2–11   The original (7, 4) Hamming code is a *nonsystematic* code with

$$H = \begin{bmatrix} 1 & 0 & 1 & 0 & 1 & 1 & 1 \\ 0 & 1 & 1 & 0 & 0 & 1 & 1 \\ 0 & 0 & 0 & 1 & 1 & 0 & 1 \end{bmatrix}$$

(a) Construct the lookup table and explain the rationale for this nonsystematic form. (b) Write the equations for s_1, s_2, and s_3 taking $Y = X = (x_1 \; x_2 \; \cdots \; x_7)$. Then determine which are the message and check bits in X, and obtain the check-bit equations.

13.2–12   The (7, 4) Hamming code can be *extended* to form an (8, 4) code by appending a fourth check bit chosen such that all codewords have even parity. (a) Apply this extension process to Table 13.2–1 to show that $d_{min} = 4$. (b) Find the corresponding equation for c_4 in terms of the message bits. (c) Explain how the decoding system in Fig. 13.2–2 should be modified to perform double-error detection as well as single-error correction.

13.2–13*   Consider a systematic (7, 3) cyclic code generated by $G(p) = p^4 + p^3 + p^2 + 0 + 1$. Find $Q_M(p)$, $C(p)$, and X when $M = (1 \; 0 \; 1)$. Then take $Y = X'$ and confirm that $S(p) = 0$. You may carry out the divisions using binary words rather than polynomials if you wish.

13.2–14   Do Prob. 13.2–13 with $G(p) = p^4 + 0 + p^2 + p + 1$.

13.2–15   Given $G(p)$, the equivalent generator matrix of a systematic cyclic code can be found from the polynomials $R_i(p) = \text{rem}\left[p^{n-i}/G(p)\right]$, $1 \le i \le k$, which correspond to the rows of the P submatrix. Use this method to obtain P for a (7, 4) Hamming code generated by $G(p) = p^3 + 0 + p + 1$. (You may carry out the divisions using binary words rather than polynomials if you wish.) Compare your result with the G matrix in Example 13.2–1.

13.2–16   Do Prob. 13.2–15 with $G(p) = p^3 + p^2 + 0 + 1$.

13.2–17   Figure P13.2–17 is a shift-register circuit that divides an arbitrary mth-order polynomial $Z(p)$ by a fixed polynomial $G(p) = p^q + g_{q-1}p^{q-1} + \cdots + g_1p + 1$. If the register has been cleared before $Z(p)$ is shifted in, then the output equals the quotient and the remainder appears in the register after m shift cycles. Confirm the division operation by constructing a table similar to Fig. 13.2–4b, taking $Z(p) = p^3M(p)$ and $G(p)$ as in Example 13.2–3.

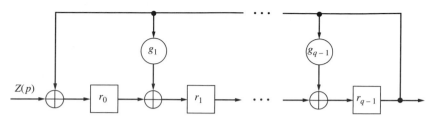

Figure P13.2–17

13.2–18   Use the shift-register polynomial divider in Prob. 13.2–17 to diagram a circuit that produces the syndrome $S(p)$ from $Y(p)$ for the code in Example 13.2–3. Then construct a table like Fig. 13.2–4b to show that $S(p) = 0$ when $Y = (1\ 1\ 0\ 0\ 0\ 1\ 0)$.

13.2–19*   Suppose an uncoded system has a $P_{be} = 10^{-5}$. What is P_{be} and α for the following Hamming codes: (a) (7, 4), (b) (15, 11), (c) (31, 26)?

13.2–20   For a fixed value of γ_c show that a (7, 4) code has a lower P_{be} than a (31, 26) code. Why is this the case, and what is the advantage of the (31, 26) code?

13.2–21   Suppose an uncoded system has a $P_{be} = 10^{-5}$. What would be the overall P_{be} if the data were sent three times but at three times the data rate with the receiver deciding the bit value based on a majority vote? Compare your results to an equivalent system that uses (7, 4) and (15, 11) FEC coding.

13.2–22   Determine M given the following Y vectors that were generated via a (7, 4) Hamming code. Each one may have a single error. (a) $Y = [0100101]$, (b) $Y = [0111111]$, (c) $Y = [1010111]$, (d) $Y = [1101000]$.

13.2–23*   Determine X for ASCII letter "E" and CRC-12 polynomial.

13.2–24   For the transmitted codeword in Prob. 13.2–23, show that the following errors will be detected: errors in the last two digits.

13.2–25   A channel that transmits 7-bit ASCII information at 9600 bits per second often has noise bursts lasting 125 ms, but with at least 125 ms between bursts. Using a (63, 45) BCH code, design an interleaving system to eliminate the effects of these error bursts. What is the maximum delay time between interleaving and deinterleaving?

13.3–1   Diagram the encoder for a systematic (3, 2, 3) convolutional code. Label the input and output rates and the current input and state at an arbitrary time.

13.3–2   Diagram the encoder for a systematic (4, 3, 1) convolutional code. Label the input and output rates and the current input and state at an arbitrary time.

13.3–3   A (3, 1, 2) encoder achieves maximum free distance when

$$x'_j = m_{j-2} \oplus m_j \quad x''_j = x'''_j = m_{j-1} \oplus x'_j$$

(*a*) Construct the code trellis and state diagram. Then find the state and output sequence produced by the input sequence 1011001111. (*b*) Construct the modified state diagram like Fig. 13.1–8*a*, identify the minimum-weight path or paths, and determine the values of d_f and $M(d_f)$.

13.3–4   Do Prob. 13.3–3 for a (2, 1, 3) encoder with

$$x'_j = m_{j-3} \oplus m_{j-1} \oplus m_j \quad x''_j = m_{j-2} \oplus x'_j$$

which achieves maximum free distance. Use the pattern in Fig. P13.3–4 for your state diagram.

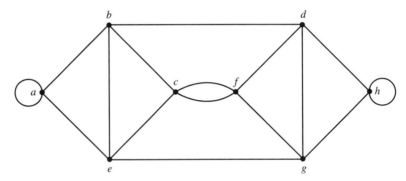

Figure P13.3–4

13.3–5   Determine the output from the encoder of Fig. P13.3–5 for message input of $M = (1101011101110000 \ldots)$.

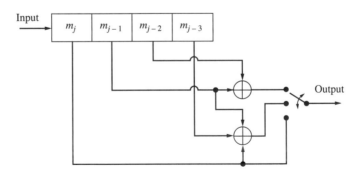

Figure P13.3–5

13.3–6*   Determine the output from the (3, 2, 1) encoder of Fig. 13.3–5 for message input of $M = (1101011101110000 \ldots)$.

13.3–7   Determine the output from the (3, 2, 1) encoder of Fig. P13.3–7 for message input of $M = (1101011101110000\ldots)$.

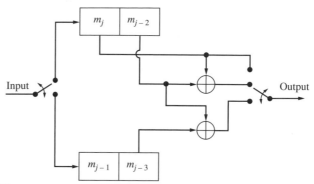

Figure P13.3–7

13.3–8   Determine if the codes produced by the encoders of Figs. 13.3–2, P13.3–5, P13.3–7, and P13.3–8 are catastrophic.

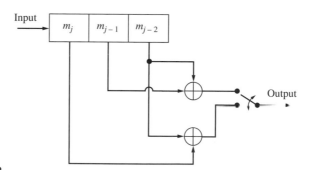

Figure P13.3–8

13.3–9   Use block-diagram reduction of Fig. 13.3–8b to get $T(D, I)$ as in Eq. (7a).

13.3–10   Derive Eq. (10) from Eqs. (8) and (9).

13.3–11*   Consider a (2, 1, 1) code with $x_j' = m_j$ and $x_j'' = m_{j-1} \oplus m_j$. (a) Construct the code trellis and modified state diagram. Then identify the minimum-weight path or paths and find the values of d_f and $M(d_f)$. (b) Use block-diagram reduction to obtain $T(D, I)$ in the form of Eq. (7a). (c) Find and compare P_{be} from Eqs. (9) and (10), assuming $\alpha \ll 1$.

13.3–12   Do Prob. 13.3–11 for a (3, 1, 2) code with $x_j' = m_j$, $x_j'' = m_{j-2} \oplus m_j$, and $x_j''' = m_{j-2} \oplus m_{j-1} \oplus m_j$.

13.3–13*   Use the Viterbi algorithm and the code trellis in Fig. 13.3–4 to find the sequences $Y + \hat{E}$ and $\hat{M}$ when $Y = 01\ 11\ 01\ 01\ 11\ 01\ 10\ 11$. If two paths arriving at a given node have equal running metrics, arbitrarily keep the upper path.

13.3–14   Construct the code trellis for a (2, 1, 2) code with $x_j' = m_{j-1} \oplus m_j$ and $x_j'' = m_{j-2} \oplus m_{j-1}$. Then apply the Viterbi algorithm to find the sequences $Y + \hat{E}$ and $\hat{M}$ when $Y = 10\ 11\ 01\ 01\ 10\ 01\ 10\ 11$. If two paths arriving at a given node have equal running metrics, arbitrarily keep the upper path.

13.3–15   A systematic (2, 1, 4) code intended for feedback decoding with majority logic has $x_j'' = m_{j-4} \oplus m_{j-3} \oplus m_j$. Construct a parity-check table like Fig. 13.3–14 showing all the error bits that appear in the check sums s_{j-4} through s_j. Then find an orthogonal set of three check sums, and draw a complete diagram of the decoder.

13.3–16   Do Prob. 13.3–15 with $x_j'' = m_{j-4} \oplus m_{j-1} \oplus m_j$.

13.3–17   Show the block diagrams for the following RSC encoders: (a) (35, 21), (b) (34, 23).

13.4–1*   Encrypt and decrypt sequence $x = 8, 27, 51$ using the RSA algorithm with: (a) $p = 5$, $q = 11$, $e = 7$; (b) $p = 11$, $q = 37$, $e = 7$; (c) $p = 13$, $q = 37$, $d = 173$.

13.4–2   List all the private/public keys for an RSA system with $p = 11$ and $q = 31$.

chapter
14

Bandpass Digital Transmission

CHAPTER OUTLINE

ong-haul digital transmission usually requires CW modulation to generate a bandpass signal suited to the transmission medium—be it radio, cable, telephone lines (for personal computer Internet connection), or whatever. Just as there are a multitude of modulation methods for analog signals, there are many ways of impressing digital information upon a carrier wave. This chapter applies concepts of baseband digital transmission and CW modulation to the study of bandpass digital transmission.

We begin with waveforms and spectral analysis of digital CW modulation for binary and M-ary modulating signals. Then we focus on the demodulation of binary signals in noise to bring out the distinction between *coherent* (synchronous) detection and *noncoherent* (envelope) detection. Next we deal with quadrature-carrier M-ary systems, leading to a comparison of modulation methods with regard to spectral efficiency, hardware complexity, and system performance in the face of corrupting noise. We then conclude with a discussion of trellis-coded modulation (TCM), where we combine convolutional coding and M-ary digital modulation.

OBJECTIVES

After studying this chapter and working the exercises, you should be able to do each of the following:

1. Identify the format of binary ASK, FSK, PSK, and DSB with baseband pulse-shaping waveforms (Sect. 14.1).
2. State the distinctions between the various ASK, PSK, and FSK methods (Sect. 14.1).
3. Calculate the error probabilities of binary and M-ary modulation systems. Or, for a specified error probability and noise level, specify the transmitter or receiver power (Sects. 14.2, 14.3, and 14.4).
4. Describe the operation of the correlation receiver (Sect. 14.2).
5. Predict error probabilities given conditional PDFs (Sect. 14.2).
6. Specify the appropriate detector(s) for each of the binary and M-ary modulation systems (Sects. 14.2, 14.3, and 14.4).
7. Explain trellis-coded modulation and predict the corresponding coding gains over conventional digital modulation methods (Sect. 14.5).

14.1   DIGITAL CW MODULATION

A digital signal can modulate the amplitude, frequency, or phase of a sinusoidal carrier wave. If the modulating waveform consists of NRZ rectangular pulses, then the modulated parameter will be switched or *keyed* from one discrete value to another. Figure 14.1–1 illustrates binary **amplitude-shift keying** (ASK), **frequency-shift keying** (FSK), and **phase-shift keying** (PSK). Also shown, for contrast, is the waveform produced by DSB modulation with Nyquist pulse shaping at baseband. Other modulation techniques combine amplitude and phase modulation, with or without baseband pulse shaping.

In this section we'll define specific types of digital modulation in terms of mathematical models and/or transmitter diagrams. We'll also examine their power spectra and estimate therefrom the transmission bandwidth required for a given digital signaling rate. As preparation, we first develop a technique for spectral analysis of bandpass digital signals.

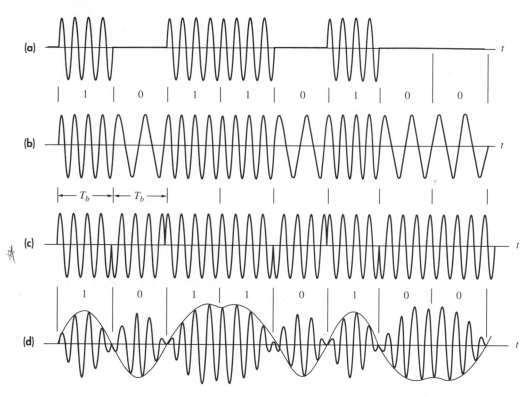

Figure 14.1–1   Binary modulated waveforms. (a) ASK; (b) FSK; (c) PSK; (d) DSB with baseband pulse shaping.

Spectral Analysis of Bandpass Digital Signals

Any modulated bandpass signal may be expressed in the quadrature-carrier form

$$x_c(t) = A_c[x_i(t) \cos(\omega_c t + \theta) - x_q(t) \sin(\omega_c t + \theta)] \qquad [1]$$

The carrier frequency f_c, amplitude A_c, and phase θ are constant, while the time-varying i (in-phase) and q (quadrature) components contain the message. Spectral analysis of $x_c(t)$ is relatively easy when the i and q components are statistically independent signals and at least one has zero mean. Then, from the superposition and modulation relations in Sect. 9.2, the power spectrum of $x_c(t)$ becomes

$$G_c(f) = \frac{A_c^2}{4}[G_i(f - f_c) + G_i(f + f_c) + G_q(f - f_c) + G_q(f + f_c)]$$

where $G_i(f)$ and $G_q(f)$ are the power spectra of the i and q components. For a more compact expression, we define the *equivalent lowpass spectrum*

$$G_{\ell p}(f) \triangleq G_i(f) + G_q(f) \qquad [2]$$

so that

$$G_c(f) = \frac{A_c^2}{4}[G_{\ell p}(f - f_c) + G_{\ell p}(f + f_c)] \qquad [3]$$

Thus, the bandpass spectrum is obtained from the equivalent lowpass spectrum by simple frequency translation.

Now suppose that the i component is an M-ary digital signal, say

$$x_i(t) = \sum_k a_k p(t - kD) \qquad [4a]$$

where a_k represents a sequence of source digits with rate $r = 1/D$. We assume throughout that source digits are equiprobable, statistically independent, and uncorrelated. Consequently, Eq. (11), Sect. 11.1, applies here and

$$G_i(f) = \sigma_a^2 r |P(f)|^2 + (m_a r)^2 \sum_{n=-\infty}^{\infty} |P(nr)|^2 \delta(f - nr) \qquad [4b]$$

Similar expressions hold when the q component is another digital waveform.

The pulse shape $p(t)$ in Eq. (4a) depends on the baseband filtering, if any, and on the type of modulation. Keyed modulation involves NRZ rectangular pulses, and we'll find it convenient to work with pulses that start at $t = kD$, rather than being centered at $t = kD$ as in Chap. 11. Accordingly, let

$$p_D(t) \triangleq u(t) - u(t - D) = \begin{cases} 1 & 0 < t < D \\ 0 & \text{otherwise} \end{cases} \qquad [5a]$$

whose Fourier transform yields

$$|P_D(f)|^2 = D^2 \operatorname{sinc}^2 fD = \frac{1}{r^2} \operatorname{sinc}^2 \frac{f}{r} \qquad [5b]$$

If $p(t) = p_D(t)$ in Eq. (4a), then the continuous spectral term in Eq. (4b) will be proportional to $|P_D(f)|^2$. Since $\operatorname{sinc}^2(f/r)$ is not bandlimited, we conclude from Eqs. (2) and (3) that keyed modulation requires $f_c \gg r$ in order to produce a *bandpass* signal.

Amplitude Modulation Methods

The binary ASK waveform illustrated in Fig. 14.1–1a could be generated simply by turning the carrier on and off, a process described as **on-off keying** (OOK). In general, an M-ary ASK waveform has $M - 1$ discrete "on" amplitudes as well as the "off" state. Since there are no phase reversals or other variations, we can set the q component of $x_c(t)$ equal to zero and take the i component to be a *unipolar* NRZ signal, namely

$$x_i(t) = \sum_k a_k p_D(t - kD) \qquad a_k = 0, 1, \ldots, M - 1 \qquad [6a]$$

The mean and variance of the digital sequence are

$$m_a = \overline{a_k} = \frac{M-1}{2} \qquad \sigma_a^2 = \overline{a_k^2} - m_a^2 = \frac{M^2-1}{12} \qquad \text{[6b]}$$

Hence, the equivalent lowpass spectrum is

$$G_{\ell p}(f) = G_i(f) = \frac{M^2-1}{12\,r}\; \text{sinc}^2 \frac{f}{r} + \frac{(M-1)^2}{4}\,\delta(f) \qquad \text{[7]}$$

obtained with the help of Eqs. (2), (4b), and (5b).

Figure 14.1–2 shows the resulting bandpass spectrum $G_c(f)$ for $f > 0$. Most of the signal power is contained within the range $f_c \pm r/2$, and the spectrum has a *second-order rolloff* proportional to $|f - f_c|^{-2}$ away from the carrier frequency. These considerations suggest the estimated transmission bandwidth to be $B_T \approx r$. If an M-ary ASK signal represents binary data at rate $r_b = r \log_2 M$, then $B_T \approx r_b/\log_2 M$ or

$$r_b/B_T \approx \log_2 M \qquad \text{bps/Hz} \qquad \text{[8]}$$

This ratio of bit rate to transmission bandwidth serves as our measure of modulation "speed" or spectral efficiency. Binary OOK has the poorest spectral efficiency since $r_b/B_T \approx 1$ bps/Hz when $M = 2$.

Drawing upon the principle of quadrature-carrier multiplexing, **quadrature-carrier AM (QAM)** achieves twice the modulation speed of binary ASK. Figure 14.1–3a depicts the functional blocks of a binary QAM transmitter with a *polar* binary input at rate r_b. The serial-to-parallel converter divides the input into two streams consisting of alternate bits at rate $r = r_b/2$. Thus, the i and q modulating signals are represented by

$$x_i(t) = \sum_k a_{2k}\, p_D(t - kD) \qquad x_q(t) = \sum_k a_{2k+1}\, p_D(t - kD)$$

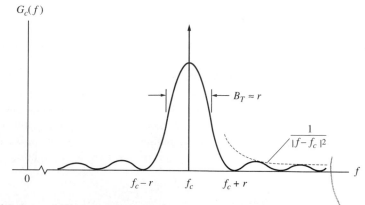

Figure 14.1–2   ASK power spectrum.

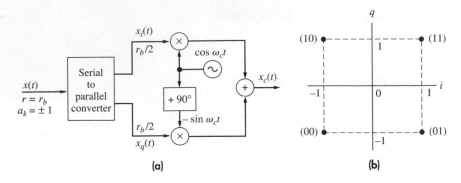

Figure 14.1–3   Binary QAM. (a) Transmitter; (b) signal constellation.

where $D = 1/r = 2T_b$ and $a_k = \pm 1$. The peak modulating values are $x_i = x_q = \pm 1$ during an arbitrary interval $kD < t < (k + 1)D$. Figure 14.1–3b conveys this information as a two-dimensional **signal constellation.** The four signal points have been labeled with the corresponding pairs of source bits, known as **dibits.**

Summing the modulated carriers finally yields the QAM signal in the form of Eq. (1). The i and q components are independent but they have the same pulse shape and the same statistical values, namely, $m_a = 0$ and $\sigma_a^2 = 1$. Thus,

$$G_{\ell p}(f) = 2 \times r|P_D(f)|^2 = \frac{4}{r_b} \operatorname{sinc}^2 \frac{2f}{r_b} \qquad [9]$$

where we've used Eqs. (4b) and (5b) with $r = r_b/2$. Binary QAM achieves $r_b/B_T \approx 2$ bps/Hz because the dibit rate equals one-half of the input bit rate, reducing the transmission bandwidth to $B_T \approx r_b/2$.

Keep in mind, however, that ASK and QAM spectra actually extend beyond the estimated transmission bandwidth. Such spectral "spillover" outside B_T becomes an important concern in radio transmission and frequency-division multiplexing systems when it creates interference with other signal channels. Bandpass filtering at the output of the modulator controls spillover, but heavy filtering introduces ISI in the modulated signal and should be avoided.

Spectral efficiency without spillover is achieved by the **vestigial-sideband** modulator diagrammed in Fig. 14.1–4a. This VSB method applies Nyquist pulse shaping to a polar input signal, as covered in Sect. 11.3, producing a bandlimited modulating signal with $B = (r/2) + \beta_N$. The VSB filter then removes all but a vestige of width β_V from one sideband, so $G_c(f)$ looks something like Fig. 14.4b—a bandlimited spectrum with $B_T = (r/2) + \beta_N + \beta_V$. Therefore, if $r = r_b/\log_2 M$, then

$$r_b/B_T \leq 2 \log_2 M \qquad [10]$$

and the upper bound holds when $\beta_N \ll r$ and $\beta_V \ll r$.

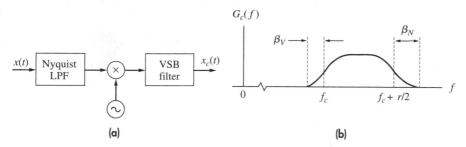

Figure 14.1–4   Digital VSB. (a) Transmitter; (b) power spectrum.

Binary data is to be transmitted on a 1-MHz carrier. Spillover is not a concern, but B_T must satisfy the fractional bandwidth constraint $B_T/f_c \leq 0.1$. Estimate the maximum possible bit rate r_b when the modulation is: (a) OOK, (b) binary QAM, (c) VSB with $M = 8$.

EXERCISE 14.1–1

Phase Modulation Methods

The binary PSK waveform back in Fig. 14.1–1c contains phase shifts of $\pm\pi$ radians, often described as **binary phase-shift keying** (BPSK) or **phase-reversal keying** (PRK). An M-ary PSK signal has phase shift ϕ_k in the time interval $kD < t < (k + 1)D$, expressed in general by

$$x_c(t) = A_c \sum_k \cos\left(\omega_c t + \theta + \phi_k\right) p_D(t - kD) \qquad [11]$$

Trigonometric expansion of the cosine function yields our desired quadrature-carrier form with

$$x_i(t) = \sum_k I_k p_D(t - kD) \qquad x_q(t) = \sum_k Q_k p_D(t - kD) \qquad [12a]$$

where

$$I_k = \cos\phi_k \qquad Q_k = \sin\phi_k \qquad [12b]$$

To ensure the largest possible phase modulation for a given value of M, we'll take the relationship between ϕ_k and a_k to be

$$\phi_k = \pi(2a_k + N)/M \qquad a_k = 0, 1, \ldots, M - 1 \qquad [13]$$

in which N is an integer, usually 0 or 1.

Examples of PSK signal constellations are shown in Fig. 14.1–5, including the corresponding binary words in Gray code. The binary words for adjacent signal

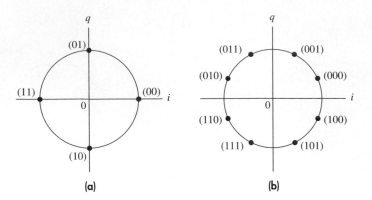

Figure 14.1–5   PSK signal constellations. (a) $M = 4$; (b) $M = 8$.

points therefore differ by just one bit. The PSK signal with $M = 4$ and $N = 0$ represented in Fig. 14.1–5a is designated **quaternary** or **quadriphase** PSK (QPSK). Had we taken QPSK with $N = 1$, the signal points would have been identical to binary QAM (Fig. 14.1–3b). Indeed, you can think of binary QAM as two BPSK signals on quadrature carriers. M-ary PSK differs from M-ary ASK, of course, since an ideal PSK waveform always has a *constant envelope*.

PSK spectral analysis becomes a routine task after you note from Eqs. (12b) and (13) that

$$\overline{I}_k = \overline{Q}_k = 0 \qquad \overline{I_k^2} = \overline{Q_k^2} = 1/2 \qquad \overline{I_k Q_j} = 0$$

Hence, the i and q components are statistically independent, and

$$G_{\ell p}(f) = 2 \times \frac{r}{2}|P_D(f)|^2 = \frac{1}{r}\operatorname{sinc}^2\frac{f}{r} \qquad \text{[14]}$$

Comparison with Eq. (7) reveals that $G_c(f)$ will have the same shape as an ASK spectrum (Fig. 14.1–2) without the carrier-frequency impulse. The absence of a discrete carrier component means that PSK has better power efficiency, but the spectral efficiency is the same as ASK.

Some PSK transmitters include a BPF to control spillover. However, bandpass filtering produces *envelope variations* via the FM-to-AM conversion effect discussed in Sect. 5.2. (Remember that a stepwise phase shift is equivalent to an FM impulse.) The typical nonlinear amplifier used at microwave carrier frequencies will flatten out these envelope variations and restore spillover—largely negating the function of the BPF. A special form of QPSK called **staggered** or **offset-keyed** QPSK (OQPSK) has been devised to combat this problem. The OQPSK transmitter diagrammed in Fig. 14.1–6 delays the quadrature signal such that modulated phase shifts occur every $D/2 = T_b$ seconds but they never exceed $\pm\pi/2$ radians. Cutting the maximum phase shift in half results in much smaller envelope variations after bandpass filtering.

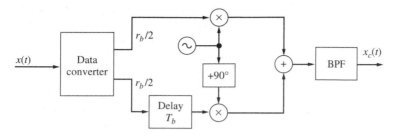

Figure 14.1–6   Offset-keyed QPSK transmitter.

When envelope variations are allowable, combined **amplitude-phase keying** (APK) is an attractive family of modulation methods. APK has essentially the same spectral efficiency as PSK, but it can achieve better performance with respect to noise and errors. Further discussion is postponed to Sect. 14.4.

Draw the signal constellation for binary PSK with $\phi_k = \pi(2a_k - 1)/4$ and $a_k = 0, 1$. Then determine the lowpass equivalent spectrum and sketch $G_c(f)$.

EXERCISE 14.1–2

Frequency Modulation Methods

There are two basic methods for digital frequency modulation. **Frequency-shift keying** (FSK) is represented conceptually by Fig. 14.1–7a, where the digital signal $x(t)$ controls a switch that selects the modulated frequency from a bank of M oscillators. The modulated signal is discontinuous at every switching instant $t = kD$. Unless the amplitude, frequency, and phase of each oscillator has been carefully adjusted, the resultant output spectrum will contain relatively large *sidelobes* which don't carry any additional information and thus waste bandwidth. Discontinuities are avoided in **continuous-phase FSK** (CPFSK) represented in Fig. 14.1–7b, where $x(t)$ modulates the frequency of a single oscillator. Both forms of digital frequency modulation pose significant difficulty for spectral analysis, so we'll limit our consideration to some selected cases.

First, consider M-ary FSK. Let all oscillators in Fig. 14.1–7a have the same amplitude A_c and phase θ, and let their frequencies be related to a_k by

$$f_k = f_c + f_d a_k \qquad a_k = \pm 1, \pm 3, \ldots, \pm(M-1) \qquad \text{[15a]}$$

which assumes that M is even, Then

$$x_c(t) = A_c \sum_k \cos\left(\omega_c t + \theta + \omega_d a_k t\right) p_D(t - kD) \qquad \text{[15b]}$$

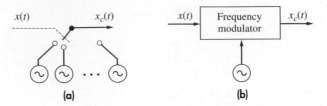

Figure 14.1–7   Digital frequency modulation. (a) FSK; (b) continuous-phase FSK.

where $\omega_d = 2\pi f_d$. The parameter f_d equals the frequency shift away from f_c when $a_k = \pm 1$, and adjacent frequencies are spaced by $2f_d$. Continuity of $x_c(t)$ at $t = kD$ is assured if $2\omega_d D = 2\pi N$ where N is an integer.

We'll analyze a version of binary FSK known as **Sunde's** FSK, defined by the foregoing relations with $M = 2$, $D = T_b = 1/r_b$, and $N = 1$. Then $p_D(t) = u(t) - u(t - kT_b)$ and

$$f_d = r_b/2 \tag{16}$$

After trigonometric expansion of $x_c(t)$, we use the fact that $a_k = \pm 1$ to write

$$\cos \omega_d a_k t = \cos \omega_d t \qquad \sin \omega_d a_k t = a_k \sin \omega_d t$$

The i component thereby reduces to

$$x_i(t) = \cos \pi r_b t \tag{17a}$$

independent of a_k. The q component contains a_k in the form

$$x_q(t) = \sum_k a_k \sin (\pi r_b t)[u(t - kT_b) - u(t - kT_b - T_b)] \tag{17b}$$

$$= \sum_k Q_k p(t - kT_b) \qquad Q_k = (-1)^k a_k$$

where

$$p(t) = \sin (\pi r_b t)[u(t) - u(t - T_b)] \tag{17c}$$

The intervening manipulations are left to you as an instructive exercise.

Once again, we have independent i and q components. The i component, being a sinusoid, just contributes spectral impulses at $\pm r_b/2$ in the equivalent lowpass spectrum. The power spectrum of the q component contains no impulses since $\overline{Q_k} = 0$, whereas $\overline{Q_k^2} = \overline{a_k^2} = 1$. Thus,

$$G_{\ell p}(f) = \frac{1}{4}\left[\delta\left(f - \frac{r_b}{2}\right) + \delta\left(f + \frac{r_b}{2}\right)\right] + r_b|P(f)|^2 \tag{18a}$$

where

$$|P(f)|^2 = \frac{1}{4\,r_b^2}\left[\operatorname{sinc}\frac{f-(r_b/2)}{r_b} + \operatorname{sinc}\frac{f+(r_b/2)}{r_b}\right]^2 \qquad \textbf{[18b]}$$

$$= \frac{4}{\pi^2 r_b^2}\left[\frac{\cos(\pi f/r_b)}{(2f/r_b)^2 - 1}\right]^2$$

The resulting bandpass spectrum is shown in Fig. 14.1–8.

Observe that the impulses correspond to the keyed frequencies $f_c \pm f_d = f_c \pm r_b/2$, and that the spectrum has a fourth-order rolloff. This rapid rolloff means that Sunde's FSK has very little spillover for $|f - f_c| > r_b$. We therefore take $B_T \approx r_b$, even though the central lobe of $G_c(f)$ is 50% wider than the central lobe of a binary ASK or PSK spectrum.

Another special case is **M-ary orthogonal** *FSK*, in which the M keyed frequencies are equispaced by $2f_d = 1/2D = r/2$. Without attempting the spectral analysis, we can surmise that $B_T \geq M \times 2f_d = Mr/2 = Mr_b/(2\log_2 M)$. Therefore,

$$r_b/B_T \leq (2\log_2 M)/M \qquad \textbf{[19]}$$

and the modulation speed is less than M-ary ASK or PSK for $M \geq 4$. In other words, orthogonal FSK is a *wideband* modulation method.

CPFSK may be wideband or narrowband depending on the frequency deviation. Let $x(t)$ in Fig. 14.1–7b start at $t = 0$, so

$$x(t) = \sum_{k=0}^{\infty} a_k p_D(t - kD) \qquad a_k = \pm 1, \pm 2, \ldots, \pm(M - 1)$$

and frequency modulation produces the CPFSK signal

$$x_c(t) = A_c \cos\left[\omega_c t + 0 + \omega_d \int_0^t x(\lambda)\,d\lambda\right] \qquad t \geq 0$$

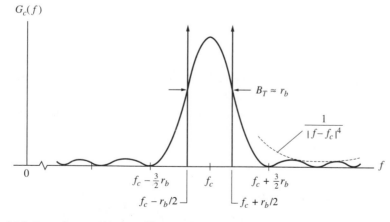

Figure 14.1–8   Power spectrum of binary FSK with $f_d = r_b/2$.

To bring out the difference between CPFSK and FSK, consider the integral

$$\int_0^t x(\lambda)\, d\lambda = \sum_{k=0}^{\infty} a_k \int_0^t P_D(\lambda - kD)\, d\lambda$$

in which $p_D(\lambda - kD) = 0$ except for $kD < \lambda < (k+1)D$ when $p_D(\lambda - kD) = 1$. Piecewise integration yields

$$\int_0^t x(\lambda)\, d\lambda = a_0 t \qquad\qquad\qquad 0 < t < D$$

$$= a_0 D + a_1(t - D) \qquad\qquad D < t < 2D$$

$$= \left(\sum_{j=0}^{k-1} a_j \right) D + a_k(t - kD) \qquad kD < t < (k+1)D$$

Now we can express $x_c(t)$ in the summation form

$$x_c(t) = A_c \sum_{k=0}^{\infty} \cos\left[\omega_c t + \theta + \phi_k + \omega_d a_k(t - kD) \right] p_D(t - kD) \qquad [20a]$$

where $t \geq 0$ and

$$\phi_k \overset{\Delta}{=} \omega_d D \sum_{j=0}^{k-1} a_j \qquad\qquad\qquad [20b]$$

with the understanding that $\phi_k = 0$ for $k = 0$.

Equation (20) shows that CFPSK has a frequency shift $f_d a_k$ in the interval $kD < t < (k+1)D$, just like FSK. But it also has a phase shift ϕ_k that depends on the *previous* digits. This phase shift results from the frequency-modulation process and ensures phase continuity for all t. Unfortunately, the past history embodied in ϕ_k greatly complicates CPFSK spectral analysis. Proakis (2001, Chap. 4) gives further details and plots of $G_c(f)$ for various values of f_d when $M = 2$, 4, and 8. To conclude this section, we'll examine an important special case of binary CPFSK called **minimum-shift keying** (MSK).

EXERCISE 14.1–3   Carry out the details omitted in the derivation of Eqs. (17a)–(17c). *Hint:* Show that $\sin \omega_d t = \sin\left[\omega_d(t - kT_b) + k\pi \right] = \cos(k\pi) \times \sin\left[\omega_d(t - kT_b) \right]$.

Minimum-Shift Keying★

Minimum-shift keying, also known as **fast** FSK, is binary CPFSK with

$$f_d = \frac{r_b}{4} \qquad a_k = \pm 1 \qquad \phi_k = \frac{\pi}{2} \sum_{j=0}^{k-1} a_j \qquad\qquad [21]$$

Notice that the frequency spacing $2f_d = r_b/2$ is half that of Sunde's FSK. This fact, together with the continuous-phase property, results in a more compact spectrum, free of impulses. Subsequent analysis will prove that $G_i(f) = G_q(f)$ and

$$G_{\ell p}(f) = \frac{1}{r_b} \left[\text{sinc} \frac{f - (r_b/4)}{(r_b/2)} + \text{sinc} \frac{f + (r_b/4)}{(r_b/2)} \right]^2 \qquad [22]$$

$$= \frac{16}{\pi^2 r_b} \left[\frac{\cos(2\pi f/r_b)}{(4 f/r_b)^2 - 1} \right]^2$$

The bandpass spectrum $G_c(f)$ plotted in Fig. 14.1–9 has minuscule spillover beyond the central lobe of width $3r_b/2$. The rapid rolloff justifies taking $B_T \approx r_b/2$, so

$$r_b/B_T \approx 2 \text{ bps/Hz}$$

which is twice the modulation speed of Sunde's FSK and accounts for the name "fast" FSK.

Our investigation of MSK starts with the usual trigonometric expansion to put $x_c(t)$ in quadrature-carrier form with

$$x_i(t) = \sum_{k=0}^{\infty} \cos(\phi_k + a_k c_k) p_{T_b}(t - kT_b)$$

$$x_q(t) = \sum_{k=0}^{\infty} \sin(\phi_k + a_k c_k) p_{T_b}(t - kT_b)$$

where

$$c_k \triangleq \frac{\pi r_b}{2}(t - kT_b) \qquad p_{T_b}(t) = u(t) - u(t - kT_b)$$

We'll also draw upon the behavior of ϕ_k versus k as displayed in the trellis pattern of Fig. 14.1–10. This pattern clearly reveals that $\phi_k = 0, \pm\pi, \pm2\pi, \dots$, for even values of k while $\phi_k = \pm\pi/2, \pm3\pi/2, \dots$, for odd values of k.

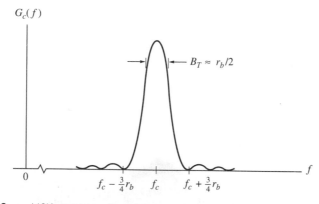

Figure 14.1–9   MSK power spectrum.

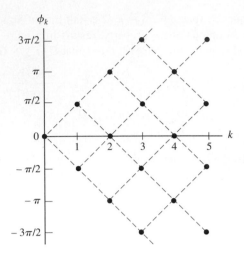

Figure 14.1–10   MSK phase trellis.

As a specific example, let the input message sequence be 100010111. The resulting phase path ϕ_k is shown in Fig. 14.1–11a, taking $a_k = +1$ for input bit 1 and $a_k = -1$ for input bit 0. The corresponding i and q waveforms calculated from the foregoing expressions are sketched in Fig. 14.1–11b. We see that both waveforms have zeros spaced by $2T_b$, but staggered such that the zeros of $x_i(t)$ coincide with the peaks of $x_q(t)$, and vice versa. These observations will guide our subsequent work.

Consider an arbitrary time interval between adjacent zeros of the i component, i.e.,

$$(k - 1)T_b < t < (k + 1)T_b$$

with k being *even*. During this interval,

$$x_i(t) = \cos(\phi_{k-1} + a_{k-1}c_{k-1})p_{T_b}[t - (k - 1)T_b]$$
$$+ \cos(\phi_k + a_k c_k)p_{T_b}(t - kT_b)$$

which we seek to combine into a single term. Since k is even, $\sin \phi_k = 0$, and routine trigonometric manipulations yield

$$\cos(\phi_k + a_k c_k) = \cos \phi_k \cos(a_k c_k) = \cos \phi_k \cos c_k$$

Likewise, using

$$\cos \phi_{k-1} = 0 \qquad \phi_{k-1} = \phi_k - a_{k-1}\pi/2 \qquad c_{k-1} = c_k + \pi/2$$

we get

$$\cos(\phi_{k-1} + a_{k-1}c_{k-1}) = -\sin \phi_{k-1} \sin(a_{k-1}c_{k-1})$$
$$= a_{k-1}^2 \cos \phi_k \cos c_k = \cos \phi_k \cos c_k$$

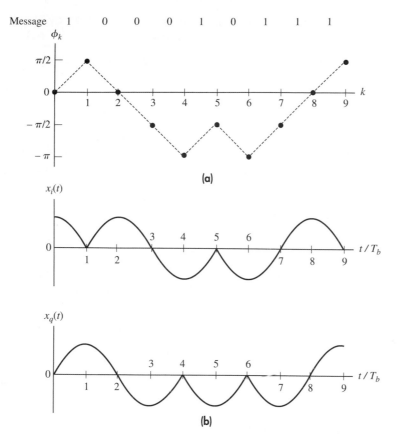

Figure 14.1-11   Illustration of MSK. (a) Phase path; (b) i and q waveforms.

Thus, for the interval in question,

$$x_i(t) = \cos \phi_k \cos c_k \{p_{T_b}[t - (k-1)T_b] + p_{T_b}(t - kT_b)\}$$

$$= \cos \phi_k \cos [(\pi r_b/2)(t - kT_b)][u(t - kT_b + T_b) - u(t - kT_b - T_b)]$$

Summing intervals to encompass all $t \geq 0$ finally yields

$$x_i(t) = \sum_{k \text{ even}} I_k p(t - kT_b) \qquad I_k = \cos \phi_k \qquad \text{[23]}$$

where

$$p(t) = \cos (\pi r_b t/2)[u(t + T_b) - u(t - T_b)] \qquad \text{[24]}$$

This result checks out against the waveform in Fig. 14.1–11b since $I_k = \cos \phi_k = \pm 1$ when k is even.

Now, for the q component, we consider the interval $(k - 1)T_b < t < (k + 1)T_b$ with k *odd*. Similar manipulations as before lead to

$$x_q(t) = \sin \phi_k \cos c_k\{p_{T_b}[t - (k - 1)T_b] + p_{T_b}(t - kT_b)\}$$

Thus, for all $t \geq 0$,

$$x_q(t) = \sum_{k \text{ odd}} Q_k p(t - kT_b) \qquad Q_k = \sin \phi_k \qquad \text{[25]}$$

which also agrees with Fig. 14.1–11*b*. Equation (22) follows from Eqs. (23)–(25) since the i and q components are independent, with $\overline{I_k} = \overline{Q_k} = 0$ and $\overline{I_k^2} = \overline{Q_k^2} = 1$.

14.2   COHERENT BINARY SYSTEMS

Coherent bandpass digital systems employ information about the carrier frequency and phase at the receiver to detect the message—like synchronous analog detection. **Noncoherent** systems don't require synchronization with the carrier phase, but they fall short of the optimum performance made possible by coherent detection.

This section examines coherent binary transmission, starting with a general treatment of optimum binary detection in the presence of additive white gaussian noise (AWGN). The results are then applied to assess the performance of specific binary modulation systems. We'll focus throughout on keyed modulation (OOK, PRK, and FSK), without baseband filtering or transmission distortion that might produce ISI in the modulated signal.

Optimum Binary Detection

Any bandpass binary signal with keyed modulation can be expressed in the general quadrature-carrier form

$$x_c(t) = A_c\left\{\left[\sum_k I_k p_i(t - kT_b)\right] \cos (\omega_c t + \theta) - \left[\sum_k Q_k p_q(t - kT_b)\right] \sin (\omega_c t + \theta)\right\}$$

For practical coherent systems, the carrier wave should be synchronized with the digital modulation. Accordingly, we'll take $\theta = 0$ and impose the condition

$$f_c = N_c/T_b = N_c r_b \qquad \text{[1]}$$

where N_c is an integer—usually a very large integer. Then

$$x_c(t) = A_c \sum_k [I_k p_i(t - kT_b) \cos \omega_c(t - kT_b) - Q_k p_q(t - kT_b) \sin \omega_c(t - kT_b)]$$

and we can concentrate on a single bit interval by writing

$$x_c(t) = s_m(t - kT_b) \qquad kT_b < t < (k + 1)T_b \qquad \text{[2]}$$

with

$$s_m(t) \triangleq A_c[I_k p_i(t) \cos \omega_c t - Q_k p_q(t) \sin \omega_c t]$$

Here, $s_m(t)$ stands for either of two *signaling waveforms*, $s_0(t)$ and $s_1(t)$, representing the message bits $m = 0$ and $m = 1$.

Now consider the received signal $x_c(t)$ corrupted by white gaussian noise. We showed in Sect. 11.2 that an optimum baseband receiver minimizes error probability with the help of a filter matched to the baseband pulse shape. But binary CW modulation involves two different signaling waveforms, as in Eq. (2), rather than one pulse shape with two different amplitudes. Consequently, we must redo our previous analysis in terms of $s_0(t)$ and $s_1(t)$.

Figure 14.2–1 shows the proposed receiver structure labeled with the relevant signals and noise for the interval under consideration. This bandpass receiver is just like a baseband receiver with a BPF in place of an LPF. The filtered signal plus noise $y(t)$ is sampled at $t_k = (k + 1)T_b$, the end of the bit interval, and compared with a threshold level to regenerate the most likely message bit $\hat{m}$. We seek the BPF impulse response $h(t)$ and threshold level V for optimum binary detection, resulting in the smallest average regeneration error probability.

As in Sect. 11.2 let H_1 and H_0 denote the hypotheses that $m = 1$ and $m = 0$, respectively. The receiver decides between H_1 and H_0 according to the observed value of the random variable

$$Y = y(t_k) = z_m + n$$

where

$$z_m \triangleq z_m(t_k) = \left[s_m(t - kT_b) * h(t) \right]\Big|_{t=t_k} \tag{3}$$

$$= \int_{kT_b}^{(k+1)T_b} s_m(\lambda - kT_b) h(t_k - \lambda) \, d\lambda$$

$$= \int_0^{T_b} s_m(\lambda) h(T_b - \lambda) \, d\lambda$$

The noise sample $n = n(t_k)$ is a gaussian RV with zero mean and variance σ^2, so the conditional PDFs of Y given H_1 or H_0 will be gaussian curves centered at z_1 or z_0,

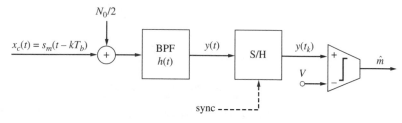

Figure 14.2–1   Bandpass binary receiver.

portrayed by Fig. 14.2–2. With the usual assumption of equally likely zeros and ones, the optimum threshold is at the intersection point, i.e.,

$$V_{opt} = \frac{1}{2}(z_1 + z_0)$$

Then, from the symmetry of the PDFs, $P_{e_1} = P_{e_0}$ and

$$P_e = Q(|z_1 - z_0|/2\sigma)$$

in which the absolute-value notation $|z_1 - z_0|$ includes the case of $z_1 < z_0$.

But what BPF impulse response $h(t)$ maximizes the ratio $|z_1 - z_0|/2\sigma$ or, equivalently, $|z_1 - z_0|^2/4\sigma^2$? To answer this question, we note from Eq. (3) that

$$|z_1 - z_0|^2 = \left| \int_{-\infty}^{\infty} [s_1(\lambda) - s_0(\lambda)]h(T_b - \lambda)\, d\lambda \right|^2 \qquad [4a]$$

where the infinite limits are allowed since $s_m(t) = 0$ outside of $0 < t < T_b$. We also note that

$$\sigma^2 = \frac{N_0}{2}\int_{-\infty}^{\infty} |h(t)|^2\, dt = \frac{N_0}{2}\int_{-\infty}^{\infty} |h(T_b - \lambda)|^2\, d\lambda \qquad [4b]$$

Application of Schwartz's inequality now yields

$$\frac{|z_1 - z_0|^2}{4\sigma^2} \le \frac{1}{2N_0}\int_{-\infty}^{\infty} [s_1(t) - s_0(t)]^2\, dt \qquad [5]$$

and the ratio is maximum if $h(T_b - t) = K[s_1(t) - s_0(t)]$. Thus,

$$h_{opt}(t) = K[s_1(T_b - t) - s_0(T_b - t)] \qquad [6]$$

with K being an arbitrary constant.

Equation (6) says that:

The filter for optimum binary detection should be matched to the *difference* between the two signaling waveforms.

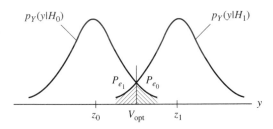

Figure 14.2–2   Conditional PDFs.

Alternatively, you could use two matched filters with $h_1(t) = Ks_1(T_b - t)$ and $h_0(t) = Ks_0(T_b - t)$ arranged in parallel per Fig. 14.2–3a; subtracting the output of the lower branch from the upper branch yields the same optimum response. In either case, any stored energy in the filters must be discharged after each sampling instant to prevent ISI in subsequent bit intervals.

Another alternative, with built-in discharge, is based on the observation that the sampled signal value from the upper branch in Fig. 14.2–3a is

$$z_{m1}(t_k) = \int_0^{T_b} s_m(\lambda)h_1(T_b - \lambda) \, d\lambda$$
$$= \int_{kT_b}^{(k+1)T_b} s_m(t - kT_b)Ks_1(t - kT_b) \, dt$$

and likewise for $z_{m0}(t_k)$. Hence, optimum filtering can be implemented by the system diagrammed in Fig. 14.2–3b, which requires two multipliers, two integrators, and stored copies of $s_0(t)$ and $s_1(t)$.

This system is called a **correlation detector** because it correlates the received signal plus noise with noise-free copies of the signaling waveforms. Note that correlation detection is a generalization of the integrate-and-dump technique for matched filtering. It should also be noted that the matched filter and correlation detector are equivalent only at the sample time t_k.

Regardless of the particular implementation method, the error probability with optimum binary detection depends upon the ratio maximized in Eq. (5). This ratio,

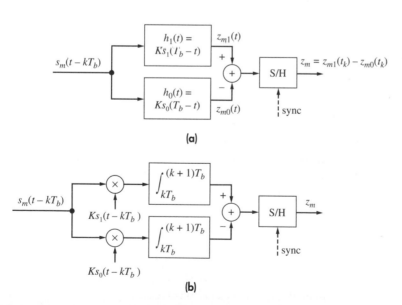

(a)

(b)

Figure 14.2–3   Optimum binary detection. (a) Parallel matched filters; (b) correlation detector.

in turn, depends on the signal energy per bit and on the similarity of the signaling waveforms. To pursue this point, consider the expansion

$$\int_0^{T_b} [s_1(t) - s_0(t)]^2 \, dt = E_1 + E_0 - 2E_{10}$$

where

$$E_1 \triangleq \int_0^{T_b} s_1^2(t) \, dt \qquad E_0 \triangleq \int_0^{T_b} s_0^2(t) \, dt \qquad \text{[7]}$$

$$E_{10} \triangleq \int_0^{T_b} s_1(t)s_0(t) \, dt$$

We identify E_1 and E_0 as the respective energies of $s_1(t)$ and $s_0(t)$, while E_{10} is proportional to their correlation coefficient. We define this correlation coefficient as

$$\rho \triangleq \frac{1}{\sqrt{E_1 E_0}} \int_0^{T_b} s_1(t)s_0(t) \, dt \qquad \text{[8]}$$

Since zeros and ones are equally likely, the average signal energy per bit is

$$E_b = \frac{1}{2}(E_1 + E_0)$$

Therefore,

$$\left(\frac{z_1 - z_0}{2\sigma}\right)^2_{\max} = \frac{E_1 + E_0 - 2E_{10}}{2N_0} = \frac{E_b - E_{10}}{N_0} \qquad \text{[9a]}$$

and

$$P_e = Q\left[\sqrt{(E_b - E_{10})/N_0}\right] \qquad \text{[9b]}$$

or, if equal signal energies,

$$P_e = Q\left[\sqrt{E_b(1 - \rho)/N_0}\right] \qquad \text{[9c]}$$

Equation (9) brings out the importance of E_{10} relative to system performance when E_b and N_0 are fixed and how system performance depends on the correlation coefficient of the two signals.

Finally, substituting Eq. (6) into Eq. (3) yields $z_1 = K(E_1 - E_{10})$ and $z_0 = K(E_{10} - E_0)$, so

$$V_{\text{opt}} = \frac{1}{2}(z_1 + z_0) = \frac{K}{2}(E_1 - E_0) \qquad \text{[10]}$$

Note that the optimum threshold does not involve E_{10}.

Derive Eqs. (5) and (6) from Eqs. (4a) and (4b). Use Eq. 17, Sect. 3.6, written in the   **EXERCISE 14.2–1**
form

$$\frac{\left|\int_{-\infty}^{\infty} V(\lambda)W^*(\lambda)\,d\lambda\right|^2}{\int_{-\infty}^{\infty} |W(\lambda)|^2\,d\lambda} \leq \int_{-\infty}^{\infty} |V(\lambda)|^2\,d\lambda$$

and recall that the equality holds when $V(\lambda)$ and $W(\lambda)$ are proportional functions.

Coherent OOK, BPSK, and FSK

Although the crude nature of ASK hardly warrants sophisticated system design, a brief look at coherent on-off keying helps clarify optimum detection concepts. The OOK signaling waveforms are just

$$s_1(t) = A_c\, p_{T_b}(t)\, \cos \omega_c t \qquad s_0(t) = 0 \tag{11}$$

Our carrier-frequency condition $f_c = N_c/T_b$ means that $s_1(t - kT_b) = A_c \cos \omega_c t$ for any bit interval while, of course, $s_0(t - kT_b) = 0$. Thus, a receiver with correlation detection simplifies to the form of Fig. 14.2–4, in which a local oscillator synchronized with the carrier provides the stored copy of $s_1(t)$. The bit sync signal actuates the sample-and-hold unit and resets the integrator. Both sync signals may be derived from a single source, thanks to the harmonic relationship between f_c and r_b.

Now we use Eqs. (7) and (11) to obtain $E_0 = E_{10} = 0$ and

$$E_1 = A_c^2 \int_0^{T_b} \cos^2 \omega_c t\, dt - \frac{A_c^2 T_b}{2}\left[1 + \text{sinc}\,\frac{4f_c}{r_b}\right] = \frac{A_c^2 T_b}{2}$$

so $E_b = E_1/2 = A_c^2 T_b/4$. Setting the threshold at $V = K(E_1 - E_0)/2 = KE_b$ yields the minimum average error probability given by Eq. (9), namely

$$P_e = Q\left(\sqrt{E_b/N_0}\right) = Q\left(\sqrt{\gamma_b}\right) \tag{12}$$

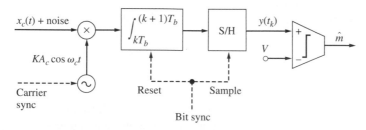

Figure 14.2–4   Correlation receiver for OOK or BPSK.

Not surprisingly, the performance of coherent OOK is identical to unipolar base-band transmission.

Better performance is achieved by coherent phase-reversal keying. Let the two phase shifts be 0 and π radians, so

$$s_1(t) = A_c p_{T_b}(t) \cos \omega_c t \qquad s_0(t) = -s_1(t) \tag{13}$$

The relation $s_0(t) = -s_1(t)$ defines *antipodal* signaling, analogous to polar base-band transmission. It quickly follows that

$$E_b = E_1 = E_0 = A_c^2 T_b/2 \qquad E_{10} = -E_b$$

so $E_b - E_{10} = 2E_b$ and

$$P_e = Q\left(\sqrt{2E_b/N_0}\right) = Q\left(\sqrt{2\gamma_b}\right) \tag{14}$$

BPSK therefore gets by with 3 dB less signal energy than OOK, other factors being equal.

Since $s_0(t) = -s_1(t)$, a coherent BPSK receiver requires only one matched fil-ter or correlator, just like OOK. But now $V = 0$ since $E_1 = E_0$, so the BPSK thresh-old level need not be readjusted if the received signal undergoes fading. Further-more, the constant envelope of BPSK makes it relatively invulnerable to nonlinear distortion. BPSK is therefore superior to OOK on several counts, and has the same spectral efficiency. We'll see next that BPSK is also superior to binary FSK.

Consider binary FSK with frequency shift $\pm f_d$ and signaling waveforms

$$s_1(t) = A_c p_{T_b}(t) \cos 2\pi(f_c + f_d)t \tag{15}$$

$$s_0(t) = A_c p_{T_b}(t) \cos 2\pi(f_c - f_d)t$$

When $f_c \pm f_d \gg r_b$, $E_b \approx A_c^2 T_b/2$, whereas

$$E_{10} = E_b \operatorname{sinc}(4 f_d/r_b) \tag{16}$$

which depends on the frequency shift. If $f_d = r_b/2$, corresponding to Sunde's FSK, then $E_{10} = 0$ and the error probability is the same as OOK.

Some improvement is possible when phase discontinuities are allowed in $x_c(t)$, but $E_b - E_{10} \le 1.22E_b$ for any choice of f_d. Hence, binary FSK does not provide any significant wideband noise reduction, and BPSK has an energy advantage of at least $10 \log (2/1.22) \approx 2$ dB. Additionally, an optimum FSK receiver is more com-plicated than Fig. 14.2–4.

EXERCISE 14.2–2   Suppose the optimum receiver for Sunde's FSK is implemented in the form of Fig. 14.2–3a. Find and sketch the amplitude response of the two filters.

Timing and Synchronization

Finally, we should give some attention to the timing and synchronization problems associated with optimum coherent detection. For this purpose, consider the band-pass signaling waveform and matched filter

$$s(t) = A_c p_{T_b}(t) \cos \omega_c t \qquad f_c T_b = N_c \gg 1$$

$$h(t) = Ks(T_b - t) = KA_c p_{T_b}(t) \cos \omega_c t$$

When $s(t)$ is applied to its matched filter, the resulting response is

$$z(t) = s(t) * h(t) \approx KE\Lambda \left(\frac{t - T_b}{T_b} \right) \cos \omega_c t \qquad \text{[17]}$$

where $E = A_c^2 T_b / 2$. The sketch of $z(t)$ in Fig. 14.2–5 shows the expected maximum value $z(T_b) = KE$, and the response for $t > T_b$ would be eliminated by discharging the filter after the sampling instant.

But suppose there's a small timing error such that sampling actually occurs at $t_k = T_b(1 \pm \epsilon)$. Then

$$z(t_k) \approx KE \cos \theta_\epsilon \qquad \theta_\epsilon = \omega_c T_b \epsilon = 2\pi N_c \epsilon$$

so the timing error reduces the effective signal level by the factor $\cos \theta_\epsilon$. Since $|z_1 - z_0|^2$ will be reduced by $\cos^2 \theta_\epsilon$, while σ^2 remains unchanged, the error probability becomes

$$P_e = Q \left(\sqrt{\frac{E_b - E_{10}}{N_0}} \cos^2 \theta_\epsilon \right) \qquad \text{[18]}$$

which follows from Eq. (9). As an example of the magnitude of this problem, take BPSK with $\gamma_b = 8$, $r_b = 2$ kbps, and $f_c = 100$ kHz; perfect timing gives $P_e = Q(\sqrt{16}) \approx 3 \times 10^{-5}$, while an error of just 0.3 percent of the bit interval results in $\theta_\epsilon = 2\pi(100/2) \times 0.003 = 54°$ and $P_e = Q(\sqrt{16 \cos^2 54°}) \approx 10^{-2}$. These numbers illustrate why a bandpass matched filter is not a practical method for coherent detection.

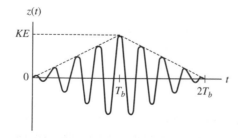

Figure 14.2–5   Response of bandpass matched filter.

A correlation detector like Fig. 14.2–4 has much less sensitivity to timing error, since the integrated output does not oscillate at the carrier frequency. Correlation detection is therefore used in most coherent binary systems. However, the local oscillator must be synchronized accurately with the carrier, and a phase synchronization error θ_ϵ again reduces the effective signal level by the factor $\cos \theta_\epsilon$.

In the case of BPSK, the carrier sync signal can be derived from $x_c(t)$ using techniques such as the Costas PLL system back in Fig. 7.3–4. Another approach known as *phase-comparison detection* is discussed in the next section, along with noncoherent detection of OOK and FSK.

14.3   NONCOHERENT BINARY SYSTEMS

Optimum coherent detection may not be essential if the signal is strong enough for adequate reliability with a less sophisticated receiver. A prime example of this situation is digital transmission over voice telephone channels, which have relatively large signal-to-noise ratios dictated by analog performance standards. There are also applications in which it would be very difficult and expensive to carry out coherent detection. For instance, the propagation delay on some radio channels changes too rapidly to permit accurate tracking of the carrier phase at the receiver, and unsynchronized or noncoherent detection becomes the only viable recourse.

Here we examine the suboptimum performance of noncoherent OOK and FSK systems that employ envelope detection to bypass the synchronization problems of coherent detection. We'll also look at *differentially coherent* PSK systems with phase-comparison detection. For all three cases we must first analyze the envelope of a sinusoid plus bandpass noise.

Envelope of a Sinusoid Plus Bandpass Noise

Consider the sinusoid $A_c \cos (\omega_c t + \theta)$ plus gaussian bandpass noise $n(t)$ with zero mean and variance σ^2. Using the quadrature-carrier expression

$$n(t) = n_i(t) \cos (\omega_c t + \theta) - n_q(t) \sin (\omega_c t + \theta)$$

we write the sum as

$$A_c \cos (\omega_c t + \theta) + n(t) = A(t) \cos [\omega_c t + \theta + \phi(t)]$$

where, at any instant t,

$$A = \sqrt{(A_c + n_i)^2 + n_q^2} \qquad \phi = \arctan \frac{n_q}{A_c + n_i} \qquad [1]$$

We recall from Sect. 10.1 that the i and q noise components are independent RVs having the same distribution as $n(t)$. Now we seek the PDF of the envelope A.

Before plunging into the analysis, let's speculate on the nature of A under extreme conditions. If $A_c = 0$, then A reduces to the noise envelope A_n, with the

Rayleigh distribution

$$p_{A_n}(A_n) = \frac{A_n}{\sigma^2} e^{-A_n^2/2\sigma^2} \qquad A_n \geq 0 \qquad\qquad [2]$$

At the other extreme, if $A_c \gg \sigma$, then A_c will be large compared to the noise components most of the time, so

$$A = A_c\sqrt{1 + (2\,n_i/A_c) + (n_i^2 + n_q^2)/A_c^2} \approx A_c + n_i$$

which implies that A will be approximately gaussian.

For an arbitrary value of A_c, we must perform a rectangular-to-polar conversion following the procedure that led to Eq. (10), Sect. 8.4. The joint PDF of A and ϕ then becomes

$$p_{A\phi}(A, \phi) = \frac{A}{2\pi\sigma^2} \exp\left(-\frac{A^2 - 2A_cA\cos\phi + A_c^2}{2\sigma^2}\right) \qquad [3]$$

for $A \geq 0$ and $|\phi| \leq \pi$. The term $A\cos\phi$ in the exponent prevents us from factoring Eq. (3) as a product of the form $p_A(A)p_\phi(\phi)$, meaning that A and ϕ are not statistically independent. The envelope PDF must therefore be found by integrating the joint PDF over the range of ϕ, so

$$p_A(A) = \frac{A}{2\pi\sigma^2} \exp\left(-\frac{A^2 + A_c^2}{2\sigma^2}\right) \int_{-\pi}^{\pi} \exp\left(\frac{A_cA\cos\phi}{\sigma^2}\right) d\phi$$

Now we introduce the *modified Bessel function* of the first kind and order zero, defined by

$$I_0(v) \triangleq \frac{1}{2\pi} \int_{-\pi}^{\pi} \exp\left(v\cos\phi\right) d\phi \qquad\qquad [4a]$$

with the properties

$$I_0(v) \approx \begin{cases} e^{v^2/4} & v \ll 1 \\[2mm] \dfrac{e^v}{\sqrt{2\pi v}} & v \gg 1 \end{cases} \qquad\qquad [4b]$$

We then have

$$p_A(A) = \frac{A}{\sigma^2} e^{-(A^2 + A_c^2)/2\sigma^2} I_0\left(\frac{A_cA}{\sigma^2}\right) \qquad A \geq 0 \qquad [5]$$

which is called the **Rician distribution.**

Although Eq. (5) has a formidable appearance, it easily simplifies under large-signal conditions to

$$p_A(A) \approx \sqrt{\frac{A}{2\pi A_c\sigma^2}}\, e^{-(A - A_c)^2/2\sigma^2} \qquad A_c \gg \sigma \qquad [6]$$

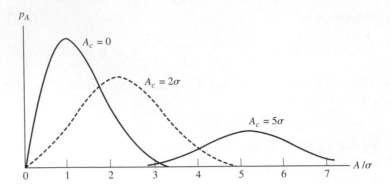

Figure 14.3–1   PDFs for the envelope of a sinusoid plus bandpass noise.

obtained from the large-v approximation in Eq. (4b). Since the exponential term dominates in Eq. (6), we have confirmed that the envelope PDF is essentially a gaussian curve with variance σ^2 centered at $\overline{A} \approx A_c$. Figure 14.3–1 illustrates the transition of the envelope PDF from a Rayleigh curve to a gaussian curve as A_c becomes large compared to σ.

Noncoherent OOK

Noncoherent on-off keying is intended to be a simple system. Usually the carrier and data are unsynchronized so, for an arbitrary bit interval $kT_b < t < (k + 1)T_b$, we write

$$x_c(t) = A_c a_k p_{T_b}(t - kT_b) \cos{(\omega_c t + \theta)} \qquad a_k = 0, 1 \qquad [7]$$

The signaling energies are $E_0 = 0$ and

$$E_1 = \frac{A_c^2 T_b}{2}\left[1 + \frac{\sin{(2\omega_c T_b + 2\theta)} - \sin{2\theta}}{2\omega_c T_b}\right] \approx \frac{A_c^2 T_b}{2}$$

where we've assumed that $f_c \gg r_b$. The average signal energy per bit is then $E_b = E_1/2 \approx A_c^2 T_b/4$ since we'll continue to assume that 1s and 0s are equally likely.

The OOK receiver diagrammed in Fig. 14.3–2 consists of a BPF followed by an envelope detector and regenerator. The BPF is a matched filter with

$$h(t) = KA_c p_{T_b}(t) \cos{\omega_c t} \qquad [8]$$

which ignores the carrier phase θ. The envelope detector eliminates dependence on θ by tracing out the dashed line back in Fig. 14.2–5. Thus, when $a_k = 1$, the peak signal component of the envelope $y(t)$ is $A_1 = KE_1$. Let's take $K = A_c/E_1$ for convenience, so that $A_1 = A_c$. Then

$$A_c^2/\sigma^2 = 4E_b/N_0 = 4\gamma_b \qquad [9]$$

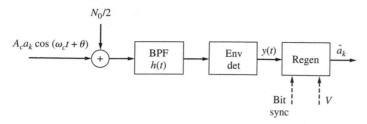

Figure 14.3-2   Noncoherent OOK receiver.

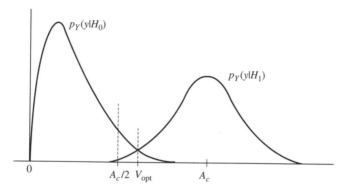

Figure 14.3-3   Conditional PDFs for noncoherent OOK.

where σ^2 is the variance of the bandpass noise at the input to the envelope detector, calculated from $h(t)$ using Eq. (4b), Sect. 14.2.

Now consider the conditional PDFs of the random variable $Y = y(t_k)$. When $a_k = 0$, we have a sample value of the envelope of the noise alone; hence, $p_Y(y|H_0)$ is the Rayleigh function $p_{A_n}(y)$. When $a_k = 1$, we have a sample value of the envelope of a sinusoid plus noise; hence, $p_Y(y|H_1)$ is the Rician function $p_A(y)$. Figure 14.3–3 shows these two curves for the case of $\gamma_b \gg 1$, so the Rician PDF has a nearly gaussian shape. The intersection point defines the optimum threshold, which turns out to be

$$V_{\text{opt}} \approx \frac{A_c}{2}\sqrt{1 + \frac{2}{\gamma_b}} \approx \frac{A_c}{2} \qquad \gamma_b \gg 1$$

But we no longer have symmetry with respect to the threshold and, consequently, $P_{e_1} \neq P_{e_0}$ when P_e is minimum.

Noncoherent OOK systems require $\gamma_b \gg 1$ for reasonable performance, and the threshold is normally set at $A_c/2$. The resulting error probabilities are

$$P_{e_0} = \int_{A_c/2}^{\infty} p_{A_n}(y)\, dy = e^{-A_c^2/8\sigma^2} = e^{-\gamma_b/2} \qquad \text{[10a]}$$

$$P_{e_1} = \int_0^{A_c/2} p_A(y)\, dy \approx Q\left(\frac{A_c}{2\sigma}\right) = Q\left(\sqrt{\gamma_b}\right)$$

$$\approx \frac{1}{\sqrt{2\pi\gamma_b}}\, e^{-\gamma_b/2} \quad \gamma_b \gg 1 \qquad\qquad [10b]$$

where we've introduced the asymptotic approximation for $Q(\sqrt{\gamma_b})$ to bring out the fact that $P_{e_1} \ll P_{e_0}$ when $\gamma_b \gg 1$. Finally,

$$P_e = \tfrac{1}{2}(P_{e_0} + P_{e_1}) = \tfrac{1}{2}\left[e^{-\gamma_b/2} + Q\left(\sqrt{\gamma_b}\right)\right] \qquad\qquad [11]$$

$$\approx \tfrac{1}{2} e^{-\gamma_b/2} \quad \gamma_b \gg 1$$

which is plotted versus γ_b in Fig. 14.3–4 along with curves for other binary systems.

EXERCISE 14.3–1   Consider the BPF output $z(t) = x_c(t) * h(t)$ when $x_c(t) = A_c p_{T_b}(t)\cos(\omega_c t + \theta)$ and $K = 2/A_c T_b$. Show that, for $0 < t < T_b$,

$$z(t) = \frac{A_c t}{T_b}\left[\cos\theta\,\cos\omega_c t - \left(\sin\theta - \frac{\cos\theta}{\omega_c t}\right)\sin\omega_c t\right]$$

Then find and sketch the envelope of $z(t)$ assuming $f_c \gg r_b$.

Noncoherent FSK

Although envelope detection seems an unlikely method for FSK, a reexamination of the waveform back in Fig. 14.1–1b reveals that binary FSK consists of two interleaved OOK signals with the same amplitude A_c but different carrier frequencies, $f_1 = f_c + f_d$ and $f_0 = f_c - f_d$. Accordingly, noncoherent detection can be implemented with a pair of bandpass filters and envelope detectors, arranged per Fig. 14.3–5 where

$$h_1(t) = KA_c p_{T_b}(t)\cos\omega_1 t \qquad h_0(t) = KA_c p_{T_b}(t)\cos\omega_0 t \qquad [12]$$

We'll take $K = A_c/E_b$, noting that $E_b = E_1 = E_0 \approx A_c^2 T_b/2$. Then

$$A_c^2/\sigma^2 = 2E_b/N_0 = 2\gamma_b \qquad\qquad [13]$$

where σ^2 is the noise variance at the output of either filter.

We'll also take the frequency spacing $f_1 - f_0 = 2f_d$ to be an integer multiple of r_b, as in Sunde's FSK. This condition ensures that the BPFs effectively separate the two frequencies, and that the two bandpass noise waveforms are uncorrelated at the sampling instants. Thus, when $a_k = 1$, the sampled output $y_1(t_k)$ at the upper branch has the signal component $A_1 = KE_1 = A_c$ and a Rician distribution, whereas $y_0(t_k)$ at the lower branch has a Rayleigh distribution—and vice versa when $a_k = 0$.

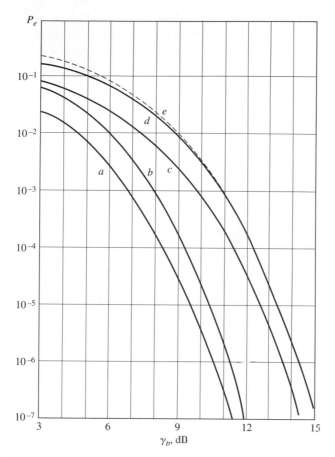

Figure 14.3-4 Binary error probability curves. (*a*) Coherent BPSK; (*b*) DPSK; (*c*) coherent OOK or FSK; (*d*) noncoherent FSK; (*e*) noncoherent OOK.

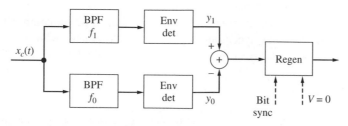

Figure 14.3-5 Noncoherent detection of binary FSK.

Regeneration is based on the envelope difference $Y_1 - Y_0 = y_1(t_k) - y_0(t_k)$. Without resorting to conditional PDFs, we conclude from the symmetry of the receiver that the threshold should be set at $V = 0$, regardless of A_c. It then follows that $P_{e_1} = P(Y_1 - Y_0 < 0|H_1)$ and $P_{e_0} = P_{e_1} = P_e$. Therefore,

$$P_e = P(Y_0 > Y_1|H_1)$$

$$= \int_0^\infty p_{Y_1}(y_1|H_1)\left[\int_{y_1}^\infty p_{Y_0}(y_0|H_1)\,dy_0\right]dy_1$$

where the inner integral is the probability of the event $Y_0 > Y_1$ for a *fixed* value of y_1. Inserting the PDFs $p_{Y_0}(y_0|H_1) = p_{A_n}(y_0)$ and $p_{Y_1}(y_1|H_1) = p_A(y_1)$ and performing the inner integration yields

$$P_e = \int_0^\infty \frac{y_1}{\sigma^2}e^{-(2y_1^2+A_c^2)/2\sigma^2}I_0\left(\frac{A_c y_1}{\sigma^2}\right)dy_1$$

Rather amazingly, this integral can be evaluated in closed form by letting $\lambda = \sqrt{2}y_1$ and $\alpha = A_c/\sqrt{2}$ so that

$$P_e = \frac{1}{2}e^{-A_c^2/4\sigma^2}\int_0^\infty \frac{\lambda}{\sigma^2}e^{-(\lambda^2+\alpha^2)/2\sigma^2}I_0\left(\frac{\alpha\lambda}{\sigma^2}\right)d\lambda$$

The integrand is now exactly the same function as the Rician PDF in Eq. (5), whose total area equals unity. Hence, our final result simply becomes

$$P_e = \frac{1}{2}e^{-A_c^2/4\sigma^2} = \frac{1}{2}e^{-\gamma_b/2} \qquad \text{[14]}$$

having used Eq. (13).

A comparison of the performance curves for noncoherent FSK and OOK plotted in Fig. 14.3–4 reveals little difference except at small values of γ_b. However, FSK does have three advantages over OOK: constant modulated signal envelope, equal digit error probabilities, and fixed threshold level $V = 0$. These advantages usually justify the extra hardware needed for the FSK receiver.

Differentially Coherent PSK

Noncoherent detection of binary PSK would be impossible since the message information resides in the phase. Instead, the clever technique of **phase-comparison detection** gets around the phase synchronization problems associated with coherent BPSK and provides much better performance than noncoherent OOK or FSK. The phase-comparison detector in Fig. 14.3–6 looks something like a correlation detector except that the local oscillator signal is replaced by the BPSK signal itself after a delay of T_b. A BPF at the front end prevents excess noise from swamping the detector.

Successful operation requires f_c to be an integer multiple of r_b, as in coherent BPSK. We therefore write

$$x_c(t) = A_c p_{T_b}(t - kT_b)\cos(\omega_c t + \theta + a_k\pi) \qquad \text{[15]}$$

$$a_k = 0, 1 \qquad kT_b < t < (k+1)T_b$$

In the absense of noise, the phase-comparison product for the kth bit interval is

$$x_c(t) \times 2\,x_c(t - T_b) = 2A_c^2 \cos\left(\omega_c t + \theta + a_k \pi\right)$$

$$\times \cos\left[\omega_c(t - T_b) + \theta + a_{k-1}\pi\right]$$

$$= A_c^2\{\cos\left[(a_k - a_{k-1})\pi\right]$$

$$+ \cos\left[2\omega_c t + 2\theta + (a_k + a_{k-1})\pi\right]\}$$

where we've used the fact that $\omega_c T_b = 2\pi N_c$. Lowpass filtering then yields

$$z(t_k) = \begin{cases} +A_c^2 & a_k = a_{k-1} \\ -A_c^2 & a_k \neq a_{k-1} \end{cases} \tag{16}$$

so we have polar symmetry and the threshold should be set at $V = 0$.

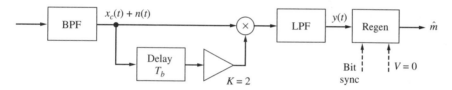

Figure 14.3–6   Differentially coherent receiver for binary PSK.

Since $z(t_k)$ only tells you whether a_k differs from a_{k-1}, a BPSK system with phase-comparison detection is called **differentially coherent** PSK (DPSK). Such systems generally include *differential encoding* at the transmitter, which makes it possible to regenerate the message bits directly from $z(t_k)$. Differential encoding starts with an arbitrary initial bit, say $a_0 = 1$. Subsequent bits are determined by the message sequence m_k according to the rule: $a_k = a_{k-1}$ if $m_k = 1$, $a_k \neq a_{k-1}$ if $m_k = 0$. Thus, $z(t_k) = +A_c^2$ means that $m_k = 1$ and $z(t_k) = -A_c^2$ means that $m_k = 0$. Figure 14.3–7 shows a logic circuit for differential encoding; this circuit implements the logic equation

$$a_k = a_{k-1} m_k \oplus \bar{a}_{k-1}\, \bar{m}_k \tag{17}$$

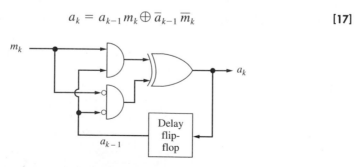

Figure 14.3–7   Logic circuit for differential encoding.

where the overbar stands for logical inversion. An example of differential encoding and phase-comparison detection (without noise) is given in Table 14.3–1.

To analyze the performance of DPSK with noise, we'll assume that the BPF performs most of the noise filtering, like the BPFs in an FSK receiver. Hence, the carrier amplitude and noise variance at the BPF output are related by

$$A_c^2/\sigma^2 = 2E_b/N_0 = 2\gamma_b$$

We'll also exploit the symmetry and focus on the case when $a_k = a_{k-1} = 0$, so an error occurs if $y(t_k) < 0$.

Now let the delayed i and q noise components be denoted by $n_i'(t) = n_i(t - T_b)$ and $n_q'(t) = n_q(t - T_b)$. The inputs to the multiplier during the kth bit interval are $x_c(t) + n(t) = [A_c + n_i(t)] \cos(\omega_c t + \theta) - n_q(t) \sin(\omega_c t + \theta)$ and $2[x_c(t - T_b) + n(t) - T_b)] = 2[A_c + n_i'(t)] \cos(\omega_c t + \theta) - 2 n_q'(t) \sin(\omega_c t + \theta)$. The LPF then removes the high-frequency terms from the product, leaving

$$Y = y(t_k) = (A_c + n_i)(A_c + n_i') + n_q n_q' \qquad [18]$$

where all four noise components are independent gaussian RVs with zero mean and variance σ^2.

Equation (18) has a *quadratic form* that can be simplified by a diagonalization process, resulting in

$$Y = \alpha^2 - \beta^2 \qquad [19a]$$

with

$$\alpha^2 = (A_c + \alpha_i)^2 + \alpha_q^2 \qquad \beta^2 = \beta_i^2 + \beta_q^2 \qquad [19b]$$

and

$$\alpha_i \triangleq \frac{1}{2}(n_i + n_i') \qquad \beta_i \triangleq \frac{1}{2}(n_i - n_i') \qquad [19c]$$

$$\alpha_q \triangleq \frac{1}{2}(n_q + n_q') \qquad \beta_q \triangleq \frac{1}{2}(n_q - n_q')$$

Note that α_i is a zero-mean gaussian RV with variance $\overline{\alpha_i^2} = (\overline{n_i^2} + \overline{n_i'^2})/4 = 2\sigma^2/4 = \sigma^2/2$; identical conclusions hold for the other i and q components of α and β. Therefore, α has a *Rician* PDF given by Eq. (5) with $\sigma^2/2$ in place of σ^2, while β has a *Rayleigh* PDF given by Eq. (2) with $\sigma^2/2$ in place of σ^2.

Table 14.3–1

Input message	1	0	1	1	0	1	0	0	
Encoded message	1	1	0	0	0	1	1	0	1
Transmitted phase	π	π	0	0	0	π	π	0	π
Phase-comparison sign		+	−	+	+	−	+	−	−
Regenerated message		1	0	1	1	0	1	0	0

Lastly, since α and β are nonnegative, we can write the average error probability as

$$P_e = P(Y < 0 | a_k = a_{k-1}) = P(\alpha^2 < \beta^2) = P(\beta > \alpha)$$

and we've arrived at an expression equivalent to the one previously solved for noncoherent FSK. Substituting $\sigma^2/2$ for σ^2 in Eq. (14) now gives our DPSK result

$$P_e = \frac{1}{2}e^{-A_c^2/2\sigma^2} = \frac{1}{2}e^{-\gamma_b} \qquad [20]$$

The performance curves in Fig. 14.3–4 now show that DPSK has a 3-dB energy advantage over noncoherent binary systems and a penalty of less than 1 dB compared to coherent BPSK at $P_e \leq 10^{-4}$.

DPSK does not require the carrier phase synchronization essential for coherent PRK, but it does involve somewhat more hardware than noncoherent OOK or FSK—including differential encoding and carrier-frequency synchronization with r_b at the transmitter. A minor annoyance is that DPSK errors tend to occur in groups of two (why?).

Binary data is to be sent at the rate $r_b = 100$ kbps over a channel with 60-dB transmission loss and noise density $N_0 = 10^{-12}$ W/Hz at the receiver. What transmitted power S_T is needed to get $P_e = 10^{-3}$ for various types of modulation and detection?

EXAMPLE 14.3–1

To answer this question, we first write the received signal power as $S_R = E_b r_b = N_0 \gamma_b r_b = S_T/L$ with $L = 10^6$. Thus,

$$S_T = L N_0 \gamma_b r_b = 0.1\gamma_b$$

Next, using the curves in Fig. 14.3–4 or our previous formulas for P_e, we find the value of γ_b corresponding to the specified error probability and calculate S_T therefrom.

Table 14.3–2 summarizes the results. The systems have been listed here in order of increasing difficulty of implementation, bringing out the trade-off between signal power and hardware complexity.

Table 14.3–2

System	S_T, W
Noncoherent OOK or FSK	1.26
Differentially coherent PSK	0.62
Coherent BPSK	0.48

EXERCISE 14.3–2   Suppose the system in the previous example has a limitation on the *peak envelope power*, such that $L A_c^2 \leq 2$ watts at the transmitter. Find the resulting minimum error probability for noncoherent OOK and FSK and for DPSK.

14.4   QUADRATURE-CARRIER AND *M*-ARY SYSTEMS

This section investigates the performance of *M*-ary modulation systems with coherent or phase-comparison detection, usually in a quadrature-carrier configuration. Our primary motivation here is the increased modulation speed afforded by QAM and related quadrature-carrier methods, and by *M*-ary PSK and *M*-ary QAM modulation. These are the modulation types best suited to digital transmission on telephone lines and other bandwidth-limited channels.

As in previous sections, we continue to assume independent equiprobable symbols and AWGN contamination. We also assume that *M* is a power of two, consistent with binary to *M*-ary data conversion. This assumption allows a practical comparison of binary and *M*-ary systems.

Quadrature-Carrier Systems

We pointed out in Sect. 14.1 that both quadriphase PSK and keyed polar QAM are equivalent to the sum of two BPSK signals impressed on quadrature carriers. Here we'll adopt that viewpoint to analyze the performance of QPSK/QAM with coherent detection. Accordingly, let the source information be grouped into *dibits* represented by $I_k Q_k$. Each dibit corresponds to one symbol from a quaternary ($M = 4$) source or two successive bits from a binary source. In the latter case, which occurs more often in practice, the dibit rate is $r = r_b/2$ and $D = 1/r = 2T_b$.

Coherent quadrature-carrier detection requires synchronized modulation, as discussed in Sect. 14.2. Thus, for the kth dibit interval $kD < t < (k + 1)D$, we write

$$x_c(t) = s_i(t - kD) - s_q(t - kD) \qquad [1a]$$

with

$$s_i(t) = A_c I_k p_D(t) \cos \omega_c t \qquad I_k = \pm 1 \qquad [1b]$$

$$s_q(t) = A_c Q_k p_D(t) \sin \omega_c t \qquad Q_k = \pm 1$$

Since f_c must be harmonically related to $r = 1/D$, the signaling energy is

$$\int_{kD}^{(k+1)D} x_c^2(t) \, dt = \frac{1}{2} A_c^2 (I_k^2 + Q_k^2) D = A_c^2 D$$

and we have

$$E = 2E_b \qquad E_b = A_c^2 D/2 \qquad [2]$$

where E is the energy per dibit or quaternary symbol.

From Eq. (1) and our prior study of coherent BPSK, it follows that the optimum quadrature-carrier receiver can be implemented with two correlation detectors arranged as in Fig. 14.4–1. Each correlator performs coherent binary detection, independent of the other. Hence, the average error probability *per bit* is

$$P_{be} = Q(\sqrt{2E_b/N_0}) = Q(\sqrt{2\gamma_b}) \tag{3}$$

where the function $Q(\sqrt{2\gamma_b})$ denotes the area under the gaussian tail—not to be confused with Q symbolizing quadrature modulation.

We see from Eq. (3) that coherent QPSK/QAM achieves the same bit error probability as coherent BPSK. But recall that the transmission bandwidth for QPSK/QAM is

$$B_T \approx r_b/2$$

whereas BPSK requires $B_T \approx r_b$. This means that the additional quadrature-carrier hardware allows you to cut the transmission bandwidth in half for a given bit rate or to double the bit rate for a given transmission bandwidth. The error probability remains unchanged in either case.

Equation (3) and the bandwidth/hardware trade-off also hold for *minimum-shift keying*, whose *i* and *q* components illustrated back in Fig. 14.1–11*b* suggest quadrature-carrier detection. An MSK receiver has a structure like Fig. 14.4–1 modified in accordance with the pulse shaping and staggering of the *i* and *q* components. There are only two significant differences between MSK and QPSK: (1) the MSK spectrum has a broader main lobe but smaller side lobes than the spectrum of QPSK with the same bit rate; (2) MSK is inherently binary frequency modulation, whereas QPSK can be viewed as either binary or quaternary phase modulation.

When QPSK/QAM is used to transmit quaternary data, the output converter in Fig. 14.4–1 reconstructs quaternary symbols from the regenerated dibits. Since bit errors are independent, the probability of obtaining a *correct* symbol is

$$P_c = (1 - P_{be})^2$$

The average error probability *per symbol* thus becomes

$$P_e = 1 - P_c = 2Q(\sqrt{E/N_0}) - Q^2(\sqrt{E/N_0}) \tag{4}$$

$$\approx 2Q(\sqrt{E/N_0}) \qquad E/N_0 \gg 1$$

where $E = 2E_b$ represents the average symbol energy.

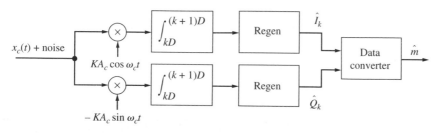

Figure 14.4–1   Quadrature-carrier receiver with correlation detectors.

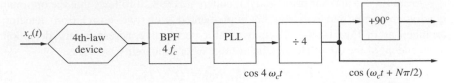

Figure 14.4–2   PLL system for carrier synchronization in a quadrature-carrier receiver.

Various methods have been devised to generate the carrier sync signals necessary for coherent detection in quadrature-carrier receivers. Figure 14.4–2 shows a simple PLL system based on the fact that the fourth power of $x_c(t)$ contains a discrete frequency component at $4f_c$. However, since $\cos 4\omega_c t = \cos(4\omega_c t + 2\pi N)$, fourfold frequency division produces $\cos(\omega_c t + N\pi/2)$ so the output has a fixed phase error of $N\pi/2$ with N being an integer whose value depends on the lock-in transient. A known preamble may be transmitted at the start of the message to permit phase adjustment, or differential encoding may be used to nullify the phase error effects. Another carrier sync system will be described in conjunction with M-ary PSK; additional methods are covered by Lindsey (1972).

Phase-comparison detection is also possible in quadrature-carrier systems with differential encoding. From our study of DPSK in Sect. 14.3, you may correctly infer that differentially coherent QPSK (DQPSK) requires somewhat more signal energy than coherent QPSK to get a specified error probability. The difference turns out to be about 2.3 dB.

EXERCISE 14.4–1   Consider a QPSK signal like Eq. (1) written as $x_c(t) = A_c \cos(\omega_c t + \phi_k)$ with $\phi_k = \pi/4, 3\pi/4, 5\pi/4, 7\pi/4$. Show that $x_c^4(t)$ includes an unmodulated component at $4f_c$.

M-ary PSK Systems

Now let's extend our investigation of coherent quadrature-carrier detection to encompass M-ary PSK. The carrier is again synchronized with the modulation, and f_c is harmonically related to the symbol rate r. We write the modulated signal for a given symbol interval as

$$x_c(t) = s_i(t - kD) - s_q(t - kD) \qquad [5a]$$

with

$$s_i(t) = A_c \cos \phi_k p_D(t) \cos \omega_c t \qquad [5b]$$

$$s_q(t) = A_c \sin \phi_k p_D(t) \sin \omega_c t$$

where

$$\phi_k = 2\pi a_k/M \qquad a_k = 0, 1, \ldots, M - 1$$

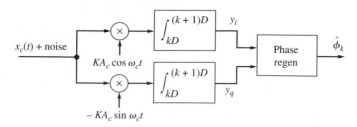

Figure 14.4–3   Coherent *M*-ary PSK receiver.

from Eq. (13), Sect. 14.1, taking $N = 0$. The signaling energy per symbol then becomes

$$E = \frac{1}{2}A_c^2(\cos^2\phi_k + \sin^2\phi_k)D = \frac{1}{2}A_c^2 D \qquad [6]$$

equivalent to $E_b = E/\log_2 M$ if each symbol represents $\log_2 M$ binary digits. The transmission bandwidth requirement is $B_T \approx r = r_b/\log_2 M$, from our spectral analysis in Sect. 14.1.

An optimum receiver for *M*-ary PSK can be modeled in the form of Fig. 14.4–3. We'll let $K = A_c/E$ so, in absence of noise, the quadrature correlators produce $z_i(t_k) = A_c \cos\phi_k$ and $z_q(t_k) = A_c \sin\phi_k$ from which $\phi_k = \arctan z_q/z_i$.

When $x_c(t)$ is contaminated by noise, message symbol regeneration is based on the noisy samples

$$y_i = A_c \cos\phi_k + n_i \qquad y_q = A_c \sin\phi_k + n_q$$

in which the i and q noise components are independent gaussian RVs with zero mean and variance

$$\sigma^2 = K^2 E N_0/2 = A_c^2 N_0/2E = N_0 r \qquad [7]$$

The generator has *M angular* thresholds equispaced by $2\pi/M$, as illustrated in Fig. 14.4–4, and it selects the point from the signal constellation whose angle is closest to $\arctan y_q/y_i$.

The circular symmetry of Fig. 14.4–4, together with the symmetry of the noise PDFs, means that all phase angles have the same error probability. We'll therefore focus on the case of $\phi_k = 0$, so

$$\arctan \frac{y_q}{y_i} = \arctan \frac{n_q}{A_c + n_i} = \phi$$

and we recognize ϕ as the phase of a sinusoid plus bandpass noise. Since no error results if $|\phi| < \pi/M$, the symbol error probability can be calculated using

$$P_e = P(|\phi| > \pi/M) = 1 - \int_{-\pi/M}^{\pi/M} p_\phi(\phi)\, d\phi \qquad [8]$$

for which we need the PDF of the phase ϕ.

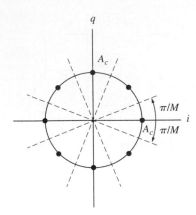

Figure 14.4–4   Decision thresholds for *M*-ary PSK.

The joint PDF for the envelope and phase of a sinusoid plus bandpass noise was given in Eq. (3), Sect. 14.3. The PDF of the phase alone is found by integrating the joint PDF over $0 \le A < \infty$. A few manipulations lead to the awesome-looking expression

$$p_\phi(\phi) = \frac{1}{2\pi} e^{-A_c^2/2\sigma^2} + \frac{A_c \cos \phi}{\sqrt{2\pi\sigma^2}} \exp\left(-\frac{A_c^2 \sin^2 \phi}{2\sigma^2}\right)\left[1 - Q\left(\frac{A_c \cos \phi}{\sigma}\right)\right] \quad [9]$$

for $-\pi < \phi < \pi$. Under the large-signal condition $A_c \gg \sigma$, Eq. (9) simplifies to

$$p_\phi(\phi) \approx \frac{A_c \cos \phi}{\sqrt{2\pi\sigma^2}} e^{-(A_c \sin \phi)^2/2\sigma^2} \quad |\phi| < \frac{\pi}{2} \quad [10]$$

which, for small values of ϕ, approximates a *gaussian* with $\overline{\phi} = 0$ and $\overline{\phi^2} = \sigma^2/A_c^2$. Equation (10) is invalid for $|\phi| > \pi/2$, but the probability of that event is small if $A_c \gg \sigma$. Figure 14.4–5 depicts the transition of $p_\phi(\phi)$ from a uniform distribution when $A_c = 0$ to a gaussian curve when A_c becomes large compared to σ. (See Fig. 14.3–1 for the corresponding transition of the envelope PDF.)

We'll assume that $A_c \gg \sigma$ so we can use Eq. (10) to obtain the error probability of coherent *M*-ary PSK with $M > 4$. (We already have the results for $M = 2$ and 4.) Inserting Eq. (10) with $A_c^2/\sigma^2 = 2E/N_0$ into Eq. (8) gives

$$P_e \approx 1 - \frac{1}{\sqrt{2\pi}} \int_{-\pi/M}^{\pi/M} \sqrt{\frac{2E}{N_0}} \cos \phi \, e^{-(2E/N_0)(\sin \phi)^2/2} \, d\phi \quad [11]$$

$$\approx 1 - \frac{2}{\sqrt{2\pi}} \int_0^L e^{-\lambda^2/2} \, d\lambda$$

where we've noted the even symmetry and made the change of variable $\lambda = \sqrt{2E/N_0} \sin \phi$ so $L = \sqrt{2E/N_0} \sin (\pi/M)$. But the integrand in Eq. (11) is a gaussian function, so $P_e \approx 1 - [1 - 2Q(L)] = 2Q(L)$. Hence,

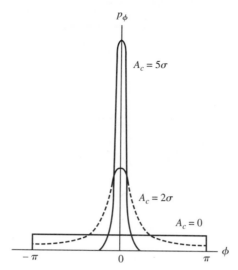

P_ϕ

$A_c = 5\sigma$

$A_c = 2\sigma$

$A_c = 0$

$-\pi$ ⟶ 0 ⟶ π ⟶ ϕ

Figure 14.4–5   PDFs for the phase of a sinusoid plus bandpass noise.

$$P_e \approx 2Q\!\left(\sqrt{\frac{2E}{N_0}\sin^2\frac{\pi}{M}}\,\right) \qquad \text{[12]}$$

which is our final result for the symbol error probability with $M > 4$. We'll discuss the equivalent bit error probability in our comparisons at the end of the chapter.

Returning to the receiver in Fig. 14.4–3, the carrier sync signals can be derived from the *M*th power of $x_c(t)$ using a modified version of Fig. 14.4–2. The more sophisticated *decision-feedback* PLL system in Fig. 14.4–6 uses the estimated phase $\hat{\phi}_k$ to generate a control signal $v(t)$ that corrects any VCO phase error. The two delayors here simply account for the fact that $\hat{\phi}_k$ is obtained at the end of the *k*th symbol interval.

If accurate carrier synchronization proves to be impractical, then differentially coherent detection may be used instead. The noise analysis is quite complicated, but Lindsey and Simon (1973) have obtained the simple approximation

$$P_e \approx 2Q\!\left(\sqrt{\frac{4E}{N_0}\sin^2\frac{\pi}{2M}}\,\right) \qquad \text{[13]}$$

which holds for $E/N_0 \gg 1$ with $M \geq 4$. We see from Eqs. (12) and (13) that *M*-ary DPSK achieves the same error probability as coherent PSK when the energy is increased by the factor

$$\Gamma = \frac{\sin^2(\pi/M)}{2\sin^2(\pi/2M)}$$

This factor equals 2.3 dB for DQPSK ($M = 4$), as previously asserted, and it approaches 3 dB for $M \gg 1$.

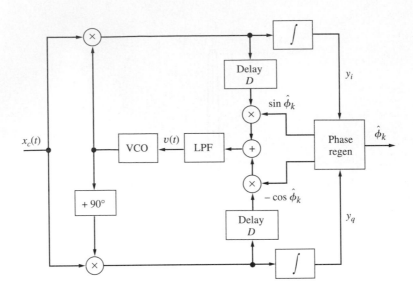

Figure 14.4–6   M-ary PSK receiver with decision-feedback system for carrier synchronization.

EXERCISE 14.4–2   Derive Eq. (7) by replacing one of the correlation detectors in Fig. 14.4–3 with an equivalent BPF, as in Fig. 14.2–3.

M-ary QAM Systems

We can represent the source symbols by combining amplitude and phase modulation to form **M-ary QAM.** *M*-ary QAM is also called *M*-ary amplitude-phase keying (APK). It is useful for channels having limited bandwidth and provides lower error rates than other *M*-ary systems with keyed modulation operating at the same symbol rate. Here we'll study the class of *M*-ary QAM systems defined by square signal constellations, after a preliminary treatment of suppressed-carrier *M*-ary ASK.

Consider *M*-ary ASK with synchronized modulation and suppressed carrier. Carrier-suppression is readily accomplished by applying a polar modulating signal. Thus, for the *k*th symbol interval, we write

$$x_c(t) = A_c I_k p_D(t - kD) \cos \omega_c t \qquad \text{[14a]}$$

where

$$I_k = \pm 1, \pm 3, \ldots, \pm(M - 1) \qquad \text{[14b]}$$

The transmission bandwidth is $B_T \approx r$, the same as *M*-ary PSK.

An optimum coherent receiver consists of just one correlation detector, since there's no quadrature component, and regeneration is based on the noisy samples

$$y_i = A_c I_k + n_i$$

The noise component is a zero-mean gaussian RV with variance $\sigma^2 = N_0 r$, as in Eq. (7). Figure 14.4–7 shows the one-dimensional signal constellation and the corre-

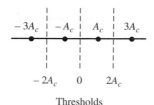

Figure 14.4–7   Decision thresholds for ASK with $M = 4$.

sponding $M - 1$ equispaced thresholds when $M = 4$. The symbol error probability for any even value of M is

$$P_e = 2\left(1 - \frac{1}{M}\right)Q\left(\frac{A_c}{\sqrt{N_0 r}}\right) \qquad [15]$$

obtained by the same analysis used for polar *M*-ary baseband transmission in Sect. 11.2.

Suppose that two of these ASK signals are transmitted on the same channel via quadrature-carrier multiplexing, which requires no more bandwidth than one signal. Let the information come from an *M*-ary source with $M = \mu^2$ so the message can be converted into two μ-ary digit streams, each having the same rate r. The performance of *M*-ary QAM fundamentally depends upon the μ-ary error rate and therefore will be superior to direct *M*-ary modulation with $M > \mu$.

Figure 14.4–8*a* diagrams the structure of our *M*-ary QAM transmitter. The output signal for the *k*th symbol interval is

$$x_c(t) = s_i(t - kD) - s_q(t - kD) \qquad [16a]$$

with

$$s_i(t) = A_c I_k p_D(t) \cos \omega_c t \qquad I_k = \pm 1, \pm 3, \dots, \pm(\mu - 1) \qquad [16b]$$

$$s_q(t) = A_c Q_k p_D(t) \sin \omega_c t \qquad Q_k = \pm 1, \pm 3, \dots, \pm(\mu - 1)$$

The average energy per *M*-ary symbol is

$$E = \frac{1}{2}A_c^2(\overline{I_k^2} + \overline{Q_k^2})D = \frac{1}{3}A_c^2(\mu^2 - 1)D \qquad [17]$$

since $\overline{I_k^2} = \overline{Q_k^2} = (\mu^2 - 1)/3$.

Coherent QAM detection is performed by the receiver in Fig. 14.4–8*b*, whose quadrature correlators produce the sample values

$$y_i = A_c I_k + n_i \qquad y_q = A_c Q_k + n_q$$

We then have a *square* signal constellation and threshold pattern, illustrated in Fig. 14.4–8*c* taking $M = 4^2 = 16$. Now let P denote the probability of error for I_k or Q_k, as given by Eq. (15) with M replaced by $\mu = \sqrt{M}$. The error probability per *M*-ary symbol is $P_e = 1 - (1 - P)^2$ and $P_e \approx 2P$ when $P \ll 1$. Therefore,

$$P_e \approx 4\left(1 - \frac{1}{\sqrt{M}}\right)Q\left[\sqrt{\frac{3E}{(M - 1)N_0}}\right] \qquad [18]$$

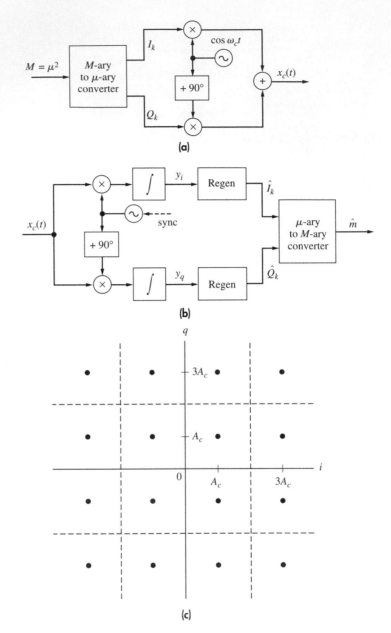

Figure 14.4–8   M-ary QAM system. (a) Transmitter; (b) receiver; (c) square signal constellation and thresholds with $M = 16$.

in which we've inserted the average symbol energy from Eq. (17).

Calculations using this result confirm the superior performance of *M*-ary QAM. By way of example, if $M = 16$ and $E/N_0 = 100$, then $P_e \approx 4 \times 3/4 \times Q(\sqrt{20})$ $= 1.2 \times 10^{-5}$, whereas an equivalent PSK system with $M = 16$ would have $P_e \approx 2Q(\sqrt{7.6}) = 6 \times 10^{-3}$.

Comparison of Digital Modulation Systems

A performance comparison of digital modulation systems should consider several factors, including: error probability, transmission bandwidth, spectral spillover, hardware requirements, and the differences between binary and *M*-ary signaling. To establish an equitable basis for comparison, we'll make the realistic assumption that the information comes from a binary source with bit rate r_b. This allows us to compare systems in terms of the modulation speed r_b/B_T and the energy-to-noise ratio γ_b needed to get a specified error probability per bit.

Our previous results for binary modulation systems apply directly to the comparison at hand, especially the error probability curves back in Fig. 14.3–4. Table 14.4–1 serves as a more abbreviated summary when γ_b is large enough to justify the applicable approximations. (Thus, in the case of noncoherent OOK, almost all the errors correspond to the carrier "off" state.) The table omits coherent OOK and FSK, which have little practical value, but it includes QAM and QPSK viewed as binary rather than quaternary modulation. This listing emphasizes the fact that doubled modulation speed goes hand-in-hand with coherent quadrature-carrier detection. Also recall that minimizing spectral spillover requires staggered keyed modulation (MSK or OQPSK) or additional pulse shaping.

Now consider *M*-ary transmission with symbol rate r and energy E per symbol. We'll take $M = 2^K$ and introduce the data-conversion factor

$$K = \log_2 M$$

which equals the number of bits per *M*-ary symbol. The equivalent bit rate and energy are $r_b = Kr$ and $E_b = E/K$, so

$$\gamma_b = E/KN_0$$

The modulation speed of *M*-ary PSK or *M*-ary QAM is

$$r_b/B_T \approx K \qquad\qquad [19]$$

Table 14.4–1   Summary of binary modulation systems

Modulation	Detection	r_b/B_T	P_{be}
OOK or FSK ($f_d = r_b/2$)	Envelope	1	$\frac{1}{2}e^{-\gamma_b/2}$
DPSK	Phase-comparison	1	$\frac{1}{2}e^{-\gamma_b}$
BPSK	Coherent	1	$Q(\sqrt{2\gamma_b})$
MSK, QAM, or QPSK	Coherent quadrature	2	$Q(\sqrt{2\gamma_b})$

since $B_T \approx r = r_b/K$. The error probability per bit is given by

$$P_{be} \approx P_e/K$$

providing that the data converter employs a Gray code, as discussed in conjunction with Eq. (24), Sect. 11.2. After incorporating these adjustments in our previous expressions, we get the comparative results listed in Table 14.4–2. The quantity r_b/B_T is often referred to as **bandwidth efficiency.**

Table 14.4–2   Summary of M-ary modulation systems with $r_b/B_T = K = \log_2 M$

Modulation	Detection	P_{be}
DPSK ($M \geq 4$)	Phase-comparison quadrature	$\dfrac{2}{K} Q\left(\sqrt{4K\gamma_b \sin^2 \dfrac{\pi}{2M}} \right)$
PSK ($M \geq 8$)	Coherent quadrature	$\dfrac{2}{K} Q\left(\sqrt{2K\gamma_b \sin^2 \dfrac{\pi}{M}} \right)$
QAM (K even)	Coherent quadrature	$\dfrac{4}{K}\left(1 - \dfrac{1}{\sqrt{M}} \right) Q\left(\sqrt{\dfrac{3K}{M-1}\gamma_b} \right)$

All of the quadrature-carrier and M-ary systems increase modulation speed at the expense of error probability or signal energy. Suppose, for example, that you want to keep the error probability fixed at $P_{be} \approx 10^{-4}$—a common standard for comparison purposes. The value of γ_b needed for different modulation systems with various modulation speeds then can be calculated from our tabulated expressions. Figure 14.4–9 depicts the results as plots of r_b/B_T versus γ_b in dB, and each point is labeled with the corresponding value of M. Clearly, you would choose QAM over PSK for $r_b/B_T \geq 4$ with coherent detection. M-ary DPSK eliminates the carrier-synchronization problems of coherent detection, but it requires at least 7 dB more energy than QAM for $r_b/B_T \geq 4$.

As our final comparison, Table 14.4–3 combines M-ary data from Fig. 14.4–9 and calculates values for binary systems with the same error probability. The various systems are listed here in order of increasing complexity to bring out the trade-offs between modulation speed, signal energy, and hardware expense.

You should keep two points in mind when you examine this table. First, the numerical values correspond to *ideal* systems. The modulation speed of an actual system is typically about 80 percent of the theoretical value, and the required energy is at least 1–2 dB higher. Second, the characteristics of specific transmission channels may impose additional considerations. In particular, rapidly changing transmission delay prohibits coherent detection, while transmission nonlinearities dictate against the envelope modulation of OOK and M-ary QAM.

Other factors not covered here include the effects of interference, fading, and delay distortion. These are discussed in an excellent paper by Oetting (1979), which also contains an extensive list of references.

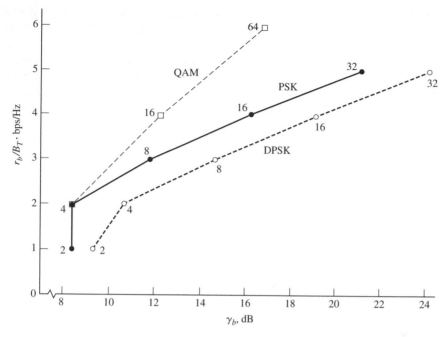

Figure 14.4–9   Performance comparison of *M*-ary modulation systems with $P_{be} = 10^{-4}$.

Table 14.4–3   Comparison of digital modulation systems with $P_{be} = 10^{-4}$

Modulation	Detection	r_b/B_T	γ_b, dB
OOK or FSK $(f_d = r_b/2)$	Envelope	1	12.3
DPSK $(M = 2)$	Phase-comparison	1	9.3
DQPSK	Phase-comparison quadrature	2	10.7
BPSK	Coherent	1	8.4
MSK, QAM, or QPSK	Coherent quadrature	2	8.4
DPSK $(M = 8)$	Phase-comparison quadrature	3	14.6
PSK $(M = 8)$	Coherent quadrature	3	11.8
PSK $(M = 16)$	Coherent quadrature	4	16.2
QAM $(M = 16)$	Coherent quadrature	4	12.2

14.5   TRELLIS-CODED MODULATION

Let's consider a standard voice-telephone line with a 3.2 kHz bandwidth and a *S/N* of 35 dB used in conjunction with a modem to transmit a digital message. According to the Hartley-Shannon law, discussed in Sect. 16.3, the line's capacity, or message rate, should be $C = 37.2$ kbps. However, for a fixed P_{be}, practical modems using conventional digital modulation methods such as *M*-ary PSK and QAM have only achieved rates of up to 9.6 kbps—nowhere near the Hartley-Shannon limit. The rate could be increased without requiring additional bandwidth by simply adding more points in the signal constellation. But since the *euclidean* distances between these points are decreased, P_{be} increases. Conventional error correction codes could be employed, but the added redundancy reduces the overall message rate. The newly developed turbo codes would also allow us to approach the Hartley-Shannon limit, but due to their iterative nature they have a relatively long *latency time* which may not be acceptable.

In order to achieve message rates closer to the Hartley-Shannon limit on *band-limited* channels such as telephone and cable TV lines, we now consider **trellis-coded modulation** (TCM) which was developed in the early 1980s by Gottfried Ungerboeck of IBM Zurich Research Laboratory (Ungerboeck, 1982, 1987).

> TCM is a scheme that combines convolutional coding and modulation.

TCM enables coding gains of at least 7 dB without bandwidth expansion and has enabled much higher baud rates for telephone modems. The cost incurred for this improvement is slightly increased decoder complexity.

TCM Basics

We first consider the *M*-ary PSK system whose constellation is shown in Fig. 14.5–1. Recall,

$$P_{be} = \frac{2}{K} Q\left(\sqrt{\frac{2E}{N_0} \sin^2 \frac{\pi}{M}} \right)$$ [1]

with errors most likely to occur when adjacent points are confused. Thus,

> Minimizing errors requires that we maximize the euclidean distance between adjacent points.

In this case, the distance squared between adjacent points is

$$d_{\min}^2 = 4 \sin^2 \frac{\pi}{M}$$ [2]

and thus we can express Eq. (1) as

$$P_{be} = \frac{2}{K} Q\left(\sqrt{\frac{E}{2N_0} d_{\min}^2} \right)$$ [3]

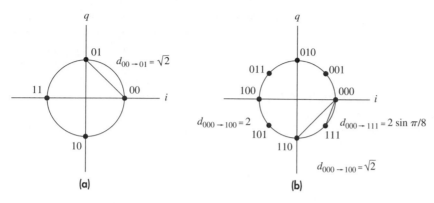

Figure 14.5–1   PSK signal constellation. (a) $M = 4$; (b) $M = 8$.

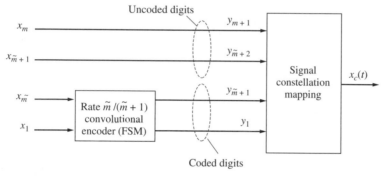

Figure 14.5–2   Generic $m/(m + 1)$ encoder-modulator for TCM.

To explain TCM, we first look at a QPSK system that has two bits per symbol and a constellation diagram as shown in Fig. 14.5–1a. In this case $x_1x_2 = 01$ is mapped to $\pi/2$ radians, $x_1x_2 = 10$ is mapped to $3\pi/2$ radians, and so on. As Fig. 14.5–1 and Eq. (2) show, after mapping into the signal constellation, the minimum distance squared between adjacent symbols is $d_{\min}^2 = d_{00\to01}^2 = 2$. We use this as a reference for M-ary PSK TCM systems.

The generic TCM system of Fig. 14.5–2 expands m message inputs $x_1, x_2, \ldots, x_m$ to generate $m + 1$ signal outputs $y_1, y_2, \ldots, y_{m+1}$ giving us $M = 2^{m+1}$ channel symbols that are then mapped to an M-ary signal constellation. The overall encoding rate is thus $R = m/(m + 1)$. The TCM structure includes using an $\widetilde{m}/(\widetilde{m} + 1)$ rate convolutional encoder, where $\widetilde{m}$ is equal to number of coded message bits and $\widetilde{m} \leq m$. This encoder expands $\widetilde{m}$ message bits into $\widetilde{m} + 1$ signal outputs.

The system of Fig. 14.5–3a, uses an $\widetilde{m}/(\widetilde{m} + 1) = 1/2$, 4-state, convolutional encoder to encode message bit x_1 into two bits, y_1 and y_2, while the system of Fig. 14.5–3b, uses an $\widetilde{m}/(\widetilde{m} + 1) = 2/3$, 8-state convolutional encoder to encode the two message bits x_1, x_2 into three bits y_1, y_2, and y_3. In both systems, bits $y_1, y_2,$

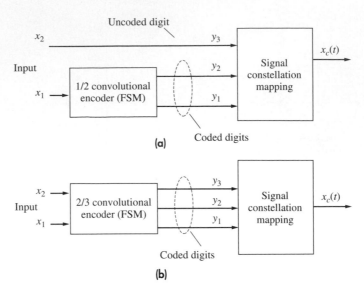

Figure 14.5–3   8-ary PSK encoder for TCM. (a) 4-state, $m = 2$, $\tilde{m} = 1$ convolutional encoder; (b) 8-state, $m = 2$, $\tilde{m} = 2$ convolutional encoder.

and y_3 are then mapped into an 8-ary PSK signal constellation of Fig. 14.5–1b with $y_3 y_2 y_1 = 001$ mapped to $\pi/4$ radians, $y_3 y_2 y_1 = 010$ mapped to $\pi/2$, and so on.

> The basic idea of TCM is to maximize the *euclidean distance* between message symbols most likely to be confused by using *set partitioning*. This amounts to maximizing the *free distance* between different message sequences.

This is in contrast to conventional error control coding schemes that seek to maximize *Hamming distances*. An example of an 8-PSK partitioning process is shown in Fig. 14.5–4.

Consider the system of 14.5–3a and its corresponding trellis diagram of Fig. 14.5–5. From each node, we have M branches corresponding to M outputs. Note that the parallel transitions are caused by the uncoded message bits. Let's assume the message is a successive sequence of $x_2 x_1 = 00$ inputs which produces a corresponding sequence of $y_3 y_2 y_1 = 000$ outputs as shown in the dashed line of Fig. 14.5–6. At the destination, the signal is decoded using a similar trellis structure and the *Viterbi algorithm* to determine the trellis path that most likely correlates with the received sequence. Therefore to minimize errors, we want to maximize the *free distance* between alternate sequences.

Let's say due to noise, an error occurs so that during decoding the received sequence diverges from the correct path. However, inherent in the TCM design, only

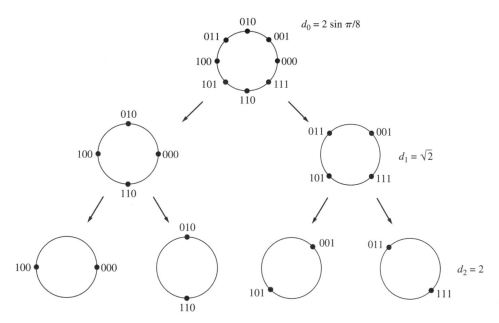

Figure 14.5–4   Partitioning of an 8-PSK signal set.
                      SOURCE: Ungerbroeck, 1982.

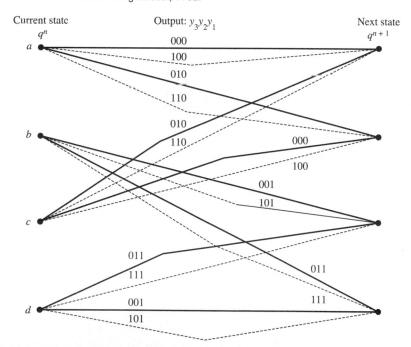

Figure 14.5–5   Trellis diagram for the encoder of Figure 14.5–3a.
                      SOURCE: Ungerbroeck, 1982.

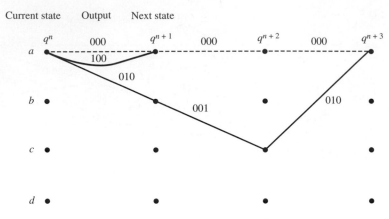

Figure 14.5–6   Two possible error events for the encoder of Figure 14.5–3a. Dashed line is transmitted sequence; solid lines are alternate received sequences; bold line is most likely error.

certain deviations are allowed; therefore it may take several transmission intervals until the signal path eventually remerges back to the correct node. This serves to increase the free distance between the correct and alternate paths.

Two different error events are shown in the solid lines of Fig. 14.5–6. Deviations from the correct path are measured by their respective euclidean distances away from 000 on the signal constellation. The shorter error path has a distance squared of $d_1^2 = d_{000 \to 100}^2 = 4$. Another possible error path sequence has a distance squared of $d_2^2 = d_{000 \to 010}^2 + d_{000 \to 001}^2 + d_{000 \to 010}^2 = 2 + 4 \sin^2 \pi/8 + 2 = 4.586$. Therefore, when an error event does occur, its distance will be at least $d_{min}^2 = 4$. There are other possible error paths, but these will have the same or greater euclidean distances, and thus we select $d_{min}^2 = d_{free}^2 = \min(d_1^2, d_2^2 \dots)$.

For TCM, we define coding gain as

$$g \triangleq \frac{(d_{min}^2/E')_{coded}}{(d_{min}^2/E)_{uncoded}} \qquad [4]$$

with E' and E being the energy of the coded and uncoded signals respectively. If both the coded and uncoded signals have the same normalized energy levels then

$$g = \frac{(d_{min}^2)_{coded}}{(d_{min}^2)_{uncoded}} \qquad [5]$$

and the *coding gain* is the same as the *distance gain* squared. With uncoded QPSK as our reference, the coding gain for the system of Fig. 14.5–3a is $g = 4/2 = 2$, or 3 dB.

For increased coding gain, we go to the 8-state, $\tilde{m} = 2$ TCM system of Fig. 14.5–3b with its corresponding trellis diagram of Fig. 14.5–7. Because all the message bits are coded, there are no parallel transitions to the next state. As before, let's assume the correct signal is a sequence of $y_3 y_2 y_1 = 000$'s as shown in the

State Output symbols*   Output: $y_3y_2y_1$

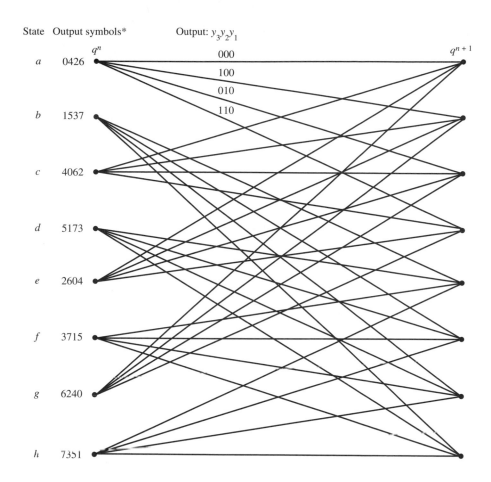

*Decimal equivalent encoder output

Figure 14.5–7   Trellis diagram for the encoder of Figure 14.5–3*b*.
                     SOURCE: Ungerbroeck, 1982.

dashed line of Fig. 14.5–8 with the minimum distance error event shown in the solid
lines. Again, when an error does occur causing the signal to stray from the correct
path, because only certain transitions are allowed, the signal may take several transitions before it remerges to the correct node. As we also stated before, other error
paths exist, but they will have the same or larger distances than this one. For the signal constellation distances of Fig. 14.5–1*b*, we get

$$d_{\min}^2 = (d_{000\to110}^2 + d_{000\to111}^2 + d_{000\to110}^2) = 2 + 4\sin^2\frac{\pi}{8} + 2 = 4.586$$

and the coding gain becomes $g = 4.586/2 = 2.293$ or 3.6 dB.

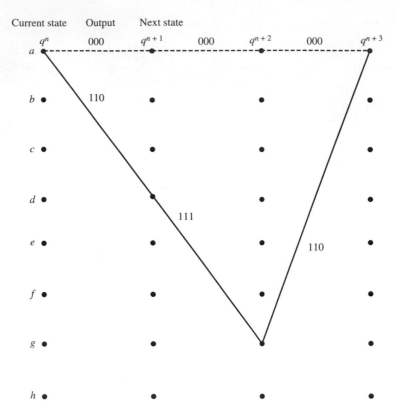

Figure 14.5–8   Error event for the encoder of Figure 14.5–3b. Dashed line is transmitted sequence; solid line is received sequence containing an error.

As we already stated, we achieve maximum coding gains by maximizing the free distance between alternate sequences. This is done by careful encoding and definition of allowable transitions in the trellis diagram. Ungerbroeck (1982) has stated the following rules for optimal signal assignment:

1. All signals should occur with equal frequency.

2. All parallel transitions in the trellis diagram should be with signals that have the maximum euclidean distance. (This is why in the case of the TCM system of Figs. 14.5–3a and 14.5–5, transitions occur with signals such as $y_3 y_2 y_1 = 000$ and 100 that have a euclidean distance of $d_{000 \to 100} = 2$ instead of signals $y_3 y_2 y_1 = 000$ and 001 that only have a euclidean distance of $d_{000 \to 001} = 2 \sin \pi/8 = 1.414$.)

3. All other signals entering or leaving a given state should have the next maximum possible euclidean distance.

Table 14.5-1   Coding gains for PSK-TCM systems

States	$\tilde{m}$	$g_{8-PSK/QPSK}$, dB $m = 2$	$N_{min}(m \to \infty)$
4	1	3.01	1.0
8	2	3.60	2.0
16	2	4.13	≈ 2.3
32	2	4.59	4.0
64	2	5.01	≈ 5.3
128	2	5.10	≈ 0.5
256	2	5.75	≈ 1.5

SOURCE: Ungerbroeck, 1987.

Table 14.5-2   Coding gains for QAM TCM systems

States	$\tilde{m}$	$g_{16-QAM/8-PSK}$, dB $m = 3$	$^1 g_{32-QAM/16-QAM}$, dB $m = 4$	$N_{min}(m \to \infty)$
4	1	4.36	3.01	4
8	2	5.33	3.98	16
16	2	6.12	4.77	56
32	2	6.12	4.77	16
64	2	6.79	5.44	56
128	2	7.37	6.02	344
256	2	7.37	6.02	44
512	2	7.37	6.02	4

SOURCE: Ungerbroeck, 1987.
NOTE: [1]The 32-QAM has a "cross"-shaped constellation pattern and is referred to by Ungerbroeck as 32CR.

Tables 14.5–1 and 14.5–2 show the potential coding gains versus number of states. Again, as stated, this assumes optimal encoding and partitioning of the signal set.

TCM can also be implemented with M-ary QAM. The potential coding gains of coded 16-QAM and 32-QAM, as compared to uncoded 8-PSK and 16-QAM respectively, are shown in Table 14.5–2.

Error Probability for Uncoded QPSK Versus 8-PSK TCM   **EXAMPLE 14.5-1**

For high signal-to-noise ratios and unity signal energy, an approximate expression for the *lower bound* for the error probability per M-ary symbol is given by (Ungerbroeck, 1982)

$$P_e \approx N_{min} Q\left(\sqrt{d_{min}^2/2 N_0}\right) \qquad [6]$$

where $N_{\min}$ is the average number of sequences whose distance is $d_{\min}$ from the correct sequence. Note the factor of $N_{\min}$ in Eq. (6) assumes only the likely error paths will significantly affect P_e. It does not take into account other longer sequences. For uncoded QPSK, we observed from Fig. 15.5–1a, that each signal point has two adjacent points that are spaced at $d_{\min} = \sqrt{2}$ and thus $N_{\min} = 2$. For $E_b/N_0 = 9$ dB $= 7.94$, the error probability is

$$P_e = 2Q\left(\sqrt{\frac{2}{2/7.94}}\right) = 5 \times 10^{-3}$$

Previously, it was determined for TCM with 4-states and 8-PSK that $d_{\min} = 2$. The corresponding trellis diagram of Fig. 14.5–6 shows that, at any given state, for a set of $y_3 y_2 y_1 = 000$ sequences, there is only one path, whose distance is $d_{\min}$, that deviates from the correct sequence and therefore $N_{\min} = 1$. Thus, we get

$$P_e = Q\left(\sqrt{\frac{4}{2/7.94}}\right) = 3 \times 10^{-5}$$

We said that $N_{\min}$ is an average. This is because with some signal constellations the number of nearest neighbors to a given point depends on its location. For example, in square signal constellations, interior points will have more neighbors than the exterior points.

Hard Versus Soft Decisions

As stated earlier, the coding gains we obtained with TCM come at a cost of increased decoder complexity. In conventional binary systems such as the one in Fig. 14.2–1, the demodulator outputs either $\hat{m} = 0$ or $\hat{m} = 1$ so that the decoder makes a **hard decision** based on two possible values of $\hat{m}$. Thus, a significant amount of information may be lost prior to decoding. On the other hand, TCM uses a demodulator that outputs a value of $\hat{m}$ that is typically 3 bits in length; thus the TCM decoder makes a **soft decision** based on $2^3 = 8$ levels of $\hat{m}$. These additional levels give the decoder more information about the signal than possible with only 2 levels of $\hat{m}$. Soft decision systems, TCM or otherwise, where the demodulator outputs 8 quantization levels will have gains of 2 dB over hard decision systems. Extending this to an analog system with an infinite number of quantization levels, the gain would be only 2.2 dB. Therefore,

There is little advantage in soft decisions based on more than 3 bits.

For more information on TCM, see Ungerbroeck (1982), Biglieri et al. (1991), and Schlegal (1997).

Table 14.5-3   Selected telephone line data modems

Model	Bit Rate	Modulation	Transmitting Frequencies (Hz)
Bell 103A[2]	300 bps	FSK	$(1{,}070–1{,}270)/(2{,}025–2{,}225)$[1]
Bell 212A[3]	300/1200 bps	DPSK	1200/2400[1]
V.32	9600 bps	16-QAM or 32-QAM with TCM	1800
V.32bis	14.4 kbps	128-QAM with TCM	1800
V.34	28.8 kbps	960-QAM with TCM	1800
V.34.bis[4]	33.6 kbps	1664-QAM with TCM	1800
V.90	56/33.6 kbps[5]	PCM	—

SOURCE: Lewart, 1998; Forney et al., 1996.

NOTES: [1]Originating/answering. [2]Similar to V.21. [3]Similar to V.22. [4]Similar to V.34-1996 or V.34+.
[5]56 kbps downstream (server to user), 33.6 kbps upstream; requires conditioned line.

Modems

A common application of bandpass digital modulation is the voice telephone **modem** (modulator/demodulator). This device modulates baseband digital signal from computer or fax to be put on a voice telephone line and then vice versa with the demodulator. As we discussed in Sect. 12.2, modems are an alternative to DSLs. Table 14.5–3 displays a selected list of Bell and the ITU (International Telecommunications Union) modem standards along with their respective rates and modulation methods. As we stated earlier, the Shannon limit for standard voice grade telephone lines is 37 kbps. Improvements in modulation and coding have enabled modems so they can get relatively close to this limit. Most commercial modems also have a fall-back option, so that at the initial connection and even during the session the modem will test the telephone line's *S/N* to set or adjust the modem's data rate. Thus in the case of a V.34 modem, if the line's *S/N* is severely degraded, the 28.8 kpbs data rate drops to 14.4 or 9.6 kbps.

Other types of modems for computer communication include cable modems for communication via the cable-TV network, LAN modems, wireless modems, and cellular telephone modems. Cable-TV systems with their bandwidths of 300 MHz promise data rates in the Gbps range.

14.6   PROBLEMS

14.1–1*   Find from Eq. (7) the average power $\overline{x_c^2}$ and the carrier-frequency power P_c of an *M*-ary ASK signal. Then form the ratio $P_c/\overline{x_c^2}$ and simplify for $M = 2$ and $M \gg 1$.

14.1–2   Suppose a binary ASK signal consists of RZ rectangular pulses with duration $T_b/2$, where $r_b = 1/T_b \ll f_c$. (a) Find the equivalent lowpass spectrum, and sketch and label $G_c(f)$ for $f > 0$. (*b*) Sketch the signal representing the sequence 010110. Then find the ratio of the carrier-frequency power P_c to the average power $\overline{x_c^2}$.

14.1–3   Consider a binary ASK signal with *raised-cosine pulse shaping* so, from Example 2.5–2

$$p(t) = \frac{1}{2}\left[1 + \cos\left(\frac{\pi t}{\tau}\right)\right]\Pi\left(\frac{t}{2\tau}\right) \qquad P(f) = \frac{\tau \operatorname{sinc} 2f\tau}{1 - (2f\tau)^2}$$

(*a*) Sketch the signal representing the sequence 010110 when $\tau = T_b/2$. Then find the equivalent lowpass spectrum, and sketch and label $G_c(f)$ for $f > 0$.

(*b*) Redo part *a* with $\tau = T_b$.

14.1–4   The *envelope* and *phase* variations of a QAM signal are

$$A(t) = A_c[x_i^2(t) + x_q^2(t)]^{1/2} \qquad \phi(t) = \arctan\left[x_q(t)/x_i(t)\right]$$

(*a*) By considering the time interval $kD < t < (k + 1)D$, obtain expressions for $A(t)$ and $\phi(t)$ with a rectangular pulse shape $p_D(t)$.

(*b*) Redo part *a* with an arbitrary pulse shape $p(t)$ whose duration does not exceed D.

14.1–5   Let a polar *M*-ary VSB signal have Nyquist pulse shaping per Eq. (6), Sect. 11.3. Find the equivalent lowpass spectrum before VSB filtering. Then sketch and label $G_c(f)$ for $f > 0$ when the filter has $\beta_V \ll r$.

14.1–6   Before bandpass filtering, the *i* and *q* components of the OQPSK signal generated in Fig. 14.1–6 can be written as

$$x_i(t) = \sum_k a_{2k}p(t - 2kT_b) \qquad x_q(t) = \sum_k a_{2k+1}p(t - 2kT_b - T_b)$$

where $a_k = (2A_k - 1)$ is the polar sequence corresponding to the message bit sequence A_k, and $p(t) = \Pi(t/2T_b)$ for NRZ rectangular pulse shaping.

(*a*) Sketch $x_i(t)$ and $x_q(t)$ for the bit sequence 10011100. Use your sketch to draw the signal constellation and to confirm that the phase $\phi(t) = \arctan\left[x_q(t)/x_i(t)\right]$ never changes by more than $\pm\pi/2$ rad.

(*b*) Find the equivalent lowpass spectrum.

14.1–7‡   Let $A_k B_k C_k$ denote the Gray-code binary words for the eight-phase PSK constellation in Fig. 14.1–5b. Construct a table listing $A_k B_k C_k$ and the corresponding values of I_k and Q_k expressed in terms of $\alpha = \cos \pi/8$ and $\beta = \sin \pi/8$. Then write *algebraic* expressions for I_k and Q_k as functions of A_k, B_k, C_k, α, and β. Devise from these expressions the diagram of a quadrature-carrier transmitter to generate the PSK signal, given a serial-to-parallel converter that supplies the binary words in the inverted polar form $a_k = 1 - 2A_k$, etc.

14.1–8   Suppose a binary FSK signal with discontinuous phase is generated by switching between two oscillators with outputs $A_c \cos (2\pi f_0 t + \theta_0)$ and $A_c \cos (2\pi f_1 t + \theta_1)$. Since the oscillators are unsynchronized, the FSK signal may be viewed as the interleaved sum of two independent binary ASK signals. Use this approach to find, sketch, and label $G_c(f)$ for $f > 0$ when $f_0 = f_c - r_b/2$ and $f_1 = f_c + r_b/2$ with $f_c \gg r_b$. Estimate B_T by comparing your sketch with Figs. 14.1–2 and 14.1–8.

14.1–9 Starting with $p(t)$ in Eq. (17c), obtain both forms of $|P(f)|^2$ as given in Eq. (18b).

14.1–10‡ Consider a binary FSK signal defined by Eq. (15) with $M = 2, D = T_b$, and $\omega_d = N/T_b$, where N is an integer. Modify the procedure used in the text and the hint given in Exercise 14.1–3 to obtain $x_i(t)$ and $x_q(t)$. Then show that

$$G_i(f) = \frac{1}{4}\left[\delta\left(f - \frac{Nr_b}{2}\right) + \delta\left(f + \frac{Nr_b}{2}\right)\right]$$

$$G_q(f) = \frac{1}{4r_b}\left[j^{N-1}\,\text{sinc}\,\frac{f - Nr_b/2}{r_b} + (-j)^{N-1}\,\text{sinc}\,\frac{f + Nr_b/2}{r_b}\right]^2$$

which reduce to Sunde's FSK when $N = 1$.

14.1–11 Use the FSK spectral expressions given in Prob. 14.1–10 to sketch and label $G_c(f)$ for $f > 0$ when $N = 2$ and $N = 3$. Compare with Fig. 14.1–8.

14.1–12 An OQPSK signal with *cosine pulse shaping* has many similarities to MSK. In particular, let the i and q components be as given in Prob. 14.1–6 but take $p(t) = \cos(\pi r_b t/2)\Pi(t/2T_b)$.

(a) Sketch $x_i(t)$ and $x_q(t)$ for the bit sequence 100010111. Use your sketch to draw the signal constellation and to find the phase $\phi(t) = \arctan\left[x_q(t)/x_i(t)\right]$ at $t = kT_b$, $0 \le k \le 7$. Compare these with Fig. 14.1–11.

(b) By considering an arbitrary interval $2kT_b < t < (2k + 1)T_b$, confirm that the envelope $A(t) = A_c[x_q^2(t) + x_i^2(t)]^{1/2}$ is constant for all t.

(c) Justify the assertion that $G_{lp}(f)$ is identical to an MSK spectrum.

14.1–13 Derive the q component of an MSK signal as given in Eq. (25).

14.1–14* Consider a BPSK system for a bandlimited channel with $B_T = 3000$ Hz where the spectral envelope must be at least 30 dB below the maximum outside the channel. What is the maximum data rate r_b to achieve this objective?

14.1–15 Repeat Prob. 14.1–14 for (a) FSK, (b) MSK.

14.2–1 Draw and label the block diagram of an optimum coherent BPSK receiver with matched filtering.

14.2–2 Suppose an OOK signal has raised-cosine pulse shaping so that

$$s_1(t) = A_c \sin^2(\pi t/T_b)p_{T_b}(t)\cos\omega_c t$$

Draw and label the diagram of an optimum coherent receiver using: (a) matched filtering; (b) correlation detection.

14.2–3 Obtain an *exact* expression for E_b when a binary FSK signal has $f_c = N_c r_b$ but f_d is arbitrary. Simplify your result when $N_c - f_d T_b \gg 1$.

14.2–4* Use Table T.4 to show that taking $f_d \approx 0.35r_b$ yields the lowest possible error probability for binary FSK with AWGN. Write the corresponding expression for P_e in terms of γ_b.

14.2–5‡ Draw the complete block diagram of an optimum receiver for Sunde's FSK. Use correlation detection and just *one* local oscillator whose frequency equals r_b. Assume that a bit-sync signal has been extracted from $x_c(t)$ and that N_c is known.

14.2–6* With perfect synchronization, a certain BPSK system would have $P_e = 10^{-5}$. Use Eq. (18) to find the condition on θ_ϵ so that $P_e < 10^{-4}$.

14.2–7 Consider a BPSK receiver in the form of Fig. 14.2–4 with local-oscillator output $K A_c \cos{(\omega_c t + \theta_\epsilon)}$, where θ_ϵ is a synchronization error. Show that the signal component of $y(T_b)$ is reduced in magnitude by the factor $\cos{\theta_\epsilon}$.

14.2–8 Find the exact expression for $z(t)$ in Eq. (17). Then take $N_c \gg 1$ to obtain the given approximation.

14.2–9 Consider a BPSK signal with pilot carrier added for synchronization purposes, resulting in

$$s_1(t) = [A_c \cos \omega_c t + \alpha A_c \cos{(\omega_c t + \theta)}]p_{T_b}(t)$$

$$s_0(t) = [-A_c \cos \omega_c t + \alpha A_c \cos{(\omega_c t + \theta)}]p_{T_b}(t)$$

Take $\theta = 0$ and show that an optimum coherent receiver with AWGN yields $P_e = Q\left[\sqrt{2\gamma_b/(1 + \alpha^2)}\,\right]$.

14.2–10 Do Prob. 14.2–9 with $\theta = -\pi/2$.

14.2–11‡ When the noise in a coherent binary system is gaussian but has a *nonwhite* power spectrum $G_n(f)$, the noise can be "whitened" by inserting at the front end of the receiver a filter with transfer function $H_w(f)$ such that $|H_w(f)|^2 G_n(f) = N_0/2$. The rest of the receiver must then be matched to the distorted signaling waveforms $\tilde{s}_1(t)$ and $\tilde{s}_0(t)$ at the output of the whitening filter. Furthermore, the duration of the unfiltered waveforms $s_1(t)$ and $s_0(t)$ must be reduced to ensure that the whitening filter does not introduce appreciable ISI. Apply these conditions to show from Eq. (9a) that

$$\left(\frac{z_1 - z_0}{2\sigma}\right)^2_{\max} = \int_{-\infty}^{\infty} \frac{|S_1(f) - S_0(f)|^2}{4G_n(f)}\, df$$

where $S_1(f) = \mathcal{F}[s_1(t)]$, etc. *Hint*: Recall that if $v(t)$ and $w(t)$ are real, then

$$\int_{-\infty}^{\infty} v(t)w(t)\, dt = \int_{-\infty}^{\infty} V(f)W^*(f)\, df = \int_{-\infty}^{\infty} V^*(f)W(f)\, df$$

14.2–12 For coherent binary FSK, show that $\rho = \text{sinc}\,(4f_d/r_b)$.

14.2–13 Determine ρ that minimizes P_e for a coherent binary FSK system.

14.2–14* Determine A_c required to achieve a $P_e = 10^{-5}$ for a channel with $N_0 = 10^{-11}$ W/Hz, and BPSK with (a) $r_b = 9.6$ kpbs, (b) for $r_b = 28.8$ kpbs.

14.2–15   Repeat Prob. 14.2–14 using coherent FSK.

14.3–1*   A noncoherent OOK system is to have $P_e < 10^{-3}$. Obtain the corresponding bounds on γ_b and P_{e_1}.

14.3–2   Do Prob. 14.3–1 with $P_e < 10^{-5}$.

14.3–3   Repeat Prob. 14.2–14 using noncoherent FSK.

14.3–4*   Determine P_e for a channel using Sunde's FSK with $(S/N) = 12$ dB, $r_b = 14.4$ kbps, and (*a*) coherent detection, (*b*) noncoherent detection.

14.3–5   Obtain Eq. (9) from Eq. (8) with $K = A_c/E_1$.

14.3–6   Suppose the OOK receiver in Fig. 14.3–2 has a simple BPF with $H(f) = [1 + j\,2(f - f_c)/B]^{-1}$ for $f > 0$, where $2r_b \leq B \ll f_c$. Assuming that $f_c \gg r_b$ and $\gamma_b \gg 1$, show that the signal energy must be increased by at least 5 dB get the same error probability as an incoherent receiver with a matched filter.

14.3–7   Consider a noncoherent system with a *trinary* ASK signal defined by Eq. (7) with $a_k = 0, 1, 2$. Let E be the average energy per symbol. Develop an expression similar to Eq. (11) for the error probability.

14.3–8*   A binary transmission system with $S_T = 200$ mW, $L = 90$ dB, and $N_0 = 10^{-15}$ W/Hz is to have $P_e \leq 10^{-4}$. Find the maximum allowable bit rate using: (a) noncoherent FSK; (*b*) DPSK; (*c*) coherent BPSK.

14.3–9   Do Prob. 14.3–8 with $P_e \leq 10^{-5}$.

14.3–10   A binary transmission system with phase modulation is to have $P_e \leq 10^{-4}$. Use Eq. (18), Sect. 14.2, to find the condition on the synchronization error θ_ϵ such that BPSK will require less signal energy than DPSK.

14.3–11   Do Prob. 14.3–10 with $P_e \leq 10^{-6}$.

14.3–12   Derive the joint PDF in Eq. (3) by rectangular-to-polar conversion, starting with $x = A_c + n_i$ and $y = n_4$.

14.4–1*   Binary data is to be transmitted at the rate $r_b = 500$ kbps on a radio channel having 400-kHz bandwidth.

（*a*) Specify the modulation method that minimizes signal energy, and calculate γ_b in dB needed to get $P_{be} \leq 10^{-6}$.

(*b*) Repeat part *a* with the additional constraint that coherent detection is not practical for the channel in question.

14.4–2   Do Prob. 14.4–1 with $r_b = 1$ Mbps.

14.4–3*   Binary data is to be transmitted at the rate $r_b = 800$ kbps on a radio channel having 250-kHz bandwidth.

(*a*) Specify the modulation method that minimizes signal energy, and calculate γ_b in dB needed to get $P_{be} \leq 10^{-6}$.

(b) Repeat part *a* with the additional constraint that channel nonlinearities call for a constant envelope signal.

14.4–4   Do Prob. 14.4–3 with $r_b = 1.2$ Mbps.

14.4–5   Let the VCO output in Fig. 14.4–6 be $2 \cos{(\omega_c t + \theta_\epsilon)}$. In absence of noise, show that the control voltage $v(t)$ will be proportional to $\sin{\theta_\epsilon}$.

14.4–6*   Suppose an *M*-ary QAM system with $M = 16$ is converted to DPSK to allow phase-comparison detection. By what factor must the symbol energy be increased to keep the error probability essentially unchanged?

14.4–7   Suppose a PSK system with $M \gg 1$ is converted to *M*-ary QAM. By what factor can the symbol energy be reduced, while keeping the error probability essentially unchanged?

14.4–8‡   Obtain the phase PDF given in Eq. (9) from the joint PDF in Eq. (3), Sect. 14.3. *Hint*: Use the change of variable $\lambda = (A - A_c \cos{\phi})/\sigma$.

14.4–9   Using the technique given in Fig. 14.4–2, design a system that will create the carrier reference needed for a *M*-ary PSK receiver.

14.4–10   Generalize the design of Fig. 14.4–2 to enable the creation of *M*-reference signals for an *M*-ary PSK receiver.

14.4–11*   For $\gamma_b = 13$ dB, calculate P_e for (a) FSK (noncoherent), (b) BPSK, (c) 64-PSK, (d) 64-QAM.

14.5–1*   A QPSK system has a $P_e = 10^{-5}$. What is the new P_e if we employed TCM with $m = \tilde{m} = 2$ and 8 states? With TCM, does the output symbol rate change?

14.5–2   Partition a 16-QAM signal constellation in a similar way that was done for the 8-PSK constellation of Fig. 14.5–4 to maximize the distance between signal points. If the original minimum distances between adjacent points are unity, show the new minimum distances for each successive partition.

14.5–3*   Given the system of Figs. 14.5–3b and 14.5–7, with initial state of *a*, determine the output sequence $y_3 \, y_2 \, y_1$ for an input sequence of $x_2 \, x_1 = 00\ 01\ 10\ 01\ 11\ 00$.

14.5–4   For the system of Fig. 14.5–7, what is the distance between paths (0, 2, 4, 2) and (6, 1, 3, 0)?

14.5–5   Ungerboeck (1982) increases the coding gain of the $m = \tilde{m} = 2$, 8-PSK TCM system of Fig. 14.5–7 by adding the following new states and corresponding output symbols: (*i*: 4062, *j*: 5173, *k*: 0426, *l*: 1537, m: 6240, *n*: 7351, *o*: 2604, *p*: 3715). Following the pattern of Fig. 14.5–7, construct the new trellis diagram and show that $g_{8-PSK/QPSK} = 4.13$ dB.

chapter
15

Spread Spectrum Systems

CHAPTER OUTLINE

671

J ust prior to World War II, Hedy Lamarr, a well-known actress and political refugee from Austria, struck up a conversation with music composer George Antheil that led to a scheme to control armed torpedoes over long distances. The technique was immune to enemy jamming and detection. Instead of a conventional guidance system consisting of a single frequency signal that could easily be detected or jammed, their signal would hop from one frequency to another in a pseudorandom fashion known only to an authorized receiver (i.e., the torpedo). This would cause the transmitted spectrum to be spread over a range much greater than the message bandwidth. Thus, *frequency-hopping spread spectrum* (FH-SS) was born and eventually patented by Lamarr and Antheil.

Spread spectrum (SS) is similar to angle modulation in that special techniques spread the transmitted signal over a frequency range much greater than the message bandwidth. The spreading combats strong interference and prevents casual eavesdropping by unauthorized receivers. In addition to FH-SS, there is also *direct-sequence spread-spectrum* (DSS) based on a *direct spreading* technique in which the message spectrum is spread by multiplying the signal by a wideband *pseudonoise* (PN) sequence.

We begin our study of spread-spectrum systems by defining *direct sequence* and *frequency hopping* systems and then examining their properties in the presence of broadband noise, single- and multiple-tone jammers as well as other SS signals. We'll then consider the generation of PN codes that have high values of autocorrelation between identical codes (so authorized users can easily communicate) and low values of crosscorrelation between different codes (to minimize interference by outsiders). Next we examine the method of *code-division-multiple-access* (CDMA) in which several users have different PN codes but share a single RF channel. Finally, we discuss synchronization and wireless telephone systems.

OBJECTIVES

After studying this chapter and working the exercises, you should be able to do each of the following:

1. Describe the operation of DSS and FH-SS systems (Sects. 15.1 and 15.2).

2. Calculate probability of error for DSS systems under single-tone jamming, broadband noise, and multiple-user conditions (Sect. 15.1).

3. Calculate probability of error for FH-SS for single- and multiple-tone jamming, narrowband and wideband noise conditions, and muliple-user conditions (Sect. 15.2).

4. Design and analyze code generators that produce spreading codes with high autocorrelation and low crosscorrelation values (Sect. 15.3).

5. Describe how a SS can be used for distance measurement (Sect. 15.3).

6. Describe SS receiver synchronization and calculate the average time it takes to achieve synchronization (Sect. 15.4).

7. Describe the differences between conventional cellular phone and the personal communications systems (PCS) (Sect. 15.5).

15.1   DIRECT SEQUENCE SPREAD SPECTRUM

DSS is similar to FM in that the modulation scheme causes the transmitted message's frequency content to be greatly spread out over the spectrum. The difference is that with FM, the message causes the spectrum spreading, whereas with DSS, a pseudorandom number generator causes the spreading.

DSS Signals

A DSS system and its associated spectra are illustrated in Fig. 15.1–1, where the message $x(t)$ is multiplied by a wideband PN waveform $c(t)$ prior to modulation resulting in

$$\tilde{x}(t) = x(t)c(t) \tag{1}$$

Multiplying by $c(t)$ effectively masks the message and spreads the spectrum of the modulated signal. The spread signal $\tilde{x}(t)$ can then be modulated by a balanced modulator (or a multiplier) to produce a DSB signal. If $x(t)$ had values of ±1 that represented a digital message, the output from the DSB modulator would be a BPSK (PRK) signal.

Let's look at this more closely. The PN generator produces a pseudorandom binary wave $c(t)$, illustrated in more detail in Fig. 15.1–2, consisting of rectangular pulses called **chips** (CPS). Each chip has a duration of T_c and an amplitude of ±1 so that $c^2(t) = 1$—an essential condition for message recovery. To facilitate analysis, we'll assume that $c(t)$ has the same properties as the random digital wave in Example 9.1–3 when $D = T_c$, $a_k = \pm1$ and $\sigma^2 = \overline{c^2} = 1$. Thus, from our previous studies,

$$R_c(\tau) = \Lambda(t/T_c) \qquad \text{and} \qquad G_c(f) = T_c \, \text{sinc}^2 \frac{f}{W_c} \tag{2}$$

which are sketched in Fig. 15.1–2. The parameter

$$W_c \triangleq \frac{1}{T_c}$$

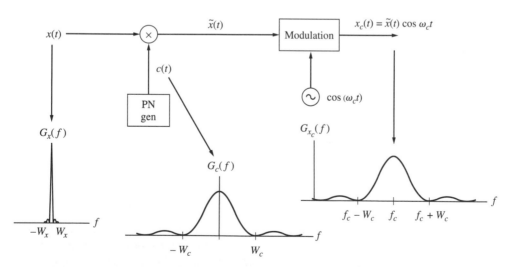

Figure 15.1–1   DSS transmitter system and spectra.

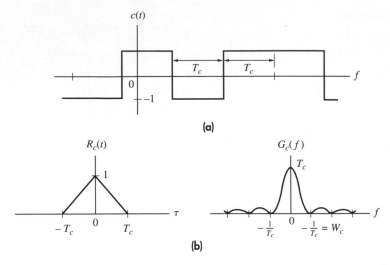

Figure 15.1–2   Pseudrandom binary wave. (a) Waveform; (b) autocorrelation and power spectrum.

serves as a measure of the PN bandwidth. Next consider the "chipped message" $\tilde{x}(t) = x(t)c(t)$. Treating $x(t)$ as the output of an ergodic information process independent of $c(t)$ we have

$$\overline{\tilde{x}^2} = E[x^2(t)c^2(t)] = \overline{x^2} = S_x$$

Recall further that multiplying independent random signals corresponds to multiplying their autocorrelation functions and *convolving* their power spectra. For clarity we denote the message bandwidth by W_x such that $G_x(f) \approx 0$ for $|f| > W_x$ and

$$G_{\tilde{x}}(f) = G_x(f)*G_c(f) = \int_{-W_x}^{W_x} G_x(\lambda)G_c(f - \lambda)\, d\lambda$$

But effective spectral spreading calls for $W_c \gg W_x$ in which case $G_c(f - \lambda) \approx G_c(f)$ over $|\lambda| \le W_x$. Therefore

$$G_{\tilde{x}}(f) \approx \left[\int_{-W_x}^{W_x} G_x(\lambda)\, d\lambda \right] G_c(f) = S_x G_c(f) \tag{3}$$

and we conclude from Eq. (3) and Fig. 15.1–1 that $x(t)$ has a spread spectrum whose bandwidth essentially equals the PN bandwidth W_c. With practical systems, the **bandwidth expansion factor** W_c/W_x can range from 10 to 10,000 (10 to 40 dB). As will be shown later, the higher this ratio the better the system's immunity to interference.

DSB or DPSK modulation produces a transmitted signal proportional to $\tilde{x}(t)\ \cos \omega_c t$ requiring a bandwidth $B_T \gg W_x$.

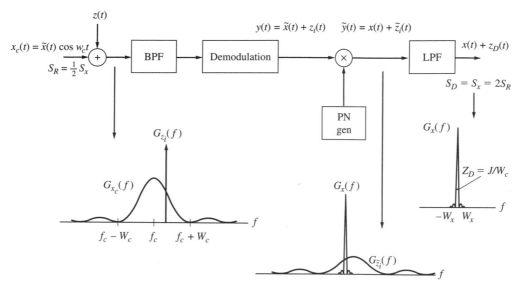

Figure 15.1–3   DSS receiver system and the effects of a single-tone jammer.

We'll analyze the system's performance taking unit-amplitude carrier at the receiver so $S_R = \frac{1}{2}\overline{\tilde{x}^2} = \frac{1}{2}S_x$ and we'll let $z(t)$ stand for additive noise or interference with in-phase component $z_i(t)$. Synchronous detection after bandpass filtering yields $y(t) = \tilde{x}(t) + z_i(t)$ which is multiplied by a locally generated PN wave to get

$$\tilde{y}(t) = [\tilde{x}(t) + z_i(t)]c(t) \qquad\qquad\qquad [4]$$

$$= x(t)c^2(t) + z_i(t)c(t) = x(t) + \tilde{z}_i(t)$$

Notice that this multiplication spreads the spectrum of $z_i(t)$ but de-spreads $\tilde{x}(t)$ and recovers $x(t)$ assuming near perfect synchronization of the local PN generator. Final lowpass filtering removes the out-of-band portion of the $\tilde{z}_i(t)$, leaving $y_D(t) = x(t) + z_D(t)$ with output signal power $S_D = S_x = 2S_R$. Our next step is to find the contaminating power $\overline{z_D^2}$ at the output.

When $z(t)$ stands for white noise $n(t)$, the in-phase component $n_i(t)$ has the lowpass power spectrum back in Fig. 10.1–3 and

$$R_{n_i}(\tau) = \mathcal{F}[G_{n_i}(f)] \approx N_0 B_T \, \text{sinc}\,(B_T \tau)$$

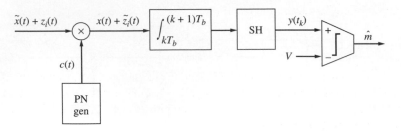

Figure 15.1–4   DSS correlation receiver for BPSK.

The autocorrelation of the chipped lowpass noise $\tilde{n}_i(t)$ equals the product $R_{n_i}(\tau)R_c(\tau)$. Since $R_{n_i}(\tau)$ becomes quite small for $|\tau| \geq 1/B_T \ll 1/W_c$ whereas $R_c(\tau) \approx 1$ for $|\tau| \ll T_c = 1/W_c$, we conclude that $R_{\tilde{n}_i}(\tau) \approx R_{n_i}(\tau)$ and hence

$$G_{\tilde{n}_i}(f) \approx G_{n_i}(f) \approx N_0 \Pi\,(f/B_T)$$

The output noise power from the lowpass filter is then $N_D = 2N_0 W_x$, so

$$(S/N)_D = 2S_R/2N_0W_x = S_R/N_0W_x \tag{5}$$

Comparing this result with Eq. (4b), Sect. 10.2, confirms that spread spectrum with synchronous detection has the same noise performance as a conventional DSB system.

Similarly, if our message is digital and sent via BPSK we can use the correlation detector of Fig. 15.1–4. Thus, in the presence of white noise, the probability of error would be

$$P_e = Q\big(\sqrt{2E_b/N_0}\big) \tag{6}$$

DSS Performance in Presence of Interference

Fig. 15.1–3 also illustrates the effects of a single-tone jammer with DSS system. Let $z(t)$ stand for an interfering sinusoid or CW jamming signal, say

$$z(t) = \sqrt{2J}\,\cos\big[(\omega_c + \omega_z)t + \theta\big]$$

with average power $\overline{z^2} = J$ at frequency $f_c + f_z$. Then the in-phase component is $z_i(t) = \sqrt{2J}\,\cos\,(\omega_z t + \theta)$ so that

$$G_{z_i}(f) = \frac{J}{2}[\delta(f - f_z) + \delta(f + f_z)] \tag{7}$$

Multiplication by $c(t)$ spreads this spectrum so that $G_{\tilde{z}_i}(f) = G_{z_i}(f)*G_c(f)$, a routine convolution with impulses. Because $c(t)$ is relatively broadband, $\tilde{z}_i(t)$ approximates another broadband noise source with power spectral density J/W_c. If $|f_z| \ll W_c$, then the corresponding upper bound on the output jamming power is

$$\overline{z_D^2} = \int_{-W_x}^{W_x} G_{\tilde{z}_i}(f)\,df \leq 2W_x\frac{J}{W_c} \tag{8a}$$

and the signal-to-jamming ratio becomes

$$\left(\frac{S}{J}\right)_D \geq \frac{W_c}{W_x}\frac{S_R}{J}$$ [8b]

The single-tone jammer spectra has been spread and thus, relative to the output signal power, has been reduced by a factor $W_x/W_c \ll 1$. This reduction is illustrated in the output spectra of Fig. 15.1–3. The bandwidth expansion ratio W_c/W_x is also called the **process gain** (Pg) or

$$Pg = W_c/W_x$$ [9]

which is a measure of a system's immunity to interference. It should be observed that the larger the Pg, the better the system's immunity to interference.

We can then treat the jammer as just another source of noise, and in the case of digital information sent via BPSK the probability of error would then be

$$P_e = Q\left(\sqrt{2E_b/N_J}\right)$$ [10]

where $N_J = J/W_c$. If the channel is corrupted by both broadband white noise and a CW jammer then

$$P_e = Q\left[\sqrt{2E_b/(N_0 + N_J)}\right]$$ [11]

If we substitute the received power and jammer power terms of $S_R = E_b r_b$ and $N_J = J/W_c$ respectively into Eq. (10) we have

$$P_e = Q\left(\sqrt{\frac{2W_c/r_b}{J/S_R}}\right)$$ [12]

Substituting $r_b = W_x$, and Eq. (9), Eq. (12) becomes

$$P_e = Q\left(\sqrt{\frac{2\,Pg}{J/S_R}}\right)$$ [13]

Now let's say we specify a minimum P_e value for Eq. (10). Combining with Eq. [13] and converting to dB, we have

$$10 \log (J/S_R) = 10 \log (Pg) - 10 \log (E_b/N_J)$$ [14]

The term $10 \log (J/S_R)$ is called the **jamming margin** and is used as a measure of a system's ability to operate in the presence of interference. If, in a given system, we specify a minimum P_e or minimum E_b/N_J ratio and a relatively large Pg, then the system will have margin against interference.

A DSS-BPSK system has $r_b = 3$ kbps, $N_0 = 10^{-12}$, and is received with $P_e = 10^{-9}$. **EXAMPLE 15.1–1**
Let's calculate the Pg needed for the system to achieve $P_e = 10^{-8}$ in the presence of a single-tone jammer whose received power is 10 times larger than the correct signal. Using Table T.6 and Eq. (6) we get $2E_b/N_0 = 36$ or $E_b = 1.8 \times 10^{-11}$. With

$S_R = E_b r_b = 5.4 \times 10^{-8}$, then $J = 10 S_R = 5.4 \times 10^{-7}$ W. If with a jammer, $P_{e_{min}} = 10^{-8}$ then, using Eq. (11) and Table T.6, $N_J = 1.48 \times 10^{-13}$ and $N_J = J/W_c$, $W_c = 3.6 \times 10^6$. If $W_x = r_b$ then $Pg = W_c/W_x = 1200$.

EXERCISE 15.1–1   Given the above problem specifications, what is the jamming margin if $Pg = 10,000$?

Multiple Access

Finally, if we are sharing the channel with $M - 1$ other spread spectrum users—as is the case with *code division multiple access* (CDMA)—each one would have their own unique spreading code and arrival times at the receiver and thus the interference term becomes,

$$z(t) = \sum_{m=1}^{M-1} A_m x_m(t - t_m) c_m(t - t_m) \cos \theta_m \tag{15}$$

where A_m, $c_m(t)$, t_m, and θ_m denote the signal amplitude, spreading code, time delay, and phase, respectively, of the mth user. Thus

$$y(t) = \tilde{x}(t) + \sum_{m=1}^{M-1} A_m x_m(t - t_m) c_m(t - t_m) \cos \theta_m \tag{16}$$

If we assume each of the other users has identical signal strengths of unit value, then after despreading, Eq. (16) becomes,

$$\tilde{y}(t) = x(t) + \left[\sum_{m=1}^{M-1} x_m(t - t_m) c_m(t - t_m) \cos \theta_m \right] c(t) \tag{17}$$

In the case of BPSK, the output of the correlation receiver would be

$$x(t_k) + \sum_{m=1}^{M-1} \left[\cos \theta_m \int_{kT_b}^{(k+1)T_b} x_m(t - t_m) c_m(t - t_m) c(t)\, dt \right] \tag{18a}$$

$$= x(t_k) + z(t_k) \tag{18b}$$

where $z(t_k)$ is the cumulative interference of the additional $M - 1$ CDMA users. Notice that since $x_m(t) = \pm 1$, the integration term in Eq. (15a) becomes the crosscorrelation between the desired and the interferer's PN codes. Therefore, minimizing the crosscorrelation between spreading codes minimizes the interference between CDMA users. Ideally, each PN code would be chosen to be orthogonal to the other, thereby making $z(t_k) = 0$. Unfortunately, as will be shown later, with practical systems this is not possible.

Pursley (1977) and Lathi (1998) derive an analytical expression for the bit error probability for M users of a CDMA channel corrupted by white noise. These show that if each of the M users have identical signal strengths, then

$$P_e = Q\big(1/\sqrt{(M-1)/(3Pg) + N_0/2E_b}\big) \qquad [19]$$

Note that if the channel has only one user (i.e., $M = 1$), Eq. (19) reduces to Eq. (6). Conversely, if the channel is noiseless but contains other users, the error probability would be

$$P_e = Q\big(\sqrt{3Pg/(M-1)}\big) \qquad [20]$$

Thus, even if the channel is noiseless, the error probability is still nonzero if it contains other users.

As Eqs. (6), (10) and (19) show, all interfering signals appear to the system as broadband noise. Therefore, unlike TDMA and FDMA, in which additional users manifest themselves as crosstalk, each CDMA user merely adds to the ambient noise floor.

This has regulatory implications. In the past, FCC regulations regarding maximum power levels were dictated by crosstalk considerations. In today's SS world, additional users merely decrease the other's signal-to-noise ratio.

A single-user DSS-BPSK system has $P_e = 10^{-7}$ and $Pg = 30$ dB. How many additional users can this system support if we allow $P_e \lesssim 10^{-5}$?

EXERCISE 15.1–2

15.2   FREQUENCY HOP SPREAD SPECTRUM

As previously noted, the wider the spread spectrum bandwidth, the more the jammer has to increase its power to be effective. However, the practical limitations of PN sequence generation hardware imposes a constraint on bandwidth spreading and therefore the processing gain. To enable even larger processing gains, the PN generator can drive a frequency synthesizer that produces a wideband sequence of frequencies that can cause the data-modulated carrier to hop from one frequency to another. This process is called *frequency hopping spread spectrum* (FH-SS). Because the message is spread out over numerous carrier frequencies, the jammer has a reduced probability of hitting any one in particular. Or the jammer has to spread its power over a wider frequency range in order to be effective.

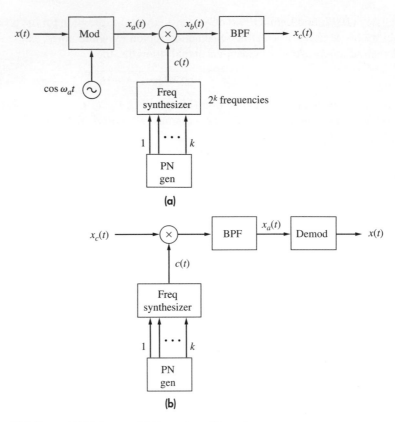

Figure 15.2–1   FH-SS System. (a) Transmitter; (b) receiver.

FH-SS Signals

Frequency hopping SS is shown in Fig. 15.2–1 and works as follows. The message is usually M-ary FSK modulated to some carrier frequency f_c, although some systems use BPSK. The modulated message is then mixed with the output of a frequency synthesizer. The frequency synthesizer's output is one of $Y = 2^k$ values, where k equals the number of outputs from the PN generator. The BPF selects the sum term from the mixer for transmission on the channel. The receiver in part (b) of the figure is the reverse of this process. Because of practical difficulties in maintaining phase coherence, most systems use noncoherent detection such as an envelope detector.

There are two types of FH-SS systems. In **slow hop SS** one or more message symbols are transmitted per hop: in **fast hop SS** there are two or more frequency hops per message symbol. With slow hop SS, the receiver demodulates the signal like any other M-ary FSK signal. However, with fast hop SS there are several hops per symbol, so the detector determines the value based on either a majority vote or some decision rule such as maximum likelihood.

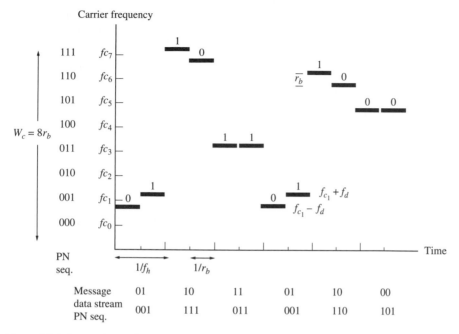

Figure 15.2–2   Output frequency versus data input for slow hop FH-SS system of Example 15.2–1.

Consider a slow hop SS system with binary FSK that transmits two symbols per frequency hop and has a PN generator with $k = 3$ outputs. For a given binary message sequence, the spectral output is illustrated in Fig. 15.2–2. With $k = 3$ the system can hop to $Y = 2^k = 8$ different carrier frequencies. For each hop the output frequency is $f_{c_i} + f_d$ where f_{c_i} is the carrier frequency at hop i and f_d is the frequency shift from a particular carrier frequency. The number of frequencies clustered around each f_{c_i} is equal to the number of symbols/hop. If the bit rate is r_b, then from Eq. (16) of Sect. 14.1 we have $f_d = r_b/2$. With a maximum frequency of $f_{c_7} + f_d$ and a minimum frequency of $f_{c_0} - f_d$ then the amount of spreading or transmission bandwidth is $B_T = W_c = 8r_b$. From Sect. 15.1 we defined process gain as $Pg = W_c/W_x$. If $W_x = r_b$ then with FH-SS we have

$$Pg = 2^k \qquad [1]$$

EXAMPLE 15.2–1

Consider a fast hop SS system with binary FSK, two hops per symbol, and a PN generator with outputs with the same binary message of Example 15.2–1. The message is transmitted using the following PN sequence: {010, 110, 101, 100, 000, 101,

EXERCISE 15.2–1

$011, 001, 001, 111, 011, 001, 110, 101, 101, 001, 110, 001, 011, 111, 100, 000, 110,$ 110}. In a format similar to that done in Example 15.2–1, plot the output frequencies for the input message.

FH-SS Performance in the Presence of Interference

We now want to consider the performance of a FH-SS system with respect to bit errors. There are several types of interference or noise to deal with: white noise of barrage jamming, single-tone jamming, partial-band jamming, or interference due to multiple FH-SS users on the same band as occurs with CDMA systems. Several of these are illustrated in Fig. 15.2–3.

Slow hop SS is the most susceptible to jamming because one or more symbols are being transmitted at a particular frequency. However, if the hop period is shorter than the transit time between the jammer and user's transmitter/receiver, then by the time the jammer has decided which frequency to jam, the transmitter has already hopped to another frequency. With M-ary FSK and noncoherent detection, the bit error probability in the presence of white noise is

$$P_e = \tfrac{1}{2}e^{-E_b/2N_0} \qquad\qquad [2]$$

If a jammer has the power spectral density of Fig. 15.2-3(a) with N_J the power level over the entire bandpass of the system, the jammer appears as white noise and Eq. (2) becomes

$$P_e = \tfrac{1}{2}e^{-E_b/[2(N_0+N_J)]} \qquad\qquad [3]$$

where $N_J = J/W_c$.

Figure 15.2–3(b) illustrates partial-band jamming with Δ the fraction of the band being jammed. The quantity N_J/Δ is equal to the jammer's PSD and has a bandwidth of ΔW_c. With Δ also equal to the probability of being jammed and $1 - \Delta$ equal to the probability of not being jammed then, using the chain rule, we have

$$P_e = P(e|jammed^c)P(jammed^c) + P(e|jammed)P(jammed)$$

The probability of error for partial-band jamming of FH-SS would then be

$$P_e = \frac{1 - \Delta}{2} e^{-E_b/2N_0} + \frac{\Delta}{2} e^{-E_b/[2(N_0+N_J/\Delta)]} \qquad\qquad [4]$$

In systems in which there are a relatively large number of frequency slots to hop to and the jammer's distance is large compared to the transmitter-receiver distance, the probability of a single tone causing an error is quite small.

Finally, let's consider a CDMA system with M users and M-1 potential interferers, as is shown in Fig. 15.2–3(d). Let the probability of one user causing an error to another, given that they are using the same frequency slot (thus causing a collision c), be 1/2, or $P(error|c) = 1/2$. Then the probability of an error given

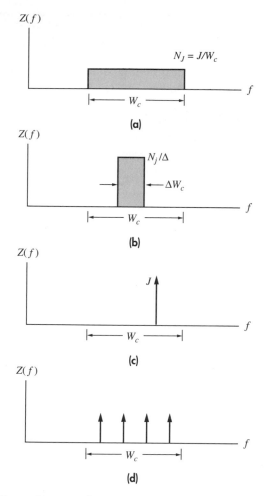

$Z(f)$

$N_J = J/W_c$

W_c

(a)

$Z(f)$

N_j/Δ

ΔW_c

W_c

(b)

$Z(f)$

J

W_c

(c)

$Z(f)$

W_c

(d)

Figure 15.2–3   Types of jamming for FH-SS systems. (a) Barrage; (b) partial-band; (c) single-tone; (d) multiple-tone.

that there is no collision is the same as what occurs when there is only noise, given by $P(error|c^c) = \frac{1}{2}e^{-E_b/2N_0}$. If there are $Y = 2^k$ frequencies in the CDMA channel, then the probability of an intereferer being in a given slot is $1/Y$ (or the probability of not being in that slot is $(1 - 1/Y)$). Therefore, the probability that $M - 1$ users will not interfere is $(1 - 1/Y)^{M-1}$. Consequently, the probability that $M - 1$ users could collide with the Mth user is $P_c = 1 - \left(1 - \frac{1}{Y}\right)^{M-1} \approx \frac{M-1}{Y}$ if $M \ll Y$. Again using the chain rule, we have $P_e = P(e|c)P(c) + P(e|c^c)P(c^c)$ so

$$P_e \approx \frac{1}{2}\left(\frac{M-1}{Y}\right) + \frac{1}{2}e^{-E_b/2N_0}\left(1 - \frac{M-1}{Y}\right)$$ [5]

EXAMPLE 15.2–2

Let's calculate B_T and the error probabilities for a SH-SS system with a message rate of 3 kbps, $S_R = 5.4 \times 10^{-8}$ W, given that the PN generator has $k = 10$ outputs and the system has partial band jamming with $\Delta = 100\%$, 10%, and 0.1%. The jammer power at the receiver is 10 times the signal power and the PSD of the random noise is $N_0 = 10^{-12}$ W/Hz. If $k = 10$, then $Pg = 2^{10} = 1024 = W_c/W_x$. If $r_b = W_x = 3000$ then $W_c = 3.07 \times 10^6$. With $E_b = S_R/r_b$, then $E_b = 5.4 \times 10^{-8}/3000 = 1.80 \times 10^{-11}$ and $N_J = J/W_c = 5.4 \times 10^{-7}/3.07 \times 10^6 = 1.76 \times 10^{-13}$. Substituting these values into Eq. (4) for $\Delta = 1, 0.1, 0.001$, we get P_e values of 2.37×10^{-4}, 0.0020, and 5.36×10^{-4} respectively. You will note from these that there will be an optimum value of Δ that minimizes P_e.

EXERCISE 15.2–2

A single-user FH-SS system has $P_e = 10^{-7}$. For the same number of users as the system of Exercise 15.1–2, what is the minimum value of Pg for $P_e \le 10^{-5}$?

15.3   CODING

The PN generator used to generate $c(t)$ and thereby spread the signal is of the same type and properties described in Sect. 11.4. The reader should refer back to Sect. 11.4 for the discussion of PN generators.

> In order for a receiver to properly recognize a transmitter's signal and, in particular, prevent false synchronization, it is important that the PN code's autocorrelation function have the largest peak possible at $\tau = 0$ and $\tau = \pm kT_c$ and as low as possible everywhere else.

This objective is met if the PN generator produces an ml sequence.

To minimize jamming and/or casual eavesdropping, the PN sequence should be as long as possible; the longer the sequence, the more effort it takes for an unauthorized listener to determine the PN sequence. However, with linear codes, an eavesdropper only requires the knowledge of 2^n chips to determine the shift register connections. Therefore vulnerability to unauthorized listening can be reduced by either making frequent changes in the PN sequence during transmission or by using some nonlinear scheme for the feedback connections. PN codes that are both secure and have desirable correlation characteristics are generally difficult to find, and thus if the goal is secure communications, then the message should be encrypted separately.

With CDMA systems, in which each user has a unique PN code, the receiver should be able to reject other interfering SS signals and/or prevent false correlations. As stated by Eqs. (17) and (18), Sect. 15.1,

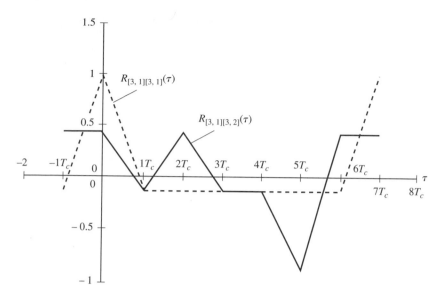

Figure 15.3–1   Auto- and crosscorrelation of [3, 1] and [3, 2] PN sequences.

Minimizing interference requires the upper bound on the crosscorrelation between different PN sequences to be as small as possible.

Let's consider the crosscorrelation between two different ml sequences. From Example 11.4–1, the [3, 1] shift register configuration produces the autocorrelation function plotted in Fig. 15.3–1. Note the single periodic correlation peak at $\tau = 0, 7T_c$. It is easily shown that a [3, 2] register configuration produces an ml sequence of 1110010 and has the same autocorrelation function as the [3,1] register configuration. Now if we calculate the crosscorrelation of these two sequences using a shortcut similar to that that we used in Example 11.4–1, and we compare the output from the [3, 1] configuration with shifted versions of the [3, 2] configuration, we get the results shown in Table 15.3–1 and Fig. 15.3–1. Remember that just as with Example 11.4–1, these calculations are based on the PN sequence being a polar NRZ signal. Note that there is relatively little difference between the autocorrelation and crosscorrelation values, and therefore these particular shift-register configurations would be unsuitable for CDMA systems because we cannot easily discriminate between signals that use these PN codes.

A PN sequence for CDMA applications should have the following characteristics: (a) $R_{ss}(0)_{max}$ relative to $R_{ss}(\tau \neq 0)$ should be as large as possible; (b) if s and t represent different PN sequences, then $|R_{st}(\tau)|_{max}$ should be as small as possible for

Table 15.3–1

τ	[3, 1]/[3, 2]	$v(\tau)$	$R_{[3,1][3,2]}(\tau) = v(\tau)/N$
0	1110100	$5 - 2 = 3$	$3/7 = 0.43$
	1110010		
1	1110100	-1	-0.14
	0111001		
2	1110100	3	0.43
	1011100		
3	1110100	-1	-0.14
	0101110		
4	1110100	-1	-0.14
	0010111		
5	1110100	-5	-0.71
	1001011		
6	1110100	3	0.43
	1100101		
7	1110100	3	0.43
	1110010		

all values of τ; and therefore (c) we want to minimize the ratio $|R_{st}(\tau)|_{max}/R_{ss}(0)$. Table 15.3–2 shows the most optimum ratios of $|R_{st}(\tau)|_{max}/R_{ss}(0)$ for various register lengths. This table also gives a partial listing of the possible feedback connections that enable the generation of ml sequences.

Unfortunately, as we observe in the plot of Fig. 15.3–1 and the values in Table 15.3–2, ml sequences generated from a single shift register do not have good cross-correlation properties, making them unsuitable for CDMA systems. Also, for a single shift register to generate a different output sequence requires changing the feedback connections; therefore, a given shift register length will give us relatively few unique output sequences. To overcome both of these limitations, Gold codes are

Table 15.3–2   Crosscorrelation ratios for various ml sequence lengths and feedback connections (produced from a single shift register). Dixon, 1994.

n	$N = 2^n - 1$	$|R_{st}(\tau)|_{max}/R_{ss}(0)$	Feedback Taps
3	7	0.71	[3, 1], [3, 2]
5	31	0.35	[5, 2], [5, 4, 3, 2], [5, 4, 2, 1]
8	255	0.37	[8, 5, 3, 1], [8, 6, 5, 1], [8, 7, 6, 1]
9	511	0.22	[9, 4], [9, 6, 4, 3], [9, 8, 6, 5], [9, 8, 7, 6, 5, 3]
12	4095	0.34	[12, 6, 4, 1], [12, 10, 9, 8, 6, 2], [12, 11, 10, 5, 2, 1]

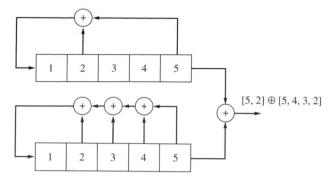

Figure 15.3–2   Gold code generator.

generated by the mod-2 combination of the output of two or more registers as illustrated in Fig. 15.3–2. For certain combinations of n-bit registers and feedback connections which we call **preferred pairs,** Gold (1967, 1968) states that the maximum value of crosscorrelation between $N = 2^n - 1$ length sequences is bounded by

$$R_{st} = |\phi|/N \qquad \text{[1a]}$$

where

$$\phi = \begin{cases} (2^{(n+1)/2} + 1 & n \text{ odd} \\ (2^{(n+2)/2} + 1 & n \text{ even} \end{cases} \qquad \text{[1b]}$$

or in some cases

$$R_{st} - \cdot 1/N \qquad \text{[1c]}$$

Table 15.3–3 lists some preferred pairs of registers that will generate Gold codes and their respective crosscorrelation upper bound. To see the superior crosscorrelation properties of Gold codes versus single register ml sequences, we note from Table 15.3–3 that the sequences produced from a 12-bit Gold code will have $R_{st_{max}} = 0.031$, whereas from Table 15.3–2 the best of the sequences produced from a single 12-shift register will have $R_{st_{max}} = 0.34$.

Table 15.3–3   Some preferred pairs for Gold Code generation and their auto/crosscorrelation ratios. Dixon, 1994.

n	Preferred Pair	$|R_{st}(\tau)|_{max}/R_{ss}(0)$
5	[5, 2] [5, 4, 3, 2]	0.290
7	[7, 3] [7, 3, 2, 1]	0.134
8	[8, 7, 6, 5, 2, 1] [8, 7, 6, 1]	0.129
10	[10, 8, 5, 1] [10, 7, 6, 4, 2, 1]	0.064
12	[12, 9, 8, 5, 4, 3] [12, 7, 6, 4]	0.031

The autocorrelation function for the Gold code sequence has an off-peak autocorrelation value that is also bounded by Eq. (1) (Proakis, 2001). Therefore, unlike the ml sequence produced by a single shift register that has an off-peak autocorrelation value bounded by $-1/N$, the off-peak autocorrelation value from a Gold code sequence could be as high as $|\phi|/N$. Thus with Gold codes, we obtain a greatly reduced set of crosscorrelation values at a cost of a slight increase in the off-peak autocorrelation values.

For each set of initial conditions in the registers (or register offsets), we get a different periodic output sequence. Therefore, two n-bit preferred register pairs will give us $N = 2^n - 1$ unique output sequences. If we also include the two original sequences, our register pair can then produce a total of $N + 2$ sequences. This is in contrast to what was stated earlier, where a given single register configuration can only give us one unique periodic output sequence.

Dixon (1994) has a more extensive list of preferred pairs for Gold code generation and also describes an algorithm for the proper selection of preferred pairs used to generate Gold codes.

> We conclude that Gold codes are attractive for CDMA applications because for a given shift register configuration we can create numerous unique code sequences with desirable crosscorrelation properties.

EXAMPLE 15.3–1   **Ranging using DSS**

A common application of DSS is for distance measurement, and is based on measuring the phase difference or time delay in chips between transmitter and receiver correlation functions. Let's say we have a transmitter that produces a signal $v(t)$ with a PN sequence whose autocorrelation is $R_{vv}(\tau)$ with its maximum value at $\tau = 0$. The transmitted signal is reflected off a target at some distance d back to the transmitter. Let the reflected signal received back at the transmitter location be $w(t) = v(t - t_d)$ where $t_d = 2\,d/c$, with c being the speed of light. If we correlate the received signal with the transmitter PN sequence we get $R_{vw}(\tau) = R_{vv}(\tau - \tau_d)$ where τ_d is kT_c chips. The relative phase difference between the two peaks (k chips or $\tau_d = kT_c$ seconds) is the transit time between the transmitted and received signals, and is a measure of the distance. Let's say we have a DSS transmitter that sends to a target a PN ml sequence of length $N = 31$ and clock period 10 ns. The sequence is received back such that the difference in correlation peak locations is 20 chips. The distance between the transmitter and target would then be

$$d = kT_c c/2 \tag{2}$$

$$= 20 \text{ chips} \times 10 \times 10^{-9} \text{ seconds/chip} \times 2.99 \times 10^8 \text{ m/s} \times 1/2$$

$$= 29.90 \text{ meters}$$

Thus if DSS is used for distance measurement the resolution in meters would be

$$\Delta d = cT_c \qquad\qquad [3]$$

An important application of DSS that uses this property of distance measurement and triangulation is the **global positioning system** (GPS). GPS can be used to determine the time and a person's precise latitude, longitude, and altitude on the earth. It consists of 24 satellites such that at any one time at least four are visible from anywhere on earth. Each one is 22,200 km above the earth and circles it every 12 hours.

Show that for any nonzero initial condition, a 5-bit shift register with a [5, 2] configuration will produce only one unique PN sequence. Generalize your result for any single n-bit shift register PN generator.

EXERCISE 15.3–1

15.4   SYNCHRONIZATION

Up to now, we have assumed that we know the transmitted carrier frequency and phase and that there is perfect alignment between the transmitted and receiver PN codes. Unfortunately, achieving these objectives is not easy and, in fact, is one of the most difficult problems to solve in spread spectrum systems. The initial objective is to achieve carrier frequency and phase synchronization. This is made somewhat simpler if we use noncoherent detection. We then have to align the transmitter and receiver's PN codes and to maintain synchronization by overcoming frequency drift in the transmitted carrier or PN clock. With mobile or satellite SS systems, the carrier frequency and code clock phases can change because of Doppler frequency shift[†] caused by the relative motion between the transmitter and receiver.

Tracking the incoming carrier frequency and phase is done in the same way as with any other digital or analog modulation systems. Specific techniques are discussed in Sects. 7.3 and and 14.4. Synchronizing or aligning the receiver PN code to the incoming PN code is done in two steps. The first step is called **acquisition,** where there is initial acquisition and a coarse alignment to within a half a chip between the two codes. The second step is the ongoing and fine alignment called **tracking.** Both acquisition and tracking involve a feedback loop, with tracking done using PLL techniques.

Acquisition

Figure 15.4–1 is a block diagram showing a serial search acquisition system for a DSS system, and Fig. 15.4–2 shows one for a FH-SS system. Both work in a similar manner. The transmitted PN code is contained in $y(t)$, and the receiver generates a

---

[†]Doppler frequency shift is a change in the perceived received frequency caused by the relative motion between transmitter and receiver. Doppler frequency shift equals $\Delta f = \pm v f_0/c$ where v is the relative velocity between transmitter and receiver, f_0 is the nominal frequency, and c is the speed of light.

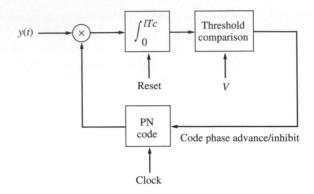

Figure 15.4–1   DS serial search acquisition.

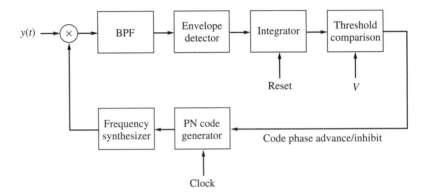

Figure 15.4–2   Frequency hopping serial search acquisition.

replica code, but likely is offset by some phase difference of N_c which is an integer number of chips. N_c is also a random variable whose maximum value is the PN period in chips minus one. The goal is that these two PN codes be aligned to within a half chip. Then the tracking phase takes over. For initial acquisition, some systems use an abbreviated version of the PN code called a **preamble.**

Acquisition works as follows. The received signal $y(t)$, which contains l PN chips, is multiplied by its replica that contains the same number of chips. The product is then integrated from 0 to lT_c seconds where T_c is the PN clock period. If there is alignment, the signal will have been de-spread and the integrator's output will be relatively high compared to the threshold V, and therefore the PN code phase is not changed. If there isn't alignment, the threshold circuitry causes the PN code generator to increment its phase by half a chip, and the entire process is repeated until alignment is achieved. Assuming the system properly recognizes the correct PN sequence, the maximum time for acquisition would be

$$T_{acq} = 2N_c lT_c \qquad [1]$$

where the factor 2 occurs because we are correlating in increments of half chips. In systems that contain noise, there is a possibility of an incorrect synchronization. Equation (1) is modified to obtain the average acquisition time of

$$\overline{T_{acq}} = \frac{2 - P_D}{P_D}(1 + \alpha P_{FA})N_c lT_c \qquad [2]$$

where P_D is the probability of a correct detection, P_{FA} is the probability of a false synchronization, and α is a factor that reflects a penalty due to a false alarm with $\alpha \gg 1$. The variance in acquisition time is

$$\sigma_{T_{acq}}^2 = (2N_c lT_c)^2(1 + \alpha P_{FA})^2\left(\frac{1}{12} + \frac{1}{P_D^2} - \frac{1}{P_D}\right) \qquad [3]$$

Eqs. (2) and (3) are derived by Simon et al. (1994).

To decrease the acquisition time for DS-SS, we could modify the system of Fig. 15.4–1 to become a **sliding-correlator** where instead of waiting for the integration of the entire lT_c sequence, we do a continual integration and feed the result to the comparator such that when the output is above threshold, we stop the process because alignment has been achieved. Alternatively, we can speed up the process by a factor of $2N_c$ by going to a parallel structure and adding $2(N_c - 1)$ correlators to the structure of Fig. 15.4–1.

Tracking

Once we have achieved course synchronization, tracking or fine synchronization begins. For a DSS signal, we can use a *delay-locked loop* (DLL) as shown in Fig. 15.4–3. The received signal $y(t)$ is fed to two multipliers. Each version of $y(t)$ is multiplied by a retarded or advanced version of the PN code generator output, that is $c(t \pm \delta)$ where δ is a small fraction of the clock period T_c. The output of each multiplier is then bandpass filtered and envelope detected. The two outputs are summed and fed to a loop filter that controls the VCO (voltage controlled oscillator). If there is synchronization error, one of the envelope detectors will have a larger value than the other, causing a jump in the summer output, which in turn will cause the VCO to advance or retard its output and thereby causes $c(t)$ to straddle between the values of $c(t + \delta)$ and $c(t - \delta)$.

A simpler alternative to the DLL is the Tau-Dither loop of Fig. 15.4–4. Here only one loop is required, so we don't have to worry about two loops having identical gains. It works in a similar manner to the DLL except that one loop is shared by the advance and retard signal. When signal $q(t)$ is $+1$, the input signal $y(t)$ is multiplied by $c(t + \delta)$. This product is then detected, multiplied by $q = 1$ and fed to the loop filter. At the next cycle $q(t) = -1$, the input signal $y(t)$ is multiplied by $c(t - \delta)$. It too is enveloped detected, scaled by -1 and fed to the loop filter. If both signals are the same, the VCO output remains unchanged. But, like the DLL, if one signal is larger than the other, then the VCO output changes causing the PN code generator to change accordingly.

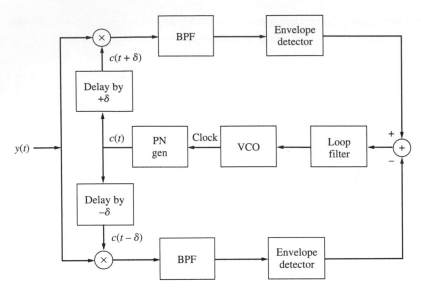

Figure 15.4–3   Delay-locked loop (DLL).

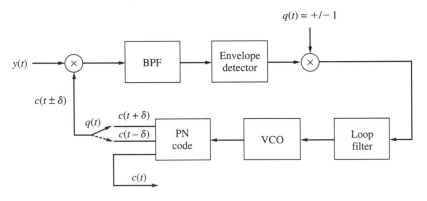

Figure 15.4–4   Tau-Dither synchronization.

15.5   WIRELESS TELEPHONE SYSTEMS

In the last 15 years, wireless telephone systems have undergone a tremendous transformation from being a luxury to almost a necessity. In fact, formerly theater patrons were requested not to smoke; now they are requested to turn off their cell phones! In some cases wireless systems have even supplanted *land-line* phone systems. For example, some nations are considering not upgrading their copper-wire telephone system but instead replacing them with wireless telephones. Wireless telephones have enabled everybody to be reachable by telephone at any time and location. In this section we want to consider the two primary wireless systems: *cellular telephones* and *personal communications systems* or *services* (PCS).

Cellular Telephone Systems

Cellular telephones, by definition, are FDMA FM systems that operate in the 800-MHz band with $W = 3$ kHz and $B_T = 30$ kHz. They work in the following manner. A given service area is divided up into cells; each cell has a base station we call a **mobile telephone switching office** (MTSO), consisting of a transmitter, receiver, and tower for the antenna. The MTSO in turn is connected to a telephone central office via telephone lines and thereby interfaces the cell-phone user to the rest of the telephone system. When a call is placed, the MTSO authenticates the cell-phone user based on the cell phone serial number and assigned phone number and then assigns a set of available transmit and receive frequencies to the cell phone. During the call, the MTSO monitors the cell phone's signal strength so that if the user moves, the MTSO switches the call to a different cell station and thereby maintains signal strength. The process is seamless. Increasing the density of cell towers/stations in a given area, while not necessarily aesthetically pleasing, enables the cell phones and MTSO transmitters to be low power. In turn, this means their signals will not propagate much beyond a given cell area, thus enabling much greater reuse of available frequencies in other cell areas without interference. Low power also means cell phones with smaller physical size and reduced battery requirements.

Personal Communications Systems

Recent government telecommunication deregulation has made available the 900 MHz and 1.8 GHz (1.9 GHz in Europe) frequencies for portable digital phones using either TDMA or DSS-CDMA digital technology. To differentiate themselves from traditional analog cellular phone systems, the service providers have called this new system **personal communications systems** (PCS) or **wireless** phones—wireless being synonymous with the new digital technology. Not only are there more frequencies, but digital technology has expanded the number of available services. These include Internet for e-mail and World-Wide-Web access, navigation aides, messaging, biomedical telesensors, and so on. As we stated earlier, the inherent nature of CDMA is that cross talk is virtually nonexistent if the spread codes are mutually orthogonal and we eliminate, or at least reduce, the need for frequency allocation. More users only raise the noise floor, and thus there is no hard limit on the number in a given area. Unlike cell phone technology, we don't have to be as strict with respect to maintaining minimum physical distances between systems that use the same frequencies. Furthermore, the FCC has also allowed DSS-CDMA communication to overlay the existing 800 MHz cell-phone band. These additional users do not pose a serious interference problem to existing cell-phone users since a DSS signal received at any unauthorized receiver sounds like random noise regardless of the type of detector used.

More than other terrestrial radio systems, wireless phone systems suffer from the *near-far problem.* This is where you have unequal signal powers arriving at the receiver, and the stronger one drowns out the weaker one. Overcoming this problem requires receiver amplifiers that have a wide dynamic range, so strong signals will not overload the front-end amplifiers and other adaptive gain control techniques. For more information on wireless systems see Rappaport (1996).

15.6   PROBLEMS

15.1–1*   A DSS-BPSK system has an information rate of 3 kbps, and is operating in an environment where, at the receiver, there is a single-tone interferring signal five times more powerful than the received signal and $N_0 = 10^{-21}$ W/Hz. Without the jammer, the $(S/N)_R = 60$ dB. If the required $P_e = 10^{-7}$, calculate (a) the chip rate, (b) B_T.

15.1–2   A DSS-BPSK system has $Pg = 30$ dB, and in the presence of a jammer the $P_e = 10^{-7}$. What is the jamming margin?

15.1–3   A DSS-CDMA system is used for telephone transmitting information at a rate of 6 kbps. The system uses a code rate of 10 MHz and $(S/N)_R = 20$ dB with just one signal transmitted. The required $P_e = 10^{-7}$, and each signal is received with equal power. (a) How many users can share the channel? (b) If each user reduces power by 6 dB, how many can share the channel?

15.1–4   What is the minimum chip rate required to overcome multipath interference if the multipath length signal is 500 meters longer than the direct signal?

15.1–5*   Ten DSS-BPSK users, with each one transmitting information at a rate of 6 kbps, want to share a channel via CDMA. Assuming additive noise can be ignored, what is the minimum chip rate in order to achieve $P_e = 10^{-7}$?

15.1–6   Repeat Prob. 15.1–5 assuming the channel is corrupted by white gaussian noise such that with a single user, $P_e = 10^{-9}$.

15.1–7   A DSS system transmits information at a rate of 9 kbps in the presence of a single-tone jammer whose power is 30 dB greater than the desired signal. What is the processing gain required in order to achieve $P_e < 10^{-7}$?

15.1–8   How many customers of equal power can share a CDMA system if each one is using DS-BPSK where the processing gain is 30 dB and with $P_e < 10^{-7}$?

15.1–9   A DSS-BPSK user transmits information at a rate of 6 kbps, using a chip rate of 10×10^6 and the received signal has a $P_e = 10^{-10}$. How many users can share the channel if the P_e requirement was reduced to 10^{-5} and each user reduces their power by 3 dB?

15.2–1*   A binary FSK SH-SS system has an information rate of 3 kbps, and is operating in a barrage jamming environment where the entire channel is being jammed with a power level five times greater than the received signal. Without the jammer, the $(S/N)_R = 60$ dB and $N_0 = 10^{-21}$ W/Hz. If the required $P_e = 10^{-7}$, determine the minimum Pg and corresponding B_T.

15.2–2   Determine the minimum Pg and corresponding B_T for a binary FSK system that has an information rate of 3 kbps, 10 users, and $P_e \leq 10^{-5}$. It has been found for one user that for one user that $P_e = 10^{-10}$.

15.2–3   Given the system in Prob. 15.2–1, what would be Pg and B_T for a DSS that enables the same P_e?

15.2–4*   What is P_e for a SH-SS system in the presence of partial band jamming with $\gamma = 10\%$, $J = 6$ mW, $E_b = 2 \times 10^{-11}$, $N_0 = 10^{-12}$, $r_b = 6$ kbps, and the PN generator has $k = 10$ outputs?

15.2–5   Calculate the minimum hop rate for a FH-SS system that will prevent a jammer from operating five miles away from the receiver.

15.3–1   Given a four-stage shift register with [4, 1] configuration, where the initial state is 0100, and is driven by a 10 MHz clock, what is (*a*) its output sequence, and (*b*) PN sequence period? Plot the autocorrelation function produced by this shift register.

15.3–2   Repeat Prob. 15.3–1 for a four-stage shift register with [4, 2] configuration.

15.3–3   Show that the output sequence for the [4, 1] shift register has the properties of a ml sequence as described in Sect. 11.4.

15.3–4   What is the output sequence and its corresponding length from a 5-bit shift register PN generator with [5, 4] feedback connections with the initial state 11111? Is it a ml sequence?

15.3–5*   Given a Gold code generator consisting of preferred pair [5, 2] and [5, 4, 3, 2], where both registers initially have all 1s, what is its (*a*) output sequence, (*b*) cross-correlation bound?

15.3–6   What chip rate is required for a DSS distance measuring system to achieve a resolution of 0.01 miles?

15.3–7   What is the distance resolution for a DSS system with a 20-MHz code clock rate?

15.4–1   A 10-MHz code generator drifts 5 Hz/hour. What is the chip uncertainity after one day?

15.4–2   A DSS system operating at a carrier frequency of 900 MHz with a 10-MHz code clock is traveling toward us at 500 mph. What is the maximum shift in carrier frequency and chip uncertainty due to Doppler?

15.4–3*   What is the average acquisition time and its standard deviation for a DSS-BPSK, with chip rate of 10 MHz, that uses a serial search of a 2048-bit preamble and has a $P_D = 0.9$, $P_{FA} = 0.01$, and $\alpha = 100$? Assume the maximum initial phase difference between the input signal's PN sequence and receiver's PN generator.

15.4–4   A DSS-BPSK system uses a code clock of 50 MHz and a 12-stage shift register set up for maximal length. Assuming the phase difference between the received signal's PN sequence and the receiver PN generator is half a PN period, calculate the average acquisition time and its standard deviation if a serial search is done with $P_D = 0.9$, $P_{FA} = 0.01$, and $\alpha = 10$.

chapter

16

Information and Detection Theory

CHAPTER OUTLINE

Throughout this book we have studied electrical communication primarily in terms of *signals*—desired information-bearing signals corrupted by noise and interference signals. Although signal theory has proved to be a valuable tool, it does not come to grips with the fundamental communication process of *information transfer*. Recognizing the need for a broader viewpoint, Claude Shannon drew upon the earlier work of Nyquist and Hartley and the concurrent investigations of Wiener to develop and set forth in 1948 a radically new approach he called "A Mathematical Theory of Communication."

Shannon's paper isolated the central task of communication engineering in this question: Given a message-producing source, not of our choosing, how should the messages be represented for reliable transmission over a communication channel with its inherent physical limitations? To address that question, Shannon concentrated on the message information per se rather than on the signals. His approach was soon renamed **information theory,** and it has subsequently evolved into a hybrid mathematical and engineering discipline.

Information theory deals with three basic concepts: the measure of source information, the information capacity of a channel, and coding as a means of utilizing channel capacity for information transfer. The term *coding* is taken here in the broadest sense of message representation, including both discrete and continuous waveforms. The three concepts of information theory are tied together through a series of theorems that boil down to the following:

> If the rate of information from a source does not exceed the capacity of a communication channel, then there exists a coding technique such that the information can be transmitted over the channel with an arbitrarily small frequency of errors, despite the presence of noise.

The surprising, almost incredible aspect of this statement is its promise of *error-free* transmission on a noisy channel, a condition achieved with the help of coding.

Optimum coding matches the source and channel for maximum reliable information transfer, roughly analogous to impedance matching for maximum power transfer. The coding process generally involves two distinct encoding/decoding operations, portrayed diagrammatically by Fig. 16.0. The *channel* encoder/decoder units perform the task of error-control coding as discussed in Sects. 13.1–13.3 where we saw how the effects of channel noise could be reduced. Information theory goes a step further, asserting that optimum channel coding yields an equivalent *noiseless* channel with a well-defined capacity for information transmission. The *source* encoder/decoder units then match the source to the equivalent noiseless channel, provided that the source information rate falls within the channel capacity.

This chapter starts with the case of digital or *discrete* information, including information measure, source coding, information transmission, and discrete channel capacity. Many concepts and conclusions gained from the study of discrete information carry over to the more realistic case of information transmission on a *continuous* channel, where messages take the form of time-varying signals. The *Hartley-Shannon law* defines continuous channel

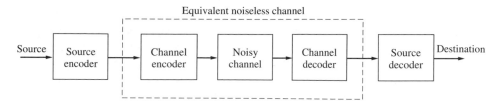

Equivalent noiseless channel

Figure 16.0   Communication system with source and channel coding.

capacity in terms of bandwidth and signal-to-noise ratio, and thereby serves as a benchmark for comparing the performance of communication systems in the light of information theory.

The chapter closes with an application of theory to optimum digital detection. We'll first introduce signal space as a means of representing and quantifying signals. Then the principles of optimum detection will be applied to digital communication.

OBJECTIVES

After studying this chapter and working the exercises, you should be able to do each of the following:

1. Define Shannon's measure of information (Sect. 16.1).
2. Calculate the information rate of a discrete memoryless source (Sect. 16.1).
3. Analyze a discrete memoryless channel, given the source and transition probabilities (Sect. 16.2).
4. State and apply Shannon's fundamental theorem for information transmission on a noisy channel (Sects. 16.2 and 16.3).
5. State and apply the Hartley-Shannon law for a continuous channel (Sect. 16.3).
6. Draw the vector representation of two or more signals (Sect. 16.4).
7. Construct the decision regions for MAP detection in a two-dimensional space (Sect. 16.5).
8. Find the decision functions for a MAP receiver, and evaluate the symbol error probabilities when there are rectangular decision regions (Sect. 16.5).

16.1   INFORMATION MEASURE AND SOURCE ENCODING

Quite logically, we begin our study of information theory with the measure of information. Then we apply information measure to determine the information rate of discrete sources. Particular attention will be given to binary coding for discrete memoryless sources, followed by a brief look at predictive coding for sources with memory.

Information Measure

The crux of information theory is the measure of information. Here we use *information* as a technical term, not to be confused with "knowledge" or "meaning"—concepts that defy precise definition and quantitative measurement. In the context of communication, information is simply the commodity produced by the source for transfer to some user at the destination. This implies that the information was previously unavailable at the destination; otherwise, the transfer would be zero.

Suppose, for instance, that you're planning a trip to a distant city. To determine what clothes to pack, you might hear one of the following forecasts:

- The sun will rise.
- There will be scattered rainstorms.
- There will be a tornado.

The first message conveys virtually no information since you are quite sure in advance that the sun will rise. The forecast of rain, however, provides information not previously available to you. The third forecast gives you more information, tornadoes being rare and unexpected events; you might even decide to cancel the trip!

Notice that the messages have been listed in order of decreasing likelihood and increasing information. The less likely the message, the more information it conveys. We thus conclude that information measure must be related to *uncertainty,* the uncertainty of the user as to what the message will be. Alternatively, we can say that information measures the *freedom of choice* exercised by the source in selecting a message. When a source freely chooses from many different messages, the user is highly uncertain as to which message will be selected.

Whether you prefer the source or user viewpoint, it should be evident that information measure involves the *probability* of a message. If x_i denotes an arbitrary message and $P(x_i) = P_i$ is the probability of the event that x_i is selected for transmission, then the amount of information associated with x_i should be some function of P_i. Specifically, Shannon defined information measure by the logarithmic function

$$I_i \triangleq -\log_b P_i = \log_b \frac{1}{P_i} \qquad [1]$$

where b is the logarithmic base. The quantity I_i is called the **self-information** of message x_i. The value of I_i depends only on P_i, irrespective of the actual message content or possible interpretations. For suppose you received a weather forecast that said

- The sun will rain tornadoes

It will convey lots of information—being very improbable—despite the lack of substance.

Although not immediately obvious, the definition in Eq. (1) has several important and meaningful consequences, including

$$I_i \geq 0 \quad \text{for} \quad 0 \leq P_i \leq 1 \tag{2a}$$

$$I_i \to 0 \quad \text{for} \quad P_i \to 1 \tag{2b}$$

$$I_i > I_j \quad \text{for} \quad P_i < P_j \tag{2c}$$

Hence, I_i is a nonnegative quantity with the properties that $I_i = 0$ if $P_i = 1$ (no uncertainty) and $I_i > I_j$ if $P_i < P_j$ (information increases with uncertainty). Furthermore, suppose a source produces two successive and independent messages, x_i and x_j, with joint probability $P(x_i x_j) = P_i P_j$; then

$$I_{ij} = \log_b \frac{1}{P_i P_j} = \log_b \frac{1}{P_i} + \log_b \frac{1}{P_j} = I_i + I_j \tag{3}$$

so the total information equals the sum of the individual message contributions. Shannon's information measure $\log_b (1/P_i)$ is the *only* function that satisfies all of the properties in Eqs. (2) and (3).

Specifying the logarithmic base b determines the *unit* of information. The standard convention of information theory takes $b = 2$ and the corresponding unit is the **bit,** a name coined by J. W. Tukey as the contraction for "binary digit." Equation (1) thus becomes $I_i = \log_2 (1/P_i)$ bits. This convention normalizes information measure relative to the most elementary source, a source that selects from just two equiprobable messages. For if $P(x_1) = P(x_2) = 1/2$, then $I_1 = I_2 = \log_2 2 = 1$ bit. In other words, 1 bit is the amount of information needed to choose between two equally likely alternatives.

Binary digits enter the picture simply because any two things can be represented by the two binary digits. However, you must carefully distinguish information bits from binary digits per se—especially since a binary digit may convey more or less than one bit of information, depending upon its probability. To prevent possible misinterpretation, the abbreviation *binits* is sometimes used for binary digits as message or code elements.

When necessary, you can convert base-2 to natural or common logarithms via

$$\log_2 v = \frac{\ln v}{\ln 2} = \frac{\log_{10} v}{\log_{10} 2} \tag{4}$$

If $P_i = 1/10$, for instance, then $I_i = (\log_{10} 10)/0.301 = 3.32$ bits. Hereafter, all logarithms will be base-2 unless otherwise indicated.

Entropy and Information Rate

Now consider an information source that emits a sequence of symbols selected from an alphabet of M different symbols, i.e., an M-ary alphabet. Let X denote the entire set of symbols $x_1, x_2, \ldots, x_M$. We can treat each symbol x_i as a message that occurs with probability P_i and conveys the self-information I_i. The set of symbol probabilities, of course, must satisfy

$$\sum_{i=1}^{M} P_i = 1 \qquad \text{[5]}$$

We'll assume that the source is stationary, so the probabilities remain constant over time. We'll also assume that successive symbols are statistically independent and come from the source at an average rate of r symbols per second. These properties define the model of a **discrete memoryless source.**

The amount of information produced by the source during an arbitrary symbol interval is a *discrete random variable* having the possible values $I_1, I_2, \ldots, I_M$. The expected information per symbol is then given by the statistical average

$$H(X) \triangleq \sum_{i=1}^{M} P_i I_i = \sum_{i=1}^{M} P_i \log \frac{1}{P_i} \quad \text{bits/symbol} \qquad \text{[6]}$$

which is called the source **entropy.** Shannon borrowed the name and notation H from a similar expression in statistical mechanics. Subsequently, various physical and philosophical arguments have been put forth relating thermodynamic entropy to communication entropy (or *comentropy*).

But we'll interpret Eq. (6) from the more pragmatic observation that when the source emits a sequence of $n \gg 1$ symbols, the total information to be transferred is about $nH(X)$ bits. Since the source produces r symbols per second on average, the time duration of this sequence is about n/r. The information must therefore be transferred at the average rate $nH(X)/(n/r) = rH(X)$ bits per second. Formally, we define the source **information rate**

$$R \triangleq rH(X) \quad \text{bits/sec} \qquad \text{[7]}$$

a critical quantity relative to transmission. Shannon asserted that information from any discrete memoryless source can be encoded as binary digits and transmitted over a noiseless channel at the signaling rate $r_b \geq R$ binits/sec. As preparation for the study of source coding, we need to know more about entropy.

The value of $H(X)$ for a given source depends upon the symbol probabilities P_i and the alphabet size M. Nonetheless, the source entropy always falls within the limits

$$0 \leq H(X) \leq \log M \qquad \text{[8]}$$

The lower bound here corresponds to no uncertainty or freedom of choice, which occurs when one symbol has probability $P_j = 1$ while $P_i = 0$ for $i \neq j$—so the source almost always emits the same symbol. The upper bound corresponds to maximum uncertainty or freedom of choice, which occurs when $P_i = 1/M$ for all i—so the symbols are equally likely.

To illustrate the variation of $H(X)$ between these extremes, take the special but important case of a binary source $(M = 2)$ with

$$P_1 = p \qquad P_2 = 1 - p$$

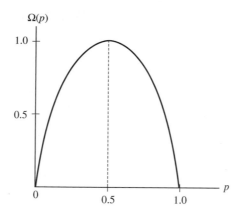

Figure 16.1–1   Binary entropy function.

Substituting these probabilities into Eq. (6) yields the *binary entropy*

$$H(X) = \Omega(p) \triangleq p \log \frac{1}{p} + (1 - p) \log \frac{1}{1 - p} \qquad \text{[9]}$$

in which we've introduced the "horseshoe" function $\Omega(p)$. The plot of $\Omega(p)$ in Fig. 16.1–1 displays a rather broad maximum centered at $p = 1 - p = 1/2$ where $H(X) = \log 2 = 1$ bit/symbol; $H(X)$ then decreases monotonically to zero as $p \rightarrow 1$ or $1 - p \rightarrow 1$.

Proving the lower bound in Eq. (8) with arbitrary M is easily done once you note that $v \log (1/v) \rightarrow 0$ as $v \rightarrow 0$. The proof of the upper bound $H(X) \leq \log M$ involves a few more steps but deserves the effort.

First, we introduce another set of probabilities $Q_1, Q_2, \ldots, Q_M$, and replace $\log (1/P_i)$ in Eq. (6) with $\log (Q_i/P_i)$. Conversion from base-2 to natural logarithms gives the quantity

$$\sum_i P_i \log \frac{Q_i}{P_i} = \frac{1}{\ln 2} \sum_i P_i \ln \frac{Q_i}{P_i}$$

where it's understood that all sums range from $i = 1$ to M. Second, we invoke the inequality $\ln v \leq v - 1$, which becomes an equality only if $v = 1$, as seen in Fig. 16.1–2. Thus, letting $v = P_i/Q_i$ and using Eq. (5),

$$\sum_i P_i \ln \frac{Q_i}{P_i} \leq \sum_i P_i \left(\frac{Q_i}{P_i} - 1 \right) = \left(\sum_i Q_i \right) - 1$$

Third, we impose the condition

$$\sum_{i=1}^{M} Q_i = 1 \qquad \text{[10a]}$$

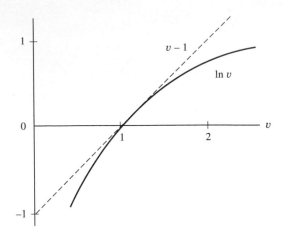

Figure 16.1–2

so it follows that

$$\sum_{i=1}^{M} P_i \log \frac{Q_i}{P_i} \leq 1 - 1 = 0 \tag{10b}$$

Finally, taking $Q_i = 1/M$ we have

$$\sum_i P_i \log \frac{1}{P_i M} = \sum_i P_i \log \frac{1}{P_i} - \sum_i P_i \log M = H(X) - \log M \leq 0$$

thereby confirming that $H(X) \leq \log M$. The equality holds only in the equally likely case $P_i = 1/M$ so $v = Q_i/P_i = 1/Mp_i = 1$ for all i.

EXAMPLE 16.1–1   Suppose a source emits $r = 2000$ symbols/sec selected from an alphabet of size $M = 4$ with symbol probabilities and self information listed in Table 16.1–1. Equation (6) gives the source entropy

$$H(X) = \tfrac{1}{2} \times 1 + \tfrac{1}{4} \times 2 + \tfrac{1}{8} \times 3 + \tfrac{1}{8} \times 3 = 1.75 \text{ bits/symbol}$$

which falls somewhat below the maximum value $\log M = 2$. The information rate is

$$R = 2000 \times 1.75 = 3500 \text{ bits/sec}$$

and appropriate coding should make it possible to transmit the source information at the binary signaling rate $r_b \geq 3500$ binits/sec.

Table 16.1–1

x_i	P_i	I_i
A	1/2	1
B	1/4	2
C	1/8	3
D	1/8	3

EXERCISE 16.1–1

Suppose a source has $M = 3$ symbols with probabilities $P_1 = p$ and $P_2 = P_3$. Show that $H(X) = \Omega(p) + 1 - p$. Then evaluate $H(X)\big|_{\max}$ and the corresponding value of p.

Coding for a Discrete Memoryless Channel

When a discrete memoryless source produces M equally likely symbols, so $R = r \log M$, all symbols convey the same amount of information and efficient transmission can take the form of M-ary signaling with a signaling rate equal to the symbol rate r. But when the symbols have different probabilities, so $R = rH(X) < r \log M$, efficient transmission requires an encoding process that takes account of the variable amount of information per symbol. Here, we'll investigate source encoding with a *binary* encoder. Equivalent results for nonbinary encoding just require changing the logarithmic base.

The binary encoder in Fig. 16.1–3 converts incoming source symbols to codewords consisting of binary digits produced at some fixed rate r_b. Viewed from its output, the encoder looks like a binary source with entropy $\Omega(p)$ and information rate $r_b\Omega(p) \le r_b \log 2 = r_b$. Coding obviously does not generate additional information, nor does it destroy information providing that the code is **uniquely decipherable.** Thus, equating the encoder's input and output information rates, we conclude that $R = rH(X) = r_b\Omega(p) \le r_b$ or $r_b/r \ge H(X)$.

The quantity $\overline{N} = r_b/r$ is an important parameter called the **average code length.** Physically, $\overline{N}$ corresponds to the average number of binary digits per source symbol. Mathematically, we write the statistical average

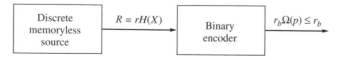

Figure 16.1–3

$$\bar{N} = \sum_{i=1}^{M} P_i N_i \tag{11}$$

where N_i represents the length of the codeword for the ith symbol.

Shannon's **source coding theorem** states that the minimum value of $\bar{N}$ is bounded by

$$H(X) \leq \bar{N} < H(X) + \epsilon \tag{12}$$

with ϵ being a positive quantity. The value of ϵ can be made arbitrarily small, in theory, and optimum source encoding achieves the lower bound $\bar{N} = H(X)$. In practice, we often settle for suboptimum coding with $\bar{N} > H(X)$ if the code has reasonably good efficiency. The ratio $R/r_b = H(X)/\bar{N} \leq 1$ serves as the measure of efficiency for suboptimum codes.

The source coding theorem presumes a uniquely decipherable code to ensure that no information is lost. This requirement imposes an additional but indirect constraint on N. Specifically, as a necessary and sufficient condition for a uniquely decipherable binary code, the word lengths N_i must be such that

$$K = \sum_{i=1}^{M} 2^{-N_i} \leq 1 \tag{13}$$

which is the **Kraft inequality.** The simplest encoding process generates a *fixed-length* code, all codewords having the same length $N_i = \bar{N}$. The Kraft inequality in Eq. (13) then becomes $K = M2^{-\bar{N}} \leq 1$, so decipherability requires $\bar{N} \geq \log M$ and the resulting efficiency is $H(X)/\bar{N} \leq H(X)/\log M$. When $H(X) < \log M$, higher efficiency calls for *variable-length* coding to reduce the average code length $\bar{N}$. An example should help clarify these concepts.

EXAMPLE 16.1–2

Table 16.1–2 lists some potential codes for the source in Example 16.1–1. Code I is a fixed-length code with $\bar{N} = \log M = 2$ binits/symbol, as compared to $H(X) = 1.75$ bits/symbol. The efficiency is $H(X)/\bar{N} = 1.75/2 \approx 88\%$—not bad, but we can do better with a variable-length code.

Application of Eqs. (11) and (13) to Code II gives

$$\bar{N} = \tfrac{1}{2} \times 1 + \tfrac{1}{4} \times 1 + \tfrac{1}{8} \times 2 + \tfrac{1}{8} \times 2 = 1.25 < H(X)$$

$$K = 2^{-1} + 2^{-1} + 2^{-2} + 2^{-2} = 1.5 > 1$$

The result $\bar{N} < H(X)$ is meaningless because $K > 1$, which tells us that this code is not uniquely decipherable. For instance, the code sequence 10011 could be decoded as BAABB or CABB or CAD, and so on. Hence, Code II effectively destroys source information and cannot be used.

Code III, known as a **comma code,** has $K < 1$ and ensures decipherability by marking the start of each word with the binary digit 0. However, the extra comma digits result in $\bar{N} = 1.875 > H(X)$.

Code IV is a **tree code** with the property that no codeword appears as the *prefix* in another codeword. Thus, for example, you can check that the code sequence 110010111 unambiguously represents the message CABD. This code is *optimum* for the source in question since it has $\overline{N} = 1.75 = H(X)$ as well as $K = 1$.

Table 16.1–2   Illustrative source codes

x_i	P_i	Code I	Code II	Code III	Code IV
A	1/2	00	0	0	0
B	1/4	01	1	01	10
C	1/8	10	10	011	110
D	1/8	11	11	0111	111
	$\overline{N}$	2.0	1.25	1.875	1.75
	K	1.0	1.5	0.9375	1.0

Having demonstrated optimum coding by example, we turn to a general proof of the source coding theorem. For this purpose we start with Eq. (10b), taking $Q_i = 2^{-N_i}/K$ where K is the same as in Eq. (13), which satisfies Eq. (10a). Then

$$\sum_i P_i \log \frac{Q_i}{P_i} = \sum_i P_i \left(\log \frac{1}{P_i} - N_i - \log K \right) = H(X) - \overline{N} - \log K \leq 0$$

or, since $\log K \leq 0$, we have

$$\overline{N} \geq H(X)$$

which establishes the lower bound in Eq. (12). The equality holds when $K = 1$ and $P_i = Q_i$. Optimum source encoding with $\overline{N} = H(X)$ therefore requires $K = 1$ and symbol probabilities of the form

$$P_i = 2^{-N_i} \qquad i = 1, 2, \ldots, M \qquad\qquad \textbf{[14]}$$

so that $N_i = -\log P_i = I_i$.

An optimum code must also have equally likely 1s and 0s in order to maximize the binary entropy at $\Omega(p) = 1$. Code IV in Table 16.1–2 exemplifies these optimal properties.

Granted, we can't expect every source to obey Eq. (14), and we certainly can't control the statistics of an information source. Even so, Eq. (14) contains a significant implication for practical source coding, to wit:

Symbols that occur with high probability should be assigned shorter codewords than symbols that occur with low probability.

Long before Shannon, Samuel Morse applied this commonsense principle to his telegraph code for English letters. (Incidentally, Morse estimated letter probabilities by counting the distribution of type in a printer's font.) We'll invoke this principle to establish the upper bound in the source coding theorem.

Let the length of the ith codeword be an integer N_i that falls within $I_i \le N_i < I_i + 1$ or

$$\log \frac{1}{P_i} \le N_i < \log \frac{1}{P_i} + 1 \qquad [15]$$

You can easily confirm that this relationship satisfies the Kraft inequality. Multiplying Eq. (15) by P_i and summing over i then yields

$$H(X) \le \overline{N} < H(X) + 1 \qquad [16]$$

Hence, a code constructed in agreement with Eq. (15) will be reasonably efficient if either $H(X) \gg 1$ or $N_i \approx \log(1/P_i)$ for all i. If neither condition holds, then we must resort to the process called **extension coding.**

For an extension code, n successive source symbols are grouped into blocks and the encoder operates on the blocks rather than on individual symbols. Since each block consists of n statistically independent symbols, the block entropy is just $nH(X)$. Thus, when the coding rule in Eq. (15) is modified and applied to the extension code, Eq. (16) becomes $nH(X) \le n\overline{N} < nH(X) + 1$, where $n\overline{N}$ is the average number of binary digits per block. Upon dividing by n we get

$$H(X) \le \overline{N} < H(X) + \frac{1}{n} \qquad [17]$$

which is our final result.

Equation (17) restates the source coding theorem with $\epsilon = 1/n$. It also shows that $\overline{N} \to H(X)$ as $n \to \infty$, regardless of the source statistics. We've thereby proved that an nth-extension code comes arbitrarily close to optimum source coding. What we haven't addressed is the technique of actual code construction. Systematic algorithms for efficient and practical source coding are presented in various texts. The following example serves as an illustration of one technique.

EXAMPLE 16.1–3   **Shannon-Fano coding**

Shannon-Fano coding generates an efficient code in which the word lengths increase as the symbol probabilities decrease, but not necessarily in strict accordance with Eq. (15). The algorithm provides a tree-code structure to ensure unique decipherability. We'll apply this algorithm to the source with $M = 8$ and $H(X) = 2.15$ whose statistics are listed in Table 16.1–3.

Table 16.1–3   Shannon-Fano coding

		Coding steps						
x_i	P_i	1	2	3	4	5	6	Codeword
A	0.50	0						0
B	0.15	1	0	0				100
C	0.15	1	0	1				101
D	0.08	1	1	0				110
E	0.08	1	1	1	0			1110
F	0.02	1	1	1	1	0		11110
G	0.01	1	1	1	1	1	0	111110
H	0.01	1	1	1	1	1	1	111111

The Shannon-Fano algorithm involves a succession of divide-and-conquer steps. For the first step, you draw a line that divides the symbols into two groups such that the group probabilities are as nearly equal as possible; then you assign the digit 0 to each symbol in the group above the line and the digit 1 to each symbol in the group below the line. For all subsequent steps, you subdivide each group into subgroups and again assign digits by the previous rule. Whenever a group contains just one symbol, as happens in the first and third steps in the table, no further subdivision is possible and the codeword for that symbol is complete. When all groups have been reduced to one symbol, the codewords are given by the assigned digits reading from left to right. A careful examination of Table 16.1–3 should clarify this algorithm.

The resulting Shannon-Fano code in this case has $\overline{N} = 2.18$ so the efficiency is $2.15/2.18 \approx 99\%$. Thus, if the symbol rate is $r = 1000$, then $r_b = \overline{N}r = 2180$ binits/sec—slightly greater than $R = rH(X) = 2150$ bits/sec. As comparison, a fixed-length code would require $\overline{N} = \log 8 = 3$ and $r_b = 3000$ binits/sec.

Apply the Shannon-Fano algorithm to the source in Table 16.1–2 (p. 707). Your result should be identical to Code IV. Then confirm that this code has the optimum property that $N_i = I_i$ and that 0s and 1s are equally likely.

EXERCISE 16.1–2

Predictive Coding for Sources with Memory

Up to this point we've assumed a memoryless source whose successive symbols are statistically independent. But many information sources have *memory* in the sense that the symbol probabilities depend on one or more previous symbols. Written language, being governed by rules of spelling and grammar, provides a good illustration of a source with memory. For instance, the letter *U* (capitalized or not) occurs in English text with probability $P(U) = 0.02$ on the basis of relative frequency; but if the previous letter is Q, then the conditional probability becomes $P(U\,|\,Q) \approx 1$. Clearly, memory effect reduces uncertainty and thereby results in a lower value of entropy than would be calculated using absolute rather than conditional source statistics.

Suppose that a source has a **first-order memory,** so it "remembers" just one previous symbol. To formulate the entropy, let P_{ij} be the conditional probability that symbol x_i is chosen after symbol x_j. Substituting P_{ij} for P_i in Eq. (6), we have the **conditional entropy**

$$H(X\,|\,x_j) \triangleq \sum_i P_{ij} \log \frac{1}{P_{ij}} \tag{18a}$$

which represents the average information per symbol given that the previous symbol was x_j. Averaging over all possible previous symbols then yields

$$H(X) = \sum_j P_j H(X\,|\,x_j) \tag{18b}$$

An equivalent expression applies to the general case of a qth-order memory. However, the notation gets cumbersome because x_j must be replaced by the state of the source defined in terms of the previous q symbols, and there are M^q possible states to consider.

A source with memory is said to be **redundant** when the conditional probabilities significantly reduce $H(X)$ compared to the upper bound $\log M$. The redundancy of English text has been estimated at about 50 percent, meaning that roughly half the symbols in a long passage are not essential to convey the information. By way of example, yu shd babl t read ths evntho sevrl ltrs r msng. It likewise follows that if uncertainty is reduced by memory effect, then *predictability* is increased. Coding for efficient transmission can then be based on some prediction method. Here we'll analyze the scheme known as **predictive run encoding** for a discrete source with memory.

Consider the encoding system in Fig. 16.1–4a, where source symbols are first converted to a binary sequence with digit rate r. Let $x(i)$ denote the ith digit in the binary sequence, and let $\tilde{x}(i)$ be the corresponding digit generated by a predictor. Mod-2 addition of $x(i)$ and $\tilde{x}(i)$ yields the binary **error sequence** $\epsilon(i)$ such that $\epsilon(i) = 0$ when $\tilde{x}(i) = x(i)$ and $\epsilon(i) = 1$ when $\tilde{x}(i) \neq x(i)$. The error sequence is fed back to the predictor in order to update the prediction process. The decoding system in Fig. 16.1–4b employs an identical predictor to reconstruct the source symbols from the mod-2 sum $x(i) = \tilde{x}(i) \oplus \epsilon(i)$. We'll assume the source is sufficiently predictable that a predictor can be devised whose probability of a *correct* prediction is

$$p = P[\epsilon(i) = 0] > 1/2$$

We'll also assume that prediction errors are statistically independent. Hence the error sequence entropy is $\Omega(p) < 1$ and appropriate encoding should allow information transmission at the rate $r_b < r$.

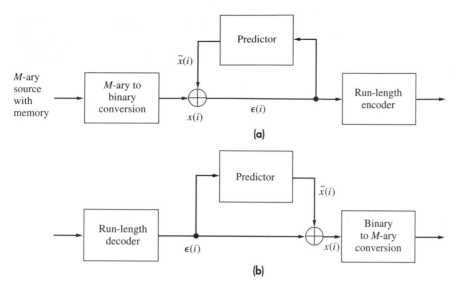

Figure 16.1–4   Predictive coding system for a source with memory. (a) Encoder;
(b) decoder.

Predictive run encoding exploits the fact that the error sequence consists of more 0s than 1s—the better the prediction, the higher the percentage of 0s. We define a **run of n** to be a string of n successive 0s followed by a 1. Instead of transmitting these *strings* of $n + 1$ digits, the encoder represents the successive values of n as k-digit binary codewords. A single codeword can represent any value of n in the range $0 \leq n \leq m - 1$, where

$$m = 2^k - 1$$

When a run has $n \geq m$, the encoder sends a k-digit word consisting entirely of 1s, which tells the decoder to wait for the next codeword to determine n. Table 16.1–4 summarizes the encoding and decoding algorithm.

Table 16.1–4   Run encoding algorithm

n	Encoder Output (k digits/codeword)	Decoder Output ($\leq m$ digits/word)
0	00 ... 000	1
1	00 ... 001	01
2	00 ... 010	001
$\vdots$	$\vdots$	$\vdots$
$m - 1$	11 ... 110	000 ... 01
$\geq m$	11 ... 111	000 ... 00

Efficient transmission results if, on the average, the codewords have fewer digits than the runs they represent. In particular, let $\overline{E}$ be the average number of error digits per run and let $\overline{N}$ be the average number of codeword digits per run, so the encoded digit rate is $r_b = (\overline{N}/\overline{E})r$. If $\overline{N}/\overline{E} < 1$ then $r_b < r$ and we say that predictive run encoding achieves **rate reduction** or **data compression.** The compression ratio $\overline{N}/\overline{E}$ is found from the following calculations.

First, note that a run consists of $E = n + 1$ error digits, including the 1 at the end, so the probability of a run of n is

$$P(n) = p^n(1 - p)$$

Hence,

$$\overline{E} = \sum_{n=0}^{\infty} (n + 1)P(n) = (1 - p)\sum_{n=0}^{\infty} (n + 1)p^n = \frac{1}{1 - p} \qquad [19]$$

where we've evaluated the summation using the series expansion

$$(1 - v)^{-2} = 1 + 2v + 3v^2 + \cdots = \sum_{n=0}^{\infty} (n + 1)v^n \qquad [20]$$

which holds when $v^2 < 1$.

Next, since long runs must be represented by more than one k-digit codeword, we write

$$N = Lk \qquad L = 1, 2, 3, \ldots$$

where L is related to n via $(L - 1)m \leq n \leq Lm - 1$. The average number of code digits per run is then given by

$$\overline{N} = k\sum_{n=0}^{m-1} P(n) + 2k\sum_{n=m}^{2m-1} P(n) + \cdots = k(1 - p)\left(\sum_{n=0}^{m-1} p^n\right)\sum_{i=0}^{\infty} (i + 1)(p^m)^i$$

The first summation is a standard form and the second can be evaluated using Eq. (20), with the final result that

$$\overline{N} = k(1 - p)\frac{p^m - 1}{p - 1}(1 - p^m)^{-2} = \frac{k}{1 - p^m} = \frac{k}{1 - p^{2k-1}} \qquad [21]$$

Since $\overline{N}$ depends on k and p, there is an optimum value of k that minimizes $\overline{N}$ for a given predictor.

Table 16.1–5 lists a few values of p and the corresponding optimum values of k. Also listed are the resulting values of $\overline{N}$, $\overline{E}$, and the compression ratio $\overline{N}/\overline{E} = r_b/r$. This table shows that predictive run encoding achieves substantial compression when we know enough about the source to build a predictor with $p > 0.8$.

Table 16.1–5   Performance of predictive run encoding

p	k_{opt}	$\overline{N}$	$\overline{E}$	$\overline{N}/\overline{E}$
0.6	1	2.50	2.5	1.00
0.8	3	3.80	5.0	0.76
0.9	4	5.04	10.0	0.50
0.95	6	6.25	20.0	0.31

In facsimile transmission a text is scanned and sample values are converted to 0s and 1s for white and black, respectively. Since there are usually many more 0s than 1s, the source clearly has memory. Suppose a predictor can be built with $p = 0.9$. Estimate an upper bound on the source entropy less than $\log 2 = 1$ bit/sample.

EXERCISE 16.1–3

16.2   INFORMATION TRANSMISSION ON DISCRETE CHANNELS

This section applies information theory to the study of information transmission. We'll assume that both the source and the transmission channel are discrete, so we can measure the amount of information transferred and define the *channel capacity*. Shannon asserted that, with appropriate coding, nearly errorless information transmission is possible on a noisy channel if the rate does not exceed the channel capacity. This fundamental theorem will be examined in conjunction with the *binary symmetric channel*, an important channel model for the analysis of digital communication systems.

Mutual Information

Consider the information transmission system represented by Fig. 16.2–1. A discrete source selects symbols from an alphabet X for transmission over the channel. Ideally, the transmission channel should reproduce at the destination the symbols

Figure 16.2–1   Discrete information transmission system.

emitted by the source. However, noise and other transmission impairments alter the source symbols, resulting in a different symbol alphabet Y at the destination. We want to measure the information transferred in this case.

Several types of symbol probabilities will be needed to deal with the two alphabets here, and we'll use the notation defined as follows:

* $P(x_i)$ is the probability that the source selects symbol x_i for transmission;
* $P(y_j)$ is the probability that symbol y_j is received at the destination;
* $P(x_i\, y_j)$ is the joint probability that x_i is transmitted and y_j is received;
* $P(x_i\,|\,y_j)$ is the conditional probability that x_i was transmitted given that y_j is received;
* $P(y_j\,|\,x_i)$ is the conditional probability that y_j is received given that x_i was transmitted.

We'll assume, for simplicity, that the channel is *time-invariant* and *memoryless,* so the conditional probabilities are independent of time and previous symbol transmissions. The conditional probabilities $P(y_j\,|\,x_i)$ then have special significance as the channel's **forward transition probabilities.**

By way of example, Fig. 16.2–2 depicts the forward transitions for a noisy channel with two source symbols and three destination symbols. If this system is intended to deliver $y_j = y_1$ when $x_i = x_1$ and $y_j = y_2$ when $x_i = x_2$, then the symbol *error probabilities* are given by $P(y_j\,|\,x_i)$ for $j \neq i$.

Our quantitative description of information transfer on a discrete memoryless channel begins with the **mutual information**

$$I(x_i; y_j) \triangleq \log \frac{P(x_i \mid y_j)}{P(x_i)} \quad \text{bits} \tag{1}$$

which measures the amount of information transferred when x_i is transmitted and y_j is received. To lend support to this definition, we'll look at two extreme cases. On the one hand, suppose we happen to have an ideal noiseless channel such that each

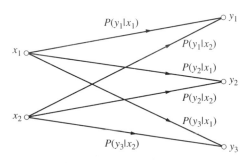

Figure 16.2–2   Forward transition probabilities for a noisy discrete channel.

y_j uniquely identifies a particular x_i; then $P(x_i | y_j) = 1$ and $I(x_i; y_j) = \log [1/P(x_i)]$ —so the transferred information equals the self-information of x_i. On the other hand, suppose the channel noise has such a large effect that y_j is totally unrelated to x_i; then $P(x_i | y_j) = P(x_i)$ and $I(x_i; y_j) = \log 1 = 0$—so no information is transferred. These extreme cases make sense intuitively.

Most transmission channels fall somewhere between the extremes of perfect transfer and zero transfer. To analyze the general case, we define the **average mutual information**

$$I(X; Y) \triangleq \sum_{x,y} P(x_iy_j)I(x_i; y_j) = \sum_{x,y} P(x_iy_j) \log \frac{P(x_i | y_j)}{P(x_i)} \text{ bits/symbol} \quad [2]$$

where the summation subscripts indicate that the statistical average is taken over both alphabets. The quantity $I(X; Y)$ represents the average amount of source information gained per *received* symbol, as distinguished from the average information per source symbol represented by the source entropy $H(X)$.

Several different but equivalent expressions for the mutual information can be derived using the probability relationships

$$P(x_iy_j) = P(x_i | y_j)P(y_j) = P(y_j | x_i)P(x_i) \quad [3a]$$

$$P(x_i) = \sum_y P(x_iy_j) \qquad P(y_j) = \sum_x P(x_iy_j) \quad [3b]$$

In particular, let's expand Eq. (2) as

$$I(X; Y) = \sum_{x,y} P(x_iy_j) \log \frac{1}{P(x_i)} - \sum_{x,y} P(x_iy_j) \log \frac{1}{P(x_i | y_j)}$$

so the first term simplifies to

$$\sum_x \left[\sum_y P(x_iy_j) \right] \log \frac{1}{P(x_i)} = \sum_x P(x_i) \log \frac{1}{P(x_i)} = H(X)$$

Hence,

$$I(X; Y) = H(X) - H(X|Y) \quad [4]$$

where we've introduced the **equivocation**

$$H(X | Y) \triangleq \sum_{x,y} P(x_iy_j) \log \frac{1}{P(x_i | y_j)} \quad [5]$$

Equation (4) says that the average information transfer per symbol equals the source entropy minus the equivocation. Correspondingly, the equivocation represents the information lost in the noisy channel.

For another perspective on information transfer, we return to Eq. (2) and note from Eq. (3a) that $P(x_i | y_j)/P(x_i) = P(y_j | x_i)/P(y_j)$ so $I(X; Y) = I(Y; X)$. Therefore, upon interchanging X and Y in Eq. (4), we have

$$I(X; Y) = H(Y) - H(Y \mid X) \qquad [6]$$

with

$$H(Y) = \sum_y P(y_j) \log \frac{1}{P(y_j)} \qquad H(Y \mid X) = \sum_{x,y} P(x_i y_j) \log \frac{1}{P(y_j \mid x_i)} \qquad [7]$$

Equation (6) says that the information transferred equals the *destination* entropy $H(Y)$ minus the **noise entropy** $H(Y \mid X)$ added by the channel. The interpretation of $H(Y \mid X)$ as noise entropy follows from our previous observation that the set of forward transition probabilities $P(y_j \mid x_i)$ includes the symbol error probabilities.

EXAMPLE 16.2–1

The binary symmetric channel

Figure 16.2–3 depicts the model of a **binary symmetric channel** (BSC). There are two source symbols with probabilities

$$P(x_1) = p \qquad P(x_2) = 1 - p$$

and two destination symbols with forward transition probabilities

$$P(y_1 \mid x_2) = P(y_2 \mid x_1) = \alpha \qquad P(y_1 \mid x_1) = P(y_2 \mid x_2) = 1 - \alpha$$

This model could represent any binary transmission system in which errors are statistically independent and the error probabilities are the same for both symbols, so the average error probability is

$$P_e = P(x_1)P(y_2 \mid x_1) + P(x_2)P(y_1 \mid x_2) = p\alpha + (1 - p)\alpha = \alpha$$

Since we know the forward transition probabilities, we'll use $I(X; Y) = H(Y) - H(Y \mid X)$ to calculate the mutual information in terms of p and α.

The destination entropy $H(Y)$ is easily found by treating the output of the channel as a binary source with symbol probabilities $P(y_1)$ and $P(y_2) = 1 - P(y_1)$. We therefore write $H(Y) = \Omega[P(y_1)]$ where $\Omega(\ )$ is the binary entropy function defined in Eq. (9), Sect. 16.1 (p. 703), and

$$P(y_1) = P(y_1 \mid x_1)P(x_1) + P(y_1 \mid x_2)P(x_2) = \alpha + p - 2\alpha p$$

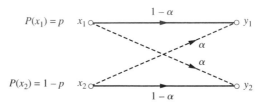

Figure 16.2–3   Binary symmetric channel (BSC).

obtained with the help of Eqs. (3*a*) and (3*b*). For the noise entropy $H(Y|X)$, we substitute $P(x_i\,y_j) = P(y_j\,|\,x_i)P(x_i)$ into Eq. (7) to get

$$H(Y\mid X) = \sum_x P(x_i)\left[\sum_y P(y_j\mid x_i)\,\log\frac{1}{P(y_j\mid x_i)}\right]$$

which then reduces to $H(Y|X) = \Omega(\alpha)$. The channel's symmetry causes this result to be independent of p.

Putting the foregoing expressions together, we finally have

$$I(X;\,Y) = \Omega(\alpha + p - 2\alpha p) - \Omega(\alpha) \qquad [8]$$

so the information transfer over a BSC depends on both the error probability α and the source probability p. If the noise is small, then $\alpha \ll 1$ and $I(X;\,Y) \approx \Omega(p) = H(X)$; if the noise is very large, then $\alpha = 1/2$ and $I(X;\,Y) = 0$.

Confirm that $H(Y|X) = \Omega(\alpha)$ for a BSC. **EXERCISE 16.2–1**

Consider a channel with the property that x_i and y_j are statistically independent for all i and j. Show that $H(X|Y) = H(X)$ and $I(X;\,Y) = 0$. **EXERCISE 16.2–2**

Discrete Channel Capacity

We've seen that discrete memoryless channels transfer a definite amount of information $I(X;\,Y)$, despite corrupting noise. A given channel usually has fixed source and destination alphabets and fixed forward transition probabilities, so the only variable quantities in $I(X;\,Y)$ are the *source* probabilities $P(x_i)$. Consequently, *maximum information transfer* requires specific source statistics—obtained, perhaps, through source encoding. Let the resulting maximum value of $I(X;\,Y)$ be denoted by

$$C_s \triangleq \max_{P(x_i)} I(X;\,Y) \qquad \text{bits/symbol} \qquad [9]$$

This quantity represents the maximum amount of information transferred per channel symbol (on the average) and is called the **channel capacity.** We also measure capacity in terms of information rate. Specifically, if s stands for the maximum symbol rate allowed by the channel, then the capacity per unit time is

$$C = sC_s \qquad \text{bits/sec} \qquad [10]$$

which represents the maximum rate of information transfer.

The significance of channel capacity becomes most evident in the light of Shannon's **fundamental theorem for a noisy channel,** stated as follows:

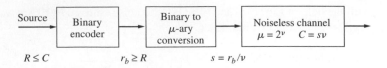

$$R \le C \qquad\qquad r_b \ge R \qquad\qquad s = r_b/v$$

Figure 16.2–4   Encoding system for a noiseless discrete channel.

If a channel has capacity C and a source has information rate $R < C$, then there exists a coding system such that the output of the source can be transmitted over the channel with an arbitrarily small frequency of errors. Conversely, if $R > C$, then it is not possible to transmit the information without errors.

A general proof of this theorem goes well beyond our scope, but we'll attempt to make it plausible by considering two particular cases.

First, suppose we have an ideal noiseless channel with $\mu = 2^v$ symbols. Then $I(X; Y) = H(X)$, which is maximized if $P(x_i) = 1/\mu$ for all i. Thus,

$$C_s = \max_{P(x_i)} H(X) = \log \mu = v \qquad C = sv \qquad\qquad \text{[11]}$$

Errorless transmission rests in this case on the fact that the channel is noiseless. However, we still need a coding system like the one diagrammed in Fig. 16.2–4 to match the source and channel. The binary source encoder generates binary digits at the rate $r_b \ge R$ for conversion to μ-ary channel symbols at the rate $s = r_b/\log \mu = r_b/v$. Hence,

$$R \le r_b = sv = C \qquad \text{bits/sec}$$

and optimum source encoding achieves maximum information transfer with $R = r_b = C$. Transmission at $R > C$ would require a coding system that violates the Kraft inequality; consequently, decoding errors would occur even though the channel is noiseless.

A more realistic case, including channel noise, is the binary symmetric channel from Example 16.2–1 (p. 716). We previously found that $I(X; Y) = \Omega(\alpha + p - 2\alpha p) - \Omega(\alpha)$, with $\Omega(\alpha)$ being constant for a fixed error probability α. But $\Omega(\alpha + p - 2\alpha p)$ varies with the source probability p and reaches a maximum value of unity when $\alpha + p - 2\alpha p = 1/2$, which is satisfied for any α if $p = 1/2$, that is, equally likely binary input symbols. Thus, the capacity of a BSC is

$$C_s = 1 - \Omega(\alpha) \qquad\qquad \text{[12]}$$

The plot of C_s versus α in Fig. 16.2–5 shows that $C_s \approx 1$ for $\alpha \ll 1$, but the capacity rapidly drops to zero as $\alpha \to 0.5$. The same curve applies for $0.5 \le \alpha \le 1$ if you replace α with $1 - \alpha$, equivalent to interchanging the two output symbols.

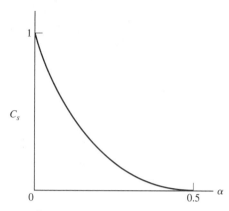

Figure 16.2–5   Capacity of a BSC.

Reliable transmission on a BSC requires channel coding for *error control* in addition to source coding. Our study of error-control coding in Chap. 13 demonstrated that the error probability per binary message digit could, in fact, be made much smaller than the transmission error probability α. For instance, the (15, 11) Hamming code has code rate $R_c = 11/15$ and output error probability $P_{be} \approx 14\alpha^2$ per message digit; if the BSC has $\alpha = 10^{-3}$ and symbol rate s, then the Hamming code yields $P_{be} \approx 10^{-5}$ at the message digit rate $r_b = R_c s \approx 0.73s$. Shannon's theorem asserts that a better coding system would yield virtually errorless transmission at the rate $r_b = R \leq C = s[1 - \Omega(\alpha)] \approx s$. We'll conclude this section with an outline of the argument Shannon used to prove the existence of such a coding system for the BSC.

Coding for the Binary Symmetric Channel ★

Consider the encoding system diagrammed in Fig. 16.2–6. The BSC has symbol rate s and capacity $C_s = 1 - \Omega(\alpha)$, with $\alpha < 1/2$. The source produces information at rate R. To maximize the mutual information, an optimum binary source encoder operates on the source symbols and emits equiprobable binary digits. Channel coding for error control is then carried out in two steps. First, a converter transforms

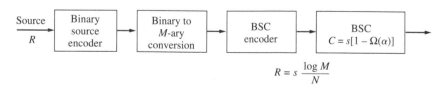

$$R = s\,\frac{\log M}{N}$$

Figure 16.2–6   Encoding system for a BSC.

blocks of binary digits into equiprobably M-ary symbols conveying $\log M$ bits/symbol. Second, the BSC encoder represents each M-ary symbol by a *channel codeword* consisting of N binary channel symbols, hereafter called binits. (Don't confuse the channel wordlength N with the average code length $\overline{N}$ at the output of the source encoder.)

The average information per binit is $(\log M)/N$, and binits are generated at the channel symbol rate s. Hence, we can express the source information rate as

$$R = s(\log M)/N$$

where M and N are parameters of the channel encoder. Shannon's theorem requires $R \leq C = sC_s$, equivalent here to $(\log M)/N \leq C$. The parameters M and N must therefore be related by

$$M = 2^{N(C_s - \epsilon)} \tag{13}$$

with $0 \leq \epsilon < C_s$. We'll show that ϵ can be arbitrarily small, so $R \to C_s$, and that appropriate channel coding makes it possible to recover the M-ary symbols at the destination with arbitrarily low probability of error—providing that the wordlength N is *very large*. In fact, we'll eventually let $N \to \infty$ to ensure errorless information transfer. Ideal channel coding for the BSC thus involves *infinite time delay.* Practical coding systems with finite time delay and finite wordlength will fall short of ideal performance.

The reasoning behind large wordlength for error control comes from the vector picture of binary codewords introduced in Sect. 13.1. Specifically, recall that all of 2^N possible words of length N can be visualized as vectors in an N-dimensional space where distance is measured in terms of *Hamming distance.* Let the vector V in Fig. 16.2–7 represent one of the M channel codewords, and let V' be the received codeword with n erroneous binits caused by transmission errors. The Hamming distance between V' and V equals n, a random variable ranging from 0 to N. However, when N is very large, V' almost always falls within a **Hamming sphere** of radius $d < N$ around the tip of V. The decoder correctly decides that V was transmitted if V' is received within the sphere and if none of the remaining $M - 1$ code vectors fall within this sphere.

But Shannon did not propose an explicit algorithm for codeword selection. Instead, he said that *randomly* selected codewords would satisfy the requirement

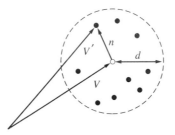

Figure 16.2–7   Vector representation of codewords.

when $N \to \infty$. Although the random coding approach met with considerable criticism at first, it has subsequently been recognized as a powerful method in coding and information theory. We'll adopt random codeword selection and we'll write the probability of a decoding error as

$$P_e = P_{ne} + P_{ce}$$

Here, P_{ne} is the probability of a noise error, so V' falls outside the Hamming sphere, and P_{ce} is the probability of a "code error" in the sense that two or more selected codewords fall within the same Hamming sphere.

A noise error corresponds to the event $n \geq d$. Since transmission errors are statistically independent and occur with probability $\alpha < 1/2$, n is governed by the *binomial distribution* (p. 338) with mean and variance

$$\bar{n} = N\alpha \qquad \sigma^2 = N\alpha(1 - \alpha)$$

If we take the Hamming radius as

$$d = N\beta \qquad \alpha < \beta < 1/2 \qquad\qquad \textbf{[14a]}$$

then Chebeyshev's inequality yields the upper bound

$$P_{ne} = P(n \geq d) \leq \left(\frac{\sigma}{d - \bar{n}} \right)^2 = \frac{\alpha(1 - \alpha)}{N(\beta - \alpha)^2} \qquad\qquad \textbf{[14b]}$$

Thus, for any $\beta > \alpha$, the probability of a noise error becomes vanishingly small as $N \to \infty$.

To formulate the probability of a "code error," let the vector V in Fig. 16.2–7 represent one of the selected codewords, and let m denote the number of vectors within the Hamming sphere. The remaining $M - 1$ codewords are chosen randomly from the entire set of 2^N vectors, and the probability of selecting one of the m vectors simply equals $m/2^N$. Hence,

$$P_{ce} = (M - 1)m2^{-N} < Mm2^{-N} = m2^{-N[\Omega(\alpha) - \epsilon]} \qquad\qquad \textbf{[15]}$$

where we've inserted Eq. (13) for M and written $C_s = 1 - \Omega(\alpha)$. Now we need an upper bound on m as $N \to \infty$.

All of the m vectors within a Hamming sphere of radius d represent binary words of length N that differ in no more than d places. The number of these words that differ in i places equals a binomial coefficient, so

$$m = \sum_{i=0}^{d} \binom{N}{i} = \binom{N}{0} + \binom{N}{1} + \cdots + \binom{N}{d}$$

There are $d + 1$ terms in this sum and the last term is the largest since $d = N/\beta < N/2$; hence,

$$m \leq (d + 1)\binom{N}{d} = (d + 1)\frac{N!}{d!(N - d)!}$$

Then, for the factorial of each of the large numbers N, $d = N/\beta$, and $N - d = N(1 - \beta)$, we apply **Stirling's approximation**

$$k! \approx \sqrt{2\pi k}\, k^k\, e^{-k} \qquad k \gg 1 \tag{16}$$

and a few manipulations lead to the upper-bound expression

$$m \leq \frac{N\beta + 1}{\sqrt{2\pi N\beta(1 - \beta)}}\, 2^{N\Omega(\beta)} \tag{17}$$

Combining Eqs. (15) and (17) then gives

$$P_{ce} < \frac{N\beta + 1}{\sqrt{2\pi N\beta(1 - \beta)}}\, 2^{-N[\epsilon + \Omega(\alpha) - \Omega(\beta)]} \tag{18}$$

which is our final result.

Equation (18) shows that the probability of a decoding error caused by random codeword selection goes to zero as $N \to \infty$, providing that $\epsilon > \Omega(\beta) - \Omega(\alpha)$. For a given value of the channel parameter α, we can take $\beta > \alpha$ to satisfy Eq. (14) and still have an arbitrarily small value of $\epsilon > \Omega(\beta) - \Omega(\alpha)$. Under these conditions, we achieve errorless information transfer on the BSC with $R \to C$ as $N \to \infty$.

16.3   CONTINUOUS CHANNELS AND SYSTEM COMPARISONS

Having developed concepts of information transmission for the simplified discrete case, we're now ready to tackle the more realistic case of a continuous source and channel.

We'll begin with the measure of information from a source that emits a continuous signal. The material may seem heavy going at first, but we'll then make some reasonable assumptions about the transmission of continuous signals to express channel capacity in terms of *bandwidth* and *signal-to-noise ratio,* a result known as the *Hartley-Shannon law.* This result leads us to the definition of an ideal communication system, which serves as a standard for system comparisons and a guide for the design of improved communication systems.

Continuous Information

A continuous information source produces a time-varying *signal x(t)*. We'll treat the set of possible signals as an ensemble of waveforms generated by some random process, assumed to be ergodic. We'll further assume that the process has a finite bandwidth, meaning that $x(t)$ is completely characterized in terms of periodic *sample values.* Thus, at any sampling instant, the collection of possible sample values constitutes a *continuous random variable X* described by its probability density function $p(x)$.

The average amount of information per sample value of $x(t)$ is measured by the **entropy function**

$$H(X) \triangleq \int_{-\infty}^{\infty} p(x) \log \frac{1}{p(x)} \, dx \qquad [1]$$

where, as before, the logarithmic base is $b = 2$. This expression has obvious similarities to the definition of entropy for a discrete source. However, Eq. (1) turns out to be a *relative* measure of information rather than an absolute measure. The *absolute* entropy of a continuous source can, in principle, be defined from the following limiting operation.

Let the continuous RV X be approximated by a discrete RV with values $x_i = i \, \Delta x$ for $i = 0, \pm 1, \pm 2, \ldots$, and probabilities $P(x_i) \approx p(x_i) \, \Delta x$. Then, based on the formula for discrete entropy, let the absolute entropy be

$$H_{\text{abs}}(X) = \lim_{\Delta x \to 0} \sum_i p(x_i) \Delta x \log \frac{1}{p(x_i) \, \Delta x}$$

$$= \lim_{\Delta x \to 0} \sum_i \left[p(x_i) \log \frac{1}{p(x_i)} - p(x_i) \log \Delta x \right] \Delta x$$

Passing from summation to integration yields

$$H_{\text{abs}}(X) = H(X) + H_0(X) \qquad [2a]$$

where

$$H_0(X) = -\lim_{\Delta x \to 0} \log \Delta x \int_{-\infty}^{\infty} p(x) \, dx = -\lim_{\Delta x \to 0} \log \Delta x \qquad [2b]$$

which is the *reference* for the relative entropy $H(X)$. Since $H_0(X) = -\log 0 = \infty$, the absolute entropy of a continuous source is always infinite—a useless but reasonable result in view of the fact that X is a continuous RV with an uncountable number of possible values.

Relative entropy, being finite, serves as a useful measure of information from continuous sources if you avoid misleading comparisons involving different references. In particular, consider two information signals $x(t)$ and $z(t)$. For a valid comparison of their entropy functions we write

$$H_{\text{abs}}(Z) - H_{\text{abs}}(X) = H(Z) - H(X) + [H_0(Z) - H_0(X)]$$

where

$$H_0(Z) - H_0(X) = -\lim_{\Delta z \to 0} \log \Delta z + \lim_{\Delta x \to 0} \log \Delta x \qquad [3]$$

$$= -\lim_{\Delta z, \Delta x \to 0} \log \left| \frac{\Delta z}{\Delta x} \right| = -\log \left| \frac{dz}{dx} \right|$$

If the signals are related in some manner and if $|dz/dx| = 1$, then the reference values are equal and $H(Z)$ and $H(X)$ are directly comparable.

The inherent nature of relative entropy precludes absolute bounds on $H(X)$. In fact, the value of $H(X)$ can be positive, zero, or negative, depending upon the source PDF. Nonetheless, a reasonable question to ask is: What $p(x)$ maximizes $H(X)$ for a given source? This question is significant because a given source usually has specific signal constraints, such as fixed peak value or average power, that limit the possible PDFs.

Stating the problem in more general terms, we seek the function $p = p(x)$ that maximizes an integral of the form

$$I = \int_a^b F(x, p)\, dx$$

whose integrand $F(x, p)$ is a specified function of x and p. The variable function p is subject to a set of k constraints given by

$$\int_a^b F_k(x, p)\, dx = c_k$$

with the c_k being constants. A theorem from the calculus of variations says that I is maximum (or minimum) when p satisfies the equation

$$\frac{\partial F}{\partial p} + \sum_k \lambda_k \frac{\partial F_k}{\partial p} = 0 \tag{4}$$

where the λ_k are **Lagrange's undetermined multipliers.** The values of λ_k are found by substituting the solution of Eq. (4) into the constraint equations.

In the problem at hand, we wish to maximize $I = H(X)$ as defined by Eq. (1). Thus, we take

$$F(x, p) = p \log (1/p) = -p (\ln p)/\ln 2$$

Furthermore, $p(x)$ must obey the essential PDF property

$$\int_{-\infty}^{\infty} p(x)\, dx = 1 \tag{5}$$

so we always have the constraint function and constant

$$F_1(x, p) = p \qquad c_1 = 1$$

Additional constraints come from the particular source limitations, as illustrated in the important example that follows.

EXAMPLE 16.3–1   **Source entropy with fixed average power**

Consider the case of a source with *fixed average power* defined in terms of $p(x)$ by

$$S \triangleq \overline{x^2} = \int_{-\infty}^{\infty} x^2 p(x)\, dx \tag{6}$$

which imposes the additional constraint function $F_2 = x^2 p$ and $c_2 = S$. Inserting F, F_1, and F_2 into Eq. (4) yields

$$-(\ln p + 1) + \lambda_1 + \lambda_2 x^2 = 0$$

where ln 2 has been absorbed in λ_1 and λ_2. Thus,

$$p(x) = e^{\lambda_1 - 1} e^{\lambda_2 x^2}$$

After using Eqs. (5) and (6) to evaluate the multipliers, we get

$$p(x) = \frac{1}{\sqrt{2\pi S}} e^{-x^2/2S} \qquad [7]$$

a *gaussian* function with zero mean and variance $\sigma^2 = S$.

The corresponding maximum entropy is calculated from Eq. (1) by writing $\log 1/p(x) = \log \sqrt{2\pi S} + (x^2/2S) \log e$. Therefore,

$$H(X) = \log \sqrt{2\pi S} + \tfrac{1}{2} \log e = \tfrac{1}{2} \log 2\pi e S \qquad [8]$$

obtained with the help of Eqs. (5) and (6). Note that this relative entropy has a negative value when $2\pi e S < 1$. Even so, for any fixed average power S, we've established the important result that

$$H(X) \leq \tfrac{1}{2} \log 2\pi e S$$

and that the entropy is maximum when $p(x)$ is a zero-mean gaussian PDF. Other source constraints of course lead to different results.

Suppose a source has a peak-value limitation, such that $-M \leq x(t) \leq M$. (a) By finding the PDF that maximizes $H(X)$, show that $H(X) = \log 2M$. (b) Let the source waveform be amplified to produce $z(t) = Kx(t)$. Find $H(Z)$ and confirm that the absolute entropy remains unchanged even though $H(Z) \neq H(X)$.

EXERCISE 16.3–1

Continuous Channel Capacity

Information transfer on a continuous channel takes the form of signal transmission. The source emits a signal $x(t)$ which, after corruption by transmission noise, appears at the destination as another signal $y(t)$. The **average mutual information** is defined by analogy with the discrete case to be

$$I(X; Y) \triangleq \int\!\!\int_{-\infty}^{\infty} p_{XY}(x, y) \log \frac{p_X(x \mid y)}{p_X(x)} \, dx \, dy \qquad [9]$$

where $p_X(x)$ is the source PDF, $p_{XY}(x, y)$ is the joint PDF, and so on. Averaging with respect to both X and Y removes the potential ambiguities of relative entropy. Thus, $I(X; Y)$ measures the *absolute* information transfer per sample values of $y(t)$ at the destination. It can be shown from Eq. (9) that $I(X; Y) \geq 0$ and that $I(X; Y) = 0$ when the noise is so great that $y(t)$ is unrelated to $x(t)$.

Usually, we know the forward transition PDF $p_Y(y \mid x)$ rather than $p_X(x \mid y)$. We then calculate $I(X; Y)$ from the equivalent expression

$$I(X; Y) = H(Y) - H(Y \mid X) \qquad [10]$$

in which $H(Y)$ is the destination entropy and $H(Y \mid X)$ is the noise entropy given by

$$H(Y \mid X) = \int \int_{-\infty}^{\infty} p_X(x) p_Y(y \mid x) \log \frac{1}{p_Y(y \mid x)} \, dx \, dy$$

If the channel has *independent additive noise* such that $y(t) = x(t) + n(t)$, then $p_Y(y \mid x) = p_Y(x + n) = p_N(y - x)$, where $p_N(n)$ is the noise PDF. Consequently, $H(Y \mid X)$ reduces to

$$H(Y \mid X) = \int_{-\infty}^{\infty} p_N(n) \log \frac{1}{p_N(n)} \, dn \qquad [11]$$

independent of $p_X(x)$.

Now consider a channel with fixed forward transition PDF, so the maximum information transfer per sample value of $y(t)$ is

$$C_s \overset{\triangle}{=} \max_{p_X(x)} I(X; Y) \qquad \text{bits/sample} \qquad [12]$$

If the channel also has a fixed *bandwidth B*, then $y(t)$ is a bandlimited signal completely defined by sample values taken at the Nyquist rate $f_s = 2B$—identical to the maximum allowable signaling rate for a given bandwidth B. (Samples taken at greater than the Nyquist rate would not be independent and carry no additional information.) The maximum rate of information transfer then becomes

$$C \overset{\triangle}{=} 2BC_s \qquad \text{bits/sec} \qquad [13]$$

which defines the *capacity* of a bandlimited continuous channel. Shannon's fundamental theorem for a noisy channel applies here in the sense that errorless transmission is theoretically possible at any information rate $R \leq C$.

The continuous channel model of greatest interest is known as the **additive white gaussian noise** (AWGN) **channel,** defined by the following properties:

1. The channel provides distortion-free transmission over some bandwidth B, and any transmission loss is compensated by amplification.

2. The channel constrains the input from the source to be a bandlimited signal $x(t)$ with fixed average power $S = \overline{x^2}$.

3. The signal received at the destination is contaminated by the addition of bandlimited white gaussian noise $n(t)$ with zero mean and average noise power $N = \overline{n^2} = N_0 B$.

4. The signal and noise are independent so that $y(t) = x(t) + n(t)$ and

$$\overline{y^2} = \overline{x^2} + \overline{n^2} = S + N$$

These properties agree with the assumptions made in earlier chapters to obtain upper bounds on the performance of analog systems. Here, we'll determine the corresponding channel capacity.

First, we find the noise entropy $H(Y|X)$ via Eq. (11). Since $p_N(n)$ is a zero-mean gaussian function with variance $\sigma^2 = N$, it follows from Eqs. (7) and (8) that $H(Y|X) = \frac{1}{2} \log 2\pi eN$. Second, noting that $H(Y|X)$ does not depend on the source PDF, we use Eqs. (10) and (12) to write

$$C_s = \max_{p_X(x)}[H(Y) - H(Y|X)] = \left[\max_{p_X(x)} H(Y)\right] - \frac{1}{2} \log 2\pi eN$$

But the destination signal $y(t)$ has fixed average power $\overline{y^2} = S + N$, so $H(Y) \le \frac{1}{2} \log 2\pi e(S + N)$. If $p_X(x)$ is a zero-mean gaussian function, then $y = x + n$ has a gaussian PDF and $H(Y)$ is maximized. Hence,

$$C_s = \frac{1}{2} \log 2\pi e(S + N) - \frac{1}{2} \log 2\pi eN = \frac{1}{2} \log\left(\frac{S + N}{N}\right)$$

Finally, substituting for C_s in Eq. (13) yields the simple result

$$C = B \log (1 + S/N) \qquad\qquad \textbf{[14]}$$

where we recognize S/N as the *signal-to-noise ratio* at the destination.

Equation (14) is the famous **Hartley-Shannon law.** When coupled with the fundamental theorem, it establishes an upper limit for reliable information transmission on a bandlimited AWGN channel, namely, $R \le B \log (1 + S/N)$ bits/sec. Additionally, since bandwidth and signal-to-noise ratio are basic transmission parameters, Eq. (14) establishes a general performance standard for the comparison of communication systems. We therefore devote the remainder of this section to implications of the Hartley-Shannon law.

Ideal Communication Systems

Virtually every realistic communication system is capable of transmitting continuous signals, subject to some power and bandwidth limitations. Furthermore, if we exclude radio transmission with fading, careful system design can largely eliminate all other contaminations except unavoidable thermal noise. The AWGN channel thus serves as a reasonable model under these conditions, and the Hartley-Shannon law gives the maximum rate for reliable communication. Therefore, we here define

An *ideal* communication system is one that achieves nearly error-free information transmission at a rate approaching $R = B \log (1 + S/N)$.

Shannon described the operation of such an ideal system as follows.

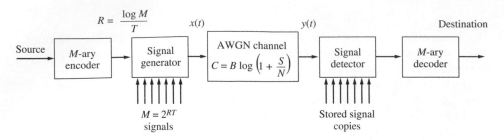

$$R = \frac{\log M}{T}$$

Figure 16.3–1   Ideal communication system with AWGN channel.

The ideal system has the structure diagrammed in Fig. 16.3–1. Information from the source is observed for intervals of T seconds and encoded as equiprobable M-ary symbols, such that $(\log M)/T = R$. For each symbol, the signal generator emits a unique waveform $x(t)$ selected from a set of $M = 2^{RT}$ signals. These signals are generated by a bandlimited white gaussian process, and stored copies are available at the destination. The signal detector compares the received signal $y(t) = x(t) + n(t)$ with the stored copies and chooses the most probable one for decoding. The operations of encoding at the source and signal detection at the destination result in a total delay of $2T$ seconds, called the **coding delay.**

The number of different signals required depends on the desired error probability P_e, and Shannon showed that

$$\lim_{P_e \to 0} \lim_{T \to \infty} \frac{\log M}{T} = B \log\left(1 + \frac{S}{N}\right)$$

Hence, $R \to C$ in the limit as $T \to \infty$—which means that the number of signals and the coding delay becomes infinitely large, so the ideal system is physically *unrealizable*. However, real systems having large but finite M and T can be designed to come as close as we wish to ideal performance. (One such system is described in Sect. 16.5.) With this thought in mind, let's further consider the properties of a hypothetical ideal system.

The capacity relation $C = B \log (1 + S/N)$ underscores the fundamental role of bandwidth and signal-to-noise ratio in communication. It also shows that we can exchange increased bandwidth for decreased signal power, a feature previously observed in wideband noise-reduction systems such as FM and PCM. The Hartley-Shannon law specifies the *optimum* bandwidth-power exchange and, moreover, suggests the possibility of *bandwidth compression*.

Keeping in mind that noise power varies with bandwidth as $N = N_0 B$, we explore the trade-off between bandwidth and signal power by writing

$$C = B \log\left(1 + \frac{S}{N_0 B}\right) \qquad [15]$$

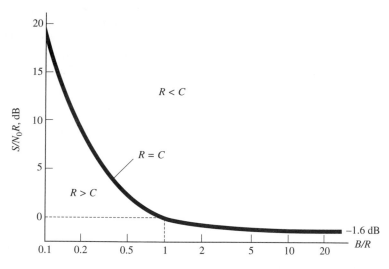

Figure 16.3-2   Trade-off between bandwidth and signal power with an ideal system.

Thus, if N_0 and R have fixed values, information transmission at the rate $R \le C$ requires

$$\frac{S}{N_0 R} \ge \frac{B}{R}\left(2^{R/B} - 1\right) \qquad [16]$$

which becomes an equality when $R = C$. Figure 16.3–2 shows the resulting plot of $S/N_0 R$ in dB versus B/R. The region on the lower left corresponds to $R > C$, a forbidden condition for reliable communication. This plot reveals that **bandwidth compression** $(B/R < 1)$ demands a dramatic increase of signal power, while **bandwidth expansion** $(B/R > 1)$ reduces $S/N_0 R$ asymptotically toward a distinct limiting value of about -1.6 dB as $B/R \to \infty$.

In fact, an ideal system with *infinite bandwidth* has finite channel capacity given by

$$C_\infty \triangleq \lim_{B\to\infty} B \log\left(1 + \frac{S}{N_0 B}\right) = \frac{S}{N_0 \ln 2} \approx 1.44 \frac{S}{N_0} \qquad [17]$$

Equation (17) is derived from Eq. (15) written in the form

$$C = \frac{S}{N_0 \lambda} \frac{\ln(1 + \lambda)}{\ln 2} \qquad \lambda = \frac{S}{N_0 B}$$

The series expansion $\ln(1 + \lambda) = \lambda - \frac{1}{2}\lambda^2 + \cdots$ then shows that $[\ln(1 + \lambda)]/\lambda \to 1$ as $\lambda \to 0$, corresponding to $B \to \infty$. Note that C_∞ is the maximum capacity for fixed S and N_0, so

$$S/N_0 R \ge S/N_0 C_\infty = \ln 2 \approx -1.6 \text{ dB}$$

From the shape of the curve in Fig. 16.3–2, we conclude that $C \approx C_\infty$ for $B/R > 10$.

It must be stressed that these results pertain to an ideal but unrealizable system. Consequently, they may be used to estimate upper limits on the performance of any real system whose transmission channel approximates the AWGN model. An example should help illustrate the value of such estimates.

EXAMPLE 16.3–2   **TV pictures from Mars**

A roving vehicle equipped with a monochrome TV camera is proposed to continue the exploration of the surface of Mars. The TV pictures will be digitized for transmission back to Earth. We want to estimate the time required to transmit one picture, given the following information.

A digitized picture will consist of $n_p = 400 \times 300$ pixels (picture elements), each pixel having one of 16 possible brightness levels. Hence, the information rate R and picture transmission time T are related by

$$RT = n_p \log 16 = 480,000 \text{ bits}$$

assuming equally likely brightness levels.

The Mars-to-Earth link is a microwave radio system with carrier frequency $f_c = 2$ GHz and path length $\ell = 3 \times 10^8$ km. The transmitter on the vehicle delivers $S_T = 20$ W signal power to a dish antenna one meter in diameter. The Earth station has a 30-m dish antenna and a low-noise receiver with noise temperature $\mathcal{T}_N = 58$ K. The signal power S at the receiver is calculated using $L = 268$ dB, $g_T = 26$ dB, and $g_R = 56$ dB, so

$$S = \frac{g_T \, g_R}{L} S_T = 5 \times 10^{-18} \text{ W}$$

The noise density at the receiver is

$$N_0 = k\mathcal{T}_N = 8 \times 10^{-22} \text{ W/Hz}$$

obtained from Eq. (6), Sect. 9.4.

Since no transmission bandwidth was specified, let's assume that $B/R > 10$. An ideal system would then have

$$R \leq C \approx C_\infty = 1.44 S/N_0 \approx 9000 \text{ bits/sec}$$

and the corresponding bandwidth must be $B > 10R \geq 90$ kHz. Therefore, the transmission time per picture is

$$T \geq \frac{480,000 \text{ bits}}{9000 \text{ bits/sec}} \approx 53 \text{ sec}$$

A real system, of course, would require more time for picture transmission. The point here is that no system with the same assumed specifications can achieve a smaller transmission time—unless some predictive source encoding reduces RT.

EXERCISE 16.3–2

A keyboard machine with 64 different symbols is connected to a voice telephone channel having $B = 3$ kHz and $S/N_0B = 30$ dB.   (a) Calculate the maximum possible symbol rate for errorless transmission.   (b) Assuming B can be changed while other parameters are fixed, find the symbol rate with $B = 1$ kHz and $B \to \infty$.

System Comparisons

Having determined the properties of an ideal communication system, we're ready to reexamine various systems from previous chapters and compare their performance in the light of information theory. Such comparisons provide important guidelines for the development of new or improved communication systems, both digital and analog. For all comparisons, we'll assume an AWGN channel with transmission bandwidth B_T and signal-to-noise ratio $(S/N)_R = S_R/N_0B_T$ at the receiver.

Consider first the case of binary baseband transmission. Previously, we found the maximum signaling rate $2B_T$ and minimum transmission error probability $Q[\sqrt{(S/N)_R}]$, which requires the use of polar sinc-pulse signals and matched filtering. Now we'll view this system as a *binary symmetric channel* with symbol rate $s = 2B_T$ and transmission error probability $\alpha = Q[\sqrt{(S/N)_R}]$. Hence, the BSC system capacity is

$$C = 2B_T[1 - \Omega(\alpha)] \qquad \alpha = Q[\sqrt{(S/N)_R}] \qquad [18]$$

where $\Omega(\alpha)$ is the binary entropy function from Eq. (9), Sect. 16.1. With sufficiently elaborate error-control coding, we can obtain nearly error-free information transfer at $R \le C$.

Figure 16.3–3 shows C/B_T versus $(S/N)_R$ for a BSC and for an ideal continuous system with $C/B_T = \log [1 + (S/N)_R]$. The BSC curve flattens off at $C/B_T = 2$ as $(S/N)_R \to \infty$ because the maximum entropy of a noise-free binary signal is one bit per symbol. Hence, we must use M-ary signaling to get closer to ideal performance when $(S/N)_R \gg 1$. At lower signal-to-noise ratios, where $C/B_T < 2$, the gap between the BSC and ideal curves suggests that there should be a way of extracting more information from a noisy binary signal. This observation has led to the sophisticated technique known as **soft-decision decoding** in which binary codewords are decoded from sample values of the continuous signal-plus-noise waveform, rather than from a regenerated (hard-decision) binary waveform.

Now let's examine the performance of M-ary digital communication systems without error-control coding. We'll assume a binary source that emits equiprobable symbols at rate r_b. If the error probability is small, say $P_{be} \le 10^{-4}$, then the information rate is $R \approx r_b$ since almost all of the binary digits correctly convey information to the destination. For transmission purposes, let the binary data be converted to a Gray-code M-ary signal with $M = 2^K$. An M-ary baseband system with polar sinc pulses and matched filtering then has

$$\frac{r_b}{B_T} = 2K \qquad P_{be} \approx \frac{2}{K}\left(1 - \frac{1}{M}\right)Q\left(\sqrt{\frac{6K}{M^2 - 1}\,\gamma_b}\right) \qquad [19]$$

where $\gamma_b = E_b/N_0$ and $E_b = S_R/r_b$, which is the average energy per binary digit.

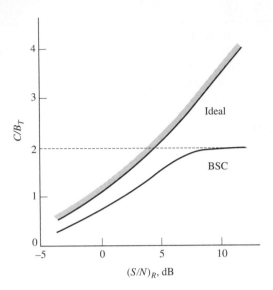

Figure 16.3–3

Upon solving Eq. (19) for the value of γ_b that gives $P_{be} \approx 10^{-4}$, we obtain the plot of r_b/B_T versus γ_b in Fig. 16.3–4. This same curve also applies for bandpass transmission via APK modulation with M replaced by $\sqrt{M}$, while other modulation methods have poorer performance. The comparison curve for an ideal system is calculated from Eq. (16) taking $R = r_b$ and $S/N_0R = S_R/N_0r_b = \gamma_b$. We thus see that real digital systems with $r_b/B_T \geq 2$ and small but nonzero error probability require at least 6–7 dB more signal energy than an ideal system. Error-control coding would reduce the energy needed for a specified error probability, thereby shifting the system performance curve closer to the ideal.

Finally, we come to *analog* communication in which bandwidth and signal-to-noise ratio are the crucial performance factors, rather than information rate and error probability. Consider, therefore, the analog modulation system represented by Fig. 16.3–5, where the analog signal at the destination has bandwidth W and signal-to-noise ratio $(S/N)_D$. The maximum output information rate is $R = W \log [1 + (S/N)_D]$, and it cannot exceed the transmission channel capacity $C = B_T \log [1 + (S_R/N_0B_T)]$. Setting $R \leq C$ and solving for $(S/N)_D$ yields

$$\left(\frac{S}{N}\right)_D \leq \left(1 + \frac{S_R}{N_0B_T}\right)^{B_T/W} - 1 = \left(1 + \frac{\gamma}{b}\right)^b - 1 \qquad [20]$$

where the normalized parameters b and γ are defined by

$$b = B_T/W \qquad \gamma = S_R/N_0W$$

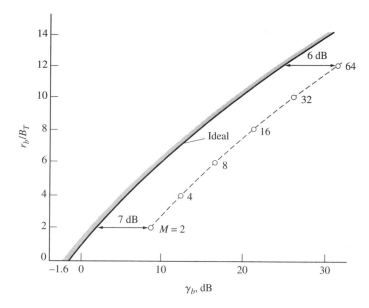

Figure 16.3–4   Performance comparison of ideal system and *M*-ary system with $P_{be} = 10^{-4}$.

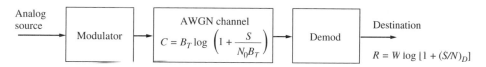

Figure 16.3–5   Analog modulation system.

Equation (20) becomes an equality for an ideal system operating at full capacity, so we have a standard for the comparison of all other analog communication systems. In particular, we'll compare the analog performance curves given by $(S/N)_D$ versus γ with fixed bandwidth ratio b, and the power-bandwidth exchange given by γ versus b with fixed $(S/N)_D$.

Figure 16.3–6 repeats some of our previous analog performance curves for $b = 6$, together with the curve for an ideal system. The heavy dots mark the threshold points, and we see that wideband noise-reduction systems (FM, PCM, and PPM) fall short of ideal performance primarily as a result of threshold limitations. Threshold extension techniques may improve performance by a few dB, but threshold effect must *always* occur in wideband analog modulation; otherwise, the curves would cross into the forbidden region to the left.

Figure 16.3–7 depicts the power-bandwidth exchange of analog modulation systems, taking $(S/N)_D = 50$ dB and $S_x = \overline{x^2} = 1/2$. SSB and DSB appear here as single

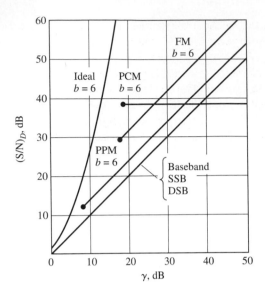

Figure 16.3–6   Performance comparison of analog modulation systems.

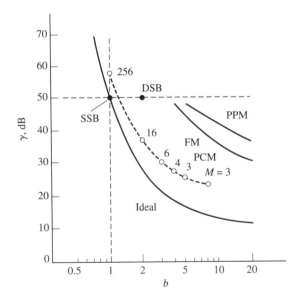

Figure 16.3–7   Power-bandwidth exchange of analog modulation systems with $(S/N)_D = 50$ dB.

points since b is fixed, and SSB is equivalent to an ideal system with $b = 1$—meaning no wideband improvement. The FM and PPM curves are calculated from previous equations, and the PCM curve corresponds to operation just above threshold. Observe that PCM comes closest to the ideal when we want $b > 1$ for wideband improvement.

Based on these two comparison figures, we conclude that *digital* transmission via PCM makes better use of the capacity of a continuous channel than conventional analog modulation. This difference between digital and analog transmission of analog information stems from the fundamental fact that transmission noise prevents *exact* reproduction of an analog waveform. The PCM quantization process takes advantage of this fact and provides a better match of signal to channel than does analog modulation.

Shannon went further in this direction by proposing that the information rate of a continuous source be measured in terms of some specified *fidelity* or *distortion* criterion. Appropriate digital source coding then allows efficient information transmission over a channel whose capacity equals the rate in question. Shannon's proposal has subsequently been developed into the field known as **rate distortion theory.** An introduction to this theory is given by Proakis and Salehi (1994, Chap. 4).

16.4   SIGNAL SPACE

As we have seen, *digital* transmission is the most efficient form of communication in the light of information theory. Thus we should give some attention to the theory of *optimum digital detection*. First, however, we need a method of representing signals, which is the purpose of this section. In particular, we'll introduce the important technique of representing signals as *vectors,* a technique originated by Kotel'nikov in his theory of optimum noise immunity.

Signals as Vectors

A **signal space** is a multidimensional space wherein vectors represent a specified set of signals. By describing signals as vectors, the familiar relations and insights of geometry can be brought to bear on signal analysis and communication system design. Here we introduce those aspects of signal-space theory relevant to digital detection, restricting our attention to the case of *real energy signals.*

Consider any two real energy signals $v(t)$ and $w(t)$. If α and β are finite real constants, then the linear combination $z(t) = \alpha v(t) + \beta w(t)$ is also an energy signal. Formally, we say that the set of all real energy signals is **closed under linear combination,** which is the essential condition for establishing a **linear space.** We thus define the signal space Γ to be a linear space wherein vectors such as v and w represent energy signals $v(t)$ and $w(t)$, and any linear combination such as $z = \alpha v + \beta w$ is another vector in Γ.

Vector addition and scalar multiplication in Γ obey the usual rules, and there exists a zero element such that $v + 0 = v$ and $v - v = 0$. However, to complete the geometric structure of Γ, we need appropriate measures of vector length and

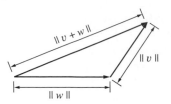

Figure 16.4–1

angle expressed in terms of signal properties. (We can't use Hamming distance here because we're not necessarily working with binary sequences.) Since we're dealing with energy signals, we'll define the length or **norm** of a signal vector v to be the square root of the signal energy, written as

$$\|v\| \triangleq \sqrt{E_v} = \left[\int_{-\infty}^{\infty} v^2(t)\, dt \right]^{1/2} \tag{1}$$

This vector norm is a nonnegative quantity, and $\|v\| = 0$ only if $v(t) = 0$. We use the square root of the signal energy in Eq. (1) so that the norm of the sum $z(t) = v(t) + w(t)$ satisfies the **triangle inequality**

$$\|v + w\| \leq \|v\| + \|w\| \tag{2}$$

Figure 16.4–1 shows the corresponding picture as a triangle composed of the vectors v, w, and $z = v + w$.

To measure the angle of v relative to some other vector w, we first recall that the crosscorrelation function measures the similarity between the shifted signal $v(t + \tau)$ and $w(t)$. Now, setting $\tau = 0$, we define the **scalar product**

$$(v, w) \triangleq R_{vw}(0) = \int_{-\infty}^{\infty} v(t)w(t)\, dt \tag{3}$$

which has the properties

$$(v, v) = \|v\|^2 \tag{4}$$

$$|(v, w)| \leq \|v\|\|w\| \tag{5}$$

Equation (5) simply restates *Schwarz's inequality,* and the equality holds when $v(t)$ and $w(t)$ are proportional signals.

Justification of the scalar product for angular measurement comes from the expansion

$$\|v + w\|^2 = (v + w, v + w) = (v, v) + (v, w) + (w, v) + (w, w)$$

$$= \|v\|^2 + 2(v, w) + \|w\|^2 \tag{6}$$

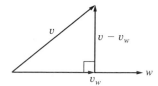

Figure 16.4–2   Projection of v on w.

On the one hand, if $v(t)$ and $w(t)$ are fully correlated in the sense that $v(t) = \pm\alpha w(t)$, then $(v, w) = \pm \|v\|\|w\|$ and $\|v + w\| = \|v\| \pm \|w\|$. Figure 16.4–1 thus reduces to *colinear* vectors, with v parallel to w. On the other hand, if $v(t)$ and $w(t)$ are uncorrelated such that $(v, w) = 0$, then $\|v + w\|^2 = \|v\|^2 + \|w\|^2$ and Fig. 16.4–1 becomes a right triangle with *perpendicular* vectors v and w. Accordingly, we say that

$v(t)$ and $w(t)$ are **orthogonal signals** when $(v, w) = 0$

Orthogonality in signal space implies *superposition of energy,* since $\|v\|^2 + \|w\|^2 = E_v + E_w$.

For the intermediate case of $0 < |(v, w)| < \|v\|\|w\|$, we can decompose v into the two orthogonal (perpendicular) components shown in Fig. 16.4–2. The **projection** of v on w is defined by the colinear and orthogonal conditions $v_w = \alpha w$ and $(v - v_w, w) = 0$, and solving for α yields

$$v_w = \frac{(v, w)}{\|w\|^2} w \qquad [7]$$

You can confirm from Schwarz's inequality that $\|v_w\| \leq \|v\|$ as expected.

Next, as further preparation for the study of optimum detection, suppose we have a set of M energy signals denoted by $\{s_i(t); i = 1, 2, \ldots, M\}$. Each of these signals can be represented by a vector written in the form

$$s_i = \alpha_{i1}\phi_1 + \alpha_{i2}\phi_2 + \cdots + \alpha_{iK}\phi_K = \sum_{k=1}^{K} \alpha_{ik}\phi_k \qquad K \leq M \qquad [8a]$$

where the ϕ_k are vectors and the α_{ik} are scalar coefficients calculated from

$$\alpha_{ik} = (s_i, \phi_k) \qquad [8b]$$

Two important aspects of this representation should be stressed. First, Eq. (8a) is an *exact* expression, rather than a series approximation. Second, each signal is completely specified by the scalars in Eq. (8b), so we can work with a set of K coefficients α_{ik} instead of the entire waveform $s_i(t)$. These points will become clearer from the signal-space interpretation.

The vectors ϕ_k in Eq. (8) represent a set of **orthonormal basis functions $\phi_k(t)$** with the properties

$$\|\phi_k\|^2 = (\phi_k, \phi_k) = 1 \qquad k = 1, 2, \ldots, K \tag{9}$$

$$(\phi_k, \phi_j) = 0 \qquad j \neq k$$

These basis functions are mutually orthogonal and each one has unit norm—analogous to the *unit vectors* of ordinary two- and three-dimensional spaces. By extrapolation, we say that the ϕ_k span a *K-dimensional subspace* Γ_K of the signal space Γ. If Γ_K contains the signal vector s_i, its projection on the basis vector ϕ_k is

$$(s_i, \phi_k)\phi_k/\|\phi_k\|^2 = (s_i, \phi_k)\phi_k = \alpha_{ik}\phi_k$$

where we've used Eqs. (7), (9), and (8b). Therefore, the coefficients α_{ik} are the **coordinates** of the vector s_i with respect to the ϕ_k basis.

The subspace Γ_K and basis ϕ_k depend upon the specific set of signals we wish to represent. In fact, we usually generate the basis functions directly from the set of signals $\{s_i(t)\}$, thereby ensuring that Γ_K contains all of the signal vectors in question. Although a given set of signals defines a unique subspace, we have considerable flexibility in the choice of basis functions as long as they span Γ_K and satisfy Eq. (9).

Having generated an orthonormal basis for a set of signals, we can compute the vector coordinates and express the norms and scalar products in terms of the coordinates. The scalar product of any two signals is found using Eqs. (8) and (9), with the result that

$$(s_i, s_m) = \left(\sum_k \alpha_{ik}\phi_k, \sum_j \alpha_{mj}\phi_j \right) = \sum_{k=1}^{K} \alpha_{ik}\alpha_{mk} \tag{10a}$$

Hence, letting $m = i$ we have

$$\|s_i\|^2 = (s_i, s_i) = \sum_{k=1}^{K} \alpha_{ik}^2 \tag{10b}$$

which is a *K*-dimensional statement of the *pythagorean theorem*.

The Gram-Schmidt Procedure

A sequential method for constructing an orthonormal basis is the **Gram-Schmidt procedure,** as follows:

1. Select from the signal set any $s_1(t)$ having nonzero norm, and take $\phi_1 = s_1/\|s_1\|$ so $\|\phi_1\| = 1$.

2. Select another signal $s_2(t)$, compute the coordinate $\alpha_{21} = (s_2, \phi_1)$, and define the auxiliary vector $g_2 = s_2 - \alpha_{21}\phi_1$ in Fig. 16.4–3a. Since g_2 is orthogonal to ϕ_1, we can take $\phi_2 = g_2/\|g_2\|$.

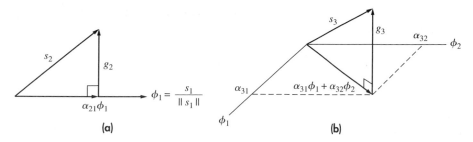

Figure 16.4–3   Vector constructions for the Gram-Schmidt procedure.

3. Select another signal $s_3(t)$, compute the coordinates α_{31} and α_{32} to define $g_3 = s_3 - [\alpha_{31}\phi_1 + \alpha_{32}\phi_2]$ in Fig. 16.4–3b. Then take $\phi_3 = g_3/\|g_3\|$ so ϕ_3 has unit norm and is orthogonal to ϕ_1 and ϕ_2.

4. Continue in this manner until all M signals have been considered.
    The basis resulting from this procedure can be expressed compactly by

$$\phi_k = g_k/\|g_k\| \qquad\qquad [11a]$$

with $g_1 = s_1$ and

$$g_k = s_k - \sum_{j=1}^{k-1} \alpha_{kj}\phi_j \qquad 2 \leq k \leq K \qquad\qquad [11b]$$

Equation (11b) yields $g_k = 0$ whenever s_k has zero norm or equals a linear combination of previous signal vectors. Such cases are omitted from the basis, and the dimensionality will be $K < M$.

Let's apply the Gram-Schmidt procedure to obtain an orthonormal basis for the three signals in Fig. 16.4–4a. First, noting that $\|s_1\|^2 = E_1 = 6^2$, we take $\phi_1(t) = s_1(t)/6$ as plotted in Fig. 16.4–4b. Then

EXAMPLE 16.4–1

$$\alpha_{21} = \int_{-\infty}^{\infty} s_2(t)\phi_1(t)\, dt = 3$$

so $g_2(t) = s_2(t) - 3\phi_1(t)$, and a sketch of $g_2(t)$ quickly yields $\|g_2\|^2 = 3^2$, and we take $\phi_2(t) = g_2(t)/3$ in Fig. 16.4–4c. A third basis function is not needed since we can write the linear combination $s_3(t) = s_1(t) - s_2(t)$.

By calculating the remaining coordinates, or by inspection of the waveforms, we obtain the vector representations

$$s_1 = 6\phi_1 \qquad s_2 = 3\phi_1 + 3\phi_2 \qquad s_3 = 3\phi_1 - 3\phi_2$$

which are diagrammed as vectors in Fig. 16.4–4d. This diagram brings out the fact that s_2 and s_3 are orthogonal and their sum equals s_1.

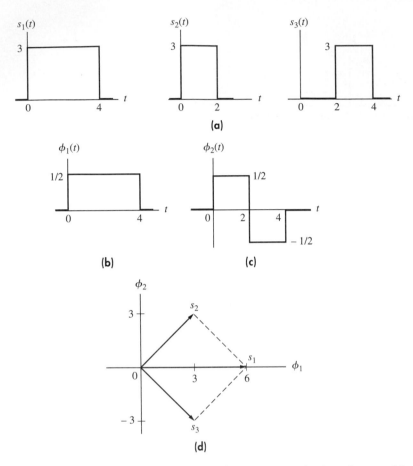

Figure 16.4–4   (a) A signal set; (b) one basis function; (c) another basis function; (d) vector diagram.

EXERCISE 16.4–1   Let another set of signals be defined from Fig. 16.4–4a by taking $s_1'(t) = \sqrt{2}\, s_2(t)$, $s_2'(t) = \sqrt{2}\, s_1(t)$, and $s_3'(t) = -\sqrt{2}\, s_3(t)$. (a) Apply the Gram-Schmidt procedure and draw the resulting vector diagram. (b) Calculate $\|s_2'\|^2$ using Eq. (10b) and compare with the result from Eq. (1).

16.5   OPTIMUM DIGITAL DETECTION

The signal/vector concepts developed in the previous section are used here to design optimum receivers for digital signals with AWGN contamination. Vector concepts also facilitate the calculation of error probabilities and the selection of signals for efficient digital communication.

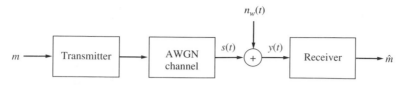

Figure 16.5–1   *M*-ary digital communication system.

Optimum Detection and MAP Receivers

Consider the *M*-ary digital communication system in Fig. 16.5–1. The source selects a symbol *m* from the set $\{m_i; i = 1, 2, \dots, M\}$. The transmitter generates a signal that appears at the destination as $s(t)$. This signal belongs to the set $\{s_i(t); i = 1, 2, \dots, M\}$, in one-to-one correspondence with the symbol set. The received signal is corrupted by independent additive white gaussian noise $n_w(t)$ with zero mean and power spectral density $G(f) = N_0/2$. The receiver produces the *estimated* symbol $\hat{m}$ by comparing $y(t) = s(t) + n_w(t)$ with stored copies of the uncorrupted signals. We seek an optimum detection strategy that minimizes the average error probability $P_e = P(\hat{m} \neq m)$.

To formulate the optimization criterion more precisely, let $P(m \mid y)$ be the conditional probability that *m* was selected by the source given that $y(t)$ has been observed at the destination. Clearly, if

$$P(m_j \mid y) > P(m_i \mid y) \qquad \text{all } i \neq j \tag{1}$$

then the receiver should chose $\hat{m} = m_j$ to minimize $P(\hat{m} \neq m)$. The quantities in Eq. (1) are known as **a posteriori** probabilities, meaning "after observation," and Eq. (1) defines a **maximum a posteriori** (MAP) receiver. In the unlikely event of a tie, the receiver chooses any one of the m_i that maximizes $P(m_i \mid y)$. We'll assume equiprobable symbols, in which case the MAP criterion reduces to **maximum likelihood** detection.

Our statement of the optimum detection problem contains only two critical assumptions: independent additive gaussian noise and stored copies at the receiver. The assumption of equiprobable symbols is not critical since this condition can be ensured by source encoding, as described in Sect. 16.1. Nor is the assumption of a white noise spectrum critical since a nonwhite noise spectrum can be equalized by a *prewhitening filter* at the front end of the receiver. The assumption that stored copies are available at the receiver essentially means that we're working with a *coherent* system, but we have no preconceptions about the structure of the receiver itself. Signal-space analysis eventually determines specific receiver implementations.

All of the *M* possible signal waveforms will be represented by vectors in a subspace Γ_K of dimensionality $K \leq M$ and spanned by some orthonormal basis ϕ_k. However, the received vector $y = s + n_w$ does not fall entirely within Γ_K. We therefore decompose n_w as $n_w = n + n'$ shown in Fig. 16.5–2 where, for convenience,

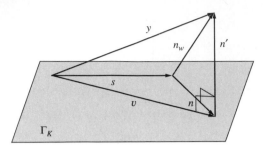

Figure 16.5–2   Projection of $y = s + n_w$ on Γ_K.

Γ_K is drawn as a plane. The component n', called the **irrelevant noise,** is orthogonal to Γ_K in the sense that

$$(n', \phi_k) = 0 \qquad k = 1, 2, \dots, K$$

The remaining noise vector n is contained in Γ_K, so we can write

$$n = \sum_{k=1}^{K} \beta_k \phi_k \qquad \qquad \textbf{[2a]}$$

where the noise coordinates β_k are given by

$$\beta_k = (n, \phi_k) = (n_w - n, \phi_k) = (n_w, \phi_k) \qquad \textbf{[2b]}$$

Finally, we define the **relevant data vector**

$$v \overset{\triangle}{=} y - n' = s + n \qquad \qquad \textbf{[3]}$$

which is the *projection* of y on Γ_K since $y = v + n'$ and n' is orthogonal to every vector in Γ_K.

Figure 16.5–3 illustrates the vectors in Γ_K for the case of $K = 2$ and $M = 3$. The three dots mark the tips of the uncorrupted signal vectors, and v is a typical relevant data vector when $s = s_2$. Intuitively, this diagram suggests that the receiver should choose the signal vector whose tip is closest to the tip of v. Subsequent analysis indeed confirms the **maximum-likelihood detection rule:**

Given an observed vector v, choose $\hat{m} = m_j$ with j such that $\|v - s_j\| < \|v - s_i\|$ for all $i \neq j$.

This rule defines M **decision regions** in Γ_K, one region for each signal point. In Fig. 16.5–3, for instance, the three decision regions are bounded by dashed lines formed by constructing *perpendicular bisectors* of lines connecting the signal points. These decision-region boundaries correspond to the decision thresholds introduced in Chap. 11.

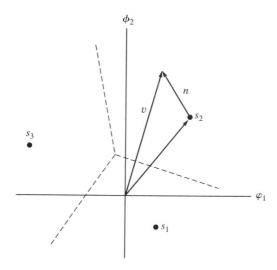

Figure 16.5–3   Signal vectors and decision regions in Γ_K with $K = 2$ and $M = 3$.

The foregoing detection rule involves vector norms in Γ_K. We know the signal norms $\|s_i\| = \sqrt{E_i}$, but the noise n is a *random* vector whose properties require further study. From Eq. (2a) and Eq. (10b), Sect. 16.4, we have

$$\|n\|^2 = \sum_{k=1}^{K} \beta_k^2$$

where $\beta_k = (n_w, \phi_k)$, which is a linear operation on a gaussian RV. Hence, the β_k are also gaussian with mean value $\overline{\beta_k} = (\overline{n_w}, \phi_k) = 0$. We turn then to the variance and the question of statistical independence by considering the expectation

$$E[\beta_k \beta_i] = E\left[\int_{-\infty}^{\infty} n_w(t)\phi_k(t)\, dt \int_{-\infty}^{\infty} n_w(\lambda)\phi_i(\lambda)\, d\lambda \right]$$

$$= \int\!\!\int_{-\infty}^{\infty} E[n_w(t)n_w(\lambda)]\phi_k(t)\phi_i(\lambda)\, dt\, d\lambda$$

But, from the autocorrelation function of white noise, $E[n_w(t)n_w(\lambda)] = \dfrac{N_0}{2}\delta(t - \lambda)$, so we get

$$E[\beta_k \beta_i] = \begin{cases} N_0/2 & i = k \\ 0 & i \neq k \end{cases}$$

which means that the β_k are *independent* gaussian RV's with variance $\sigma^2 = N_0/2$.

Therefore, the PDF of any noise coordinate $\beta_k = \beta$ is the zero-mean gaussian function

$$p_\beta(\beta) = \frac{1}{\sqrt{2\pi\sigma^2}} e^{-\beta^2/2\sigma^2} \qquad \sigma^2 = \frac{N_0}{2} \qquad \text{[4a]}$$

The joint PDF of all K coordinates is then $p_n(\beta_1, \beta_2, \ldots, \beta_K) = p_\beta(\beta_1)p_\beta(\beta_2)\cdots p_\beta(\beta_K)$ since the coordinates are mutually independent. Accordingly, we express the vector PDF as

$$p_n(n) = p_n(\beta_1, \beta_2, \ldots, \beta_K) = (\pi N_0)^{-K/2} e^{-\|n\|^2/N_0} \qquad \text{[4b]}$$

which has *spherical symmetry* in Γ_K.

Now we're prepared to analyze optimum detection from the criterion in Eq. (1), noting that $P(m \mid y) = P(m \mid v)$. Since $P(m \mid v)p_v(v)\, dv = P(m)p_v(v \mid m)\, dv$, we can write $P(m \mid v) = P(m)p_v(v \mid m)/p_v(v)$ where the unconditional PDF $p_v(v)$ does not depend on m. Thus, $P(m_j \mid y) > P(m_i \mid y)$ is equivalent to the vector PDF condition

$$P(m_j)p_v(v \mid m_j) > P(m_i)p_v(v \mid m_i) \qquad \text{[5]}$$

But when $m = m_i$, $s = s_i$ so $v = s_i + n$ and

$$p_v(v \mid m_i) = p_n(v - s_i) = (\pi N_0)^{-K/2} e^{-\|v-s_i\|^2/N_0}$$

Substituting this expression into Eq. (5) yields, after simplification the **MAP detection rule**

$$\|v - s_j\|^2 - \ln P(m_j) < \|v - s_i\|^2 - \ln P(m_i) \qquad \text{[6a]}$$

When the symbols are equiprobable, $P(m_i) = 1/M$ for all i and Eq. (6a) simplifies to the **maximum-likelihood detection rule**

$$\|v - s_j\|^2 < \|v - s_i\|^2 \qquad \text{[6b]}$$

which is the same as the decision region condition $\|v - s_j\| < \|v - s_i\|$.

Our next task is to devise a receiver structure that implements the detection rule. Recognizing that the receiver must operate on the waveform $y(t)$ instead of the vector v, we first write $\|v - s_i\|^2 = \|v\|^2 - 2(v, s_i) + \|s_i\|^2$ where $\|s_i\|^2 = E_i$ and $(v, s_i) = (y - n', s_i) = (y, s_i)$ since n' is orthogonal to any vector in Γ_K. The term $\|v\|^2$ will appear on both sides of Eq. (6) and cancel out, so we define the observable **decision function**

$$z_i \overset{\triangle}{=} (y, s_i) - c_i \qquad \text{[7a]}$$

in which

$$(y, s_i) = \int_{-\infty}^{\infty} y(t)s_i(t)\, dt \qquad \text{[7b]}$$

$$c_i = \begin{cases} \frac{1}{2} E_i & \text{equiprobable } m_i \\[2mm] \frac{1}{2}[E_i - N_0 \ln P(m_i)] & \text{otherwise} \end{cases} \qquad \textbf{[7c]}$$

The detection rule thus becomes:

Choose $\hat{m} = m_j$ with j such that $z_j > z_i$ for all $i \neq j$.

The scalar products (y, s_i) needed for optimum detection can be calculated at the receiver using the stored signal copies and a bank of **correlation detectors,** each correlator having the form of Fig. 16.5–4a. If the signals are timelimited so that $s_i(t) = 0$ outside of $0 \leq t \leq D$, then correlation detection reduces to *matched filtering* and *sampling* as diagrammed in Fig. 16.5–4b. In either case, the operation (y, s_i) projects y along the direction of s_i and strips off the irrelevant noise n'.

Figure 16.5–5 shows a complete MAP receiver with a bank of correlators or matched filters that operate in parallel on $y(t)$ to produce (y, s_i), $i = 1, 2, \ldots, M$.

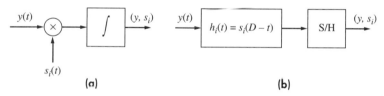

Figure 16.5–4   Scalar product calculation. (a) Correlation detector; (b) matched filter.

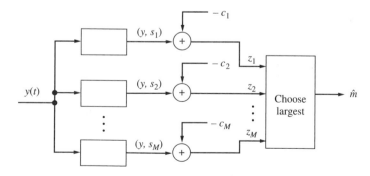

Figure 16.5–5   MAP receiver with bank of correlators or matched filters.

Subtracting the bias terms c_i yields the set of decision functions $\{z_i\}$. The receiver compares the decision functions to find the largest one, say z_j, and chooses the optimum estimate $\hat{m} = m_j$.

Various simplifications of this structure are possible, depending upon the specific set of signals. When the symbols are equiprobable and all signals have the same energy, the bias terms are independent of i and may be omitted so that $z_i = (y, s_i)$. When the signals are such that the dimensionality of Γ_K is $K < M$, we can use Eq. (8a), Sect. 16.4, to write

$$(y, s_i) = \sum_{k=1}^{K} \alpha_{ik}(y, \phi_k) \qquad [8]$$

and the scalar products can be calculated from stored copies of the K basis functions $\phi_k(t)$ rather than the M signals. Only for this implementation do we need to know the basis functions explicitly. Modifications such as these account for the simplified receivers shown in previous chapters.

EXAMPLE 16.5–1   Suppose a set if $M = 8$ signals are generated by combined amplitude and phase modulation (QAM) such that

$$s_1(t) = -s_5(t) = A_c \cos \omega_c t$$

$$s_2(t) = -s_6(t) = \sqrt{2}\, A_c \cos (\omega_c t + \pi/4) = A_c(\cos \omega_c t - \sin \omega_c t)$$

$$s_3(t) = -s_7(t) = A_c \cos (\omega_c t + \pi/2) = -A_c \sin \omega_c t$$

$$s_4(t) = -s_8(t) = \sqrt{2}\, A_c \cos (\omega_c t + 3\pi/4) = -A_c(\cos \omega_c t + \sin \omega_c t)$$

All signals have duration D, and the carrier frequency f_c is an integer multiple of $1/D$.

Observing that each signal can be expressed as a linear combination of $\cos \omega_c t$ and $\sin \omega_c t$, we take the basis functions to be

$$\phi_1(t) = \sqrt{2/D} \cos \omega_c t \qquad \phi_2(t) = -\sqrt{2/D} \sin \omega_c t$$

We thus obtain the two-dimensional signal set in Fig. 16.5–6a, where $E_1 = A_c^2 D/2$ and $E_2 = 2E_1$. Figure 16.5–6b shows the corresponding decision boundaries for maximum-likelihood detection. (These diagrams differ from the class of QAM signals discussed in Sect. 14.4 which have $M = \mu^2$ points in a square array.)

Since $K < M$ and the basis functions are known, Eq. (8) leads to the simplified optimum receiver portrayed in Fig. 16.5–6c. The computational unit uses the inputs (y, ϕ_1) and (y, ϕ_2) to calculate the decision functions

$$z_i = [\alpha_{i1}(y, \phi_1) + \alpha_{i2}(y, \phi_2)] - E_i/2 \qquad i = 1, 2, \ldots, 8$$

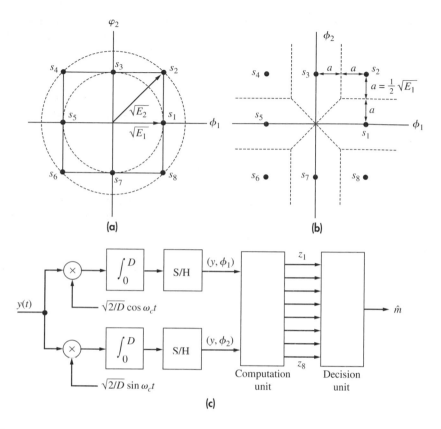

Figure 16.5–6   (a) A two-dimensional signal set; (b) decision regions; (c) optimum receiver.

in accordance with Eqs. (7) and (8). When the source produces a sequence of symbols, the integration and decision process is repeated over each successive symbol interval $kD \leq t \leq (k + 1)D$.

Error Probabilities

But the structure of an optimum receiver is not much good without knowledge of the error probability. So here we formulate general and specific expressions for the average error probability in an M-ary system with equiprobable symbols and optimum detection.

As a starting point, take an arbitrary symbol m_j and let $P(e \,|\, m_j)$ stand for the probability of erroneous detection of the corresponding signal $s_j(t)$. Averaging over the M equiprobable symbols gives

$$P_e = \sum_{j=1}^{M} P(m_j)P(e \mid m_j) = \frac{1}{M}\sum_{j=1}^{M} P(e \mid m_j) \qquad \text{[9a]}$$

Alternatively, we can write

$$P_e = 1 - P_c \qquad P_c = \frac{1}{M}\sum_{j=1}^{M} P(c \mid m_j) \qquad \text{[9b]}$$

where $P(c \mid m_j)$ stands for the probability of *correct* detection. We'll use geometric vector arguments to determine $P(e \mid m_j)$ or $P(c \mid m_j)$.

Figure 16.5–7a represents the situation in Γ_K when $s_j(t)$ has been corrupted by noise $n_w(t)$, resulting in the vector $v = s_j + n$. Also shown is another signal vector s_i whose tip-to-tip "distance" from s_j is $d_{ij} \triangleq \|s_i - s_j\|$. A detection error occurs if the projection of n along the direction of $s_i - s_j$ exceeds $d_{ij}/2$, so v falls on the wrong side of the decision boundary halfway between s_i and s_j. This error event will be denoted by e_{ij}. To find the probability $P(e_{ij})$, let the vectors be translated and/or rotated until $s_i - s_j$ becomes colinear with an arbitrary basis vector ϕ_k as shown in Fig. 16.5–7b. Any translation or rotation of signal points and decision boundaries does not effect error-probability calculations because $p_n(n)$ has spherical symmetry in Γ_K. But now we more readily see that $P(e_{ij})$ equals the probability that $\beta_k > d_{ij}/2$. Hence,

$$P(e_{ij}) = \int_{d_{ij}/2}^{\infty} p_\beta(\beta_k)\, d\beta_k = Q\!\left(\frac{d_{ij}/2}{\sqrt{N_0/2}}\right) = Q\!\left(\frac{d_{ij}}{\sqrt{2N_0}}\right) = Q\!\left(\frac{\|s_i - s_j\|}{\sqrt{2N_0}}\right) \qquad \text{[10]}$$

which follows from Eq. (4a).

Equation (10) immediately applies to the *binary* case since there are only two signal vectors, which we label s_0 and s_1 to be consistent with previous notation. Clearly, $P(e \mid m_0) = P(e_{10})$ and $P(e \mid m_1) = P(e_{01}) = P(e_{10})$, so maximum-likelihood detection of a binary signal yields the average error probability

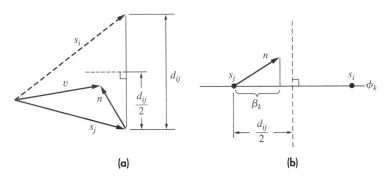

(a)   (b)

Figure 16.5–7   (a) Signal-plus-noise vector; (b) noise vector projection obtained by translation and/or rotation of signal points.

$$P_e = Q\left(\frac{\|s_1 - s_0\|}{\sqrt{2N_0}}\right) \qquad [11]$$

A more familiar expression emerges when we note that $\|s_1 - s_0\|^2 = E_1 - 2(s_1, s_0) + E_0 = 2[E_b - (s_1, s_0)]$, where $E_b = (E_1 + E_0)/2$ is the average energy per bit. If the signals are polar, so $s_0(t) = -s_1(t)$, then $(s_1, s_0) = -E_b$ and $\left(\|s_1 - s_0\|/\sqrt{2N_0}\right)^2 = 4E_b/2N_0 = 2\gamma_b$. Equation (11) is therefore identical to the result we previously obtained for polar baseband transmission with matched filtering.

In the general *M*-ary case, Eq. (10) accounts for just one of the boundaries between s_j and the other $M - 1$ signal points. A complete expression for $P(e \mid m_j)$ or $P(c \mid m_j)$ necessarily involves the specific geometry of the signal set, and it may or may not be easy to obtain. If the signal points form a *rectangular* array, then translation and rotation yields rectangular decision regions like Fig. 16.5–8 where we see that correct detection of m_j requires $-a_1 < \beta_1 < b_1$ and $-a_2 < \beta_2 < b_2$. Since these orthogonal noise coordinates are statistically independent, we have

$$P(c \mid m_j) = \int_{-a_1}^{b_1} p_\beta(\beta_1)\, d\beta_1 \int_{-a_2}^{b_2} p_\beta(\beta_2)\, d\beta_2 \qquad [12a]$$

Each integral in Eq. [12a] can be expanded in the form

$$\int_{-a}^{b} p_\beta(\beta)\, d\beta = 1 - Q\left(\frac{a}{\sqrt{N_0/2}}\right) - Q\left(\frac{b}{\sqrt{N_0/2}}\right) \qquad [12b]$$

obtained from the graphical interpretation of the Q function for a gaussian PDF with zero mean and variance $N_0/2$.

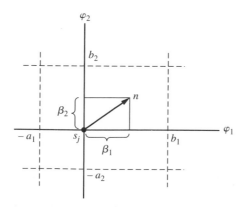

Figure 16.5–8   Noise vector coordinates relative to rectangular decision regions.

When the geometry of the signal set makes exact analysis difficult, we may settle instead for an upper bound on the average error probability. To develop an upper bound, first consider the case of $M = 3$; $P(e \mid m_1)$ then equals the probability of the union of the error events e_{21} and e_{31}, so we write $P(e \mid m_1) = P(e_{21} + e_{31}) = P(e_{21}) + P(e_{31}) - P(e_{21}, e_{31}) \leq P(e_{21}) + P(e_{31})$. Extrapolating to arbitrary M and m_j gives the **union bound**

$$P(e \mid m_j) \leq \sum_{i=1}^{M} P(e_{ij}) \qquad i \neq j$$

with $P(e_{ij})$ as in Eq. (10). There are $M - 1$ terms in this sum, and a simpler but looser bound is $P(e \mid m_j) \leq (M - 1)Q(d_j / \sqrt{2N_0})$, where d_j stands for the "distance" between s_j and its nearest neighbor, i.e.,

$$d_j \stackrel{\triangle}{=} \min_{i \neq j} \| s_i - s_j \| \tag{13a}$$

Then, using Eq. (9a), we have

$$P_e \leq \frac{M - 1}{M} \sum_{j=1}^{M} Q\left(\frac{d_j}{\sqrt{2N_0}}\right) \tag{13b}$$

which is our final result.

EXAMPLE 16.5–2

Consider the signal set back in Fig. 16.5–6b. The decision regions have a rather complicated pattern, but the nearest-neighbor distance is $d_j = 2a = \sqrt{E_1}$ for all eight points. We also observe that $\| s_i \|^2 = E_1 = (2a)^2$ for even i, while $\| s_i \|^2 = E_2 = 2E_1$ for odd i. The average energy per symbol is then

$$E = \frac{1}{8}[4 \times (2a)^2 + 4 \times 2(2a)^2] = 6a^2$$

Hence, from Eq. (13),

$$P_e \leq \frac{7}{8} 8 Q\left(\frac{2a}{\sqrt{2N_0}}\right) = 7Q\left(\sqrt{\frac{E}{3N_0}}\right)$$

which is an upper bound on the error probability in terms of the average signal energy.

Closer examination of Fig. 16.5–6b reveals that the corner points have rectangular boundaries, equivalent to Fig. 16.5–8 with $a_1 = a_2 = a$ and $b_1 = b_2 = \infty$. Therefore, for even values of j, Eq. (12) gives the *exact* result

$$P(c \mid m_j) = (1 - q)^2 \qquad q = Q(a/\sqrt{N_0/2}) = Q(\sqrt{E/3N_0})$$

Furthermore, for odd values of j, we can omit the triangular area and write

$$P(c \mid m_j) > \int_{-a}^{\infty} p_\beta(\beta_1) d\beta_1 \int_{-a}^{a} p_\beta(\beta_2) \, d\beta_2 = (1 - q)(1 - 2q)$$

Thus, substitution in Eq. (9b) yields

$$P_c > \frac{1}{8} \left[4(1 - q)^2 + 4(1 - q)(1 - 2q) \right] \qquad P_e = 1 - P_c < \frac{5}{2} q - \frac{3}{2} q^2$$

which is a more accurate result than the union bound.

Let a set of $M = 6$ signals be defined by $s_1 = -s_2 = a\phi_1$, $s_3 = -s_4 = a\phi_1 + 2a\phi_2$, and $s_5 = -s_6 = a\phi_1 - 2a\phi_2$. Construct the decision boundaries and use Eq. (12) to show that $P_e = (7q - 4q^2)/3$, where $q = Q(\sqrt{6E/11N_0})$ with E being the average signal energy.

EXERCISE 16.5–1

Signal Selection and Orthogonal Signaling ★

Having learned how to implement optimum detection given a set of signals, we come at last to the important design task of *signal selection* for digital communication. We continue to assume equiprobable symbols and gaussian noise, but we now add a constraint on the average signal energy E. In this context, we say that:

> An *optimum signal set* achieves the lowest error probability for a specified value of E or, equivalently, achieves a specified error probability with the lowest value of E.

Our vector interpretation suggests that the corresponding signal points should be arranged spherically around the origin to minimize vector length and signal energy, with the largest possible spacing between points to maximize separation and minimize error probability. These optimal properties define the so called **simplex set** in a subspace with $K = M - 1$ dimensions.

The simplex signal points form the vertices of a K-dimensional pyramid whose center of gravity coincides with the origin. The pyramid reduces to an equilateral triangle when $M = 3$, and to antipodal points $(s_2 = -s_1)$ when $M = 2$. When M is large, there's little difference between the optimal simplex set and a set of **equal-energy orthogonal signals.** Since orthogonal signals are easier to generate and analyze than simplex signals, we'll focus attention on the nearly optimum case of digital communication via M-ary orthogonal signals.

Specifically, let an M-dimensional subspace be spanned by a set of orthonormal basis functions, and let

$$s_i(t) = \sqrt{E}\, \phi_i(t) \qquad i = 1, 2, \ldots, M \qquad \text{[14a]}$$

so that

$$(s_i, s_j) = \begin{cases} E & j = i \\ 0 & j \neq i \end{cases} \qquad \text{[14b]}$$

$$d_{ij}^2 = \|s_i - s_j\|^2 = 2E \qquad j \neq i \qquad \text{[14c]}$$

These relations define a set of mutually orthogonal signals with average signal energy E, as illustrated in Fig. 16.5–9 for $M = 3$. We'll also impose the timelimited condition that $s_i(t) = 0$ for $t < 0$ and $t > D$. The values of D and M are related by

$$r_b = \frac{\log_2 M}{D}$$

where r_b is the equivalent bit rate of the information source.

There are many possible sets of timelimited orthogonal signals. At baseband, for instance, the signals could be nonoverlapping pulses of duration $\tau \leq D/M$—i.e., digital pulse-position modulation (PPM). Or, for bandpass transmission, orthogonal signaling may take the form of frequency-shift keying (FSK) with frequency spacing $1/2D$. Regardless of the particular implementation, M-ary orthogonal signaling is a *wideband* method requiring transmission bandwidth

$$B_T \geq \frac{M}{2D} = \frac{Mr_b}{2 \log_2 M} \geq r_b \qquad \text{[15]}$$

We'll see that this method trades increased bandwidth for decreased error probability, similar to wideband noise-reduction in analog communication.

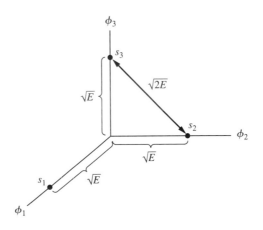

Figure 16.5–9   Orthogonal signal set with $M = 3$.

First note from symmetry that $P(c \mid m_j)$ is the same for all M signals, so

$$P_c = P(c|m_j) = P(z_j > z_i, \text{ all } i \neq j)$$

where z_i is the decision function in Eq. (7). Since the bias term c_i is independent of i, we can drop it and write $z_i = (y, s_i) = (v, s_i)$. When $m = m_j$, $v = s_j + n = \sqrt{E}\varphi_j + n$ and

$$z_i = \left(\sqrt{E}\,\phi_j + n, \sqrt{E}\,\phi_i\right) = E(\phi_j, \phi_i) + \sqrt{E}(n, \varphi_i) = \begin{cases} E + \sqrt{E}\beta_j & i = j \\ \sqrt{E}\beta_i & i \neq j \end{cases}$$

in which we've introduced the noise coordinates $\beta_i = (n, \phi_i)$.

Next, consider some particular value of β_j so, for any $i \neq j$, the probability of the event $z_j > z_i$ given β_j is

$$P(z_j > z_i|\beta_j) = P(E + \sqrt{E}\beta_j > \sqrt{E}\beta_i) = P(\beta_i < \sqrt{E} + \beta_j)$$

$$= \int_{-\infty}^{\sqrt{E}+\beta_j} p_\beta(\beta_i)\, d\beta_i$$

Then, since the noise coordinates are independent RVs and there are $M - 1$ coordinates with $i \neq j$, the probability of correct detection given β_j is

$$P(c \mid \beta_j) = \left[\int_{-\infty}^{\sqrt{E}+\beta_j} p_\beta(\beta_i)\, d\beta_i \right]^{M-1}$$

Averaging $P(c \mid \beta_j)$ over all possible values of β_j finally yields

$$P_c = \int_{-\infty}^{\infty} P(c \mid \beta_j) p_\beta(\beta_j)\, d\beta_j = \pi^{-M/2} \int_{-\infty}^{\infty} \left[\int_{-\infty}^{\sqrt{E/N_0}+\lambda} e^{-\mu^2}\, d\mu \right]^{M-1} e^{-\lambda^2}\, d\lambda \quad \textbf{[16]}$$

where $\mu = \beta_i/\sqrt{N_0}$ and $\lambda = \beta_j/\sqrt{N_0}$.

The formidable expression in Eq. (16) has been evaluated numerically by Viterbi (1966), and plots of $P_e = 1 - P_c$ are shown in Fig. 16.5–10 for selected values of M. These curves are plotted versus

$$\frac{E}{N_0 \log_2 M} = \frac{SD}{N_0 \log_2 M} = \frac{S}{N_0 r_b} \qquad \textbf{[17]}$$

where $S = E/D$ is the average signal power. We see that when $S/N_0 r_b$ has a fixed value, the error probability can be made as small as desired by increasing M. In fact, Viterbi proves analytically that

$$\lim_{M \to \infty} P_e = \begin{cases} 0 & S/N_0 r_b > \ln 2 \\ 1 & S/N_0 r_b < \ln 2 \end{cases}$$

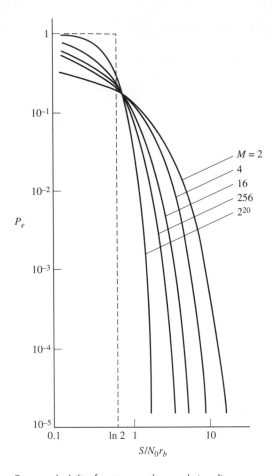

Figure 16.5–10   Error probability for *M*-ary orthogonal signaling.

represented by the dashed line in Fig. 16.5–10. Hence, if $M \to \infty$, then orthogonal signaling with optimum detection approaches *errorless* transmission at any bit rate

$$r_b < \frac{S}{N_0 \ln 2} = C_\infty$$

where C_∞ is the maximum capacity of an AWGN channel as predicted from information theory.

But keep in mind that increasing *M* means increasing both the bandwidth and the receiver complexity, which is proportional to *M*. Furthermore, Fig. 16.5–10 reveals that orthogonal signaling with fixed $M \gg 1$ has a *threshold effect* in the sense that P_e rises abruptly if $S/N_0 r_b$ should happen to decrease somewhat. These observations point to the important conclusion that reliable, efficient, and practical digital communication usually involves some form of *error-control coding* as well as careful signal selection and optimum receiver design.

16.6   PROBLEMS

16.1–1* Suppose that equal numbers of the letter grades A, B, C, D, and F are given in a certain course. How much information in bits have you received when the instructor tells you that your grade is not F? How much more information do you need to determine your grade?

16.1–2 A card is drawn at random from an ordinary deck of 52 playing cards.   (a) Find the information in bits that you receive when you're told that the card is a heart; a face card; a heart face card.   (b) If you're told that the card is a red face card, then how much more information do you need to identify the specific card?

16.1–3 Calculate the amount of information needed to open a lock whose combination consists of three integers, each ranging from 00 to 99.

16.1–4* Calculate $H(X)$ for a discrete memoryless source having six symbols with probabilities

$$P_A = 1/2 \quad P_B = 1/4 \quad P_C = 1/8 \quad P_D = P_E = 1/20 \quad P_F = 1/40$$

Then find the amount of information contained in the messages *ABABBA* and *FDDFDF* and compare with the expected amount of information in a six-symbol message.

16.1–5 Do Prob. 16.1–4 with

$$P_A = 0.4 \quad P_B = 0.2 \quad P_C = 0.12 \quad P_D = P_E = 0.1 \quad P_F = 0.08$$

16.1–6 A certain source has eight symbols and emits data in blocks of three symbols at the rate of 1000 blocks per second. The first symbol in each block is always the same, for synchronization purposes; the remaining two places are filled by any of the eight symbols with equal probability. Find the source information rate.

16.1–7 A certain data source has 16 equiprobable symbols, each 1 ms long. The symbols are produce in blocks of 15, separated by 5-ms spaces. Find the source information rate.

16.1–8 Calculate the information rate of a telegraph source having two symbols: dot and dash. The dot duration is 0.2 sec. The dash is twice as long as the dot and half as probable.

16.1–9* Consider a source with $M = 3$. Find $H(X)$ as a function of p when $P_1 = 1/3$ and $P_2 = p$. Also evaluate $H(X)$ when $p = 0$ and $p = 2/3$.

16.1–10 Consider a source with $M > 2$. One symbol has probability $\alpha \ll 1/M$, and the remaining symbols are equally likely. Show that $H(X) \approx \log(M - 1) + \alpha \log(1/\alpha)$.

16.1–11 Obtain the bounds on K in Eq. (13) (p. 706) when N_i satisfies Eq. (15).

16.1–12 Obtain the Shannon-Fano code for the source in Prob. 16.1–4, and calculate the efficiency.

16.1–13 Show that there are two possible Shannon-Fano codes for the source whose symbol probabilities are given in Prob. 16.1–5. Then calculate the efficiency of the better code.

16.1–14   Consider a source with $M = 3$ and symbol probabilities 0.5, 0.4, and 0.1. (*a*) Obtain the Shannon-Fano code and calculate its efficiency.   (*b*)   Repeat part *a* for the second-extension code, grouping the symbols in blocks of two.

16.1–15   A binary source has symbol probabilities 0.8 and 0.2.   (*a*) Group the symbols in blocks of two, obtain the corresponding second-extension Shannon-Fano code, and calculate its efficiency.   (*b*) Repeat part *a* with blocks of three for a third-extension code.

16.1–16* A binary source has $P_0 = P_1 = 1/2$ and first-order memory such that $P(0\,|\,1) = P(1\,|\,0) = 3/4$. Calculate the resulting conditional entropy.

16.2–1   The **joint entropy** of a discrete system is defined as

$$H(X, Y) = \sum_{x, y} P(x_i y_j) \log \frac{1}{P(x_i y_j)}$$

Show that $H(X,Y) = H(Y) + H(X\,|\,Y)$.

16.2–2   Expand Eq. (2) (p. 715) to obtain $I(X; Y) = H(X) + H(Y) - H(X, Y)$, where $H(X, Y)$ is the joint entropy defined in Prob. 16.2–1.

16.2–3   Consider a noiseless system with μ source symbols, μ destination symbols, and forward transition probabilities

$$P(y_j \mid x_i) = \begin{cases} 1 & j = i \\ 0 & j \neq i \end{cases}$$

Show that $H(Y) = H(X)$, $H(Y\,|\,X) = 0$, and therefore $I(X; Y) = H(X)$. *Hint:* Write the sums with indices $0 \leq i \leq \mu$ and $0 \leq j \leq \mu$.

16.2–4   Figure P16.2–4 represents a *nonsymmetric* binary channel. Follow the method used in Example 16.2–1 (p. 716) to obtain

$$I(X; Y) = \Omega[\beta + (1 - \alpha - \beta)p] - p\Omega(\alpha) - (1 - p)\Omega(\beta)$$

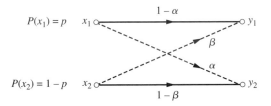

Figure P16.2–4

16.2–5   The channel in Prob. 16.2–4 is said to be **useless** if $\beta = 1 - \alpha$. Justify this name from intuitive and analytical arguments.

16.2–6   When a and b are constants, show that

$$\frac{d}{dp}\Omega(a + bp) = b \log \frac{1 - a - bp}{a + bp}$$

Then apply this relation to confirm that $I(X; Y)$ in Eq. (8) (p. 717) is maximum when $p = 1/2$.

16.2–7   The channel in Prob. 16.2–4 is called a **Z channel** when $\beta = 0$. Use the relation in Prob. 16.2–6 to show that if $\alpha = 1/2$, then $I(X; Y)$ is maximum when $p = 2/5$. Then evaluate C_s.

16.2–8*   Let the channel in Prob. 16.2–4 have $\alpha = 1/4$ and $\beta = 0$. Use the relation in Prob. 16.2–6 to find C_s.

16.2–9   The **binary erasure channel** has two source symbols, 0 and 1, and three destination symbols, 0, 1, and E, where E denotes a detected but uncorrectable error. The forward transition possibilities are

$$P(0\,|\,0) = 1 - \alpha \quad P(E\,|\,0) = \alpha \quad P(1\,|\,0) = 0$$

$$P(0\,|\,1) = 0 \quad\quad P(E\,|\,1) = \alpha \quad P(1\,|\,1) = 1 - \alpha$$

It follows from symmetry that $I(X; Y)$ is maximum when the source symbols are equiprobable. Find C_s in terms of α.

16.2–10   Derive the bound on $P(n \geq d)$ given in Eq. (14b) (p. 721).

16.3–1*   Find $H(X)$ in terms of $S = \overline{x^2}$ and confirm that $H(X) < \frac{1}{2} \log 2\pi eS$ when $p(x)$ is a uniform PDF over $|x| \leq a$.

16.3–2   Find $H(X)$ in terms of $S = \overline{x^2}$ and confirm that $H(X) < \frac{1}{2} \log 2\pi eS$ when $p(x)$ is a Laplace PDF in Example 8.3–1.

16.3–3   Find $H(X)$ in terms of $S = \overline{x^2}$ and confirm that $H(X) < \frac{1}{2} \log 2\pi eS$ when $p(x) = ae^{-ax}u(x)$.

16.3–4   Express $H(Z)$ in terms of $H(X)$ when $Z = aX + b$.

16.3–5*   Find $p(x)$ that maximizes $H(X)$ and determine the resulting $H(X)$, given the signal constraints $x(t) \geq 0$ and $\overline{x(t)} = m$.

16.3–6   Find $p(x)$ that maximizes $H(X)$ and determine the resulting $H(X)$, given the signal constraints $x(t) \geq 0$ and $\overline{x^2(t)} = S$.

16.3–7   Write $p_x(x\,|\,y)/p_x(x) = [p_x(x)p_y(y)/p_{xy}(x, y)]^{-1}$ and use $\ln v \leq v - 1$ to show that $I(X; Y) \geq 0$.

16.3–8   Consider an AWGN channel with $S/N_0 = 10^4$. Find the maximum rate for reliable information transmission when $B = 1$ kHz, 10 kHz, and 100 kHz.

16.3–9   Consider an AWGN channel with $B = 3$ kHz. Find the minimum value of S/N in dB for reliable information transmission at $R = 2400$, 4800, and 9600 bits/sec.

16.3–10* Consider an AWGN channel with $B = 1$ kHz and $N_0 = 1\ \mu$W/Hz. Find the minimum value of S in milliwatts for reliable information transmission at $R = 100$, 1000, and 10,000 bits/sec.

16.3–11 An engineer claims to have designed an analog communication system that has $(S/N)_D = 60$ dB when $S_R/N_0B_T = 5$ dB and $B_T = 10W$. Do you believe this claim?

16.3–12 An ideal system has $(S/N)_D = 40$ dB when $B_T = 4W$. Find $(S/N)_D$ when B_T is tripled while all other parameters are held fixed.

16.3–13 Do Prob. 16.3–12 with B_T reduced to $W/2$ instead of being tripled.

16.3–14* A communication system has $B_T = 12$ kHz, $L = 60$ dB, and $N_0 = $ pW/Hz. Find the minimum value of S_T needed to get $(S/N)_D = 30$ dB when $W = 3$ kHz.

16.3–15 Do Prob. 16.3–14 with $W = 6$ kHz.

16.4–1 Consider the signals $v(t)$, $w(t)$, and $z(t) = v(t) + w(t)$ with energy $E_v = E_w = 4$ and $E_z = 8$. Construct a vector diagram based on this information, and use it to find E_x when $x(t) = 3v(t) - 0.5w(t)$.

16.4–2 Do Prob. 16.4–1 with $E_v = 4$, $E_w = 12$, and $E_z = 16$.

16.4–3 Prove the triangle inequality in Eq. (2) (p. 736) from Eqs. (5) and (6).

16.4–4 Derive Eq. (7) (p. 737) and show that $\|v_w\| \leq \|v\|$.

16.4–5* Apply the Gram-Schmidt procedure to obtain an orthonormal basis for the signals

$$s_1(t) = 1 \quad s_2(t) = t \quad s_3(t) = t^2$$

defined over $-1 \leq t \leq 1$. Then write the signal vectors in terms of the basis functions.

16.4–6 Apply the Gram-Schmidt procedure to obtain an orthonormal basis for the signals

$$s_1(t) = 1 \quad s_2(t) = \cos \pi t/2 \quad s_3(t) = \sin \pi t$$

defined over $-1 \leq t \leq 1$. Then write the signal vectors in terms of the basis functions.

16.5–1* Consider the following three sets of binary signals defined in terms of orthonormal basis functions ϕ_1 and ϕ_2:

(a) $s_1 = a\phi_1 \quad s_2 = a\phi_2$

(b) $s_1 = a\phi_1 \quad s_2 = -a\phi_2$

(c) $s_1 = a\phi_1 \quad s_2 = 0$

Construct the maximum-likelihood decision regions and write the decision functions for each signal set. Then simplify and diagram optimum receivers using correlation detection.

16.5–2   Consider the following two sets of quaternary signals $(M = 4)$ defined in terms of orthonormal basis functions ϕ_1 and ϕ_2:

(a) $s_1 = a\phi_1$ $s_2 = a\phi_2$ $s_3 = -a\phi_1$ $s_4 = -a\phi_2$

(b) $s_1 = a\phi_1$ $s_2 = a\phi_2$ $s_3 = a(\phi_1 + \phi_2)$ $s_4 = 0$

Construct the maximum-likelihood decision regions and write the decision functions for each signal set. Then simplify and diagram optimum receivers using correlation detection.

16.5–3   Consider the following set of quaternary signals $(M = 4)$ defined in terms of orthonormal basis functions ϕ_1 and ϕ_2:

$$s_1 = -s_3 = a(\phi_1 + \phi_2)   s_2 = -s_4 = a(\phi_1 - \phi_2)$$

Construct the maximum-likelihood decision regions and find the error probability with AWGN, expressing your result in terms of the average signal energy E.

16.5–4   Suppose the signal $s_5 = 0$ is added to the set in Prob. 16.5–3. Construct the maximum-likelihood decision regions and obtain an upper bound on the error probability with AWGN, expressing your result in terms of the average signal energy E.

16.5–5*   Suppose the signal $s_9 = 0$ is added to the set in Fig. 16.5–6a (p. 747). Construct the maximum-likelihood decision regions and find the error probability with AWGN, expressing your result in terms of the average signal energy E. Compare with the upper bound obtained from Eq. (13) (p. 750).

Appendix

Circuit and System Noise

OUTLINE

Transmission loss usually results in a very feeble signal at the input of a communication receiver. Consequently, to obtain an adequate signal level for further processing, the "front end" of a typical receiver includes several stages of amplification. But high-gain amplifiers amplify any noise that accompanies the received signal, and they also add their own internally generated noise. An accurate assessment of system performance must therefore take into account amplifier noise.

This appendix starts with circuit and device noise, as preparation for the description of amplifiers and other noisy two-port networks. Then we analyze noisy two-ports connected in cascade, a configuration that pertains to repeater systems as well as to receivers. Our analysis brings out the critical factors in receiver noise and the design principles for low-noise systems.

OBJECTIVES

After studying this appendix and working the exercises, you should be able to do each of the following:

1. Calculate the mean square noise voltage or current and the available power density for a passive network with internal resistance noise.

2. Use the concepts of available power gain, effective noise temperature, and noise figure to analyze the performance of a noisy two-port network.

3. Calculate the overall noise figure and noise temperature for a cascade system, and use them to find $(S/N)_o$ given $(S/N)_s$.

Circuit and Device Noise

Noise generated within an electrical circuit may come from numerous sources involving several different physical phenomena. Here we'll describe major types of device noise found in communication systems, and we'll develop appropriate circuit models and analysis methods.

The assumption of stationarity is reasonable for most electrical noise processes, thereby allowing us to represent noise sources in the frequency domain. For convenience, we'll adopt the common practice of working entirely with *positive frequency* and *one-sided frequency functions*—as distinguished from the two-sided functions used in Sect. 9.3. To clarify this distinction, let $G(f)$ be the two-sided available power spectrum of some noise source. Since $G(f)$ has even symmetry, the corresponding **one-sided available power density** will be defined by

$$\eta(f) \triangleq 2G(f) \quad f \geq 0 \tag{1a}$$

and the *total available noise power* is

$$N = \int_0^\infty \eta(f)\, df \tag{1b}$$

Likewise, the one-sided mean square voltage or current densities equal twice the two-sided frequency functions for $f \geq 0$.

Figure A–1a shows the frequency-domain Thévenin circuit model of a stationary but otherwise arbitrary noise source with noiseless internal impedance $Z(f)$.

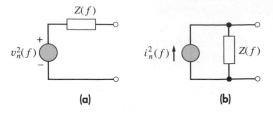

Figure A–1   Frequency-domain models of a noise source. (a) Thévenin circuit; (b) Norton circuit.

The function $v_n^2(f)$ represents the **open-circuit mean square voltage density,** defined such that

$$\eta(f) = \frac{v_n^2(f)}{4 \operatorname{Re}\,[Z(f)]} \qquad [2]$$

Alternatively, an ideal voltmeter with bandwidth df centered at some frequency f would measure the rms value $\sqrt{v_n^2(f)}\, df$ across the open terminals of this source. Converting Fig. A–1a to a Norton equivalent circuit gives the circuit model in Fig. A–1b, with

$$i_n^2(f) = \frac{v_n^2(f)}{|Z(f)|^2} \qquad [3]$$

which represents the **short-circuit mean square current density.** Equation (3) also expresses Ohm's law in the form needed for circuit noise analysis.

Suppose the source in question happens to be a *thermal resistance R* at temperature $\mathcal{T}$, so $G(f) = k\mathcal{T}/2$ from Eq. (5), Sect. 9.3. Then

$$\eta(f) = k\mathcal{T} \qquad [4a]$$

and setting $Z(f) = R$ yields

$$v_n^2(f) = 4\,Rk\mathcal{T} \quad i_n^2(f) = 4k\mathcal{T}/R \qquad [4b]$$

These constant densities correspond to *white* noise, at least up to infrared frequencies. For computational purposes, the quantity $k\mathcal{T}$ can be rewritten as

$$k\mathcal{T} = k\mathcal{T}_0(\mathcal{T}/\mathcal{T}_0) \approx 4 \times 10^{-21}(\mathcal{T}/\mathcal{T}_0)$$

where we've inserted the standard room temperature $\mathcal{T}_0 = 290$ K.

Resistance noise occurs in almost all circuits, but reactive elements may alter the frequency density. In particular, let Fig. A–2a be a one-port (two-terminal) network containing only resistance, capacitance, and inductance, and having the equivalent impedance $Z(f) = R(f) + jX(f)$. When the resistances are in thermal equilibrium at temperature $\mathcal{T}$, **Nyquist's formula** states that

$$v_n^2(f) = 4\,R(f)k\mathcal{T} \qquad [5]$$

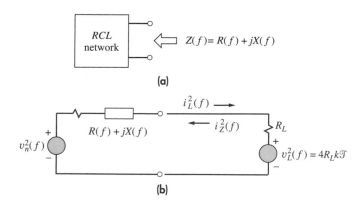

Figure A–2

Hence, the mean square voltage density takes the shape of the equivalent resistance $R(f)$. Equation (5) includes the special case of an all-resistive network whose equivalent resistance will be independent of f.

We prove Nyquist's formula by considering the complete circuit in Fig. A–2b, where the source has been connected to a load resistance R_L at the same temperature $\mathcal{T}$. This load resistance, of course, generates thermal noise with $v_L^2(f) = 4 R_L k \mathcal{T}$. Since the two noise sources are physically independent, superposition applies and we can calculate the mean square current densities in each direction, namely,

$$i_L^2(f) = \frac{v_n^2(f)}{|Z(f) + R_L|^2} \qquad i_Z^2(f) = \frac{4 R_L k \mathcal{T}}{|Z(f) + R_L|^2}$$

The average power delivered from the source to the load in any frequency band df is $R_L i_L^2(f)\, df$, while the load delivers $R(f) i_Z^2(f)\, df$ back to the source. But the net power transfer must be zero at every frequency for the circuit to be in thermal equilibrium. Therefore,

$$\frac{R_L v_n^2(f)\, df}{|Z(f) + R_L|^2} = \frac{R(f) 4 R_L k \mathcal{T}\, df}{|Z(f) + R_L|^2}$$

and Eq. (5) follows after cancellation.

Nyquist's formula does not hold when the resistances are at different temperatures or the network contains nonthermal sources. However, such cases are easily analyzed provided that the sources are independent. You simply use superposition and sum the mean square values to find the resulting frequency density.

The most common type of nonthermal noise in most electrical circuits is **shot noise.** This phenomenon occurs whenever charged particles cross a potential barrier—as in semiconductor junctions or vacuum tubes. Small variations of kinetic energy among the individual particles cause random fluctuations of the total current, roughly analogous to the sound of a stream of buckshot falling on a drum. Schottky first

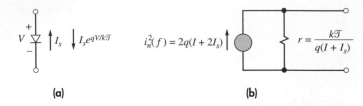

Figure A–3   (a) Semiconductor junction diode; (b) noise model.

studied this effect in a *vacuum-tube diode* operated under *temperature-limited* conditions. He found that shot noise could be represented by a current source with

$$i_n^2(f) = 2qI \tag{6}$$

in which q is the electronic charge (1.6×10^{-19} coulombs) and I is the dc current. Since Eq. (6) is independent of both frequency and temperature, shot noise is *nonthermal white noise*. Subsequent investigations have shown that $i_n^2(f)$ actually decreases at frequencies above $f \approx 1/\tau$, where τ denotes the average particle transit time.

Schottky's result also holds for the **semiconductor junction diode** in Fig. A–3a. The net dc current consists of two components, given by the diode equation

$$I = I_s e^{qV/kT} - I_s \tag{7a}$$

where V is the junction voltage and I_s is the reverse saturation current. The two current components produce statistically independent shot noise, so the total mean square noise current density becomes

$$i_n^2(f) = 2qI_s e^{qV/kT} + 2qI_s = 2q(I + 2I_s) \tag{7b}$$

Figure A–3b shows the complete noise source model, including the diode's *dynamic resistance*

$$r = \frac{1}{dI/dV} = \frac{kT}{q(I + I_s)} \tag{7c}$$

Unlike ohmic resistance, dynamic resistance is *noiseless* since it does not correspond to any power dissipation.

Now consider the **junction field-effect transistor** (JFET) in Fig. A–4a, which has a reverse-biased junction between the gate terminal G and the semiconductor channel from the source terminal S to the drain terminal D. Figure A–4b shows a simplified noise equivalent circuit with current sources representing the gate shot noise and the channel terminal noise, given by

$$i_g^2(f) = 2qI_g \quad i_d^2(f) = 4kT \times \tfrac{2}{3}g_m$$

where g_m is the transconductance. When external circuitry connects to the gate and source terminals, the shot noise is amplified via the controlled current generator $g_m V_{gs}$ and adds to the output noise at the drain terminal.

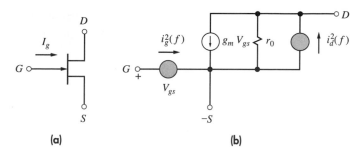

Figure A–4   (a) Junction field-effect transistor; (b) noise model.

An FET with an insulated gate structure avoids junction shot noise. But a **bipolar junction transistor** (BJT) has *two* semiconductor junctions and two sources of shot noise in addition to thermal noise from internal ohmic resistance. Besides thermal and shot noise, all transistors generate certain types of *nonwhite* noise that may or may not be significant, depending upon the particular device and the application.

Transistors, vacuum tubes, and other devices exhibit a low-frequency phenomenon known as **flicker noise**—often called "one-over-f" noise because the mean square density is proportional to $1/f^\nu$ with $\nu \approx 1$. Some semiconductor devices also produce **burst** or **"popcorn" noise,** whose waveform resembles the random telegraph wave in Fig. 9.2–3a, (p. 366). Hence, the mean square density eventually falls off as $1/f^2$. Figure A–5 illustrates the frequency variation that results when $i_n^2(f)$

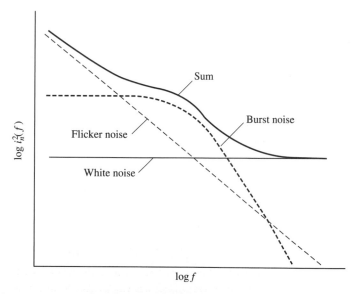

Figure A–5   Frequency variation of semiconductor noise with flicker and burst noise.

consists of flicker, burst, and white noise. Flicker and burst noise pose serious problems for low-frequency applications, but they usually can be ignored at frequencies above a few kilohertz. At much higher frequencies, capacitive coupling and various other effects tend to increase noise in electronic devices.

EXAMPLE A–1

Vacuum diode noise generator

Noise generators are used in the lab to study noise characteristics, and Fig. A–6a gives the schematic diagram of a noise generator employing a temperature-limited vacuum diode. The choke coil and blocking capacitor serve to separate the dc and and shot-noise currents, as indicated.

The corresponding noise circuit model in Fig. A–6b includes the diode's noiseless dynamic resistance r and the noisy thermal resistance R, presumed to be at room temperature $\mathcal{T}_0$. The equivalent impedance is the parallel combination

$$Z(f) = \frac{R(r + 1/j\omega C)}{R + (r + 1/j\omega C)}$$

but the nonthermal shot noise precludes the use of Nyquist's formula. Instead, we convert the shot-noise current and dynamic resistance into Thévenin form shown in

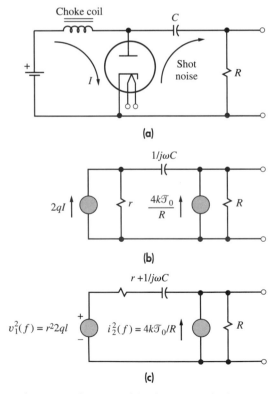

Figure A–6   Circuit diagram and noise models of a vacuum diode noise generator.

Fig. A–6c. Now, applying superposition, the total short-circuit mean square current density is clearly seen to be

$$i_n^2(f) = \frac{r^2 2qI}{|r + 1/j\omega C|^2} + \frac{4k\mathcal{T}_0}{R}$$

and therefore

$$v_n^2(f) = |Z(f)|^2 i_n^2(f) = \frac{R^2(r^2 2qI)}{|R + r + 1/j\omega C|^2} + \frac{R^2 |r + 1/j\omega C|^2}{|R + r + 1/j\omega C|^2} \frac{4k\mathcal{T}_0}{R}$$

Note that the first term is the open-circuit mean square voltage density produced by $v_1^2(f)$ when $i_2^2(f) = 0$, while the second term is produced by $i_2^2(f)$ when $v_1^2(f) = 0$.

In normal operation of the noise generator, we're not concerned with the noise at low frequencies (which would probably be dominated by flicker noise), and the dynamic resistance r is much greater than R. Under these conditions, our results simplify to

$$Z(f) \approx R \quad v_n^2(f) \approx 2R^2 qI + 4\,Rk\mathcal{T}_0$$

over the frequency range

$$1/2\pi rC \ll f < 1/\tau$$

Tubes designed for this application have very short transit times, typically $\tau \le 1$ ns, so the shot noise remains constant up to about 1 GHz. Finally, from Eq. (2), the available noise power density is

$$\eta(f) \approx \tfrac{1}{2} qIR + k\mathcal{T}_0 = k(\mathcal{T}_x + \mathcal{T}_0)$$

where

$$\mathcal{T}_x = qIR/2k$$

which we interpret as the equivalent *noise temperature* of the shot noise. The voltage applied to the filament of the tube controls I and thus controls $\mathcal{T}_x$, whose value can be determined from an ammeter reading.

Consider the network in Fig. A–7, where R_1 and R_2 are thermal resistances at different temperatures. Let $R_2 = R_1 = R$ and $L = R/2\pi$. Obtain expressions for $v_n^2(f)$ and $\eta(f)$. Then simplify your results when $\mathcal{T}_1 = \mathcal{T}_2 = \mathcal{T}$.

EXERCISE A–1

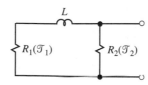

Figure A–7

Amplifier Noise

A detailed circuit model showing all the individual noise sources within an amplifier would be very complicated and of little practical value to a communications engineer. Consequently, alternative methods have been devised for the analysis of noise in amplifiers. Two particularly useful measures of amplifier noise are the *effective noise temperature* and the *noise figure*. Both of these measures involve the concept of *available power gain*, which we introduce first.

Let Fig. A–8 be the circuit model of a *noiseless* amplifier inserted between a source and a load. (For simplicity, we've omitted any reactances that might be associated with the source, amplifier, or load impedances.) The amplifier itself is characterized by an input resistance r_i, output resistance r_o, and voltage transfer function $H(f)$. The source generates a mean square voltage density $v_s^2(f)$—representing noise or an information signal or both—and the available power density from the source is $\eta_s(f) = v_s^2(f)/4\,R_s$. The available power density at the output of the amplifier is

$$\eta_o(f) = \frac{v_o^2(f)}{4r_o} = \frac{|H(f)|^2 v_i^2(f)}{4r_o} = \frac{|H(f)|^2}{4r_o}\left(\frac{r_i}{R_s + r_i}\right)^2 v_s^2(f)$$

We define the amplifier's **available power gain** $g_a(f)$ as the ratio of these available power densities, i.e.,

$$g_a(f) \triangleq \frac{\eta_o(f)}{\eta_s(f)} = \frac{v_o^2(f)R_s}{v_s^2(f)r_o} = \left(\frac{|H(f)|\,r_i}{R_s + r_i}\right)^2 \frac{R_s}{r_o} \qquad [8]$$

The actual power gain of an amplifier equals the available power gain when the impedances are *matched* to obtain maximum power transfer at input and output.

We'll assume hereinafter that the source generates white noise, thermal or nonthermal, with noise temperature $\mathcal{T}_s$. Then $\eta_s(f) = k\mathcal{T}_s$ and the available noise power at the output of a *noiseless* amplifier would be

$$\eta_o(f) = g_a(f)\eta_s(f) = g_a(f)k\mathcal{T}_s$$

But a *noisy* amplifier contributes additional internally generated noise. Since the internal noise is independent of the source noise, we write

$$\eta_o(f) = g_a(f)k\mathcal{T}_s + \eta_{\text{int}}(f) \qquad [9]$$

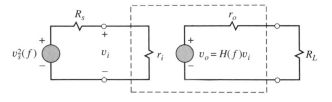

Figure A–8   Circuit model of a noiseless amplifier.

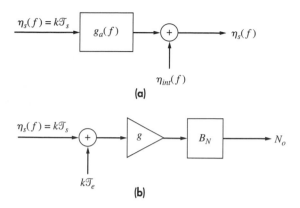

Figure A–9   Block-diagram representation of a noisy amplifier.

where $\eta_{\text{int}}(f)$ stands for the available power density of the internal noise seen at the output. Figure A–9*a* depicts Eq. (9) in the form of a block diagram. Integration then yields the total available output noise power

$$N_o = \int_0^\infty \eta_o(f)\,df = k\mathcal{T}_s \int_0^\infty g_a(f)\,df + \int_0^\infty \eta_{\text{int}}(f)\,df$$

an expression that calls for some simplifications.

Most amplifiers in a communication system have a frequency-selective response, with *maximum power gain g* and *noise equivalent bandwidth B_N*. These parameters are related to $g_a(f)$ by

$$gB_N = \int_0^\infty g_a(f)\,df \tag{10}$$

so the first term of N_o reduces to $k\mathcal{T}_s g B_N$. Next, to simplify the second term, we define the **effective noise temperature** of the amplifier to be

$$\mathcal{T}_e \triangleq \frac{1}{gkB_N} \int_0^\infty \eta_{\text{int}}(f)\,df \tag{11}$$

Hence, the total output noise power becomes

$$N_o = k\mathcal{T}_s g B_N + gkB_N\mathcal{T}_e = gk(\mathcal{T}_s + \mathcal{T}_e)B_N \tag{12}$$

diagrammatically portrayed by Fig. A–9*b*. This diagram brings out the fact that $\mathcal{T}_e$ represents the internal noise *referred to the input* and thereby expedites calculations of signal-to-noise ratios.

Now let Fig. A–10 represent a noisy amplifier with signal plus white noise at the input. The available signal power from the source is S_s, and the signal spectrum

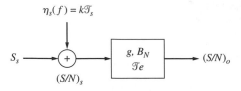

Figure A–10   Noisy amplifier with input signal and noise.

falls within the passband of the amplifier, so the available signal power at the output will be $S_o = gS_s$. Thus, using Eq. (12), the output signal-to-noise ratio is

$$\left(\frac{S}{N}\right)_o = \frac{gS_s}{N_o} = \frac{S_s}{k(\mathcal{T}_s + \mathcal{T}_e)B_N} \qquad \text{[13]}$$

where the gain g has cancelled out in numerator and denominator. Although the source noise does not necessarily have a defined bandwidth, the source signal-to-noise ratio is taken by convention to be

$$\left(\frac{S}{N}\right)_s \triangleq \frac{S_s}{k\mathcal{T}_s B_N}$$

which just corresponds to the signal-to-noise ratio produced by an ideal noiseless filter with unit gain and bandwidth B_N. However, from Eq. (13), the noisy amplifier produces

$$\left(\frac{S}{N}\right)_o = \frac{S_s}{(1 + \mathcal{T}_e/\mathcal{T}_s)k\mathcal{T}_s B_N} = \frac{1}{1 + \mathcal{T}_e/\mathcal{T}_s}\left(\frac{S}{N}\right)_s \qquad \text{[14]}$$

and we see that $(S/N)_o < (S/N)_s$.

We also see that the degradation of signal-to-noise ratio due to a noisy amplifier depends on the value of effective noise temperature *relative* to the source noise temperature. In particular, if $\mathcal{T}_e \ll \mathcal{T}_s$ then $(S/N)_o \approx (S/N)_s$; under this condition the internal noise has little effect, and the amplifier appears to be noiseless. This condition holds primarily at carrier frequencies below about 30 MHz, whereas $\mathcal{T}_e$ becomes significant at higher frequencies. Such considerations often affect the design of receivers and repeaters.

When impedances are not matched at the input or output all signal and noise powers will be less than the available powers. Nonetheless, Eqs. (13) and (14) are still valid because they express power *ratios* measured at specific points, so the impedance mismatch factor cancels out along with the gain. The effective noise temperature is therefore a significant parameter, irrespective of impedance matching.

Another measure of amplifier noise is the **noise figure** F. (This is also called the *noise factor* and symbolized by n_F or NF. Some authors reserve the term "noise figure" for the value of the noise factor expressed in dB.) The noise figure is defined such that

$$\left(\frac{S}{N}\right)_o = \frac{1}{F}\left(\frac{S}{N}\right)_s \quad \text{when} \quad \mathcal{T}_s = \mathcal{T}_0 \qquad [15]$$

Since $S_o = gS_s$ and $(S/N)_s = S_s/k\mathcal{T}_0 B_N$ when $\mathcal{T}_s = \mathcal{T}_0$,

$$F = \frac{N_o}{gk\mathcal{T}_0 B_N} = 1 + \frac{\mathcal{T}_e}{\mathcal{T}_0} \qquad [16]$$

and conversely

$$\mathcal{T}_e = (F - 1)\mathcal{T}_0 \qquad [17]$$

A very noisy amplifier has $\mathcal{T}_e \gg \mathcal{T}_0$ and $F \gg 1$—in which case we usually express the value of F in decibels. Then, if the source noise is at room temperature, Eq. (15) says that $(S/N)_o$ in dB equals $(S/N)_s$ in dB minus F in dB. A low-noise amplifier has $\mathcal{T}_e < \mathcal{T}_0$ and $1 < F < 2$—in which case we usually work with effective noise temperature and calculate $(S/N)_o$ from Eq. (13).

Table A–1 lists typical values of effective noise temperature, noise figure, and maximum power gain for various types of high-frequency amplifiers. Many low-noise amplifiers have cryogenic cooling systems to reduce the physical temperature and thus reduce internal thermal noise. Other amplifiers operate at room temperature, but the internal noise comes from nonthermal sources that may result in $\mathcal{T}_e > \mathcal{T}_0$.

Equation (16) defines the *average* or *integrated* noise figure in the sense that N_o involves the integral of $\eta_o(f)$ over all frequency. But sometimes we need to know

Table A–1   Noise parameters of typical amplifiers

Type	Frequency	$\mathcal{T}_e$, K	F, dB	g, dB
Maser	9 GHz	4	0.06	20–30
Parametric Amplifier				
Room temperature	9 GHz	130	1.61	10–20
Cooled with liquid N_2	6 GHz	50	0.69	
Cooled with liquid He	4 GHz	9	0.13	
FET Amplifier				
GaAs	9 GHz	330	3.3	6
	6 GHz	170	2.0	10
	1 GHz	110	1.4	12
Silicon	400 MHz	420	3.9	13
	100 MHz	226	2.5	26
Integrated Circuit	10.0 MHz	1160	7.0	50
	4.5 MHz	1860	8.7	75

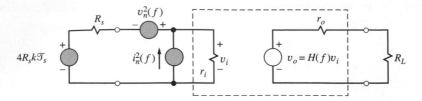

Figure A–11   Two-generator circuit model of amplifier noise.

how the internal noise varies with frequency. The **spot noise figure** $F(f)$ contains this information in the form

$$F(f) \triangleq \frac{\eta_o(f)}{kT_s g_a(f)} \quad \text{when} \quad T_s = T_0 \qquad [18]$$

The value of $F(f)$ at a particular frequency serves as an estimate of the noise figure when a wideband amplifier is used for narrowband amplification with the help of additional filtering.

Finally, to relate system and circuit models of amplifier noise, we should mention the **two-generator model** diagrammed in Fig. A–11. This circuit model represents the internal noise in terms of fictitious voltage and current sources at the input terminals of an equivalent noiseless amplifier. The total mean square voltage density at the open-circuited output is

$$v_o^2(f) = \left(\frac{|H(f)| \, r_i}{R_s + r_i} \right)^2 [4 \, R_s kT_s + v_n^2(f) + R_s^2 i_n^2(f)]$$

provided that $v_n^2(f)$ and $i_n^2(f)$ are uncorrelated. Using Eqs. (8) and (18) with $\eta_o(f) = v_o^2(f)/4r_o$, we find that

$$F(f) = 1 + \frac{v_n^2(f) + R_s^2 i_n^2(f)}{4 \, R_s kT_0} \qquad [19]$$

which shows that the spot noise figure depends in part upon the external source resistance. Hence, optimizing $F(f)$ at a particular frequency often requires a transformer to obtain the optimum source resistance.

EXAMPLE A–2   **Amplifier Noise Measurement**

Measuring *absolute* noise power is a difficult chore, so clever techniques have been developed for amplifier noise measurement with a *relative* power meter connected at the output. One technique utilizes a calibrated source of white noise, such as a diode noise generator, impedance-matched to the input of the amplifier. The procedure goes as follows.

First, set the source noise temperature at $\mathcal{T}_s = \mathcal{T}_0$ and record the output meter reading N_1. From Eq. (12), this value corresponds to

$$N_1 = CN_o = Cgk(\mathcal{T}_0 + \mathcal{T}_e)B_N$$

where the proportionality constant C includes any impedance mismatch factor at the output. Second, increase the source temperature to $\mathcal{T}_s = \mathcal{T}_x + \mathcal{T}_0$ such that the meter reading has *doubled* and

$$N_2 = Cgk[(\mathcal{T}_x + \mathcal{T}_0) + \mathcal{T}_e]B_N = 2N_1$$

Then $N_2/N_1 = (\mathcal{T}_x + \mathcal{T}_0 + \mathcal{T}_e)/(\mathcal{T}_0 + \mathcal{T}_e) = 2$, so

$$\mathcal{T}_e = \mathcal{T}_x - \mathcal{T}_0 \qquad F = \mathcal{T}_x/\mathcal{T}_0$$

Note that we don't need to know g, B_N, or the constant C.

An amplifier with $g = 60$ dB and $B_N = 2$ MHz has $N_o = 40$ nW when the source noise is at room temperature. (a) Find the effective noise temperature and noise figure. (b) Calculate the source temperature needed for the second step of the measurement procedure in Example A–2.

EXERCISE A–2

System Noise Calculations

Here we take up the analysis of cascade-connected systems that include amplifiers and other noisy two-port networks. Our objective is to develop expressions for the *overall* performance of the system in terms of the parameters of the individual stages.

First, we must give consideration to *lossy* two-port networks such as transmission lines and connecting cables. Power loss implies dissipation by internal resistance. Consequently, the internal noise is *thermal noise* at the ambient temperature $\mathcal{T}_{amb}$, and $\eta_{int}(f) = k\mathcal{T}_{amb}$. However, we cannot use the model in Fig. A–9a because lossy two-ports are *bilateral*, meaning that a portion of the internal noise flows back to the input. When impedances are matched, a bilateral two-port has constant gain $g < 1$ in both directions, so $g\eta_{int}(f)$ flows back to the input while $(1 - g)\eta_{int}(f)$ goes to the output. The total available noise power in bandwidth B_N at the output thus becomes

$$N_o = gk\mathcal{T}_s B_N + (1 - g)k\mathcal{T}_{amb}B_N = gk[\mathcal{T}_s + (L - 1)\mathcal{T}_{amb}]B_N$$

where

$$L = 1/g$$

which represents the transmission loss or attenuation. Comparing our expression for N_o with Eq. (12), we obtain the effective noise temperature

$$\mathcal{T}_e = (L - 1)\mathcal{T}_{amb} \qquad [20a]$$

Figure A–12   Cascade of two noisy two-ports.

and Eq. (16) gives the noise figure

$$F = 1 + (L - 1)(\mathcal{T}_{amb}/\mathcal{T}_0) \qquad [20b]$$

If a lossy two-port is at room temperature, then $\mathcal{T}_{amb} = \mathcal{T}_0$ and Eq. (20b) reduces to $F = L$.

Next, consider the cascade of two noisy two-ports in Fig. A–12, where subscripts identify the maximum power gain, noise bandwidth, and effective noise temperature of each stage. We reasonably assume that both stages are linear and time invariant (LTI). We further assume that the passband of the second stage falls within the passband of the first stage, so $B_2 \le B_1$ and the overall noise bandwidth is $B_N \approx B_2$. This condition reflects the sensible strategy of designing the last stage to mop up any remaining noise that falls outside the signal band. The overall power gain then equals the product

$$g = g_1 g_2$$

since the first stage amplifies everything passed by the second stage.

The total output noise power consists of three terms:

1. Source noise amplified by both stages;

2. Internal noise from the first stage, amplified by the second stage;

3. Internal noise from the second stage.

Thus,

$$N_o = gk\mathcal{T}_s B_N + g_2(g_1 k\,\mathcal{T}_1 B_N) + g_2 k\,\mathcal{T}_2 B_N = gk\left(\mathcal{T}_s + \mathcal{T}_1 + \frac{\mathcal{T}_2}{g_1}\right)B_N$$

and the overall effective noise temperature is

$$\mathcal{T}_e = \mathcal{T}_1 + \frac{\mathcal{T}_2}{g_1}$$

The overall noise figure is

$$F = 1 + \frac{\mathcal{T}_1}{\mathcal{T}_0} + \frac{\mathcal{T}_2}{g_1\mathcal{T}_0} = F_1 + \frac{F_2 - 1}{g_1}$$

which follows from the general relationship $F = 1 + \mathcal{T}_e/\mathcal{T}_0$.

The foregoing analysis readily generalizes to the case of three or more cascaded LTI two-ports. The overall effective noise temperature is given by **Friis' formula** as

$$\mathcal{T}_e = \mathcal{T}_1 + \frac{\mathcal{T}_2}{g_1} + \frac{\mathcal{T}_3}{g_1 g_2} + \cdots \qquad [21]$$

and the overall noise figure is

$$F = F_1 + \frac{F_2 - 1}{g_1} + \frac{F_3 - 1}{g_1 g_2} + \cdots \qquad \text{[22]}$$

Both expressions bring out the fact that

> The *first stage* plays a critical role and must be given careful attention in system design.

On the one hand, suppose the first stage is a *preamplifier* with sufficiently large gain g_1 that Eq. (21) reduces to $\mathcal{T}_e \approx \mathcal{T}_1$. The system noise is then determined primarily by the preamplifier. The remaining stages provide additional amplification and filtering, amplifying the signal and noise without appreciably changing the signal-to-noise ratio. The design of low-noise receivers is usually based on this preamplification principle.

But, on the other hand, suppose the first stage happens to be a connecting cable or any other *lossy* two-port. From the noise viewpoint, the attenuation is twice cursed since $g_1 = 1/L_1 < 1$ and $\mathcal{T}_1 = (L_1 - 1)\mathcal{T}_{\text{amb}}$. Equation (21) thus becomes

$$\mathcal{T}_e = (L_1 - 1)\mathcal{T}_{\text{amb}} + L_1 \mathcal{T}_2 + L_1 \mathcal{T}_3 / g_2 + \cdots$$

which shows that L_1 multiplies the noise temperatures of *all* subsequent stages.

Now consider a complete communications receiver as drawn in Fig. A–13a. The receiver has been divided into two major parts: a predetection unit followed by a detector. The detector processes the amplified signal plus noise and carries out a *nonlinear* operation, i.e., analog demodulation or digital regeneration. These operations are analyzed in Chaps. 10, 11, and 14, making the reasonable assumption that the detector introduces negligible noise compared to the amplified noise coming from the predetection unit. We're concerned here with the *predetection* signal-to-noise ratio denoted by $(S/N)_R$.

The predetection portion of a receiver is a cascade of noisy amplifiers and other functional blocks that act as LTI two-ports under the usual small-signal conditions. Hence, as indicated in Fig. A–13a, the entire predetection unit can be characterized by its overall effective noise temperature calculated from Eq. (21). (When the predetection unit includes a frequency converter, as in a superheterodyne receiver, its *conversion gain* takes the place of available power gain.) For a well-designed receiver, the predetection noise bandwidth essentially equals the transmission bandwidth B_T required for the signal. If the available signal power at the receiver input is S_R and the accompanying noise has temperature $\mathcal{T}_R$, then Eq. (13) becomes

$$(S/N)_R = S_R / N_0 B_T \qquad \text{[23a]}$$

with

$$N_0 \triangleq k(\mathcal{T}_R + \mathcal{T}_e) = k\mathcal{T}_N \qquad \text{[23b]}$$

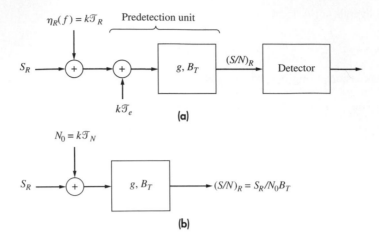

Figure A-13   (a) Communication receiver; (b) noise model of predetection unit.

The sum $\mathcal{T}_N = \mathcal{T}_R + \mathcal{T}_e$ is called the **system noise temperature,** and N_0 represents the total noise power density referred to the input of an equivalent noiseless receiver— corresponding to the diagram in Fig. A–13b.

EXAMPLE A–3   **Satellite ground station**

The signal received at a satellite ground station is extremely weak. Fortunately, the accompanying noise comes primarily from "cold" atmospheric phenomena and has a very low temperature. Minimizing the receiver noise is therefore both essential and justifiable. (In contrast, a receiving antenna pointed at or below the horizon picks up blackbody radiation from the "hot" earth; then $\mathcal{T}_R \approx \mathcal{T}_0$ and the receiver noise will be relatively less important.)

   Figure A–14 depicts an illustrative low-noise microwave receiver for a satellite signal with frequency modulation. The waveguide is part of the antenna feed structure and introduces a small loss; the corresponding effective noise temperature is $\mathcal{T}_1 = (1.05 - 1)290 = 14.5$ K, from Eq. (20a). Two preamplifiers are employed to mitigate the noise of the high-gain FM receiver. Inserting numerical values into Eq. (21), we get the overall effective noise temperature

$$\mathcal{T}_e = 14.5 + 1.05 \times 9 + \frac{1.05 \times 170}{100} + \frac{1.05 \times 1860}{100 \times 10}$$

$$= 14.5 + 9.5 + 1.8 + 2.0 = 27.8 \text{ K}$$

Notice that the waveguide loss accounts for half of $\mathcal{T}_e$, whereas the noise from the FM receiver has been nearly washed out by the preamplification gain.

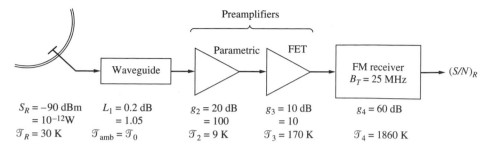

Figure A–14   Satellite ground station.

The system noise temperature is $\mathscr{T}_N = \mathscr{T}_R + \mathscr{T}_e = 57.8 \text{ K} \approx 0.2\mathscr{T}_0$. Therefore, using Eq. (23),

$$\left(\frac{S}{N}\right)_R = \frac{10^{-12}}{4 \times 10^{-21} \times 0.2 \times 25 \times 10^6} = 50 = 17 \text{ dB}$$

This small signal-to-noise ratio would be insufficient for analog communication, were it not for the further improvement afforded by FM demodulation.

Suppose the parametric amplifier in Fig. A–14 could be mounted directly on the antenna, ahead of the waveguide. Find $\mathscr{T}_N$ with and without the FET preamplifier.

EXERCISE A–3

Cable Repeater Systems

The adverse noise effect of lossy two-ports obviously cannot be avoided in cable transmission systems. However, we previously asserted that inserting *repeater amplifiers* improves performance compared to cable transmission without repeaters. Now we have the tools needed to analyze the noise performance of such repeater systems.

A repeater system normally consists of m identical cable/repeater sections like Fig. A–15. The cable has loss L_c and ambient temperature $\mathscr{T}_{\text{amb}} \approx \mathscr{T}_0$, so $F_c = L_c$. The repeater amplifier has noise figure F_r and gain $g_r = L_c$ to compensate for the cable loss. We'll treat each section as a single unit with power gain

$$g_{cr} = (1/L_c)g_r = 1$$

and noise figure

$$F_{cr} = F_c + \frac{F_r - 1}{(1/L_c)} = L_c + L_c(F_r - 1) = L_c F_r$$

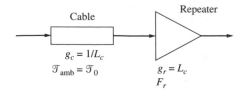

Figure A–15   One section of a cable repeater system.

calculated from Eq. (22). The overall noise figure for m cascaded sections then becomes

$$F = mF_{cr} - (m - 1) \qquad [24a]$$

$$\approx mF_{cr} \qquad [24b]$$

The approximation in Eq. (24b) assumes $F_{cr} \gg 1$, consistent with $L_c \gg 1$. This result explains the rule-of-thumb saying that "doubling the number of repeaters increases the noise figure by 3 dB."

Most systems do have $F_{cr} \gg 1$, so the effective noise temperature is

$$\mathcal{T}_e = (F - 1)\mathcal{T}_0 \approx mF_{cr}\mathcal{T}_0$$

Furthermore, we can reasonably presume that the noise temperature at the transmitter will be small compared to $\mathcal{T}_e$. Under these conditions, the transmitted signal power S_T yields at the destination the predetection signal-to-noise ratio

$$\left(\frac{S}{N}\right)_R \approx \frac{S_T}{k\mathcal{T}_e B_N} \approx \frac{S_T}{m(kF_{cr}\mathcal{T}_0 B_T)} = \frac{1}{m}\left(\frac{S}{N}\right)_1$$

where $(S/N)_1 = S_T/(kF_{cr}\mathcal{T}_0 B_T)$, which corresponds to the signal-to-noise ratio at the output of the *first* repeater.

Problems

A–1* Obtain expressions for $v_n^2(f)$ and $i_n^2(f)$ when resistance R_1 at temperature $\mathcal{T}_1$ is connected in series with R_2 at temperature $\mathcal{T}_2$. Check your result by taking $\mathcal{T}_1 = \mathcal{T}_2 = \mathcal{T}$.

A–2 Do Prob. A–1 for a parallel connection.

A–3 Find $v_n^2(f)$ when the circuit in Fig. A–7 has $\mathcal{T}_1 = \mathcal{T}_2 = \mathcal{T}$, $R_1 = 1$, $R_2 = 9$, and $L = 1/2\pi$.

A–4 Let the inductance in Fig. A–7 be replaced by capacitance C. Find $v_n^2(f)$ when $\mathcal{T}_1 = \mathcal{T}_2 = \mathcal{T}$ and $R_1 = R_2 = R$.

A–5* Let the voltage in Fig. A–3 be $V \gg k\mathcal{T}/q$. Write $i_n^2(f)$ in terms of r and explain why junction shot noise is sometimes called "half-thermal noise."

A–6 An amplifier with $r_i = r_o = 50\ \Omega$ is connected to a room-temperature source with

$R_s = 50 \ \Omega$. The amplifier has $|H(f)| = 200 \ \Pi[(f - f_c)/B]$ for $f \geq 0$ and $\eta_{int}(f) = 2 \times 10^{-16}\Pi[(f - f_c)/B]$, where $f_c = 100$ MHz and $B = 1$ MHz. Find gB_N, $k\mathcal{T}_e$, and N_o.

A–7   An amplifier with $g = 50$ dB and $B_N = 20$ kHz is found to have $N_o = 80$ pW when $\mathcal{T}_s = \mathcal{T}_0$. Find $\mathcal{T}_e$ and F, and calculate N_o when $\mathcal{T}_s = 2\mathcal{T}_0$.

A–8   When the noise temperature at the input to a certain amplifier changes from $\mathcal{T}_0$ to $2\mathcal{T}_0$, the output noise increases by one-third. Find $\mathcal{T}_e$ and F.

A–9   A sinusoidal oscillator may be used in place of a noise source for the measurement process in Example A–2 (p. 772). The oscillator is connected but turned off for the first measurement, so its internal resistance provides the source noise. The oscillator is then turned on, and its signal power S is adjusted to double the meter reading. Obtain an expression for F in terms of S, and discuss the disadvantages of this method.

A–10*   Impedance matching between a 300-Ω antenna and a 50-Ω receiver is sometimes accomplished by putting a 300-Ω resistor in series with the antenna and a 50-Ω resistor across the receiver's input terminals. Find the noise figure of this resistive two-port network by calculating its power gain with a 300-Ω source resistance and a 50-Ω load resistance.

A–11   Two cascaded amplifiers have the following properties: $\mathcal{T}_1 = 3\mathcal{T}_0$, $g_1 = 10$ dB, $F_2 = 13.2$ dB, $g_2 = 50$ dB, $B_1 > B_2 = 100$ kHz. What input signal power is required to get $(S/N)_o = 30$ dB when $\mathcal{T}_s = 10\mathcal{T}_0$?

A–12   A system consists of a cable whose loss is 2 dB/km followed by an amplifier with $F = 7$ dB. If $\mathcal{T}_s = \mathcal{T}_0$, then what's the maximum path length for $(S/N)_o \geq 0.05(S/N)_s$?

A–13   A receiver system consists of a preamplifier with $F = 3$ dB and $g = 20$ dB; a cable with $L = 6$ dB and $\mathcal{T}_{amb} = \mathcal{T}_0$; and a receiver with $F = 13$ dB. (*a*) Calculate the system noise figure in dB. (*b*) Repeat part *a* with the preamplifier between the cable and the receiver.

A–14*   A receiver system consists of a waveguide with $L_1 = 1.5$ dB; a preamplifier with $g_2 = 20$ dB and $\mathcal{T}_2 = 50$ K; and a receiver with $F_3 = 10$ dB. To what temperature must the waveguide be cooled so that the system has $\mathcal{T}_e \leq 150$ K?

A–15   The **Haus-Adler noise measure** for an amplifier is defined as $M = (F - 1)/(1 - 1/g)$. Show from noise considerations that in a cascade of two different amplifiers, the first amplifier should have the lower value of M. *Hint:* Write $F_{12} - F_{21}$ for the two possible configurations.

Table T.1

Fourier Transforms

Definitions

Transform
$$V(f) = \mathcal{F}[v(t)] = \int_{-\infty}^{\infty} v(t)e^{-j2\pi ft}\, dt$$

Inverse transform
$$v(t) = \mathcal{F}^{-1}[V(f)] = \int_{-\infty}^{\infty} V(f)e^{j2\pi ft}\, df$$

Integral theorem

$$\int_{-\infty}^{\infty} v(t)w^*(t)\, dt = \int_{-\infty}^{\infty} V(f)W^*(f)\, df$$

Theorems

Operation	Function	Transform
Superposition	$a_1v_1(t) + a_2v_2(t)$	$a_1V_1(f) + a_2V_2(f)$
Time delay	$v(t - t_d)$	$V(f)e^{-j\omega t_d}$
Scale change	$v(\alpha t)$	$\dfrac{1}{\lvert\alpha\rvert} V\left(\dfrac{f}{\alpha}\right)$
Conjugation	$v^*(t)$	$V^*(-f)$
Duality	$V(t)$	$v(-f)$
Frequent translation	$v(t)e^{j\omega_c t}$	$V(f - f_c)$
Modulation	$v(t)\cos(\omega_c t + \phi)$	$\frac{1}{2}[V(f - f_c)e^{j\phi} + V(f + f_c)e^{-j\phi}]$
Differentiation	$\dfrac{d^n v(t)}{dt^n}$	$(j2\pi f)^n V(f)$
Integration	$\displaystyle\int_{-\infty}^{t} v(\lambda)\, d\lambda$	$\dfrac{1}{j2\pi f} V(f) + \frac{1}{2} V(0)\,\delta(f)$
Convolution	$v * w(t)$	$V(f)W(f)$
Multiplication	$v(t)w(t)$	$V * W(f)$
Multiplication by t^n	$t^n v(t)$	$(-j2\pi)^{-n}\dfrac{d^n V(f)}{df^n}$

Transforms

Function	$v(t)$	$V(f)$
Rectangular	$\Pi\left(\dfrac{t}{\tau}\right)$	$\tau \operatorname{sinc} f\tau$
Triangular	$\Lambda\left(\dfrac{t}{\tau}\right)$	$\tau \operatorname{sinc}^2 f\tau$
Gaussian	$e^{-\pi(bt)^2}$	$(1/b)\, e^{-\pi(f/b)^2}$
Causal exponential	$e^{-bt}u(t)$	$\dfrac{1}{b + j2\pi f}$
Symmetric exponential	$e^{-b|t|}$	$\dfrac{2b}{b^2 + (2\pi f)^2}$
Sinc	$\operatorname{sinc} 2Wt$	$\dfrac{1}{2W}\,\Pi\left(\dfrac{f}{2W}\right)$
Sinc squared	$\operatorname{sinc}^2 2Wt$	$\dfrac{1}{2W}\,\Lambda\left(\dfrac{f}{2W}\right)$
Constant	1	$\delta(f)$
Phasor	$e^{j(\omega_c t + \phi)}$	$e^{j\phi}\,\delta(f - f_c)$
Sinusoid	$\cos(\omega_c t + \phi)$	$\frac{1}{2}[e^{j\phi}\,\delta(f - f_c) + e^{-j\phi}\delta(f + f_c)]$
Impulse	$\delta(t - t_d)$	$e^{-j\omega t_d}$
Sampling	$\displaystyle\sum_{k=-\infty}^{\infty} \delta(t - kT_s)$	$\displaystyle f_s \sum_{n=-\infty}^{\infty} \delta(f - nf_s)$
Signum	$\operatorname{sgn} t$	$1/j\pi f$
Step	$u(t)$	$\dfrac{1}{j2\pi f} + \frac{1}{2}\delta(f)$

Table T.2

Fourier Series

Definitions

If $v(f)$ is a periodic with fundamental period

$$T_0 = 1/f_0 = 2\pi/\omega_0$$

then it can be written as the exponential Fourier series

$$v(t) = \sum_{n=-\infty}^{\infty} c_n e^{jn\omega_0 t}$$

The series coefficients are

$$c_n = \frac{1}{T_0} \int_{t_1}^{t_1 + T_0} v(t) e^{-jn\omega_0 t}\, dt$$

where t_1 is arbitrary.

When $v(t)$ is real, its Fourier series may be expressed in the trigonometric forms

$$v(t) = c_o + \sum_{n=1}^{\infty} |2c_n| \cos(n\omega_0 t + \arg c_n)$$

$$= c_o + \sum_{n=1}^{\infty} (a_n \cos n\omega_0 t + b_n \sin n\omega_0 t)$$

where

$$a_n = 2\,\text{Re}\,[c_n] \qquad b_n = -2\,\text{Im}\,[c_n]$$

Coefficient calculations

If a single period of $v(t)$ has the known Fourier transform

$$Z(f) = \mathcal{F}\left[v(t)\Pi\left(\frac{t - t_1}{T_0} \right) \right]$$

then

$$c_n = \frac{1}{T_0} Z(n f_0)$$

The following relations may be used to obtain the exponential series coefficients of a waveform that can be expressed in terms of another waveform $v(t)$ whose coefficients are known and denoted by $c_v(n)$.

Waveforms	Coefficients	
	$Ac_v(0) + B$	$n = 0$
$Av(t) + B$	$Ac_v(n)$	$n \neq 0$
$v(t - t_d)$	$c_v(n)e^{-jn\omega_0 t_d}$	
$dv(t)/dt$	$(j2\pi n f_0)c_v(n)$	
$v(t) \cos(m\omega_0 t + \phi)$	$\frac{1}{2}c_v(n-m)e^{j\phi} + \frac{1}{2}c_v(n+m)e^{-j\phi}$	

Series coefficients for selected waveforms

The waveforms in the following list are periodic and are defined for one period, either $0 < t < T_0$ or $|t| < T_0/2$, as indicated. This listing may be used in conjunction with the foregoing relationships to obtain the exponential series coefficients for other waveforms.

Waveform		Coefficients	
Impulse train			
$\delta(t)$	$|t| < T_0/2$	$1/T_0$	
Rectangular pulse train			
$\Pi(t/\tau)$	$|t| < T_0/2$	$(\tau/T_0)\,\text{sinc}\, nf_0\tau$	
Square wave (odd symmetry)			
$+1$	$0 < t < T_0/2$	0	n even
-1	$-T_0/2 < t < 0$	$-j2/\pi n$	n odd
Triangular wave (even symmetry			
$1 - \dfrac{4|t|}{T_0}$	$\begin{array}{l}0 \\ |t| < T_0/2\end{array}$		n even
		$(2/\pi n)^2$	n odd
Sawtooth wave			
t/T_0	$0 < t < T_0$	$1/2$	$n = 0$
		$j/2\pi n$	$n \neq 0$
Half-rectified sine wave			
$\sin \omega_0 t$	$0 < t < T_0/2$	$\dfrac{1}{\pi(1-n^2)}$	n even
0	$T_0/2 < t < T_0$	$-jn/4$	$n = \pm 1$
		0	otherwise

Table T.3

Mathematical Relations

Certain of the mathematical relationship encountered in this text are listed below for convenient reference. However, this table is not intended as a substitute for more comprehensive handbooks.

Trigonometric identities

$e^{\pm j\theta} = \cos\theta \pm j\sin\theta$

$e^{j2\alpha} + e^{j2\beta} = 2\cos(\alpha - \beta)e^{j(\alpha+\beta)}$

$e^{j2\alpha} - e^{j2\beta} = j2\sin(\alpha - \beta)e^{j(\alpha+\beta)}$

$\cos\theta = \dfrac{1}{2}(e^{j\theta} + e^{-j\theta}) = \sin(\theta + 90°)$

$\sin\theta = \dfrac{1}{2j}(e^{j\theta} - e^{-j\theta}) = \cos(\theta - 90°)$

$\sin^2\theta + \cos^2\theta = 1$

$\cos^2\theta - \sin^2\theta = \cos 2\theta$

$\cos^2\theta = \frac{1}{2}(1 + \cos 2\theta)$

$\cos^3\theta = \frac{1}{4}(3\cos\theta + \cos 3\theta)$

$\sin^2\theta = \frac{1}{2}(1 - \cos 2\theta)$

$\sin^3\theta = \frac{1}{4}(3\sin\theta - \sin 3\theta)$

$\sin(\alpha \pm \beta) = \sin\alpha\cos\beta \pm \cos\alpha\sin\beta$

$\cos(\alpha \pm \beta) = \cos\alpha\cos\beta \mp \sin\alpha\sin\beta$

$\tan(\alpha \pm \beta) = \dfrac{\tan\alpha \pm \tan\beta}{1 \mp \tan\alpha\tan\beta}$

$\sin\alpha\sin\beta = \frac{1}{2}\cos(\alpha - \beta) - \frac{1}{2}\cos(\alpha + \beta)$

$\cos\alpha\cos\beta = \frac{1}{2}\cos(\alpha - \beta) + \frac{1}{2}\cos(\alpha + \beta)$

$\sin\alpha\cos\beta = \frac{1}{2}\sin(\alpha - \beta) + \frac{1}{2}\sin(\alpha + \beta)$

$A\cos(\theta + \alpha) + B\cos(\theta + \beta) = C\cos\theta - S\sin\theta = R\cos(\theta + \phi)$

where

$\quad C = A\cos\alpha + B\cos\beta$

$\quad S = A\sin\alpha + B\sin\beta$

$\quad R = \sqrt{C^2 + S^2} = \sqrt{A^2 + B^2 + 2AB\cos(\alpha - \beta)}$

$\quad \phi = \arctan\dfrac{S}{C} = \arctan\dfrac{A\sin\alpha + B\sin\beta}{A\cos\alpha + B\cos\beta}$

Series expansions and approximations

$$(1 + x)^n = 1 + nx + \frac{n(n-1)}{2!}x^2 + \cdots \qquad |nx| < 1$$

$$e^x = 1 + x + \frac{1}{2!}x^2 + \cdots$$

$$a^x = 1 + x \ln a + \frac{1}{2!}(x \ln a)^2 + \cdots$$

$$\ln(1 + x) = x - \tfrac{1}{2}x^2 + \tfrac{1}{3}x^3 + \cdots$$

$$\sin x = x - \frac{1}{3!}x^3 + \frac{1}{5!}x^5 - \cdots$$

$$\cos x = 1 - \frac{1}{2!}x^2 + \frac{1}{4!}x^4 - \cdots$$

$$\tan x = x + \tfrac{1}{3}x^3 + \tfrac{2}{15}x^5 + \cdots$$

$$\arcsin x = x + \tfrac{1}{6}x^3 + \tfrac{3}{40}x^5 + \cdots$$

$$\arctan x = \begin{cases} x - \tfrac{1}{3}x^3 + \tfrac{1}{5}x^5 - \cdots & |x| < 1 \\ \dfrac{\pi}{2} - \dfrac{1}{x} + \dfrac{1}{3x^3} - \cdots & x > 1 \end{cases}$$

$$\operatorname{sinc} x = 1 - \frac{1}{3!}(\pi x)^2 + \frac{1}{5!}(\pi x)^4 - \cdots$$

$$J_n(x) = \frac{1}{n!}\left(\frac{x}{2}\right)^n - \frac{1}{(n+1)!}\left(\frac{x}{2}\right)^{n+2} + \frac{1}{2!(n+2)!}\left(\frac{x}{2}\right)^{n+4} - \cdots$$

$$J_n(x) \approx \sqrt{\frac{2}{\pi x}}\cos\left(x - \frac{\pi}{4} - \frac{n\pi}{2}\right) \qquad x \gg 1$$

$$I_0(x) \approx \begin{cases} e^{x^2/4} & x^2 \ll 1 \\ e^x/\sqrt{2\pi x} & x \gg 1 \end{cases}$$

Summations

$$\sum_{m=1}^{M} m = \frac{M(M+1)}{2}$$

$$\sum_{m=1}^{M} m^2 = \frac{M(M+1)(2M+1)}{6}$$

$$\sum_{m=1}^{M} m^3 = \frac{M^2(M+1)^2}{4}$$

$$\sum_{m=0}^{M} x^m = \frac{x^{M+1} - 1}{x - 1}$$

Definite integrals

$$\int_0^\infty \frac{x^{m-1}}{1+x^n}\,dx = \frac{\pi/n}{\sin(m\pi/n)} \qquad n > m > 0$$

$$\int_0^\infty \sin x^2\,dx = \int_0^\infty \cos x^2\,dx = \frac{1}{2}\sqrt{\frac{\pi}{2}}$$

$$\int_0^\infty \frac{\sin x}{x}\,dx = \int_0^\infty \frac{\tan x}{x}\,dx = \frac{\pi}{2}$$

$$\int_0^\infty x^n e^{-ax}\,dx = \frac{n!}{a^{n+1}} \qquad n \geq 1, a > 0$$

$$\int_0^\infty \frac{\sin x \cos ax}{x}\,dx = \begin{cases} \dfrac{\pi}{2} & a^2 < 1 \\[2mm] \dfrac{\pi}{4} & a^2 = 1 \\[2mm] 0 & a^2 > 1 \end{cases}$$

$$\int_0^\infty e^{-a^2 x^2}\,dx = \frac{1}{2a}\sqrt{\pi} \qquad a > 0$$

$$\int_0^\infty x^2 e^{-x^2}\,dx = \tfrac{1}{4}\sqrt{\pi}$$

$$\int_0^\infty \frac{\sin^2 x}{x^2}\,dx = \frac{\pi}{2}$$

$$\int_0^\infty e^{-ax}\cos x\,dx = \frac{a}{1+a^2} \qquad a > 0$$

$$\int_0^\infty \frac{\cos nx}{1+x^2}\,dx = \frac{\pi}{2}e^{-|n|}$$

$$\int_0^\infty e^{-ax}\sin x\,dx = \frac{1}{1+a^2} \qquad a > 0$$

$$\int_0^\infty \mathrm{sinc}\,x\,dx = \int_0^\infty \mathrm{sinc}^2 x\,dx = \tfrac{1}{2}$$

$$\int_0^\infty e^{-a^2 x^2}\cos bx\,dx = \frac{1}{2a}\sqrt{\pi}e^{-(b/2a)^2}$$

Schwarz's inequality

$$\left|\int_a^b v(\lambda)w(\lambda)\,d\lambda\right|^2 \leq \int_a^b |v(\lambda)|^2\,d\lambda \int_a^b |w(\lambda)|^2\,d\lambda$$

The equality holds if $v(\lambda) = Kw(\lambda)$ where K is a constant.

Poisson's sum formula

$$\sum_{n=-\infty}^\infty e^{\pm j2\pi n\lambda/L} = L\sum_{m=-\infty}^\infty \delta(\lambda - mL)$$

Table T.4

The Sinc Function

Numerical values of sinc x $= (\sin \pi x)/\pi x$ and its square are tabulated below for x from 0 to 3.9 in increments of 0.1.

x	sinc x	sinc$^2 x$	x	sinc x	sinc$^2 x$
0.0	1.000	1.000	2.0	0.000	0.000
0.1	0.984	0.968	2.1	0.047	0.002
0.2	0.935	0.875	2.2	0.085	0.007
0.3	0.858	0.737	2.3	0.112	0.013
0.4	0.757	0.573	2.4	0.126	0.016
0.5	0.637	0.405	2.5	0.127	0.016
0.6	0.505	0.255	2.6	0.116	0.014
0.7	0.368	0.135	2.7	0.095	0.009
0.8	0.234	0.055	2.8	0.067	0.004
0.9	0.109	0.012	2.9	0.034	0.001
1.0	0.000	0.000	3.0	0.000	0.000
1.1	−0.089	0.008	3.1	−0.032	0.001
1.2	−0.156	0.024	3.2	−0.058	0.003
1.3	−0.198	0.039	3.3	−0.078	0.006
1.4	−0.216	0.047	3.4	−0.089	0.008
1.5	−0.212	0.045	3.5	−0.091	0.008
1.6	−0.189	0.036	3.6	−0.084	0.007
1.7	−0.151	0.023	3.7	−0.070	0.005
1.8	−0.104	0.011	3.8	−0.049	0.002
1.9	−0.052	0.003	3.9	−0.025	0.001

Table T.5

Probability Functions

Binomial distribution

Let the discrete RV I be the number of times an event A occurs in n independent trials. If $P(A) = \alpha$, then

$$P_I(i) = \binom{n}{i} \alpha^i (1 - \alpha)^{n-i} \qquad i = 0, 1, \ldots, n$$

$$E[I] = n\alpha \qquad \sigma_I^2 = n\alpha(1 - \alpha)$$

If $n \gg 1$, $\alpha \ll 1$, and $m = n\alpha$ remains finite, then

$$P_I(i) \approx e^{-m} \frac{m^i}{i!}$$

Poisson distribution

Let the discrete RV I be the number of times an event A occurs in time T. If $P(A) = \mu \Delta T \ll 1$ in a small interval ΔT, and if multiple occurrences are statistically independent, then

$$P_I(i) = e^{-\mu T} \frac{(\mu T)^i}{i!} \qquad E[I] = \mu T$$

Uniform distribution

If the continuous RV X is equally likely to be observed anywhere in a finite range, and nowhere else, then

$$p_X(x) = \frac{1}{b - a} \quad a \le x \le b$$

$$E[X] = \frac{1}{2}(a + b) \qquad \sigma_X^2 = \frac{1}{12}(b - a)^2$$

Sinusoidal distribution

If X has a uniform distribution with $b - a = 2\pi$ and $Z = A \cos(X + \theta)$, where A and θ are constants, then

$$p_Z(z) = \frac{1}{\pi\sqrt{A^2 - z^2}} \qquad |z| \leq A$$

$$E[Z] = 0 \qquad \sigma_Z^2 = \frac{1}{2}A^2$$

Laplace distribution

If the continuous RV X is governed by

$$p_X(x) = \frac{a}{2}e^{-a|x|}$$

then

$$E[X] = 0 \qquad \sigma_X^2 = \frac{2}{a^2}$$

Gaussian (normal) distribution

If X represents the sum of a large number of independent random components, and if each component makes only a small contribution to the sum, then

$$p_X(x) = \frac{1}{\sqrt{2\pi\sigma^2}}e^{-(x-m)^2/2\sigma^2}$$

$$E[X] = m \qquad \sigma_X^2 = \sigma^2$$

(See Table T.6 for gaussian probabilities.)

Rayleigh distribution

If $R^2 = X^2 + Y^2$, where X and Y are independent gaussian RVs with zero mean and variance σ^2, then

$$p_R(r) = \frac{r}{\sigma^2}e^{-r^2/2\sigma^2} \qquad r \geq 0$$

$$E[R] = \sqrt{\frac{\pi}{2}}\,\sigma \qquad E[R^2] = 2\sigma^2$$

Table T.6

Gaussian Probabilities

The probability that a gaussian random variable with mean m and variance σ^2 will have an observed value greater than $m + k\sigma$ is given by the function

$$Q(k) \triangleq \frac{1}{\sqrt{2\pi}} \int_k^{\infty} e^{-\lambda^2/2}\, d\lambda$$

called the area under the gaussian tail. Thus

$$P(X > m + k\sigma) = P(X \le m - k\sigma) = Q(k)$$

$$P(|X - m| > k\sigma) = 2Q(k)$$

$$P(m < X \le m + k\sigma) = P(m - k\sigma < X \le m) = \tfrac{1}{2} - Q(k)$$

$$P(|X - m| \le k\sigma) = 1 - 2Q(k)$$

$$P(m - k_1\sigma < X \le m + k_2\sigma) = 1 - Q(k_1) - Q(k_2)$$

Other functions related to $Q(k)$ are as follows:

$$\operatorname{erf} k \triangleq \frac{2}{\sqrt{\pi}} \int_0^k e^{-\lambda^2}\, d\lambda = 1 - 2Q(\sqrt{2}\,k)$$

$$\operatorname{erfc} k \triangleq \frac{2}{\sqrt{\pi}} \int_k^{\infty} e^{-\lambda^2}\, d\lambda = 1 - \operatorname{erf} k = 2Q(\sqrt{2}\,k)$$

$$\Phi(k) \triangleq \frac{1}{\sqrt{2\pi}} \int_0^k e^{-\lambda^2/2}\, d\lambda = \tfrac{1}{2} - Q(k)$$

All of the foregoing relations are for $k \ge 0$. If $k < 0$, then

$$Q(-|k|) = 1 - Q(|k|)$$

Numerical values of $Q(k)$ are plotted below for $0 \leq k \leq 7.0$. For larger values of k, $Q(k)$ may be approximated by

$$Q(k) \approx \frac{1}{\sqrt{2\pi k}} e^{-k^2/2}$$

which is quite accurate for $k > 3$.

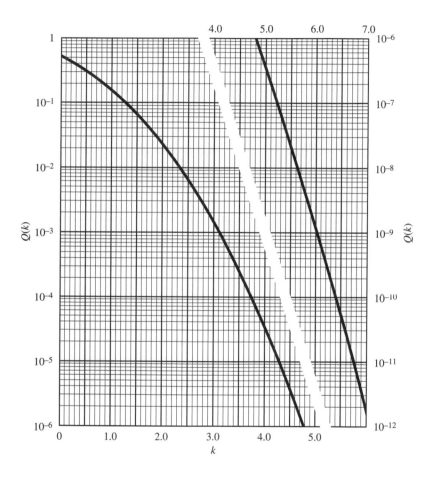

Table T.7

Glossary of Notation

Operations

z*	Complex conjugate
Re $[z]$, Im $[z]$	Real and imaginary parts
$|z|$	Magnitude or absolute value
$\arg z = \arctan \dfrac{\text{Im } [z]}{\text{Re } [z]}$	Angle of a complex quantity
$\langle v(t) \rangle = \lim\limits_{T\to\infty} \dfrac{1}{T} \displaystyle\int_{-T/2}^{T/2} v(t)\, dt$	Time average
$\mathcal{F}[v(t)] = \displaystyle\int_{-\infty}^{\infty} v(t) e^{-j2\pi ft}\, dt$	Fourier transform
$\mathcal{F}^{-1}[V(f)] = \displaystyle\int_{-\infty}^{\infty} V(f) e^{j2\pi ft}\, df$	Inverse Fourier transform
$v * w(t) = \displaystyle\int_{-\infty}^{\infty} v(\lambda) w(t - \lambda)\, d\lambda$	Convolution
$\hat{v}(t) = \dfrac{1}{\pi} \displaystyle\int_{-\infty}^{\infty} \dfrac{v(\lambda)}{t - \lambda}\, d\lambda$	Hilbert transform
$R_{vw}(\tau)$	Crosscorrelation
$R_v(\tau) = R_{vv}(\tau)$	Autocorrelation
$G_v(f) = \mathcal{F}_\tau[R_v(\tau)]$	Spectral density or power spectrum
$E[X]$	Mean or expected value
$E[v(t)]$	Ensemble average

Functions

$$Q(k) = \frac{1}{\sqrt{2\pi}} \int_k^{\infty} e^{-\lambda^2/2}\, d\lambda \qquad\qquad \text{Gaussian probability}$$

$$\exp t = e^t \qquad\qquad \text{Exponential}$$

$$\operatorname{sinc} t = \frac{\sin \pi t}{\pi t} \qquad\qquad \text{Sinc}$$

$$\operatorname{sgn} t = \begin{cases} 1 & t > 0 \\ -1 & t < 0 \end{cases} \qquad\qquad \text{Sign}$$

$$u(t) = \begin{cases} 1 & t > 0 \\ 0 & t < 0 \end{cases} \qquad\qquad \text{Step}$$

$$\Pi\!\left(\frac{t}{\tau}\right) = \begin{cases} 1 & |t| < \dfrac{\tau}{2} \\[2mm] 0 & |t| > \dfrac{\tau}{2} \end{cases} \qquad\qquad \text{Rectangle}$$

$$\Lambda\!\left(\frac{t}{\tau}\right) = \begin{cases} 1 - \dfrac{|t|}{\tau} & |t| < \tau \\[2mm] 0 & |t| > \tau \end{cases} \qquad\qquad \text{Triangle}$$

Miscellaneous symbols

$$\triangleq \qquad\qquad \text{“Equals by definition”}$$

$$\approx \qquad\qquad \text{“Approximately equals”}$$

$$\int_T \qquad \int_{t_1}^{t_1+T} \quad \text{where } t_1 \text{ is arbitrary}$$

$$\leftrightarrow \qquad\qquad \text{Denoting a Fourier transform pair}$$

$$\binom{n}{i} = \frac{n!}{i!(n-i)!} \qquad\qquad \text{Binomial coefficient}$$

$$\bigstar \qquad\qquad \text{Text material that may be omitted}$$

Solutions to Exercises

2.1-1 $v(t) = 3\cos(2\pi 0t \pm 180°) + 4\cos(2\pi 15t - 90° \pm 180°)$

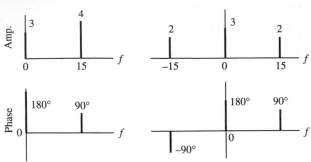

2.1-2

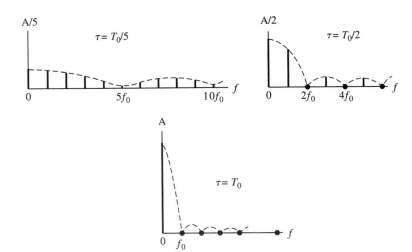

2.1-3 $P = 7^2 + 2 \times 5^2 + 2 \times 2^2 = 107$

2.2-1 $V(f) = 2\int_0^\infty Ae^{-bt}\cos\omega t\, dt = \dfrac{2A}{\omega}\dfrac{b/\omega}{1+(b/\omega)^2} = \dfrac{2Ab}{b^2+(2\pi f)^2}$

$|V(f)| \geq \dfrac{1}{2}\left(\dfrac{2A}{b}\right) \Rightarrow |f| \leq \dfrac{b}{2\pi}$

2.2-2 $\displaystyle\int_{-\infty}^\infty |V(f)|^2\, df = \dfrac{2A^2}{b^2}\int_0^\infty \dfrac{df}{1+(2\pi f/b)^2} = \dfrac{A^2}{\pi b}\dfrac{\pi/2}{\sin \pi/2} = \dfrac{A^2}{2b}$

$\displaystyle\int_{-\infty}^\infty |v(t)|^2\, dt = A^2\int_0^\infty e^{-2bt}\, dt = \dfrac{A^2}{2b}$

2.2-3 $z(t) = V(t)$ with $b = 1$ and $2A = B$, so $Z(f) = Ae^{-b|-f|} = \dfrac{B}{2}e^{-|f|}$

2.3–1 $\mathscr{F}[v(-t)] = \dfrac{1}{|-1|}\,[V_e(-f) + jV_o(-f)] = V_e(f) - jV_o(f)$

$Z(f) = a_1[V_e(f) + jV_o(f)] + a_2[V_e(f) - jV_o(f)]$
$= (a_1 + a_2)V_e(f) + j(a_1 - a_2)V_o(f)$

2.3–2 $\dfrac{d}{df}\left[\displaystyle\int_{-\infty}^{\infty} v(t)e^{-j2\pi ft}\,dt\right] = \int_{-\infty}^{\infty} v(t)(-j2\pi t)e^{-j2\pi ft}\,dt = -j2\pi\mathscr{F}[tv(t)]$

Thus, $tv(t) \leftrightarrow \dfrac{1}{-j2\pi}\dfrac{d}{df}V(f)$

2.4–1 $(A \operatorname{sinc} 2Wt)^2 \leftrightarrow \dfrac{A}{2W}\,\Pi\!\left(\dfrac{f}{2W}\right) * \dfrac{A}{2W}\,\Pi\!\left(\dfrac{f}{2W}\right) = \begin{cases} \dfrac{A^2}{2W}\,\Lambda\!\left(\dfrac{f}{2W}\right) \\ 0 \qquad |f| > 2W \end{cases}$

2.5–1 (a) $\displaystyle\int_{-\infty}^{\infty} v(t)\delta(t + 4)\,dt = v(-4) = 49,$

(b) $v(t) * \delta(t + 4) = v(t + 4) = (t + 1)^2$
(c) $v(t)\delta(t + 4) = v(-4)\delta(t + 4) = 49\delta(t + 4)$
(d) $v(t) * \delta(-t/4) = |-4|v(t) * \delta(t) = 4(t - 3)^2$

2.5–2 $\mathscr{F}[Au(t) \cos \omega_c t] = \dfrac{A}{2}\left[\dfrac{1}{j2\pi(f - f_c)} + \tfrac{1}{2}\delta(f - f_c)\right.$

$\left.+ \dfrac{1}{j2\pi(f + f_c)} + \tfrac{1}{2}\delta(f + f_c)\right]$

2.5–3 $\dfrac{dv(t)}{dt} = \dfrac{2A}{\tau}\,\Pi\!\left(\dfrac{t}{\tau}\right) - A\delta\!\left(t + \dfrac{\tau}{2}\right) - A\delta\!\left(t - \dfrac{\tau}{2}\right)$

$j2\pi fV(f) = 2A \operatorname{sinc} f\tau - Ae^{j\pi f\tau} - Ae^{-j\pi f\tau}$

$V(f) = \dfrac{jA}{\pi f}\,(\cos \pi f\tau - \operatorname{sinc} f\tau)$

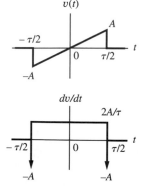

3.1–1 $g(t) = e^{-t/RC}u(t)$

$$h(t) = e^{-t/RC}\frac{du}{dt} + \frac{d}{dt}(e^{-t/RC})u(t) = \delta(t) - \frac{1}{RC}e^{-t/RC}u(t)$$

3.1–2 $H(f) = \dfrac{j2\pi fL}{R + j2\pi fL} = \dfrac{jf}{f_l + jf}$

3.1–3 $H(f) = T \operatorname{sinc} fT\, e^{-j2\pi fT}$, $X(f) = A\tau \operatorname{sinc} f\tau$

$\tau \ll T$, $Y(f) \approx A\tau H(f)$, $y(t) \approx A\tau h(t)$

$\tau = T$, $Y(f) = AT^2 \operatorname{sinc}^2 fT\, e^{-j2\pi fT}$, $y(t) = AT\Lambda\left(\dfrac{t-T}{T}\right)$

$\tau \gg T$, $Y(f) \approx TX(f)$, $y(t) \approx Tx(t)$

3.2–1 $t_d(f) = \begin{cases} -\dfrac{1}{2\pi f}\left(-\dfrac{\pi}{2}\,\text{rad}\right)\dfrac{f}{30\text{ kHz}} = \dfrac{1}{120}\text{ ms} & |f| < 30\text{ kHz} \\[3mm] -\dfrac{1}{2\pi f}\left(-\dfrac{\pi}{2}\,\text{rad}\right) = \dfrac{1}{4f} & |f| > 30\text{ kHz} \end{cases}$

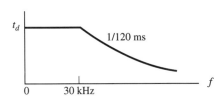

3.2–2 $|H_{eq}(f)| = 1/4\,|H(f)|$

$$= \begin{cases} \dfrac{1}{4}\dfrac{20\text{ kHz}}{f} & |f| < 20\text{ kHz} \\[3mm] \tfrac{1}{4} & |f| > 20\text{ kHz} \end{cases}$$

$$\arg H_{eq}(f) = -2\pi f\,\frac{10^{-3}}{120} - \arg H(f)$$

$$= \begin{cases} 0 & |f| < 30\text{ kHz} \\[3mm] -2\pi f\dfrac{10^{-3}}{120} + \dfrac{\pi}{2} & |f| > 30\text{ kHz} \end{cases}$$

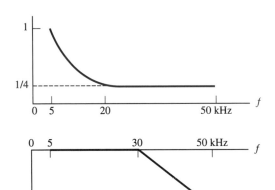

3.3–1 (*a*)

$$P_{\mathrm{dBm}} = 10 \log_{10}\left(\frac{P}{1 \text{ mW}} \times \frac{10^3 \text{ mW}}{1 \text{ W}}\right)$$

$$= 10 \log_{10}\frac{P}{1 \text{ W}} + 10 \log_{10} 10^3 = P_{\mathrm{dBW}} + 30 \text{ dB}$$

(*b*)

$$|H(f)|^2 = 10^{(-3 \text{ dB}/10)} = 10^{-0.3} = 0.501 \Rightarrow |H(f)| \approx \frac{1}{\sqrt{2}}$$

3.3–2 (*a*) 33 dBm $-$ 24 $\times$ 2.5 dB $= -27$ dBm $= 10^{-2.7}$ mW $\approx 2 \ \mu$W

(*b*) -27 dBm $+ 64$ dB $- (40 - 24) \times 2.5$ dB

$$= -3 \text{ dBm} = 10^{-0.3} \text{ mW} \approx 0.5 \text{ mW}$$

3.4–1 Let $f_c = f_l + B/2 = (f_l + f_u)/2$ and let $V(f) = 2K\Pi(f/B)$, so
$H(f) = \frac{1}{2}\left[V(f - f_c) + V(f + f_c)\right]e^{-j\omega t_d}$
$h(t) = v(t - t_d)\cos\omega_c(t - t_d)$ where $v(t) = 2BK \text{ sinc } Bt$

3.4–2 $|H(f)|_{\mathrm{dB}} = 10 \log_{10}\dfrac{1}{1 + (f/B)^{2n}} \approx 10 \log_{10}\left(\dfrac{f}{B}\right)^{-2n}$

$$= -20n \log_{10}\left(\frac{f}{B}\right) \quad \text{for } f > B$$

$|H(2B)|_{\mathrm{dB}} \approx -20n \log_{10} 2 = -6.0n \le -20 \text{ dB} \Rightarrow n \ge \dfrac{20}{6},\ n_{\min} = 4$

3.4–3 $\tau_{\min} = 10 \ \mu$s, but the minimum pulse spacing is 30 μs $- \tau_{\max} = 5 \ \mu$s, so

$$B \ge \frac{1}{2 \times 5 \ \mu\text{s}} = 100 \text{ kHz}, \quad t_r \approx \frac{1}{2B} = 5 \ \mu\text{s}$$

3.5–1 $\mathscr{F}[\hat{x}(t)] = (-j \text{ sgn} f)X(f)$ and

$$\mathscr{F}\left[-\frac{1}{\pi t}\right] = -H_Q(f) = +j \text{ sgn} f, \text{ so}$$

$$\mathscr{F}\left[\hat{x}(t) * \left(-\frac{1}{\pi t}\right)\right] = (\text{sgn} f)^2 X(f)$$

$$= X(f) \Rightarrow \hat{x}(t) * \left(-\frac{1}{\pi t}\right) = x(t)$$

3.6–1 Let $z(t) = v(t) + w(t)$ where $v(t) = \frac{A}{2} e^{j\phi}e^{j\omega_0 t}$,
$w(t) = \frac{A}{2} e^{-j\phi}e^{-j\omega_0 t}$
then $R_{vw}(\tau) = 0$ since $\omega_w \neq \omega_v$, so
$R_z(\tau) = R_v(\tau) + R_w(\tau) = \left|\frac{A}{2} e^{j\phi}\right|^2 e^{j\omega_0 t}$
$\quad + \left|\frac{A}{2} e^{-j\phi}\right|^2 e^{-j\omega_0 t} = \frac{A^2}{2} \cos \omega_0 \tau$

3.6–2 $z(t) = w * (-t) = w(-t),$

$$R_{vw}(\tau) = \int_{-\infty}^{\infty} v(\lambda)z(\tau - \lambda)\, d\lambda$$

$E_v = E_w = A^2D$

$|R_{vw}(\tau)|^2_{\max} = (A^2D)^2 = E_v E_w$ at

$\tau = -t_d$

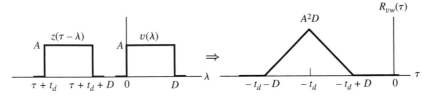

3.6–3 $\mathcal{F}_\tau[v * (-\tau)] = \int_{-\infty}^{\infty} v * (-\tau)e^{-j\omega\tau}\, d\tau$

$$= \int_{-\infty}^{\infty} v * (\lambda)e^{-j\omega\lambda}\, d\lambda$$

$$= \left[\int_{-\infty}^{\infty} v(\lambda)e^{-j\omega\lambda}\, d\lambda\right]^* = V*(f) \text{ so}$$

$$G_v(f) = \mathcal{F}_\tau[R_v(\tau)] = \mathcal{F}_\tau[v(\tau) * v * (-\tau)]$$
$$= V(f)V*(f) = |V(f)|^2$$

4.1–1 $v_{bp}(t) = z(t) + z * (t),\ z(t) = v_{lp}(t)e^{j\omega_c t}$

$V_{bp}(f) = \mathcal{F}[z(t)] + \mathcal{F}[z * (t)]$ where

$\mathcal{F}[z(t)] = V_{lp}(f - f_c)$

and $\mathcal{F}[z * (t)] = Z * (-f) = V_{lp}^*(-f - f_c)$

so $V_{bp}(f) = V_{lp}(f - f_c) + V_{lp}^*(-f - f_c)$

4.1–2 $H_{lp}(f) = K_0 + K_1 f/f_c,\quad f_l - f_c < f < f_u - f_c$

$$Y_{lp}(f) = K_0 X_{lp}(f) + \frac{K_1}{j2\pi f_c}\,[j2\pi f X_{lp}(f)] \text{ where}$$

$x_{bp}(t) = A_x(t)\cos\omega_c t \Rightarrow x_{lp}(t) = \tfrac{1}{2}A_x(t)$

so $y_{lp}(t) = \tfrac{1}{2}K_0 A_x(t) + j\tfrac{1}{2}\left[\dfrac{-K_1}{2\pi f_c}\dfrac{dA_x(t)}{dt}\right]$ and

$y_i(t) = K_0 A_x(t),\ y_q(t) = \dfrac{-K_1}{2\pi f_c}\dfrac{dA_x(t)}{dt}$

4.2–1

AM, $\mu = 0.5$

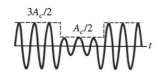

AM, $\mu = 1$

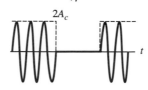

DSB

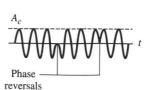

Phase
reversals

4.2–2 DSB: $S_T = 2P_{sb} = 20$ W, $A^2_{max} = \dfrac{P_{sb}}{S_x/4} = 200$ W

AM: $P_c = \dfrac{P_{sb}}{\frac{1}{2}\mu^2 S_x} = 100$ W $\Rightarrow S_T = P_c + 2P_{sb} = 120$ W and

$A^2_{max} = \dfrac{P_{sb}}{S_x/16} = 800$ W

4.2–3 $x_c(t) = \dfrac{A_c}{2}\cos(\omega_c - \omega_m)t + \dfrac{A_c}{2}\cos(\omega_c + \omega_m)t$

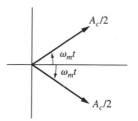

$v_i = \dfrac{A_c}{2}\cos\omega_m t + \dfrac{A_c}{4}\cos\omega_m t = \dfrac{3A_c}{4}\cos\omega_m t$

$v_q = \dfrac{A_c}{2}\sin\omega_m t - \dfrac{A_c}{4}\sin\omega_m t = \dfrac{A_c}{4}\sin\omega_m t$

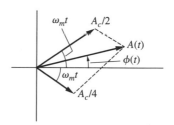

$A(t) = \sqrt{\left(\tfrac{3}{4}A_c\cos\omega_m t\right)^2 + \left(\tfrac{1}{4}A_c\sin\omega_m t\right)^2} = \dfrac{A_c}{4}\sqrt{9\cos^2\omega_m t + \sin^2\omega_m t}$

$ = \dfrac{A_c}{4}\sqrt{8\cos^2\omega_m t + 1} = \dfrac{A_c}{4}\sqrt{5 + 4\cos 2\omega_m t}$

$\phi(t) = \arctan\dfrac{A_c/4\,\sin\omega_m t}{3A_c/4\,\cos\omega_m t} = \arctan\left(\dfrac{\tan\omega_m t}{3}\right)$

4.3–1 Expanding $\cos^3\theta = \tfrac{3}{4}\cos\theta + \tfrac{1}{4}\cos 3\theta$,

$v_{out\pm} = a_1\underbrace{\left(A_c\cos\omega_c t \pm \tfrac{1}{2}x\right)} + a_2\underbrace{\left(A_c^2\cos^2\omega_c t \pm 2\tfrac{x}{2}A_c\cos\omega_c t + \tfrac{x^2}{4}\right)}$

$\phantom{v_{out\pm} =} + a_3\underbrace{\left(A_c^3\tfrac{3}{4}\cos\omega_c t + A_c^3\tfrac{1}{4}\cos 3\omega_c t \pm 3\tfrac{x}{2}A_c^2\cos^2\omega_c t + 3\tfrac{x^2}{4}A_c\cos\omega_c t \pm \tfrac{x^3}{8}\right)}$

Only underlined terms are passed by BPFs, so

$$x_c(t) = v_{out+} - v_{out-} = 2a_2x(t)A_c \cos \omega_c t = (2a_2A_c)x(t) \cos \omega_c t$$

4.4-1 $x_c(t) = \frac{1}{2}A_cA_m(\cos \omega_m t \cos \omega_c t \mp \sin \omega_m t \sin \omega_c t)$

$$= \frac{1}{4}A_cA_m[\cos (\omega_c - \omega_m)t + \cos (\omega_c + \omega_m)t$$
$$\mp \cos (\omega_c - \omega_m)t \pm \cos (\omega_c + \omega_m)t]$$
$$= \frac{1}{2}A_cA_m \cos (\omega_c \pm \omega_m)t$$

$$A(t) = \frac{1}{2}A_c\sqrt{A_m^2 \cos^2 \omega_m t + A_m^2 \sin^2 \omega_m t} = \frac{1}{2}A_cA_m$$

4.4-2 $x(t) = \cos \omega_m t$ $A_c/2\, x(t) = \cos \omega_c t = A_c/4\, [\cos(\omega_c - \omega_m)t + \cos(\omega_c + \omega_m)t]$

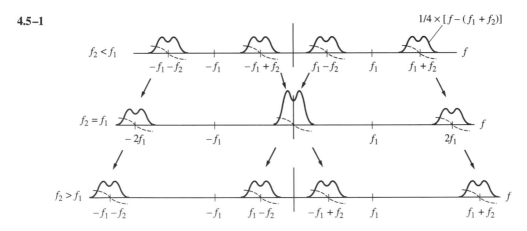

$\hat{x}(t) = \cos (\omega_m t - 90°)$ $\dfrac{A_c}{2}\, \hat{x}(t)\cos (\omega_c t - 90°) =$

$$\frac{A_c}{4} \{\cos (\omega_c - \omega_m)t + \cos [(\omega_c + \omega_m)t - 180°]\}$$

4.5-1

4.5-2 Let $a = A_cA_m/2$, so

$$A^2(t) = (A_{LO} \cos \phi' + a \cos \omega_m t)^2 + (A_{LO} \sin \phi' \pm a \sin \omega_m t)^2$$
$$= A_{LO}^2 + a^2 + 2A_{LO}a\underbrace{(\cos \omega_m t \cos \phi' \pm \sin \omega_m t \sin \phi')}_{\cos(\omega_m t \mp \phi')}\quad \text{and}$$

$$A(t) = A_{LO} \sqrt{1 + \left(\frac{a}{A_{LO}}\right)^2 + \frac{2a}{A_{LO}} \cos\left(\omega_m t \mp \phi'\right)}$$

$$\approx A_{LO} + \tfrac{1}{2} A_c A_m \cos\left(\omega_m t \mp \phi'\right)$$

5.1–1 $x_c(t) = A_c \cos\left[\omega_c t + \omega_c \mu x(t) t\right] \Rightarrow \theta_c(t) = 2\pi\left[f_c t + f_c \mu x(t) t\right]$

$f(t) = \frac{1}{2\pi} \dot{\theta}_c(t) = f_c + f_c \mu x(t) + f_c \mu \dot{x}(t) t$

$= f_c\left[1 + \mu \cos \omega_m t - \mu \omega_m t \sin \omega_m t\right]$

so $f(t) \approx -f_c \mu \omega_m t \sin \omega_m t$ for $\mu \omega_m t \gg 1$ and $|f(t)| \to \infty$ as $t \to \infty$.

5.1–2 $\mathcal{F}[\phi(t)] = \frac{\phi_\Delta}{2W} \Pi\left(\frac{f}{2W}\right)$, $\mathcal{F}[\phi^2(t)] = \frac{\phi_\Delta}{2W} \Lambda\left(\frac{f}{2W}\right)$

$$X_c(f) = \frac{1}{2} A_c \left\{\delta(f - f_c) + \frac{j\phi_\Delta}{2W} \Pi\left(\frac{f - f_c}{2W}\right) - \frac{\phi_\Delta}{4W} \Lambda\left(\frac{f - f_c}{2W}\right)\right\}, f \geq 0$$

5.1–3 $\beta = 8 \text{ kHz}/4 \text{ kHz} = 2$

$f_c = 30 \text{ kHz}$

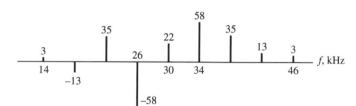

$f_c = 11 \text{ kHz}$

Note "folded" terms at

$|11 - 12| = 1 \text{ kHz}$

$|11 - 16| - 5 \text{ kHz}$

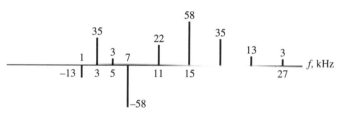

5.2–1

D	2M(D)	Approximation
0.3	3.0	$2(D + 1) = 2.6$
3.0	10	$2(D + 2) = 10$
30	$\cdots$	$2(D + 1) = 62$

5.2–2 Since $H(f + f_c) = e^{-j2\pi t_1 f}$, we have $K_0 = 1$, $K_1 = 0$,

and $t_0 = 0$ in Eq. (12), so $A(t) = A_c$,

$\phi(t - t_1) = \beta \sin \omega_m(t - t_1)$

$= \beta(\cos \omega_m t_1 \sin \omega_m t - \sin \omega_m t_1 \cos \omega_m t)$

$\approx \beta(\sin \omega_m t - \omega_m t_1 \cos \omega_m t) \quad \omega_m t_1 \ll \pi$

and $y_c(t) \approx A_c \cos\left(\omega_c t + \beta \sin \omega_m t - \beta \omega_m t_1 \cos \omega_m t\right)$

For Eq. (14), $|H(f_c)| = 1$ and $f(t) = f_c + \beta f_m \cos \omega_m t$, so

$\arg H[f(t)] = -2\pi t_1[f(t) - f_c] = -\beta \omega_m t_1 \cos \omega_m t$ and

$y_c(t) = A_c \cos\left(\omega_c t + \beta \sin \omega_m t - \beta \omega_m t_1 \cos \omega_m t\right)$

5.3–1 $x_c(t) = A_c \cos \omega_c t - A_c \phi_\Delta x \sin \omega_c t = A_c \sqrt{1 + (\phi_\Delta x)^2} \cos \left[\omega_c t + \arctan (\phi_\Delta x) \right]$

Thus, $\phi(t) = \arctan (\phi_\Delta x) = \phi_\Delta x(t) - \frac{1}{3} \phi_\Delta^3 x^3(t) + \frac{1}{5} \phi_\Delta^5 x^5(t) + \cdots$

5.3–2 $x_c(t) - x_c(t - t_0) \approx t_0 \dot{x}_c(t) = t_0 2\pi A_c [f_c + f_\Delta x(t)] \sin \left[\theta_c(t) \pm 180° \right]$

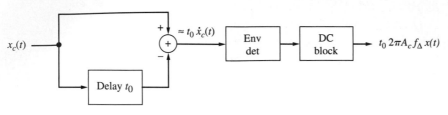

5.4–1 $1 + \cos \theta_i = 2 \cos^2 \dfrac{\theta_i}{2}$ so

$$A_v(t) = A_c \sqrt{2 + 2 \cos \theta_i} = A_c \sqrt{4 \cos^2 \frac{\theta_i}{2}} = 2A_c \left| \cos \frac{\omega_i t}{2} \right|$$

$$\frac{\sin \theta_i}{1 + \cos \theta_i} = \frac{2 \sin \dfrac{\theta_i}{2} \cos \dfrac{\theta_i}{2}}{2 \cos \dfrac{\theta_i}{2}} = \tan \frac{\theta_i}{2} \text{ so}$$

$$\phi_v(t) = \arctan \left(\tan \frac{\theta_i}{2} \right) = \frac{\omega_i t}{2}$$

Envelope detector

Phase detector

Frequency detector

5.4–2 $A_{mpe} = A_m \sqrt{1 + (f/B_{de})^2} \leq \dfrac{1 \text{ kHz}}{15 \text{ kHz}} \sqrt{1 + 7.5^2} \approx 0.5$

$\beta = 0.5 \times 75 \text{ kHz}/15 \text{ kHz} = 2.5, \ M(\beta) \approx 4.5$

$B \approx 2 \times 4.5 \times 15 \text{ kHz} = 135 \text{ kHz} < B_T$

6.1–1 $s_p(t) = \sum\limits_{n=-\infty}^{\infty} c_n e^{-jn\omega_s t}$, $p(t) = s_p(t)\Pi\left(\dfrac{t}{T_s}\right) \Rightarrow c_n = \dfrac{1}{T_s} P(nf_s)$

Thus, $S_p(f) = \mathcal{F}\left[\sum\limits_{n} \dfrac{1}{T_s} P(nf_s)e^{-jn\omega_s t}\right] = f_s \sum\limits_{n=-\infty}^{\infty} P(nf_s)\delta(f - nf_s)$

6.1–2 Sample values are identical, so the reconstructed waveforms will be the same for both signals.

6.2–1 $\dfrac{1}{\tau} = \dfrac{1}{0.1T_s} = 10f_s = 80$ kHz, $B_T \geq \dfrac{1}{2\tau} = 40$ kHz

6.3–1 $c_n = f_s\tau \,\text{sinc}\, nf_s\tau = \dfrac{1}{\pi n}\sin \pi nf_s\tau$

$x_p(t) = A\left[f_s\tau + \sum\limits_{n=1}^{\infty} \dfrac{2}{\pi n}\sin \pi nf_s\tau \cos n\omega_s t\right]$ $\quad \tau = \tau_0[1 + \mu x(t)]$

$= Af_s\tau_0[1 + \mu x(t)] + \sum\limits_{n=1}^{\infty} \dfrac{2A}{\pi n}\sin\{n\pi f_s\tau_0[1 + \mu x(t)]\}\cos n\omega_s t$

7.1–1 $f_{IF} = 7.0$ and $10 < f_{LO} < 10.5$ with $f_c' - 10.5 = 7 \Rightarrow 17 < f_c' < 17.5$

$f_{IF} = 7.0$ and $30 < f_{LO} < 31.5$ with $31.5 - f_c'' = 7 \Rightarrow 23 < f_c'' < 24.5$

$f_{IF} = 7.0$ and $30 < f_{LO} < 31.5$ with $f_c''' - 31.5 = 7 \Rightarrow 37 < f_c''' < 38.5$

With 1st order Butterworth LPF, spurious rejection is

$\left[20 \log \dfrac{1}{\sqrt{1 + (f/4)^2}}\right]_{f=17,\,23,\,37 \text{ MHz}} = -12.8$ dB, -15.3 dB, and

-19.4 dB

7.1–2 $H_{RF}(f_c) = 1$, $H_{RF}(f_c') = \left[1 + jQ\left(x - \dfrac{1}{x}\right)\right]^{-1}$ where $x = \dfrac{f_c'}{f_c}$

$RR = 1 + 50^2\left(x - \dfrac{1}{x}\right)^2 = 10^6 \Rightarrow x \approx 20$ or $\dfrac{1}{20}$

But $\dfrac{f_c'}{f_c} = 1 + \dfrac{2f_{IF}}{f_c} > 1$ so take $\dfrac{f_c'}{f_c} \approx 20$ and $f_{IF} \approx 9.5f_c$

7.2–1 $(v_2 \cos \omega_2 t)^2 v_1 \cos \omega_1 t = \tfrac{1}{2}v_2^2(1 + \cos 2\omega_2 t)v_1 \cos \omega_1 t$

$= \tfrac{1}{2}v_1 v_2^2 \cos \omega_1 t + $ components at $|2f_2 \pm f_1|$

AM: $v_1 v_2^2 = 1 + x_1(t) + \underline{2x_2(t)} + \underline{2x_1(t)x_2(t) + x_1(t)x_2^2(t) + x_2^2(t)}$

                           intelligible   unintelligible

DSB: $v_1 v_2^2 = x_1(t)x_2^2(t)$ unintelligible

7.2–2 $\tau = \frac{1}{2} \times \frac{1}{8} \times \frac{1}{750}$, $B_T \geq \frac{1}{2\tau} = 8 \times 750 = 6$ kHz

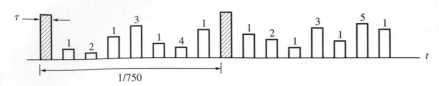

7.2–3 $T_g = (-60)/(-54.5 \times 4 \times 10^5) \approx 2.8$ μs,

$T_s/M = 1/(10 \times 8 \times 10^3) = 12.5$ μs,

$\tau = 12.5/5 = 2.5$ μs, $t_0 \leq \frac{1}{2}(12.5 - 2.5 - 2.8) \approx 3.6$ μs

7.3–1

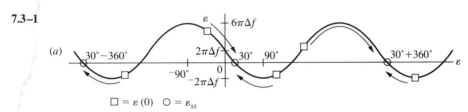

(a)

$\square = \varepsilon(0)$ $\bigcirc = \varepsilon_{ss}$

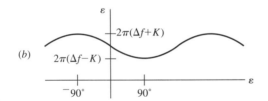

(b)

$\varepsilon > 0$ for all ε
so $\varepsilon(t)$ continually increases
and ε_{ss} does not exist

7.3–2

$$f_v = \frac{nf_c}{m} - \Delta f, \quad K \geq |\Delta f| = \left| f_v - \frac{nf_c}{m} \right|$$

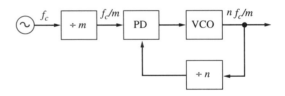

7.4–1 $n_p = (37 \text{ cm} \times 40 \text{ lines/cm})(59 \text{ cm} \times 40 \text{ lines/cm})$

$\approx 3.5 \times 10^6$

$$T_{frame} = \frac{1}{3.2 \times 10^3} \times \frac{0.714 n_p}{1 \times 1} = 781 \text{ sec} \approx 13 \text{ min}$$

7.4–2 $-\sin \omega_{cv} t = \cos(\omega_{cv} t + 90°)$

$A_{ca} \ll A_{cv}$, $|\mu x| < 1$, and $|\mu x_q| \ll 1$

Thus, $A_y \approx A_{cv}(1 + \mu x) + A_{ca}\cos(\omega_a t + \phi)$

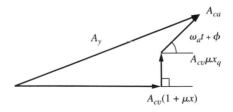

8.1–1 $M = 9$ equally likely outcomes

$P(A) = 4/9$

$P(B) = 3/9 = 1/3$

$P(AB) = P(GG + RR) = 2/9$

$P(A + B) = 5/9$

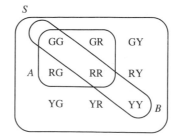

8.1–2 $P(D) = 4/8$, $P(BD) = 1/8$,

$P(B|D) = (1/8)/(4/8) = 2/8 = P(B)$

$P(D|B) = (1/8)/(2/8) = 4/8 = P(D)$,

$P(B)P(D) = (2/8)(4/8) = 1/8 = P(BD)$

8.2–1

Outcome	GG	GR	GY	RG	RR	RY	YG	YR	YY
Weights	2,2	2,−1	2,0	−1,2	−1,−1	−1,0	0,2	0,−1	0,0
X	2.0	0.5	1.0	0.5	−1.0	−0.5	1.0	−0.5	0.0

x_i	−1.0	−0.5	0.0	0.5	1.0	2.0
$P_X(x_i)$	1/9	2/9	1/9	2/9	2/9	1/9
$F_X(x_i)$	1/9	3/9	4/9	6/9	8/9	9/9

$P(-1.0 < X \le 1.0) = F_X(1.0) - F_X(-1.0) = 7/9$

8.2–2 $P(\pi < X < 3\pi/2) = \displaystyle\int_{\pi}^{3\pi/2} \frac{1}{2\pi}\, dx = \frac{1}{4}$,

$P(X > 3\pi/2) = \displaystyle\int_{3\pi/2}^{2\pi} \frac{1}{2\pi}\, dx = \frac{1}{4}$

$P(\pi < Z \le 3\pi/2) = \displaystyle\int_{\pi^+}^{3\pi/2} \frac{1}{2\pi}\, dz = \frac{1}{4}$,

$P(\pi \le Z \le 3\pi/2) = \displaystyle\int_{\pi^-}^{3\pi/2} \left[\frac{1}{2}\delta(z + \pi) + \frac{1}{2\pi} \right] dz = \frac{3}{4}$

8.2–3 $p_X(x) = 1/4$ for $0 < x \le 4$, $g^{-1}(z) = z^2 \Rightarrow dg^{-1}(z)/dz = 2z$

$p_Z(z) = z/2 \quad 0 < z \le 2$

$ = 0 \quad$ otherwise

8.3–1 $m_X = \displaystyle\int_0^{2\pi} x\frac{1}{2\pi}\, dx = \pi \quad \overline{X^2} = \displaystyle\int_0^{2\pi} x^2\frac{1}{2\pi}\, dx = \frac{4\pi^2}{3}$

$\sigma_X = \sqrt{(4\pi^2/3) - \pi^2} = \pi/\sqrt{3}$,

$P(|X - m_X| < 2\sigma_X)$

$\quad = P(\pi - 2\pi/\sqrt{3} < X < \pi + 2\pi/\sqrt{3}) = 1$

8.3–2 $E[X + Y] = \displaystyle\int_{-\infty}^{\infty} \int_{-\infty}^{\infty} (x + y)p_{XY}(x, y)\, dx\, dy$

$\qquad\qquad = \displaystyle\int_{-\infty}^{\infty} x\left[\int_{-\infty}^{\infty} p_{XY}(x, y)\, dy\right] dx$

$\qquad\qquad\quad + \displaystyle\int_{-\infty}^{\infty} y\left[\int_{-\infty}^{\infty} p_{XY}(x, y)\, dx\right] dy$

$\qquad\qquad = \displaystyle\int_{-\infty}^{\infty} x p_X(x)\, dx + \int_{-\infty}^{\infty} y p_Y(y)\, dy = \overline{X} + \overline{Y}$

8.3–3 $\Phi_X(2\pi t) = \mathscr{F}^{-1}[a^{-1}\Pi(f/a)] = \operatorname{sinc} at$, so

$\qquad \Phi_X(\nu) = \operatorname{sinc} at|_{t = \nu/2\pi} = \operatorname{sinc}(a\nu/2\pi)$

8.4–1 $m = 10^4 \times 5 \times 10^{-5} = 0.5$, $P_I(i) \approx e^{-0.5}(0.5^i/i!)$

$\qquad F_I(2) \approx e^{-0.5}\left(\dfrac{0.5^0}{0!} + \dfrac{0.5^1}{1!} + \dfrac{0.5^2}{2!}\right) = 0.986$

8.4–2 $\sigma = 8$ so $9 = m + 0.5\sigma$, $25 = m + 2.5\sigma$

$\qquad P(9 < X \le 25) = P(X > 9) - P(X \ge 25)$

$\qquad\qquad\qquad\qquad = P(X - m > 0.5\sigma) - P(X - m \ge 2.5\sigma)$

$\qquad\qquad\qquad\qquad = Q(0.5) - Q(2.5) \approx 0.31 - 0.06 \approx 0.30$

8.4–3 $F_R(r) = \displaystyle\int_{-\infty}^{r} p_R(\lambda)\, d\lambda = \int_0^r \left(\dfrac{\lambda}{\sigma^2}\right) e^{-\lambda^2/2\sigma^2}\, d\lambda \quad r \ge 0$

$\qquad$ Let $\alpha = \lambda^2/2\sigma^2$ so

$\qquad F_R(r) = \displaystyle\int_0^{r^2/2\sigma^2} e^{-\alpha}\, d\alpha = 1 - e^{-r^2/2\sigma^2} \quad r \ge 0$

9.1–1 $\overline{v(t)} = E[X + 3t] = \overline{X} + 3t = 3t$

$\qquad R_v(t_1, t_2) = E[X^2 + 3(t_1 + t_2)X + 9t_1 t_2]$

$\qquad\qquad\qquad = \overline{X^2} + 3(t_1 + t_2)\overline{X} + 9t_1 t_2 = 5 + 9t_1 t_2$

$\qquad \overline{v^2(t)} = R_v(t, t) = 5 + 9t^2$

9.1–2 $E[z^2(t_1, t_2)] = E[v^2(t_1) + v^2(t_2) \pm 2v(t_1)v(t_2)]$

$\qquad\qquad\qquad = \overline{v^2(t_1)} + \overline{v^2(t_2)} \pm 2R_v(t_1, t_2) \ge 0$

$\qquad$ Since $\overline{v^2(t)} = R_v(0)$ for all t,

$\qquad |R_v(\tau)| = |R_v(t, t - \tau)|$

$\qquad\qquad\qquad \le \tfrac{1}{2}[\overline{v^2(t)} + \overline{v^2(t - \tau)}] = R_v(0)$

9.1–3 Being produced by a linear operation on a gaussian process, $w(t)$ is another gaussian process with

$$R_w(t_1, t_2) = E[4v(t_1)v(t_2) - 16v(t_1) - 16v(t_2) + 64]$$

$$= 4R_v(t_1, t_2) - 16[\overline{v(t_1)} + \overline{v(t_2)}] + 64$$

$$= 36e^{-5|t_1 - t_2|} + 64$$

Thus, $R_w(\tau) = 36e^{-5|\tau|} + 64$ and

$\overline{w^2} = R_w(0) = 100, \quad m_w = \sqrt{R_w(\pm\infty)} = 8,$

$\sigma_w = \sqrt{100 - 8^2} = 6$

Hence, $w(t)$ is stationary and ergodic.

9.2–1 $R_z(\tau) = E[v(t)v(t - \tau) - m_V v(t - \tau) - m_V v(t) + m_V^2]$

$\qquad = R_v(\tau) - m_V^2 - m_V^2 + m_V^2$

Thus, $R_v(\tau) = R_z(\tau) + m_V^2 \Rightarrow G_v(f) = G_z(f) + m_V^2 \delta(f)$

9.2–2 Let $w(t)$ be a randomly phased sinusoid with $A = 1$, so

$G_w(f) = \frac{1}{4}[\delta(f - f_c) + \delta(f + f_c)]$ and

$G_z(f) = G_v(f) * G_w(f) = \frac{1}{4}[G_v(f - f_c) + G_v(f + f_c)]$

9.2–3 $G_x(f) = \sigma^2 D \, \text{sinc}^2 fD \approx \sigma^2 D$ for $|f| \ll 1/D$. Thus, if $B \ll 1/D$,

$G_y(f) \approx \dfrac{1}{1 + (f/B)^2}\, \sigma^2 D \quad \text{and} \quad R_y(\tau) \approx \sigma^2 D \pi B e^{-\pi B |\tau|}$

9.3–1 $\overline{v^2} = \dfrac{2(\pi 4 \times 10^{-22})^2}{3 \times 6.62 \times 10^{-34}} \times 1000 = 1.6 \times 10^{-6}\ \text{V}^2,$

$\sigma_V \approx 1.26\ \text{mV} \quad h/2k\mathcal{T} \approx 8 \times 10^{-13} \text{ so}$

$h|f|/2k\mathcal{T} \ll 1 \text{ for } |f| \le 10^9$

$\displaystyle\int_{-10^9}^{10^9} G_v(f)\, df \approx 2 \times 10^9 G_v(0) = 1.6 \times 10^{-9}\ \text{V}^2, \text{ and}$

$\dfrac{1.6 \times 10^{-9}}{1.6 \times 10^{-6}} = 0.1\%$

9.3–2 $|H(f)|^2 = 1/[1 + (f/B)^{2n}]$ and $g = |H(0)|^2 = 1$ so

$B_N = \displaystyle\int_0^\infty \dfrac{df}{1 + (f/B)^{2n}} = B \int_0^\infty \dfrac{d\lambda}{1 + \lambda^{2n}}$

$\qquad = B\,\dfrac{\pi/2n}{\sin(\pi/2n)} = \dfrac{\pi B}{2n \sin(\pi/2n)}$

and $[\sin(\pi/2n)]/(\pi/2n) = \text{sinc}\,(1/2n) \to 1$ as $n \to \infty$

9.4–1 $S_{R_{dBm}} + 174 - 10 \log_{10} (5 \times 4.2 \times 10^6) \geq 50 \text{ dB}$

$$\Rightarrow S_R \geq -51 \text{ dBm} = 8.4 \times 10^{-6} \text{ mW}$$

$S_T \geq 10^{14} S_R = 840 \text{ kW without repeater}$

$S_T \geq 840 \text{ kW}/(5 \times 10^6) = 168 \text{ mW with repeater}$

9.5–1 (a) $\sigma_A/A = \sqrt{N_0/2E_p} = 0.1,$

$\sigma_t/\tau = \sqrt{N_0/4BE_p\tau} = \sqrt{N_0/2E_p} = 0.1$

(b) $\sigma_A/A = \sqrt{N_0 B_T/A^2} = \sqrt{N_0 B_T \tau/E_p} = 0.4,$

$\sigma_t/\tau = \sqrt{N_0/4B_T E_p\tau} = 0.025$

9.5–2 $h_{opt}(t) = (2K/N_0)[u(t_d - t) - u(t_d - t - \tau)]$

$$= \begin{cases} 1 & t_d - \tau < t < t_d \\ 0 & \text{otherwise} \end{cases}$$

so, with $2K/N_0 = 1$, $h_{opt}(t) = u(t - t_d + \tau) - u(t - t_d)$

Realizability requires $h_{opt}(t) = 0$ for $t < 0$, so take $t_d \geq \tau$.

$h_{opt}(t) * x_R(t) = A\Lambda\left(\dfrac{t - t_d}{\tau}\right)$ where, at $t = t_d$,

$$A = \int_{t_d-\tau}^{t_d} A_p \, d\lambda = A_p\tau.$$

10.1–1 $G_n(f)$ for $f > 0$

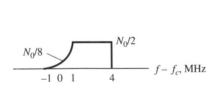
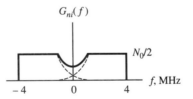

$$G_{n_i}(0) = 2 \times N_0/8 = N_0/4$$

10.1–2 $\overline{A_n} = \sqrt{\pi \times \dfrac{10^{-6}}{2}} \approx 1.3 \text{ mV and } \overline{A_n^2} = 2 \times 10^{-6}$

so $\sigma_{A_n} = \sqrt{2 - \dfrac{\pi}{2}} \times 10^{-3} = 0.655 \text{ mV}$

Let $a^2 = (2\overline{A_n})^2 = 2\pi N_R$ so $P(A_n > a) = e^{-\pi} = 0.043$

10.2–1 $G_n(f)$

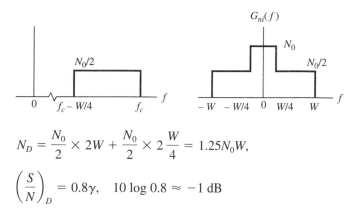

$$N_D = \frac{N_0}{2} \times 2W + \frac{N_0}{2} \times 2\frac{W}{4} = 1.25N_0W,$$

$$\left(\frac{S}{N}\right)_D = 0.8\gamma, \quad 10 \log 0.8 \approx -1 \text{ dB}$$

10.2–2 $S_x = 1 \Rightarrow A_c^2 = S_R, \qquad P(A_c \geq A_n) = 0.99 \Rightarrow P(A_n > A_c) = 0.01$

Thus, $e^{-A_c^2/2N_R} = e^{-S_R/2N_R} = 0.01$ and $\left(\dfrac{S}{N}\right)_{R_{th}} = 2 \ln \dfrac{1}{0.01} = 4 \ln 10 = 9.2$

10.3–1 $|H_{de}(f)|^2 G_\xi(f) = \dfrac{N_0}{2S_R} \dfrac{f^2}{1 + (f/B_{de})^2} \Pi\left(\dfrac{f}{B_T}\right) \approx \dfrac{N_0}{2S_R} B_{de}^2$

for $B_{de} < |f| < B_T/2$

$$N_D \approx \frac{N_0}{2S_R} B_{de}^2 \times 2\frac{B_T}{2} = \frac{N_0 B_{de}^2 B_T}{2S_R} - \left(\frac{B_T}{2W}\right)\frac{N_0 B_{de}^2 W}{S_R}$$

10.3–2 $\left(\dfrac{f_\Delta}{B_{de}}\right)^2 S_x \dfrac{S_R(FM)}{N_0W} = \phi_\Delta^2 S_x \dfrac{S_R(PM)}{N_0W}$ where $\phi_\Delta \leq \pi$ and $S_T(FM) = 1$ W

$$\frac{S_T(PM)}{S_T(FM)} = \left(\frac{f_\Delta}{\phi_\Delta B_{de}}\right)^2 \geq \left(\frac{7.5}{\pi \times 2.1}\right)^2 \Rightarrow S_T(PM) \geq 130 \text{ W}$$

10.3–3 $B_T = 5W \Rightarrow \gamma_{th} = 10 \times 5 = 50$

so $\left(\dfrac{S}{N}\right)_D \geq \left(\dfrac{10B_{de}}{B_{de}}\right)^2 \times \dfrac{1}{2} \times 50 = 2500 \approx 34$ dB

10.6–1 $\tau_{max} = \tau_0(1 + \mu) \leq T_s$ and $\tau_{min} = \tau_0(1 - \mu) \geq 0$ so

$\tau_{max} - \tau_{min} = 2\mu\tau_0 \leq T_s \Rightarrow \mu\tau_0 \leq T_s/2 = 1/4W$

$$\left(\frac{S}{N}\right)_D = 4(\mu\tau_0)^2 B_T\left(\frac{W}{f_s\tau_0}\right)S_x\gamma = 4\mu^2\tau_0 B_T \frac{W}{f_s} S_x\gamma \leq \frac{1}{2}\frac{B_T}{W}S_x\gamma$$

since $\mu \leq 1$, $\tau_0 \leq T_s/2 = 1/2f_s$, and $f_s \geq 2W$

11.1–1 $\sin c^2 \, at = \begin{cases} 1 & t = 0 \\ 0 & t = \pm\frac{1}{a}, \pm\frac{2}{a}, \ldots \end{cases}$ so take $r = a$

$\mathcal{F}[\sin c^2 \, at] = \dfrac{1}{a} \Lambda\left(\dfrac{f}{a}\right) = 0$ for $|f| > a$ so $B \geq a \Rightarrow r \leq B$

11.1–2 $P(f) = \dfrac{1}{r_b} \sin c\left(\dfrac{f}{r_b}\right) = 0$ for $f = \pm r_b, \pm 2r_b, \ldots$

Thus, $G_x(f) = \dfrac{A^2}{4r_b} \sin c^2 \dfrac{f}{r_b} + \dfrac{A^2}{4} \delta(f)$

$\overline{x^2} = A^2/2$ by inspection of $x(t)$ or integration of $G_x(f)$

11.1–3 $P(f) = \dfrac{1}{r_b} \Pi\left(\dfrac{f}{r_b}\right) = 0$ for $|f| > \dfrac{r_b}{2}$

Thus, $G_x(f) = \dfrac{A^2}{r_b} \sin^2 \dfrac{\pi f}{r_b} \Pi\left(\dfrac{f}{r_b}\right)$, $\quad \overline{x^2} = \frac{1}{2}\dfrac{A^2}{r_b} \times 2\dfrac{r_b}{2} = \dfrac{A^2}{2}$

11.2–1 $A/\sigma = 2\sqrt{\frac{1}{2} \times 50} = 10$

$P_{e_0} = Q(0.4 \times 10) \approx 3.4 \times 10^{-5}$, $P_{e_1} = Q(0.6 \times 10) \approx 1.2 \times 10^{-9}$

$P_e = \frac{1}{2}(P_{e_0} + P_{e_1}) \approx 1.7 \times 10^{-5}$

whereas $P_{e_{\min}} = Q(0.5 \times 10) \approx 3 \times 10^{-7}$

11.2–2   (b) $S_R = \frac{1}{4} A^2$, $\tau = \frac{1}{2} T_b = \dfrac{1}{2r_b}$, $\sigma^2 = \dfrac{N_0}{2\tau} = N_0 r_b$

$(A/2\sigma)^2 = A^2/4\sigma^2 = 4S_R/4N_0r_b = S_R/N_0r_b = \gamma_b$

11.2–3 $P_e = \frac{1}{2} \times 2Q(A/2\sigma) + 2 \times \frac{1}{4}Q(A/2\sigma) = \frac{3}{2}Q(A/2\sigma)$

$S_R = \frac{1}{4}A^2 + \frac{1}{4}(-A)^2 + \frac{1}{2}0 = \dfrac{A^2}{2}$, $\left(\dfrac{A}{2\sigma}\right)^2 = \dfrac{2S_R}{4N_R} \leq \dfrac{S_R}{2N_0r_b/2} = \gamma_b$

so $P_e = \frac{3}{2}Q(\sqrt{\gamma_b})$

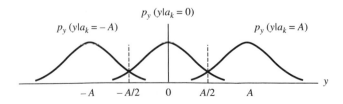

11.3–1   Note that $P(f)$ has even symmetry, so consider only $f > 0$.

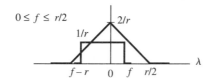

$$P(f) = \frac{1}{r}\left[\int_{f-r}^{0} \frac{2}{r}\left(1 + \frac{2\lambda}{r}\right)d\lambda + \int_{0}^{f} \frac{2}{r}\left(1 - \frac{2\lambda}{r}\right)d\lambda\right] = \frac{1}{r}\left(1 - \frac{2f^2}{r^2}\right)$$

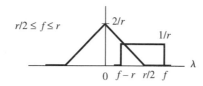

$$P(f) = \frac{1}{r}\int_{f-r}^{r/2} \frac{2}{r}\left(1 - \frac{2\lambda}{r}\right)d\lambda = \frac{2}{r}\left(1 - \frac{f}{r}\right)^2$$

Thus, $P(f) = \begin{cases} \dfrac{1}{r}\left(1 - \dfrac{2f^2}{r^2}\right) & |f| \le \dfrac{r}{2} \\[3mm] \dfrac{2}{r}\left(1 - \dfrac{|f|}{r}\right)2 & \dfrac{r}{2} \le |f| \le r \end{cases}$

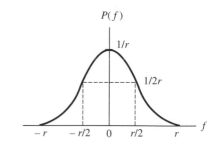

$$p_\beta(t) = \operatorname{sinc}^2\frac{rt}{2}, \quad p(t) = \operatorname{sinc}^2\frac{rt}{2}\operatorname{sinc} rt$$

No additional zero-crossings, but $|p(t)| < 0.01$ for $|t| > 2D$.

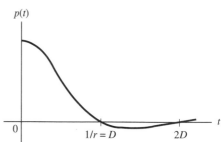

11.3–2   Let $I_{HR} = \displaystyle\int_{-\infty}^{\infty} |V(f)|^2\, df \int_{-\infty}^{\infty} |W(f)|^2\, df$ with

$$V(f) = |H_R(f)|\,\sqrt{G_n(f)}\,, \quad W(f) = \frac{|P(f)|}{|H_c(f)|\,|H_R(f)|}$$

Then I_{HR} is minimized when $V(f) = gW(f)$, so

$$|H_R(f)| \sqrt{G_n(f)} = g \frac{|P(f)|}{|H_c(f)||H_R(f)|} \Rightarrow |H_R(f)|^2 = \frac{g|P(f)|}{\sqrt{G_n(f)} \; |H_c(f)|},$$

and $|H_T(f)|^2 = \dfrac{|P(f)|^2}{|P_x(f)H_c(f)H_R(f)|^2} = \dfrac{|P(f)| \sqrt{G_n(f)}}{g|P_x(f)|^2|H_c(f)|}$

11.3–3

m_k	m'_{k-2}	m'_k	$m'_k - m'_{k-2}$
0	0	0	0
0	1	1	0
1	0	1	1
1	1	0	-1

$$y(t_k) = (m'_k - m'_{k-2})A = \begin{cases} 0 & m_k = 0, m'_{k-2} = 0 \\ 0 & m_k = 0, m'_{k-2} = 1 \\ A & m_k = 1, m'_{k-2} = 0 \\ -A & m_k = 1, m'_{k-2} = 1 \end{cases}$$

11.4–1 $m_1 = m_2 + m_3 + m_4 + m_5$ and output $= m_5$

shift	m_1	m_2	m_3	m_4	m_5	shift	m_1	m_2	m_3	m_4	m_5
0	1	1	1	1	1	16	0	1	1	0	1
1	0	1	1	1	1	17	1	0	1	1	0
2	0	0	1	1	1	18	0	1	0	1	1
3	1	0	0	1	1	19	1	0	1	0	1
4	0	1	0	0	1	20	0	1	0	1	0
5	0	0	1	0	0	21	0	0	1	0	1
6	1	0	0	1	0	22	0	0	0	1	0
7	1	1	0	0	1	23	1	0	0	0	1
8	0	1	1	0	0	24	1	1	0	0	0
9	0	0	1	1	0	25	1	1	1	0	0
10	0	0	0	1	1	26	0	1	1	1	0
11	0	0	0	0	1	27	1	0	1	1	1
12	1	0	0	0	0	28	1	1	0	1	1
13	0	1	0	0	0	29	1	1	1	0	1
14	1	0	1	0	0	30	1	1	1	1	0
15	1	1	0	1	0	31	1	1	1	1	1

12.1–1 $\nu f_s \le 36{,}000$ and $f_s \ge 2W = 6400$ $\nu \le \dfrac{36{,}000}{6400} = 5.6 \Rightarrow \nu = 5$

so $q = 2^5 = 32$, $f_s = r/\nu = 7.2$ kHz

12.1–2 (a) $4.8 + 6\nu \ge 50$ dB $\Rightarrow \nu = 8$, $r = \nu f_s = 80$ Mbps

(b) $(S/N)_D = 4.8 + 6\nu - 10 \ge 50$ dB $\Rightarrow \nu = 11$, $r = 110$ Mbps

12.1–3 3 − bit quantizer $\Rightarrow$ 8 levels, with $x_{max} = 8.75$ V $\Rightarrow$ step size $= 2.5$ V.

For an input of 0.6 V $\Rightarrow x_q = 1.25$ V $\Rightarrow \varepsilon_q = 1.25 - 0.60 = 0.65$ V.

With companding: $z(x) = 8.75\left(\dfrac{\ln(1 + 255 \times 0.6/8.75)}{\ln(1 + 255)}\right) = 4.60$

4.601 feeds to a quantizer $\Rightarrow x_q' = 3.75$ V.

x_q' is then expanded using Eq. (13):

$$\hat{x} = \frac{8.75}{255}\left[(1 + 255)^{3.75/8.75} - 1\right] = 0.34$$

$\varepsilon_q' = 0.60 - 0.34 = 0.26$ (with companding) versus

$\varepsilon_q = 0.65$ (without companding)

12.2–1 $1 + 4q^2 P_e = 10^{0.1} = 1.259 \Rightarrow P_e = 0.065/q^2 \approx 10^{-6}$

$M = 2$, $P_e = Q[\sqrt{(S/N)_R}] = 10^{-6} \Rightarrow (S/N)_R \approx 4.76^2 = 13.6$ dB

Eq. (5) gives $(S/N)_{R_{th}} = 6(2^2 - 1) = 12.6$ dB

12.2–2 $\gamma_{th} \approx 6\dfrac{B_T}{W}(M^2 - 1) \Rightarrow M_{th}^2 = 1 + \dfrac{W}{6B_T}\gamma_{th} = 1 + \dfrac{\gamma_{th}}{6b}$

Thus, $(S/N)_{D_{th}} = 3M_{th}^{2b}S_x = 3\left(1 + \dfrac{\gamma_{th}}{6b}\right)^b S_x$

For WBFM, $(S/N)_{D_{th}} = 3(b/2)^2 S_x \gamma_{th} = \frac{3}{4}b^2 \gamma_{th} S_x$

12.3–1 $W_{rms}^2 = \dfrac{1}{S_x}\displaystyle\int_{-W}^{W} f^2 \dfrac{S_x}{2W}\, df = \dfrac{W^2}{3} \Rightarrow W_{rms} = \dfrac{W}{\sqrt{3}}$

$s = \dfrac{f_s \Delta \sqrt{3}}{2\pi\sigma W} = \dfrac{\Delta \sqrt{3}\, b}{\pi \sqrt{S_x}}$,

$s_{opt} \approx \ln 2b \Rightarrow \Delta_{opt} = \dfrac{\pi \sqrt{S_x}}{\sqrt{3b}}\ln 2b = 0.393\sqrt{S_x}$

12.3–2 PCM: $(S/N)_D = 4.8 + 6.0\nu + 10\log_{10} S_x$ dB

DPCM: $(S/N)_D = G_{p_{dB}} + 4.8 + 6.0\nu' + 10\log_{10} S_x$ dB

If $G_p = 6$ dB, then $6 + 6.0\nu' = 6.0\nu \Rightarrow \nu' = \nu - 1$

12.4–1 One frame has a total of 588 bits consisting of 33 symbols and 17 bits/symbol.
But, of 17 bits, only 8 are info, so 8 info bits × 33 symbols frame = 264 info bits/frame.

Output is 4.3218 Mbits/sec × one frame/588 bits = 7350 frames/sec.

12.5–1 Voice PCM bits/frame = 30 channels × 8 bits/channel = 240 bits,
$T_{frame} = 1/(8 \text{ kHz}) = 125 \ \mu s$

$$r = \frac{240 + n}{125 \ \mu s} = 2.048 \text{ Mbps} \Rightarrow n = 256 - 240 = 16 \text{ bits/frame}$$

13.1–1 $\alpha = 10^{-3}, n = 15$

$$P(0, n) = (1 - \alpha)^{15} = 0.985, \ P(1, n) = 15\alpha(1 - \alpha)^{14} = 0.0148$$

$$P(2, n) = \frac{15 \times 14}{2} \alpha^2 (1 - \alpha)^{13} = 1.04 \times 10^{-4}$$

$$P(3, n) = \frac{15 \times 14 \times 13}{3 \times 2} \alpha^3 (1 - \alpha)^{12} = 4.50 \times 10^{-7}$$

We see that $P(2, n) \gg P(3, n)$, and $P(4, n)$ will be even smaller, etc.

Hence, $\displaystyle\sum_{i=2}^{n} P(i, n) \approx P(2, n)$

13.1–2 We want $R_c' = r_b/r \geq 0.5$, given $2t_d r_b/k = 2.2$ and $p \approx 10\alpha = 0.011$

Go-back-N: $R_c' \leq \dfrac{9}{10} \dfrac{0.989}{0.989 + 2.2 \times 0.011} = 0.879$ OK

Stop-and-wait: $R_c' \leq \dfrac{9}{10} \dfrac{0.989}{1 + 2.2} = 0.278$ Unacceptable

13.2–1 $(c_1 \quad c_2 \quad c_3) = (m_1 \quad m_2 \quad m_3) \begin{bmatrix} 1 & 1 & 0 \\ 0 & 1 & 1 \\ 1 & 0 & 1 \end{bmatrix}$

$c_1 = m_1 \oplus 0 \oplus m_3, \ c_2 = m_1 \oplus m_2 \oplus 0, \ c_3 = 0 \oplus m_2 \oplus m_3$

$m_1 m_2 m_3$	$c_1 c_2 c_3$	W
0 0 0	0 0 0	0
0 0 1	1 0 1	3
0 1 0	0 1 1	3
0 1 1	1 1 0	4
1 0 0	1 1 0	3
1 0 1	0 1 1	4
1 1 0	1 0 1	4
1 1 1	0 0 0	3

13.2–2 $S = Y\left[P^T \mid I_q^T\right] = (y_1 \quad y_2 \quad \cdots \quad y_n)\begin{bmatrix} p_{11} & p_{21} & \cdots & p_{k1} & | & 1 & 0 & \cdots & 0 \\ p_{12} & p_{22} & \cdots & p_{k2} & | & 0 & 1 & \cdots & 0 \\ \vdots & & & \vdots & | & \vdots & & & \vdots \\ p_{1q} & p_{2q} & \cdots & p_{kq} & | & 0 & 0 & \cdots & 1 \end{bmatrix}^T$

$$s_j = y_1 p_{1j} \oplus y_2 p_{2j} \oplus \cdots \oplus y_k p_{kj} \underbrace{\oplus 0 \oplus 0 \oplus \cdots}_{j-1 \text{ terms}} \oplus y_{k+j} \underbrace{\oplus 0 \oplus 0 \oplus \cdots \oplus 0}_{g-j \text{ terms}}$$

For $P^T = \begin{bmatrix} 1 & 1 & 1 & 0 \\ 0 & 1 & 1 & 1 \\ 1 & 1 & 0 & 1 \end{bmatrix}$

$s_1 = y_1 \oplus y_2 \oplus y_3 \oplus 0 \oplus y_5$, $s_2 = 0 \oplus y_2 \oplus y_3 \oplus y_4 \oplus y_6$,
$s_3 = y_1 \oplus y_2 \oplus 0 \oplus y_4 \oplus y_7$

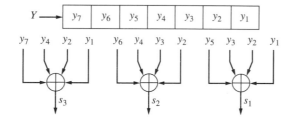

13.2–3 "J" $= 1\,0\,0\,1\,0\,1\,0 \Rightarrow Q_m(p) = p^6 + p^3 + p$
CRC-8: $G(p) = p^8 + p^2 + p + 1$
$X(p) = Q_m(p)G(p) = p^{14} + p^{11} + p^9 + p^8 + p^7 + p^6 + p^5 + p^4 + p^2 + p$
Y is received version of X with errors in first two digits, so
$Y(p) = p^{13} + p^{11} + p^9 + p^8 + p^7 + p^6 + p^5 + p^4 + p^2 + p$

$$\frac{Y(p)}{G(p)} = p^8 + p^2 + p + 1\!\!\overline{\Big)p^{13} + 0 + p^{11} + 0 + p^9 + p^8 + p^7 + p^6 + p^5 + p^4 + 0 + p^2 + p}$$

$$
\begin{array}{l}
\underline{p^{13} \hspace{9.5em} + p^7 + p^6 + p^5} \\
\hspace{3em} p^{11} \hspace{3em} + p^9 + p^8 \hspace{5em} + p^4 \hspace{3em} + p^2 + p \\
\underline{\hspace{3em} p^{11} \hspace{11em} + p^5 + p^4 + p^3} \\
\hspace{5.5em} p^9 + p^8 \hspace{6em} + p^5 \hspace{3em} + p^3 + p^2 + p \\
\underline{\hspace{5.5em} p^9 \hspace{11.5em} + p^3 + p^2 + p} \\
\hspace{8em} p^8 \hspace{6em} + p^5 \\
\underline{\hspace{8em} p^8 \hspace{10.5em} + p^2 + p + 1} \\
\hspace{10.5em} p^5 \hspace{6em} + p^2 + p + 1
\end{array}
$$

$S(p) = \text{rem}\left[\dfrac{Y(p)}{G(p)}\right] = p^5 + p^2 + p + 1 \neq 0 \Rightarrow$ an error has occurred

13.2–4 $n = 63, k = 15 \Rightarrow t = \dfrac{63 - 15}{2} = 24$ errors can be corrected

13.3–1

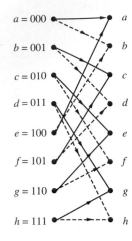

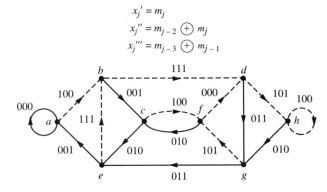

$$x_j' = m_j$$
$$x_j'' = m_{j-2} \oplus m_j$$
$$x_j''' = m_{j-3} \oplus m_{j-1}$$

13.3–2   *(a)*

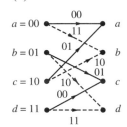

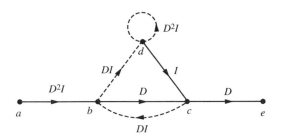

Minimum-weight paths: $\left. \begin{array}{l} abce = D^4I \\ abdce = D^4I^2 \end{array} \right\}$ $d_f = 4, \quad M(d_f) = 1 + 2 = 3$

(b) $\alpha \approx \dfrac{1}{\sqrt{20\pi}} e^{-5} = 8.5 \times 10^{-4} \Rightarrow P_{be} = 3 \times 2^4 \times \alpha^2 = 3.5 \times 10^{-5}$

$P_{ube} \approx \dfrac{1}{\sqrt{40\pi}} e^{-10} = 4.1 \times 10^{-6} < P_{be}$

Coding *increases* error probability when $R_c d_f/2 = 1$.

14.1–1 $B_T \leq 0.1 f_c = 100$ kHz, $r_b \leq (r_b/B_T) \times 100$ kHz
(a) $r_b/B_T \approx 1$ so $r_b \leq 100$ kbps   *(b)* $r_b/B_T \approx 2$ so $r_b \leq 200$ kbps
(c) $r_b/B_T \approx 2 \log_2 8 = 6$ so $r_b \leq 600$ kbps

14.1–2 $\phi_k = \pm \dfrac{\pi}{4} \Rightarrow I_k = \cos\phi_k = \dfrac{1}{\sqrt{2}}, \; Q_k = \sin\phi_k = \pm\dfrac{1}{\sqrt{2}}$

$$x_i(t) = \sum_k \frac{1}{\sqrt{2}} p_{T_b}(t - kT_b) = \frac{1}{\sqrt{2}} \Rightarrow G_i(f) = \frac{1}{2}\delta(f)$$

$$\overline{Q_k} = 0, \; \overline{Q_k^2} = \frac{1}{2} \Rightarrow G_q(f) = \frac{1}{2} r_b |P_{T_b}(f)|^2 = \frac{1}{2r_b} \text{sinc}^2 \frac{f}{r_b}$$

Thus, $G_{lp}(f) = \dfrac{1}{2}\delta(f) + \dfrac{1}{2r_b}\text{sinc}^2\dfrac{f}{r_b}$

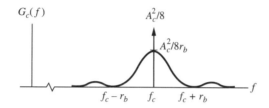

14.1–3 $x_c(t) = A_c \sum_k [\cos(\omega_d a_k t)\cos(\omega_c t + \theta) - \sin(\omega_d a_k t)\sin(\omega_c t + \theta)]p_{T_b}(t - kT_b)$

where $a_k = \pm 1$, $p_{T_b}(t) = u(t) - u(t - T_b)$, $\omega_d = \dfrac{\pi}{T_b} = \pi r_b$

so $\cos(\omega_d a_k t) = \cos\omega_d t$, $\sin(\omega_d a_k t) = a_k \sin\omega_d t$. Thus,

$$x_i(t) = \sum_k \cos(\omega_d a_k t)p_{T_b}(t - kT_b) = \sum_k \cos\omega_d t \; p_{T_b}(t - kT_b) = \cos\omega_d t, \text{ and}$$

$$x_q(t) = \sum_k \sin(\omega_d a_k t)p_{T_b}(t - kT_b) = \sum_k a_k \sin\omega_d t \; p_{T_b}(t - kT_b). \text{ But}$$

$$\sin\omega_d t = \sin\frac{\pi t}{T_b} = \sin\left[\tfrac{\pi}{T_b}(t - kT_b) + k\pi\right]$$

$$= \cos k\pi \sin\left[\tfrac{\pi}{T_b}(t - kT_b)\right] = (-1)^k \sin\left[\pi r_b(t - kT_b)\right]$$

so $x_q(t) = \sum_k \underbrace{(-1)^k a_k}_{Q_k} \underbrace{\sin\left[\pi r_b(t - kT_b)\right]p_{T_b}(t - kT_b)}_{p(T - kT_b)}$

14.2–1   Let $V(\lambda) = s_1(\lambda) - s_0(\lambda)$ and $W*(\lambda) = h(T_b - \lambda)$, so

$$\frac{|z_1 - z_0|^2}{4\sigma^2} = \frac{\left|\int_{-\infty}^{\infty} V(\lambda)W*(\lambda)\,d\lambda\right|^2}{4\frac{N_0}{2}\int_{-\infty}^{\infty}|W(\lambda)|^2\,d\lambda} \leq \frac{1}{2N_0}\int_{-\infty}^{\infty}|V(\lambda)|^2\,d\lambda$$

$$= \frac{1}{2N_0}\int_{-\infty}^{\infty}|s_1(\lambda) - s_0(\lambda)|^2\,d\lambda$$

The equality holds when $W(\lambda) = KV(\lambda)$, so

$$h(T_b - \lambda) = K[s_1(\lambda) - s_0(\lambda)] \Rightarrow h_{opt}(t) = K[s_1(T_b - t) - s_0(T_b - t)]$$

14.2–2 $h(t) = A_c p_{T_b}(T_b - t) \cos\left[2\pi(f_c \pm f_d)(T_b - t)\right]$ with $f_c T_b = N_c, f_d T_b = \frac{1}{2}$

$$= A_c[u(T_b - t) - u(-t)]\cos\left[2\pi(N_c \pm \tfrac{1}{2}) - 2\pi(f_c \pm f_d)t\right]$$

$$= -A_c \cos\left[2\pi(f_c \pm f_d)t\right][u(t) - u(t - T_b)], \quad f_d = \frac{r_b}{2}$$

since $[u(T_b - t) - u(-t)] = [u(t) - u(t - T_b)] = \Pi\left(\dfrac{t - \dfrac{T_b}{2}}{T_b}\right)$

Thus, $|H(f)| = \dfrac{A_c T_b}{2}\left|\text{sinc}\left[\dfrac{f - \left(f_c \pm \dfrac{r_b}{2}\right)}{r_b}\right] + \text{sinc}\left[\dfrac{f + \left(f_c \pm \dfrac{r_b}{2}\right)}{r_b}\right]\right|$

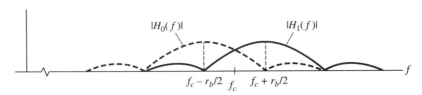

14.3–1 $z(t) = \int_0^t A_c \cos(\omega_c\lambda + \theta) KA_c \cos(\omega_c t - \omega_c\lambda)\,d\lambda \quad KA_c = \dfrac{2}{T_b}$

$$= \frac{2A_c}{T_b}\frac{1}{2}\left[\int_0^t \cos(\omega_c t + \theta)\,d\lambda + \int_0^t \cos(2\omega_c\lambda - \omega_c t + \theta)\,d\lambda\right]$$

$$= \frac{A_c}{T_b}\left[t\cos(\omega_c t + \theta) + \frac{\sin(\omega_c t + \theta) + \sin(\omega_c t - \theta)}{2\omega_c}\right] \quad 0 < t < T_b$$

where $\cos(\omega_c t + \theta) = \cos\omega_c t \cos\theta - \sin\omega_c t \sin\theta$ and

$$\sin(\omega_c t + \theta) + \sin(\omega_c t - \theta) = 2\sin\omega_c t \cos\theta$$

Thus, $z(t) = \dfrac{A_c t}{T_b} \left[\cos\theta \cos\omega_c t - \left(\sin\theta - \dfrac{\cos\theta}{\omega_c t} \right) \sin\omega_c t \right]$ and

$$A_z(t) = \dfrac{A_c t}{T_b} \sqrt{ \cos^2\theta + \left(\sin\theta - \dfrac{\cos\theta}{\omega_c t} \right)^2 }$$

$$= \dfrac{A_c t}{T_b} \sqrt{ 1 - \dfrac{2\sin\theta\cos\theta}{\omega_c t} + \left(\dfrac{\cos\theta}{\omega_c t} \right)^2 }$$

$$\approx \dfrac{A_c t}{T_b} \qquad \omega_c t \gg 1$$

14.3–2 $A_c^2 \leq 2 \times 10^{-6} \text{W}, \quad \gamma_b = \dfrac{A_c^2}{N_0} \dfrac{E_b}{A_c^2} \leq 2 \times 10^{-6} \dfrac{E_b}{A_c^2}$

OOK: $\dfrac{E_b}{A_c^2} = \dfrac{1}{4r_b} = 2.5 \times 10^{-6}$ so $\gamma_b \leq 5, \; P_e \geq \frac{1}{2}\left[e^{-2.5} + Q(\sqrt{5}) \right] \approx 5 \times 10^{-2}$

FSK: $\dfrac{E_b}{A_c^2} = \dfrac{1}{2r_b} = 5 \times 10^{-6}$ so $\gamma_b \leq 10, \; P_e \geq \frac{1}{2} e^{-5} \approx 3 \times 10^{-3}$

DPSK: $\dfrac{E_b}{A_c^2} = \dfrac{1}{2r_b} = 5 \times 10^{-6}$ so $\gamma_b \leq 10, \; P_e \geq \frac{1}{2} e^{-10} \approx 2 \times 10^{-5}$

14.4–1   Let $\psi = \omega_c t + \phi_k$ so

$$x_c^4 = A_c^4 \cos^3\psi \cos\psi = \dfrac{A_c^4}{4} (3\cos\psi + \cos 3\psi)\cos\psi$$

$$= \dfrac{A_c^4}{8}(3 + 4\cos 2\psi + \cos 4\psi)$$

where $\cos 4\psi = \cos(4\omega_c t + 4\phi_k) = -\cos 4\omega_c t$ since $4\phi_k - \pi, 3\pi, 7\pi$

14.4–2   For correlation detection

$$y(t_k) = \int_{kD}^{(k+1)D} x_c(\lambda)\, KA_c \cos\omega_c\lambda \, d\lambda, \quad t_k = (k+1)D$$

For filter detection $y(t_k) = \displaystyle\int_{-\infty}^{\infty} x_c(\lambda)\, h(t_k - \lambda)\, d\lambda$

Thus, $h[(k+1)D - \lambda] = \begin{cases} KA_c \cos\omega_c\lambda & kD \leq \lambda \leq (k+1)D \\ 0 & \text{otherwise} \end{cases}$

So $h(t) = KA_c \cos[(k+1)\omega_c D - \omega_c t] \quad 0 \leq t \leq D$
$\qquad\quad = KA_c \cos\omega_c t \, p_D(t) \quad \text{since} \quad \omega_c D = 2\pi N_c$

$E = \frac{1}{2} A_c^2 D = \dfrac{A_c^2}{2r} \Rightarrow K = \dfrac{A_c}{E} = \dfrac{2r}{A_c}$

$\sigma^2 = \dfrac{N_0}{2} \displaystyle\int_{-\infty}^{\infty} |h(t)|^2 \, dt = \dfrac{N_0}{2} \int_0^D (KA_c)^2 \cos^2\omega_c t \, dt$

$\quad = \dfrac{N_0}{4}(KA_c)^2 D = \dfrac{A_c^2 N_0}{2E} = N_0 r$

15.1–1 $10 \log (J/S_R) = 10 \log (Pg) - 10 \log (E_b/N_J)$
$$= 10 \log (10{,}000) - 10 \log (1.80 \times 10^{-11}/1.48 \times 10^{-13})$$
$$= 19.2 \text{ dB}$$

15.1–2 $P_e = Q(\sqrt{2E_b/N_0}) = 1 \times 10^{-7} \Rightarrow 2E_b/N_0 = 27.04$

Multiple users:

$$P_e = Q\left(\cfrac{1}{\sqrt{(M-1)/3Pg + N_0/2E_b}}\right) = 1 \times 10^{-5}$$

$$= Q\left(\cfrac{1}{\sqrt{(M-1)/3000 + 1/27.04}}\right) \Rightarrow (M-1) = 54$$

$\Rightarrow 54$ total users

15.2–1

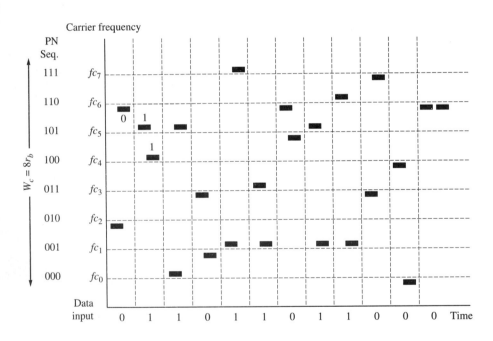

PN Seq.   010 110   101 100   000 101   011 001   001 111   011 001   110 101   101 001   110 001   011 111   100 000   110 110

15.2–2   Multiple users: $P_e = \dfrac{1}{2}\left(\dfrac{M-1}{Y}\right) + \dfrac{1}{2}e^{-E_b/2N_0}\left(1 - \dfrac{M-1}{Y}\right)$

Assume 2nd term does not significantly contribute to the overall error so that with $M = 54$ users, we have

$$10^{-5} = \frac{1}{2}\left(\frac{54-1}{Y}\right) \Rightarrow Y = 2650.$$

But with FH-SS, $Pg = 2^k \Rightarrow k = 12 \Rightarrow Y = 4096$.

Comparing to Exercise 15.1–2, for $M = 54$, and the same P_e, we require $Pg = 1000$.

15.3–1 $m_1 = m_2 + m_5$ and output $= m_5$

shift	m_1	m_2	m_3	m_4	m_5	shift	m_1	m_2	m_3	m_4	m_5
0	1	1	1	1	1	8	0	0	1	0	1
1	0	1	1	1	1	9	1	0	0	1	0
2	0	0	1	1	1	10	0	1	0	0	1
3	1	0	0	1	1	11	0	0	1	0	0
4	1	1	0	0	1	12	0	0	0	1	0
5	0	1	1	0	0	13	0	0	0	0	1
6	1	0	1	1	0	14	1	0	0	0	0
7	0	1	0	1	1	15	0	1	0	0	0

shift	m_1	m_2	m_3	m_4	m_5	shift	m_1	m_2	m_3	m_4	m_5
16	1	0	1	0	0	24	0	1	1	0	1
17	0	1	0	1	0	25	0	0	1	1	0
18	1	0	1	0	1	26	0	0	0	1	1
19	1	1	0	1	0	27	1	0	0	0	1
20	1	1	1	0	1	28	1	1	0	0	0
21	0	1	1	1	0	29	1	1	1	0	0
22	1	0	1	1	1	30	1	1	1	1	0
23	1	1	0	1	1	31	1	1	1	1	1

The above output occurs with all 1s as initial conditions. Any other set of nonzero initial conditions will produce a delayed version of the above output. Therefore, this register configuration only produces one unique sequence. Any n-bit register configured to produce a ml sequence will only have one unique output sequence regardless of initial conditions.

16.1–1 $P_2 + P_3 = 1 - P_1$ so $2P_2 = 2P_3 = 1 - p$ and

$$H(X) = p \log \frac{1}{p} + 2 \frac{1 - p}{2} \log \frac{2}{1 - p} = p \log \frac{1}{p} + (1 - p) \left[\log \frac{1}{1 - p} + \log 2 \right]$$

$$= p \log \frac{1}{p} + (1 - p) \log \frac{1}{1 - p} + (1 - p) = \Omega(p) + 1 - p$$

$$H(X)\big|_{\max} = \log M = \log 3 = 1.58 \text{ at } p = 1/M = 1/3$$

16.1–2

x_i	P_i	1	2	3	Codeword	N_i	I_i
A	1/2	0			0	1	1
B	1/4	1	0		10	2	2
C	1/8	1	1	0	110	3	3
D	1/8	1	1	1	111	3	3

$$N_0 = \tfrac{1}{2} \times 1 + \tfrac{1}{2} \times 1 + \tfrac{1}{8} \times 1 = \tfrac{7}{8} = \overline{N}/2$$

$$N_1 = \tfrac{1}{4} \times 1 + \tfrac{1}{8} \times 2 + \tfrac{1}{8} \times 3 = \tfrac{7}{8} = \overline{N}/2$$

16.1–3   From Table 16.1–5 with $p = 0.9$ we have the data compression $\overline{N}/\overline{E} = 0.50$ so $r_b/r = 0.50$. But $R = rH(X) \le r_b$, so $H(X) \le r_b/r = 0.50$ bits/sample.

16.2–1 $H(Y \mid X) = P(x_1)\left[P(y_1 \mid x_1) \log \dfrac{1}{P(y_1 \mid x_1)} + P(y_2 \mid x_1) \log \dfrac{1}{P(y_2 \mid x_1)} \right]$

$$+ P(x_2)\left[P(y_1 \mid x_1) \log \dfrac{1}{P(y_1 \mid x_2)} + P(y_2 \mid x_2) \log \dfrac{1}{P(y_2 \mid x_2)} \right]$$

$$= p\left[(1 - \alpha) \log \dfrac{1}{1 - \alpha} + \alpha \log \dfrac{1}{\alpha} \right]$$

$$+ (1 - p)\left[\alpha \log \dfrac{1}{\alpha} + (1 - \alpha) \log \dfrac{1}{1 - \alpha} \right]$$

$$= \alpha \log \dfrac{1}{\alpha} + (1 - \alpha) \log \dfrac{1}{1 - \alpha} = \Omega(\alpha)$$

16.2–2 $P(x_i y_j) = P(x_i)P(y_j) \quad P(x_i \mid y_j) = P(x_i y_j)/P(y_j) = P(x_i)$

Thus, $H(X \mid Y) = \displaystyle\sum_{x,y} P(x_i)P(y_j) \log \dfrac{1}{P(x_i)}$

$$= \left[\sum_y P(y_j) \right]\left[\sum_x P(x_i) \log \dfrac{1}{P(x_i)} \right] = 1 \times H(X)$$

so $I(X; Y) = H(X) - H(X \mid Y) = 0$

16.3–1   (a) $p(x) = 0$ for $|x| > M$ so

$$I = \int_{-M}^{M} p(x) \log \dfrac{1}{p(x)} \, dx \quad \text{and} \quad \int_{-M}^{M} p(x) \, dx = 1 \Rightarrow F_1 = p, c_1 = 1$$

$$-\dfrac{(\ln p + 1)}{\ln 2} + \lambda_1 = 0 \Rightarrow \ln p = \lambda_1 \ln 2 - 1 \Rightarrow p = e^{(\lambda_1 \ln 2 - 1)} = \text{constant}$$

Thus, $p(x) = \dfrac{1}{2M}$ for $-M < x < M$, and $H(X) = \displaystyle\int_{-M}^{M} \dfrac{1}{2M} \log 2M \, dx$

$$= \log 2M$$

SOLUTIONS TO EXERCISES **823**

(b) $p(z) = 1/2KM$ for $-KM < z < KM$ so $H(Z) = \log 2KM$

But $dz/dx = K$ so $H_0(Z) - H_0(X) = -\log K$ and

$H_{abs}(Z) - H_{abs}(X) = \log 2KM - \log 2M - \log K$
$$= \log\left[2KM/(2M \times K)\right] = 0$$

16.3-2 (a) $R = r \log 64 \le B \log (1 + S/N) \Rightarrow r \le (3 \times 10^3 \log 1001)/6$

$$= 5000 \text{ symbols/sec}$$

(b) $S/N_0 B = 10^3 \Rightarrow S/N_0 = 3 \times 10^3 \times 10^3 = 3 \times 10^6$

$B = 1$ kHz:

$C = 10^3 \log (1 + 3 \times 10^6/10^3) \approx 1.2 \times 10^4 \Rightarrow r \le 1.2 \times 10^4/6 = 2000$

$B \to \infty$:

$C_\infty = 1.44 \times 3 \times 10^6 = 4.32 \times 10^6 \Rightarrow r \le 4.32 \times 10^6/6 = 720{,}000$

16.4-1 (a) $\|s_1'\|^2 = \left(3\sqrt{2}\right)^2 \times 2 = 36 \quad \phi_1 = s_1'/6$

$$\alpha_{21} = \int s_2'\phi_1 \, dt = 3\int_0^2 dt = 6 \quad g_2 = s_2' - 6\phi_1$$

$$\|g_2\|^2 = 36 \quad \phi_2 = g_2/6$$

(b) $\left\|s_2'\right\|^2 = 6^2 + 6^2 = 72 \quad \left\|s_2'\right\|^2 = \int_0^4 \left(3\sqrt{2}\right)^2 dt = 18 \times 4 = 72$

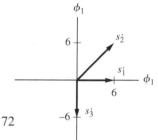

16.5-1 $E_i = a^2 \qquad i = 1, 2$
$$= a^2 + (2a)^2 \qquad i = 3, 4, 5, 6$$
$$E = [2 \times a^2 + 4 \times 5a^2]/6 = 11a^2/3$$

$$\frac{a}{\sqrt{N_0/2}} = \sqrt{\frac{6E}{11N_0}} \text{ so let } q = Q\left(\sqrt{\frac{6E}{11N_0}}\right)$$

For $i = 1, 2$

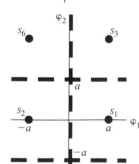

$$P(c \mid m_i) = \int_{-a}^{a} p_\beta(\beta_1) \, d\beta_1 \int_{-a}^{\infty} p_\beta(\beta_2) \, d\beta_2 = (1 - 2q)(1 - q)$$

For $i = 3, 4, 5, 6 \quad P(c \mid m_i) = \int_{-a}^{\infty} p_\beta(\beta_1) \, d\beta_1 \int_{-a}^{\infty} p_\beta(\beta_2) \, d\beta_2 = (1 - q)^2$

$P_c = \frac{1}{6}\left[2(1 - 2q)(1 - q) + 4(1 - q)^2\right] = \frac{1}{3}(3 - 7q + 4q^2)$

Thus, $P_e = 1 - P_c = \frac{1}{3}(7q - 4q^2)$

A–1

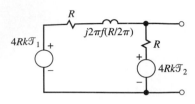

$$i_n^2(f) = \frac{4Rk\mathcal{T}_1}{|R + jfR|^2} + \frac{4Rk\mathcal{T}_2}{R^2} = \frac{4k}{R}\left(\frac{\mathcal{T}_1}{1 + f^2} + \mathcal{T}_2\right)$$

$$Z(f) = \frac{R(R + jfR)}{R + R + jfR} = \frac{R(1 + jf)}{2 + jf}, \quad |Z(f)|^2 = R^2\frac{1 + f^2}{4 + f^2},$$

$$\mathrm{Re}[Z(f)] = R\frac{2 + f^2}{4 + f^2}$$

$$v_n^2(f) = |Z(f)|^2 i_n^2(f) = 4kR\frac{\mathcal{T}_1 + (1 + f^2)\mathcal{T}_2}{4 + f^2}$$

$$\eta(f) = \frac{v_n^2(f)}{4\,\mathrm{Re}[Z(f)]} = k\frac{\mathcal{T}_1 + (1 + f^2)\mathcal{T}_2}{2 + f^2}, \text{ If } \mathcal{T}_1 = \mathcal{T}_2 = \mathcal{T}, \text{ then}$$

$$\eta(f) = k\mathcal{T}.$$

A–2   (a) $N_o = 10^6 k(\mathcal{T}_0 + \mathcal{T}_e) \times 2 \times 10^6$

$$= 2 \times 10^{12} \times 4 \times 10^{-21}\frac{\mathcal{T}_0 + \mathcal{T}_e}{\mathcal{T}_0} = 40 \times 10^{-9}$$

so $(\mathcal{T}_0 + \mathcal{T}_e)/\mathcal{T}_0 = 5 \Rightarrow \mathcal{T}_e = 4\mathcal{T}_0, F = 1 + 4\mathcal{T}_0/\mathcal{T}_0 = 5$

(b) $F = \mathcal{T}_x/\mathcal{T}_0 = 5, \mathcal{T}_i = \mathcal{T}_0 + \mathcal{T}_x = 6\mathcal{T}_0 = 1740$ K

A–3   With FET: $\mathcal{T}_e = 9 + \dfrac{14.5}{100} + 1.8 + 2.0 = 12.9, \quad \mathcal{T}_N = 42.9$ K

Without: $\mathcal{T}_e = 9 + \dfrac{14.5}{100} + \dfrac{1.05 \times 1860}{100} = 28.7, \quad \mathcal{T}_N = 58.7$ K

Note that FET increase $(S/N)_R$ by $58.7/42.9 = 1.37 \approx 1.4$ dB

Answers to Selected Problems

Selected answers are given here for problems marked with an asterisk (*).

2.1–8 $0.23A^2, 0.24A^2, 0.21A^2$

2.2–4 $-j(A/\pi f)[\text{sinc } 2f\tau - \cos 2\pi f\tau]$

2.2–6 50%, 84%

2.2–10 $AW[\text{sinc } (2Wt - \frac{1}{2}) - \text{sinc } (2Wt + \frac{1}{2})]$

2.3–1 $2A\tau \text{ sinc } f\tau \cos 2\pi fT$

2.3–6 $(1/|a|)V(f/a)e^{-j\omega t_d/a}$

2.3–13 $(j4Abf)/[b^2 + (2\pi f)^2]^2$

2.4–2 $\begin{aligned} y(t) &= 0 & t < 0, t > 5 \\ &= At^2/2 & 0 < t < 2 \\ &= 2A & 2 < t < 3 \\ &= (A/2)[4 - (t-3)^2] & 3 < t < 5 \end{aligned}$

2.4–7 $\begin{aligned} y(t) &= 0 & t < 0 \\ &= [ab/(a-b)][e^{-bt} - e^{-at}] & t > 0 \end{aligned}$

2.4–14 $y(t) = \frac{1}{2} \text{sinc}\left(\frac{t}{2}\right)$ (*Hint:* Use duality)

2.5–5 $2A\tau \text{ sinc } 2f\tau \, e^{-j2\pi f\tau}$

2.5–11 $2A\tau \text{ sinc } f\tau(1 + \cos 4\pi fT)$

2.5–13 4

3.1–3 $AH^2(f)e^{-j2\pi ft_d}$

3.1–9 $y(t) \approx 2\pi B \displaystyle\int_{-\infty}^{t} x(\lambda) \, d\lambda$

3.1–14 $h(t) = \dfrac{1}{K}\delta(t) - \dfrac{1}{K^2}e^{-t/K}u(t), \; g(t) = \dfrac{1}{K}e^{-t/K}u(t)$

3.2–3 $\begin{aligned} y(t) &= 1.28 \cos (\omega_0 t + 72°) + 0.31 \cos (3\omega_0 t + 45°) \\ &\quad + 0.14 \cos (5\omega_0 t + 31°) \end{aligned}$

3.2–8 $y(t) \approx \dfrac{\alpha}{2}x(t) + x(t - T) - \dfrac{\alpha}{2}x(t - 2T)$

3.3–1 $\ell_1 = 22$ km, $g_2 = 56$ dB, $g_4 = 34$ dB

3.3–6 $r = 0.9$ m

3.4–2 $h(t) = K\delta(t - t_d) - 2BK \text{ sinc } B(t - t_d) \cos \omega_c(t - t_d)$

3.4–8 $g(t) = 1 - e^{-bt}(\sin bt + \cos bt)$ for $t \geq 0$, $t_r \approx 1/2.8B$

3.5–2 $\hat{v}(t) = \lim_{t\to\infty}(A/\pi)\ln |(2t + \tau)/(2t - \tau)| = 0$

3.5–5 $\hat{x}(t) = 4 \sin \omega_0 t + \frac{4}{9} \sin 3\omega_0 t + \frac{4}{25} \sin 5\omega_0 t$

3.6–5 $G_v(f) = \left(\dfrac{A}{2W}\right)^2 \Pi\left(\dfrac{f}{2W}\right), R_v(\tau) = \left(\dfrac{A^2}{2W}\right)\text{sinc } 2W\tau, E_v = \dfrac{A^2}{2W}$

3.6–10 $R_v(\tau) = A^2/2, P_v = A^2/2, G_v(f) = (A^2/2)\delta(f)$

4.1–4 $v_{lp}(t) = 400 \operatorname{sinc} 400t \, e^{-j2\pi 100t}$, $v_i(t) = 800 \operatorname{sinc} 400t \cos 2\pi 100t$,
$v_q(t) = -800 \operatorname{sinc} 400t \sin 2\pi 100t$

4.1–10 $y_{bp}(t) = a(1 - e^{-\pi Bt}) \cos \omega_c t \, u(t)$

4.2–3 AM: $B_T = 400$, $S_T = 68$; DSB: $B_T = 400$, $S_T = 50$

4.2–9 $5 < f_c < 50$

4.3–2 $K = \sqrt{b/a}$

4.3–7 $f_c > 200$

4.4–7 $S_T = 7A_c^2/4$, $B_T = 400$

4.4–12 $x_c(t) = (A_c/2)[\cos \omega_m t \cos \omega_c t - 2a \sin \omega_m t \sin \omega_c t]$

4.5–5 $a = \frac{1}{2}$ no distortion; $a = 1$ or $a = 0$ produces SSB (max distortion)

5.1–4 $\theta_c(t) = 2\pi[f_1 t + (f_2 - f_1)t^2/T]$

5.1–10 $f(t) = 40 + 40 \cos 2\pi 20t$; $S_T = 6441.5$

5.2–5 $B_T = 10^{13}$, $f_\Delta = 5 \times 10^{12}$

5.2–9 $y_c(t) = A_c[\cos \omega_c t - \phi_\Delta(\pi f_c/Q)\tilde{x}(t) \sin \omega_c t]$;
$\phi(t) = \arctan[\phi_\Delta(\pi f_c/Q)\tilde{x}(t)]$

5.2–15 $f_\Delta < (f_c - 2W)/9$

5.3–5 Signal has low frequencies boosted.

5.3–11 $K_1 = A_c/2 \sqrt{2} f_c$, $K_2 = A_c/8 \sqrt{2} f_c^2$; $f_\Delta/f_c < 0.08$

5.4–5 S_x will increase dramatically and will have a larger impact on DSB.
$P_{sb}/A_{max}^2 = S_x/4$ for DSB however $P_{sb}/A_{max}^2 = S_x/16$ for AM.

5.4–10 $y_D(t) = \{(1 + \rho \cos[\phi(t) - \theta_i(t)])\dot{\phi}(t)/2\pi$
$+ (\rho + \cos[\phi(t) - \theta_i(t)])\rho f_i\}/\{1 + \rho^2 + 2\rho \cos[\phi(t) - \theta_i(t)]\}$

6.1–4 $K_1 = 2$ and $K_2 = \pi/2$

6.1–15 $f_s \le 12$ MHz, $W_{presampling} \le 137.5$ MHz

6.1–17 (a) 1.64%, (b) −14.3%

6.1–20 (a) 100 Hz, (b) 200 Hz, (c) 400 kHz

6.2–3 (a) $H(f) = \operatorname{sinc} f\tau \, e^{-j\omega\tau/2}$, $X_p(f) = P(f)\overline{X}_\delta(f)$ where $\overline{X}_\delta(f) = f_s \sum_n H(f - nf_s)X(f - nf_s)$

(b) $H_{eq}(f) = Ke^{-j\omega(t_d - \tau/2)}/\operatorname{sinc}^2 f\tau$, $|f| \le W$

6.3–1 $B_T \ge 400$ kHz

7.1–1 $f_{IF} \ge 532.5$ kHz, $f_{LO} = 1072.5$ to 2132.5 kHz, 10 kHz $< B_{RF} <$ 1065 kHz

7.1–3 $C = 6$ to 25.6 nF, 19.3 to 3,506 nF

7.1–9 (a) 50–54 MHz, 50.910 to 54.910 MHz
(b) 50–54 MHz, 64 to 68 MHz

7.1–12 (a) 2 MHz at 0 dB, 2.9 MHz at −10 dB, 4.550 at −23.1 dB, 5.365 MHz at −25.4 dB

7.1–14 (a) −24 dB, (b) −35.6 dB

7.2–4 $B_g \geq 0.76W, B_T \geq 17W$

7.2–11 $r = 150 \text{ kHz}, \tau = 2 \ \mu s, B_T \geq 250 \text{ kHz}$

7.2–13   (a) $B_T \geq 250 \text{ kHz}$,   (b) $B_T \approx 250 \text{ kHz}$

7.2–18 $M = 28$

7.3–5 $\cos \theta_v(t) = \cos \left[(\omega_c + \omega_1)t + \phi_0 + \phi_1 + 90^0 - \varepsilon_{ss} \right]$

7.3–8 $f_v = 5 \text{ kHz}, 50 \text{ kHz}$

7.3–13   (a) $A_\varepsilon = \dfrac{f_\Delta}{K} \dfrac{A_m}{\sqrt{1 + (f_m/K)^2}}$,   (b) $K \geq 2f_\Delta$

7.4–4 $n_p = 25.921, B = 805 \text{ kHz}$

7.4–6 $n_p = 2.18 \times 10^5, T_{line} = 64 \ \mu s, B = 4.99 \text{ MHz}$

8.1–1 $P(A^c B) = 2/12$

8.1–5 $P(C) = P(A) + P(B) - 2P(AB)$

8.1–11 $P(\text{all tails}) = 11/24$

8.2–1 $P(X \geq 2) = 0.4$

8.2–5 $K = 0.01$

8.2–9 $p_Z(z) = e^{-(z+5)}u(z + 5)$

8.2–15 $p_Y(y) = e^{-y}u(y)$

8.3–1 $m_X = 1/a$

8.3–5 $m_X = \frac{K-1}{2} a$

8.3–9 $\beta = -m_X$

8.3–15 $\Phi_X(\nu) = \left(1 - j\dfrac{\nu}{a} \right)^{-1}$

8.4–1 $P(i < 3) = 56/1024$

8.4–5 $P(I > 1) = 0.264$

8.4–9 $P(X > 20) = Q(0.5) \approx 0.31$

8.4–17 $E[Y] = e^{\sigma_X^2/2} e^{m_X}$

8.4–24 $m_Z = 0, \sigma_Z^2 = 100$

9.1–1 $\overline{v(t)} = \frac{3}{t} (e^{2t} - 1)$

9.1–7 $R_{vw}(t_1, t_2) = \sigma^2 \sin \omega_0(t_1 - t_2)$

9.1–10 $\overline{z(t)} = 0$

9.2–1 $\langle v(t) \rangle = \pm 3$

9.2–12 $R_{yx}(\tau) = \mathscr{F}_\tau^{-1}[(j2\pi f)G_x(f)]$

9.3–3 $\overline{y^2} = N_0 T/2$

9.3–7 $\overline{y^2} = N_{0v}R/4L$

9.3–10 $\overline{z} = \sqrt{\frac{2}{\pi}} \sigma \approx 16 \ \mu v$

9.3–14 $B_N = \sqrt{\pi}/2 \sqrt{2} \ a$

9.4–1 $(S/N)_{D_{dB}} = 61$ dB

9.4–5 $(S/N)_D = 2S_T/3LN_0W$

9.4–9 $m_{min} = 5$

9.5–1 $(\sigma_A/A)^2 = k\mathcal{T}_N B_N \tau/E_p = 0.4$

9.5–5 $E_p \geq 5 \times 10^{-8}$

9.5–8 $h_{opt}(t) = \dfrac{2K\tau}{N_0} \Lambda\left(\dfrac{t - t_d}{\tau}\right)$

10.1–3 $G_n(f) \div N_0 = 0.1$ at $f/f_c = \pm 0.5$

10.1–7 $\sigma_Y = \sqrt{8 - 2\pi} = 1.3$

10.1–10 $R_n(\tau) = N_0 B_T \text{ sinc } B_T\tau \cos \omega_c\tau$

10.1–16 $R_{n_i n_q}(\tau) = -\pi N_0 B_T^2 \tau \text{ sinc}^2 B_T\tau$

10.2–1 $(S/N)_D = 50$ dB

10.2–5 $(S/N)_D = \gamma/2$ for both channels

10.2–11 $g_T = \gamma/\gamma_{th} \approx 1500 = 32$ dB

10.3–1   FM:$N_D = N_0 W^3/3S_R = 1.67 \times 10^5$

10.3–6 $(S/N)_D = 290 = 24.6$ dB

10.3–10 $(S/N)_{D_{th}} \leq 2030 \approx 33$ dB

10.4–1   (c) $S_T \approx 200$ W

10.4–5 $(S/N)_D = 90 = 19.5$ dB

10.6–1 $(S/N)_D = 2.78 \times 10^5 = 54.4$ dB

11.1–7   (a) $B \geq 160$ kHz,   (b) $M \geq 8$

11.1–11 $r \approx 0.7B$

11.2–1 $(S/N)_R = 19.2, P_e = 6 \times 10^{-6}$ (polar)

11.2–4   (a) $P_e = \dfrac{1}{2} Q\left(\dfrac{A - 2\alpha}{2\sigma}\right) + \dfrac{1}{2} Q\left(\dfrac{A + 2\alpha}{2\sigma}\right)$, $P_e = 3.7 \times 10^{-4}$ and 3.4×10^{-5} (if $\epsilon = 0$)

11.2–8 $P_e = 1.5 \times 10^{-22}$ (regenerative), $P_e = 1.2 \times 10^{-2}$ (nonregenerative)

11.2–13 $M_{min} = 16, S_R \geq 670$ pW

11.3–3   (a) 3 kpbs,   (b) 4 kpbs,   (c) 4.8 kbps

11.4–1   scrambled output: 010010111001100, dc value: 0.47 (scrambled), 0.80 (unscrambled)

12.1–1 $\nu = 3, f_s \leq 33.3$ kHz, $n = 3$

12.1–5 $M = 3, \nu = 5, f_s \leq 6.4$ kHz

12.1–10 $q = 4096, \Delta = 0.488$ mV, $(S/N)_D = 74$ dB

12.1–13   24 Mbits

12.2–1 $M = 8, \nu = 2, q = 64, S_R = 56.7$ mW

12.2–4 $\gamma \approx 22.8$ dB (PCM), $\gamma = 50$ dB (analog), advantage: 27.2 dB

12.3–2 0.372

12.3–7 $W_{rms} = 1.3$ kHz, $K = \dfrac{f_0 S_x}{2 \arctan (W/f_0)}$

12.3–10 $n = 1$: $c_1 = \rho_1$, $G_\rho = 10.1$ dB, $n = 2$: $c_1 = 0.9744$, $c_2 = -0.0256$, $G_\rho = 10.1$ dB

12.4–1 5.9 Gbits

12.5–1 $N = 4$, efficiency $= 7.8\%$

12.5–4 247 secs/page (both B channels used)

13.1–1 $P(\text{no errors}) = 0.6561$, $P(\text{detected error}) = 0.2916$, $P(\text{undetected error}) = 0.0523$

13.1–5 $(31, 26)$: $\gamma_b = 7.0$ dB, $(31, 21)$: $\gamma_b = 8.2$ dB, $(31, 16)$: $\gamma_b = 8.3$ dB, uncoded: $\gamma_b = 10.5$ dB

13.1–10 $P_{be} \approx 10^{-6}$, $N \geq 10$, $r_b \leq 269$ kbps

13.2–13 $Q_M(p) = p^2 + p + 1$, $C(p) = 0 + 0 + p + 1$, $X = (1\,0\,1\,0\,0\,1\,1)$

13.2–19 (a) $\alpha = 5 \times 10^{-4}$, $P_{be} = 1.5 \times 10^{-6}$

13.2–23 $X = (1\,0\,0\,0\,1\,0\,1\,1\,0\,1\,0\,1\,0\,1\,0\,0\,1\,1\,1)$

13.3–6 $X = (101\quad 100\quad 010\quad 000\quad 100\quad 000\quad 011\quad 000)$

13.3–11 (a) minimum weight path: $abc = D^3 I^1$, $d_f = 3$, $M(d_f) = 1$

(b) $T(D, I) = \dfrac{D^3 I}{1 - DI}$ (c) Eq. (9): $P_{be} = 8\alpha^{3/2}$, Eq. (10): $P_{be} = 8\alpha^{3/2}$

13.3–13 $Y + \hat{E} = 00\quad 11\quad 01\quad 01\quad 11\quad 11\quad 10\quad 11$
$\hat{M} = 0\quad 1\quad 1\quad 0\quad 0\quad 1\quad 0\quad 0$

13.4–1 (a) $x = \{8, 27, 51\} \rightarrow y = \{2, 3, 6\}$ with $pq = 55$, $e = 7$, $d = 23$

14.1–1 $\overline{x_c^2} = \dfrac{A_c^2}{12} (M - 1)(2M - 1)$, $\dfrac{P_c}{\overline{x_c^2}} = 1/2\ (M = 2)$, $\dfrac{P_c}{\overline{x_c^2}} = 3/4\ (M \gg 1)$

14.1–14 $r_b = 385$ bps

14.2–4 $P_e = Q(\sqrt{1.216\gamma_b})$

14.2–6 $\theta_\varepsilon < \arccos (3.74/4.27) \approx 29°$

14.2–14 (a) $A_c = 0.00133$,   (b) $A_c = 0.0023$

14.3–1 $\gamma_b > 10.9$ dB, $P_{e1} < 3 \times 10^{-4}$

14.3–4 (a) $P_e = 3.4 \times 10^{-5}$,   (b) $P_e = 1.76 \times 10^{-4}$

14.3–8 (a) 11.8 kbps   (b) 23.5 kbps   (c) 28.4 kbps

14.4–1 (a) $\gamma_b \geq 10.5$ dB,   (b) $\gamma_b \geq 12.8$ dB

14.4–3 (a) 16-QAM with $\gamma_b \geq 14.4$ dB,   (b) 16-PSK with $\gamma_b \geq 18.4$ dB

14.4–6 5.2

14.4–11 (a) $P_e = 2.32 \times 10^{-5}$, (b) $P_e = 1.8 \times 10^{-10}$, (c) $P_e = 0.4$, (d) $P_e = 0.017$

14.5–1 $P_e = 1.6 \times 10^{-11}$, output symbol rate does not change

14.5–3 $x_2 x_1 = \{00, \quad 01, \quad 10, \quad 01, \quad 11, \quad 00\} \rightarrow$
              $y_3 y_2 y_1 = \{000, \quad 100, \quad 011, \quad 010, \quad 111, \quad 111\}$

15.1–1   (a) $W_c = 203$ kcps,   (b) $B_T = 0.41$ MHz

15.1–5 $W_c > 487$ kcps

15.2–1 $Pg = 2^7 = 256, B_T = 768$ kHz

15.2–4 $P_e = 4.95 \times 10^{-2}$

15.3–5 $\{0000 \quad 0000 \quad 1001 \quad 0100 \quad 1001 \quad 1110 \quad 1010 \quad 110\}, |R_{st}(\tau)|_{max} = 0.29$

15.4–3 $\bar{T}_{acq} = 0.51$ secs, $\sigma_{acq} = 0.38$ secs

16.1–1 $I(\text{not } F) = \log 5/4 = 0.322$ bits

16.1–4 $H(X) = 1.94$ bits

16.1–9 $H(X) = \dfrac{1}{3} \log 3 + p \log \dfrac{1}{p} + \left(\dfrac{2}{3} - p\right)\log \dfrac{1}{\dfrac{2}{3} - p}$

16.1–16 $H(X) = 0.811$ bits

16.2–8 $C_s = 0.577$ bits/symbol

16.3–1 $H(X) = \log \sqrt{12S}$

16.3–5 $p(x) = \dfrac{1}{m} e^{-x/m} u(x) \quad H(X) = \log em$

16.3–10 $S \geq 10^{-3}(2^{R/1000} - 1)$

16.3–14 $S_T = \gamma L N_0 W$

16.4–5 $s_1 = \sqrt{2}\,\phi_1 \quad s_2 = \sqrt{\dfrac{2}{3}}\,\phi_2 \quad s_3 = \dfrac{\sqrt{2}}{3}\,\phi_1 + \sqrt{\dfrac{8}{45}}\,\phi_3$

16.5–1   (a) $z_1 = (y, s_1) \quad z_2 = (y, s_2)$

16.5–5 $P_e = \dfrac{20}{9} Q\left(\sqrt{\dfrac{9E}{24N_0}}\right) - \dfrac{16}{9} Q^2\left(\sqrt{\dfrac{9E}{24N_0}}\right)$

A–1 $v_n^2(f) = 4R_1 k\mathcal{T}_1 + 4R_2 k\mathcal{T}_2$

A–5 $i_n^2(f) \approx 4(r/2)k\mathcal{T}, \quad r \approx k\mathcal{T}/qI$

A–10 $F = 26$

A–14 $\mathcal{T} \leq 103$ K

SUPPLEMENTARY READING

Listed below are books and papers that provide expanded coverage or alternative treatments of particular topics. Complete citations are given in the References.

Communication Systems

The following texts present about the same general scope of communication systems as this book: Schwartz (1990), Ziemer and Tranter (1995), Couch (1995), Roden (1996), Lathi (1998), and Haykin (2001). A somewhat more advanced treatment is provided by Proakis and Salehi (1994). See also Kamen and Heck (1997) or Proakis and Salehi (1998) for additional MATLAB material.

Bellamy (1991) provides details of digital telephony. Optical systems are discussed by Gagliardi and Karp (1995) and by Palais (1998).

Fourier Signal Analysis

Expanded presentations of signal analysis and Fourier methods are contained in Lathi (1998) and Stuart (1966). Two graduate texts dealing entirely with Fourier transforms and applications are Bracewell (1986), which features a pictorial dictionary of transform pairs, and Papoulis (1962), which strikes a nice balance between rigor and lucidity.

Advanced theoretical treatments will be found in Lighthill (1958) and Franks (1969). The article by Slepian (1976) expounds on the concept of bandwidth.

Probability and Random Signals

Probably the best general reference on probability and random signals is Leon-Garcia (1994). Other texts in order of increasing sophistication are Drake (1967), Beckmann (1967), Peebles (1987a), Papoulis (1984), and Breipohl (1970).

The classic reference papers on noise analysis are Rice (1944) and Rice (1948). Bennett (1956) is an excellent tutorial article.

CW Modulation and Phase-Lock Loops

Goldman (1948), one of the earliest books on CW modulation, has numerous examples of spectral analysis. More recent texts that include chapters on this subject are Stremler (1990), Ziemer and Tranter (1995), and Haykin (2001).

Detailed analysis of FM transmission is found in Panter (1965), a valuable reference work. Taub and Schilling (1986) gives clear discussions of FM noise and threshold extension. The original papers on FM by Carson (1922) and Armstrong (1936) remain informative reading.

The theory and applications of phase-lock loops are examined in depth in Brennan (1996), which also includes a discussion of noise in PLLs.

Sampling and Coded Pulse Modulation

Shenoi (1995) presents sampling and digital signal processing with emphasis for telecommunications. Oppenheim, Schafer, and Buck (1999) is a classic book that covers sampling and digital signal processing. Ifeachor and Jervis (1993) presents many of the practical aspects of sampling. Other discussions of sampling are in papers by Linden (1959) and Jerri (1977).

Oliver, Pearce, and Shannon (1948) is a landmark article on the philosophy of PCM while Reeves (1965) recounts the history of his invention. The book by Cattermole (1969) is entirely devoted to PCM. Jayant and Noll (1984) covers the full range of digital encoding methods for analog signals.

Digital Communication and Transmission Methods

Undergraduate-level texts that cover digital communication are Gibson (1993), Couch (2001), and Haykin (2001). Graduate-level treatments are given by Proakis (2001), Sklar (2001), and Lindsey and Simon (1973). Also see the books listed under Detection Theory.

The following references deal with specific aspects of digital transmission: Fehr (1981) on microwave radio; Mitola (2000) on software radio; Spilker (1977) and Sklar (2001) on satellite systems; Dixon (1994) and Peterson, Ziemer, and Borth (1995) on spread spectrum; Tannenbaum (1989) and Stallings (2000) on computer networks; Lewart (1998) on modems; Ungerboeck (1982), Biglieri, Divsalar, McLane, and Simon (1991) and Schlegel (1997) on trellis coded modulation systems; and Rappaport (1996) on wireless communications.

Of the many papers that could be mentioned here, the following have special merit: Arthurs and Dym (1962) on optimum detection; Lender (1963) on duobinary signaling; Lucky (1965) on adaptive equalization; Gronemeyer and McBride (1976) on MSK and OAPSK; Oetting (1979) on digital radio.

Coding and Information Theory

Abramson (1963), Hamming (1986), and Wells (1999) provide very readable introductions to both coding and information theory. Also see Chaps. 4–6 of Wilson (1996) and Chaps. 4, 7, and 8 of Lafrance (1990). Mathematically advanced treatments are given by Gallager (1968) and McElice (1977).

Texts devoted to error-control coding and applications are Wiggert (1978), Lin (1970), and Adámek (1991) at the undergraduate level, and Berlekamp (1968), Peterson and Weldon (1972), Lin and Costello (1983), Sweeney (1991) and Wicker (1995) at the graduate level. A history of coding and a treatment of code-breakers and the present encryption systems is given in Singh (1999).

Introductions to information theory are given by Blahut (1987) and Cover and Thomas (1991). The classic papers on the subject are Nyquist (1924, 1928a),

Hartley (1928), and Shannon (1948, 1949). Especially recommended is Shannon (1949), which contained the first exposition of sampling theory applied to communication. A fascinating nontechnical book on information theory by Pierce (1961) discusses implications to art, literature, music, and psychology.

Detection Theory

The concise monograph by Selin (1965) outlines the concepts and principles of detection theory. Applications to optimum receivers for analog and digital communication are developed in Sakrison (1968), Van Trees (1968), and Wozencraft and Jacobs (1965). The latter includes a clear and definitive presentation of vector models. Viterbi (1966) emphasizes phase-coherent detection. Tutorial introductions to matched filters are given in the papers by Turin (1960, 1976).

Electrical Noise

There are relatively few texts devoted to electrical noise. Perhaps the best general reference is Pettai (1984). Useful sections on system noise are found in Freeman (1997), Johns and Martin (1997), and Ludwig and Bretchko (2000). Noise in microwave systems is described by Siegman (1961) using the informative transmission-line approach. Electronic device noise is treated by Ambrózy (1982) and Van der Ziel (1986).

Communication Circuits and Electronics

The design and implementation of filter circuits are detailed in Hilburn and Johnson (1973) and Van Valkenburg (1982). Recent introductory treatments of communication electronics are found in texts such as van der Puije (1992), Tomasi (1998), and Miller (1999). More advanced details are given by Clarke and Hess (1971), Krauss, Bostian, and Raab (1980), Smith (1986), Freeman (1997), Johns and Martin (1997), and Ludwig and Bretchko (2000).

REFERENCES

Abate, J. E. (1967), "Linear and Adaptive Delta Modulation," *Proc. IEEE,* vol. 55, pp. 298–308.

Abramson, N. (1963), *Information Theory and Coding,* McGraw-Hill, New York.

Adámek, J. (1991), *Foundations of Coding,* Wiley, New York.

Ambrózy, A. (1982), *Electronic Noise,* McGraw-Hill, New York.

Armstrong, E. H. (1936), "A Method of Reducing Disturbances in Radio Signaling by a System of Frequency Modulation," *Proc. IRE,* vol. 24, pp. 689–740.

Arthurs, E., and H. Dym (1962), "On the Optimum Detection of Digital Signals in the Presence of White Gaussian Noise—A Geometric Interpretation and a Study of Three Basic Data Transmission Systems," *IRE Trans. Commun. Systems,* vol. CS-10, pp. 336–372.

ATSC (1995), *Guide to the Use of the ATSC Digital Television Standard,* Advanced Television System Committee, Washington, DC, doc. A/54, Oct. 4, 1995.

Baert, L., L. Theunissen, J. M. Bergult, and J. Arts (1998), *Digital Audio and Compact Disc Technology,* 3d ed., Focal Press, Oxford, Great Britain.

Bahl, L. R., J. Cocke, F. Jelinek, and J. Raviv, (1972), "Optimal decoding of linear codes for minimizing symbol error rate," *Abstracts of Papers, Int. Symp. Inform. Theory,* Asilomar, CA, p. 90.

Balmer, L. (1991), *Signals and Systems: An Introduction,* Prentice-Hall International, Upper Saddle River, NJ.

Beckmann, P. (1967), *Probability in Communication Engineering,* Harcourt, Brace, & World, New York.

Bellamy, J. C. (1991), *Digital Telephony,* 2d ed., Wiley, New York.

Benedetto, S., and Montorsi, G. (1996). "Unveiling Turbo Codes: Some Results on Parallel Concatenated Coding Schemes," *IEEE Trans. Inform. Theory,* vol. IT-42, pp. 409–428.

Bennett, W. R. (1956), "Methods of Solving Noise Problems," *Proc. IRE,* vol. 44, pp. 609–638.

Berlekamp, E. R. (1968), *Algebraic Coding Theory,* McGraw-Hill, New York.

Berrou, C., and A. Glavieux (1996), "Near Optimum Error Correcting Coding and Decoding: Turbo Codes," *IEEE Trans. Commun.,* vol. 44, pp. 1261–1271.

Berrou, C., A. Glavieux, and P. Thitmajshima (1993), "Near Shannon Limit Error-Correcting Coding and Decoding: Turbo Codes," *Proc. IEEE Int. Conf. Commun.,* pp. 1064–1070, Geneva, Switzerland.

Biglieri, E., D. Divsalar, P. J. McLane, and M. K. Simon (1991), *Introduction to Trellis-Coded Modulation,* Macmillan, New York.

Blahut, R. E. (1987), *Principles and Practice of Information Theory,* Addison-Wesley, Reading, MA.

Blanchard, A. (1976), *Phase-Locked Loops,* Wiley, New York.

Bracewell, R. (1986), *The Fourier Transform and Its Applications,* 2d ed., McGraw-Hill, New York.

Breipohl, A. M. (1970), *Probabilistic Systems Analysis,* Wiley, New York.

Brennan, P. V. (1996), *Phase-Locked Loops: Principles and Practice,* McGraw-Hill, New York.

Campbell, R. (1993), "High-Performance, Single Signal Direct-Conversion Receivers," *QST,* pp. 32–40.

Carson, J. R. (1922), "Notes on the Theory of Modulation," reprinted in *Proc. IEEE,* vol. 51, pp. 893–896.

Cattermole, K. W. (1969), *Principles of Pulse Code Modulation,* American Elsevier, New York.

Chaffee, J. G. (1939), "The Application of Negative Feedback to Frequency-Modulation Systems," *Proc. IRE,* vol. 27, pp. 317–331.

Clarke, K. K., and D. T. Hess (1971), *Communication Circuits: Analysis and Design,* Addison-Wesley, Reading, MA.

Couch, L. W., II (1995), *Modern Communication Systems,* Prentice-Hall, Upper Saddle River, NJ.

Couch, L. W., II (2001), *Digital and Analog Communication Systems,* 6th ed., Prentice-Hall, Upper Saddle River, NJ.

Cover, T. M., and J. A. Thomas (1991), *Elements of Information Theory,* Wiley, New York.

Diffie, W., and M. E. Hellman (1976), "New Directions in Cryptography," *IEEE Trans. Inf. Theory,* vol. IT-22, pp. 644–654.

Diffie, W., and M. E. Hellman (1979), "Privacy and Authentication: An Introduction to Cryptography," *Proc. IEEE,* vol. 67, pp. 397–427.

Dixon, R. (1994), *Spread Spectrum Systems with Commercial Applications,* 3d ed., Wiley, New York.

Drake, A. W. (1967), *Fundamentals of Applied Probability Theory,* McGraw-Hill, New York.

Dutta-Roy, A. (2000), "A Second Wind for Wiring," *IEEE Spectrum,* vol. 36, pp. 52–60.

Feher, K. (1981), *Digital Communications,* Prentice-Hall, Englewood Cliffs, NJ.

FIPS-46 National Institute of Standards and Technology, "Data Encryption Standard (DES)," *Federal Information Processing Standards,* Publication no. 46, Dec. 1993.

Forney, G. D., L. Brown, M. V. Eyuboglu, and J. L. Moran (1996), "The V.34 High- Speed Modem Standard," *IEEE Communications Magazine,* vol. 34, pp. 28–33.

Franks, L. E. (1969), *Signal Theory,* Prentice-Hall, Englewood Cliffs, NJ.

Freeman, R. L. (1997), *Radio Systems Design for Telecommunications,* 2d ed., Wiley, New York.

Gagliardi, R., and S. Karp (1995), *Optical Communications,* 2d ed., Wiley, New York.

Gallager, R. H. (1968), *Information Theory and Reliable Communication,* Wiley, New York.

Gardner, F. M. (1979), *Phaselock Techniques,* 2d ed., Wiley, New York.

Gibson, J. D. (1993), *Principles of Digital and Analog Communications,* 2d ed., Prentice-Hall, Upper Saddle River, NJ.

Glisic, S., and B. Vucetic (1997), *Spread Spectrum CDMA Systems for Wireless Communications,* Artech House, Boston.

Gold, R. (1967), "Optimum Binary Sequences for Spread Spectrum Multiplexing," *IEEE Trans. Inform. Theory,* vol. IT-13, pp. 619–621.

Gold, R. (1968), "Maximal Recursive Sequences with 3-valued Recursive Cross Correlation Functions," *IEEE Trans. Inform. Theory,* vol. IT-14, pp. 154–156.

Goldman, S. (1948), *Frequency Analysis, Modulation, and Noise,* McGraw-Hill, New York.

Gonzalez, R., and R. Woods (1992), *Digital Image Processing,* Addison-Wesley, Reading, MA.

Gronemeyer, S. A., and A. L. McBride (1976), "MSK and Offset QPSK Modulation," *IEEE Trans. Commun.,* vol. COM-24, pp. 809–820.

Hagenauer, J., E. Offer, and L. Papke (1996). "Iterative Decoding of Binary Block and Convolutional Codes," *IEEE Trans. Inform. Theory,* vol. IT-42, pp. 429–445.

Hamming, R. W. (1950), "Error Detecting and Error Correcting Codes," *Bell System Tech. J.,* vol. 29, pp. 147–160.

Hamming, R. W. (1986), *Coding and Information Theory,* 2d ed., Prentice-Hall, Englewood Cliffs, NJ.

Hartley, R. V. (1928), "Transmission of Information," *Bell System Tech. J.,* vol. 7, pp. 535–563.

Haykin, S. (2001), *Communication Systems,* 4th ed., Wiley, New York.

Hilburn, J. L., and D. E. Johnson (1973), *Manual of Active Filter Design,* McGraw-Hill, New York.

http://www331.jpl.nasa.gov/public/tcodes-bib.html (List of papers that cover turbo codes).

http://www.webproforum.com/sonet (Information on SONET system).

Ifeachor, E. C., and B. W. Jervis (1993), *Digital Signal Processing, A Practical Approach,* Addison-Wesley, New York.

Jayant, N. S., and P. Noll (1984), *Digital Coding of Waveforms: Principles and Applications to Speech and Video,* Prentice-Hall, Englewood Cliffs, NJ.

Jerri, A. J. (1977), "The Shannon Sampling Theorem—Its Various Extensions and Applications: A Tutorial Review," *Proc. IEEE,* vol. 65, pp. 1565–1569.

Johannesson, R., and K. Sh. Zigangirov (1999), *Fundamentals of Convolutional Coding,* IEEE Press, New York.

Johns, D. A., and K. Martin (1997), *Analog Integrated Circuit Design,* Wiley, New York.

Johnson, J. B. (1928), "Thermal Agitation of Electricity in Conductors," *Phys. Rev.,* vol. 32, pp. 97–109.

Kamen, E. W., and B. S. Heck (1997), *Fundamentals of Signals and Systems Using MATLAB,* Prentice-Hall, Upper Saddle River, NJ.

Kotel'nikov, V. A. (1959), *The Theory of Optimum Noise Immunity,* McGraw-Hill, New York.

Krauss, H. L., C. W. Bostian, and F. H. Raab (1980), *Solid State Radio Engineering,* Wiley, New York.

Lafrance, P. (1990), *Fundamental Concepts in Communication,* Prentice-Hall, Englewood Cliffs, NJ.

Lathi, B. P. (1998), *Modern Digital and Analog Communication Systems,* 3d. ed., Oxford, New York.

Lender, A. (1963), "The Duobinary Technique for High Speed Data Transmission," *IEEE Trans. Commun. and Electron.,* vol. 82, pp. 214–218.

Leon-Garcia, A. (1994), *Probability and Random Processes for Electrical Engineering,* 2d ed., Addison-Wesley, Reading, MA.

Lewart, C. R. (1998), *The Ultimate Modem Handbook: Your Guide to Selection, Installation, Troubleshooting and Optimization,* Prentice-Hall, Upper Saddle River, NJ.

Lighthill, M. J. (1958), *An Introduction to Fourier Analysis and Generalized Functions,* Cambridge, New York

Lin, S. (1970), *An Introduction to Error-Correcting Codes,* Prentice-Hall, Englewood Cliffs, NJ.

Lin, S., and D. J. Costello, Jr. (1983), *Error Control Coding: Fundamentals and Applications,* Prentice-Hall, Englewood Cliffs, NJ.

Linden, D. A. (1959), "A Discussion of Sampling Theorems," *Proc. IRE,* vol. 47, pp. 1219–1226.

Lindsey, W. C. (1972), *Synchronization Systems in Communication and Control,* Prentice-Hall, Englewood Cliffs, NJ.

Lindsey, W. C., and M. K. Simon (1973), *Telecommunication Systems Engineering,* Prentice-Hall, Upper Saddle River, NJ. Reprinted by Dover Press, NY, 1991.

Lucky, R. W. (1965), "Automatic Equalization for Digital Communication," *Bell System Tech. J.,* vol. 44, pp. 547–588.

Ludwig, R., and P. Bretchko (2000), *RF Circuit Design,* Prentice-Hall, Upper Saddle River, NJ.

McElice, R. J. (1977), *The Theory of Information and Coding,* Addison-Wesley, Reading, MA.

Mengali, U., and A. N. D'Andrea (1997), *Synchronization Techniques for Digital Receivers,* Plenum Press, New York.

Merkle, R. C., and M. E. Hellman (1978), "Hiding Information and Signatures in Trap-Door Knapsacks," *IEEE Trans. Inf. Theory,* vol. IT-24, pp. 525–530.

Meyr, H., and G. Ascheid (1990), *Synchronization in Digital Communications,* vol. 1, Wiley, New York.

Miller, G. M. (1999), *Modern Electronic Communication,* 6th ed., Prentice-Hall, Upper Saddle River, NJ

Miller, M. (2000), *Data & Network Communications,* Delmar, Thomson Learning, Albany, NY.

Mitola, J. (2000), *Software Radio Architecture: Object Oriented Approaches to Wireless Systems Engineering,* Wiley, New York.

National Institute of Standards and Technology (1993), "Data Encryption Standard (DES)," *Federal Information Processing Standards (FIPS),* publication no. 46, Dec. 1993.

Nellist, J. G. (1992), *Understanding Telecommunications and Lightwave Systems: An Entry Level Guide,* IEEE Press, New York.

Nyquist, H. (1924), "Certain Factors Affecting Telegraph Speed," *Bell System Tech. J.,* vol. 3, pp. 324–346.

Nyquist, H. (1928a), "Certain Topics in Telegraph Transmission Theory," *Trans. AIEE,* vol. 47, pp. 617–644.

Nyquist, H. (1928b), "Thermal Agitation of Electric Charge in Conductors," *Phys. Rev.,* vol. 32, pp. 110–113.

Oetting, J. D. (1979), "A Comparison of Modulation Techniques for Digital Radio," *IEEE Trans. Commun.,* vol. COM-27, pp. 1752–1762.

Oliver, B. M., J. R. Pierce, and C. E. Shannon (1948), "The Philosophy of PCM," *Proc. IRE,* vol. 36, pp. 1324–1332.

Oppenheim, A. V., R. W. Shafer, and R. W. Buck (1999), *Discrete-Time Signal Processing,* 2d ed., Prentice-Hall, Upper Saddle River, NJ.

Palais, J. C. (1998), *Fiber Optic Communications,* 4th ed., Prentice-Hall, Upper Saddle River, NJ

Panter, P. F. (1965), *Modulation, Noise, and Spectral Analysis,* McGraw Hill, New York.

Papoulis, A. (1962), *The Fourier Integral and Its Applications,* McGraw Hill, New York.

Papoulis, A. (1984), *Probability, Random Variables, and Stochastic Processes,* 2d ed., McGraw-Hill, New York.

Peebles, P. Z., Jr. (1987*a*), *Probability, Random Variables, and Random Signal Principles,* 2d ed., McGraw-Hill, New York.

Peebles, P. Z., Jr. (1987*b*), *Digital Communication Systems,* Prentice-Hall, Englewood Cliffs, NJ.

Peterson, L. L., and B. S. Davie (2000), *Computer Networks,* Morgan Kaufman Publishers, San Francisco, CA.

Peterson, W. W., and D. T. Brown (1961), "Cyclic Codes for Error Detection," *Proc. IRE,* vol. 49, pp. 228–235.

Peterson, W. W., and E. J. Weldon, Jr. (1972), *Error Correcting Codes,* 2d ed., MIT Press, Cambridge, MA.

Peterson, R. L., R. E. Ziemer, and D. E. Borth (1995), *Introduction to Spread Spectrum Communications,* Prentice-Hall, Upper Saddle River, NJ.

Pettai, R. (1984), *Noise in Receiving Systems,* Wiley, New York.

Pierce, J. R. (1961), *Symbols, Signals, and Noise,* Harper & Row, New York.

Proakis, J. G. (2001), *Digital Communications,* 4th ed., McGraw-Hill, New York.

Proakis, J. G., and M. Salehi (1994), *Communication Systems Engineering,* Prentice-Hall, Englewood Cliffs, NJ.

Proakis, J. G., and M. Salehi (1998), *Contemporary Communication Systems Using MATLAB,* PWS, Pacific Grove, CA.

Pursley, M. B. (1977), "Performance Evaluation for Phase-Coded Spread Spectrum Multiple-Access Communications—Part I: System Analysis," *IEEE Trans. Commun.,* vol. COM-25, pp. 795–799.

Rabiner, L. R., and B. Gold (1975), *Theory and Application of Digital Signal Processing,* Prentice-Hall, Englewood Cliffs, NJ.

Rabiner, L. R., and R. W. Schafer (1978), *Digital Processing of Speech Signals,* Prentice-Hall, Englewood Cliffs, NJ.

Rappaport, T. S. (1996), *Wireless Communications: Principles and Practice,* Prentice-Hall, Upper Saddle River, NJ.

Reeves, A. H. (1965), "The Past, Present, and Future of PCM," *IEEE Spectrum,* vol. 2, pp. 58–63.

Rice, S. O. (1944), "Mathematical Analysis of Random Noise," *Bell System Tech. J.,* vol. 23, pp. 282–332 and vol. 24, pp. 46–156.

Rice, S. O. (1948), "Statistical Properties of a Sine-Wave plus Random Noise," *Bell System Tech. J.,* vol. 27, pp. 109–157.

Rivest, R. L., A. Shamir, and L. Adleman (1978), "A Method for Obtaining Digital Signatures and Public Key Cryptosystems," *Communications of the Association for Computing Machinery,* vol. 21, pp. 120–126.

Roden, M. S. (1996), *Analog and Digital Communication Systems,* 4th ed., Prentice-Hall, Upper Saddle River, NJ.

Sakrison, D. J. (1968), *Communication Theory: Transmission of Waveforms and Digital Information,* Wiley, New York.

Schlegel, C. (1997), *Trellis Coding,* IEEE Press, NY.

Schmidt, H. (1970), *Analog/Digital Conversion,* Van Nostrand Reinhold, Princeton, NJ.

Schwartz, M. (1990), *Information Transmission, Modulation and Noise,* 4th ed., McGraw-Hill, New York.

Selin, L. (1965), *Detection Theory,* Princeton University Press, Princeton, NJ.

Shannon, C. E. (1948), "A Mathematical Theory of Communication," *Bell System Tech. J.,* vol. 27, pp. 379–423 and 623–656.

Shannon, C. E. (1949), "Communication in the Presence of Noise," *Proc. IRE,* vol. 37, pp. 10–21.

Shenoi, K. (1995), *Digital Signal Processing in Telecommunications,* Prentice-Hall, Upper Saddle River, NJ.

Siegman, A. E. (1961), "Thermal Noise in Microwave Systems," *Microwave J.,* vol. 4, pp. 66–73 and 93–104.

Simon, M. K., J. K. Omura, R. A. Scholtz, and B. K. Levitt (1994), *Spread Spectrum Communications Handbook,* 2d ed., McGraw-Hill, New York.

Singh, S. (1999), *The Code Book,* Doubleday, New York.

Sklar, B. (2001), *Digital Communications: Fundamentals and Applications,* 2d ed., Prentice-Hall, Upper Saddle River, NJ.

Slepian, D. (1976), "On Bandwidth," *Proc. IEEE,* vol 64, pp. 292–300.

Smith, J. R. (1986), *Modern Communication Circuits,* McGraw-Hill, New York.

Spilker, J. J., Jr. (1977), *Digital Communications by Satellite,* Prentice-Hall, Englewood Cliffs, NJ.

Stallings, W. (2000), *Data and Computer Communications,* 6th ed., Prentice-Hall, Upper Saddle River, NJ.

Stark, H., F. B. Tuteur, and J. B. Anderson (1988), *Modern Electrical Communications: Analog, Digital and Optical Systems,* 2d ed., Prentice-Hall, Upper Saddle River, NJ.

Stremler, F. G. (1990), *Introduction to Communication Systems,* 3d ed., Addison-Wesley, Reading, MA.

Stuart, R. D. (1966), *An Introduction to Fourier Analysis,* Methuen, London.

Stuber, G. (1996), *Principles of Mobile Communications,* Kluwer Academic Publishers, Boston.

Stumpers, F. L. (1948), "Theory of Frequency-Modulation Noise," *Proc. IRE,* vol. 36, pp. 1081–1092.

Sweeney, P. (1991), *Error Control Coding,* Prentice-Hall, Upper Saddle River, NJ.

Tannenbaum, A. S. (1989), *Computer Networks,* 2d ed., Prentice-Hall, Upper Saddle River, NJ.

Taub, H., and D. L. Schilling (1986), *Principles of Communication Systems,* 2d ed., McGraw-Hill, New York.

Tomasi, W. (1998), *Electronic Communication Systems,* 2d ed., Prentice-Hall, Englewood Cliffs, NJ.

Turin, G. (1960), "An Introduction to Matched Filters," *IRE Trans. Inform. Theory,* vol. IT-6, pp. 311–329.

Turin, G. (1976), "An Introduction to Digital Matched Filters," *Proc. IEEE,* vol. 64, pp.1092–1112.

Ungerboeck, G. (1982), "Channel Coding with Multilevel/Phase Signals," *IEEE Trans. on Information Theory,* vol. IT-28, pp. 55–66.

Ungerboeck, G. (1987), "Trellis-Coded Modulation with Redundant Signal Sets, Parts I and II," *IEEE Communications Magazine,* vol. 25, pp. 5–21. van der Puije, P. A. (1992), *Telecommunication Circuit Design,* Wiley, New York.

Van der Ziel, A. (1986), *Noise in Solid State Devices and Circuits,* Wiley, New York.

Van Trees, H.L. (1968), *Detection, Estimation, and Modulation Theory,* Part I, Wiley, New York.

Van Valkenburgh, M. E. (1982), *Analog Filter Design,* Holt, Rinehart, and Winston, New York.

Viterbi, A. J. (1966), *Principles of Coherent Communication,* McGraw-Hill, New York.

Wells, R. B. (1999), *Applied Coding and Information Theory for Engineers,* Prentice-Hall, Upper Saddle River, NJ.

Whitaker, J. C. (1999), *HDTV—The Revolution in Digital Video,* McGraw-Hill, New York.

Wicker, S. B., (1995), *Error Control Systems,* Prentice-Hall, Upper Saddle River, NJ.

Wiggert, D. (1978), *Error-Control Coding and Applications,* Artech House, Dedham, MA.

Wilson, S. G. (1996), *Digital Modulation and Coding,* Prentice-Hall, Upper Saddle River, NJ.

Wozencraft, J. M., and I. M. Jacobs (1965), *Principles of Communication Engineering,* Wiley, New York.

Ziemer, R. E., and W. H. Tranter (1995), *Principles of Communications,* Houghton Mifflin, Boston.

Ziemer, R. E., W. H. Tranter, and D. R. Fanin (1998), *Signals and Systems,* 4th ed, Prentice-Hall, Upper Saddle River, NJ.

INDEX